Advances in
Occupational, Social, and
Organizational Ergonomics

Advances in Human Factors and Ergonomics Series

Series Editors

Gavriel Salvendy
Professor Emeritus
Purdue University
West Lafayette, Indiana

Chair Professor & Head
Tsinghua University
Beijing, People's Republic of China

Waldemar Karwowski
Professor & Chair
University of Central Florida
Orlando, Florida, U.S.A.

Advances in Human Factors and Ergonomics in Healthcare
V. Duffy

Advances in Applied Digital Human Modeling
V. Duffy

Advances in Cross-Cultural Decision Making
D. Schmorrow and D. Nicholson

Advances in Cognitive Ergonomics
D. Kaber and G. Boy

Advances in Occupational, Social, and Organizational Ergonomics
P. Vink and J. Kantola

Advances in Human Factors, Ergonomics, and Safety in Manufacturing
and Service Industries
W. Karwowski and G. Salvendy

Advances in Ergonomics Modeling and Usability Evaluation
H. Khalid, A. Hedge, and T. Ahram

Advances in Understanding Human Performance: Neuroergonomics,
Human Factors Design, and Special Populations
T. Marek, W. Karwowski, and V. Rice

Advances in
Occupational, Social, and
Organizational Ergonomics

Edited by
Peter Vink
Jussi Kantola

CRC Press
Taylor & Francis Group
Boca Raton   London   New York

CRC Press is an imprint of the
Taylor & Francis Group, an **informa** business

CRC Press
Taylor & Francis Group
6000 Broken Sound Parkway NW, Suite 300
Boca Raton, FL 33487-2742

First issued in paperback 2017

© 2011 by Taylor and Francis Group, LLC
CRC Press is an imprint of Taylor & Francis Group, an Informa business

No claim to original U.S. Government works

ISBN-13: 978-1-4398-3507-4 (hbk)
ISBN-13: 978-1-138-11763-1 (pbk)

Visit the Taylor & Francis Web site at
http://www.taylorandfrancis.com

and the CRC Press Web site at
http://www.crcpress.com

Table of Contents

Section I: Participation and Collaboration

Section II: Human Performance

Section III: Working and Working Environment

Section IV: Health and Well-Being

Section V: Environment and Living Environment

Section VI: Virtual Environment

Section VII: Macro-Ergonomic Aspects

Preface

Worldwide the attention for health, innovation and productivity is increasing. In all situations humans interact with their environment, which is the field of ergonomics. The need for knowledge and its applications is large and this book contributes to knowledge development as well as its application. The content varies from the effect of a complete new office interior to gloves, form interaction to work place design and from vibration to the latest virtual reality applications.

This book is concerned with issues in Occupational, Social and Organizational ergonomics. The book contains a total of 90 articles. The authors of the articles represent 24 countries on five continents. We would like to thank all the authors for their hard work contributing to this book. Each of the chapters of the book were either reviewed by the members of Editorial Board or germinated by them. For these our sincere thanks and appreciation goes to the Board members listed below.

This book contains articles that deal with ergonomic and human factor issues ranging from individual perspective to multi-organizational perspective in many different settings. Explicitly, the articles were organized according to the following themes:

I: Participation and Collaboration
II: Human Performance
III: Health and Well-being

We hope that you find the articles in this book interesting and helpful to you and your work.

April 2010

Peter Vink
The Netherlands

Jussi Kantola
Daejeon, Republic of Korea

Editors

Participatory Ergonomics in a Mobile Factory: Ergonomic Device to Decrease Neck Pain

Symone A. Miguez[1,2], Peter Vink[2,3], M. Susan Hallbeck[4]

[1]Ergosys Consulting, Brasil

[2]Delft University of Technology, Industrial Design Engineering, Delft, The Netherlands

[3]TNO, Work and Employment, Hoofddorp, The Netherlands

[4]University of Nebraska-Lincoln, Industrial and Management Systems Engineering, Lincoln, NE, USA

ABSTRACT

To reduce neck complaints in a cell phone assembly task a participatory ergonomics approach is used. Five stages of a participatory ergonomics program were followed (initiative, problem identification, selection of solution, implementation and evolution). Twenty-eight women, all operators on an assembly line of cell phone boards, voluntarily participated in the design and evaluation of a device before implementing the device to all 215 employees performing that job. Prior to the intervention, RULA, comfort experiences and interviews were used. These same metrics were employed after the intervention, as well. After the intervention, an ergonomic aid which is an adjustable angled small counter, these metrics showed both a posture improvement and comfort improvement. It also showed that the initial prototype needed to be modified to reduce sharp edges/compression points for the forearm. This project shows the importance of iterative testing and that the initiative should come from workers.

Keywords: Participatory ergonomics; neck pain; cell phone; assembly line; comfort; design; RULA; ergonomic device.

INTRODUCTION

The cell phone industry currently faces challenges such as high quality manufacturing, market competition and constant technological innovation. These challenges, in turn, force the sector to continuously alter its production process. These changes are demanding because they also need to incorporate the quality of life of workers in the company. Implementation of ergonomic improvements can be difficult, but an approach showing successes is participatory ergonomics (Vink et al, 2008). Therefore participatory ergonomics was the strategy used for the development of an ergonomic intervention in cell phone assembly. Core of the approach is the involvement of people from different company areas for the fostering of general acceptance and direct participation in solving problems.

According to Vink et al. (2006), participatory ergonomics implies adapting the environment to the human through the involvement of the people who will benefit throughout the proposed design process. Also, involvement of management, engineers and health and safety is advised by participatory ergonomics experts (e.g. Noro & Imada, 1992). Therefore, this study involved participation from the maintenance technician, cell phone manufacturing workers and had the support from both the area management and the department of Environment, Health and Labor Security from this company.

The demand for this study (neck pain) arose from the workers themselves during the ergonomic evaluation carried out by the company's ergonomist. Work-related musculoskeletal disorders (WRMD), such as neck pain, have been widespread research objects, mainly in the manufacturing industry. The literature indicates that even low levels of muscular activity of the trapezius muscle for long periods, in other words, for longer than 8 minutes, over successive years can increase the risk of neck pain (Østensvik et al., 2009). According to a Dutch study (Borghouts et.al, 1999), the direct costs of medical expenses due to problems in the cervical region (neck) were about 160 million Euros (10 times the number of inhabitants in the Netherlands), and the indirect costs related to loss of productivity were estimated at 527 million Euros.

Due to the need by the workers and the potential cost saving ways of preventing musculoskeletal disorders are relevant. This preliminary study aimed to develop an ergonomic device or intervention, which we called a "small counter" ("bancadinha"), which was based on the user's need and improved iteratively using participatory ergonomic processes. This process resulted in the development of two prototypes before the final version was designed. The final version will be deployed throughout the company.

METHODS

Subjects: 28 female operators at the assembly line of cell phone boards, between 20 and 37 years of age, between 5'0.6'' and 5'8'' (153.7 to 173 cm) tall and with middle school or better education voluntarily participated in this study. The subjects work in fixed shifts of 8 hours a day, from Monday through Friday.

Instruments: The ergonomic analysis of the task was done through direct observation of postures, unstructured interviews with the workers, and pictures. Comparisons of data before and after the ergonomic intervention were done through RULA (Rapid Upper Limb Assessment), which is "a screening tool that assesses biomechanical and postural loading on the whole body with particular attention to the neck, trunk and upper limbs." (McAtamney & Corlett, 1993). RULA values range from 1 to 7 and they define the action level to be taken, as is shown in table 1:

Table 1: RULA's action level. (McAtamney & Corlett, 1993)

Action Level	Results
Action Level 1	A score of *one* or *two* indicates that posture is acceptable if it is not maintained or repeated for long periods of time.
Action Level 2	A score of *three* or *four* indicates that further investigation is needed and changes may be required.
Action Level 3	A score of *five* or *six* indicates investigation and changes are required soon.
Action Level 4	A score of *seven* or more indicates investigation and changes are required immediately.

Description of the activity before intervention: Cell phone board assembling consists of visual inspection, manual insertion of components on the board, board positioning on the jig and welding of the components using a tin-welding tool. The work is performed on a horizontal counter which is height adjustable, and the worker statically stands on an anti-fatigue mat. The cycle time of the activity varies according to the cell phone model and the demand of production of the day, alternating between short cycles (less than 30 seconds) and long cycles (greater than 30 seconds) over an 8-hour day.

The production layout consists of parallel workstations, where each worker is responsible for finalizing the cell phone board and putting it on the belt that runs between the counters (Figure 1). Although the type of layout described here suggests that the tasks are less monotonous, in this case they are monotonous because there are very few technical actions to be performed.

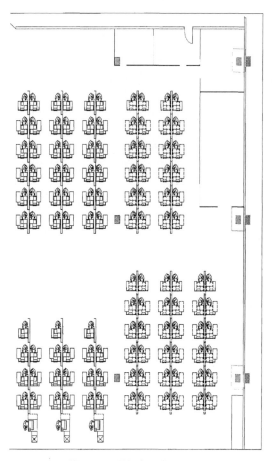

Figure 1: Workstations Layout

APPROACH AND RESULTS:

The participatory ergonomics program chosen in this paper consisted of 5 stages: 1. Initiative; 2. Problem identification; 3. Selection of solution; 4. Implementation and 5. Evolution (Vink et al., 2006). This interventional study followed the participatory ergonomic stages in the following way:

Stage 1 - Initiative: According to Patry et al. (1991), the initiative may arise from the occupational health department of the company. According to Vink et al. (2008), the initiative may also come from the workers themselves or from their union. In the present study, the initiative was generated by the workers themselves during unstructured interviews with the ergonomist during an ergonomic evaluation of their workstations. It should be noted that these ergonomic evaluations and interviews are standard because they are part of the Ergonomics Program at the company participating in this study.

Stage 2 - Problem identification: This phase is deemed crucial for specifying and understanding the problem (Howard et al., 2008; Niku, 2009). The problem was identified after the ergonomic evaluation which, at first, found three different situations that could be triggering the neck pain:

a) Many of the workers examined do not adjust the height of their work/task counters; b) neck flexion occurs in varying degrees among individual operators, and this means that some workers flex their necks more than others, related to the demands of the activity, to the modus operandi and the non-adjustment of the height of the counter; c) Although the counters do meet the various anthropometric dimensions of workers, there extra counter support is needed to facilitate precision tasks when using the screwdriver and welding tool, as shown in Figures 2 and 3.

 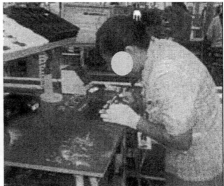

Figure 2 - Posture when screwing   **Figure 3** - Posture when welding
(before ergonomic device)   (before ergonomic device)

Prior to the intervention, RULA was used in order to quantify the problem using direct observation of the postures, pictures and unstructured interviews with workers. This resulted in RULA scores of:

Table 2 - RULA Scores prior to intervention

Action Level 3 – for the posture adopted when screwing (Figure 2)	Score of *5* for the right side assessment, indicating the need to introduce changes soon.
Action Level 3 - for the posture adopted when welding (Figure 3)	Score of *6* for the right side assessment, indicating the need to introduce changes fairly soon. (McAtamney & Corlett, 1993)

Stage 3 –Selection of a Solution: After identifying the problem in stage 2, an ergonomic device ("small counter") was developed. Its purpose was to reduce neck flexion in the tasks of screwing and welding. It was found that the screwing tasks were best performed on a horizontal surface and the welding task would benefit from a slope on the counter, as this slope would be encourage more upright neck postures and more neutral positions of shoulders, arms and wrists. Taking these facts into account plus the fact that there were few financial resources for the development of an ergonomic device, this device – the first prototype of the small counter - was produced with pieces of PVC pipe and MDF boards (all of which could be found in the company waste). Prototype 1 allows the worker to adjust the inclination of the small counter by pushing a lever to position it horizontally (Figure 4) or in a sloped position (Figure 5).

Figure 4 - Horizontal small counter   **Figure 5** – Sloped version

Prototype 1 solved the problem of the neck angle. However, in any intervention, it is extremely important to test any device in real working situations (Vink et al., 2008) and to re-evaluate the outcome. As suggested in the literature, an evaluation of prototype 1 was carried out through unstructured interviews with 28 female workers who used the ergonomic device (prototype 1) for at least 2 hours a week on the cell phone board assembly line. All 28 workers reported an improvement in the posture of the neck, but discomfort in the forearm from resting them on the new counter, due to a raised edge at the front, which was necessary to prevent the jig from slipping down when the small counter was inclined, as shown in Figure 6.

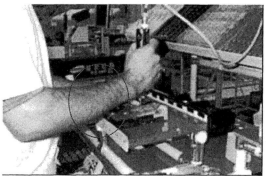

Figure 6 - Prototype 1, with raised edge at the front causing forearm discomfort.

Even though the concept of comfort is subjective and there is no universally accepted definition of comfort (Van der Linden, 2007), one should never disregard the opinion of workers in a participatory approach developing an ergonomic device. Additionally, compression points may create new ergonomic hazards for the workers. Therefore, the reports from the workers who tested the device were taken into consideration and prototype 2 was developed. The difference in prototype 2 from prototype 1 was in the front cut, whose raised edge had been removed from where the forearms rest to avoid discomfort and compression points as shown in Figure 7. The development cost of the small counter was U$9.63 dollars.

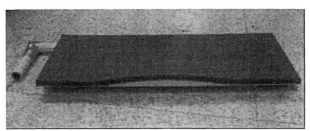

Figure 7 - Small counter without raised edge on forearm rest.

Stage 4 – Implementation: The workers were given individual guidance by the ergonomist on how to use the small counter. Prototype 2 has been tested and accepted by the 28 operators and will be implemented in all workstations in the assembly line of the company. Approximately 215 professionals will now have the benefits of an improved workstation. The postures adopted while using the new prototype 2 small counter device are better than prior to the intervention, as shown in Figures 8 and 9.

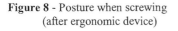

Figure 8 - Posture when screwing
(after ergonomic device)

Figure 9 - Posture when welding
(after ergonomic device)

Improvement in the worker's postures was verified by means of applying RULA to assess a worker using the small counter using prototype 2. The results were:

Table 3 – RULA Scores after intervention

Action Level 1 – for the posture adopted when screwing (Figure 8)	Score of *2* for the right side assessment, indicating that the posture is acceptable if not maintained for long periods of time.
Action Level 2 - for the posture adopted when welding (Figure 9)	Score of *3* indicates that new studies are needed and it may be necessary to introduce changes. (McAtamney & Corlett, 1993)

Stage 5 – Evaluation. According to Niku (2009), feedback gained from each phase is essential for correcting and modifying the design process, if necessary. Changes in the original ergonomic device (prototype1) were performed and resulted in prototype 2. This tested intervention will be implemented for the entire cell phone board assembly workforce. New interviews will be conducted with workers who have not yet tested the device in order for the evolution of results to keep on taking place with potential updates to the design of the intervention.

DISCUSSION

This study is an example of how participatory ergonomics can contribute to practical, inexpensive solutions that meet the needs of the worker and increase productivity. This productivity increase is due, in part to an expected reduction in sick leave caused by neck pain of the 215 employees, as well as a better working posture which is likely to increase the quality of the cell phone boards.

There were no changes in the description of the activity after the intervention. Nevertheless, as it can be seen from the pictures, the small counter intervention allowed workers to adopt better working postures during the screwing and welding tasks. RULA also confirms the visible difference in postures by decreasing the scores for screwing and welding from 5 and 6 before intervention to 2 and 3 after intervention, respectively (McAtamney & Corlett, 1993). Workers who tested Prototype 2 gave positive feedback and have encouraged us to implement the project to benefit other operators.

Support came from management, from the Environment, Health and Labor Security, from the employees who voluntarily tested the prototypes and in particular from the maintenance technician who developed the device with us was crucial for this study to happen. However, we agree with Vink et al. (2006), and believe the most important factor for the success of the implementation of this device because the initiative to seek solutions for an ergonomic problem came from the workers.

REFERENCES

Borghouts, J., Koes, B., Vondeling. H, Bouter L. (1994), "Cost of illness of neck pain in the Netherlands in 1996". *Pain*, 80, 629-636.

Howard, T.J., Culley, S.J.,& Dekoninck, E. (2008), "Describing the creative design process by the integration of engineering design and cognitive psychology literature". *Design Studies*, 29, 160-180.

McAtamney, L. and Corlett, E.N. (1993), "RULA: A survey method for the investigation of work-related upper limb disorders". *Applied Ergonomics*, 24 (2), 91-99.

Niku,S.B. (2009), *Creative Design of Products and Systems*. Hoboken, NJ: Wiley.

Noro, K. and Imada, A. (1992), *Participatory Ergonomics*, London: Taylor & Francis.

Patry, L., Cote, M. and Kuorinka, I. (1991), "Participatory ergonomics of two work stations : Warehouse employees and police officers. " In: Y. Quéinnec and F. Daniellou, eds., *Designing for Everyone: Proceedings of the 11th Congress of the International Ergonomics Association*, Vol. II, London: Taylor & Francis, 1747-1749.

Østensvik, T, Veierste,K.B., and Nilsen, P. (2009), "Association between numbers of long periods with sustained low-level trapezius muscle activity and neck pain". *Ergonomics*, 52 (12), 1556-1567.

Van der Linden, J. (2007), *Ergonomia e Design: prazer, conforto e risco no uso de produtos*. Porto Alegre: Editora UniRitter.

Vink, P., Imada, A.S, Zenk, K.J. (2008), "Defining stakeholder involvement in participatory design processes". *Applied Ergonomics*, 39, 519-526.

Vink, P., Koningsveld E.A.P., Molenbroek J.F. (2006), "Positive outcomes of participatory ergonomics in terms of greater comfort and higher productivity". *Applied Ergonomics*, 37, 537-546.

ACKNOWLEDGEMENTS

We would like to thank the workers who participated in this research, Mr. Sérgio Henrique Avelar and his team, who developed prototypes, the management and the Environment, Health and Labor Security departments of the company.

CHAPTER 2

Reflection-For-Action: A Collaborative Approach to the Design of a Future Factory

Åsa Wikberg Nilsson, Stina Johansson

Department of Human Work Science
Luleå University of Technology
Sweden

ABSTRACT

This paper describes a collaborative process of designing a concept of a future factory, in a three-year research project called "the Future Factory". In the paper we aim at presenting our approach and some results from the same. We argue for a resumed discussion on future work organizations, and do so through the project aim of designing a conceptual future factory. The Future Factory project may be considered somewhat different from traditional research projects, since it includes a design team consisting entirely of women. The reason for this is that Swedish manufacturing industry is heavily male-dominated in numbers, and solutions based only on their ideas thus risk being incomplete or unilateral. The approach of the project encompasses multiple perspectives of values of industrial organizations. The main method has been participatory workshops sessions, in which we have explored contents of future factories together with our stakeholders. Our joint proposal includes aspects of gender awareness, participation in change processes, control of work situations; continuous learning and development in order to become both an attractive and efficient future factory.

Keywords: Participatory design, Reflection-for-action, the Future factory, Collaborative approach

INTRODUCTION

There is a reason to argue that something needs to be done to resume a discussion on future management and work organizations (Isaksson, 2008). This paper describes the methodological approach of our contribution to such a discussion; a proposal of a future factory designed by both employers and employees and other stakeholders. The approach is applied in an interdisciplinary three-year project, called "the Future Factory". This paper addresses the contributions from the field of design to the project approach, including workshop sessions that are inquiries into a future situation of use. The inquiry implies a search of participants' understanding of values and needs. The objective of the approach is that ideas grounded in the participants own life- and work contexts, provides better conditions for an outcome that will fit in its context.

We draw on a tradition of a Scandinavian human-centered approach, initially focused on people's use of technology but that today has a broadened scope of everyday life issues (Westerlund, 2009). One example from the early 80s is the UTOPIA project that dealt with design of, and training in, technology and work organization (e.g. Ehn, 1988; Bødker et al., 2000; Sundblad, 2000). An additional example is Volvo's development of the Swedish Kalmar and Uddevalla plant (e.g. Sandberg, 2007).

Within the automotive industry it is common to develop a concept car, as a way of marketing business and test new ideas. Such a concept car is seldom produced (at least without changes), the purpose being to draw attention to the brand. The idea behind the Future Factory project is based on similar ideas; to develop a concept factory in order to draw attention to, not a particular brand but, to industrial organizations and their design.

THE FUTURE FACTORY PROJECT

The Future Factory project is carried out by an interdisciplinary research group at the department of Human Work Science, Luleå University of Technology, Sweden. The team consists of seven researchers representing expertise in the areas of production systems, systems design, gender, organizational design, ergonomics, and industrial design. The practical aspect of the project is to explore the possibility to design a sustainable and at the same time efficient concept of a "factory". The overall purpose of the project is to explore new approaches to change management and production design as well as to develop theories of organizational design in an industrial context. We base our approach on the notion that change management and developmental work must continuously be upgraded as the social context and the conditions of production change. Furthermore, from our perspective there is an emerging trend to encompass multiple perspectives within the field of ergonomics, (e.g. Charytonowicz, 2009) something we find to be at least to some extent addressed in this project.

The Future Factory is a triennial project expected to be completed by the end of 2010 and that is performed in three phases; 1) knowledge overview and mapping of relevant areas such as inter alia contemporary manufacturing industry, new trends within production systems, organizational design, and change management etc., 2) exploratory study of social and organizational aspects together with the interest groups, and 3) design work, which includes working with a design team consisting of participating women production engineers, systems designers, human resource managers, CEO's, industrial designers, architects, students, and researchers.

AIM

The specific aim of this paper is to describe our collaborative approach based on a design team consisting entirely of women and to present some of the outcomes that this has resulted in. Moreover, from our perspective there is a difficulty in transferring research discussions on work organizations to the people that have the ability to implement them in their organizations. Our approach includes participating stakeholders, and we thus want to contribute to transferring knowledge on these issues. However, we also see the participants' knowledge as something that will contribute to refining our approach and the methods used. The expected outcome of the project is thus a proposal of a future factory that includes needs and preferences of *both* women and men needs. Therefore, there is no particular problem, solution, organizational design or production system to explore. Research projects can explore how things are; the Future Factory project instead aims at exploring how things might be.

OUR APPROACH

A design process can be performed in many different ways. The design process of the Future factory project focuses on a future vision of a factory, including ideas on organizational design, production systems, and information technologies. In extent research there is claimed that this makes the process more complex than problem-solving, meaning that requirements cannot be fully stated beforehand (Westerlund, 2005). Furthermore, such problems are proposed as "wicked"; as problems that cannot be definitively described and that involves aspects that cannot be solved neither correctly nor falsely (Rittel & Webber, 1973). As a result, our driver to start a design process has not been problems, but instead to develop a future factory based on new perspectives, possibilities and visions.

A design process is often described as a linear process starting with a problem and ending with a solution to that problem. Such a process is often based on various disciplines contributing with their expertise at different stages during the road. For example, the marketing department performs a mapping of needs and then forwards them to a designer. However, Westerlund suggests that this is a waste of understanding since not all relevant knowledge can be represented this way. In contrast, there is an emerging trend to engage people from different disciplines in the design process, an approach also applied in the Future Factory project. Henderson (2005) suggests the approach of interacting needs of technology, user value, business value, and strategic value. Moreover, he emphasizes the importance of addressing all these values at the same time.

Given that we in this project are trying to include all dimensions, we consider it important to address all these values in our activities as well. Our solution to this has been to form a multidisciplinary design team, as previously mentioned composed of production engineers, systems designers, human resource managers, CEO's, industrial designers, architects, students and researchers. What makes this project different, and from our perspective exciting, is the fact that we have chosen to work with a design team consisting exclusively of women. The reason for this is due to a problem faced by Swedish manufacturing industry; e.g. the dilemma of how to attract young people and women. Thus we want to explore a possible constitution of a future factory particularly with these stakeholders. In addition to this, Swedish industry as a whole is heavily male dominated in numbers, meaning that women are excluded from organizational design processes. Solutions based solely on men's ideas and visions are not general and their solutions often are not all inclusive, and thus risk being incomplete or limited.

The expected outcome of the design process is a proposal for a future factory, i.e. a visualized model of a future factory including ideas based on women's preferences for a; production systems, organizational design, and change management.

Krippendorff (1995) argues for a development process that has a network of stakeholders, since he believes design to be a social process that relies on stakeholders with different and potentially conflicting interests. An approach based on this perspective is sometimes described as an inquiry into a future situation of use (Gedenryd, 1998). From our perspective such an inquiry builds on people's own experiences and provides resources for them to be able to act. For us, such an inquiry includes an understanding of the situated work practice and human factors as a ground for technology and system development, as suggested by e.g. Heath and Luff (1991), Suchman (1997) and Bannon (1998). A proposal is that if stakeholders are involved in the process, it ensures that its outcome will be better received and will be better adapted to their needs and preferences (e.g. Saschkin, 1984; Müller & Kuhn, 1993; Bødker et al., 2000). By reflecting on how manufacturing industries conceive the concepts of efficiency and attractiveness together with our participants, we therefore believe that it is possible to design a proposal for a future factory that benefits both organizational and individual perspectives. This means that the approach of the project builds on a model of design based on the notion that the interest is in how things out to be rather than how things are (Simon, 1969).

REFLECTION

The concept of reflection was introduced by Dewey (1998/1933) and further developed by Schön (1995), who initiated reflection-in-action and reflection-on-action. The latter includes a process of identifying and developing knowledge of a particular context or workplace and the actions taken there. The process of reflection is according to this concept, claimed to deal with learning and thinking, since we are said to reflect in order to learn and learn as a result of reflection. Ghaye (2007) includes the analysis of what people do in the concept and consider it dealing with important work-placed based and profession-specific knowledge. Furthermore, he adds *reflection-for-action*, occurring for a reason, a particular purpose, and including elements of thinking of the practice in order to improve it, which is: plan for action. As mentioned, in this project there is no particular practice or context to reflect upon, however, there is a particular purpose upon which we might reflect, i.e. the design of a future factory. It is, for example, rare in Swedish industries to include all groups of employees in change projects (Bellgran & Safsten, 2005), even though it is described as a successful approach that would contribute to improvements (Vink, Imada & Zink, 2008). This might result in changes that are not based on the knowledge and experience of the organizations members, which might lead to restorative actions (e.g. Abrahamsson, 2002; Charytonowicz, 2009) and/or outcomes that do not work in their contexts (Ullmark, 1996).

To reflect is also described as looking forward into *what we want to achieve* (Ghaye). Such future scenario-making is also proposed as a more solution-oriented back-casting approach (Edeholt). The assumption behind this is that when thinking of a practice in order to improve it, a plan for action takes place. If planning is based on a future practice, *what might be*, and thinking of actions that have to take place in order to implement the solutions, the solution-driven back-casting approach is used.

PARTICIPATION

Participatory design is defined as researchers and people at stake working together to define project goals and design new technologies, attending to implications for new ways of working (Blomberg, 1999). Participation is proposed as both a mean and as an end; as a mean to achieve, maintain or develop a purpose (e.g. Sashkin), or as an ideological manifest to achieve a democratic, progressive, and humanistic approach (e.g. Ehn). Already mentioned is the idea that participation better ensures a satisfactory outcome of the process. Additionally, participation is in itself claimed to increase the chance of successful implementation, regarding both productivity and job satisfaction (e.g. Vink et al.).

However, participation is also said to engage people in reflecting on their own knowledge, their understandings, skills and values, as well as the ways they interpret themselves and their actions (Kemmis & McTaggart, 2000). Since there is no particular context or practice to change in this project, this is what we emphasize.

METHOD

As previously mentioned, we draw on a human-centered approach,involving an understanding of the world as not true, fixed or stable, but instead socially constructed in collaboration with others , as suggested by inter alia Krippendorff (2006). As a basis for the project, we therefore performed a comprehensive survey of Swedish manufacturing industries including collaborative inquiries within the *research* team. The inquiries within the research team help us understand the issues we are dealing with, the stakeholders we are concerned with, as well as help us discover possibilities through communicating different perspectives on the issues in an interdisciplinary group.

We are especially dedicated to highlighting and developing gender awareness in the social and organizational design, which is why our *design* team consists entirely of women. However, in the beginning of the project we performed activities with young people, trade unions, industrial employers and employees in various companies to explore meaning and context with them. In total, about 140 people have participated so far in the project (see the contribution of Johansson & Wikberg Nilsson to this conference).

These initial activities became the basis for developing a number of personas (see Wikberg Nilsson, Fältholm & Abrahamsson, 2009). Personas are fictive characters used to describe and communicate stakeholders to others (Cooper, 1999). The aim of using personas is to overcome difficulties in communicating and understanding meaning and needs. We have also found inspiration in the Critical Incident Technique (CIT), as proposed by Flanagan (1954).The main method, though, has been participatory workshops, with the design team, supported by techniques as personas, scenarios, and CIT.

PARTICIPATORY WORKSHOPS

In the project, rather than workplace actions, we explore the participants' knowledge through participatory workshops. Such workshops are proposed to support co-operative learning and understanding of what meaning is to the participants (Westerlund).

In the approach to our workshops we draw on inspiration from the Future Workshop technique as introduced by Jungk (1989). The aim of such a workshop is suggested to support participants in a change process to reflect on the past and present, future possibilities and visions, and plan for action. The future workshop approach is facilitated by specific rules of communication, claimed to create a dialogue on equal terms (ibid.).The proposed structure consists of three phases of reflection on the present in 1) *the critique phase*, creating visions of the future in 2) the *utopia phase,* and finally the discussion of how to move from the present to the desired future in 3) the *implementation phase*. Such workshops are described to nurture shared understanding, compromises and shared commitment (Muller, 2003).

Furthermore, the proposition is that the setting of a workshop ideally should promote each actor to evolve as a function of other actors' behaviors in order for a synergetic interaction to occur (Latour, 1996).

An important aspect of a workshop is of course the participants. Westerlund demonstrates that participants that act only as representatives for stakeholders do not commit to the issue. For that reason, the suggestion is that the participants *should have personal experience and knowledge that they contribute within the workshop and thereby participate in the proposals* (p. 73). In our activities, this has meant that each participant represents her/his own point of view, in contrast to representing "all engineers", or their organizations perspectives, or something else.

We have held most of our workshops in our DesignLab, where there is material and space for a lot of people. So far, we have carried out four workshops with the design team, varying between eight and seventeen participants at a time.

It is alleged that the discussions during a workshop should not be understood as represented in language, but instead constituted *through* the dialogue (Westerlund). From our understanding this means to carefully explore the dialogues in terms of meaning, both with the participants and within the research team after each activity.

DISCUSSION

In this paper we have presented parts of our methodological approach, its background, and incentives. In this final discussion, we take up a few of the aspects of a future factory we believe this approach have contributed to. As mentioned, our task in the project is to design a proposal for a future factory that includes preferences and needs of both women and men. For that reason we involved different stakeholders in the pre-study and introduced this initial mapping of needs to the Co-design team of women for further development. It is however important to note, that the aim of the project, and thereby its solutions, were never intended to be solely directed at women. Instead the aim of the design team of women is to include their ideas and visions into a further discussion of design of industrial organizations.

In our activities, we have aimed at *encompassing multiple values*, introduced as concepts of attractive and efficient factories, and explored what these might consist of; i.e. what is an attractive workplace, and what is needed to make an attractive workplace also competitive and effective.

The collaborative approach of this project has turned out as also part of the solution proposed for a future factory. Thus, there should be a broad participation of all stakeholders that might contribute to the process of change. *Gender awareness* in recruitment, workplace culture, group interactions and participation is something we jointly emphasize as well. Several of our participants have told us about incidents when they felt discriminated on the basis of their gender. Moreover, we see the synergetic effects of addressing all resources simultaneously; people and technology as well as business and strategic values. Based on our understanding, if the participants find the process meaningful, they are more likely to adapt and agree upon the outcome of the process. Additionally, we find the inquiry into the future, as a *positive proactive approach* that we would like to propose as a solution for the future factory's management as well.

Nevertheless, we also see a need of *continuous learning*, including *learning to participate*, because many of our stakeholders have not been accustomed to such processes. Based on our inquiries, our participants have found it difficult to address all needs at the same time. However, this may also be due to the project task being "conceptual", and thus not having any real context to change or improve.

Furthermore, a design of a future factory should consider the incentives for people to work there, and their needs and preferences for *confirmation, diversity, control of work tasks* and *participation in change processes*. We understand incentives to be different for different people, so diversity should be given space and be affirmed. Furthermore, our participants have recognized that atmosphere and how people interact with each other are of great importance. Thus, we propose for a future factory to embrace a *positive workplace culture*, also recognized within management literature as a driving force for development.

FUTURE WORK

In the project, there are still some activities to be completed, including finalizing a model, which is intended to address all aspects found as a proposal for a future factory. Our intention is then to use this model as a basis for further discussion with our stakeholders about future industrial organizations. The Future Factory project will not have a direct impact on Swedish manufacturing industry, but we hope to contribute to a resumed discussion on future management and work organizations and a growing recognition of the need to consider multiple perspectives in the development and design of the same.

ACKNOWLEDGMENTS

We would like to express our gratitude to all those who has participated in the activities within the project and have contributed with their experience and expertise, without them the project could not have been conducted. The project is funded by AFA Insurance, FAS (Swedish Council for Working Life and Social Research) and the European Regional Development Fund.

18

REFERENCES

Abrahamsson, L. (2002) "Restoring the order: gender segregation as an obstacle to organisational development". *Applied Ergonomics 33 (2002) pp. 549-557*

Bannon, L. (1998) "CSCW: Towards a Social Ergonomics?" *Proceedings of RTO HFM Symposium on "Collaborative Crew Performance in Complex Operational Systems". Edinburgh, UK*

Bellgran, M. & Safsten, K. (2005) *Produktionsutveckling: utveckling och drift av produktionssystem.* (In Swedish). Lund: Studentlitteratur

Blomberg, J. (1999) On Participation and Service Innovation. In T. Binder, J. Löwgren, & L. Malmborg's (Eds) *(Re) Searching the Digital Bauhaus.* London: Springer-Verlag

Bødker, S. Ehn, P. Sjogren, D, & Sundblad, Y. (2000) "Co-operative Design – perspectives on 20 years with the"Scandinavian IT Design Model"". *Proceedings of NordCHI 2000*

Charytonowicz, J. (2009) Evolutionary Changes in the Traditional Ergonomics. In C. Stephanidis (Ed) *Universal Access in HCI, Part 1, HCII 2009, LNCS 5614, pp. 450-459.* Berlin, Heidelberg: Springer-Verlag

Cooper, A. (1999) *The Inmates Are Running the Asylum: Why High-Tech Products Drive Us Crazy and How to Restore the Sanity.* Indianapolis: SAMS

Dewey. J. (1998/1933) *How we think: a restatement of the relation of reflective thinking to the educative process.* Boston: Houghton Mifflin

Ehn, P. (1988) Work-oriented Design of Computer Artifacts. Stockholm: Arbetslivscentrum

Edeholt, H. (2007) Design och innovationer. In S. Ilstedt-Hjelm (Ed) *Under Ytan: En antologi om designforskning. (In Swedish), pp. 222-235* Stockholm: Raster förlag

Flanagan (1954) "The Critical Incident Technique". *Psychological Bulletin, Vol. 51, No. 4, July 1954, pp. 327-357*

Gedenryd, H. (1998) *How Designers Work: Making Sense of Authentic Cognitive Activities.* Diss. Lund: Lund University

Ghaye, T. (2007) *Building the Reflective Health Care Organization.* Oxford: Blackwell Publishing

Heath, C. & Luff, P. (1991) "Work, Interaction and Technology: Empirical Studies of Social Ergonomics". *Proceedings of the 11th International Conference on Ergonomics. Paris, France*

Henderson, A. (2005) "The Innovation Pipeline. Design Collaborations between Research and Development". *Interactions, January + February 2005*

Isaksson, P. (2008) *Leading Companies in a Global Age – Managing the Swedish Way.* VINNOVA Report VR 2008:14

Jungk, R. (1989) *Håndbog i fremtidsværksteder.* (2nd Ed.) (In Danish). Kobenhaven: Politisk revy

Kemmis, S. & McTaggart, R. (2000) Participatory Action Research. In Denzin, K. & Lincoln, Y.S. (Eds) *The Handbook of Qualitative Research* (2nd ed) Sage Publication, Inc

Krippendorff, K. (1995) Redesigning design: An Invitation to a Responsible Future. In P. Tahkokallio & S. Vihma's (Eds) *Design: Pleasure or Responsibility?*. Helsinki: University of Art and Design

Krippendorff, K. (2006) *The Semantic Turn: A New Foundation for Design*. Boca Raton, FL: Taylor & Francis

Latour, B. (1996)" On Interobjectivity". *Mind, Culture, and Activity, Vol. 3, No. 4 (1996), pp. 228-245*

Muller, M. J. & Kuhn, S. (1993) "Participatory Design". *Communication of the ACM, Vol. 36(6), pp 24-28*

Muller, M. J. (2003) Participatory Design: The Third Space in HCI. In J. A. Jacko & A. Sears' (Eds) *The Human-Computer Interaction Handbook: Fundamentals, Evolving Technologies, and Emerging Applications.* Mahwah, NJ: Lawrence Erlbaum Associates Inc. p. 1051-1065

Rittel, H. W. J. & Webber, M. M. (1973) "Dilemmas in General Theory of Planning". *Policy Science 4(1973), pp. 155-169*

Sandberg, Å. (2007/1995) *Enriching Production: Perspectives on Volvo´s Uddevalla plant as an alternative to lean production.* National Institute for working life and KTH Royal Institute of Technology: Stockholm

Saschkin, M. (1984) "Participative Management Is an Ethical Imperative". *Organizational Dynamics, Vol. 12 (4), Spring 1986, pp. 62-75*

Schön, D. A. (1995) *The Reflective Practitioner: How Professionals Think in Action.* New Ed. Aldershot: Arena

Simon, H. A. (1969) *The Science of the Artificial.* Cambridge, Mass.: MIT Press

Suchman, L. (1997) Centers of Coordination: A Case and some Themes. In L. B. Resnick, R. Säljö, C. Pontecorvo, & B. Burge's (Eds) *Discourse, Tools, and Reasoning: Essays on Situated Cognition.* Berlin: Springer-Verlag

Sundblad, Y. (2000) From Utopia 1981 to Utopia 2008. In T. Binder, J. Löwgren & L. Malmborg's (Eds) *(Re)Searching the Digital Bauhaus, pp. 13-42.* London: Springer-Verlag Ltd

Ullmark, P. (1996) *Förändringsarbete. Erfarenheter från att utveckla system och organisation.* (In Swedish). Stockholm: Närings- och teknikutvecklingsverket (NUTEK)

Vink, P., Imada, A. S. & Zink, K. J. (2008) "Defining stakeholder involvement in participatory design processes". *Applied Ergonomics, 39 (2008), pp. 519-526*

Westerlund, B. (2005) "Creating shared experiences and aims among stakeholders in a design process". *In proceedings of Design for Entrepreneurship – Design for Innovation Conference, Växjö University, October 20-21 2005*

Wikberg Nilsson, Å., Fältholm, Y. & Abrahamsson, L. (2009) "The Future Factory – a concept designed by women and young people". *In proceedings of 17th World Congress on Ergonomics IEA 09, Beijing China*

Westerlund, B. (2009) *Design Space Exploration. Co-operative creation of proposals for desired interactions with future artefacts.* Diss. Stockholm: Royal Institute of Technology

Collaboration Between Design and Production by Exploiting VR/AR and PLM - Case Metso Minerals

S-P. Leino[1], M. Vehviläinen[2], S. Kiviranta[1], A. Mäkiranta[2], P. Nuutinen[2], I. Hokkanen[2], P. Multanen[2], P. Rantanen[3], J. Heikkilä[3], T. Martikainen[3]

[1]VTT Technical Research Centre of Finland
[2]Tampere University of Technology
[3]Metso Minerals

ABSTRACT

Rock crushing equipment is used in a very demanding environment. Thereby preventive and corrective maintenance work is needed. In practice all maintenance tasks are performed manually by operators. Due to heavy competition within crushing equipment industry the manufacturing strategy is getting extremely important for the companies. The real challenge is the combination: How to design products so that they can be manufactured in multiple factories, that they are suitable for mass-customization and are easy to assemble and maintain manually? Design for assembly and design for maintenance requirements conflict easily. Ergonomics of manual manufacture and maintenance should be optimized, because ergonomic and safe work is normally more productive as well. Product lifecycle management (PLM), augmented and virtual reality are potential solutions for combining conflicting requirements and optimizing the wholeness. This paper introduces preliminary results of industrial research case pre-study where goal is to investigate technology maturity and benefits of AR and VR for industrial use, and to develop PLM methods and processes in order to utilize this novel technology in collaboration between design and production and supporting for manual work.

Keywords: Augmented Reality, Virtual Reality, Product Lifecycle Management

INTRODUCTION

Many requirements coming from different product lifecycle stages to machine system design conflict easily. Product lifecycle management (PLM), augmented and virtual reality are seen as potential solutions for combining conflicting requirements and optimizing the wholeness. An EU-funded research project called ManuVAR was launched in 2009 with target on developing manual work support through system lifecycle by exploiting virtual and augmented reality.

VR/AR AND PLM

CAD software is usually ideal for mechanical design but it enables only one way interaction. VR is an artificial user environment based on 3D graphics and interaction techniques. VR tools allow the setting up of an immersive virtual environment (VE), which can be defined as a plausible artificial environment created by technical means, and allowing two way interactions between user and environment. A virtual environment system consists of software, virtual models, data communication, computers, display devices, user interface devices, and other devices and their drivers. (Mäkiranta 2009)

AR combines the virtual models or other data to the view of real world. The projection system can be e.g. a display of a PDA device or see-through glasses. Low-cost AR applications can be built by using a PC screen, a webcam and a tag. AR can be considered a suitable technology for inspections and manual work guidance whereas VR seems to be promising technology also for manual task training.

Significant advantage in using of VE is its flexibility. VE can be a basis for different kind of scenarios which require active presence of the user. Nevertheless, the cost-efficiency and easy usability of these tools is naturally still a key issue to companies. The use of the 3D CAD models in VE may require making the models lighter and converting the models to correct format. Also the data flow backwards from virtual simulation models to CAD engineering is difficult to carry out.

Product Lifecycle Management (PLM) is an integrated, holistic information-driven business approach comprised of people, processes/practices, and technology to all aspects of product's life, from its design through manufacture, deployment and maintenance. Components of PLM include the products themselves, organizational structure, working methods, processes, people, information systems and product data. (Grieves 2006, Stark 2006)

PLM is a paradigm, which can be seen as an enabler of efficient collaboration and communication between actors (e.g. engineers, managers, human factors experts and workers) in complex business environment. Virtual engineering includes set of novel technology tools of PLM: CAD for producing, VR for analyzing, AR for utilizing and PDM for managing product definition data. However, PLM implementation needs practical guidelines how for instance human factors, manual work design and VE can be integrated into company processes.

EU-MANUVAR RESEARCH PROJECT

ManuVAR "Manual work support through system lifecycle by exploiting virtual and augmented reality" is a three year European Commission project funded under the Seventh Framework (CP-IP-211-548). The ManuVAR consortium consists of 18 partners from 8 European countries representing industry, research and academia. The high share of companies in the consortium indicates that industrial cases are a key strength of the ManuVAR project. These cases are developed by clusters, one case each. The clusters are formed around partner companies with the links to the respective research institutions. (Krassi et al. 2010A)

ManuVAR Cluster 5 is built around industrial partner Metso Minerals. The main research partners in the cluster are VTT and TUT, with role of developing tools, methods and processess and providing VR/AR laboratory facilities for the study. Metso Minerals is the leading global supplier of equipment, service and process solutions (Figure 1.) to industries including among others mining and minerals processing. Rock crushing, screening and conveying equipment is used in a very demanding environment. Thereby both preventive and corrective maintenance work is needed. In practice all maintenance tasks are performed manually by one or two operators.

Figure 1. Example of quarrying application (Lokotrack – LT160) and virtual model.

Due to heavy competition within crushing equipment industry the manufacturing strategy is getting extremely important for companies. The challenge is the combination: How to design products so that they can be manufactured in multiple factories, they are suitable for mass-customization and are easy to maintain? Design for assembly and design for maintenance requirement conflicts easily.

The overall target in Metso is to improve time to market and reduce the amount of engineering changes. Business requirements and performance measures are listed in Table 1. The focus of this cluster is on developing tools, methods and processes for product development and manual assembly engineering, and for evaluation and documentation of manual maintenance tasks. The targets are on helping manual workers, designers and engineers by exploiting VR and AR.

Table 1. Business requirements and performance measures.

Business requirements	Performance measures
1.Improved collaboration between product development and manufacturing (multi-site)	
2.Improved requirement management (Design for manual assembly and maintenance)	1-5% saving in design and manufacturing costs
3.Improved product features for manual assembly and maintenance tasks	
4.Cost effective verification and documentation process for maintenance tasks	1-2% better machine utilization rate
5.Virtual techniques are used in new product productization	

Hypothesis: Benefits gained by using VR, AR and PLM include:

- Improved communication and collaboration between product development and manufacturing engineering (multi site engineering & manufacturing),
- Improved human requirement management: Design for manual work,
- Better ergonomics and safety,
- Cost effective verification and documentation process for manual work tasks,
- Quality: Less errors, proactive repairing,
- Productivity: 3D models are used more efficiently
- Faster learning and assembling mass-customized products becomes easier

Goal of the pre-study, which is the first phase of Metso case implementation. is to test the hypothesis with preliminary ideas, technology elements and methods

MATERIAL AND METHODS

Work on this research study was based on concepts and methods developed in finished VIRVO project and on-going ManuVAR project. The concepts and methods are utilized, tested and further developed in cluster case studies.

VIRVO AND MANUVAR CONCEPTS

A previous Finnish research project (VIRVO), ended in 2009 and resulted in concept and methods for effective managing of maintainability of machines with the aid of virtual environments and virtual tools. The idea behind the method was to design, test, and evaluate maintainability tasks concurrently with machine design. The analysis process of VIRVO concept consists of the following seven tasks: Initial data collection, task modeling, virtual modeling, task planning, plan inspection, task training, and documentation. The developed concept and methods were tested and redefined in two industrial case studies: Rock crusher maintainability design and elevator renovation planning. (Mäkiranta 2009, Lind et al. 2009)

ManuVAR PLM model is an adaptation of generic PLM paradigm. The model adaptation aims to close common gaps identified in manual work management during system lifecycle. The model consists of methodological and technological elements (hardware and software), application tools, and in the core of ManuVAR PLM model there is Virtual Model which contains all information related to system definition. The model aims to enable bi-directional information flow between system lifecycle stages. (Krassi et al. 2010B)

CASE STUDIES

The implementation plan can be divided into four main phases. Results in this paper are derived from the first phase "Pre-study":

Phase 1: **Pre-study**
- Interviews of problems, requirements, needs: Engineers, workers, management
- Analysis of existing processes, industrial benchmarking
- Technology maturity tests: AR, VR, Digital Human Model, PDM/PLM link
- Setting business cases: Case definitions, assessment of potential and benefits
- Feedback from potential users: Review meetings, technology demonstrations

Phase 2: Definitions for new processes and methods
Phase 3: Execution of case study (testing and fine tuning);
Phase 4: Reporting of results and action plan for future development

Implementation plan with relation to research strategy is presented in Figure 2

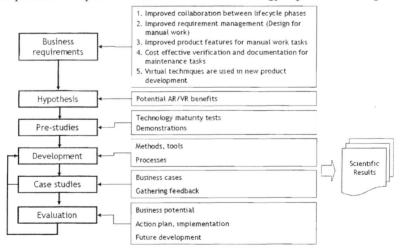

Figure 2. Research strategy.

The implementation of case studies is carried out in close co-operation between Metso and research partners. Cluster specific collaboration processes and data flows between CAD and VR/AR and other systems are defined together and tested mainly in VTT's VR-laboratory. Processes include for instance data flow from CAD to VE review meetings, and further cumulated data and information to AR-application. For detailed explanation of VE utilization concept, see Kiviranta et al. 2010.

Defined business cases

Several business cases have been identified in Cluster 5, two technology demonstrations (technology maturity tests) and one actual case study. The technology demonstrations (TD) are:

TD1) Kiosk-PC in productization: Kiosk-PC is a worker aid system which is currently used in Metso's productization (i.e. prototyping, preparation for production) department. Workers use Kiosk-PC to get working instructions for manual assembly work. Instructions include conventional drawings, but also light 3D-models and animations.

In this technology demonstration case, goal is to develop better practices for producing material for Kiosk-PC from design CAD-models and giving feedback from productization to design and engineering (PDM link). Another goal is to develop better user interface for the Kiosk-PC.

TD2) Augmented Reality demonstration: In this demonstration, an AR aided manual assembly cell of a hydraulics block on factory floor will be built. Goal is to test technology maturity of AR, gather users comments, feedback and commitment about this novel technology. Also needs for further development and most potential fields of exploitation will be studied

Actual case study is identified on higher level, and it will be defined more precisely after technology demonstrations. Case study (CS) is:

CS1) Product development project and production ramp-up: Goal is to develop methods and processes for producing 3D models and other material to Virtual and Augmented Reality and Kiosk-PC models as well as development of PDM (product data management) and PLM (product lifecycle management) links. In addition a new generation engine module will be virtually verified for serial production and new production line designed by utilizing these tools.

Developed systems will be utilized in virtual prototyping, design reviewing, manual work, designing and simulating new production line and also in ergonomics, safety and risk analyses. Potential technologies are Virtual Reality, Augmented Reality, Physics Simulation, Virtual Model, Motion Capture, Head Mounted Displays, CAD, PDM, PLM, Haptic Devices.

Metso Engine module review meeting and interviews

The review, which is pre-study task of first phase case implementation, was the first organized meeting for the new engine module concerning its manufacturability and maintainability. The meeting was arranged at the VTT's Virtual Reality laboratory premises at Tampere. The goal was to introduce VR/AR technologies to people from Metso and use virtual engineering technologies in order to review aspects mentioned earlier. It is remarkable, that the virtual environment used in the meeting, did not include all possible functionalities and user interfaces (like motion tracking and haptics), but they were explained for the participants. This was because purpose of meeting and interview was to get feedback and development ideas before implementations. VR setup consisted of interactive, functional Virtools simulation model and mono/stereo visualisation on a large screen. Participants were representatives from production, engineering, productization, maintenance, product management and project management. This ensured the comprehensive feedback from all the stakeholders involved in the new engine module development project.

The actual agenda of the review started with an introduction to VR followed by a presentation of ManuVAR project. After that participants were prepared for the actual review by showing them a questionary about aspects they were expected to pay attention during the review. The same questions formed the base for the individual feedback interviews held afterwards at Metso's premises.

The review was carried out by showing the 3D-model of the engine module on a big screen and rotating it around while people located and discussed possible problems. Virtools, the program used to create the view, also had a feature which made possible to remove all the access doors and even cover plates enabling the clear view deeper into the module. In addition, the participants were offered a chance to inspect a stereoscopic view of the module through stereo shutter goggles at the end of the review.

During the following days, all the participants were individually interviewed based on the questionary. The questionary included questions about their experiences concerning the virtual review and their thoughts of when, where and how this kind of review event should be organized. To be more specific the participants were asked about the state of the immersion experienced, interaction opportunities and handling of the virtual model. In addition they were asked to tell in what time of the product lifecycle they see fit to arrange virtual reviews, who should participate to these events and which matters should be reviewed. As the last thing participants were also asked to define prerequisites that need to be met before reviews of this nature can be permanently implemented into processes at Metso.

RESULTS

Results of this paper were made in several Metso Minerals case studies in two different research projects: ManuVAR and VIRVO. In ManuVAR project, the results are derived from the first pre-study phase of the Metso case implementation.

IDENTIFIED DEVELOPMENT TARGETS

In ManuVAR project seven gaps or common problems in system lifecycle management have been identified. In Table 2, those common gaps and specific gaps identified in prior of ManuVAR and VIRVO interviews in Metso, as well as expected solutions for closing those gaps are listed.

Table 2. Expected solutions to gaps in cluster 5.

ManuVAR Gaps	Gaps in Metso	Expected Solutions
1.Problems with communication throughout lifecycle	• Lack of feedback from production to design • No systematic approach for requirements management	• ManuVAR PLM model • Virtual and Augmented Reality
2.Poor user interfaces	• Instructions not up to date • No connection to PDM • Poor view to product data	• Improved Kiosk- PC (3D) • Virtual and Augmented Reality
3.Lack of technology acceptance	• Instructions in English • VR is considered as high tech, slow and expensive	• Animation • Training
4.Inefficient knowledge management	• Feedback is not gathered and managed systematically • Systems do not communicate	• PDM and PLM
5.Physical and cognitive stresses	• Postural loads, visibility, safety, heaviness of work • Workers have to design the equipment • Mental workload, complexity of tasks	• Posture analyses • Digital human models
6.Inflexible design process	• Design changes are difficult because a lot of people are involved • Errors are not identified in design phase	• PDM • PLM models
7.Low productivity	• Mechanical engineering is mostly in 3D-CAD but the information is not used efficiently enough • Searching and walking for components	• VR, AR, PLM • Supply chain management • Kiosk-PC and ERP integrations

FEEDBACK FROM REVIEW MEETING

The most of the feedback gathered from the interviews was rather positive. Many of the participants saw the potential of the VR technologies and generated their own visions about exploiting VR and AR in future. Following things were reported to be particularly beneficial when using VR and AR:

- VR and AR used in participatory design improves collaboration between company departments, value network actors and lifecycle phases
- Increased physical interaction with the 3D-model
- Using motion tracking system to capture workers movements while doing the manual work, and utilizing the data in ergonomics analyses
- Utilizing the collected data in generation of work instruction
- Increased productivity in terms of decreasing the amount of design failures and assembly time

DISCUSSION AND CONCLUSIONS

This paper introduces preliminary results of industrial Metso research case where goal is to study technology maturity and benefits of AR and VR for industrial use, and to develop PLM methods and processes in order to utilize this novel technology in collaboration between design and production and supporting for manual work. In Metso case identified problems were mapped to common ManuVAR gaps and our hypothesis is that solutions provided by ManuVAR platform can be utilized in closing those gaps. In the pre-study phase of the Metso case implementation, the first concepts and ideas were tested in a review meeting and interviews, where assembly and maintenance workers, design engineers, product managers and project managers had possibility to give feedback about VR and AR tools.

According to our hypothesis, utilization of VR and AR improves communication and collaboration between lifecycle stages. It also enables: Better human requirements management, better safety and ergonomics, cost effective verification and documentation process, and increased productivity. Results of the pre-study review and interviews show that our hypothesis is in right direction, but lot of further development work is still needed.

The most of the feedback gathered from the review meeting and interviews were rather positive. Basically, all gaps could have positive effect when utilizing AR and VR. Many of the participants saw the potential of the VR technologies and generated their own visions about exploiting VR and AR in future. On the negative side, the common opinion was that the review without the stereoscopic view did not different much from techniques used at Metso before. Therefore, in future reviews, it is desirable to exploit all VR/VE technology on hand, such as data gloves and haptic devices, in order of widening the scope of possibilities to participants.

What almost every interviewee told was that increased physical interaction with the model would be highly appreciated. It would enable e.g. worker to test the actual assembly work or maintenance personnel to test if consumption components are easy enough to change or clean, thus finding possible problems at an early stage of product development. Normally this type of tasks can only be done with a physical prototype, which should be then modified. With VR this step could easily be erased, or at least be made less iterative which in addition leads to improved flexibility in design process.

Many saw the possibility of using motion tracking system to capture workers movements while doing the assembly. The collected data would then be generated into an assembly instruction ready to be used by the worker assembling the real life product or an animated instruction for adjusting valves. The same method could be used in several other tasks like ones related to maintenance or customer training. The data could also be utilized in ergonomics analysis. When implemented, all the above would affect also in productivity in terms of increasing it by decreasing the amount of design failures and assembly time.

One desired feature would enable the possibility to import data to virtual model. Especially feedback, like deviations and such, from reviews should be attached within the model and somehow get back to PDM. From there the designer would easily see what needs to be improved. Two workers were a bit skeptic about this feature, because they felt that for the majority, using the computer in such a way could be overwhelming. But this is only a matter of designing the user interface to be easy-to-use and can be solved as such. By achieving this, the rate of feedback from production to design would most likely increase leading to improved quality.

It is safe to say that VR will definitely be welcome to Metso, but it needs to get a bit more mature. In addition some of the other prerequisites for the full-scale adoption stated that the technology should be reasonably priced, locate at Metso's premises and most of all be easy to operate and maintain.

In future work tools, methods and processes will be developed in order to meet the needs of the end users in engineering, production and maintenance. The results from review meeting and interviews assured that basis for the research is solid, because AR and VR technology is accepted and there is lot of expectations about it. The results also assured that the development work is on the right track and lot of good new ideas were also gathered.

ACKNOWLEDGEMENTS

The research leading to these results has received funding from the European Commission's Seventh Framework Programme FP7/2007-2013 under grant agreement 211548 "ManuVAR". The presented results are the outcome of the joint efforts of the ManuVAR and VIRVO projects. The authors are grateful to all contributors in both projects.

REFERENCES

Grieves, M., "Product Lifecycle Management – Driving the next generation of lean thinking", The McCraw-Hill Companies. New York, 2006

Kiviranta S., Poyade M., Helin K., Virtual and Augmented Reality technologies for supporting heterogeneous multidisciplinary end-users along Product Lifecycle. 3rd International Conference on Applied Human Factors and Ergonomics AHFE2010. 17-20 July 2010. Miami Florida USA

Krassi, B. (A), D'Cruz, M., Vink, P. ManuVAR: a framework for improving manual work through virtual and augmented reality.

Krassi, B. (B), Liston, P., Leino, S.-P., Kiviranta, S., Reyes-Lecuona, A., Strauchmann, M., Leva, C., Viitaniemi, J., Sääski, J. ManuVAR PLM model, methodology, architecture, and tools for manual work support throughout system lifecycle.

Lind, S.; Leino, S-P., Multanen P., Mäkiranta, A., Heikkilä, J. A virtual engineering based method for maintainability design. Maintenance Management 2009, MM 2009. Rome, 22 - 24 April 2009. Proceedings of 4th International

Conference in Maintenance and Facility Management. CNIM - Italian National Committee for Maintenance. Rome, IT (2009), 117 - 122

Mäkiranta, A V., Multanen, P., Leino, S-P S. & Lind, S M. 2009. Method and tools for machine maintainability design in VE. The Sixth International Conference on Condition Monitoring and Machinery Failure Prevention Technologies, Dublin, Ireland, 23-25 June 2009 pp. 877-888.

Stark, J., "Product Lifecycle Management – 21st Century Paradigm for Product Realisation", Springer-Verlag, London, 2006.

Chapter 4

Trust Development in Distributed Multi-Cultural Teams: Agent-Based Modeling and Simulation

Yan Liu

Department of Biomedical, Industrial, and Human Factors Engineering
Wright State University
Dayton, OH 45435

ABSTRACT

Teams that span multiple geographic and cultural boundaries have become commonplace in numerous organizations thanks to the competitive advantages they provide in human resources, products, financial means, knowledge sharing and many others. However, research suggests that trust, an important element deemed necessary to facilitate knowledge sharing and collaborative work, is rather difficult to establish in distributed multi-cultural teams.

This study uses agent-based modeling and simulation to investigate what factors and how they can affect the development of interpersonal trust relationships in distributed multi-cultural teams. Simulation experiments suggest that forming mutual trust relationships at an early stage is of particular importance in the development of long-term trust relationships not only in dyadic relationships but also in multiple-member teams. Besides, holding either a low or a high trust threshold can be ineffective in trust development and team performance, depending on the trustworthiness of the trustee agents.

Keywords: Trust, Distributed Multi-Cultural Team, Agent-Based Modeling and Simulation

INTRODUCTION

Teams that span multiple geographic and cultural boundaries have become commonplace in numerous organizations, from commercial corporations (e.g., Boeing) to military operations (e.g., coalition network-centric operations (NCO)) to nonprofit institutions (e.g., International Red Cross). According to Wright and Drewery (2006), diverse and distributed teams are becoming the norm for businesses and governments around the world thanks to the competitive advantages they provide in human resources, products, financial means, knowledge sharing, and many others. The promises of distributed teams, however, are accompanied by a number of challenges. Research suggests that trust, an important element deemed necessary to facilitate knowledge sharing and collaborative work, is rather difficult to establish in distributed teams, especially among those teams that include members with different cultural backgrounds (e.g., Jarvenpaa and Leidner, 1999; Coppola, 2004; Abuelmaati and Rezgui, 2008).

Over the years, the phenomenon of trust has been extensively explored by a variety of disciplines across communication, economics, history, psychology, sociology, and political science. As one can expect with such a diversity of scholarship, researchers have proposed many viewpoints highlighting different aspects of trust. Drawing on elements from the most-quoted definitions of trust, Dietz and Hartog (2006) maintained that trust can be broken down into three constituent parts: trust as a belief, as an intention, and as an action or behavior.

The first form of trust is a subjective, aggregated, and confident set of beliefs about the trustee's trustworthiness. Factors which may influence trust belief can be categorized as characteristics of the trustor, characteristics of the trustee, their relationships, and situational variables. A potentially decisive characteristic of the trustor is his or her propensity to trust toward others in general. Although a variety of trustee attributes have been proposed by different researchers, *ability*, *integrity*, *predictability*, and *benevolence*, or similar concepts, appear the most often in the literature and have been considered as the most salient (Mayer et al. 1995; Bews and Martins, 2002).

In addition to the individual characteristics of the trustor and trustee, some aspects of their relationships are regarded as important in influencing trust beliefs. Similarity with others positively reinforces members' own identities and contributes to their willingness to cooperate. As a result, the larger the team member diversity, the greater the likelihood that more time will be required for team members to form strong bonds (DeSanctis and Poole, 1997). Members in a team can differ at many aspects, such as in their knowledge, expertise, gender, and ethnicity. This research, however, focuses on the effect of individual cultural differences in a team. Besides, situational variables, such as institutional framework, and incentives or stakes involved, are beyond the scope of this study.

The second form of trust is an intention to place trust on the other party, through which the belief in the other party's trustworthiness is manifested. For example, trust was defined as "willingness to render oneself vulnerable" in Mayer et al.

(1995) and Rousseau et al. (1998). This intention, however, does not guarantee actual engagement of trust-informed risk-taking behaviors, the third form of trust. The most often discussed types of risk-taking behaviors in trust literature include suspending control or monitoring over important decisions, disclosing critical information, and delegating to the trustee tasks that help to achieve some of the trustor's goals.

Several attempts have been made to study trust relationships using agent-based modeling, mainly for commercial transaction applications (e.g., Gorobets and Nooteboom, 2006; Chi and De Wilde, 2008). However, the aim of most of these studies is to model intelligent software agents that serve to help people make better decisions. Few studies use agent-based modeling to investigate in interpersonal trust development. In addition, most trust studies focus on dyadic trust with a single trustor and a single trustee, without considering multiple trustors and trustees with different characteristics in teams.

AGENT-BASED MODEL OF TRUST DEVELOPMENT IN DISTRIBUTED MULTI-CULTURAL TEAMS

Figure 1 shows the conceptual diagram of the agent-based model that studies how trust develops in a team of n agents. For each task agent A_i needs to accomplish, it has two alternatives. One is to delegate (a part of) the task to another agent, in which case A_i can acquire the highest gain should the task be completed successfully with the help of the trustee agent. The other is to work on the task by itself, which will lead to less or no gain. To decide which alternative to choose, A_i will first examine whether there is any other agent in the team whose overall trustworthiness value meets its trust threshold for task delegation. If so, it will send a request of task delegation to agent A_j that it trusts the most (i.e., the agent with the highest overall trustworthiness score). If more than one agent rivals in the overall trustworthiness value, then A_i will choose the one whose cultural value is the most similar to its. If, on the other hand, no other agent in the team satisfies A_i's trust threshold, then A_i will no longer delegate tasks to another agent or accept any other agent's request of task delegation. This assumption is made because trust-building within a work relationship is a reciprocal process; it would be very difficult to build trust unilaterally if the other party is deemed untrustworthy or does not bring benefit to its partner (Six, 2005).

If A_i decides to delegate its task to A_j, after A_j receives the task delegation from A_i, it responds to the request and performs the task should it accept the request, according to its trustworthiness characteristics toward A_i. Finally, A_i receives A_j's response to the request and the result of task performance and then updates its trust in A_j's trustworthiness accordingly.

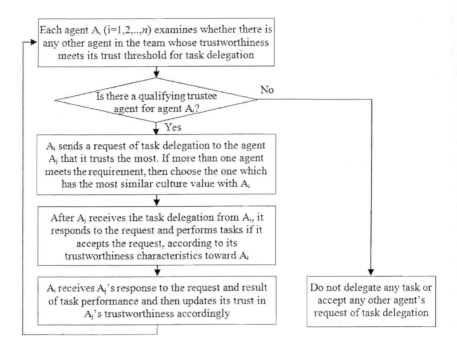

Figure 1. Conceptual Diagram of the Agent-Based Model of Trust Development in a Team of *n* Agent

Figure 2 illustrates how A_i and A_j interact and how A_i's belief of A_j's trustworthiness is updated during their interaction, in which A_i and A_j are the trustor and trustee agents, respectively. The solid arrow lines in the figure represent the interaction flows between A_i and A_j, and the dashed arrow lines represent that the elements in the pointing nodes directly impact the elements in the pointed nodes.

After A_j receives a request of task delegation from A_i, it responds to the request and performs the task according to its intended ability, integrity, and benevolence. For agent A_i, the best outcome is that A_j accepts its request and completes the task successfully, which indicates A_j's positive ability, integrity, and benevolence toward A_i. However, A_j may reject A_i's request because it is unwilling (an indication of negative benevolence) or incapable (an indication of negative ability) to help A_i. Besides, it is also possible that A_j accepts A_i's request but does not complete the task successfully, which is an indication of A_j's negative ability and negative integrity. The values of A_j's intended ability, integrity, and benevolence all range between 0 and 1 in the model.

A_i's experience with A_j's ability, integrity, and benevolence may deviate from A_j's intended values, due to the misunderstanding between A_i and A_j. Therefore, a noise component is introduced to model the variance of A_i's experience with A_j. There are various causes for the misunderstanding between A_i and A_j. This study, however, only considers the impact of cultural difference between A_i and A_j, ΔC_{ij}.

The individual cultural value of an agent in the model ranges between 0 and 1, yet it is the difference between agents' cultural values that, rather than their absolute cultural values, matters in the model. Cultural difference among teammates can negatively affect the quality of their communication and coordination (Liu and Warren, 2009). Nevertheless, more interactions between A_i and A_j may facilitate their understandings of each other and thus mitigate of the negative impact of ΔC_{ij}. The noise is modeled using an exponential decay function, as shown in Equation 1, whose initial quantity is ΔC_{ij}. The decay constant is the inverse of ΔC_{ij} to model the phenomenon that the larger the ΔC_{ij}, the more communication and coordination barriers between A_i and A_j initially and the longer it takes to overcome the barriers.

$$\text{noise} = 0.5 * \Delta C_{ij}\, e^{-0.1/\Delta C_{ij} \cdot t} \tag{1}$$

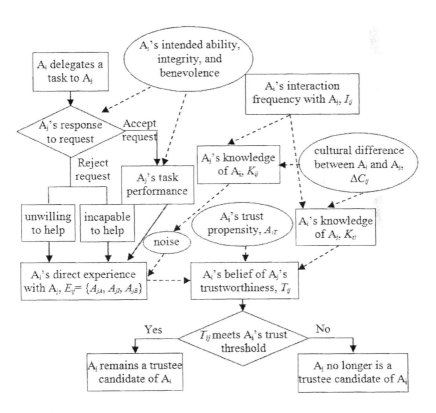

Figure 2. Conceptual Diagram of the Trust-Building Process between Trustor A_i and Trustee A_j through Their Interactions

A$_i$'s belief of A$_j$'s trustworthiness, T_{ij}, is directly affected by three factors: A$_i$'s direct experience with A$_j$ (E_{ij}), A$_i$'s knowledge A$_j$ (K_{ij}) acquired through their interaction history (I_{ij}), and A$_i$'s propensity to trust (A_{iT}). A$_i$'s direct experience with A$_j$ includes three elements, A_{jA}, A_{jI}, and A_{jB}, respectively representing A$_i$'s experience of A$_j$'s ability, integrity, and benevolence, based on A$_j$'s response to A$_i$ and its task performance described above. Trust builds up gradually and incrementally, reinforced by previous trusting behavior and positive experiences, whereas distrust is more catastrophic. This reflects the idea that, stated in trust literature, trust comes on foot but leaves on horseback (Six, 2005). Therefore, negative experiences are considered more influential than positive experiences in affecting trust development in the model.

In distributed multi-cultural teams, the tasks often need to be completed in a relatively short period of time. Therefore, members may focus more on task goals than on social/relational development. As a result, when forming trust in those teams, individuals may place more emphasis on perceived ability and integrity than on perceived benevolence.

Learning of complex tasks generally displays incremental change over time in an "S shape" (Fischer and Rose, 1994). Following the learning-curve theory, a logistic function, shown as Equation 2, is used to model K_{ij} as a function of I_{ij}. The function starts from 0 when there is no interaction history between A$_i$ and A$_j$ and asymptotically approaches 1 as the interaction frequency increases.

$$K_{ij} = -1 + \frac{2}{1+e^{-I_{ij}}} \tag{2}$$

Research suggests that the individuals' propensity to trust is especially relevant in predicting the relationship between strangers, but its impact tends to diminish over time as the partners get acquainted with each other. To model this phenomenon, T_{ij} is a function of E_{ij}, K_{ij}, and I_{ij}, as shown in Equation 3. As K_{ij} increases, the impact of E_{ij} on T_{ij} increases, whereas the effect of A_{iT} deteriorates.

$$T_{ij} = K_{ij} \cdot E_{ij} + A_{iT} \cdot (1 - K_{ij}) \tag{3}$$

If T_{ij} meets A$_i$'s trust threshold, A$_j$ remains a trustee candidate of A$_i$. Otherwise, A$_j$ is no longer a trustee candidate of A$_i$, as trust repair is beyond of scope of the study.

SIMULATION OF AGENT-BASED MODEL OF TRUST DEVELOPMENT

NetLogo 4.0.4 (http://ccl.sesp.northwestern.edu/netlogo/), a programmable modeling environment especially effective for simulating natural and social phenomena, is used to develop the agent-based model of the trust development in distributed multi-cultural teams described in the last section.

Simulation experiments with 2430 iterations have been conducted by varying individual agents' ability, integrity, benevolence, cultural values, and trust propensity in a team of 4 agents. Besides, in the experiments, negative experiences are considered 1.5 times as influential as positive ones in affecting trust development, and a trustee's perceived integrity and ability are regarded equally important but three times as important as its perceived benevolence in determining its perceived overall trust. These parameters, however, can be easily modified.

In terms of gains and losses from the tasks done by the team of agents, it is assumed that if agent A_i does not delegate a task to other agents, it does not gain any point. If A_i's delegated task is successfully completed by A_j, it gains 4 points. If A_i's delegation request is rejected by A_j, it loses 1 point. If A_j accepts A_i's request but does not complete the task successfully, A_i loses 4 points.

For simplicity, it is assumed that an agent treats all the other agents with the same intended ability, integrity, and benevolence values if it is selected by them as the trustee agent. Besides, only the cases in which all the agents have the same values of integrity, ability, benevolence, trust propensity, and trust threshold, and the values of integrity, ability, benevolence are the same are considered in the study.

Figures 4(a) and (b) respectively show the trellis displays of individual agents' gains vs. their trust threshold conditioning on the trustworthiness of the team of agents in homogenous-cultural and heterogamous-cultural teams. Each curve corresponds to one agent. In the heterogamous-cultural teams, only the extreme situation is considered – two agents have cultural values of 0 and the other two have cultural values of 1 – which results in the maximum team variation of cultural values. Each panel of the plots is a scattered plot of individual agents' gain scores vs. their trust threshold at a specific value of trustworthiness. The specific value of the trustworthiness for a panel can be found at the top of the panel. Nine possible values of trustworthiness are considered here: 0.1 – 0.9 with an interval of 0.1.

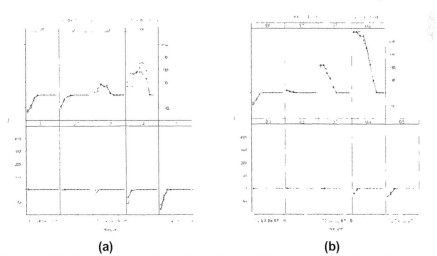

(a)   (b)

Figure 4 Trellis Displays of Individual Agents' Gains vs. Their Trust Threshold Conditioning on Trustworthiness
(a) Homogeneous-Cultural Groups
(b) Heterogeneous-Cultural Groups (Two agents have cultural values of zero and the other two have cultural values of one)

From the figures, we can see that individual agents' gains are better in the heterogamous-cultural teams than in the homogenous-cultural teams, especially when individual agents' trust threshold is relatively low and their trustworthiness is relative high (at or greater than 0.7). In addition, holding a high trust threshold is not always beneficial in terms of maximizing gains, particularly when the trustee's agent has a high value of trustworthiness (around 0.8 or above). Holding a low trust threshold, on the other hand, can be detrimental if the trustee agent has a low value of trustworthiness.

To understand what might be the cause of deteriorated performance in homogeneous-cultural teams compared to heterogamous-cultural teams, the homogeneous-cultural teams without any pair of agents that form a mutual trustor-trustee relationship initially (i.e., A_i delegates a task to A_j, and A_j also delegates a task to A_i) are compared with teams with one pair of agents that initially form a mutual trustor-trustee relationship. Figures 5(a) and (b) respectively show the trellis displays of individual agents' gains vs. their trust threshold conditioning on the trustworthiness of the team of agents in homogenous-cultural teams that do not have agents with a mutual trust relationship initially and those having one such pair of agents. These figures suggest that having initial mutual trust relationships can be significantly beneficial to increasing individual agents' gains. This phenomenon may also explain why individual agents' gains can be higher in the heterogamous-cultural teams, as shown in Figures 4(a) and (b), which consist of two pairs of agents with initial mutual trust relationships.

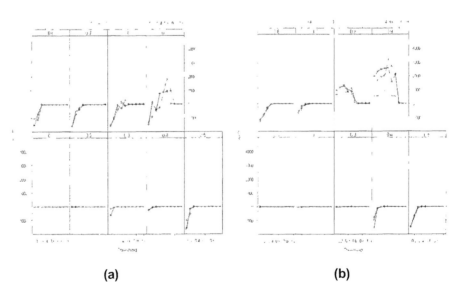

(a)   (b)

Figure 5 Trellis Displays of Individual Gains vs. Trust Threshold Conditioning on Trustworthiness in Homogeneous-Culture Groups
(a) No agents that initially form a mutual trustor-trustee relationship
(b) One pair of agents that initially form a mutual trustor-trustee relationship

CONCLUSIONS AND DISCUSSIONS

Distributed multi-cultural teams have become increasingly prevalent in all types of organizations. Merely putting people together in teams, however, does not guarantee that the teams will be effective. This is evidence that trust, an important element deemed necessary to facilitate knowledge sharing and collaborative work, is rather difficult to establish in distributed multi-cultural teams.

This study uses agent-based modeling and simulation to investigate what factors and how they can affect the development of interpersonal trust relationships in distributed multi-cultural teams. Simulation experiments suggest that forming a mutual trust relationship at an early stage is of particular importance in the development of long-term trust relationship not only in dyadic relationships but also in multiple-member teams. Besides, holding either a low or a high trust threshold can be ineffective in team performance, depending on the trustworthiness of the trustee agents which are not known in priori when a team of unfamiliar members is just formed. One strategy to cope with this issue may be to hold a low- or medium-level of trust threshold in initial interactions among team members to avoid trust breaking simply caused by miscommunications between them, and then increase trust threshold after team members have established some acquaintance.

REFERENCES

Abuelmaati, A., and Rezgui, Y. (2008), "Virtual team working: current issues and directions for the future." In: L. M. Camarinha-Matos & W. Picard (Eds.), *IFIP international federation for information processing*: *Pervasive collaborative networks* (pp. 351–360). Boston, MA: Springer.

Bews, N., and Martins, N. (2002), "An evaluation of the facilitators of trustworthiness." *South African Journal of Industrial Psychology*, 28(4), 14-19.

Chli, M., and De Wilde, P. (2008), "The emergence of knowledge exchange: an agent-based model of a software market." *IEEE Transactions on Systems, Man and Cybernetics, Part A (Systems and Humans)*. 38(5), 1056-67.

Coppola, N.W., Hiltz, S.R., and Rotter, N.G. (2004), "Building trust in virtual teams." *IEEE Transactions on Professional Communication*, 47(2), 95-104.

DeSanctis, G., and Poole, M. S. (1997). "Transitions in teamwork in new organizational forms." *Advances in Group Processes*, 14, 157-176.

Dietz, G., and Hartog, D.N. (2006). "Measuring trust inside organizations." *Personnel Review*, 35(5), 557-588.

Fischer, K.W., and Rose, S.P. (1994), "Dynamic growth cycles of brain and cognitive development." In: R. W. Thatcher, G.R. Lyon, J. Rumsey, and N. Krasnegor (Eds.), *Developmental neuroimaging: Mapping the development of brain and behavior* (pp. 263-279). San Diego, CA: Academic Press.

Gorobets, A., and Nooteboom, B. (2006), "Adaptive build-up and breakdown of trust: An agent based computational approach." *Journal of Management and Governance*, 10 (3), 277-306.

Jarvenpaa, S.L., and Leidner, D.E. (1999), "Communication and trust in global virtual teams." *Organization Science*, 10 (6), 791-815.

Liu, Y. and Warren, R. (2009), "Using fuzzy decision trees and information visualization to study the effects of cultural diversity on team planning and communication." *Proceedings of International Conference on Computational Cultural Dynamics*, pp. 45-53.

Mayer, R.C, Davis, J.H., and Schoomian, F.D. (1995), "An integrative model of organizational trust." *Academy of Management Review*, 20(3), 709-734.

Rousseau, D.M., Sitkin, S.B., Burt, R.S., and Camerer, C. (1998), "Not so different after all: A cross-discipline view of trust." *Academy of Management Review*, 23 (3), 393-404.

Six, F. (2005) *The Trouble with Trust: The Dynamics of Interpersonal Trust Development*. Edward Elgar Publishing Inc, Massachusetts, MA.

Wright, N.S, and Drewery, G.P. (2006), "Forming cohesion in culturally heterogeneous teams: Differences in Japanese, Pacific Islander, and Anglo Experience." *Cross Cultural Management*, 13, 43-53.

Chapter 5

Adaptive Coordination and Shared Leadership in High-Risk Teams: A Framework for Future Research

Gudela Grote, Michaela Kolbe

Department of Management, Technology, and Economics
ETH Zurich
Kreuzplatz 5, 8032 Zurich, Switzerland

ABSTRACT

There is a growing literature on coordination and leadership in teams, often with a special focus on teams in high-risk environments. Two concepts that have received much attention in this research are adaptive coordination and shared leadership. Adaptive coordination refers to shifts between different coordination mechanisms, such as explicit versus implicit coordination or mutual adjustment versus vertical leadership, in line with situational and task demands. Shared leadership is defined as distribution of leadership tasks across different team members, matching situational demands and individual resources and competencies. When leadership is understood as any attempt by any member of the team to influence others in order to achieve certain goals, it becomes increasingly difficult to distinguish team leadership from team coordination. This creates conceptual confusion, but also practical concerns as it becomes less clear what behaviors by whom are required. In order to guide future research and practical efforts, we propose a conceptual framework based on three constructs stemming from organization theory, namely task interdependence, coordination mechanisms, and substitutes for leadership. We suggest some relationships between different types of team coordination and leadership with team performance based on this framework and give some examples from research especially concerning high-risk teams in support of the proposed relationships.

Keywords: Coordination, leadership, high-risk teams, organization theory

INTRODUCTION

In recent years, much has been written about coordination and leadership in teams due to the increased interest in teams as the main building block of organizations and also the increased focus on team work in high-risk environments (Kozlowski & Bell, 2003). Complementing the multitude of social psychological studies on team dynamics including processes of emerging leadership, leadership researchers have begun to approach the analysis of formal and informal leadership in teams more as a group phenomenon than as an individual attribute or behavior that is brought to the team (Day, Gronn & Salas, 2004). Leadership is assumed to be reflected in any team process that is aimed at influencing others. Along with earlier work, for instance by Hackman and Walton, Zaccaro, Rittman and Marks (2001) argue that leadership should be understood in functional terms, that is, a leader needs to adaptively fulfill those functions in a team that are needed for task accomplishment and are not taken up by any of the other team members. Major functional requirements are information search and structuring, information use in problem solving, managing personnel resources, and managing material resources. Another concept stressing the process of leadership rather than the formal leadership role is shared leadership (Pearce & Conger, 2003). Shared leadership extends the argument brought forth by functional approaches by postulating that leadership functions can be fulfilled not only by the formal leader, but by any team member based on situational requirements and individual competencies.

When leadership is defined very generically as any attempt to influence others in order to promote goal attainment and additionally is not restricted to behaviors by formal leaders, but includes behavior of all team members, it becomes difficult to distinguish leadership from team coordination through mutual adjustment. This latter concept was established in organization theory as one of the main generic coordination mechanisms in organizations. Going back to Gibb, Day and colleagues (2004) state that the concept of leadership becomes diluted when studied at the group level following the notion of shared or distributed leadership. The confusion is increased further when taking into consideration the concept of adaptive coordination, which postulates the need to shift between different coordination mechanisms, for instance vertical leadership versus mutual adjustment or explicit versus implicit coordination, in line with situational and task demands (Burke et al., 2006b; Entin & Serfaty, 1999).

The difficulty of disentangling leadership and coordination, which is encountered at a conceptual level as well as empirically when trying to categorize team behaviors, is the starting point for this paper. We will discuss three constructs stemming form organization theory and integrate these into a conceptual framework which we hope will be helpful in guiding future research on team leadership and coordination. We will also use some examples of research on high-risk teams to illustrate the assumed relationships in the suggested framework.

CONSTRUCTS FROM ORGANIZATION THEORY RELEVANT TO TEAM RESEARCH

As a starting point, we propose that it is important to keep the concepts of team leadership and coordination apart because they permit complementary perspectives on team phenomena due to their different theoretical roots. Coordination from an organization theory point of view is much more than just one leadership function as it is often portrayed in the leadership literature, but leadership is also more than just "vertical personal coordination" as one finds it described in organization theory. Both leadership and coordination happen within teams, but are also part of organizational top-down processes. Rather than juxtaposing leadership and team processes as it is sometimes done (Zaccaro & Klimoski, 2002), we will look at leadership and coordination as interlinked processes at the team and organizational level.

We argue that in the study of team leadership and coordination conceptual clarity can be increased and interpretation of empirical data facilitated by more systematically exploiting organization level constructs, specifically task interdependence, coordination mechanisms, and substitutes for leadership. While there have been many calls for a multi-level perspective on group processes, few concrete proposals have been made regarding the kinds of constructs to include at the organizational level (Kozlowski & Bell, 2003). By suggesting three particular constructs stemming from organization theory, we attempt to fill this gap. While task interdependence and substitutes for leadership have been discussed in the team leadership literature, though often not in a very systematic fashion, an organization level perspective on coordination mechanisms in teams is largely missing. In the following, we provide a brief overview of the chosen constructs and illustrate how they can help to better understand leadership and coordination at the team level.

TASK INTERDEPENDENCE

Of the three constructs proposed, task interdependence is the one mentioned most in team research (Kozlowski & Bell, 2003). Originally, three types of task interdependence were distinguished in the organization theory literature, that is pooled, sequential, and reciprocal interdependence (Thompson, 1967). Pooled interdependence is present when system performance is an additive function of individual performance. Sequential interdependence is a unidirectional workflow arrangement, where individual performance depends on the proper fulfillment of prior subtasks. Reciprocal interdependence implies that information and results of work activities have to be exchanged between team members continuously. Later a fourth type of interdependence was added, which describes a particularly intensive form of reciprocal interdependence, sometimes called team interdependence (Van de Ven et al., 1976) or intensive work situations (Tesluk et al., 1997). In studies of team processes, these different forms of task interdependence are usually acknowledged as an important influencing factor, but rarely systematically compared in terms of requirements and effects for leadership and coordination. Burke et al. (2006a) performed a meta-analysis for the few studies that allowed comparing leadership in the context of high (reciprocal, team) and low (pooled, sequential) task interdependence. They found some tentative evidence that leadership is more important for team effectiveness when task interdependence is high.

COORDINATION MECHANISMS

In organization theory, task interdependence is usually linked with requirements for coordination and specific coordination mechanisms, which constitutes the second construct proposed here. The most basic distinction regarding coordination mechanisms is that between impersonal and personal forms of coordination (Van de Ven et al., 1976). Impersonal coordination happens via standardization of processes through technology or organizational rules or via cultural norms and values. Personal coordination may occur through vertical interaction between formal or informal group leaders and team members or through lateral interaction among team members often termed mutual adjustment. While pooled and sequential task interdependence should be handled mostly through impersonal forms of coordination, reciprocal and team interdependence require personal forms of coordination. The result by Burke et al. (2006a) mentioned above supports this claim regarding leadership as a form of personal coordination. Furthermore, the more uncertainties have to be handled, personal coordination should entail mutual adjustment among all team members rather than unilateral leadership (Faraj & Xiao, 2006). A similar argument is made in the literature on shared or distributed leadership which is considered particularly important in complex tasks involving high levels of uncertainty (Day et al., 2004; Pearce & Conger, 2003).

The importance of studying team coordination and leadership in view of different coordination mechanisms determined at the organizational level becomes apparent, for instance, when trying to find one's way through the maze of research on coordination in high-risk teams. Many studies have analyzed in particular how teams adapt their coordination to task and situational requirements, often distinguishing between explicit and implicit coordination (Burke et al., 2006b). Explicit coordination entails resources being spent on the task of coordinating as such, while implicit coordination occurs based on a shared mental model of the task requirements and does not require extra resources (Entin & Serfaty, 1999). In organization theory, one finds the long-standing assumption that organizational rules as a form of impersonal coordination ease the demands on personal coordination by furthering a shared understanding of the task and thereby implicit coordination (March, Schulz & Zhuo, 2000). Only very recently, this conceptual reasoning has been taken up in studies of adaptive coordination, focusing in particular on the interplay between standardization as an impersonal form of coordination and explicit and implicit personal coordination. For both medical teams and cockpit crews, it was found that there was in fact more implicit coordination in work phases with higher standardization (Grote et al., 2010; Zala-Mezö et al., 2009).

SUBSTITUTES FOR LEADERSHIP

The third construct suggested is the substitutes for leadership model presented by Kerr and Jermier (1978) several decades back. It is assumed that certain organizational, team and task characteristics like standard operating procedures, competencies and experience available in the team, or routinized tasks can act as substitutes for personal leadership in promoting team performance. An interesting new angle on the substitutes for leadership debate was opened up by the argument that creating substitutes is in itself an act of leadership (Dionne et al., 2005). This may involve, for instance, creating standards as an impersonal form of coordination, or enabling team members to coordinate tasks amongst themselves through mutual adjustment. As such activities often will be based on strategic decisions in the organization, we suggest that substitutes for leadership should be understood as an organization level construct even though the actual substitutes may concern the group or individual level, such as group cohesion or individual expertise.

To date, substitutes have mainly been considered as moderating the relationship between certain leadership behaviors and individual and team outcomes, which may be too narrow a focus, though (Dionne et al., 2005). Substitution might also entail that leadership and substitutes have parallel and independent effects. Furthermore, substitution may affect different leadership functions differently (Dionne et al., 2002). Higher levels of team member competence may require less effort by the formal leader for assuring information use in problem solving, but he or she is still needed to support information search and structuring. This behavior pattern was indeed found in well performing anesthesia teams consisting of more experienced nurses and less experienced, but formally responsible residents (Künzle et al., in press a,b).

The conceptual framework offered by the substitutes for leadership model may prove particularly helpful in disentangling leadership and coordination because it allows evaluating preconditions for team effectiveness at different levels. From the perspective of the formal leader, the ability of a team for shared or distributed leadership becomes a substitute for his or her leadership (Day, Gronn & Salas, 2006). In the same vein, the finding in Burke et al.'s meta-analysis (2006) that person-focused leadership by team leaders seems to be somewhat more important for team performance than task-focused leadership may be understood as a substitution effect stemming from high empowerment of team members. Whether formal leadership is promoted or rather substituted for by other coordination mechanisms is dependent on strategic decisions in the organization in view of a variety of contingencies that determine the coordination requirements having to be catered for.

A FRAMEWORK FOR FUTURE RESEARCH

In Figure 1, some relationships between the three discussed constructs – task interdependence, coordination mechanisms, and substitutes for leadership – and different types of team leadership and team coordination are proposed. As an initial step, it is assumed that different levels of task interdependence set different requirements for coordination. Specifically, low task interdependence can be best handled through impersonal coordination, while high task interdependence requires personal coordination through vertical leadership or mutual adjustment.

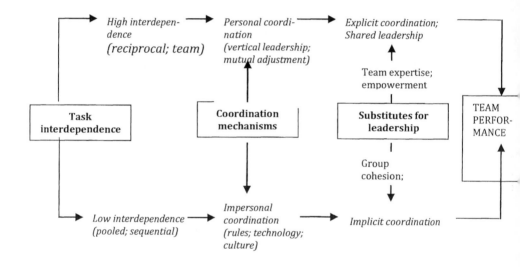

FIGURE 1.1 Relationships between the conceptual frameworks

This first assumption embedded in the framework is well founded in organization theory (e.g., Van de Ven et al., 1976). It has not been tested much in team research because usually the researched teams, especially high-risk teams, work under conditions of high task interdependence only. However, it is important to point out that implicit coordination which is frequently discussed in research on high-risk teams as a less resource-intensive and therefore desirable form of coordination does in fact imply that impersonal coordination mechanisms are in place. Only if teams have shared standards to go back to they can coordinate without resources being spent on the coordination as such. As was mentioned earlier, this reasoning has been taken up only very recently in empirical studies on coordination in high risk-teams. Evidence was provided in medical teams and cockpit crews that implicit coordination relies on higher levels of standardization (Grote et al., 2010; Zala-Mezö et al., 2009).

At the next stage in the framework, it is assumed that the actual amount and type of leadership is also influenced by the availability of substitutes, be it in the form of other coordination mechanisms like high standardization or in the form of particular team or task characteristics. Some initial evidence for the importance of standardization, team member expertise and task routinization as substitutes for leadership in anesthesia teams was found by Künzle and colleagues (in press a, b). The research by Künzle also points to the intricacies of disentangling shared leadership and substitutes for leadership. One of her findings was that leadership in better performing anesthesia teams was shared between residents and nurses, while in less well performing teams leadership was mainly exerted by residents. She also found that shared leadership did not mean that both resident and nurse were involved in the same kinds of leadership functions. The residents were more active in structuring decision-making and task execution, whereas the nurses assured information use in problem solving based on their higher level of experience. One could argue that their experience served as a substitute for leadership required by the resident, but their behavior in itself can also be considered to fulfill a leadership function.

Finally, the framework contains relationships between different types of team leadership and coordination and team performance. Again some preliminary evidence for some of these relationships has been found as mentioned above. However, to date no overall test of the suggested relationships has been undertaken. We hope that the suggested framework will be instrumental for future multi-level research on leadership and coordination in teams, be it for deriving specific testable propositions or even by spurring a much more complex study including the complete set of suggested relationships.

REFERENCES

Burke, C.S., Stagl, K.C., Klein, C., Goodwin, G.F., Salas, E. & Halpin, S.M. (2006a). What type of leadership behaviors are functional in teams? A meta-analysis. The Leadership Quarterly, 17, 288-307.

Burke, C.S., Stagl, K.C., Salas, E., Pierce, L. & Kendall, D. (2006b). Understanding team adaptation: A conceptual analysis and model. Journal of Applied Psychology, 91, 1189-1207.

Day, D.V., Gronn, P & Salas, E. (2004). Leadership capacity in teams. The Leadership Quarterly, 15, 857-880.

Day, D.V., Gronn, P & Salas, E. (2006). Leadership in team-based organizations: On the threshold of a new era. The Leadership Quarterly, 17, 211-216.

48

Dionne, S.D., Yammarino, F.J., Atwater, L.E. & James, L.R. (2002). Neutralizing substitutes for leadership theory: Leadership effects and common source bias. Journal of Applied Psychology, 69, 307-321.

Dionne, S.D., Yammarino, F.J., Howell, J.P. & Villa, J. (2005). Substitutes for leadership, or not. The Leadership Quarterly, 16, 169-193.

Entin, E. E., & Serfaty, D., 1999. Adaptive team coordination. Human Factors, 41, 312-325.

Faraj, S., & Xiao, X., 2006. Coordination in fast-response organization. Management Science, 52, 1155-1169.

Grote, G., Kolbe, M., Zala-Mezö, E., Bienefeld-Seall, N., & Künzle, B. (2010). Adaptive coordination and heedfulness make better cockpit crews. Ergonomics, 53, 211-228.

Kerr, S. & Jermier, J.M. (1978). Substitutes for leadership: Their meaning and measurement. Organizational Behavior and Human Performance, 22, 375-403.

Kozlowski, S.W.J. & Bell, B.S. (2003). Work groups and teams in organizations. In W.C. Borman, D.R. Ilgen & R.J. Klimoski (eds.), Handbook of psychology: Industrial and organizational psychology (Vol. 12, pp. 333-375). London: Wiley.

Künzle, B., Zala-Mezö, E., Kolbe, M., Wacker, J., Grote, G. (in press a). Substitutes for leadership in anaesthesia teams and their impact on leadership effectiveness. European Journal of Work and Organizational Psychology.

Künzle, B., Zala-Mezö, E., Kolbe, M., Wacker, J., Grote, G. (in press b). Leadership in anaethesia teams: The most effective leadersip is shared. Quality and Safety in Health Care.

March, J., Schulz, M. & Zhou, X. (2000). The dynamics of rules: Change in written organizational codes. Stanford, CA: Stanford University Press.

Pearce, C. L. & Conger, J. A. (eds.) (2003). Shared leadership: Reframing the how's and why's of leadership. Thousand Oaks, CA: Sage.

Tesluk, P.E., Mathieu, J.E., Zaccaro, S.J. & Marks, M. (1997). Task and aggregation issues in the analysis and assessment of team performance. In M.T. Brannick, E. Salas and C. Prince (eds.), Team performance assessment and measurement. Theory, methods, and applications (pp. 197-223). Mahwah, NJ: LEA.

Thompson, J.D. (1967). Organizations in action. New York: McGraw-Hill.

Van de Ven, A.H., Delbecq, A.L. & Koenig, R.J. (1976). Determinants of coordination modes within organizations. American Sociological Review, 41, 322-338.

Zaccaro, S.J. & Klimoski, R. (2002). The interface of leadership and team processes. Group & Organization Management, 27, 4-13.

Zaccaro, S.J., Rittman, A. & Marks, M.A. (2001). Team leadership. The Leadership Quarterly, 12, 451-483.

Zala-Mezö, E., Wacker, J., Künzle, B., Brüsch, M., & Grote, G. (2009). The influence of standardisation and task load on team coordination patterns during anaesthesia inductions. Quality and Safety in Health Care, 18, 127-130.

CHAPTER 6

Collaborative Design of Workplaces: The Role of Boundary Objects

Ole Broberg, Rikke Seim, Vibeke Andersen

DTU Management Engineering
Technical University of Denmark
Lyngby, DK 2800, DENMARK

ABSTRACT

Boundary objects may play an important role in collaborative design processes. Such objects can help participants with different background to better communicate and take part in a design process. The aim of this study is to explore the role of boundary objects in participatory ergonomics and to find out what characterizes boundary objects. The study lists eight characteristics of boundary objects and set up a framework for the conceptualization of boundary objects in collaborative design of workplaces. The framework emphasizes the need not to focus solely on the boundary object per se but also include the context of its usage and its capacity to sustain and circulate the outcomes of collaborate design processes.

Keywords: Collaborative Design, Participatory Ergonomics, Boundary Objects

INTRODUCTION

Participatory ergonomics (PE) has become a widespread field of research and practice. Haines et al. (2002) have developed a conceptual framework for understanding PE. Seven dimensions are identified each with categories explaining the range, e.g. the dimension "Focus" or topic has the categories "Designing equipment or tasks", "Designing jobs, teams or work organization", and "Formulating policies or strategies". The framework is very useful for researchers in understanding specific PE schemes. For practitioners it may serve as a reflective tool when designing PE schemes in companies. We will argue, however, that the framework may benefit from a clearer distinction between two basic topics being addressed in PE. The first topic is problem solving in a current production and workplace, and the second topic is design and planning of a completely new installation. Kuorinka (1997) makes this distinction and points to the first topic as the most usual one in PE but also acknowledges the other one. In an analysis of user-centred design Eason (1995) also widens the topics of PE to include: "…establishing design processes in which the end users themselves can influence the design so that it is compatible with their goals and beliefs, etc.". This second topic seems more widespread today. Sundin et al. (2004) claim that it is not enough to improve workplaces and production systems themselves but necessary to involve "the earlier step that affects the production system, i.e. the product design". They coin the term 'participatory ergonomics design' for such activities. In line with Sundin & Medbo (2003) we find the PE framework of Haines et al. (2002) will benefit from introducing a "Tools and methods" dimension. Such a dimension should include tools and methods for participatory ergonomics design.

The focus on design and planning of new installations or production systems as a goal for PE triggers a new question: How can workers and other workplace end users participate in setting up measures for ergonomics when the new workplace does not exist? This question is about representations of workplaces and work processes in design processes. If the ergonomist has the role of guiding the PE process he also has to consider what kind of representations are useful in the design process. Looking into studies of PE processes it seems that quite a number of different kinds of objects are used to represent features of the non-existing workplace and work process in design processes. By objects we initially refer to such different things as written documents, drawings, prototypes, and CAD systems.

The representations are means of communication and enablers of participatory design processes. For an ergonomist guiding a PE process the involvement of objects should make it easier for the workers or workplace users to participate in the design process. Hence, it can be questioned whether some objects are better than others in doing that job. In this study we will introduce the concept of 'boundary objects' originally developed within the Science and Technology Studies field. Boundary objects have the ability to translate meaning across different knowledge domains and work practices. This ability is highly wanted in PE to facilitate design processes involving workers, designers, and ergonomists.

The aim of this paper is to identify and explore the role of boundary objects in PE processes. The question in focus is: What characterizes boundary objects in PE processes?

In what follows, we first introduce the concept of boundary objects. Second, we present our empirical findings of two PE case studies in which boundary objects played and important role. Finally, based on a grounded theory approach we develop a characterization of boundary objects in PE design processes and propose a framework of how the use of boundary objects in PE processes might be conceptualized.. We will argue that the concept of boundary objects in PE processes is contextual and that the ergonomist needs to actively consider their selection and the stage in which they are used.

THE BOUNDARY OBJECT CONCEPT

The concept of boundary objects was introduced by Star & Griesemer (1989): *Boundary objects are objects which are both plastic enough to adapt to local needs and constraints of the several parties employing them, yet robust enough to maintain a common identity across sites. They are weakly structured in common use, and become strongly structured in individual-site use. They may be abstract or concrete. They have different meanings in different social worlds but their structure is common enough to more than one world to make them recognizable means of translation. The creation and management of boundary objects is key in developing and maintaining coherence across intersecting social worlds. (p. 393)*

Carlile (2002 and 2004) analyzed the concept in studies of product development and for him it describes objects that are shared and shareable across different problem solving contexts. In line with Star & Griesemer, without using the exact same terms, he operates with three categories of boundary objects in the product development setting: First, repositories, i.e. cost databases, CAD/CAM databases, and parts libraries. Second, standardized forms and methods, i.e. standards for reporting findings, problem-solving methods, or engineering change forms. Third, objects, models, and maps, i.e. sketches, assembly drawings, parts, prototype assemblies, mock-ups, computer simulation, Gantt charts, process maps. Carlile (2002) identifies three characteristics of a tool, method, or object that makes them useful in joint problem solving at a given knowledge boundary. A boundary object: 1) establishes a shared syntax or language for individuals to represent their knowledge, 2) provides a concrete means for individuals to specify and learn about their differences and dependencies across a given boundary, and 3) facilitates a process where individuals can jointly transform their knowledge. Carlile (2004) points out that depending of the type of boundary faced, boundary objects with different capacities are required.

In a broader view of engineering design Boujut & Blanco (2003) pay particular attention to the interfaces between the actors involved in collaborative design processes. They observed the features of objects, termed 'intermediary objects', as mediation, transformation or translation, and representation. They suggest that "open" objects are better than "closed" ones to facilitate the participation of various actors in the design process. A closed object can be a model built by a designer or a 2D CAD drawing. Such objects are not good at mediating collaboration in a design process because they are nearly impossible to modify. The actors cannot transform them into shared representations.

In another field of research Wenger (2000) introduces the concept of boundary objects in social learning theory. He also focuses on boundaries and boundary processes. Boundaries connect different communities of practice and they offer learning opportunities. Boundary objects can be intentionally promoted to help bridging between communities and facilitate learning. In the words of Wenger: "Some objects find their value, not just as artifacts of one practice, but mostly to the extent that they support connections between different practices". Wenger suggests three different forms of boundary objects: 1) Artifacts, 2) discourses, and 3) processes. Artifacts correspond to Carlile's second and third categories. Discourses may be the existence of a common language that allows people to communicate and negotiate meanings across boundaries. Processes include explicit routines and procedures in an organization.

While boundary objects may be part of the daily work in engineering design and product development, it is a new situation to take this concept to the field of participatory ergonomics in design and planning of new workplaces. Participatory processes in this area are often a series of discrete events with the aim to influence the overall design outcome from an ergonomics point of view. Working with boundary objects is not only about representing and transforming knowledge but also about facilitating collaborative design across work practices (Suchman 1995, Button & Harper 1996). Collaboration and coordination is not across engineers with different specializations but across workers, management, designers, and ergonomists with very different work practices.

These challenges led us to look for objects in PE processes.

OBJECTS IN DESIGN PROCESS INTERVENTIONS

The two cases are from a research program that developed and tested a framework for participatory design of workplaces in the period 2005-2008. The aim was to develop methods and tools that can be applied by ergonomists and other workplace professionals to stage direct participation in design and planning of new installations and production systems (Binder & Brandt 2008, Broberg 2010, Seim & Broberg 2010). The authors were part of this research program. In this retrospective analysis we will present a number of objects taking part in the intervention and the context in which the objects were used.

The research program was based on intervention studies in two organizations: an industrial manufacturer and a public administration office. The industrial manufacturer was to implement a new production technology. Engineering designers, workers, management, and ergonomists were involved. In the public administration three departments were merged, moving from cell offices into an open plan office, and implementing new ways of working. Architects, workers, management, human relations people and ergonomists were involved.

The interventions were aimed at facilitating a collaborative design process of the future work and workplace. It was especially emphasized to develop and test methods making it easy for the workers to articulate their concerns and wishes and to participate in design activities. The interventions had the format of workshop series in which different interactive methods and tools were applied: workbooks, layout design game, use scenarios, future workplace assessment, photo safari, and the dream office. The researchers together with ergonomists and other workplace professionals from the two organizations took the role as facilitators and workshop leaders. Other researchers in the team evaluated the process of intervention and the outcomes of the workshops. The data for the cases were collected over a 12-month period, and included interviews and observations.

We first present a brief overview of the context in which objects were used. Then we present the objects in details.

THE INTERVENTION CONTEXT

In the industrial manufacturer company we accomplished two workshop series each consisting of two workshops. The first series focused on the layout of the production facility in which the new technology was to be installed. The main object was a layout design game. As a warm-up the workers were first doing a workbook session. The first workshop series was accomplished in a meeting room in the company. The second series took departure in the layout that resulted from the first series and then focused on work processes and ergonomics when the new technology was operated in the future layout. The main objects were a 1:20 scale model and floor markings in the actual production hall. Both workshops in this second series were accomplished in the production hall in which the new technology was to be installed.

In the public administration office a similar structure of workshop series was accomplished. Here we will focus only on two objects: a photo safari, and the sketching of the dream office.

Both interventions were by all participants considered to be highly successful. The workshops in the industrial manufacturer case resulted in a new design of the layout and a detailed ergonomics specification list that was agreed by all involved actors. The public office case resulted in workplace experiments that were implemented in the existing office facilities as well as a detailed proposal for the layout of the new open space office.

OBJECTS IN THE COLLABORATIVE DESIGN PROCESSES

Workbooks (Horgen et al. 1999) are a selection of pictures from the current workplace at which the workers can make comments by colored pens. A three-color code was used to signal e.g. good solution, problematic, and problem to be solved. Problems and solutions included ergonomic topics as well as production optimization topics. Workers presented their workbooks in the first workshop. Participants included design engineers, managers, and ergonomists.

In the layout design game a game board with a blueprint of the new facility is at the middle in a workshop meeting. Moveable bricks in different colors represent machines, equipment, and installations. Each participant in turn manipulates with the bricks at the game board and explains short the idea. The game explores different possibilities for layout through "what if …" experiments at the game board and their impact on ergonomics and production issues.

Use scenarios were applied in two versions. First, the work processes and the spatial workplace were simulated by help of a 1:20 scale model of the new facility with moveable machine models and LEGO figures, and by floor markings in the actual production hall where new machines were to be installed. Different production episodes were played out as if the participants were working with the new technology in the new facility. All participants were active in reflecting upon what worked and what should be redesigned. Second, the same use scenario setup was applied for a future workplace assessment in which the ergonomist went by a virtual walk-through guided by the operators of the projected new facility. This event identified ergonomic and safety and health factors which were not recognized previously.

In a photo safari four teams of workers from the public administration office, armed with a digital camera, went to visit four organizations with open space offices. They took pictures and had short dialogues with people. Coming home each team in a workshop presented a selection of photos and explained what they liked and disliked and what could be the impact for design of their future workplace. In the dream office session teams of workers were sketching their dream office at a poster, some times using allegories when explaining the idea. The sketches included ideas of physical layout as well as work organization issues. The sketches were presented to the other workshop participants and discussed.

CHARACTERISTICS OF THE BOUNDARY OBJECTS

The five objects presented in the previous section were tested as tools in a kind of participatory design process, which in the future could be guided by an ergonomist. The objects used in the two cases had a function as boundary objects and they were by all parties considered as doing a good job. The question is what characterizes these objects and their context of use.

The five objects and their context of use were analyzed using a grounded theory approach (Corbin & Strauss 2008). The following eight characteristics emerged from this analysis.

1. The objects do not come ready-made. They need to be made by the actions of the participants. They are objects-in-the-making. The workbook has no meaning before the workers have drawn and commented on the pictures. The blank blueprint layout can not be understood before the participants have furnished it with bricks representing machines, walls, equipment and the like in a specific configuration. The use scenarios inherently are actions by the participants. The photo safari forces the participants to think what pictures from a foreign workplace are interesting and why it is so. A blank piece of paper has no meaning before a team of workers has drawn their dream office.

If we compare to some of the objects listed in Table 1 they come more as facts or at least ready-made objects. E.g. this goes for injury reports, diagrams, tables, drawings, meeting reports, and inventories. These objects may of course be discussed but they do not by themselves invite to some sort of design action.

2. The objects in the two cases have this capacity. They have built-in affordances to make a design move or to articulate what is good and what is bad design.

3. Some of the objects, the layout design game, the use scenarios and the dream office, are flexible and malleable in a way that makes it possible to go into "rapid prototyping" with the new workplace. In the layout design game, it is possible very quickly to explore "what if we place the machines in this way?". In the use scenario it is possible to illustrate and reflect upon "how about doing the work sequence in this way?". The capacity to quickly test different configurations is in contrast to CAD drawings and computer simulations, which requires time consuming programming or other actions with the software before an alternative can be presented.

4. The boundary objects are surrounded by rules and instructions for their use, some with a character of a game. The workbook: Use the red pen for problems, the green one for good design. The layout design game: Manipulating with the bricks is on a turn-basis and we need a short explanation when you move bricks. The use scenario: We are playing specific sequences of work in the new production system. We look for ergonomics in the control room. We want all participants to involve in reflections. The photo safari: You are supposed to take pictures of things you like or dislike, of designs related to how citizens are met in the office etc.

5. A facilitator of the events selected the boundary objects, developed the rules and instructions, and guided the workshops in which the objects were used.

6. The boundary objects are used in discrete events, i.e. workshops and guided by a facilitator that have developed the rules and instructions. This set up establishes a temporary learning space in the organization. The guidance and rules and instructions, to some extent, help this temporary space to dissolve the daily organizational structures, roles, and politics that participants are embedded in. The set up enables and promote a collaborative design process. All participants are set in "design mode" and not "negotiating mode" based on daily conflicts of interests. The objects become boundary objects in a learning process between different communities of practice (Wenger 2000).

7. The location of the boundary object needs attention. The layout design game played out in the company's meeting room in the administration building. This is a surrounding that is very familiar to design engineers and managers. For the workers it was to play away. In contrast, the use scenarios took place in the production hall. The workers were playing at home and the design engineers were playing away. The location issue goes further to a general consideration if the use of boundary objects can be embedded in the workplace and work practices that are to be redesigned.

8. The boundary objects become themselves an output from participatory events. They articulate a piece of design that has been materialized and then can be circulated in the organization, including the ongoing design process. When all participants agree on the new layout or a compromise has been reached it is directly

to see at the game board. Bricks are frozen at the game board in a specific configuration. Such materialized outcomes from participatory events may better support that the results are taken into the further design process. They may even have a political character in the ongoing design process. Workers can physically point to the 'frozen' boundary object and present it to other actors in the organization (Henderson 1999).

The findings of this analysis are summarized in Table 1.

Table 1 Conceptualization of boundary objects in PE design processes

Attribute	Characteristics
The boundary object	Object-in-the-making
	Built-in affordances
	Flexible and malleable
The context of use	Facilitator stages and guides
	Rules and instructions
	Temporary learning space
	Location
End product	Materialized outcome enables circulation

CONCLUSIONS

We have identified eight characteristics of boundary objects and their use that make them particular useful in PE design processes. The point is that these characteristics go beyond the object itself and hence point to a framework of boundary objects in a PE design process as indicated in Table 1.

This study has demonstrated that it is useful to address boundary objects in PE processes. Especially, this is true for PE processes in which a new workplace and work organization is designed. The selection of objects as boundary objects in PE processes is of great importance because different objects enable workers participation and collaborative design differently. The framework may serve to better understand the role of boundary objects in PE processes and to give some criteria for practitioners and intervention researchers for the selection of objects to facilitate a PE process. The framework tentatively suggests a rough categorization of objects in PE processes that corresponds with Boujut & Blanco's (2003) notion of closed and open objects. In the one end we have "passive" (closed) objects and in the other end "active" (open) objects understood as objects-in-the-making during a PE process.

The framework emphasizes the need not to focus solely on the boundary object per se but also include the context of its usage and its capacity to sustain and circulate the outcomes of PE design events, e.g. in a larger overall design process. Finally, the role of ergonomists in PE processes as the ones who initiate and guide the process is not new (Haines et al. 2002) but it may be new to reflect upon the selection of boundary objects using the framework in Table 1 in the planning of PE schemes .

REFERENCES

Binder, T., Brandt, E. (2008). The Design:Lab as platform in participatory design research. *CoDesign*, *4*(2), 115-129.

Boujut, J.-F., Blanco, E. (2003). Intermediary objects as a means to foster co-operation in engineering design. *Computer Supported Cooperative Work*, *12*, 205-219.

Broberg, O. (2010). Workspace Design: a case study applying participatory design principles of healthy workplaces in an industrial setting. *International Journal of Technology Management 51(1),* 39-56.

Button, G., Harper, R. (1996). The relevance of 'work practice' for design. *Computer Supported Cooperative Work*, *4*, 263-280.

Carlile, P. R. (2002). A pragmatic view of knowledge and boundaries: Boundary objects in new product development. *Organization Science*, *13*(4), 442-455.

Carlile, P. R. (2004). Transferring, Translating, and Transforming: An integrative framework for managing knowledge across boundaries. *Organization Science*, *15*(5), 555-568.

Corbin, J., Strauss, A. (2008). *Basics of qualitative research 3e*. Los Angeles: Sage Publications.

Eason, K. D. (1995). User-centred design: for users or by users? *Ergonomics*, *38*(8), 1667-1673.

Haines, H., Wilson, J.R., Vink, P., Koningsveld, E. (2002). Validating a framework for participatory ergonomics (the PEF). *Ergonomics*, *45*(4), 309-327.

Henderson, K. (1999). *On Line and On Paper*. Cambridge MA: The MIT Press.

Horgen, T. H., Joroff, M. L., Porter, W. L., & Schîn, D. A. (1999). *Excellence by Design: Transforming Workplace and Work Practice*. New York: John Wiley & Sons.

Kuorinka, I. (1997). Tools and means of implementing participatory ergonomics. *International Journal of Industrial Ergonomics*, *19*, 267-270.

Seim, R., Broberg, O. (2010). Participatory workspace design: A new approach for ergonomists? *International Journal of Industrial Ergonomics 40*, 25-33.

Star, S. L., Griesemer, J.R. (1989). Institutional ecology, 'translations' and boundary objects: Amateurs and professionals in Berkeley's Museum of Vertebrate Zoology. *Social Studies of Science*, *19*, 387-420.

Suchman, L. (1995). Making work visible. *Communications of the ACM, 38*(9), 56-64.

Sundin, A., Christmansson, M., Larsson, M. (2004). A different perspective in participatory ergonomics in product development improves assembly work in the automotive industry. *International Journal of Industrial Ergonomics, 33,* 1-14.

Sundin, A., Medbo, L. (2003). Computer visualization and participatory ergonomics as methods in workplace design. *Human Factors and Ergonomics in Manufacturing, 13*(1), 1-17.

Wenger, E. (2000). Communities of practice and social learning systems. *Organization, 7*(2), 225-246.

CHAPTER 7

Measuring Cross-Cultural Stimulus-Responses on the Internet

Edward P. MacKerrow, Jennifer H. Watkins

Center for the Analysis of Emerging Threats
International Applications and Technology Division
Los Alamos National Laboratory, Los Alamos, NM 87545

ABSTRACT

The Internet is now an accepted form of discourse around the world on subjects ranging from the mundane to the serious. Weblogs, chat rooms, and forums provide a new source of data on attitude diffusion and opinions to events across the world. In addition to providing valuable data for social science research, the Internet is allowing new forms of social and political behavior to occur. In this paper we present initial research on measuring the rate of Internet activity to key events over time. We present measurements of stimulus-response temporal dynamics using open-source free web analysis tools.

Keywords: blog analysis, Internet social science, attitude diffusion, opinion swarms, weblog, web analytics.

INTRODUCTION

Events can stimulate discussions on the Internet. People often provide comments and opinions on weblogs ("blogs") and web forums (Hewitt, 2005). It is expected that some events will have more responses than others. We are currently attempting to characterize the relationship between event types and their associated Internet responses. The response we measure at this point is simply whether or not a given blog discussion included mention of a specific event.

Our goal is to characterize the Internet response along the dimensions of:

- The temporal shape of the response including the rise time, response pulse width, and decay time.
- A "pulse" shaped response versus a periodic or cyclic response.
- The amplitude (*i.e.* number of blogs mentioning the search term(s)) of the response in terms of percentage of all blogs posted during the same time period.
- The geographic areas of the respondents (we discuss the current limitations in measuring geo-spatially identified responses later).

An example measurement of an Internet response pulse is shown in Figure 1. Our definitions of pulse characteristics are displayed in Figure 1.

The measurements of blog activity on a particular topic are made using keyword search tools. It is possible to accidentally register blog discussions that are not semantically relevant to the topic of search. For example, searching on the term "explosive" with the search engines can find both the phrases: "... *the explosive (noun) material ignited...*" and " *... the tennis player hit the ball with explosive (adjective) force...*" as being a positive "hit". The events we searched for had a large enough response population that these spurious situations introduced a small error. This would not necessarily be the case for obscure events having low numbers of Internet responses.

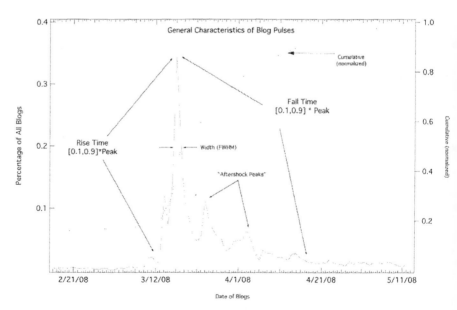

FIGURE 1. A sample blog pulse is shown here with the definitions of pulse

characteristics we measured. The left vertical axis is the amplitude of the Internet blog response and right vertical axis is the normalized cumulative rate of response over time.

TYPES OF INTERNET STIMULI

Many different types of events will trigger discussions on the Internet. For example:

- Physical events (*e.g.* hurricane, earthquake, tsunami, *etc.*)
- Statements made (normally by political elites or celebrities).
- Policies (*e.g.* new taxes, laws, regulations)

Some of these are "impulse events" that occur over a short time interval (*e.g.* a hurricane), whereas others are "ongoing events" (*e.g.* political campaigns). The Internet responses are expected to reflect the time structure of the event – discussions on old news eventually die off.

The set of events we analyzed as Internet stimuli for this initial study included:

1. During the ABC democratic debates on April 16th, 2008 Hillary Clinton stated that an attack on Israel by Iran would result in "massive retaliation" from the US (CNN, 2008).
2. The March-April 2008 collapse of The Bear Stearns Companies, Inc. investment bank (PBS, 2009).
3. Pope Benedict XVI April 2008 visit to the US (NY Times, 2008).
4. The March 2008 ABC News reports on selected excerpts taken from past sermons of Reverend Jeremiah Wright, former pastor for President Barack Obama (Ross, 2008).

Temporal measurements were made of Internet blog discussions in response to these events.

Geospatial analysis, determining where the bloggers are located, is more difficult than temporal analysis. Tools and methods do however exist to locate IP addresses (IP2Location, 2010) of the *web servers* used to host the blog websites. These tools do not provide exact location information on factual people posting to a blog. For example, a server located in New York could be hosting a weblog focused on financial markets in Asia – with the majority of bloggers living in Asia. Often there can be some correlation between the topic of discussion and the location of bloggers, since people are often more interested in news that affects them locally (Hewitt, 2005).

It is possible to measure information on the location of Internet *searches*, instead of blog postings. Geospatial information can be measured on web searches using Google™. To do this we used the Google Insights™ tool (Google, 2010). Although Internet searches (collecting information) are different than blog postings (expressing opinions), the searches can be used as a proxy for measuring interest in a particular topic or event. Search data provides no information about attitude or opinion though.

METHODOLOGY USED TO MEASURE INTERNET RESPONSES

Our initial research presented here relied on two blog search engines, Technorati™ (Technorati, 2010) and BlogPulse™ Metrics™ (Nielsen, 2010). To use these tools you merely enter search terms that you are interested in, and the tools return a time series of all blogs within a specified time period that discussed these search terms. Example measurements are shown in the next section.

In order to measure the geospatial "search interest" in a particular topic or event we utilized the Google Insights™ tool to analyze which regions of the world Internet users searched for a given term or phrase.

MEASUREMENT RESULTS

Temporal analysis of blog activity is shown in this section for the four events analyzed. In Figure 2 we show the blog activity in response to Hillary Clinton's 2008 comments expressing possible retaliation against Iran if they were to attack Israel. Clinton made multiple public comments relating to this topic during her 2008 Democratic presidential campaign. From the data it appears that Clinton's statement "obliterate Iran", that followed her initial statement of "massive retaliation", spawned a blogger response three times higher than the initial response.

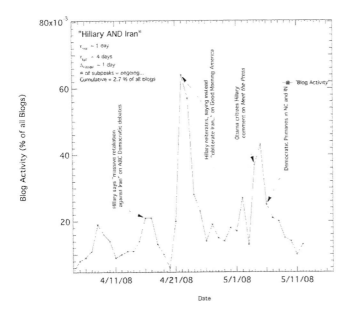

FIGURE 2. The blog activity rate is shown in this figure in response to Hillary Clinton's initial comment during the 2008 Democratic debates "That there would be massive retaliation against Iran if Iran ever attacked Israel". The blog activity increases dramatically days later when Clinton rephrased her comment to include "obliterate Iran".

The geospatial analysis of Internet *search* activity related to Hillary Clinton's comments is shown in Figure 3 where it can be seen that the US, New Zealand, South Africa, and Canada had the most Internet *search* activity related to Clinton's campaign comments.

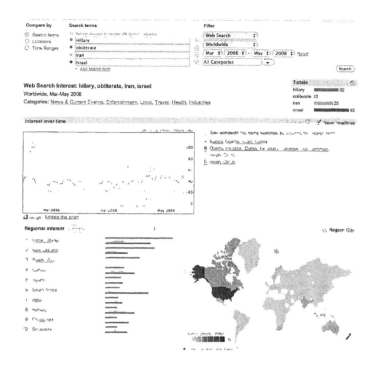

FIGURE 3. The geospatial Internet search activity for Hillary Clinton's comments on obliterating Iran if they attack Israel is shown here for the time period of March 2008 to May 2008 using the Google Insights™ tool. The time series for each independent search term is also shown.

The financial collapse of The Bear Stearns Inc. investment bank is shown in Figure 4. The initial triggering event was the Goldman Sachs announcement that they would not cover Bear Stearns derivatives. The blog response pulse to this was relatively low in amplitude as compared to the subsequent events.

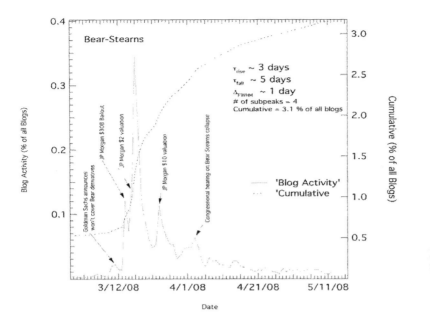

FIGURE 4. The blog response to the collapse of The Bear Stearns investment bank is shown here. It is interesting to observe the initial blog pulses that precede the huge blog storm when JP Morgan devalued Bear Stearns.

The visit of Pope Benedict XVI to the USA was also analyzed. This event differs in the others analyzed since it was not considered to be a "crisis" or a political point of contention between different political or social identity groups (see Figure 5).

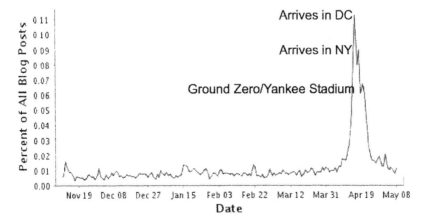

FIGURE 5. The blog activity surrounding the time period when Pope Benedict XVI visited the US in 2008 is shown. This is an example of an "impulse event".

Blogging associated with political muckraking during campaign time appears to trigger what are known as "opinion storms" – many people going onto blog sites and expressing their opinions (Hewitt, 2005). This is to be expected since blogs do provide a channel for people to express their opinions to a wide audience. An example of this type of blog activity is shown in Figure 6 for the Reverend Wright controversy during President Obama's 2008 campaign.

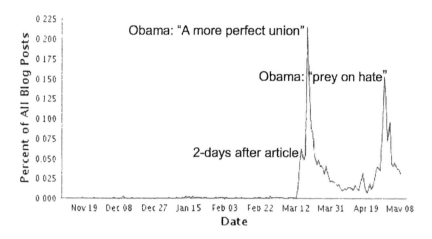

FIGURE 6. The blog activity in response to an ABC published article on excerpts of past sermons of the Reverend Wright, an early pastor of President Obama is shown here. The geospatial activity of Internet *searches* related to this controversy is shown in Figure 7.

The geospatial Internet *search* activity related to the Reverend Wright controversy is shown in Figure 7 for the USA. Heterogeneous activity levels are seen across different states in the US. These state search activity patterns can be compared to the map of red/blue election result states shown in Figure 8. There appears to be a correlation between blue states and states using the Internet to search for information on the Rev. Wright controversy.

FIGURE 7. The activity of Internet *searches* related to the Reverend Wright, Barack Obama controversy is shown here. The two temporal peaks correlate with the two peaks in Figure 6 (the 2008 election Red/Blue states are shown in Figure 8).

FIGURE 8. The 2008 US presidential election map showing the Republican (red) and Democratic (blue) majority states is shown here for comparison sake with Figure 7.

SUMMARY

The Internet can be used to measure "opinion storms" over time. We did observe well-defined response pulses to different types of events using freely available Internet analysis tools. These useful tools allow for the analysis of attitude diffusion over time.

Geospatial analysis of blog activity is currently underway and is proving to be more difficult and less accurate than temporal analysis of blog activity. The reasons are fundamental: bloggers post comments to a weblog which is hosted on a server located anywhere on the Internet. It is possible to measure the server locations based on their IP address. This does not however define where the blogger is located. A very popular blog might have posters from all over the world submitting comments and posts to it.

Estimates of geospatial Internet *search* activity are possible using the Google Insights™ analysis tool. This useful analysis tool provides search activity analysis down to the city level in the USA. The Google Insights™ tool also allows the analyst to observe the geospatial-temporal search activity unfold over time with a time-slider interactive map, allowing for visualization of opinion diffusion.

Determining *who* the individual blog respondents are is not possible with current tools, and will perhaps never be realized given the expected anonymity of bloggers and the levels of separation between the bloggers themselves and the servers hosting them.

The leading edge of the blog pulses ("rise time") was found to be very short in duration for the events analyzed – the initial blog response to a given event is very rapid. Real world events are often followed by subsequent related events (*e.g.* a politician makes one statement, the blogosphere reacts to it, then the politician makes follow-on statements). Measurements of blog pulses must take this into account – the analysis of the Bear Stearns collapse is an example where the large peak is actually composed of a sequence of closely related events and press releases over time.

The falling edge of blog pulses tended to be much slower than the rise times, however for many events we analyzed the discussions on blogs died off after approximately one week. Perhaps the majority of active bloggers may find newer events more interesting to discuss. When someone posts old news to a blog they are often "reprimanded" by the other bloggers since the topic has already been discussed. Social interactions on the Internet are important drivers to the behavioral norms of bloggers.

In order to accurately measure how different cultures respond to an event we need to improve our ability to geo-locate blogger responses. Currently we can geo-locate the regions where people are *searching* for event specific terms. At this point in our research we cannot make any conclusions as to the correlation between *blogging* on a particular topic and *searching* for a particular topic. Our current belief is that opinions should only be "measured" via blog analysis, since just because a person searches for a particular topic does not mean they have a strong opinion either way on that topic.

Since blogs provide an anonymous platform for many to express their opinions, they can capture attitude dynamics in a relatively new manner. Politicians, corporations, and influence groups can use blogs as important influence channels. The large number of Internet users available to counter/correct specific claims being broadcast on blog sites act as a collaborative editing mechanism.

REFERENCES

CNN (2008), http://www.cnn.com/2008/POLITICS/05/04/dems.election/

Google (2010), http://www.google.com/insights/search/#

Hewitt, H. (2005), "Blog, Understanding the Information Reformation That's Changing Your World", Nelson Publishing, Nashville.

IP2Location (2010), see http://www.whatismyip.com/

Nielsen (2010), see http://www.blogpulse.com/

NY Times (2008), http://www.nytimes.com/2008/04/14/us/14church.html

PBS (2009), Frontline, Feb 17, 2009, "*You've Got A Weekend to Save Yourself*", http://www.pbs.org/wgbh/pages/frontline/video/flv/generic.html?s=frol02s1f8 1q752

Ross, B., and Rehad el-Buri (2008), "*Obama's Pastor: God Damn America, U.S. to Blame for 9/11*", ABC News, March 13, 2008. http://abcnews.go.com/Blotter/story?id=4443788

Technorati (2010), see http://technorati.com/

Chapter 8

Design and Testing of a Support of Ad Hoc Communication for Tacit Workers

Floor Hickmann*, Annelise de Jong*, David Keyson*, Sander Viegers**

*Faculty Industrial Design and Engineering
Delft University of Technology
The Netherlands

**Microsoft Office Design Group
Redmond

ABSTRACT

This graduation project describes the trend of working anytime, everywhere, with anyone, which gives input to a design of an interactive wall at the workplace which supports ad hoc encounters. The problem that arises with this trend is that people have less informal encounters. It is known that a lot of things are getting done by this ad hoc communication. Research is conducted among consultants of Microsoft to understand better what actually happens during these informal encounters. The design vision is to use the serendipity of face-to-face ad hoc meetings by spontaneously showing information about contexts of related people. The final design is an interactive visualization which recognizes a person passing by and it shows contexts of people who might be interesting for that person. A test was setup at a Microsoft office (Schiphol, the Netherlands) to see how people respond to the design. Results showed that the design was perceived well amongst employees but there were several points of improvement concerning usability issues.

Keywords: Social networking for the enterprise, tacit work, serendipity in communication, interaction design, wall display, concept testing.

INTRODUCTION

The way we work is influenced by new circumstances. The society is changing with the introduction of new technologies, a new economic situation, new types of work and a changing mentality of employees.

The amount of tacit workers (also knowledge workers) increases (McKinsey, 2005) which gives more people the opportunity to work anytime, anywhere with anybody. The amount of flexible workspaces will increase and the office becomes a social environment where people go to have meetings (Microsoft, 2005).

Rapid advancements in technology continued deployment of Web 2.0 tools that allow tacit workers to mingle private life and work and tacit workers need to balance their increasingly complicated lives with work. There will be a two way blending. That is, work blending into personal and personal blending into work.

These tacit workers work more individually and there will be a shift away from more permanent, lifetime jobs toward less permanent, even nonstandard employment relationships (e.g. self-employment) and work arrangements (Microsoft, 2005) and working nomads; people who work for themselves while they travel without even having an office.

Since people are working whenever they want wherever they want, they have to stay connected to everyone to collaborate and companies will be more transparent. This asks for better technological solutions. The lack of physical contact is solved by virtual meeting devices like the roundtable and live meeting. There is little technology though to support ad hoc meetings with a colleague in the hall. It is known that this type of contact is one of the most important ways of getting things done. Informal encounters are useful means of getting people to know and like each other, of creating a common context and perspective and of supporting planning and coordination in group work (Kraut 2002).

The challenge is to understand more about the nature and value of informal and ad hoc communicative activity and seeing whether technology could be fruitfully employed to aid it.

PROBLEM EXPLORATION

What happens at an informal encounter at the workplace?

Since consultants of Microsoft Schiphol, The Netherlands, are a good example of people who work anytime, anywhere with anybody, research is done by interviews with 11 consultants and with personal booklets to understand more about the nature and value of informal and ad hoc communication. A design vision will be developed by these insights. The office where these consultants work is designed by the ideas of Bijl (2007). At this office people have the possibility to flexibly adjust the balance between their private life and work by having the freedom to work the way they want, supported by (new) virtual communication tools like Instant Messaging, Microsoft RoundTable and Microsoft Office LiveMeeting. The office becomes a(n) (ad hoc) meeting place.

This research amongst 11 consultants showed that ad hoc encounters are really important to maintain relationships, getting to know new people and to get up to date. First contact, building trust, close collaboration and expression of emotions really need to happen face to face.

The majority of the conversations are about what they are doing, what they have been doing and what they are going to do. It is also important to know how these things are going. Employees can find out what they can mean for each other and how they can help each other. They choose to have contact by going to the flexible workspace.

When people can't have this contact they miss the possibility to have easy access to others at a certain moment. By not knowing the context of a person it is hard to decide how to reach a person best and to know in which kind of mood the person is. They all use e-mail, phone and IM for ad hoc communication.

Definition of ad hoc

Kraut (2002) specified four kinds of communication at the work floor.

- *Scheduled*; one in which the initiator set out specifically to visit another party.
- *Intended*; one in which the initiator had planned to talk with other participants sometime and took advantage of a chance encounter to have the conversation.
- *Opportunistic*; one in which the initiator had planned to talk with other participants sometime and took advantage of a chance encounter to have the conversation.
- *Spontaneous*; a spontaneous interaction in which the initiator had not planned to talk with other participants.

Opportunistic and spontaneous contact could also be called 'unintended' contact. These occur more often at the workplace than intended encounters (Kraut 2002). In this paper an ad hoc meeting means an unintended encounter.

Conclusions of past research

There have been a number of experiments conducted to support remote informal communication, for example described by Heath and Luff (1992), Honda (1999), Jancke (2001), Kraut (2002), Obata (1999) and Scholl (2005). These experiments pointed out several important issues. Most of them are based on continuous visual awareness which could replace the visual aspect of normal ad hoc meetings which offers minimal costs in initiating remote conversation.

Most of these cases show that this continuous visual awareness doesn't replace the visual aspect of ad hoc communication. It touches the privacy and it has shown that the cost is still too high to start a conversation. Therefore it is decided not to continue with visual awareness as a solution space in this project.

Design Aims

It is clear that the physical contact can't be replaced. Some things need to happen face to face like having a first contact, building trust and problem solving. The goal is not to find a way to replace this contact but to support this contact.

Serendipity is an important aspect of ad hoc meetings. Since there is no visual support the same goals of general awareness and expanding of network are being used in a serendipitous way. This serendipity will surprise employees with information about different interesting contexts of others while they are not especially looking for it. Since employees have already gotten an overload of information while they are behind their computer it is decided to provide this information at the office in a spatial way.

CONCEPT

FIGURE 1. Visualizations of the concept.

A design was made to support this ad hoc contact and is designed for the office where employees work flexible; the place where people go to work or to have (ad hoc) meetings. The concept consists of a wall which recognizes a random passer-by and responds to him/her by giving adequate context information about possibly interesting colleagues, known or unknown (see Figure 1). Profile information of the social network of new SharePoint 2010 is used to match people. The passer-by has the possibility to respond to this information immediately. The aim is to create a higher general awareness of familiar colleagues and they have the possibility to expand their network by bumping into new interesting people.

TECHNOLOGY

A passer-by is recognized by the wall by a RFID chip in his/her personal security pass. The "wall" uses three inexpensive infrared sensors and a rear-mounted camera. This camera notes when something breaks through the laser line and feeds that information back to the software. The TouchWall of Microsoft Research is a proof of concept of this technology.

CONCEPT

Interaction

When an employee passes the wall he/she is recognized by the wall. Quickly different contexts appear on the wall and the contexts follow the passer-by so he/she doesn't lose time and has the possibility to glimpse at the wall. When a context seems interesting the passer-by stops and more context information is revealed together with other contexts of connections. The passer-by can choose to get an e-mail with a link to profile information or to do a coffee request. When the person of the coffee request is in the same building he/she gets an instant message and if not he/she receives an email. When there is a second passer-by the two people standing in front of the screen can choose to see how they are connected to each other by SharePoint 2010. When there is no one close to the wall random contexts are shown like a screensaver. Also at home people have access to the context information.

Determination of contexts

An overview can be made from the social network of SharePoint 2010 which shows different connections. First grade (known) or second grade (maybe unknown) connections will be shown together with random new people. The distance between people will be shown visually by the distances between context circles on the wall. In the SharePoint profiles people describe their expertise, hobbies, activities and tags. Employees are linked to each other by this information. When there is a new employee, he/she is introduced to all the other employees by showing him/her to all by the wall.

Context information

The quick contexts show a picture of the person, the name, the availability, the calendar activity and the location. Location can be determined through 802.11 networks. The existing 802.11 wireless network infrastructure (Krumm, 2004) already exists in many places.

The more extended context shows a bigger picture, name, availability, job title and project information. There are also two buttons to get an email and to do a coffee request.

TESTING

GOALS

It is important to perform a study in a real-life situation to test the functionality and user experiences of the concept. A test was setup at Microsoft Schiphol, The Netherlands. The goal of testing is to know if the goals of the design are achieved. These goals are to support ad hoc communication, to create general awareness and to expand people's network. Sub goals of the test are seeing how people respond to this concept in a real office environment and understanding what they think about the context information, the possible interaction and their privacy.

METHOD

Tools

Due to technical restrictions the concept which is tested was different than the original concept.

The test setup used a projection of a beamer at a screen to show the content on the wall since no interactive wall was available. This caused a shadow of the passer-by on the wall of the people who are standing in front of the wall and therefore the beamer is placed as high as possible to minimize shadows.

To simulate the interaction the content is built with Adobe Flash Player. With help of invisible buttons in the projected area the interaction could be simulated secretly by mouse clicks and roll-overs.

It wasn't possible to make the wall personal by RFID recognition; therefore contexts of five people of the same floor of the office were chosen to be shown where also the test took place.

The test is recorded by film inconspicuously by placing a small camera at a table nearby, hidden by surrounding objects.

Location and participants

Fourteen random participants, mostly consultants, who pass by at the 5th floor at Microsoft Schiphol, took part in the research in The Netherlands in October 2009.

FIGURE 2. (left) Test setup at Microsoft Schiphol and (right) example of projected context.

Test set up

Two navigators sit undercover on the right side of the screen (see Figure 2). They simulate the interaction secretly. When no one is near the wall, the screensaver modus turns on. The name, picture, the availability, the calendar activity and location are shown. The different contexts fade in and out slowly.

When a person enters the area in front of the screen, the navigator clicks on the screen and follows the person at the bottom of the screen with the mouse where invisible buttons are located. After passing by the contexts fade out.

When a person stops or tries to touch a context, it will grow with a mouse click. The person sees the picture, the name the current projects, the availability, the function description, some random related colleagues, and two action buttons.

When the person touches 'more info', more info will be send to the person by e-mail theoretically. The person can also choose for a coffee request.

After touching one of those two, a thanks box appears. Restricted by the use of just a few intuitive invisible buttons it wasn't possible to have more detailed feedback to each action.

After use the participants are asked to fill in a survey and a short interview is performed.

RESULTS

Observational results (first response and interaction)

It was remarkable that almost half of the people were avoiding the screen with projections; some were scared to go the screen or were too occupied to pay attention.

   In the first place all the people didn't understand that they could really do something with the screen, some just started to watch from a distance. Therefore little hands, cut out of paper, were put on the screen during the test with the text 'touch me' to let people understand that they could interact with it. Then all people started to explore the concept by touching the screen.

   Their first reaction after seeing the contexts was to touch the person to know more about him/her. Three participants wanted to see their own context first, which is possible in the design but not in this test. The majority of people wanted to touch the related people of the shown context and get more information about them. A few of them didn't understand that the different contexts were correlated. Every time that all of them considered something interesting they wanted to see more information about this item by touching it. Two buttons were interactive in the model and none of the people understood which things were interactive and which things weren't. When they explored all possibilities after a couple of minutes they walked away.

Survey and interview results

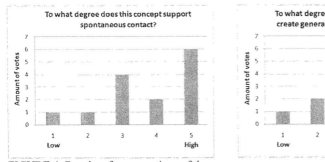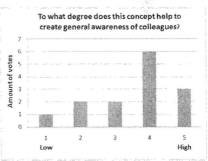

FIGURE 4. Results of two questions of the survey.

More feedback was gathered by a survey with 14 people (two examples see Figure 4) which was given right after experiencing the concept and there was asked the participants to explain themselves.

- *General opinion;* all the people liked to see the concept when passing by. The idea was considered 'out of the box' and interesting.

- *Ad hoc support;* the majority of the participants considered the concept as a useful tool to support ad hoc communication (Figure 4).
- *General awareness;* the majority of the participants considered the concept as a useful tool to create general awareness (Figure 4) and 11 people considered general awareness important.
- *Expanding network;* nine participants considered the concept as a useful tool to expand their network, which is important for 12 people.
- *Visual connections:* eleven people liked to see connections between people.
- *Privacy aspect;* only one person didn't like the idea that his context was shown on the wall. People want to have control though about which part of their context is shown at the wall. The idea of being recognized by RFID didn't bother twelve of the fourteen people.
- *Distractive at work;* ten people didn't consider the wall as too distractive unless it is situated close to the coffee corner.
- *Other;* three people emphasized that it is important to have the possibility to do something with a context immediately. Three people brought up the possibility to search for people at the wall as a browse function.

CONCLUSIONS

The idea of a social wall at the office which provides context information about colleagues to support ad hoc communication was perceived well amongst employees. Most of the participants of the test were convinced that the concept could enhance the ad hoc communication, create general awareness and expand people's network.

People liked to see the contexts on the wall and it is an interesting way of seeing what others do, without having privacy issues. It is a good way to improve the cohesion of employees physically instead of virtually.

The flow of the contexts that follows people as they walk by, the personal recognition, the possibility to dive into a context and the possible meeting actions are good aspects of the design. Nevertheless people wanted to navigate deeper into the information provided and are interested in search possibilities. This test was indicative; the system didn't function entirely but it worked to see what people's first response was. Further research of the concept could be a test with a more elaborated interaction with more advanced equipment combined with a test where the participant is recognized personally. Also the coffee request and the show of interest in others profile should be tested in a real office environment to see the effect on longer term.

The concept doesn't replace ad hoc communication. This requires the resolution of too many trade-offs. A system must provide access and openness as well as restrictions and privacy. The feeling of working together when people are separated is hard to replace by technology; in some situations face-to-face contact remains crucial. Therefore it is important to find the equilibrium between a physical minimum and a virtual optimum.

ACKNOWLEDGEMENTS

We would like to thank the Office Design Group Microsoft (Redmond) for making this project possible, Microsoft Netherlands for providing a workplace and cooperating with the explorative studies amongst consultants and test set up.

REFERENCES

Bijl (2007), *Het Nieuwe Werken.*Den Haag, Sdu Uitgevers bv.

Heath and Luff (1992) *Media Space and Communicative Asymmetries: Preliminary Observations of Video-Mediated interaction.*

Honda, S., Tomioka, H., Kimura, T., Oosawa, T., Okada, K. (1999) *A company-office system 'Valentine' providing informal communication and personal space based on 3D virtual space and avatars.*

Jancke, J.,Venolia, G.J., Grudin, J., Cadiz, J.J., Gupta, A. (2001) *Linking Public Spaces: Technical and Social Issues, CHI.*

Kraut, R.E., Fish, R.S., Root, R.W., Chalfonte, B.L., (2002) *Informal Communication in Organizations: Form, Function, and Technology.*

Krumm, J., Horvitz, E. (2004) *Locadio: Inferring Motion and Location from Wi-Fi Signal Strengths.*

McKinsey (Johnson, B.C., Manyika, J.M., Yee, L.A.) *The next revolution in interactions,* The McKinsey Quarterly (2005)

Obata, A., Sasaki, K, (1999) *Video Communication System that Facilitates Informal Communication among Distributed Offices.*

Scholl, J., (2005) *Technology for Supporting Informal Communication in Multimedia Conferencing Systems.*

Microsoft, Digital work style; *The new World of Work,* Microsoft white paper, 2005.

CHAPTER 9

The Multidisciplinarity in the Development of Work Systems Conception Projects: A Review Theoretical-Practical About the Subject

Giles Balbinotti, Leila Gontijo, Arlete Motter, Marcus Guaragni

Universidade Federal de Santa Catarina, UFSC,
Florianópolis, SC, Brasil

ABSTRACT

To the development of systems conception projects, it´s sine qua non conditions that the work in multifunctional teams through methodology that encourages social interaction looking forward to bring potential risks of a project process, for project team itself and for the customer. We did the bibliography research about this subject specially about projects steps and the action by projects actors. The outcome was satisfactory, to confirm the hypothesis that the transversal work in the project and the strong consideration of the human factors are fundamental in the achievement of the pre-defined targets and in the achievement of the social-economic-technical results.

Keywords: Quality of project; multidisciplinarity; project players; project industrial management ; social-technical questions

INTRODUCTION

We believe that through the active involvement of the specialists when the activity of conception of a process or product was tied straightly, we will reach with fewer efforts and stress the data of entry definite-daily pay.

In *L'usine of l'avenir* (Du Roy 1992) shows that the construction of an organization cannot come down simply to an evaluation of productivity of the labour and to the preparation of organization charts, it is a much richer and more complex process. One resorts then to other areas of the knowledge with sights to make rich the sciences and the technologies of the engineer through the knowledge of the expert partner - technician.

We understand that the participation of a professional with knowledge and experience as much in the technical area as the human ones, will be an efficient tool on behalf of the success of a project. (Martin 1998) wrote in his theory of doctorate " *La preoccupation dominante est d introduire tres tôt, dans le projet, les futurs exploitants en leur donnant un rôle de prescripteurs. Plusieurs raisons sont mises en avant pour faire travailler ensemble diverses categories de profissionnels dans la conception des futures situations de travail.*

For this actors' involvement with several competences, (Erdmann 1998) defines it as simultaneous engineering. This term was created in 1986, like part of a report of the it was Defense analyses Institute of the USA, where it was defined like " a systemic approach for the integrated, simultaneous design of products and his connected processes, including manufacturing and the support ". The simultaneous engineering, also called competing engineering or parallel engineering, has been defined to by some authors like simultaneous project of a product and of its process of manufacturing.

According (Ribeiro 1989), "the formation of multifunctional teams working inside a matrix scheme, and the way to begin the practice of simultaneous engineering in a correct way ". These teams, of several areas that have connection in the project, each one with its limits of autonomy and responsibility, that have with principal objective the optimal efforts and consequently the achievements of the results pre-determined

DEVELOPMENT OF A PROJECT

Stages of a Project

A project is a temporary effort undertaken to create a product, service or exclusive result. The organizations carry out a work to reach a set of objectives. In general, the work can be categorized like projects or operations, though the two occasionally are put on top. They share a great deal of the next characteristics: it is carried out by persons; restricted by resources limited and planned, executed and controlled. The projects and the operations differ principally in the fact that the operations are continuous and repetitive, while the projects are temporary and exclusive. (Guide PMBOK, 2004).

For (Slack 1999) the management of success projects has to do with some important factors: definite marks, manager of competent project, support of the superior administration, competent members of the group of project, sufficient allocation of resources, appropriate channels of communication, mechanisms of control, capacities of feedback, answers to clients, mechanisms of attack of problems and continuity of the people in the project.

According to the Guide PMBOK, the management of projects is the application of knowledge, skills, tools and techniques to the activities of the project in order to pay attention to its requisites. The management of projects is carried out through the application and the integration of the next processes of management of projects: initiation, projection, execution, follow-up and control, and ending.

The manager of projects is the person responsible for the achievement of the objectives of the project. The management of a project includes: identification of the necessities; establishment of clear and reachable objectives; balancing of the conflicting demands of quality, aim, time and range; and adaptation of the specifications, of the plans and the approach to the different preoccupations and expectations of several interested parts.

The principal stages of a project, based on (Daniellou 1989) in figure 1, where he presents the classic phases of an industrial project. The macrophases are show basic and essential phases in an industrial project.

The knowledge of the development in time of the project as well as the clear identification of the key stages are basic phases for the continuation and definition of the actions of the project.

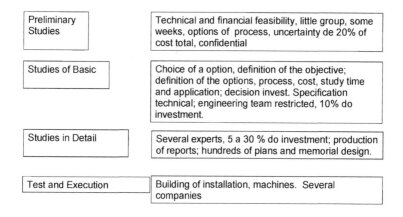

Figure 1. Classic steps of a project

For Dinsmore (1991), the potential causes of problems project pass by a projection of the insufficient time, a planning no participating, the lack of answer to the necessities of the clients, weak perception of the personal aims of team members of the of project among others

The necessity of removing, besides several potential causes of problems in a project, the causes pointed (Dinsmore 1991), show the relevance in defining and implementing clearly a method that approaches all the determinants that a winning project demands.

Conceptual Method

According to the ISO 10006:2002, management of projects means the existence of the stages of projection, organization, attendance, control besides the accounts rendered of all the aspects of a project and undoubtedly, the motivation of the persons when they were involved to reach the objectives of the project.

Garrigou (1992), defines SAC – Situation of characteristic action as being the correspondence to the set of determinants of the structuring of the activity: the objectives of the action; the involved people; the sources of information, the ways and necessary tools; the elements that delimit the action (time, environment quality criterion state of the tools, work enviroment); the relative elements to the conditions of night work, physical risks ..) and what can have consequences to the health.

Work coordinated by the sociotechnical , in the decade of 80, in order to understand and board the troubles found in the driving of many industrial projects, showed the frequent deficiencies up in the driving of project). The deficiencies from the lack of clear identification of the control of the entrepreneur (here he is the director of the factory) and of his person in charge (here he is the chief of project), up to the fragility in the definition of the objectives of the project. The weak presence of the entrepreneur and the defective association of the "users" in the project they complete the list of deficiencies. DANIELLOU (1992)

In order for quality conceptions be produced, Daniellou (*apud* Falzon (2007)) recommends some aspects: the leadership of the project: relation of entrepreneur and coordinator of project, interface between representative persistence; a clear definition of objectives social-technical; important actors involved in the project; and search of solutions for achievement of the established objectives.

The managers of projects frequently talk about a "triple restriction" — aim, time and cost in the management of conflicting necessities of the project. The quality of the project is affected by the balancing of these three factors. Projects of high quality hand over the product, service or result asked inside the aim, in the term and inside the budget. The relation between these factors take place in such a way that if some of three factors change, at least another factor probably will be affected. The managers of projects also manage projects in reply to uncertainties. A risk of the project is an event or uncertain condition that, to take place, will have a positive or negative effect in at least one objective of the project. (Guide PMBOK, 2004).

The Human Factors in Projects

The management of human resources of project includes the processes that organize and manage the team of the project. The team of the project is composed of persons with functions and responsibilities attributed for the end of the project. Though it is common to talk about functions and attributed responsibilities, the members of the team must be involved in great part of the projection and of the taking decision of the project. The involvement of the members of the team from the beginning adds specialization during the process of projection and strengthens the promise with the project. The type and the members' number of the team of the project very often can change according to the project if it develops. The members of the team of the project can be called of people of the project. (Guide PMBOK, 2004).

According to the Guide, the team of management of projects is still a subset of the team of the project and is responsible for the activities of management of projects, like projection, control and ending. This group of persons can be called of principal, executive team or leader. In less projects, the responsibilities of management in projects can be shared by the whole team or administered only by the manager of projects. The sponsor of the project works together with the team of management of projects, normally helping with questions like financial resources of the project, explaining doubts on the aim and practising influence on other persons to benefit the project.

Inside the organization, being part of the project or not, the individuals participate of social groups and it supports itself in constant social interaction. To be able to explain the human behaviour in the organizations, the Theory of the Human Relations, according to Mayo (1977) started to study this social interaction. So, they are called human Relations the actions and attitudes developed by the contacts between persons and groups.

Each individual is a differentiated personality that influences the behaviour and the attitudes of others with the one who maintains contacts and is, on the other side, equally influenced enough by others. Each individual tries to conform to other individuals and to other definite groups, claiming to be understood, to be quite accepted and to participate, in the sense of paying attention to his most immediate interests and aspirations. His behaviour is strongly influenced by the environment and for several attitudes and informal existent standards in several groups.

It is principally inside the enterprise that the opportunities of human relations appear, in view of the great number of groups and to the necessarily resultant interactions. There is exactly the understanding of the nature of these Human Relations that allows to an administrator better results of his subordinates; an understanding of the human Relations allows an atmosphere where each individual is encouraged by expressing release and health.

And in this social context, all the activities, including the work, has at least three aspects: physical, cognitive and psychological, and each one of them can determine an overload. Every individual reaches the work with his genetic capital, raising the set of his pathological history. He also brings his way of life, his personal and ethnic customs, his apprenticeships. Completely that weighs in the personal cost of the situation of work in which it is put. (WISNER 1994)

The different professional categories that appoint the " players of the conception " have not been in general by the habit of working together, and they ignore frequently the embarassment and way of work with others.

For Dinsmore (1991) two guidelines show the importance of the " human side " in the coordinating of projects of conception: persons are the causes of the problems of project and the problems of project can be resolved only by the people.

Bucciarelli (1992) emphasizes that in his studies of project, he sees many different individuals participating in a task of project, working in all the traineeships shown in his model and still, what each participant, while him and when to put a different aspect in the project gave responsibility, it needs to work together with all the other participants so that a good project happens. We understand then that the vision of project is characterized like a dialog between individuals who fulfill different papers in the institutional context. In the architecture it is similar, there is (or there should be) the interaction of an architect, the contractor, the structural engineer and the user of the design of a construction. Each one brings his different roles, languages, knowledge and interests.

To illustrate the comments of (Bucciarelli 1992) we can present the used methodology by Duarte (1999). In the phase of construction of the building of the Integrated Centre of Control of a petrochemical Brazilian industry. The evaluation of the initial project was made putting a team of three ergonomics professional, that used as a theoretical referential system Ergonomics Analysis of the Work – AET (Wisner, 1988/Guerin et alii, 1991) where from a diagnosis of the conditions of work and possible the necessities of the operation did recommendations for adaptation of the ways of work. Initially it was done a macroanalysis of the situation of the work (functioning would refine, population workers and distribution tasks between operators), after general and glimpsed observations were carried out, and simultaneously when the conditions were checked work form in other refineries, by visits to situations of reference (DANIELLOU, 1992).

Subsequently one carried out analyses of activity of the operators, next, in the definition of a new proposal of layout several meetings were carried out involving operators, engineers and *ergonomist*. The recommendations ergonomics when the enterprise was presented based between others on the experience of the integrants of the team.

Finally as Bucciarelli (1992) says, the diversity of opinions is healthy. Ambiguity, will always exist in the process of design and the different participants, with their values, beliefs, a few different ones from others will make the activity of project rich.Du Roy (1992) emphasizes: "the technical system and the social system (characteristics of the workers in terms of age, sex, education and formation, professional culture, expectations) will be considered together "

In the Sistemic School (1960) authors like Ludwig von Bertalanffy, K. Boulding and Katz and Kahn define that the organization is an integrant part of bigger system and that all the interactions are interdependent. And like consequences of this system the decisions must take into account environmental variables; and the organizations influence and are influenced by the environment (system).

At the next moment (decade of 70) the social-technical approach prescribes that the organization is a complex system where interact factors of technological order and of human/social order, besides that it is in the best interaction between these factors, that the best performance of the organization happens, endorsing what Du Roy presented.

Du Roy (1992), reinforces that the conception must be interactive, where there is interaction between the aspects social - organizational and the technical aspects; to reach this scenery, it is basic that we create alternatives and to make sure that in the conception of the project there is dialog between several actors and several types of specialized knowledge like representation of fig 3 created by Jackson (2000) and adapted by the author of this article

Figure 2 shows aspects inherent to a situation of work and that must be considered in he conception. The human factors must be considered strongly in the realization of a project, either referring to the actors who drive the project, or regarding the actors regarding whom they will act in the situations of work it predicts. And the aspects pointed in figure 2, orientate the possible creators of overloads, according to observation by Wisner (1987).

Duarte (1999) explains, why to speak in ergonomics in the projects. It is necessary to talk of ergonomics in the projects, because the limits of the corrective actions as cost and difficulties of the changes must be considered; because it is important to integrate the knowledge of the workers; also to value the experience built in the enterprise; the anticipation of the problems besides the participation of the workers in the decisions and choices of the solutions and trainning in the new technologies resulted with the new project.

Another well-adjusted representation of Jackson (2000), suggests the interaction of the players of the project in the development of the activities.

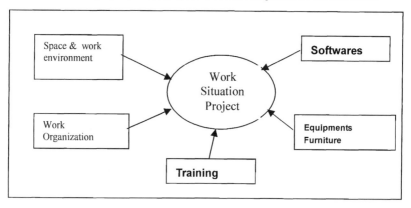

Figure 2. – Participation of ergonomic in Projects

Actions of the players responsible for the ergonomics in the projects must act together in order to reduce the future embarrassment in the factory, acting strongly together in the predictions of the future activities.

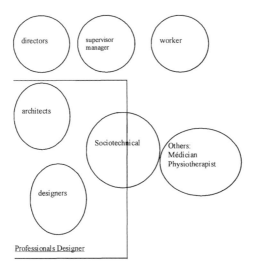

Figure 3. –Projects Players

Fundacentro proposes some actions special for these actors: - to build prediction of the future work and to integrate the principal characteristics in the project of the system and of its organization; - to introduce the beginnings ergonomics necessary; - to help to make the project rich; - eventually, to influence the structure and the management of the process of project.

THE RESULTED ON PRACTICAL CASES

The functioning of the projects in an assembly company of vehicles

Case: an assembly company of vehicles (Brazil)

In an assembly company of vehicles, there are established stages associated to the good development of a new project. A called document *"ticket d'entrée"* is characterized by the set of specific expenses to a new project. It contemplates the areas of design, engineering and manufacture.

In this context, 3 phases of development are worked: the phase "*amont*", where the preparatory studies are done and of viability of the project, then comes the phase of conception, where the studies are worked on the product and process and finally the phase of industrialization where it effectively implements the process of manufacture of the new product.

Inserted in the phases presented in the previous paragraph, the stages are (*jalons*) of project such as intention, direction, pre contract, contract, realization means of production, assembly prototypes, insertion of the product in the line of production in order to value quality criteria and decision of production.

And in this context the liveliness of the human factors inside these projects is a central point in which it refers to the efficiency of the processes and to the efficiency of the expected results. In the projects of this assembly company, important attitudes are taken like nomination of the players of the project (and also the qui fait quoi) besides agenda type of quite definite meetings and with objective and indicative of results so that significant advancement has been in the processes of the industrial project. The considerations of the human factors, specially of the future users of the systems of work central focuses and constant point of the diaries of work. The preoccupation is such that an important area is focused in the factors and human necessities , it was defined and implemented in order to secure the convergence of the objectives sociotechnical of the new projects.

The Human Factors of the point of view of a chief of project of a manufacturer of automobiles (by M Guaragni).

As for the coordination of an industrial project, we cannot forget that the coordination of a project consists basically of the coordination of persons, and that the principal challenge is lead the actions of the different persons so that the sum of the efforts agrees with the necessities that the project imposes. In individual the coordination of projects multicultural imposes a special attention to the differences that the multi cultural aspects bring, which makes it much more challenging and delicate management of the project in terms of strategic and functional aspects.

The entrepreneur has great expectations as for the execution of the project, usually invests financial sums in an idea or concept is left which has no absolute security about the concrete results that will be able to be achieved. So, the coordinator must equip oneself with different fountains of information in way to understand what the real expectations of the entrepreneur in quantitative and qualitative terms so that that one feels satisfied regarding his expectations. For the good results the relation of the binomial coordinating-enterprising debit to be based on the transparency of actions and in share of information that are made a list to the progress of the project, in what way is possible to "spin constantly the PDCA" of executed works.

If a project will have more than a representative persistence the complexity increases, when the challenge of which the same philosophical line is maintained in terms of the management was given global and of the results to be obtained. It is recommended someone that be elected as the "only" representative of the project whenever possible. In Case it is impracticable, the coordinator will have to have the insight to integrate in a method of management when the different expectations of the representatives of several levels. It increases, so, the complexity of the management of the project, demanding a profile a lot more autonomous ´, flexible and visionary on the part of the coordinator. The clear definition of the objectives social-technical is the base for the success of one undertaking, since a project is done " by persons " and " for persons ". This form, if the criteria of excellence in we will have sociotecnical professional not clearly defined it is possible for all the work " to be lost " at the moment of the final delivery of project to the organization. There will be no operational funcionality the whole effort of financial and human undertaken, will be for nothing. Man must always be the centre of attention in the management of a project, be it in the projection or in the execution. It is a question of future survival of the organization. In the whole project there are the "key" players who must be considered as the most important for the effective success of the undertaking. It is necessary to state clearly that, not always, the most important persons are not the one who have higher level hierarchically in the organization. So, it is necessary not to confuse the roles. The good manager of projects can identify well for each moment of the project what they are the players of bigger relevance, which varies from phase to the undertaking. On basis of this important identification it will have to focus his energies so that there is correct contribution of these players so that each phase is well succeeded.

The management of a project demands constant adaptability for part of all involved, obviously coordinated by the figure of manager of the project that has to direct the efforts in the correct sense for the success of the project. In this scenery it is natural that there are adaptations to be carried out so that the project continues with the good results. This search makes part of the game and it is a great engine of the manager of project.

The Function Sociotechnical as element integrator

In order to answer the preoccupation with the human factors that use the means of production in the future, there was conceived an internal proceeding that defines the mission and the responsibilities social technical for application in the stages of project. The mission of the social-technical professional of the enterprise aims to secure studies for new industrial projects in this perimeter, through internal coordination with involved correspondents, from phase pre contract until to the stage of the agreement of production, aiming at the attendance of the reduction of psychological problems and physicists connected with production; the respect to the Brazilian legislation and to the directives of the enterprise; the guarantee of the formation of the people in the new technologies; the promotion of the organizations of the work in cells of work; and the guarantee of the coordination of all the stages defined for the projects of the new vehicles.

This function has especially the responsibility to defend the physical and mental integrity of the workers of the enterprise. We list down some of them: To secure the lifting of the profile of the population of collaborators and the cartography of the existent competences (age, qualification, formation) for effectuation of future projection, together with the medical service; To establish and to follow the objectives of ergonomics of the posts of work, near the engineering and the ergonomics professional; Secures the respect to the legislation all that, to the security, hygiene in the manufacturing environment, with corresponding conditions of work; To act in the process of conception and reception of ways associated to the new projects; To participate in the construction of the plan of development of the competences and of the budget associated to the new projects ; To negotiate areas of work, which favor the development of the professionalism in the achievement of the performance; To guarantee the quality objectives, cost and term, of the chapter social-technical of the contract of project; To secure the advancement of the project of industrialization near the chief of projects of the factory; To treat the questions social-technical in the meetings of project of the departments.

The social- technical management is key to improve the financial health of the organizations, and today, the ergonomics sets herself up as a tool of management for the organizations. The challenge is to look sinergy between the technical and social systems, securing a vision *antropocentric.*

DISCUSSION

The methods of driving of projects presented by the authors quoted in this article, converge practically in all the aspects, showing that there is no secret in order that the technical stages of this process are defined.

What is debatable, that concerns the increase of the productivity and quality of the ways and of the ends (resulted) from the projects of conception, is the social question the convergence of the points of view, the cooperation between the players, the motivation to look for the best solution, these factors can take the organization to the success or the opposite.

M Guaragni, with his experience in management of teams of project, shows the condition sine qua non of the success of the expected results, when he says that the co-ordination of the persons is the principal challenge for the success and that this challenge is result also of the multicultural aspects of the actors and of the institutions involved in the process of project, this concept already worked like Antropotecnology or Macroergonomics.. Guaragny punctuates on the paper of each actor of the project saying that in projects, it is important the hierarchical questions set aside and yes to value the technical competences.

And talking in this context, the technical competence about the subjects inherent in the persons (of the project and of the users) it was presented in this article the mission and the principal activities of a professional function that takes as a principal focus the guarantee of the good management on the conditions of future work. The management social-technical, defended by Du Roy (1992) in *L'usine of l'avenir*, presents itself a model of management that collects the social and human questions in the development of conception of system of work. It is a modern and relevant concept that though little seen in the Brazilian industries, it is implemented with the purpose of reaching the social, technical and financial objectives of the enterprises.

CONCLUSIONS

The purpose of the article was reached, since it shows through the base conceptual investigated, which some daily pay-requisite are predominant for the success of the industrial projects.

Firstly, the clear definitions of the stages of a project of conception of systems of work that considers some basic activities in order that results are obtained, are financial or of quality, besides the importance of establishing these quantified objectives. As for the human factors, makes a request for central office inside the projects, they are relevant principally it raises the questions of involvement of the players connected with the project and above all the necessity of the cross work between the areas and the players.

Questions referring to the skills of the players of project, principally regarding the vision of all with the science of each part inherent in the project, come strongly to support and to drive positively the achievement of the expected results. Undoubtedly, the systemic vision of the processes of production, of the variability of the products, of the productive commitments, of the processes logistics and of engineering, as well as of the knowledge of the constraints of the system man-machine they will be of basic importance and demand so that the team of project is necessary with his paper and secures appropriate conceptions of systems of work and important performance for the people (workers) and for the organization.

The relevance of the central subject of this paper is made a list to the motivation of the persons and to the quality of life in the work that can be understood like a good was made a list to the work of an individual and the extension in which his experience of work is compensatory, satisfactory and looted of stress and other negative consequences.

The question ergonomics can be understood like the straight resultant force of the combination of several basic dimensions of the task and other dimensions not dependent straightly of the task, able to produce motivation and satisfaction in different levels, besides turning in several types of activities and conduct of the pertaining individuals to an organization.

The correspondence between productivity and quality of life is twoways and straightly proportionally, i.e. quality of high life, values of productivity also tops; low quality of life will provoke low rates of productivity.

One concludes then finally, that the preoccupation with the ergonomic question is a condition *sine qua non* of success of the enterprise for that to prosper long. We have to an effective gerenciamento, fulfilment of the relative marks to the financial health of the enterprise and the health of the workers.

This paper intends to give a relevant contribution for the engineering of production, since it boarded the development of projects of systems of work in an area, the Ergonomics / Social-Technical, which has great influence in the results of the organizations. The survival of the enterprises is a dependent-human system, in other words, the question ergonômica and sociotécnica, through the human motivation it is a base for this survival and for the consequent prosperity of the organizations.

I finished on basis of my experience of the *gemba* and through the fulfilled inquiries, which results I presented in this article, is obvious that the work together, the social interaction in the development of the activities of projects of conception is predominant for the success of the "products" resulted from this process.

AKNOWLEDGEMENTS

To my children Lucas and Adrian, for my absence. To my friends who lives in Canada Erlon, Adelia, Erlon Jr e Maria Eduarda.

REFERENCES

BALBINOTTI, Giles. A Ergonomia como Princípio e Prática nas Empresas. Curitiba: Editora Gênesis, 2003.

BANNON, Lian J; From Human Factors to Human Actors: the Role of Psychology and Human-computer Interaction Studies in System Design. 1991.

BUCCIARELLI L; Implementação dos fatores humanos na administração de amplos projetos de investimentos industriais: um ponto de vista da administração e prática de ergonomia. 1992

BUCCIARELLI, Louis L. An ethnographic perspective on engineering design. Designes Studies, [s.l.], v.9, n.3, July, 1998.

CHAPANIS, Alphonse. Ergonomics in product development: a personal view. Ergonomics, [s.l], v.38, n.8, p.1639-1660, 1995.

DANIELLOU, F.Lérgonome et lês Acteurs de la Conception. In:Actes du XXIXéme Congrès de la Société d'Ergonomie de Langue Française Eyrolles,Paris,1994, p.27-32.

DANIELLOU, F. Le statut de la pratique et des connaissances dans l'intervention ergonomique de conception. Texte d'habilitation ·à diriger des recerches. Paris: Université Toulose Le Mirail, 1992.

DINSMOREW, Paul C. Human factors in project management. New Jersey, USA: Lawrence Erlbaum, 1991.

DU ROY Olivier. L'Usine de L'Avenir – Gestão Sociotécnica do Investimento – Métoos Europeus; 1992

DUARTE F. Complementariedade entre ergonomia e engenharia em projetos industriais. 1999

FALZON, P (editor); Ergonomia; Editora Blucher; Paris. 2007

GARRIGOU, A. et al. Activy analysis in participatory design and analysis of participatory design activity. Elsevier International Journal of Industrial ergonomics, [s.l.], p.311-327, 1995.

GONTIJO, Leila. A.; SOUZA, R. J. Macoergonomia e análise do trabalho. In: II Congresso Latino Americano e VI Seminário Brasileiro de Ergonomia, Florianópolis, 1993.

GUERIN, F.; Laville, A.; DANIELLOU, f.; DURAFFOURG, j.; KERGUELLEN, A. Compreender o Trabalho para transformá-lo: A prática da Ergonomia. São Paulo. Editora Edgard Blcher. 2001.

GUIA PMBOK . Project Management Institute, Inc. Um Guia do Conjunto de Conhecimentos em Gerenciamento de Projetos. 3a edição, 2004

IIDA, I. Ergonomia - projeto e produção. São Paulo: Edgard Blucher, 1990. 465p.

JAKSON M; Processo de Ergonomia em Projetos ; 2002

LENIOR TMJ; VERHOEVEN J H M.

MALINE J. Simuler le Travail – une aide à la conduite de projet; ANACT. 1994

MARTIN Christian. La conception architecturale entre volonte politique et faisabilité technique, le positionnement de l'intervention ergonomique. Thèse de doctorat d'ergonomie.Laboratoire d'Ergonomie des Systèmes Complexes. Université Victor Segalen Bordeaux 2 – ISPED, 1998.

WISNER, A. Por dentro do trabalho. Ergonomia: método & técnica. São Paulo, Ed. FTD, 1987.

Only the intelligence of all its members may allow a company to face the turbulence and the requirements of this new environment. This is why our companies (Eastern) invest 3 or 4 times more in training its human resources than are their, this is the reason for the existence of our business in a climate of dialogue and communication so intense. (Konosuke Matsushita - Founder Group Matsushita).

"It is very important to take care of the people, so people will take care of the job". (Luis Fernando Pelaez – Director of Renault do Brasil)

Chapter 10

Using Operational Sequence Diagrams to Assess Interagency Coordination in Traffic Incident Management

Robert G. Feyen

Department of Mechanical and Industrial Engineering
University of Minnesota Duluth

ABSTRACT

One role of state-level Departments of Transportation (DOT) is traffic incident management (TIM): managing incidents that hinder traffic flow on interstate highways and requiring responses from multiple agencies (e.g., police, fire). This article reports on a portion of a larger study investigating how DOTs collect and analyze basic TIM performance measures (e.g., lane clearance times). Since evaluation of interagency coordination is one area of TIM in which little success has been attained (FHWA, 2003), process-based benchmarking methods borrowed from operations management may be useful tools. One such tool, operational sequence diagramming, was used to show how tasks and resources were allocated between the different agencies responding to a particular type of traffic incident: a disabled vehicle (no injuries or property damage). In this case, up to three agencies (state police, transportation, and towing/recovery) might interact to coordinate the safe removal of the vehicle and restore normal traffic flow. Completing these events required agency personnel to perform specific functions in specific sequences, something that did not occur consistently or following recommended guidelines. The OSDs illustrated how differences in response operations could be identified and aid decision makers in TIM in evaluating possible options to reduce variability in response protocols and, ultimately, traffic delays.

Keywords: traffic incident management, operational sequence diagrams, interagency coordination

INTRODUCTION

From a social and organizational ergonomics standpoint, a metropolitan freeway system can be considered a highly complex, tightly coupled system. The system allows people using motorized vehicles of various sizes and functions to enter the system at any time and at any access point, maneuver at relatively high speeds via a limited number of paths and exit the system at another point. The goal of this complex system is to allow rapid, unimpeded movement for all users between any access and egress points in the system, but this goal can be difficult to attain as user performance in the system is easily disrupted by weather conditions, system maintenance functions, temporal factors (e.g., time of day, day of week) and traffic incidents.

The latter includes any event creating hazardous driving conditions and/or delaying the normal flow of traffic. Within the United States, the delays alone caused by these incidents re estimated at an annual cost of $78 billion (Texas Transportation Institute, 2007) and do not include the associated costs of property damage, injuries and fatalities. Not surprisingly, these are cause for significant public concern and many departments of transportation have implemented traffic incident management (TIM) programs to avoid these costs. TIM programs utilize planned, coordinated process to detect, respond to, and remove traffic incidents while ensuring motorist safety and rapid restoration of traffic capacity (Farradyne, 2000; Balke *et al*, 2002). Although many events that significantly impede the free flow of traffic are planned (e.g., roadway construction) or expected (e.g., spectators arriving and departing sporting events), many are unplanned and unexpected – such as vehicle stalls and collisions, spilled cargo, roadway debris, and stray animals running loose on the freeway. In Minneapolis-St. Paul, incident-induced delays account for about 50% of all traffic congestion (Texas Transportation Institute, 2007); of these, at least one-half arise from unplanned and unanticipated incidents. Thus, TIM efforts are directed towards rapid detection of incidents and dispatch of necessary response personnel in order to keep these highly complex, tightly coupled systems performing well.

In fact, the entire incident response process requires effective coordination between police, fire, medical, transportation, and other public and private agencies across several jurisdictions [5, 6]. The typical phases in an unplanned incident start with the detection of an incident, followed by verification, dispatch of responders, clearance of the incident, and restoration of the roadway to normal operations (no evidence remaining of the incident). The latter two phases include many activities geared towards managing the incident scene and directing traffic around the scene to keep responders and the general public safe from harm and minimize any associated delays. However, surprisingly few areas have predetermined plans or agreements in place to guide how responders should prioritize and allocate tasks and equipment during any of these phases. Not surprisingly, 72 metro area TIM agencies previously reported that the two areas with which they have the most difficulty managing are in *interagency coordination* and *quantitative evaluation* (Balke *et al*, 2002; U.S. FHWA, 2003). Even at the TIM centers which have had some success in interagency coordination, overlaps and gaps in the incident response process often occur when multiple agencies respond to an incident (U.S. FHWA, 2003). Because each agency has different objectives (e.g., the police have an objective to maintain public safety while the transportation personnel have an objective to return traffic conditions to normal) and these objectives drive their specific response protocols, these differences often lead to conflicts between agencies during an incident. Anecdotally, the resulting overlaps and gaps contribute to prolonged incident durations and degradation in public safety, greater probability of fatalities, longer traffic delays, and increased occurrence of secondary incidents (crashes or related events caused elsewhere in the transportation network by the subsequent delays and confusion of the original incident) (Balke *et al*, 2002).

In response, recent efforts have focused on "performance measurement" of traffic incident management by the National Traffic Incident Management Coalition (NTIMC), American Association of State Highway and Transportation Officials (AASHTO) and the Federal Highway Administration (FHWA). Much of the published research and other publicly available materials related to these efforts focus on the collection and use of commonly defined metrics to facilitate external comparisons between state transportation agencies (e.g., U.S. FHWA, 2009). Only a handful of TIM studies related to interagency coordination have discussed quantitative methods beyond descriptive statistics in terms of operations *within* a state organization; these typically involve a custom simulation model with output geared towards supporting a single, targeted analysis (e.g., a cost-benefit analysis of a 'highway helper" program) (e.g.,. Minnesota DOT, 2004)

OBJECTIVES

This paper highlights a portion of an overall study to investigate quantitative analysis methods that could be used to support interagency operations within an organization or collaboration of organizations focused on TIM (Feyen and Eseonu, 2009). The overall study addressed the following research objectives:

- Identify a common interagency goal;
- Use this goal to identify quantitative performance metrics that can be used to evaluate TIM performance; and,

- Identify methods for using these metrics as feedback for decision makers in TIM interested in interagency coordination and incident response operations.

With respect to the first objective, a literature review and competitive benchmarking survey of TIM agencies in selected North American cities suggested the following common interagency goal: *"without compromising safety, minimize the time spent dealing with a traffic-related incident."* In turn, this goal suggested a set of seven time-based metrics that could effectively evaluate TIM performance across all agencies involved – meeting the second objective of this study:

- Verification time: Incident detection to dispatch
- Agency dispatch time: Dispatch to arrival
- Lane clearance time: Arrival to lane clearance
- Queue dissipation: Lane clearance to all clear
- Removal time: Arrival to all clear
- Overall incident response time: Dispatch to all clear
- Overall incident time: Detection to all clear

Finally, these metrics suggested the use of benchmarking methodologies for evaluating performance of a TIM system, the third objective of the study. Adopting the process-centered view for incident response, a pilot benchmarking study was conducted using data from a 10-mile freeway corridor in the Minneapolis-St. Paul metropolitan area. Based in part on this study, one of the methods identified from the overall project was creating visual benchmarks illustrating the task completion sequences and resource allocation between each responding agency during a major incident type.

METHOD AND RESULTS

In particular, operational sequence diagrams (OSD) were used to assess the relationships between different agencies during response to an incident. This variation on process charting has long been used for operational task analysis. In this application, the operational sequences of interest were being performed by the Minnesota State Police (MSP), Minnesota Department of Transportation (Mn/DOT) and vehicle towing/recovery services in response to detection of a stalled (disabled) vehicle. Other incident types that were considered but not analyzed at this time included vehicle fires, brush fires, crashes with property damage, crashes with minor injuries, crashes with possible fatalities, and abandoned hazardous materials (Frandrup *et al*, 2002).

In Minneapolis-St. Paul, incident response communications are recorded and archived for a year for later review. In addition, incidents on the freeway system can be monitored in real-time via an extensive camera network. Cameras are located at approximately one mile intervals and the zoom and pan of a camera can be controlled remotely by TIM personnel to verify and record incidents. In addition, the public can access (but not control) these real-time camera views of any current incident via the Minnesota 511 traveler information website. The initial OSDs were based on real-time incident response observations from the 511 website as well as recommended guidelines published in 2002 by the region's Interagency Coordination Management Team (Frandrup *et al*, 2002).

Using the generic timeline for incident response (detection, verification, dispatch, clearance and restoration of traffic flow) as the sequence profile, responses were observed and then broken down into task elements required for the process to be completed successfully. The individual task elements were categorized according to one of eight basic functions (Figure 1) and an appropriate symbol mapped to a timeline showing the flow of task sequences (Figure 2). Multiple flow diagrams were developed to show the sequence and allocation of tasks as well as any resource sharing and communications occurring between responding agencies. Decision points were incorporated to reflect flexibility in the response protocols. The resulting OSD served as an initial benchmark, illustrating the recommended sequence and allocation of tasks required for a given process.

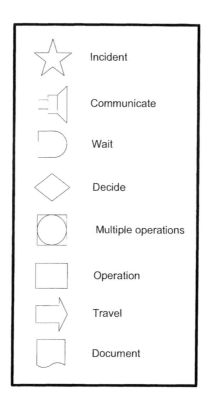

FIGURE 1 OSD symbols used to categorize traffic incident response tasks

DISCUSSION

Often for valid reasons, what actually happens in the field differs substantially from the operational guidelines. This was supported through both the real-time observations and anecdotally through conversations with various TIM personnel. However, as has been seen in other areas (e.g., manufacturing, aviation, healthcare), significant delays can result when unexpected or inconsistent methods or task sequences are used by one or more parties agencies when responding to a specific incident type. In this part of the overall study, selected incidents were studied using the process charts as a checklist to compare what responders do in the field with what they are expected to do when handling a given incident type. Differences between the field activities and operational guidelines could then be documented so that appropriate TIM personnel might review these differences and decide if changes to either current practices or the operational guidelines are needed.

In the context of TIM, several benefits of using OSDs became apparent. Some of the advantages in this particular setting include identifying what types of activities each responder was performing at a given stage in the incident process and the level of communication and resource sharing required to progress towards clearing the incident scene and returning traffic flow to normal. Differences between the actual process and guidelines indicated potential sources of variability in response times, reflected adaptations in the field to more effectively handle incidents, and highlighted coordination conflicts and resource sharing needs between agencies in handling incidents.

OSDs could also be useful when reviewing incidents with excessively long response times. In this case, the OSD becomes a template for identifying what task elements may have contributed to delays and provided clues as to how the response delays may have occurred. These analyses also point out dependencies between agencies and other resources (e.g., personnel or equipment), inefficiencies in the overall process such as redundant communications or lack of guidance in certain infrequent scenarios. Finally, when evaluating proposed interventions to improve incident response operations, the OSDs can assist in assessing the impact of changes in task duties, task ordering, agency responsibilities, resource allocation, and communications.

ACKNOWLEDGMENTS

This work was conducted in part with funding from the Northland Advanced Transportation Systems Research Laboratories (NATSRL), a cooperative research and education initiative of the Minnesota Department of Transportation (Mn/DOT), the University of Minnesota Center for Transportation Studies and its Intelligent Transportation Systems Institute, and the Swenson College of Science and Engineering at the University of Minnesota Duluth. The authors' opinions stated within do not necessarily reflect those of these supporting agencies.

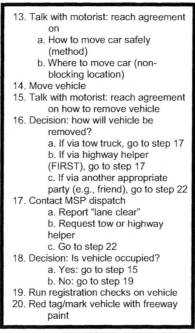

13. Talk with motorist: reach agreement on
    a. How to move car safely (method)
    b. Where to move car (non-blocking location)
14. Move vehicle
15. Talk with motorist: reach agreement on how to remove vehicle
16. Decision: how will vehicle be removed?
    a. If via tow truck, go to step 17
    b. If via highway helper (FIRST), go to step 17
    c. If via another appropriate party (e.g., friend), go to step 22
17. Contact MSP dispatch
    a. Report "lane clear"
    b. Request tow or highway helper
    c. Go to step 22
18. Decision: Is vehicle occupied?
    a. Yes: go to step 15
    b. No: go to step 19
19. Run registration checks on vehicle
20. Red tag/mark vehicle with freeway paint

Blocking   Not blocking

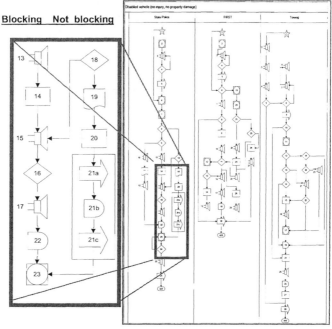

FIGURE 2 Excerpt from OSD for state police response to a disabled vehicle

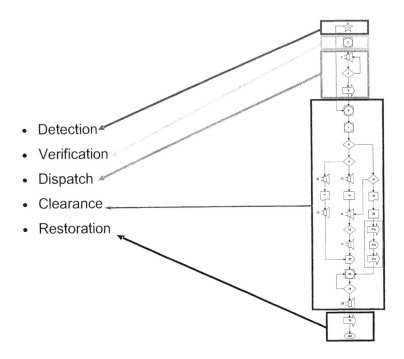

- Detection
- Verification
- Dispatch
- Clearance
- Restoration

FIGURE 3 Relationship between OSD and the incident response timeline

REFERENCES

Balke, K.N., Fenno, D.W. and Ullman, B. (2002). *Incident Management Performance Measures.* United States Federal Highway Administration: Washington D.C.

Charles, P. (2007). *Review of Current Traffic Incident Management Practices.* Austroads: Sydney, Australia.

Farradyne, P.B. (2000). *Traffic Incident Management Handbook.* United States Federal Highway Administration: Washington D.C.

Feyen, R. and Eseonu, C. (2009). *Identifying Methods and Metrics for Evaluating Interagency Coordination in Traffic Incident Management.* Report No. CTS 09-13. Center for Transportation Studies, University of Minnesota: St. Paul, Minnesota.

Frandrup, C., Groth, S., Anderson, S., Sroga, J. and Hanzalik, J. (2002). *Traffic Incident Management: Recommended Operational Guidelines.* Minnesota Department of Transportation: St. Paul, Minnesota. Retrieved March 2009 from: http://www.dot.state.mn.us/tmc/documents/Freeway Incident Management.pdf

Texas Transportation Institute (2007). *2007 Annual Urban Mobility Report: Complete Congestion Data.* Texas A&M University: College Station, Texas.

102

United States Federal Highway Administration (2003). *Traffic Incident Management (TIM) Self Assessment: National Detail Summary Report.* FHWA: Washington D.C.

United States Federal Highway Administration (2009). *Federal Highway Administration Focus States Initiative: Traffic Incident Management Performance Measures Final Report.* FHWA: Washington D.C.

Minnesota Department of Transportation. *FIRST Program Evaluation.* Mn/DOT: Minneapolis, MN, 2004.

Chapter 11

Globalizing Project Managers' Creative Tension, An Overview

Andrea Bikfalvi♣, *Pasi Porkka*♦, *Heli Aramo-Immonen*♦, *Núria Mancebo*♣,
Jussi Kantola♣, *Hannu Vanharanta*♦

♦Department of Industrial Management and Engineering
Tampere University of Technology at Pori
Finland

♣ Department of Business Administration and Product Design
University of Girona
Spain

♣ Korea Advanced Institute of Science and Technology
Republic of Korea

ABSTRACT

The present paper makes a holistic overview on project mangers' personal and social competences using the EVOLUTE methodology. The empirical evidences account for more than 2000 entries from five countries –Finland, Spain, Poland, United Kingdom and South Korea- among two clearly differentiated collectives: academic respondents and professional respondents. The results are presented accordingly to these main categories and a series of implications are discussed.

Keywords: competence evaluation, competence development, project management, international

INTRODUCTION

Competences and consequently competence evaluation has gradually become a strategic issues in academia, business and administration. Despite the conceptual ambiguity the competence approach is widely adopted. The competence framework literature provides an integrative system for human resource management. For example, competences are often conceptualized as the underlying characteristics of the individual, and as a combination of skills, knowledge and attitudes. In this paper competences refer to traits, knowledge, skills, experience and values that an individual needs to accomplish his or her tasks. Competence assessment becomes a significant instrument for predicting work-role performance, and accordingly a core element for human resource management practices.

Competences are linked to individual and organizational performance, or in a more specific level as it is the training outcome. Therefore, competences are subject to dynamic change as a result of motivation, intervention, tacit and/or explicit education and learning in a given time frame. Both at individual and organizational level learning plans represent a formal tool in competence development. The power of a learning plan enhances the ability to get results through a greater understanding of ones' own and others' competences and emotions.

Another field experimenting major transformations is Higher Education, one of the main pillars of any national/regional innovation system and one of the major suppliers of professionals working in the business field. Building a strong strategic bridge between academia and business is in the interest of the contemporary knowledge economy.

Although competences seem to be linked primordially to business environments, universities in their modern educational system complements traditional teaching/learning models with competence development. However, the competence topic in universities is still in its infancy compared to the business sector.

One major importance issue and a remaining challenge is competence evaluation and development as well as an in-depth integration of competences among the existing educational models.

Competences are highly context-dependent. Opting for a project management based competence model seems to easily fit in the complex academic setting, where students are characterized by high diversity in terms of maturity, objectives, motivation. On the other hand, indifferently of their academic background, the project management way of working is frequently habitual in any business sector they end up as professionals. Most often it is the business sector appearing as a first scenario where students meet competences, either when applying for a job, or once working with the aim of detecting training necessities.

Competence evaluation, in general, and in project management, in particular, especially in academia has received reduced attention in both theory and practice. Therefore, the main purpose of this study is to investigate and show the results of major field of development among competences and target collectives: academics and professionals. This is an explorative paper giving an overview over the topic, describing the methodology, presenting the results and drawing some conclusions.

RESEARCH METHODOLOGY

The application used, evaluates the competences of the work role of the project manager by means of 120 statements related to the individuals' every day work.

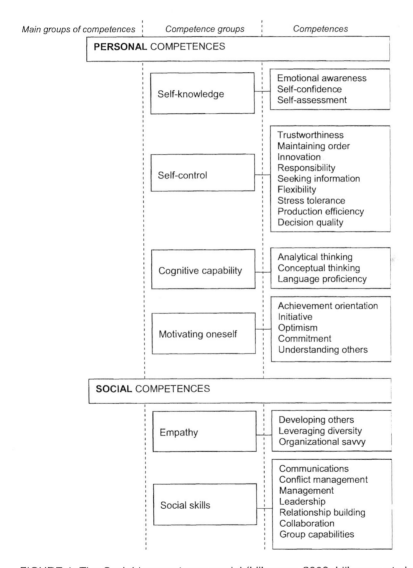

FIGURE 1: The Cycloid competence model (Liikamaa, 2006; Liikamaa et al., 2003)

It requests the self-evaluation of the current competence level and the target level desired by the respondent, identifying the possible creative tension. The responses are given on a fuzzy scale guided by standard linguistic labels (e.g., always, often, seldom, never) (Kantola, 2005; Kantola et al., 2005). The results can be visualized in detail at the level of 30 competences, grouped in 6 competence groups than can be further classified as personal or social main groups of competences (see Figure 1).

The platform also collects demographic information, which depending on a culture's/country's sensitivity towards personal data protection matters has different degrees of completeness and accuracy. According to the available information the percentage distribution of participants per country is the following: 20.5% Finland, 69.3% Spain, 4.0% Poland, 3.6% South Korea and 2.7% Great Britain. Age profile distribution shows majority of respondents younger than 31 (68.2%), 15% between 31 and 40 years old, 11.5% between 41 and 50 years old and 4.6% older than the age of 51. This is in line with the typology of collectives we manage, namely academic collective accounting for 92% of participants, and professionals 8%. Respondents also distribute through the gender variable into 68.3% men and 31.7% women. A minor comment refers to the fact that based on the type of information to fill in respondents have the right of not doing it. In that sense and as an example, age seems to be a sensitive item, since almost one third of the respondents avoid giving an answer.

RESULTS AND DISCUSSION

Since the present paper's main objective is to present an overview in the followings we show the results summarized accordingly to the different aggregation options the methodology permits: competences (30 items), groups of competences (6 groups) and main groups of competences (2 items) making difference between the two collectives considered.

The figures in Table 1 are organized by ordering the respondents' creative tension from highest to lowest values. The value *per se,* in its numeric form, has no significant interpretations, still it helps when willing to differentiate among groups of concepts, namely competences for the purpose of the present analysis. We take into account the statistical criterion of minimum significant difference (MSD). We use this technique in order to be able to mark differences between a range/list of competences. Resting the MSD value to the highest value of competence results into a cut-off point generating the upper interval of considered competences, which appear marked in dark grey in Table 1. Adding the MSD value to the lowest value of competence results into a cut-off point generating the lower interval of considered competences, which appear marked in light grey in the same table.

The interpretation of these extremes (higher end and lowest end) is the following. Those competences that appear marked in the top of the table are those for which respondents feel special interests in terms of development. We could interpret them as their main perceived weaknesses.

Table 1: Values and groupings of creative tension by collectives - 30 competences

Academic collective		Professional collective	
VALUE	MSD=1,85	VALUE	MSD=2,83
22,4	Stress tolerance	19,8	Stress tolerance
19,4	Language proficiency	18,9	Language proficiency
18,5	Decision quality	18,8	Communication
18,2	Communication	18,5	Decision quality
17,9	Flexibility	18,0	Innovativeness
17,2	Innovativeness	17,9	Relationship building
17,2	Relationship building	17,8	Group capabilities
17,1	Self-assessment	17,4	Self-assessment
17,1	Organizational savvy	17,0	Initiative
16,7	Understanding others	16,9	Production efficiency
16,5	Production efficiency	16,7	Flexibility
16,4	Seeking information	16,7	Organizational savvy
16,2	Group capabilities	16,6	Management
16,0	Leveraging diversity	16,6	Understanding others
15,9	Management	16,0	Maintaining order
15,8	Responsibility	16,0	Seeking information
15,5	Maintaining order	15,4	Leadership
15,3	Optimism	15,0	Collaboration
14,7	Conceptual thinking	14,9	Self-confidence
14,7	Achievement orientation	14,5	Leveraging diversity
14,5	Leadership	14,1	Analytical thinking
13,8	Self-confidence	14,1	Trustworthiness
13,0	Analytical thinking	13,8	Achievement orientation
13,0	Initiative	13,3	Emotional awareness
12,5	Emotional awareness	13,2	Optimism
12,4	Collaboration	13,1	Conflict management
12,4	Conflict management	12,5	Responsibility
12,3	Trustworthiness	11,0	Conceptual thinking
11,4	Commitment	11,0	Developing others
10,9	Developing others	9,5	Commitment

N = 2100	N = 191

Results statistically significant at α=0.001 level

This is the case of stress tolerance for the academic collective and it is complemented with language proficiency, communication, decision quality, innovativeness, relationship building, group capabilities, self-assessment and initiative for the case of project management professionals.

On the other hand, in the lowest part of the table we identify the competences for which respondents have a lower creative tension, meaning that they have a lowest or none interest in their development. We could interpret theses results as their perceived strengths. Concretely, emotional awareness, collaboration, conflict management, trustworthiness, commitment and developing others seem to be students' perceived strong points, while for the case of professionals these are conceptual thinking, developing others and commitment.

It is interesting to observe how the different collectives considered put emphasis on certain aspects. Only interpreting the number of items entering each group, we can deduce that the academic community has the perception of more strengthens than the professional community who, in balance, has more items on the weaknesses side. This fact might have a plausible explanation due to a certain lack of knowledge and experience in self-evaluation often translated to an over-estimation rather than under-estimation especially since that is the case of the academic community mainly characterized by young students.

Table 2: Values of creative tension by collectives - 6 groups of competences

Academic collective		Professional collective	
VALUE	MSD=0.226	VALUE	MSD=1,274
4,330	Self-control	4,037	Self-control
3,751	Cognitive capability	3,958	Social skills
3,477	Social skills	3,542	Self-knowledge
3,427	Empathy	3,497	Cognitive capability
3,165	Self-knowledge	3,241	Empathy
2,850	Motivating oneself	2,725	Motivating oneself
N = 2100		N = 191	

Results statistically significant at α=0.001 level

Grouping the 30 competences considered into 6 main groups of competences (see Table 2) we find that primordially it is self-control that participants are mostly willing to develop indifferently of the considered collective. Cognitive capability appears as second choice for the academic community and social skills for the professional one. This result is in line with the considered profile, since those in the higher education perceive more important all those abilities and capabilities that relate to the learning and cognition while those having practical experience in the working field of project management put emphasis on social and relationship relative skills. Interestingly none of the considered collectives value self-motivation as among those that should be a priority in development. It might be the case that participants consider motivation as a task that somebody else (rather than the person itself) should promote, especially teachers in the particular case of students.

Table 3: Values of creative tension by collectives - 2 main groups of competences

Academic collective		Professional collective	
VALUE	MSD=0,109	VALUE	MSD=0.143
1,527	Personal competencies	1,503	Social competencies
1,473	Social competencies	1,497	Personal competencies
N = 2100		N = 191	

Results statistically significant at α=0.001 level

One of the most summarized option of viewing and interpreting results is by main groups, differentiating between personal and social competences. Although the values are quite similar, the patterns are opposite: the academic community considers primordial for development personal competences while professionals put emphasis on social competences. One possible reason might be that those in the higher education are persons completing their studies as an important ingredient of their personal development an important issue also for the future due to their young age. On the other hand, this might be an already achieved objective for professionals who rather focus on social or relational competences.

CONCLUSIONS

Latest trends in the field of human resource management focus the attention on a variety of topics one of which is managing diversity (see two special issues of the Journal of Human Resources).

In our paper we present an overview comprising this concept from at least two perspectives: participants and concepts. It is interesting to highlight the international nature of our participants, adding strength and value to the analysis, although the results are not presented by this issue since that is beyond the objective of our present paper. Another detail worth mentioning is that the academic context, even considered a single community, is highly diverse in aims, objectives and perceptions. A further future analysis could bring valuable knowledge in the field when examining competences in the considered academic community presented accordingly to a series of different categories –age, gender, working experience, among others- that may mark important divergences in results. Further implications of such implies for academia a more target oriented training taking into account the concrete profile of students. Regarding the professional profile considered, an in-depth analysis of the available information could add to the understanding of real project management professionals' development priorities and patterns in their perceptions.

In our view we contribute by bringing valuable and recent information in a field which is far from being highly standardized. High variety in definitions and competence models characterize the field which makes difficult the use of a unique method of assessment. Our proposal is one possible choice in this direction, especially welcomed in the academic/higher education area which, in its way towards Bolognia, is approaching to the end of competence definition process searching for alternatives for a realistic assessment.

Finally, some limitations and further research concerns the breakdown of results in significant demographic groups and the use of more advanced statistical methods for a deeper analysis. Another underexplored field in the concentration of the attention of specific, strategically important single competences as it could be motivation or quality of decisions, to mention just some.

REFERENCES

Kantola, J. (2005) Ingenious Management. Industrial Engineering. Tampere, Tampere University of Technology. Doctor of Technology.

Kantola, J., Vanharanta, H., Karwowski, W. (2005) The Evolute System: A Co-Evolutionary Human Resource Development Methodology. International Encyclopedia of Human Factors and Ergonomics. W. Karwowski.

Liikamaa, K., Koskinen, K., Vanharanta, H., (2003), Project managers´ personal and social competencies, Project management: dreams, nightmares and realities, proceedings, papers and presentation, Nordnet, Oslo, Norway, September 24th-26 th

Liikamaa, K. (2006) Tacit Knowledge and Project Managers' Competences, Doctoral Thesis, Tampere University of Technology at Pori, Finland.

Chapter 12

Job-Redesign in Knowledge Work

Ole Henning Sørensen, David Holman

National Research Centre for the Working Environment
Copenhagen
Denmark

Institute of Work Psychology
University of Sheffield
Sheffield
UK

ABSTRACT

A participative job redesign intervention was used to try and improve the job design of knowledge workers in six organisations. Employees chose to alter their jobs by implementing relational initiatives (e.g., improved feedback and support) and work process initiatives (e.g., formalisation of planning and other procedures). Quantitative and qualitative evaluation methods were used. Analysis of a longitudinal survey suggested that the relational initiatives resulted in significant improvements in perceptions of leadership style, supervisory support, recognition and fairness. Work processes initiatives did not appear to result in perceived changes in work conditions, which the qualitative data indicated was due to implementation problems. The results suggest knowledge work can be made less problematic by reducing task complexity and ambiguity through enhancing social relations and through the formalisation of work process initiatives.

Keywords: Well-being, Knowledge work, Job design.

INTRODUCTION

Knowledge work jobs are an important part of the global economy (Castells 1996) and include software analysts, data analysts, consultants and designers. Although knowledge work jobs have many characteristics that are typically associated with high quality work and high employee well-being - such as high discretion, variety, responsibility and task meaningfulness – they can also be experienced as problematic, demanding and stressful. Addressing the problematic nature of knowledge work is clearly important and a means of doing this is to use a job redesign intervention to improve work design. However, there have been few, if any, job redesign interventions in knowledge work jobs. The first aim of this paper is therefore to report on an evaluation of a participative job redesign intervention in knowledge workers. Furthermore, the nature of knowledge work means that many of the typical strategies adopted in job redesign interventions, such as increasing job control and variety, may be irrelevant due to the fact that jobs are already enriched in this way, and may even be harmful as they could increase the already high levels of task ambiguity and uncertainty. The second aim of this paper is to examine the types of changes to job design adopted by knowledge workers and to use this as a means of providing insight into the nature of job design in knowledge work jobs.

KNOWLEDGE WORK

Knowledge work tasks can be broadly defined as tasks that involve the creation, manipulation, and communication of symbols (language, numbers, etc.) and ideas. These tasks can be difficult to define, highly ambiguous, have no simple solution, and it may be hard to establish when a solution is sufficient and whether work is progressing constructively (Alvesson 1993). Knowledge work tasks are often social. Problems are addressed, knowledge is exchanged and solutions are tested with others; and task are often socially distributed with different employees working on various parts of a task (Davenport 2005).

The type of task conducted by knowledge workers has implications for the characteristics of knowledge work jobs. Thus, job demands can be challenging but interesting and provide much opportunity to use and develop skills. The ambiguous nature of the task means that most knowledge work jobs provide employees with high levels of variety and discretion on how to do the job. However, the high levels of ambiguity and control can be problematic as knowledge workers not only have to craft a solution to the problem at hand but also have to continually craft the task, relational and cognitive boundaries of their work (Wrzesniewski & Dutton 2001). Allied to this, knowledge work jobs often have high workload and time pressure (Grönlund 2007). The social aspects of the task mean that relational job characteristics (e.g. interdependence, feedback, and trust) are particularly salient, and that knowledge workers have high levels of discretion over how they interact and who they interact with.

Many knowledge work jobs therefore have high levels of enriched (e.g., discretion, variety), relational and job demand characteristics (e.g., workload, ambiguity). This combination should, according to certain job design theories be beneficial for well-being (Parker & Wall 1999). But such jobs can still be experienced as problematic. Indeed, research indicates knowledge work job can be experienced as stressful (McClenahan, Giles, & Mallett 2007) and that, in Denmark, the level of stress in knowledge workers is similar to other occupational groups (Nielsen & Kristensen 2007).

Addressing the problematic nature of knowledge work is clearly important and a means of doing this is to use a job redesign intervention to improve work design. As such interventions have proven successful in other types of job, we thought it would be beneficial to run and evaluate a job redesign interventions in knowledge workers. To our knowledge there have been no previous job redesign interventions in knowledge workers. As the intervention was participative we were particularly interested as to the ways that knowledge workers would chose to improve their work design, as this can provide insight into the importance of particular job characteristics in knowledge work. For example, employees might seek to maximise control, which would be in line with the recommendation of most job design theories (Parker & Wall, 1999). But given that knowledge work jobs are highly enriched, increasing the level of job characteristics such as discretion or variety might be viewed as unnecessary or even exacerbating the ambiguity of their job. In contrast, employees might seek to reduce the level of job enrichment through greater formalisation. Indeed, it has been suggested that formalisation in ambiguous and uncertain contexts has beneficial effects on performance as it provides a structure that can aid coordination, enhance decision-making quality and improve the efficiency of work processes (Alder & Borys 1996;Juillerat 2010). Indeed, many knowledge work jobs involve formal processes and standards, such as project management techniques, software development tools, and product development tools. Other authors have, however, questioned the usefulness of formalisation when working on ambiguous and uncertain tasks (Burns & Stalker 1961;Katz & Kahn 1978).

METHOD AND RESULTS

The research project was based on a grant from the Danish Working Environment Research Fund. The main objective of the research project was to develop and test an intervention model aiming at reducing stress and increasing well-being at work. A combination of qualitative and quantitative evaluation was used.

DESIGN

A nine month intervention was conducted at six workplaces. The intervention had a strong participative element (Israel, Schurman, & House 1989;Semmer 2006) and two phases. The first phase included prioritisation and development of initiatives. A dialog workshop was conducted at each workplace where all employees in the department discussed and prioritized the main factors causing stress and wellbeing. Subsequently, the employees developed a number of suggestions for initiatives and formulated a sketch of each initiative. The managers participated in the dialog workshops. Each workshop was facilitated by two members of the research team.

The second phase was implementation, which occurred over the following six months. Each workplace was responsible for implementing the proposed initiatives. They agreed to discuss the initiatives regularly in normal meeting activities and to develop measures to assess progress. Two times during the project - after three and six months - the researchers conducted workshops to discuss activities and progress. Three months after the end of the intervention a joint workshop between the case researchers and employees at each workplace was conducted to evaluate the intervention.

The workplaces contracted participated on a voluntary basis and did not make a financial contribution. They were selected to represent knowledge work broadly. Two project organisations represented work with relatively long deadlines, two political organisations represented a mixture of short deadlines and standard administrative work tasks and two creative workplaces were dominated by short deadlines. It was not possible to establish a control group.

The two project organisations were engineering consultant departments. The first 'EngWater' was a water and environment department of a large engineering consultancy company. The second 'EngPlumb' was a 'plumbing design department' in another large engineering consultancy company. The first political organisation 'K-Union' was a negotiation department of a knowledge worker union. In the second, two different departments from a large municipal administration of social services participated. These departments had different work tasks but had virtually no contact because they were at different locations. Consequently, they are treated as separate cases. The first 'SocEco' was a department administering the economy across all social services in the municipality. The second 'SocSecr' was a political secretariat servicing politicians and handling external communication. The first creative organisation 'T-Design' was a design department in a toy company. The second was a daily newspaper but due to restructuring plans it dropped out of the study.

In total, six departments participated throughout the project. The initiatives fell in two categories: relational and work process initiatives. *Relational initiatives* included attempts to improve management feedback and support, co-worker feedback and professional dialogue. *Work process initiatives* included attempts to improve planning systems, coordination processes and working procedures.

QUALITATIVE EVALUATION

All of the workplaces appointed 'initiative leaders' among the employees to coordinate and drive the change process. Most initiative leaders volunteered. The degree to which the initiative leaders succeeded in involving other employees varied greatly. In EngWater the initiative leaders were particularly successful. Most of the employees contributed actively to the projects. The different activities were discussed and assessed regularly. Progress was made visible in public spaces such as meeting rooms and hall areas. In K-Union the initiative leaders did not make the project visible in the same way as in EngWater. However, they involved the majority of the employees and some of the initiatives were made visible through a graphs and figures. In the four remaining workplaces the initiative leaders were the main drivers of initiatives but were less successful in involving others.

All the workplaces had difficulties transforming the initiative plans made at the dialog workshop into practically workable initiatives in the subsequent implementation period. It was also difficult to keep up the enthusiasm of the dialog workshop in every day practice. The initiative leaders attributed this to a lack of time, and to competing activities. However, they also found it difficult to concretize the initiatives. The management at EngWater complained strongly and asserted that the researcher had not made it clear how much time this work required (the researchers expected the workplaces to be aware of this). EngWater and EngPlumb dealt with this problem by assigning more resources to the project and appointed a central staff representative to assist the department head and the initiative leaders. K-Union did not assign central staff to help, but got some support from the researchers in order to overcome the initial implementation problems. In T-Design a consultant from a central staff function helped to analyse an initiative. In SocEco and SocSecr the initiative leaders received some support from the department manager.

EngWater implemented both work process and relational initiatives. Relatively large changes were related to management feedback, how to handle interruptions, time registration, and work load. K-Union also implemented changes in both areas related to feedback culture between colleagues, to feedback from management, and to meeting practices. EngPlumb planned initiatives in both areas but implemented work process initiatives only. Initiatives were not adopted broadly, but some people changed their behaviour relating to open office rules, interruptions, and planning. SocEco focused on work process initiatives such as planning and interruptions. Some people adopted the initiatives but they were not adopted broadly. T-Design only worked on the 'estimation of time' initiative but it was not implemented due to changes in the department. Due to a change in management SocSecr failed to implement any initiatives until just before the end of the project.

Overall, only EngWater and K-Union implemented relational initiatives. Systematic management feedback related to performance, conflicts, and future changes was such an initiative. Systematic feedback between colleagues related to work practices was another. Especially at EngWater, the relational initiatives were perceived as a success by employees and managers. The employees at K-Union reported that the co-worker support initiative was not especially successful. All the workplaces launched work process initiatives that involved formalisation of daily procedures and social interactions such as: rules and norms about behaviour in open offices, making it legitimate to dedicate time in the calendar for concentration demanding activities, establishing visible sign to indicated need for concentration, better planning systems, and new procedures for meetings. The majority of the workplaces had difficulties implementing these bureaucratic initiatives. In some cases they were narrowly adopted and in other cases they provoked conflicts between colleagues.

QUANTITATIVE EVALUATION

A longitudinal survey was used to provide a quantitative evaluation of the intervention. The main aims were to establish whether change had occurred over time in employee perceptions of job characteristics and well-being, and whether there were differences in the extent of change between groups.

The Time 1 questionnaire was administered two weeks before the 6-month long intervention. The intervention was designed to be fully implemented within six months. The Time 2 questionnaire was administered about three months after the intervention finished.

Table 1 Response rates

| | Pre-test at Time 1 | | Post-test at Time 2 | | Both times |
	%	N	%	N	N
EngWater	79 %	34 (of 43)	69 %	36 (of 52)	19 (of 36)
K-Union	78 %	31 (of 40)	78 %	38 (of 49)	19 (of 30)
EngPlumb	74 %	28 (of 38)	66 %	23 (of 35)	16 (of 33)
SocEco	82 %	14 (of 17)	70 %	14 (of 20)	9 (of 15)
SocSecr	88 %	14 (of 16)	59 %	10 (of 17)	9 (of 13)
T-Design	88 %	35 (of 40)	81 %	30 (of 37)	26 (of 36)

At Time 1, the survey was completed by 156 of 194 potential respondents across all six cases, giving an overall response rate of 73 per cent. The exact response rates for each case are detailed in Table 1. There were 91 women and 65 men. The average age of participants was 43.8 years and their average tenure was 10.1 years. At Time 2, all employees were given the opportunity to complete the survey, including those who did not respond at Time 1 or had joined since Time 1. The survey was completed by 152 of 210 potential respondents across all six cases, giving an overall response rate of 72 per cent. At Time 2 there were 93 women and 59 men. Their average age was 41.5 years and their average tenure was 8.8 years.

The longitudinal sample, comprising those responding both Time 1 and 2, was 99 and comprised of 63 women and 36 men. Their average age at Time 1 was 41.1 years and their average tenure was 9.8 years.

All measures were based on employee perceptions and were derived from the Copenhagen Psychosocial Questionnaire (Kristensen et al. 2005). The measures were chosen to cover job design characteristics, relationship quality and well-being. The three measures of job design included: *Quantitative demands*, a four-item measure assessing the amount of work (Time 1 α=.88; Time 2 α=.89). *Work tempo*, a three-item measure of how fast work is (T1 α=.82; T2 α=.80). *Influence at work* was a four item measure concerning employees' control over their job task (T1 α=.65; T2 α=.72).

The five measures of relationship quality included: *Rewards* was a three item measure on recognition at work (T1α=.69; T2α=.65). *Leadership skills* was a four item measure on the extent to which the employee's supervisor was perceived to exhibit leadership skills, such as ability to plan and solve conflicts (T1α=.71; T2α=.74). *Supervisor support* was a three-item measure that assessed the extent to which the employee's supervisor provided social support (T1α=.70; T2α=.77). *Co-worker support* was a three-item measure that assessed the extent to which the employee's co-workers provided social support (T1α=.72; T2α=.60). *Fairness* was a four-item measure concerning the extent to which there is fairness at work. (T1α=.68; T2α=.64).

Job-related well-being was assessed using two measures derived from the Copenhagen Psychosocial Questionnaire. *Job burnout* was a four-item measure concerning the extent to which an employee felt tired and exhausted from work in the last four weeks (T1α=.84; T2α=.87). *Job Stress* was a four-item measure concerning the extent to which an employee felt anxious and irritable over the last four weeks (T1α=.79; T2α=.82). We also included a number of control variables in our analyses. They were gender, university education, post-graduate education, manager, knowledge worker (all dummy variables) and age.

When comparing between groups in interventions, the use of a control group and a treatment group allows stronger causal attributions of change to be made. We were unable to secure a control group from the outset, as all participating organisations wanted to benefit from the intervention. However, an inert-control group was created. An inert-control group receives a treatment that lacks the key active ingredient, in this case the actual implementation of changes to work (Schwartz et al. 1997).

In this study our qualitative evaluation showed that only two organisations (EngWater, K-Union) implemented relational initiatives and four did not (T-Design, EngPlumb, SocEco, SocSecr). We therefore grouped the first two into a relational implementation group and the remaining four were grouped into a relational inert-control group. To test whether the relational initiatives led to significant differences between these two groups in the perception of relevant work variables we used repeated measures ANOVA.

Our qualitative evaluation showed that four organisations (EngWater, K-Union, EngPlumb, SocEco) implemented work process initiatives and two did not (T-Design, SocSecr). We therefore grouped the first four into a work process implementation group and the remaining two were grouped into a work process inert-control group. To test whether the work process initiatives led to significant differences between these two groups in the perception of relevant work variables we also used repeated measures ANOVA.

Table 2 Results for effects of relations initiatives

Dimension	Implementation		Inert-Control		F	P
	T1	T2	T1	T2		
1. Quantitative demands	59.54	58.35	52.59	51.04	.01	n.s
2. Work tempo	70.25	65.95	69.68	68.18	.63	n.s
3. Influence	49.88	55.53	51.47	49.40	3.92	.05
4. Rewards	61.25	69.21	67.35	61.02	13.65	.01
5. Leader Skills	52.71	60.90	62.53	55.98	17.23	.01
6. Leader Support	59.46	66.13	64.14	55.37	10.24	.01
7. Co-worker Support	61.69	62.87	59.93	57.69	.92	n.s
8. Fairness	55.59	55.95	60.99	58.05	5.89	.05
9. Stress	28.95	24.21	27.80	28.47	1.59	n.s
10. Burnout	34.64	31.37	34.65	35.78	1.07	n.s

Note: Means are adjusted marginal means

The results of the RM-ANOVA that compared change in the relational implementation group and relational inert-control group (Table 2) showed significant differences in change in the expected variables: rewards, leader quality, leader support and fairness. The size of the changes reached five or even higher points, which is also clinically significant (Pejtersen, Bjørner, & Hasle 2010). All changes were positive in the relational implementation group and negative in the inert-control group. This resonates with the qualitative evaluation that indicated large improvements in management feedback and support in the treatment group. However, there was no significant difference in the level of change with regard to co-worker support. This also resonates with the qualitative evaluation as the employees the relational implementation group reported that co-worker support initiatives were not successfully implemented. Although there were no expectations that the relational initiatives would improve influence, we did find a significant difference with regard to change in the level of influence at work. This might be explained by the fact that the intervention had a large participatory element and the strongest efforts at involving employees were made at EngWater and K-Union. We also found small but non-significant changes in the well-being measures of stress and burnout.

No significant differences in the level of change were found between work process implementation group and the work process inert-control group in any of the expected work design dimension, quantitative demands and work tempo, or any other variables. This result resonates with the qualitative evaluation that indicated large problems with implementation of these initiatives.

CONCLUSIONS

One interesting outcome of the research findings is that knowledge workers in this study did not implement job design initiatives aimed at increasing job control or enlarging the job. Most of the initiatives appeared to be implemented as a means of reducing task ambiguity. The work process initiatives sought to reduce ambiguity through new planning procedure or through the introduction of rules and procedures to regulated social interaction in order to enhance meeting or to reduce interruptions and to increase possibilities to concentrate. Relational initiatives also targeted ambiguity by increasing feedback from managers to reduce uncertainty about goals, priorities etc. or feedback from colleagues to increase individual capacity to deal professionally with new and complex situations.

The evaluation showed that the workplaces had different levels of success with their efforts. The qualitative process evaluation shows that despite positive intentions at each workplace and the participative design, several workplaces were not able get a positive effect of the intervention. In several cases this was caused by insufficient effort in implement the initiatives. The quantitative analysis indicated that the relational initiatives were the most successful in inducing perceptions of change in relevant organisational measures. We did not find an effect of the work process initiatives. One reason for this may be that the work process initiatives were implemented as bureaucratic planning initiatives disregarding the complex social component. The initiative leaders may not have understood or may have overlooked the relational component of the initiatives, provoking conflicts between colleagues, and clashing with cultural values such a knowledge sharing. We also found no significant effects on well-being. This may be due to the relatively short follow-up period of three months.

This study suggests that key issues in the job design of knowledge workers are the high levels of task ambiguity and uncertainty, and the importance of formalisation and social relationships in helping the person to cope with these task attributes. This suggests that job redesign initiatives among knowledge workers should not adopt classical enrichment of job enlargement redesign strategies (e.g. increased control, variety) but should focus on reducing complexity and ambiguity and provide structures that reduce the need for a constant and demanding recrafting of the task. Based on the results of the quantitative evaluation, we recommend focusing on relational initiatives such as management feedback and support i.e. increasing leadership quality. However, the qualitative evaluation also shows that the employees desire to increase formalisation and improve bureaucracy to enhance daily work processes. They want better planning procedures and regulation of social interactions in order to increase personal task discretion. They also want to improve collective activities such as meetings and knowledge sharing. The evaluation indicated that the implementing of such initiatives may be difficult and additional research seems to be needed to understand how better formalisation can be achieved.

REFERENCES

Alder, P. S. & Borys, B. (1996), "Two Types of Bureaucracy: Enabling and Coercive." *Administrative Science Quarterly*, 41(1), 61-89.

Alvesson, M. (1993), "Organizations as Rhetoric: Knowledge-Intensive Firms and the Struggle with Ambiguity." *Journal of Management studies*, 30(6), 997-1015.

Burns, T. & Stalker, G. M. (1961), *The management of innovation*. Oxford University Press, New York.

Castells, M. (1996), *The rise of the Network Society*. Blackwell, Cambridge.

Davenport, T. H. (2005), *Thinking for a living. How to get better performance and results from knowledge workers*. Harvard Business School Press, Boston.

Grönlund, A. (2007), "Employee control in the era of flexibility: A stress buffer or a stress amplifier?" *European Societies*, 9(3), 409-428.

Israel, B. A., Schurman, S. J., & House, J. S. (1989), "Action Research on Occupational Stress - Involving Workers As Researchers." *International Journal of Health Services*, 19(1), 135-155.

Juillerat, T. L. (2010), "Friends, not foes?: Work design and formalization in the modern work context." *Journal of Organizational Behavior*, 31(2-3), 216-239.

Katz, D. & Kahn, R. L. (1978), *The social psychology of organizations*, 2 edn. John Wiley and Sons, New York.

Kristensen, T. S., Hannerz, H., Hogh, A., & Borg, V. (2005), "The Copenhagen Psychosocial Questionnaire - a tool for the assessment and improvement of the psychosocial work environment." *Scandinavian Journal of Work Environment & Health*, 31(6), 438-449.

McClenahan, C. A., Giles, M. L., & Mallett, J. (2007), "The importance of context specificity in work stress research: A test of the Demand-Control-Support model in academics." *Work and Stress*, 21(1), 85-95.

Nielsen, N. R. & Kristensen, T. S. (2007), "Stress i Danmark - hvad ved vi?".

Parker, S. K. & Wall, T. D. (1999), *Job and work design*. Sage, London.

Pejtersen, J. H., Bjørner, J. B., & Hasle, P. (2010), "Determining minimally important score differences in scales of the Copenhagen Psychosocial Questionnaire." *Scandinavian Journal of Public Health*, 38(Suppl 3), 33-41.

Schwartz, C. E., Chesney, M. A., Irvine, M. J., & Keefe, F. J. (1997), "The control group dilemma in clinical research: Applications for psychosocial and behavioral medicine trials." *Psychosomatic Medicine*, 59(4), 362-371.

Semmer, N. K. (2006), "Job stress interventions and the organization of work." *Scandinavian Journal of Work Environment & Health*, 32(6), 515-527.

Wrzesniewski, A. & Dutton, J. E. (2001), "Crafting a Job: Revisioning Employees as Active Crafters of their Work." *Academy of Management Review*, 26(2), 179-201.

Chapter 13

Human Performance Technology and Working Reliability

Oleksandr Burov[1], Volodymyr Kamyshin[1], Olena Burova[2]

[1]Institute of Gifted Child
Kiev, Ukraine

[2] Ukrainian Institute of Industrial Property
Kiev, Ukraine

ABSTRACT

Human performance (HP) control is described as a cybernetic system to improve human performance technology. The concept of the system to assess and to predict HP is discussed that explain which parameters of the psychophysiological basis of a human mental work should be measured to secure high accuracy of the prediction.

Keywords: control, cognitive ability, assessment, human performance

INTRODUCTION

Economical and social loses in aviation, road transportation, power and chemical companies are result of the personnel errors which share in accidents and incidents is 50 - 80%. According to data known, only 25 % of them fall on inadequate level of the human-operator knowledge and skills. High accurate assessment and prediction of an operator capacity and its psychophysiological basis is a necessary component of the efficiency and reliability control for the whole human-machine system. Prediction of an operator general capacity and fitness-for-duty has the maximal level of accuracy (theoretically, according to data published) 70%. In other words, the minimal level of prediction error is accepted 30% that is absolutely insufficient from practical reasons. Existing theoretical basis does not allow answering a question: what, where, why and in what way should be measured to provide a high accuracy of the human performance prediction.

IDEA

It was stated that the most fruitful approach to understand mechanisms of activity is a theory of functional systems proposed by P.Anokhin (1973) and developed by K.Sudakov (1979) and A.Navakatikyan (1981) who proposed the concept of the activity functional system that connects in one model physiological systems state, conditions of the work environment and the goal of activity. According to it a human activity is accompanied by creation and maintenance of functional systems that are activated dominant brain structures and correspond activity of one or another organism systems and are quite enough stable for particular type of the human work. But this is not enough to understand to what degree it can explain reasons of insufficient accuracy of a human performance prediction.

An operator work specific is a workload, bigger because of not so power-consuming processes, as informational ones that are, by their nature, not so discrete, but continuous (Burov 2005, 2006). The reason is that the conceptual model of activity (as result of a human psychological adaptation to the work) is expanded in the time independently on external process and an operator activity consists in discrete comparison of the information obtained from outside with the model existing constantly. I.e., conceptual model can be considered as an information stratum of professional work, and physiological chain „afferent inputs – activity acceptor – physiological control – effectors - act" is an energetic stratum. The goal of professional training is forming of the conceptual model of activity of particular type, carrying out of the particular task. It means creation of „information contour" that exists and is maintained in activated state in carrying out process for purposeful activity and embraces afferent inputs, decision making block, activity acceptor and act program, as well as the object of activity (is represented as information model in case of operator-watcher that creates the information contour together with the imagine- conceptual model).

Energetic and information stratums can be represented as two contours which partly coincide at the level of morphological structures and functions, but partly differ because of including into the information contour an activity object that does not participated in the energetic contour of the organism regulation, but is an inalienable part of the information contour. Human activity is a mediator between internal and external environment of organism, projection of structural-function specific of professional homeostasis on the operator work. Output parameters of the activity program (activity effectors) stand in the information contour as parameters of capability. In such a context, operator's activity is an activity program realized as physical and/or mental acts in external environment.

TECHNIQUES

Practical realization of these ideas is applied for the cybernetic mechanism of a human performance control. There is the set of psychophysiological parameters of human-operator **P**, which are related to forming and realization of his capacity. The set of his professionally important qualities, parameters of professional senescence and current capacity is a number of the parameters of operator professional activity **D**. The task of synthesis of the system for psychophysiological prediction of operator capacity is the task of optimum reflection **P** on **D**, that provides maximal quality of functioning SLTS **Q** subject to the system cost **C**, which does not exceed a possible level C_{lim}: [$p \in P$] **m** [**D**$^* \subset$ **D**], where **m** – operation of optimum reflection of elements **P** on the set of elements **D**; **D*** - optimum set of parameters of professional activity

$$D_0 = \arg \max_{D_0 \in D} Q(D) \text{ under } C \leq C_{lim}.$$

Notion "control" is used as a process of adduction of the set object in the state, that answers the task put. Such determination allows considering the capacity of operator as an object of control by organizational-psychological methods, and the system of assessment and prediction considering as a system of the operator capacity control. In the general case of ergonomics approach to the analysis of efficiency of the system Human-Technique-Environment (SHTE) the estimation and purpose of prognosis of functional state and operator capacity U in SHTE is to provide maximal quality of functioning of the system **Q**, which relies on the realized of operator capacity **R**, organization of the system **O**, the state of equipment **E**, interelement interface **I**, dynamics of SHTE changes in time **t**. Such task is described as:

$$\hat{U}(t) = \arg\max_{\check{R},\check{O},\check{E},\check{I}} Q(t) = \arg\max_{\check{R},\check{O},\check{E},\check{I}} f[\, R(t), O(t), E(t), I(t), t \,],$$

where $\check{R}(t) \in R(t)$, $\check{O}(t) \in O(t)$, $\check{E}(t) \in E(t)$, $\check{I}(t) \in I(t)$.

If estimation of capacity works for real $\hat{U} \boxtimes U$, quality of functioning SHTE can achieve the maximal value $Q > Q_{max}$ thanks to the use of maximal operator capacity.

RESULTS

They were:
1. Developed criteria and algorithm for determination of informative psychodynamic indices of a human-operator (HO) physiological state.

2. Proved types and parameters of psychological task tests, descriptions of their implementation that are included in the psychophysiological models of HO activity
3. Defined parameters of optimum models of HO capacity prognosis
4. Developed the methodical providing of the computer systems creation for assessment and prediction of operator professional capacity
5. Developed information technology of estimation and prognostication of capacity of operator
6. Assessed and validated prediction accuracy on the basis of the real-settings systems: 90% for the daily (pre-shift) control in power industry.

OUTCOMES

1. The P. Anokhin's theory of functional systems (FS) has received further development by ways: (a) advancement of understanding of psychophysiological mechanisms of mental performance thanks to division of energetic-information contour of regulation into two separate contours – energetic and informational, and closing of the latter outside the human organism through the object of control; (b) use of temporal pattern of the operator activity as a criterion of the functional system's steadiness.

2. Criteria and methods for creation and use of adaptive psychodynamic models of an operator capacity are differ from known ones and are based on rhythmic structure of cognitive information processing in the informational contour of FS.

3. Methodology of synthesis for systems of psychophysiological prediction of an operator capacity was developed and used above mentioned theoretical solutions that provides for the high accuracy of prediction 90% in average, i.e. error of prediction is 10%, three times less than commonly accepted.

REFERENCES

Anokhin, P.K. (1973). Principle questions of the general theory of functional system. Principles of the system organization of function. Moscow: Science, 5-61. *Russian.*

Burov O. Ergonomics, functional state and human fitness-for-duty. *Zastosowania Ergonomii.* 2005, 1-3 (57-59), 203-214.

Burov O. Development and industrial use of computer systems for operators fitness-for-duty check. *Ergonomia.* 2006. Vol. 28, # 1, p.33-45

Chapter 14

How Novice Instructor Can Learn a Stable Evaluation in Report Assignment with No Predetermined Rules

Toshiya Akasaka, Yusaku Okada

Keio University

ABSTRACT

Non-professional, novice instructors often play a leading role in training, such as teaching assistants in academic universities and experienced workers in private companies. Although they are versed in the subject to teach, they are mere novices when it comes to instructing it. One of the problems they have is that they often fail to give a stable evaluation to the report assignment that does not expect learners to follow a specific procedure and reach a certain conclusion. In this paper, we interviewed teaching assistants in a university, with the aim of identifying problems causing them to give an unstable evaluation. The result of the interview was analyzed, yielding some suggestions to prevent unstable evaluation.

Keywords : Novice Instructor, Evaluation, Report Assignment

INTRODUCTION

Training involving intellectual tasks such as report writing can be seen in several places from academic institutions to private companies. The success of a training course largely hinges on the instructor. However, instructors are not always professionals; sometimes the only criterion for selecting the instructor is their knowledge of the subject. In colleges and universities, for example, many graduate students work as teaching assistants, assisting or taking charge of an undergraduate course of their domains. In the private company, the in-house training course (as a part of Off the Job Training (Off-JT)) is often led by an experienced worker who is versed in the subject of the course. These non-professional instructors, however great their knowledge may be, are mere novices when it comes to instructing. We call these instructors NPN (Non-Professional, Novice) instructors.

One of the difficulties NPN instructors face is the evaluation of report assignments. The report assignment is often used when a training course gives a high priority to enhancing learners' skills (as opposed to teaching knowledge), such as logical thinking skill, market analysis skill, data-processing skill, etc. Due to these objectives, the report assignment does not always ask learners to follow a predetermined process, nor does it expect all learners to reach the same conclusion; processes (e.g. how to define problems, how to solve them, etc.) and conclusions are largely left up to learners. We call these report assignments TIY (Think-It-Yourself) report assignments. TIY report assignments are inevitably difficult for NPN instructors to evaluate. As a result, NPN instructors often fail to give a stable evaluation to the TIY report assignment.

However, what is important in the TIY report assignment is to give useful feedback based upon evaluation. Without the ability of giving a stable evaluation, NPN instructors have to spend much time on evaluation and there is also a concern on the validity of evaluation, preventing NPN instructors from focusing on thinking of feedback. Clearly, they first have to obtain the evaluation skill before trying to give instructive feedback. Therefore, it is important for NPN instructors to master the evaluation skill as soon as possible.

The goal of our research is to investigate a way to help NPN instructors learn quickly to give a stable evaluation to the TIY report assignment. Teaching assistants in the university can be considered typical of the NPN instructor. We studied teaching assistants in a university who took charge of evaluating the report assignments that could be considered to be TIY report assignments. As an initial investigation, we interviewed them on how they evaluated report assignments, and studied the report assignments, with the aim of finding problems causing their evaluation unstable and some framework to discuss solutions. In this paper, we present and discuss the result of the initial investigation.

INTERVIEW DETAILS

THE COURSE, TEACHING ASSISTANTS, AND REPORT ASSIGNMENT

We focused on an undergraduate course for experiments. The course is specifically dedicated to ergonomic experiments and consists of five experiments each having a different theme such as usability assessment, analysis of human behavior during VDT (Video Display Terminal) tasks, etc. Students were expected, through experiments and report writing, to foster the attitude to analyze product/task designs from the ergonomic viewpoint. Another aim was to let students enhance their skill to present arguments and to draw conclusion based upon the experiment results, as well as the report writing skill. Among the five experiments, each student underwent three experiments, each followed by one-month period for writing a report on the experiment.

In this course there were five teaching assistants, each being in charge of one of the five themes over the entire course. They were all graduate students and researching in ergonomic domains and had experience of taking this course when they had been undergraduates. Although they were familiar with the ergonomics to which their research domains belong, most of them had not had any instructing experience before becoming the teaching assistants of this course. Thus, they could be considered to be typical NPN (Non-Professional, Novice) instructors (See "INTRODUCTION" for details). Their tasks include;

- To teach students the background of the experiments they were in charge of,
- To instruct students how to carry out the experiments,
- To troubleshoot any problem happening during the experiments,
- To instruct how to deal with the report assignment,
- To respond to questions about the assignment in the report writing period,
- To evaluate submitted report assignments, and
- To instruct how to improve their reports.

Although there was a regular teacher supervising the teaching assistants, his intervention in the tasks above was limited to very few cases, such as authorizing the score of each report assignment. The teaching assistants played a leading role in completing the tasks and were given a considerable discretionary power about how to implement the tasks.

The report assignments in this course were to let students organize and analyze experiment results and draw some conclusions while taking into consideration the objective of each experiment. Students were given detailed explanation about the background of each experiment through the lecture of the teaching assistants and the textbook for this course. On the other hand, instruction about the report assignment was deliberately not specific, just indicating what kind of data to gather, what kind of charts to use, etc. Also, students were not necessarily bound to these instructions; they could neglect the proposed data gathering and instead gather different kinds of data. Thus, students did not have any specific procedure to follow and goal to achieve. These were all left up to students. Having these characteristics, the report assignments in this course could be regarded as a typical example of the TIY (Think-It-Yourself) report assignments (See "INTRODUCTION" for details). The report assignments of this course usually each have the volume of about as many as 20 pages in A4 format, adding the difficulty in evaluation. Actually, all of them did complain that their evaluation was indeed often unstable.

METHOD OF INTERVIEW

With the aim of identifying problems causing the teaching assistants to fail to give a stable evaluation, we interviewed them and investigated how they evaluated the report assignments.

We interviewed one teaching assistants at one time, holding a total of five interview sessions. By the time of the interview session, each teaching assistant had experienced evaluation of about 20 report assignments over a half-year period. Although this experience had the possibility of enabling the teaching assistants to grow beyond our assumed level (i.e. NPN instructor), all of them said that they had trouble with unstable evaluation. Thus we confirmed that it was acceptable to study their problems to achieve our goal. All interview sessions were held upon interviewees' consent. In advance of the interview session, the interviewer (i.e. the first author) studied the background and process of the experiment that the interviewee was in charge of, so that we could let them use the terminology specific to the experiment. This was important to elicit as many check points as possible.

In each interview session, the interviewer asked the interviewee to list evidences/points to look for and criteria to refer to when evaluating a report assignment. We call these evidences/points and criteria collectively "Check Points." In the beginning of an interview session, the interviewer let the interviewee list any check point they could remember in order to elicit as many check points as possible. After the interviewee appeared to finish enumerating check points, the interviewer asked further explication on some of the listed check points that seemed to deserve and have room for further elaboration. An interview session ended when the interviewee had no more check points to list and the interviewer found no room for further explication.

INTERVIEW RESULT

Through the interview described in the previous section, we gathered a number of check points (See "METHOD OF INTERVIEW" for details). Since the number of interviewees was very small (i.e. five), listing all the check points per se was not so meaningful. In order to obtain some suggestions about the causes of an unstable evaluation, we organized the check points we had gathered. To do so, we discussed with some of the teaching assistants and their supervisor who was regarded as an expert at evaluating the report assignments. We also studied the report assignments they evaluated to probe the intents of some of the check points. As a result, we found that most of the check points could be expressed by a combination of the following four elements; locus, viewpoint, focus, and spectrum.

Locus indicates the location they look into when evaluating. Example check points specifying a locus are (the locus of each example is underlined);
- How easy-to-understand the summary section is
- Whether the introductory section states the goal of the report clearly

Locus was not included in all check points. All the teaching assistants listed one or two check points specifying locus.

Viewpoint indicates a quality they focus on when evaluating. Example check points specifying a viewpoint are;

- How easy-to-read the report is (viewpoint = ease of reading)
- Whether there's any leap in logic (viewpoint = logicality)

Viewpoint indicates what quality a good report should have. Most of the check points, explicitly or implicitly, had a viewpoint.

Focus indicates any specific thing in the report that they consciously detect when evaluating. Example check points specifying a focus are (the focus of each example is underlined);

- The number of charts the report has
- Whether the analysis ends up in a mere description of data

A focus can be something objective as in the first example or situational as in the second. The situational focus is more difficult to identify than the objective one. All the teaching assistants listed some focuses, but many of them were objective.

Spectrum is a set of discrete levels about a certain level; each level indicates how much that viewpoint is satisfied, and its description needs to differentiate itself from the other levels. Among the check points we gathered, the only example specifying a spectrum had as its viewpoint "depth of analysis" with the four levels described as follows;

- Level 1: a mere description of data
- Level 2: an ergonomic analysis based upon an inappropriate data interpretation
- Level 3: an ergonomic analysis based upon an appropriate data interpretation, but having no consideration on the objective of the experiment
- Level 4: an ergonomic analysis based upon an appropriate data interpretation, having a consideration on the objective of the experiment

As you can easily recognize, the viewpoint "depth of analysis" is the most satisfied with the level 4 and the least with the level 1. A spectrum requires not only a viewpoint but also focuses. Indeed, several focuses can be seen in the examples above such as "description of data", "ergonomic analysis", "data interpretation", and "consideration on the objective of the experiment." To define a spectrum requires the most conscious effort, understanding, and experience among the four elements we propose.

A check point can be usually expressed by a combination of the four elements (A spectrum inevitably includes the combination of a viewpoint and foci) as in;

- How easy-to-understand the summary section is
  - ➤ Locus = summary section
  - ➤ Viewpoint = ease of understanding
- Whether there's any leap in logic
  - ➤ Viewpoint = logicality (= an implicit viewpoint)
  - ➤ Focus = leap in logic (= a situational focus)

On the other hand, some check points have an only one element. Examples are;

- The number of charts the report has
  - ➤ The focus "the number of charts" is the only element

- How easy-to-read the report is
  - ➢ The viewpoint "ease of reading" is the only element

We counted the number of each possible combination and figured the average frequencies per person which is shown in Figure 1. For example, the circle labeled "Locus" is completely contained by the circle "Viewpoint". This means that the

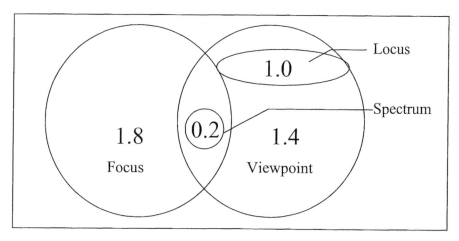

FIGURE 1. Average numbers of observed combinations. If two or more circles are overlapped, it means that the combination of the elements were observed (Although the combination of focus and viewpoint was not observed, the circles corresponding to those elements are overlapped due to a spatial reason).

element "Locus" was always accompanied by the element "Viewpoint", and that it was never observed with the element "Focus." The figure on each circle indicates the average number of check points that are represented by that combination. For example, the combination of viewpoint and locus was observed 1.0 times per person on average (i.e. 5 cases for 5 persons). Likewise, 1.4 check points had as their only element a viewpoint. We incremented the number of viewpoint only if there is a check point including a viewpoint *explicitly*, as it is the awareness of the evaluator that matters here.

Obviously it is difficult to give a stable evaluation if they only consider a viewpoint. For example, if one focuses on the viewpoint "ease of understanding", it is difficult to account for the evaluation they give, leading to an evaluation in a comparative manner, not in absolute terms. This is also the case with the combination of a viewpoint and a locus. However, these two combinations together account for more than half of the all check points. In order to give an evaluation in absolute terms, it seems necessary to form a spectrum in the evaluator's mind and to compare focal points with the description of each level of that spectrum. Furthermore, defining a spectrum requires recognizing many foci and organizing them along a certain viewpoint. The figure shows, however, that the spectrum's frequency is almost zero and that there are no check points of the viewpoint-focus combination.

DISCUSSION

As we have seen in the previous section, check points (See "METHOD OF INTERVIEW" for details) of evaluation can be put in order to some extent by focusing on locus, viewpoint, focus, and spectrum (See "INTERVIEW RESULT" for details). In this section we use these elements and discuss problems leading to an unstable evaluation and some possible solutions to the problems.

As we mentioned earlier, an evaluation based on the viewpoint or the combination of viewpoint and locus tends to be a comparative one. With the absence of any absolute criterion, the evaluator has the risk of giving different evaluations on different occasions. In other words, they tend to give an unstable evaluation. To prevent this, they have to have an absolute criterion, at least for one standard level. Forming a standard level entails having a check point with the combination of viewpoint and focus, which was not observed, at least explicitly, in any of the teaching assistants we interviewed. Thus we can say that the teaching assistants should think to what quality (i.e. viewpoint) the foci they make use of are related. Alternatively, they can decide first on the viewpoint and then on foci to define a standard level.

The apparently best way for the evaluator to prevent an unstable evaluation is to have a firm spectrum in mind. Forming a spectrum embraces defining a standard level by definition, thus being the most comprehensive approach. In order to have a spectrum, one needs to be aware of and able to recognize many foci including situational ones. However, the average number of foci the teaching assistants were aware of was very small (1.8 foci per person). Also, some of the foci seemed very difficult to recognize, such as "ergonomic analysis" and "leap in logic." The teaching assistants listed these foci said that they were somehow able to recognize the foci, but they could not mention any specific way to do so. We think that they have not been yet completely capable of recognizing these foci. Probably, what the teaching assistants need for the time being is to know as many foci as possible and to be able to recognize them.

CONCLUSION

NPN (Non-Professional, Novice) Instructors often play a leading role in training, such as teaching assistants in academic universities and experienced workers in private companies. Although they are versed in the subject to teach, they are mere novices when it comes to instructing it. One of the problems they have is that they often fail to give a stable evaluation to the TIY (Think-It-Yourself) report assignment. In this paper, we interviewed teaching assistants in a university, with the aim of identifying problems causing them to give an unstable evaluation. We organized the check points of evaluation the teaching assistants listed, by means of expressing the check points in terms of the four elements; "locus", "viewpoint", "focus", and "spectrum." As a result, we found that they needed to be aware of and able to recognize more foci, and that they should be able to organize those foci along viewpoints.

Telling them the existence of foci is a straightforward task, but enabling them to recognize the foci, especially situational ones, is a task with no clear procedure. Investigating a way to do so will be the issue that we should work on in the future.

Chapter 15

Act and Anticipate Hazard Perception Training for Young-Inexperienced Drivers

Anat Meir, Avinoam Borowsky, Tal Oron-Gilad,
Yisrael Parmet, David Shinar

The Department of Industrial Engineering and Management
Ben Gurion University of the Negev
Beer Sheva, Israel

ABSTRACT

The present research explored the development and evaluation of an innovative Hazard Perception (HP) training program – the Act and Anticipate Hazard Perception Training (AAHPT). The most important concept of the AAHPT is that young-inexperienced drivers respond to a vast array of actual hazardous situations during training so they can anticipate potential hazards during testing. Forty young-inexperienced drivers underwent one of three HP training modes ('Active', 'Instructional' or 'Hybrid') or a control group. Participants in the 'Active' mode observed 63 HP video-based traffic scenes and were asked to press a response button each time they detected a hazard. The 'Instructional' group underwent a theoretical tutorial that included written material and video-based examples concerning HP. The 'Hybrid' participants observed a concise theoretical component first, followed by a shortened active component (derived from the 'Instructional' and 'Active' modes respectively). The control participants were presented with a tutorial regarding generic road safety issues. Approximately one week later, participants performed a hazard perception test (HPT). During the HPT, participants observed 58 additional HP movies and pressed a response button each time they detected a hazard. Twenty one experienced drivers also performed the HPT and served as a gold standard for comparison. In general, the 'Active' and the 'Hybrid' programs were more aware of potential hazards relative to the 'Instructional' and control groups. The advantages of each training methodology are discussed.

Keywords: hazard perception training, hazard anticipation; young-inexperienced drivers; driving experience.

INTRODUCTION

Research has shown that young drivers - in particular during the first six months of unsupervised driving - are amongst the most vulnerable road users worldwide (e.g., Mayhew, Simpson & Pak, 2003, Pollatsek, Narayanaan, Pradhan, & Fisher., 2006). Horswill & McKenna (2004) have argued that of all driving-related skills, only hazard perception (i.e., situation awareness of hazardous traffic situations) was found to correlate with traffic crashes. According to Logan (1985) a skill is a set of goals which a person tries to obtain, and a set of constrains to which the person must adapt in order to obtain the goals. Furthermore, a more skilled performer can accomplish task goals and can better adapt to its constraints. There seems to be no limits to the degree of skill a person can attain, thus even experts can always improve their performance. Indeed A large body of evidence suggests that young-inexperienced drivers obtain poor hazard perception skills that might explain their over-involvement in traffic crashes (e.g., Pollatsek et al., 2006; McKnight & McKnight, 2003; Fisher, Pollatsek, & Pradhan, 2006). In terms of Logan's (1985) definition, young-inexperienced drivers are less aware of potential hazards (i.e. constrains) in the traffic environment while obtaining their goal (i.e. driving from point A to B).

Recently, Borowsky and his colleagues have shown that young-inexperienced drivers are less aware of potential hazards (e.g., a pedestrian walking along the curb but do not cross it; an intersection with no traffic, etc.) than experienced drivers (Borowsky, Oron-Gilad & Shinar, accepted). In their study, experienced and young-inexperienced drivers observed six short traffic scene movies and were asked to press a response button each time they detect a hazard. While no response time differences were found for actual hazards (i.e., a lead car suddenly braked) young-inexperienced drivers responded less often when potential hazards occurred (e.g., while approaching an intersection with no traffic). Moreover, young-inexperienced drivers tended to gaze straight forward when they approached a T intersection whereas experienced drivers gazed more towards the emerging road searching for hazards. Similarly, Pollatsek and his colleagues have shown that young-inexperienced drivers fixated less often than experienced drivers on areas in the traffic environment that embed potential risks (e.g., a truck that obscure a zebra crossing, or an intersection that is obscured by vegetation, etc.).

Further evidence suggests that young-inexperienced tend to categorize traffic scenes movies according to the similarity in their hazard instigators (e.g., both movies should be grouped because they both contain an episode of a lead car that suddenly brake) whereas experienced drivers tend to categorize the same movies according to the similarity in their traffic environment (e.g., these two movies should be grouped because they both contain residential traffic environments) (Borowsky, Oron-Gilad & Parmet, 2009). Borowsky and his colleagues suggested that categorizing the movies according to similar traffic environments indicates that experienced drivers are aware of potential hazards that are typically similar among similar traffic environments. Similarly, Chapman and Underwood (1998) have shown that experience drivers adopt different scanning patterns for different traffic environments whereas young-inexperienced drivers tend to scan all traffic environments in a similar fashion. Finally, Borowsky and his colleagues have argued that measuring response time difference may be useful and appropriate only if the majority of the participants responded but that usually occurs when hazards are actual and salient. Thus, utilizing only response time measurement may ignore other highly important differences between experienced and young-inexperienced drivers which are attributed to potential hazards and to traffic environment characteristics. Thus, utilizing additional measurements such as traffic scenes classification (e.g., Benda and Hoyos, 1983; Borowsky et al., 2009) response sensitivity (whether or not a driver detected an hazardous events; Borowsky et al., accepted) or eye movements (e.g., Pollatsek et al., 2006) may provide holistic understanding of hazard perception differences between experienced and young-inexperienced drivers and allow developing better training tools to enhance young-inexperienced hazard perception skills.

Ultimately, although many hazard perception training programs were developed (e.g., McKenna, F.P. & Crick, 1991; Sexton, 2000; Pollatsek et al., 2006; Regan, Deery & Triggs, 1998) and showed partial success in improving young-inexperienced hazard perception skills, there is still no agreement regarding which program is superior and which type of measurements should be applied in order to examine hazard perception performance. It is argued that a holistic training methodology that relates to more aspects of hazard perception skills has not yet developed.

The present study presents the development and evaluation of an innovative HP training program – the Act and Anticipate Hazard Perception Training (AAHPT) - which relies on Borowsky et al.'s (accepted; 2009) studies who emphasized the need to increase young-inexperienced drivers' awareness of potential hazards embedded in the traffic environment by exposing them first to a vast array of traffic situations with typical actual hazards. Specifically, it examined whether the AAHPT improves young-inexperienced sensitivity to potential hazards thus changing their behavior from reactive to proactive.

METHOD

PARTICIPANTS

Sixty one participants, 21 experienced drivers (23-29 year-olds with an average of 8.1 years of driving experience) and 40 young-inexperienced drivers (17-18 year-olds with an average of 1.74 months of driving experience) participated as paid volunteers. All participants had normal vision, with uncorrected Snellen static acuity of 6/12 or better and normal contrast sensitivity. Young-inexperienced drivers had no more than three months of driving license during which Israeli drivers are required to be accompanied by someone over 24 years of age with a valid license for at least 5 years, or someone over 30, with a valid license for over 3 years.

APPARATUS

Traffic-scene movies: training and testing databases

Eighteen hours of filmed video of driving from all around Israel were recorded utilizing two Panasonic DV 100 camcorders: one on the dashboard of a KIA carnival (i.e., from a driver's perspective) and a second on a tripod in the back of the vehicle heading towards the rear pane that was later incorporated as the rear-view mirror. Both cameras had a wide angle lenses (approx. 90 degrees) recoding 25 frames per second at a resolution of 768*576 pixels. A set of 123 daylight movie clips, 20-50 seconds long, were extracted and classified according to several dimensions such as traffic environment (i.e., residential, urban and inter-city), road type (curved, straight, mountainous etc.), hazard type (i.e. potential vs. actual), and type of hazard's instigator (e.g., vehicle, pedestrian, obstacle, etc.). These dimensions were partially based on Borowsky et al. (accepted) findings. Figure 1 presents three typical types of traffic environments.

    (1) Inter-City   (2) Residential   (3) Urban

FIGURE 1. Three typical traffic environments

The 123 traffic scenarios were divided into a training database containing 63 movie clips and a testing database containing 58 movies while the remaining 2 movies used for accommodating with the task and the system. Notably, the two databases were countered-balanced to include similar number of movies from each type of environment. However, the number of movie clips that included potential hazards was larger in the testing database than in the training database and vice versa. This manipulation reflects the essence of the AAHPT training methodology suggesting that drivers should first act and respond to actual hazards before they can anticipate potential ones (i.e., from reactive to proactive behavior).

Software and facilities

Designated software was developed to present the hazard perception movies and also recorded participants' button presses and verbal descriptions for each hazard a participant identified. A 19" LCD screen with 1024x768 pixels, connected to a Pentium 4 personal computer was used to present participants the various training methodologies. Participants sat at a distance of 70 cm from the LCD, which provided them with a visual field of 22 degrees vertically and 26 degrees horizontally In addition, a second Pentium 4 was used as the operator's computer on which eye movements were recorded and where the ASL's pan/tilt model 504 eye tracking system's software interface was operated.

Training conditions: AAHPT modes and Control

Three training methodologies were developed: (1) AAHPT Active, (2) AAHPT Instructional-based, and (3) AAHPT Hybrid. An additional group of young-inexperienced drivers used as Control. The control training condition included a PPT presentation (embedded in the software) containing contents of one of the leading Israeli websites that prepare drivers for their theoretical driving test with no specific material on hazard perception. All training methodologies lasted for approximately 1.5 hours.

AAHPT Active

The AAHPT Active mode included two training movies for accommodating with the experimental system and the training movie clips database in which participants observed 63 traffic scene movies, randomly ordered. Participants were asked to observe the movies as if they were actually driving down the road and to press a response button each time they detect a hazard. At the end of each movie they were asked to fill in blank textboxes the hazard instigator for each hazard they have identified. Notably, participants did not receive any feedback regarding their performance. This principle reflects natural driving in which the only feedback drivers receive is from the traffic environment and the consequences of their actions.

AAHPT Instructional

The AAHPT instructional mode presented participants textual explanations and definitions of hazard perception accompanied by relevant examples presented as either snapshots, or complete movies which were extracted from the hazard perception training movies database. Utilizing a Power-Point presentation, participants learned to associate various hazards with their appropriate traffic environments (e.g., a pedestrian is more likely to appear in residential areas rather than in inter-city areas). Furthermore, participants learned to recognize potential and actual hazards and then to distinguish between them (actual hazards were marked in red and potential hazards in yellow throughout training). For each traffic environment (i.e., inter-city, urban, and residential) participants observed various examples that included both potential and actual hazards that are most typical in that environment. Notably, participants were never asked to actively respond to hazardous situations as in the AAHPT Active mode. This approach was aimed at examining whether young-inexperienced drivers can improve their hazard perception skills when training is mostly based on semantic knowledge-based materials rather than, experienced based, episodic knowledge-base materials (as in the AAHPT Active)

AAHPT Hybrid

The AAHPT Hybrid mode included a combination of both Active and Instructional training modes. Participants were first acquainted with textual explanations and definitions of hazard perception accompanied by examples as previously mentioned. This part (making use of 23 of the traffic scene movies) was a reproduction of the 'Instructional' mode but here only half of the examples were used. Then, participants were asked to observe the remaining 40 traffic scene movies taken from the training database movies as if they were actually driving down the road and to press a response button each time they detect a hazard. At the end of each movie they were asked to fill in blank textboxes the hazard instigator for each hazard they have identified. Again, this part was a reproduction of the AAHPT Active. This training mode was designed to examine whether an integration of both semantic and experienced-based training may be more beneficial in enhancing young-inexperienced drivers' hazard perception skills.

Hazard Perception Testing

An additional testing database composing 58 movies used to evaluate whether participants in the various training methodologies improved their hazard perception skills. Their performance was compared to that of a control group who did not receive any hazard perception training and to a group of 21 experienced drivers who used as a goal standard for desired hazard perception performance. The hazard perception test was conducted about a week after training to allow material consolidation. The hazard perception test included two consecutive tasks: (1) hazard detection task as in training, and (2) traffic scene movies classification. The present paper reports only the results of the first task. Experienced drivers participated only in the testing session.

PROCEDURE

Participants were equally distributed to one of three training methodologies or a control: (1) AAHPT Active, (2) AAHPT Instructional, and (3) AAHPT Hybrid. Each young-inexperienced participant was invited to the Eye-Movements Laboratory to participate in two sessions: training and testing. All participants were asked to sign a consent form prior to the beginning of the experiment. For minors (under the age of 18) parents had to sign an additional consent form. Participants were paid a total amount of 120 NIS at the end of the two sessions. Furthermore, an IRB ethics committee approved the present experimental design. In the training session, each participant was seated in front of a computer screen and filled both a computerized demographical questionnaire and a sensation seeking questionnaire (Zukerman, 1990). On this basis participants were designated to one of four experimental conditions: AAHPT Active, AAHPT Instructional, AAHPT Hybrid, and Control. This selection methodology aimed at creating four matching groups according these characteristics. Once a participant was designated to a specific training program he or she underwent a short eye calibration process. Then, according to each methodology the training phase began. Data for each participant was saved and used for the testing session. Approximately a week later, participants arrived for the HP testing session.

RESULTS

The main goal of the study was to examine whether any of the AAHPT versions increased trainees' awareness of potential hazards. It was hypothesized that participants who underwent training will be more sensitive to potential hazards than the Control group and their performance will resemble that of experienced drivers. We present partial results analysis of participants response sensitivity (i.e., how likely was each group of drivers to respond for certain events; see also Borowsky et al., accepted for more details).

RESPONSE SENSITIVITY TO HAZARDOUS EVENTS

In order to examine how the various experimental conditions responded to different hazardous situations it was required to define events. A detailed explanation for how events were defined can be found in Borowsky et al., (accepted). In short, each of the 58 movies which appeared in the testing database was analyzed and Experienced-Based Events (EBEs) were defined first as situations for which at least 6 (out of 21, approximately 25%) experienced drivers responded. Note that hazardous events were not defined a priori but were rather data driven. This procedure aimed at capturing the fact that hazards are subjectively defined by the participants. Then, we compared the performance of each of training or control groups to the performance of experienced drivers. In total, 72 EBEs were recorded according to this procedure. These events were divided into five categories. Table 1 presents the distribution of EBEs across three traffic environments.

Table 1. Distribution of Experienced-Based Events (EBEs) across the traffic environments

Category	Residential	Urban	Inter-City
Pedestrian behavior	10	2	0
Obstacles on road	4	2	0
Approaching an intersection	7	8	0
Vehicles behavior	8	11	8
Limited field of view	2	4	6

Notably, the five categories emerged from the data but they also represent typical hazards in each traffic environment (e.g., pedestrians are more likely to appear in residential areas rather than inter-city). Moreover, the 'limited field of view' and 'approaching an intersection' represent pure potential hazards where all other categories might include both potential and actual hazards. Conducting one-way ANOVA on the five categories in each traffic environment revealed two main effects: (1) the experimental condition groups responded differently to 'pedestrian behavior' category in residential areas $(F(4,56=4.67,p<0.05)$, and (2) the experimental condition groups responded differently to 'Limited field of view' in residential areas $(F(3,36)= 2.88, p<0.05)$.

LSD post hoc pair-wise comparisons of the first main effect revealed that the AAHPT Hybrid (Mean button presses= 8.10, SD=1.60) pressed significantly more than experienced drivers (5.52, 2.23) and significantly more than the control group (4.80, 3.16). Also, the AAHPT Instructional responded significantly more (7.90, 1.45) than the experienced drivers (5.52, 2.23) and significantly more than the control group (4.80, 3.16). No other pair-wise comparisons were statistically significant. As a whole, participants from the AAHPT Instructional and the AAHPT Hybrid tended to press the button more often whenever they saw a pedestrian, whereas the control group and the experienced drivers' group tended to identify less hazardous situations involving pedestrians in residential areas. The AAHPT Active participants were between these two chunks in terms of response sensitivity (6.40, 2.32).

LSD post hoc pair-wise comparisons of the second main effect revealed that the AAHPT Hybrid training program pressed significantly more (mean button presses=0.70, SD=0.67) than the control group (0, 0). Moreover, the experienced drivers tended to respond significantly more (0.62, 0.8) than the control group. These results indicate that members of the control group detected less hazardous situations of the 'limited field of view' factor in residential areas. The fact that the 'limited field of view' factor relates to potential hazard instigators, suggests that HP training programs, especially the AAHPT Hybrid, may improve its trainees awareness of potential hazards.

DISCUSSION

The goal of the study was to examine and evaluate three versions of an innovative training methodology, i.e., AAHPT that aspired to increase young-inexperienced drivers' awareness of potential hazards. The most important concept of the AAHPT is that young-inexperienced drivers should respond to a vast array of actual hazardous situations during training so they can anticipate potential hazards during testing (Borowsky et al., accepted; 2009). Three modes of AAHPT were developed in order to find whether semantic knowledge based training (i.e., AAHPT Instructional) suffice or on the contrary, whether procedural, experienced-based, knowledge is superior (i.e., AAHPT Active). A third mode, i.e., AAHPT Hybrid that combines these former versions, was also considered as a contingency.

According to the results it seems that both the AAHPT Active and AAHPT Hybrid are the more promising training methodologies. Both AAHPT Active and Hybrid modes were more aware of potential hazards than the control group (they both responded more to the 'Limited field of view' category). The AAHPT 'Instructional' mode produced, in general, more responses than any other groups even on irrelevant non-hazardous situations. It seems that the absence of an experienced-based component (i.e. actively responding to hazards) did not allow its participants to distinguish between potential and actual hazards adequately and merely produced false alarms.

Next, we believe that the AAHPT Hybrid is the most promising methodology. As an example, one might notice that participants in this version responded much more on hazardous events that involved pedestrians. Since pedestrian-related hazards were emphasized during the Instructional part of training, it seems that with adequate experienced-based examples, emphasizing specific situations that are known to be more deleterious among young-inexperienced drivers may actually facilitate trainees' awareness of specific potential hazards and reduce their likelihood to be involved in a crash.

Finally, the study limitations should also be acknowledged. Although we examined a vast array of traffic situations there are many other typical situations, such as night driving or adverse weather that were not examined. Furthermore, the testing measurements, although seems to be promising, should be more confined. We are currently working on examining a group of highly experienced drivers (with more than 20 years of driving experience) so we can a-priori define hazardous events to be used as a goal standard to evaluate young-inexperienced drivers performance.

ACKNOWLEDGMENTS

This study was supported in part by the Ran Naor Foundation, Prof. Shalom Hakkert, technical monitor.

REFERENCES

Benda, H. V., & Hoyos, C. G. (1983), "Estimating hazards in traffic situations." *Accident Analysis & Prevention*, 15, 1–9.

Chapman, P.R., Underwood, G., (1998), "Visual search of driving situations: danger and experience." *Perception*, 27, 951–964.

Borowsky, A., Oron-Gilad, T., and Parmet, Y. (2009), "Age and Skill differences in classifying hazardous traffic scenes." *Transportation Research Part F*, 12, 277-288.

Borowsky, A., Shinar, D., and Oron-Gilad, T. (accepted), "Age, skill, and hazard perception in driving. *Accident Analysis and Prevention*.

Fisher, D. L., Pollatsek, A. P., and Pradhan, A. (2006), "Can novice drivers be trained to scan for information that will reduce their likelihood of a crash?" *Injury Prevention*, vol. 12, 25-29.

McKenna, F.P. and Crick, J.L. (1991), *Hazard perception in drivers: a methodology for testing and training.* Department of Transport, TRL CR313, Transport and Road Research Laboratory. Crowthorne, Berkshire.

McKnight J.A., and McKnight S.A. (2003), "Young novice drivers: Careless or clueless." *Accident Analysis and Prevention.* 35, 921–925.

Pollatsek, A., Narayanaan, V., Pradhan, A., and Fisher, D. L. (2006), "Using eye movements to evaluate a PC-based risk awareness and perception training program on a driving simulator." *Human Factors*, 48, 447–464.

Regan, M., Deery, H.A., & Triggs, T.J. (1998), *A technique for enhancing risk perception in novice car drivers.* Accepted for the Road Safety Research, Policing, Education Conference, Wellington, November 16–17.

Sexton, B. (2000), *Development of Hazard Perception Testing*, in proceedings of the DETR Novice Drivers Conference, Bristol, Available at www.dft.gov.

CHAPTER 16

Work Seats – A Short Historical Outline

Jerzy Charytonowicz

Department of Architecture
Wroclaw University of Technology
St. Prusa 53/55, 50-317 Wroclaw, Poland

ABSTRACT

Thrones can be regarded as the first "working" seats for kings and dignitaries; they symbolized power (stools, chairs and armchairs). Made of stone, wood or metal, they did not provide comfort for their stately users, however, they were an honorable equipment used for performing official state duties. Although various kinds of seats were reserved only for sovereigns, with the passing of time they were becoming more and more available for other social classes, undergoing a process of socialization. Forms of the sitting equipment have evolved through the ages. They have become adjusted to needs of various social groups, new professions and functions. Not only did sitting equipment manufacturers search for original forms, but also they refined upon the use values, sitting comfort, operating and lifetime of the equipment. The content of the paper will be a short review of specialist seats for work that developed until the end of the nineteenth century.

Keywords: ergonomics, chairs, seats for a work, desk chairs, office chairs

INTRODUCTION

Basic components of sitting workstations are seats of various types whose shapes and materials they are made of determine the comfort of use. Forms of sitting equipment reflect social diversification of people in particular historical periods, their economic and cultural status as well as the comfort care. Forms of sitting equipment reflect social diversification of people at particular historical stages, their economic and cultural status and the comfort care. Activities undertaken by the sitting equipment manufacturers included not only a quest for original forms, but also improving use values, sitting comfort, durability and manipulability of the equipment. The *throne seats*, being a symbol of power (stools, chairs and armchairs), may be deemed as the first seats used for "work" for kings and magnates. Made of wood, metal or stone, they did not provide comfort for their users, although they were honorable equipment used for performing significant state and representative functions. Various types of seating equipment initially reserved only for sovereigns were steadily undergoing a process of socialization becoming available for other social classes. During the evolution of the seating forms within history the seating equipment were becoming more and more adjusted to needs of various social groups, new professions and new functions. Specialist equipment has appeared, such as *chair – bidets, chairs for parlor games, toilet chairs, hairdresser's chairs*, etc. In the Renaissance, a *swivel chair* appeared (Luther's chair) and a *birthing chair*. The Baroque period has brought a series of new types of specialist seats such as an angle chair also called an *desk chair* since it was used for work at desks. It was a prototype of subsequent *office seats* that best embodied the ergonomic criteria.

WORK AND WORKSTATIONS

The contemporary definition says that *work* is the whole of activities, which aim is to produce specific material and cultural goods that condition the existence and development of society. Work then may have a productive (material) and non-productive character (e. g. art).

Specificity of equipment and workstations arrangement, in turn, are a consequence of a work character which, beginning from prehistory, have yielded continuous transformations.

The moment the first tools were made by people of Paleolith, *work* began. It was connected with both satisfying existential needs as well as fulfilling spiritual needs (rock drawings, glyphs, necklaces). People gradually have been realizing the need for work, but an attitude to it was diversified in subsequent historical periods. In the ancient times people held work in contempt since it was mainly slaves and common people's domain. In the middle ages, work was treated as God's imperative and a punishment for the original sin. A radical change in viewing work came in the Renaissance, in the Reformation age (Martin Luther, John Calvin). The role of human work as a significant production factor began to be appreciated not until the Industrial Revolution (Wojciech Jastrzębowski).

In the social class formation, perception of the concept of work was largely dependent on the status of specific groups of society. The physical work was subjects' responsibility, whereas a potential intellectual work including ceremonies of various kinds, social parlor games, etc. were a privilege of the ruling class.

In the past, considering the symbolic and elite character of the seating equipment it was accessible only to sovereigns.

A natural position people assumed in the distant past when performing various types of work was sitting "on the ground" squatting, kneeling or in the cross-legged position. Since the sitting persons were more or less at the same height, the need to distinguish the leader of group from its other members contributed to emergence of a custom of sitting "*on something*". Differences in sitting manners have divided the later world into areas where people can sit "*on something*" (the so called western world) and areas where people still prefer sitting "*on the ground*".

Sitting "on the ground" in any given position seems natural unlike sitting "on something", which requires adjusting a size of the seating equipment to a size of its users and positions they want to assume. It is associated with a sitting convenience popularly called comfort.

SHORT HISTORICAL OUTLINE OF SEATS FOR WORK

Some primeval furniture forms such as stumps and chopping blocks can be considered as the archetype of subsequent specialized seats for work. A stump in a vertical position could have been used as a stool from which a chair and an armchair originate; however, a trunk and a bench have their beginnings in a lying chopping block.

The first equipment for sitting were so called *natural furniture*, whose forms nature determined, and a human interference in their shapes was little (Fig. 1.1 – 1.2). As tools for wood processing developed, forms of equipment for sitting transformed.

When the class society appeared, a need for a new type of representative equipment such as *throne* occurred. It was a typical seat for work for secular and spiritual authorities. Through ages slaves and subjects could at best sit on natural stools and benches in accordance with a kind of work they performed.

In the Antiquity, unlike in the system of the primeval community, in slavery a considerable development of production powers and various crafts were a consequence of division and distribution of social work, and craftsmen came mainly from freed men.

Egyptian furniture craft reached its peak period about 2000 years BC when a quality jump occurred i. e. as a consequence of improving tools for processing wood and inventing a manual *saw* that enabled cutting wood logs, which earlier have been processed by a carpenter's, into planks. The planks then were cut into slats, which in turn were used to build furniture with a skeleton construction economical in terms of materials. Use of *arched drill bit*, in turn, enabled drilling holes in frames of seats completed next with a rope or leather plaiting ensuring sitting comfort. Wooden surfaces were smoothed away by a *grindstone* made of sandstone.

Egyptians made furniture for sitting that basically were meant for only one most important person in a given social circle. It had a ceremonial hierarchical meaning and it was a symbol of a social status and power of the equipment holder. That way the seating equipment gained a feature of thrones (Fig. 1.5 – 1.6). A specific Egyptian invention was a *folding stool* which performed a throne function during a pharaoh's journey (Fig. 1.4). Craftsmen usually used for work tripod stools with a profiled seat, which ensured a sitting comfort and freedom of movement (Fig. 1.3).

FIGURE 1 Seats for work from the prehistoric period and antiquity

1.1 – 1.2 a natural stool and chair (prehistory), 1.3 a tripod stool (Egypt), 1.4 a stool on runners, 1.5 a throne skeleton chair, 1.6 a throne chair, 1.7 a high throne chair (Assyria), King Darius's throne chair (Persia),1.8 the klismos chair (Greece), 1.10 a scissors chair difros *okladias*, 1.11 a scissors chair *sella curulis* (Rome), (author's drawings)

Assyrian seating equipment was characterized by high seats so that feet of sitting persons were easily and freely hanging when a footrest was not used (Fig. 1.7). The Persian King Darius's throne chair, in turn, had an elevated seat and a footrest and was characterized by a heavy construction made of turned elements (Fig. 1.8).

The crowning achievement of the Greek craft development fell between V and III century B.C. As a result of an advanced work division carpenters and cabinetmakers comprised separate professional groups having far better tools than the Egyptian ones. For carpenter's work *drill bits, chisels, lathes* and *files* were used. What was important for furniture carpentry was invention of a *frame saw* and *joiner's plane,* making the technology of wood processing more efficient and improving the quality of final products. The Greek invention was developing the technology of *bending wood in steam* as well as constructional *carpenter's joints* of various types. That in turn enabled to design and make the famous *klisoms* chair with a light constructions, few materials used and legs widely spaced apart towards outside (which provided stability), made with bent elements coming into stiles bent back and completed with a board as a backrest embracing a sitting person's back. Additional comfort was provided by a woven seat (Fig. 1.9). Delicate and sophisticated shapes of the *klismos* chair provide evidence for the crowning technique, artistry and a perfect sense of relations between equipment and a user's body. The vase paintings show use of chairs for different kinds of activities. Craftsmen on account of a diverse character of works use more frequently folding scissors stools *(difros okladias)* (Fig. 1.10). Functions of Greek thrones performed chairs and armchairs of a heavy construction called *thrónos.*

The Roman furniture art reached its apogee about I century B.C. The seating equipment was nearly identical to the Greek one. Still the basic seat for work were stools made of either wood or bronze. The Greek folding stool *difros okladias* had been converted into the *sella curulis stool,* being a symbol of power and performing a function of seat for work for kings, caesars and high state officials (Fig.1.11). Generally, in the Roman period any new forms of seats for work did not appear and taken from the Greek and Ethrusk culture new furniture forms were improved mainly in terms of decoration.

The medieval period (from V century B.C. to XV A.D.) was dominated by the Christian art controlled by the Church doctrine. That was manifested in a series of some new seating equipment items used in church by clergymen. They were i. e. *cathedrals* or papal or episcopal thrones and *stalls* meant for clergy and secular officials (Fig. 2.1- 2.2).

The medieval period is the beginning of a craft production organized into specialized *guilds.* Initially, craft developed in monasteries and was geared towards making furniture for churches. At that time a significant regression of furniture forms occurred relative to the ancient solutions. The equipment for seating had a heavy board or chest construction and after inventing a *mechanical water saw* (in 1322) and a *sawmill* it also had a frame and panel or transverse construction.

In the Middle Ages any special seats for work were not invented. Peasants and craftsmen used simple stools made by carpenters, but merchants and townsmen used ordinary home chairs and armchairs (Fig. 2.3 – 2.6). At that time the first *swivel stools* appeared.

In the Renaissance, a number and diversity of seating equipment grew and their forms emphasized the social status of their owners. They were characterized by clarity of forms and precise architectural structure. The renaissance carpentry craft reached a high technical and artistic level. More emphasis was put on functional aspects and the sitting comfort thanks to *upholstery nailed* to the furniture construction (the birth of modern upholstery).

In the Renaissance many specialist seats appeared such as the French chair *vertugadin* designed for women wearing crinolines or a *birthing chair* with a semicircle (in a horseshoe shape) cut out in the seat. Also, the first *swivel armchairs* belonged to that group of furniture since they were useful for various intellectual works (e. g. Luther's chair) (Fig. 2.7 – 2.8).

FIGURE 2 Seats for work from the Middle Ages and the modern era

2.1 roman stalls, 2.2. a roman chest throne chair, 2.3. a board chair (Gothic style), 2.4. a tripod chair made of turned elements, 2.5. a stool *sgabello*, 2.6. a folding transverse chair, 2.7. a revolving *Luther* chair (Renaissance), 2.8. a swivel chair, 2.9. an astride chair for reading (Baroque), 2.10. a chair for reading, 2.11. an astride chair *voyeuse*, 2.12. – 2.13. angle desk chairs (author's drawings)

The Baroque times in the furniture making are defined as the *ebeniste revolution* because they were characterized by a high innovation and ingenuity in the scope of constructing furniture for sitting in various functional types, attention to comfort, introducing a new *upholstery technology*, inventing a *loom* for weaving fabric of large sizes useful in upholstery and finally the first specialized *manufactures*. A series of baroque inventions also included some furniture decoration techniques and machines used for the purpose. Emergence of a series of specialist seats for work was a response to social needs at that time, for example a *reading* chair, a *reading armchair* and the *voyeuse chair* used when assisting at various games, on which one sat astride facing the backrest (Fig. 2.9 – 2.11). In the Baroque period, in the living space rooms for mainly office work appeared (a study), where an *angle chair* was put also called an *desk chair*, being at the same time a prototype later office seats (Fig. 2.12 – 2.13).

In the second half of the eighteenth century Classicism emerged in architecture and art. It was a style based upon simplicity and love for the ancient motifs combined with nature elements. It had numerous regional varieties which the Directoire and Empire styles belong to. Introduction simple ancient forms into the furniture making displaced the rich baroque forms made with the craft methods. The classicism period comprises the *beginnings of the machine production* in furniture making and a further development of factories. The eighteenth century was the *age of upholstery,* and a novelty of that period was *quilted upholstery*. In the Empire period furniture production was no longer only the domain of craftsmen but it became a semi-industrial production geared towards the amount of manufactured equipment and the mass consumer. In the classicism period, the *tabourets* still being the palace furniture assumed various forms adjusted also to work seats functions, whose example may be an *upholstered swivel tabouret* (Fig. 3.1). Emergence of the Vienna centre for making pianos that belonged to a group of bourgeois furniture, inspired design of a *piano stool* as a new type of a work seat.

In the first half of the nineteenth century in the German speaking areas a new bourgeois style in furniture making appeared. It was called Biedermeier and its simple forms corresponded to requirements of the industrial production. The technical novelty of the period was invention and introduction for the general use of *box springs* that increased the comfort of the sitting equipment and design of a *wood milling machine,* which allowed to obtain repeatable shapes of tenons used in furniture constructions. Some of the original furniture equipment that appeared in that period were as follows: *camp stools* that derived from rustic *support stools* used on farms in the countryside (Fig. 3.4 – 3.5).

In the second half of the century a fashion for Eclecticism came. The new movement in art and philosophy introduced diversity as a result of combining multiple styles and paradigms in the form of ornaments and designs from different historical epochs. At that time an epoch - making invention appeared that revolutionized manufacturing of the equipment for sitting. In 1841 a Viennese Michael Thonet patented a *methodology of hot bending of wood* improved relative to Samuel Gragg's invention in 1803. The technology opened a new opportunity to manufacture bent furniture assembled from prefabricated elements on an industrial scale. The example was the famous *№ 14 chair* assembled from six elements joined with a few screws used also for a comfortable seat for work (Fig. 3.6 – 3.7).

About 1877 August Thonet, Michael's son, designed, in turn, a *machine for bending seat frames* and started a mass production of seats made of plywood. In the second half of the nineteenth century, a series technical improvements appeared such as rollers fixed to the seating equipment legs, which facilitates moving. They were used in mobile chairs and armchairs manufactured by Thonet's company (Fig. 3.8 - 3.11), and also in American chairs and armchairs with a cast-iron construction shown at London Exhibition in 1851 (Fig. 3.12 – 3.13).

FIGURE 3 Seats for work from the Industrial Revolution

3.1 a stool with the seat height adjustment (classicism), 3.2. a piano stool, 3.3. a piano stool with the seat height adjustment, 3.4. a rustic support stool (biedermeier), 3.5. a folding support stool, 3.6. elements of the N°-14 chair (eclecticism), 3.7. the bent N°-14 chair, 3.8. - 3.11. M. Thonet's revolving office chairs, 3.12. - 3.13. a swivel chair and an office chair, 3.14. a chair with a lifting seat (secession), 3.15. - 3.17. desk chairs, 3.18. a chair for musicians, 3.19. a waiter chair (author's drawings)

At the end of the nineteenth century a conflict between craft and industry occurred partly against the background of aesthetic qualities of the industrial products. The technical progress and new economic and social relationships created a need for establishing new principles of manufacturing goods. They were appropriateness, functionality, usefulness and comfort. Designers of that period began searching for new solutions and ways of creativity liberated from historical tradition, accepting industry, new materials and technologies and opportunities machines provide. This way at the turn of the nineteenth and twentieth centuries Secession emerged being a response to Eclecticism. As a new approach, it tried to reconcile traditions of craft with requirements of industrial production. Considering lines waviness, floral and animal ornamentation and asymmetrical decorativeness it should be stated clearly that not all design from the period in question were suitable for serial industrial production. Undoubtedly, the precursor of soft lines in furniture making was M. Thonet whose innovative seating furniture forms was a consequence of introducing into furniture making and woodwork a technology of wood bending. In the German furniture making another new technology was introduced that based upon a *machine plywood production*. The Secession designers quested mainly for furniture forms different from those current ones. Therefore so few work seats appeared at that time that generally were not different from furniture equipment used in living interiors and public facilities. A chair with an adjustable seat height, desk chairs, an office chair, a chair for musicians and a waiter chair with a backrest supporting loins come just from that period (Fig. 3.14 – 3.19).

CONCLUSION

The history of equipment for sitting has not only reflected development of particular societies, their culture and customs, but also it illustrates the technical, technological and material development over the centuries. Each of the stages of the civilization development went down in the history of furniture design and making with a more and more sophisticated seating constructions diversified formally and functionally and a growing attention to the comfort of use.

In the period discussed, we can distinguish three stages related to the social character of seats, their availability and manufacturing methods. In the first stage, the sitting equipment had a highly elite character. The next stage is the subsequent phases of seats democratization and the beginnings of the craft manufacturing. In the third stage, however, connected with the machine manufacturing the sitting equipment became accessible to all.

A modest amount of sitting equipment designed for work until the end of the nineteenth century was mainly a result of changes in work ethics and in the attitude to work itself in subsequent stages of social development. Over the centuries work understood as a physical effort was a domain of lower social classes. Only in the nineteenth century issues of human work processes were addressed comprehensively (Wojciech Jastrzębowski).

REFERENCES

Charytonowicz J., (2007), *Evolution of sitting furniture forms. From prehistory to the machine age.* Oficyna Wydawnicza Politechniki Wroclawskiej, Wroclaw

Chapter 17

The Effects of Expectations and an Office Environment on Comfort Experience

Conne Mara Bazley[1] ,Peter Vink[2]

[1]JimConna, Inc.
Carbondale, CO 81623, USA

[2]Delft University of Technology
Faculty of Industrial Design Engineering
Delft, The Netherlands

ABSTRACT

By providing optimal internal and external environmental conditions, the concept of "comfort in an office" equates to health and wellness in the workspace and workplace to improve overall productivity. A holistic approach to comfort research includes environmental conditions, physiological, psychological, emotional levels, pre-experiences, and sustainability as comfort factors. In this study, patients were surveyed on their pre-comfort experience and comfort levels before arriving, during, and after concluding an appointment. The results confirm the time of day relationships between pre-comfort experiences and comfort levels. In addition, people (i.e., staff, clients) make an office most comfortable. Office design that considers a combination of healthy environmental conditions, guided flow, and well-being provide optimum comfort.

Keywords: comfort, holistic, office design

INTRODUCTION

Comfort is a term that is often associated with a physical state or an environmental factor.

Kolcaba and Fisher (1996) describe types of comfort: Relief – the state of having a specific comfort need addressed; Ease – the state of calm or contentment; and Transcendence – the state in which one can rise above problems or pain. Additionally, the context in which comfort occurs is Physical – pertaining to bodily sensations, homeostatic mechanisms, immune function, etc.; Psycho Spiritual – pertaining to internal awareness of self and self esteem, identity, sexuality, meaning in one's life, and understood relationship to a higher order or being; Environmental – pertaining to the external background of human experience (temperature, light, sound, odor, color, furniture, landscape, etc.) and Socio-cultural – pertaining to interpersonal, family, and societal relationships (finances, teaching, health care personnel, etc.) also pertaining to family traditions, rituals, and religious practices.

Richards (1980) stresses that comfort involves the sense of subjective well-being as the reaction a person has to an environment or situation.

A "holistic" approach to workspace and workplace design includes health, wellness, and comfort. Optimal internal and external environmental conditions include good quality air, water, climate, physical comfort and health, adequate space, smooth flow of work, general safety, and safe working conditions. Other factors to consider beyond the physical environment are the mental and emotional health that promotes growth, flexibility, creativity, and well being in the office (INQA-Buero, 2005). The integration of the outside world and our internal environment is a universal philosophy The Eastern practice of Feng Shui embraces the concept of physical and "holistic" comfort. According to Tchi (2009) a modern definition of Feng Shui is, "being in the right place at the right time".

What is Feng Shui? This simple question is difficult to answer. Feng Shui literally means "wind "and "water". It is an ancient art and science developed over 3,000 years ago in China, a complex body of knowledge, and balances the energies of the environment and assures the health and good fortune of those inhabiting that space.

The integration of the outside world and our internal environment is a universal philosophy The Eastern practice of Feng Shui embraces the concept of physical and "holistic" comfort. According to Tchi (2009) a modern definition of Feng Shui is, "being in the right place at the right time". Feng Shui brings together the vital conscious connection between person and place and heaven and earth. Feng Shui is the art of arrangement, of placement. Placing in such a way as to enhance the flow of energies and minimize dissipation.

Konieczny (2001) concluded that the changing environment of comfort at work and in service is dependent on pre-experience and attitude and considered three main elements to access pre-experiences and attitudes for flight travel: **Hardware** (airport signs, walking distance, toilets, etc.); **Software** (waiting and boarding times); and **Life ware** (staff competencies and personal support). He concluded that pre-experiences and attitude toward flight correlated with the flight outcome. Adding pre-comfort experiences to the study of comfort deepens the understanding and formation of personal comfort realities.

Taking into consideration several definitions for comfort, this paper applied a holistic approach to comfort research and combines environmental elements and psychological and emotional factors measured over time to establish changes or sustainability in comfort levels. Three questions were asked:

1. Do pre-experiences have an effect on comfort levels throughout the day?
2. Do comfort levels decline overall throughout the day?
3. What is the most important factor influencing comfort at an office?

METHOD

A One-Day Survey was administered to 51 patients at a chiropractic office. The Office staff distributed the comfort surveys to patients for a two-week period.

Participants were asked to complete a comfort survey upon arrival to the appointment and to complete the survey after receiving medical treatment. The first set of survey questions asked for a 1-5 rating (1 rated as excellent and 5 rated as very bad) on the participants waking comfort level (physical, intellectual, and emotional). In addition, participants were asked to circle a word and a color that best described the comfort level upon waking. A "yes" or "no" question was included in this set asking if the participant was looking forward to the appointment.

The next set of questions asked for comfort level (physical, intellectual, and emotional where each was rated 1-5), a word and a color to best describe comfort levels upon arriving at the medical office. Additionally, participants circled the word that best described the most significant thing they saw and most significant thing they heard on their way to their appointment.

The last part of the survey was completed after the appointment and participants were asked to rate their comfort level (physical, intellectual, and emotional where each was rated 1-5) and again circle a word and a color that best described their comfort level at that time. In addition, they circled a word that best described what made them most comfortable and a word that made them the most uncomfortable in the medical office setting.

RESULTS

A Pearson Correlation was performed on the first, second, and third set of comfort-level questions (physical, intellectual, and emotional) and the two pre-comfort experiences (most significant thing seen and heard on the way to the medical appointment). The results were most significant for the intellectual comfort level throughout the day.

A one-way repeated measures analysis of variance (ANOVA) was performed for DAYS and Time of Day (TOD) for all three-comfort levels. Significant results were found for all three-comfort levels. The physical comfort level changed the most throughout the day and the emotional comfort level was the best after the appointment (Figure 1).

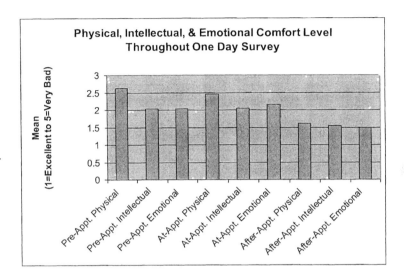

FIGURE 1 One-Day Survey physical, intellectual, and emotional comfort levels though out the day

The One Day Survey participants' age, gender, and job title (Table 1).

Table 1 Survey Participants Characteristics

Characteristic	n	%
Age Range		
22-30	8	15.7
31-40	5	9.8
41-50	6	11.8
51-60	15	29.4
61-70	17	33.3
Total	**51**	**100.0**
Gender		
Male	14	27.5
Female	37	72.5
Total	**51**	**100.0**

Job Title		
Professional	34	65.3
Homemaker	7	14.3
Retired	10	20.4
Total	**51**	**100.0**

The Pearson Correlation of the first set of comfort levels (physical, intellectual, and emotional levels upon waking) showed a statistically significant relationship between emotional and intellectual comfort levels, $r(51) = .857$, $p < .000$. This result suggests that both of these comfort levels were superior than the physical comfort level. A statistically significant relationship, $r(51) = .390$, $p < .005$, was found between physical and intellectual comfort levels suggesting that the intellectual level was superior than the physical level. A statistically significant relationship, $r(51) = .34$, $p < 0.014$, was found between physical and emotional comfort levels. This finding suggests that the emotional comfort level was better than the physical and intellectual comfort levels.

The second set of comfort levels (arrival to appointment) was correlated along with the first set and two pre-comfort experience variables were added. The first pre-comfort experiences variable was the most significant visual seen on the way to the appointment and the second variable was the most significant audio experience heard on the way to the appointment. A statistically significant relationship was found between physical and intellectual comfort levels, $r(51) = .723$, $p < 0.000$. This suggests that the pre-comfort experience variables had a positive effect on the physical and intellectual comfort levels.

The last set of comfort levels and pre-comfort experiences (after the medical treatment) showed a statistically significant relationship, $r(51) = .633$, $p < 0.000$ found between physical and intellectual comfort levels. This finding again suggests that the pre-comfort experiences had a positive effect on physical and intellectual comfort levels.

A one-way analysis of variance (ANOVA) with DAYS and TOD as factors was calculated on participants' ratings of comfort levels. There were significant results for all three comfort levels. The participants felt more physically comfortable, $F_{1, 50} = 54.1$, $P = 0.05$); than intellectually comfortable, $F_{1, 50} = 22.9$, $P = 0.05$; or emotionally comfortable, $F_{1, 50} = 37.3$, $P = 0.05$, after the medical treatment appointment.

Out of 51 participants surveyed for One-Day, all but one said "yes" that they were looking forward to their appointment. The attitude was positive. The results for the most significant visual or audio experience the participants had between waking and arriving at their appointment are as follows:

- Thirty percent observed "the roadway", 30% "people", 28% "nature", 9% "buildings", 3% the Moon and Stars. Thirty seven percent reported they heard "singing", 31% "talking", 8% "silence", 8% "nature sounds", and 16% "mechanical sounds".

- Physical comfort improved, intellectual levels were the same and emotional levels declined between the time of waking and the arrival to the appointment. When asked what word described their comfort when they arrived to the appointment, 37% were "calm", 23% "anxious", 21% "tired", 10% "relaxed", 5% "excited", and 4% "upset".
- Fourteen participants found nothing uncomfortable with the office, 8 "layout", 4 "tasks", 4 "temperature", and 3 "light". Twenty nine circled "people" for what was most comfortable in the office 8 "layout", 4 "tasks", 4 "temperature", 3 "furniture", and 3 "light".

DISCUSSION

COMFORT IS DEPENDENT ON PRE-COMFORT EXPERIENCES

A modified version for pre-experience or pre-comfort experiences for comfort studies is possible (Table 2) by exchanging hardware for physical comfort, software for intellectual comfort, and life ware for emotional comfort. The surveys used in this paper investigate pre-comfort experiences along with incremental comfort level assessments throughout the day. Adding pre-comfort experiences to the study of comfort deepens the understanding and formation of personal comfort realities.

Table 2 is a modified version of Kocienczny's (2001) table of pre-experiences for flight shows an example of elements for pre-comfort experiences. Pre-comfort experience elements may occupy more than one comfort type.

Table 2 Pre-Comfort Experiences for Flight

Comfort Type	Pre-Comfort Experiences (Elements)
Physical	Buildings
	Roadway
	People
	Sunrise
Intellectual	Looking forward to work
	Talking
	Singing
Emotional	Colors
	Upset
	Excited

Kocieczny (2001) found that comfort during airline flight correlates highest with comfort levels preceding the flight, for instance, fear of flying and ones flying attitude. This study had similar findings. Out of 51 participants surveyed for One-

Day, 98% of the respondents said "yes", they were looking forward to their appointment and their attitude was positive. The expectation that the participant had prior to an appointment was that the participant would feel better by the end of the day. Anticipation about treatment or diagnosis and attitude about the medical office can also shape the comfort outcome from an emotional standpoint. Most participants heard singing before their appointment and saw roadway signs before their appointment. Previous office visits or attitudes about the office and visits and experiences that happen earlier in the day showed to have an effect on comfort levels.

COMFORT LEVELS DECLINE DURING THE DAY

Offices designed with a consideration for comfort is important for worker longevity and productivity. The Gensler Survey (2006) found a well-designed office is the key to improving productivity. With the advent of health and wellness in the workplace, awareness to the benefits of good design and a happy workforce are evident (Sonnentag, Mojza, Binnewies, & Scholl, 2008, INQA-Buero, 2005). For therapy, treatment offices most patients feel better and more comfortable after treatments (Gionis, 2003); and a well-designed office has a positive effect on patients or clients by providing a calming and relaxed environment. However, working in an office all day as opposed to visiting an office as a client or patient and spending an hour or so can make a difference on comfort levels throughout the day. Vink, Konijn, Jongejan, & Berger (2009) found that physical comfort levels decline by the end of the day for office workers.

In this case, comfort levels did not decline during the day. After the medical treatment, all three-comfort levels improved, with the physical and emotional levels being most significant. The physical improvement results are consistent with other medical studies that show comfort level improvement after chiropractic therapy (Gionis, 2003; Winstead and Kijek, 1999; Krieger, 1975).

Patients generally do not feel well before a therapy treatment, that is why they are seeking treatment. The physical comfort level is generally the worst before an appointment for treatment therapy. After a treatment, patients usually feel more comfortable overall. Medical studies often measure pain throughout the day and throughout the treatment series (Kolcaba & Fisher, 1996). Other studies conclude that after treatment therapy a release of physical pain causes an emotional surge (Winstead and Kijek, 1999; Coulehan, 1985; Krieger, 1975).

The overall rating by participants for the office environment was very comfortable and 48 out of 51 participants found nothing that made them uncomfortable in the office setting actually they liked everything about the office. Two participants found the temperature too cold; and temperature is often the number one complaint for an office physical environment (ASHRAE, 2004; Brown, 1996; and Vink, 2005).

THE MOST IMPORTANT FACTOR INFLUENCING COMFORT AT AN OFFICE

The number one factor that made participants most comfortable in the office setting was the "people". Comments were hand written on 14 of the 51 surveys. Some of the comments included, "The people-energy in the office is invigorating and I find everything comfortable" and "This is the best office and I love the people here".

People have an influence on the office environment and can change a seemly physical environment to comfortable or uncomfortable. Comfort and discomfort includes the interaction with co-workers or clients as well as factors and experiences outside of work, such as family, friends, health, and personal business matters. Group dynamics can be comfortable or uncomfortable and somewhat dependent upon personalities, past experiences, and frames of references (Whedon, 2000).

People and their energy are constantly changing the visible and invisible flow of the physical environment. According to Heim (2001), flow is a smooth, unimpeded movement through space-time. It is an aesthetic quality of spatial movement and occurs throughout the physical world. There are three ways in which flow applies to the aesthetic of environments: atmospheric flow, the flow of words with visual images, and the flow of group dynamics.

An alternative to improving the physical environment first for comfort may be to design for flow first, flow with intention. Feng Shui establishes intention and identifies the positive "flow/energy" or "chi" of an environment. The office layout and design provide good flow, comfort, and positive energy in a space for patients, visitors, and staff. The concept of flow was incorporated into the arrangement of furniture and treatment rooms, waiting room, and equipment room. Selection of a space with adequate window light was also taken into account.

This study had limitations. Ideally surveying the participants over a month or more would have been preferable. Comparing this medical office with more medical offices would be preferable as well.

CONCLUSIONS

Participant comfort levels improved throughout the day. Pre-comfort experiences had an effect on the comfort levels The layout and office design included consideration for well-being, flow, and safety."People" bring a positive nurturing energy to the office environment and provide a comfortable space for those who occupy it. Measuring outcomes on comfort is difficult and, therefore, lends itself to the subjective. People are an open-ended system, and do not react, act, feel, or intellectualize the same way and each person has his/her own unique reality, outlook, and opinion. Further research and studies are needed in this area.

162

REFERENCES

ASHRAE (2004). ANSI/ASHRAE Satandard-55-20004: *Thermal environmental conditions for humanoccupancy.* American Society of Heating, refrigerating and Air-Conditioning Engineers, Atlanta.

Brown, D. (1996). *Alternatives to Modern Air-Conditioning Systems: Using Natural Ventilation and Other Techniques, Il,* Springfield: APT Bulletin 27(3).

Coulehan JL (1985) *Adjustment: The Hands and Healing,* Culture Med Psychiatry, 9 pp. 353-382.

The Gensler Survey (2006) (2008) Dialogue is produced by Gensler Publications,© 2010 Gensler. at dialogue@gensler.com. Retrieved on-line June 10,2009

Gionis, 2003, TA **Gionis**, E Groteke - Orthopedic technology review, Spinal Decompression, 2003 - backinactionsdc.com http://backinactionsdc.com/forms/Spinal%20Decompression%20Report.pdf

Heim, M. (2001) The Feng Shui of Virtual Reality Crossings: ejournal of Art and Technology, 1(1).Retrieved June 3, 2009, from http:crossings.tcd.ie/issues/1.1/

INQA-Buero (2005). Well-Being in the Office. Federal Institute for occupational Safety and Health (ENWHP) Dortmund-Dortsfeld, ISBN: 88261-469-9, pp.2-36

Kolcaba, K. & Fisher, E. (1996). A holistic perspective on comfort care as an advance directive. *Crit CareNurs* Q,18(4):66-7.

Konieczny G. (2001). The measurement an increase of the quality of services in the Flugzeugcabinea contribution for customer-oriented airplane development, Dissertation, TU Berlin, 2001

Krieger S., (1975) Therapeutic Touch: The imprimatur of Nursing", *American Journal of Nursing* 5 pp. 784-787.

Richards, L.G. (1980) "On the psychology of passenger comfort." In D.J. Oborne and J.A. Levis, eds., Human Factors in Transport Research, vol. 2. London: Academic Press, 15–23.

Sonnentag, S., Mojza, E., Binnewies, C., Scholl, A. (2008). Being engaged at work and detached at home: A week-level study on work engagement, psychological detachment, and affect, *Work & Stress,* 22:3, pp. 257-276.

Tchi, R. What is Feng Shui-Feng Shui Theory and Feng Shui Tools. Retrieved June 28, 2009, from http:/fengshui.about.com/od/the basics/qt/feng shui.htm

Vink, P. (2005). Comfort and design: principles and good practice, Boca Raton: CRC Press.

Vink, P., Konijn I., Jongejan B., and Berger M. (2009) Varying the Office Work Posture between Standing, Half-Standing and Sitting Results in Less Discomfort. *Ergonomics and Health Aspects,* HCII 115-120.

Whedon, C. (2000). Frames of reference that address the impact of physical environments on occupational performance, *Work* 14:165-174.

Winstead, F. and Kijek, J. (1999) Theraputic Touch, *Alternative Therapies in Health and Medicine,* 5(6) pp. 58-67.

CHAPTER 18

Office Ergonomics Interventions: Examining the Effects on Musculoskeletal Risk, Environmental Control and Comfort

Michelle M. Robertson[1], Ben Amick[2], Y.H. Huang[1], Gabriella Kernan[3]

[1]Liberty Mutual Research Institute for Safety
Hopkinton, MA, USA

[2]The University of Texas School of Public Health, Houston Texas USA and
Institute for Work & Health, Toronto, Canada

[3]University of Lowell, Lowell MA, USA

ABSTRACT

Office and computer work is associated with safety and health issues in the working age population. To address these risk factors, office ergonomics interventions need to be designed and evaluated for their effectiveness in reducing musculoskeletal and visual systems along with enhancing individual performance. Organizations show an interest in improving office workplaces; however scant research exists in longitudinal field intervention research. This paper will report on the impact of two longitudinal office ergonomics workplace and training interventions on workers' knowledge, musculoskeletal health and computing behaviors. These two study

164

results suggest that the provision of ergonomic skills, in the form of training, and workplace flexibility allows individuals to make appropriate workstation, computing behavioral changes, thus reducing musculoskeletal discomfort associated with computer use.

Keywords: Office ergonomics interventions, musculoskeletal discomfort, training, computer work

INTRODUCTION

As the dependence on computer use increases, concerns have been raised over the potential for an escalation in the incidence of Work Related Musculoskeletal Disorders (WMSDs) and visual symptoms (Aaras, et al., 2001; Bernard, et al., 1994; Bongers, et al., 1993; Tittiranonda, et al., 1999). Computer work is identified as a risk factor for WMSDs in the working age population. Systemically designed office ergonomics interventions may contribute not only to enhanced worker health and well-being, but also to organizational effectiveness (e.g., Carayon, et al., 2000; Hagberg, et al., 1995; Verbeek, 1991). When a successful office ergonomics intervention program is implemented, one of the many results is an increased ability for the worker to change his/her work environment, leading to enhanced individual effectiveness and the prevention of WMSDs and injuries (Amick et al., 2003; Robertson et al., 2006; Brisson, et al., 1999). Further, with the knowledge and ability to exert control over one's physical environment, healthy computing behaviors may emerge. Although there is a growing interest among employers to improve office workplaces, few longitudinal field studies have examined the effects of office ergonomics interventions on worker's health, comfort, and performance (Brewer, et al., 2006).

Two office ergonomics field studies will be presented where the primary aim was to investigate the effects of an office ergonomics workplace and training intervention on workers' knowledge, self-reported musculoskeletal discomfort, and computing behaviors. For both studies, it was hypothesized that the training and workplace design together would allow the worker to more effectively use their workspace through increased office ergonomics knowledge and skills and adoption of healthy computing behaviors, thus reducing WMSDs.

Both studies employed a quasi-experimental field study design and were large-scale field investigations. The first study involved two office ergonomics interventions consisting of: 1) a new flexible office space with adjustable workstations and a flexible overall facility layout, and 2) office ergonomics training regarding the use of the space that supports employee control over how the overall space is used. There were three conditions: 1) "Control" group consisting of employees who did not receive a new workspace or training; 2) "Workstation-only" employees who received the new experimental, flexible workspace, and 3) "Workstation + training" employees who received the new workspace and office ergonomics training.

METHODS

The second study consisted of an office ergonomic intervention where training was coupled with a highly adjustable chair and compared to training-only and control groups. Participants were assigned to one of three study groups: a group receiving the adjustable chair and ergonomics training (C+T), an ergonomics training-only group (T-only), and a control group. Participants in both groups were classified as knowledge workers who used a computer 4+ hours a day (Amick et al., 2003; Robertson, et al., 2009 .

Using the Instructional System Design (ISD) as a guide for both interventions, we developed an office ergonomics-training workshop with the goal of motivating workers to conduct self-evaluations and to reorganize and adjust their workspace (Gordon, 1994; Knirk & Gustafson, 1986; Kirkpatrick, D., 1979). Each of the training programs were customized to support each of the research site including, the organizational culture and the existing health and ergonomics programs and policies.

For both field sites, three methods of data collection were employed: 1) Work Environment electronic surveys, 2) Ergonomic knowledge tests and 3) worksite observations (Corlett & Bishop, 1976; Kuorinka, et al., 1987). For the first study, data were collected simultaneously from all three groups, once prior to any workplace change, and twice following the interventions. Data collection efforts for study two involved asking the participants to complete five online surveys: (2 months and 1 month prior to intervention) and (2, 6, and 12 months following intervention), over 1 16-month period. Individual workstation assessments and body posture were observed (1 pre and 1 post-intervention) along with two training knowledge tests (pre and immediate post-training).

RESULTS AND DISCUSSION

Results from study 1 revealed that following the intervention, there was a significant increase in workers' office ergonomics knowledge and awareness. Self-reported work-related musculoskeletal disorders significantly decreased for the group who had a workplace change and received ergonomic training relative to a workplace change-only group and a no intervention control group. The level of self-reported job control significantly increased for both the workplace change and training groups as compared to the control group (Robertson, et al., 2008). Significant changes in observed computing work habits and postures following the intervention were found for both the workplace change and training groups.

Results from study 2 indicated that perceived control over the physical work environment was higher for both intervention groups as compared to workers in the control group. A significant increase in overall ergonomic knowledge was observed for the intervention groups. Both intervention groups (chair + training and training-only) exhibited higher-level behavioral translation and had lower musculoskeletal risk than the control group as demonstrated by the observational results (Robertson, et al. 2009). The Chair + training group experienced significant positive improvements in chair satisfaction and comfort, whereas there was no significant difference found for the training-only group (Amick et al., 2003). This lack of training-only group effect supports implementing training in conjunction with highly adjustable office furniture and equipment to improve comfort and satisfaction.

Overall, these two study results suggest that the provision of ergonomic skills, in the form of training, and workplace flexibility allows individuals to make appropriate workstation, computing behavioral changes, thus reducing musculoskeletal discomfort associated with computer work.

REFERENCES

Aaras, A., Horgen, G., Bjorset, H. H., Ro, O., and Walsoe, H., 2001. Musculoskeletal, visual and psychosocial stress in VDU operators before and after multidisciplinary ergonomic interventions. A 6 years prospective study-Part II. Appl. Ergon. 32(6), 559-572.

Amick, B.C., Robertson, M., DeRango, K., Bazzani, L., Moore, A., Rooney, T., Harrist, R., 2003. The effect of an office ergonomics intervention on reducing musculoskeletal symptoms. Spine 28 (24), 2706-2711.

Bernard, B., Sauter, S., Fine, L., Petersen, J., and Hales, T., 1994. Job task and psychosocial risk factors for work-related musculoskeletal disorders among newspaper employees. Scand. J. Work Environ. Health 20, 417-426.

Bongers, P. M., de Winter, C. R., Kompier, M. A., and Hildebrandt, V. H., 1993. Psychosocial factors at work and musculoskeletal disease. Scand J Work Environ Health 19(5), 297-312.

Brewer, S., Van Eerd, D., Amick III, B. C., Irvin, E., Daum, K., Gerr, F., Moore, J. S., Cullen, K., and Rempel, D., 2006. Workplace interventions to prevent musculoskeletal and visual symptoms and disorders among computer users: A systematic review. J. Occup. Rehabil. 16(3).

Brisson, C., Montreuil, S., and Punnett, L., 1999. Effects of an ergonomic training program on workers with video display units. Scand. J. Work Environ. Health 25(3), 255-263.

Carayon, P., and Smith, M. J., 2000. Work organization and ergonomics. Appl. Ergon. 31(6), 649-662.

Corlett, E.N., bishop, R.P., 1976, A technique for assessing postural discomfort. Ergonomics 12 (2), 175-182.

Gordon, S. E., 1994. Systematic Training Program Design: Maximizing Effectiveness and Minimizing Liability. Prentice Hall, Englewood Cliffs, NJ.

Hagberg, M., Silverstein, B. A., Wells, R., Smith, M. J., Hendrick, H. W., Carayon, P., and Perusse, M., 1995. Work-Related Musculoskeletal Disorders (WMSDs): A Reference Book for Prevention. Taylor & Francis, London.

Kirkpatrick, D., 1979. Techniques for evaluating training programs. Train. Dev. J. 31(11), 9-12.

Knirk, F. & Gustafson, K.L.,1986. Insturcitonal technology: A systematic approach to education. New Your, NY, USA: Holt, Rinehart & Winston.

Kuorinka, I., Jonsson, B., Kilbom, A., Vinterberg, H., Biering-Sorensen, F., Andersson, G., Jorgensen, K., 1987. Standardised NORDIC quiestionnaires for the analysis of musculoskeletal sympotms. Applied Ergonomics 18 (3), 233-237.

Robertson, M.M., Amick, B.C., DeRango, K., Rooney, T., Bazzani, L., Harrist, R., Moore, A. 2009. The effects of an office ergonomics training and chair intervention on worker knowledge, behaivor and musculoskeletal risk. Applied Ergonomics, 40, 124-135.

Robertson, M.M., Huang, Y.H., Schliefier, L., 2008. Flexible workspace design and ergonomics training: impacts on the psychosocial environment, musculoskeletal health, and work effectiveness among knowledge workers, Applied Ergonomics, 39, 482-494

Robertson, M.M., Huang, Y.H., 2006. Effect of a workplace design and training intervetnion on individual performance, gorup effectiveness and collaboration: the role of enviornmetnal control. Work 27 (1), 3-12.

Tittiranonda, P., Rempel, D., Armstrong, T., and Burastero, S., 1999. Effect of four computer keyboards in computer users with upper extremity musculoskeletal disorders. Am. J. Ind. Med. 35, 647-661.

Verbeek, J., 1991. The use of adjsutable furniture: evaluation of an instruction program for office workers. Applied Ergonomics, 22, 179-184.

Chapter 19

Order-Picking in Deep Cold – A Gender-Related Analysis of Subjectively Assessed Effects

Mario Penzkofer, Karsten Kluth, Helmut Strasser

Ergonomics Division
University of Siegen
57068 Siegen, Germany

ABSTRACT

In the context of a gender-related workforce employment, an investigation of 128 employees in 24 deep cold-storage depots was carried out in order to identify the subjective experience of working in the cold. The work-physiologically oriented, structured interviews permitted examining individual effects of order-picking groceries at environmental temperatures of approximately +3°C to -24°C. By means of a standardized questionnaire, data about e.g. the physical strain, the current regulation of working time and warming-up breaks, the environmental conditions and the cold protective clothing were collected and analyzed.

In most cases gender-specific differences could not be detected. While the cold-protective clothing or the use of work equipment were assessed quite similarly by male and female subjects (Ss), the regulation of working time and warming-up breaks and the temperature sensitivity were rated rather differently.

Keywords: Working in Severe Cold, Subjective Evaluation, Gender, Cold Sensations, Workplace Design

INTRODUCTION

Order-picking systems in deep cold-storage depots at temperatures around +3°C in the chill room and -24°C or lower in the cold store represent a key element in the distribution of refrigerated or frozen food.

Despite the success in the field of automation, order-picking frozen food under these environmental conditions has largely remained "manual work" due to the problems in using technical equipment in the extreme cold. Working in a deep cold-storage depot implies that employees are exposed to temperatures of around -24°C for a considerable length of time. Transferring and stockpiling the goods with a weight up to 15 kg represents a physically high demanding task for the order-pickers in itself, which is additionally intensified by the extreme working conditions in deep cold-storage depots. In particular the working environment with extreme temperatures of -24°C or lower leads to an unusual additional strain. However, a physical "zero load" would rather increase the cold stress. Therefore, special attention should be paid to the protection of the employees working in deep cold-storage depots, particularly the group of people which is regularly exposed to these temperatures.

In case of insufficient clothing protection and extremely long exposure to the cold, certain health risks might not be excluded.

Especially for susceptible individuals possibly short-term illnesses, such as rheumatism, infections of the respiratory tract, colds and, in the long run, hypertension might not be avoidable.

Detailed interviews were conducted with 128 order-pickers out of 24 deep cold-storage depots. To identify possible impairments and health risks of the employees due to the cold exposures at +3°C or approximately -24°C, male and female order-pickers had to make statements about their cold sensations, the working conditions, the regulation of working time and warming-up breaks and possibly arising physical complaints. In addition a work-physiological field study was carried out for the objectification of the physical strain while working in the cold (Kluth et al., 2010).

METHODS

In order to gain knowledge about working in the cold and the related physical strain, a first examination was carried out measuring and evaluating the effects of order-picking work in the cold on the physiological responses of 30 subjects in whole working-day tests under real working conditions (Kluth et al., 2008/2009, Penzkofer et al., 2008/2009, 2009).

For comparison purposes the subjectively experienced stress and strain was determined through a systematic interview of 62 male and 66 female employees by means of a standardised questionnaire comprising 57 items referring to the subjectively felt effects of working in the cold. The questionnaire was not confined to the basic work in the chill room or cold store and its associated physical feelings, such as muscle and joint complaints as well as cold sensations or frostbite-symptoms, but also covered questions on the working conditions, the cold protective clothing, the work equipment, the work-rest regulation and on motivational aspects. Assessments concerning the physical strain and the entire workplace had to be carried out in some final topics. The condition for taking part in the investigation was that the order-pickers had to be employed in the cold store for a longer period of time. The investigation was carried out in form of interviews and under the supervision of a person familiar with the topic. Arising questions and uncertainties could be clarified immediately and a return rate of 100% could be guaranteed. Due to the size of the sample the results can be assumed to be reliable and valid.

Contrary to the common opinion that order-picking work in the cold is normally carried out predominantly by males, the proportion of females amounted to almost 52%.

At the time of the investigation, the age of the subjects was between 19 and 49 years. The average age was approximately 35 years, with the male Ss averaging 33 years and the female Ss – slightly older – averaging around 38 years. Table 1 gives a survey of additional specific data of the interviewees.

Table 1: Specific data of the sample of the 128 order-pickers

Characteristics	Female Ss (N=66)	Male Ss (N=62)	
Age	37.6 ± 6.8	32.8 ± 6.2	[Years]
Height	165.7 ± 5.6	180.2 ± 7.4	[cm]
Weight	68.8 ± 12.1	79.2 ± 10.5	[kg]
BMI	25.1 ± 4.4	24.4 ± 2.6	
Working Experience	7.1 ± 3.2	6.1 ± 3.3	[Years]
Order-picking in Cold-storage Depots	5.5 ± 2.9	5.0 ± 3.2	[Years]
Daily Working Time	4.3 ± 0.6	6.5 ± 1.1	[h]

The body mass index (BMI) is an index which is used to estimate a healthy body weight based on a person's height. The WHO regard a BMI of less than 18.5 as underweight, while a BMI greater than 25 is considered overweight and above 30 is considered obese. Average BMI-values of 25 for the females and of 24 for the males were calculated. In detail, 73% of males were characterized as "normal", 13% exceeded the normal weight and 14% were below the normal weight. Only 48% of the questioned females had a "normal" weight, 35% were even "underweight". Another 9% of the female order-pickers were assessed as employees with "overweight" and 8% could be assigned to obese "class 1" (cp. Fig. 1).

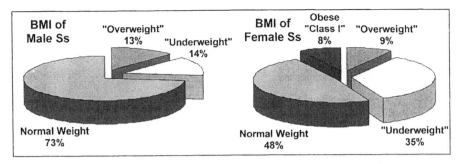

FIGURE 1. BMI-classification of male (left part) and female Ss (right part)

Figure 2 visualizes the differences in the average daily working time calculated from the weekly contracted hours of work. While males usually have full time contracts of 37.5 hours or part-time contracts of 30 hours per week, females normally are employed as part-time workers with contracts of 20 hours per week.

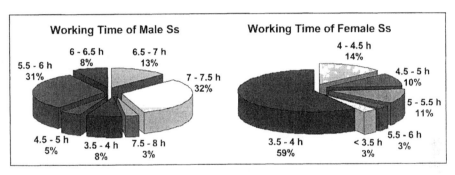

FIGURE 2. Daily working time of male (left part) and female Ss (right part)

Due to the very long experience of the employees with approximately five years of working in the cold it can be assumed that detailed and significant information is available about the working conditions and the subjectively assessed effects of the work. All ratings were carried out on a bipolar scale from -4 (very bad) to +4 (very good).

RESULTS AND DISCUSSION

EVALUATION OF THE WORKING CONDITIONS

In general, the working environment with respect to accessibility and usage of the social rooms, noise level, humidity and draft was rated "good" by both, the male (2.3 ± 2.0) and by the female employees (2.8 ± 1.3). In the deep cold-storage depots equipped with the most modern refrigeration technology, noise or distracting draft should not be expected, since the cold unit, particularly the evaporator and ventilators, is placed outside in an isolating cell, which is connected to the cooling area through cold air ducts. These cold air ducts, mounted at the ceiling, guarantee a smooth injection and distribution of the cold air to the chill room or the cold store and therefore ensure a noise-reduced and draft-minimized operating method.

The restricted effectiveness of the endogenous regulation mechanisms requires protection against the cold by means of adequate protective clothing suitable for the specific climatic conditions which have to be worn during all working phases. Therefore, besides the insulation of the clothing against cold, its comfort also was of interest. While in an overall assessment males rated the combination of these criteria as "good" (2.5 ± 1.7) the assessment of the females was again more positive (2.9 ± 1.2).

Based on previous studies of Forsthoff (1983), cold exposures also influence motor skills, especially the fingers. Therefore, the question of reliably and quickly operating the electric pallet trucks, which are supposed to make work more efficient, inevitably arose. Altogether, the electric pallet trucks were assessed "quite good" – regardless of gender.

The system of work-rest regime of the working time regulations and of warming-up breaks established in the deep cold-storage depots was also in the focus of the investigation. The minimum warming-up breaks required for the work-rest regime after corresponding, continuous cold exposures for different cold areas are listed in part 5 of the standard DIN 33 403 (1997).

In the cold store, for example, a minimum warming-up break of 30 min should follow a maximum, continuous working period of 90 min. In the chill room a 10-minute break should be granted after maximally 180 min of work.

For comparison purposes, the employees had to specify their normal regulation of working time and warming-up breaks and point out possibly arising problems caused by either too short warming-up breaks or too long working phases.

Again, gender-related differences could be established in the chill room. The males worked an average of 150 min followed by a warming-up period of 20 min. Their total working period lasted on average 23 min longer than that of their female colleagues.

Correspondingly, males estimated the work-rest regulation for this cooling area as "quite good" (1.6 ± 1.8) but considerably worse than the females (2.8 ± 1.2). In the cold store the working phases and warming-up breaks differed merely by 2 min (males: 101 min working time – 21 min warming-up break, females: 99 min working time – 23 min warming-up break). Nevertheless, the rating of the female employees was considerably more favorable than that of the males (2.9 ± 1.2 to 2.1 ± 1.8).

This more favorable evaluation of the regulation of working time and warming-up breaks by the females is probably due to their shorter daily working time (cp. Fig. 2). Fig. 3 shows the gender-related overall assessments of the different topics.

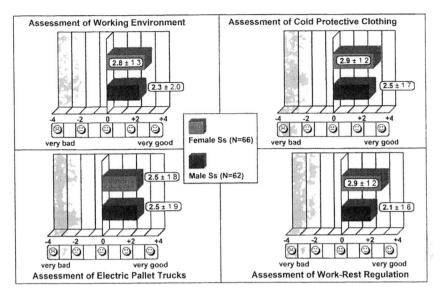

FIGURE 3. Gender-related overall assessments of the topics: working environment, protective clothing, electric pallet trucks and work-rest regulation

SUBJECTIVE EVALUATION OF THE PHYSICAL COMPLAINTS

Working in the chill room particularly in the cold store with temperatures around -24°C, however, causes considerably higher demands concerning the physical performance capacity of an order-picker than a comparable physical work at a normal surrounding temperature. In summer, for example, with more or less high environmental temperatures and a high contrast to the working temperature of +3°C or -24°C, order-picking activity leads already to additional severe physical stress.

Contrary to expectations, clearly more female Ss (71% vs. 50% of male Ss) assessed the regular shift in temperature, when moving from areas with seasonally more or less high surrounding temperatures to the cold store with its low temperatures, as rather acceptable.

According to the ratings concerning the tolerability of the permanent change between temperatures in the range of comfort and those in the cold, the question arose whether a certain adaptation to the environmental conditions occurred in the course of time. 68% of the males and 78% of the females mentioned that nowadays they felt the cold in the chill room just the same as at the beginning of their working activity. Only 24% of the male Ss and 19% of the female Ss were of the opinion that they could cope better now with the cold than at the beginning. Yet, acclimatization to the permanent temperature changes in the cold store could be observed for almost 40% of the females and even 50% of the males in the course of time. 58% of the female and 45% of the male employees, however, could not confirm this effect. At present, only 2% of the females and 5% of the males feel the cold more negatively than at the beginning of their activity.

174

Since order-pickers normally move approximately 230 items with a mean total weight of 1.6 t/h and unit weights of up to 15 kg, it seemed advisable to also collect data associated with complaints in muscles and joints. For the investigation about the physical condition of the order-pickers, special question forms were used based on a compatible visualization of all body parts, where possible muscle and joint complaints could be indicated. Statements concerning the "frequency", "duration" and "intensity" of the complaints should be made on a 4-step scale from 0 (never, none) to 4 (daily, very strong, beyond the working day). The upper part of Figure 4 shows the percentage of the employees having indicated complaints in the visualized body areas while working in the cold store.

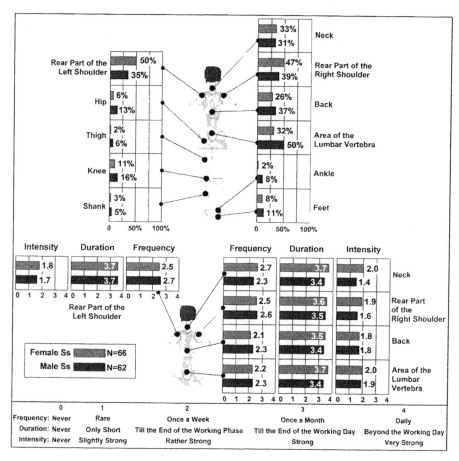

FIGURE 4. Subjective evaluation of physical complaints. Relative ratings of order-pickers who reported complaints in the visualized body regions, and details of their frequency, duration and intensity

According to Figure 4 complaints of different amounts were reported by employees in all body areas. Apparently, the rear shoulder areas on the left and on the right, the lumbar vertebra and the neck area were affected most frequently. On average, the complaints occurred once a week, usually felt beyond the working day and were classified as "rather strong" (cp. lower part of Fig. 4). Gender-related differences could be only identified regarding the frequency of occurrence of the affected body regions. Similar results were yielded in the chill room. Since the order-picking process in the two storage areas is the same, with exception of the prevailing temperature, serious differences could not be expected here.

In order to identify possible health risks due to the cold exposures, the employees were asked for details on cold sensations. Therefore, the staff members had to make statements about whether they regularly suffer from irritation of the throat, paleness, insensitivity, slight pains and slightly tingling sensations on the skin. The results of this topic are shown in Table 2.

Table 2: Health risk symptoms of the sample of the 128 order-pickers

Health Risk Symptoms	Chill Room		Cold Store	
	Female Ss	Male Ss	Female Ss	Male Ss
Irritation of the Throat	18%	8%	17%	15%
Paleness of Skin	3%	4%	8%	10%
Insensitivity of Skin	9%	0%	26%	23%
Slight Pains	6%	8%	17%	19%
Slightly Tingling Sensations	6%	12%	14%	11%

A special hazard of the cold exposure results from the cooling off of several sensitive body regions in the course of the working time. Based on experience, the acra are especially prone to such cooling off, which can be accompanied by negative impacts on the ability to work if the hands are affected (Imamura et al., 1998).

If the affected body areas are not completely re-warmed in the warming-up period, the skin temperature decreases further in the following phase of work. To gain an impression about the subjective cold experience during work in a deep cold-storage depot the order-pickers were asked to quantify the time-dependent cold sensations of all body areas at the head (forehead, nose, mouth and ears), at the upper part of the body (chest/back, arms, hands and fingers) as well as at the lower body half (abdomen/buttocks, legs, feet and toes). These subjective cold sensations were recorded in 15-minute-intervals during the time spent in the deep cold-storage depot and at the end of the warming-up phase. Remarkably, strong cold sensations occurred in both of the two cooling areas particularly at the nose, at the hands and fingers as well as at the feet and toes. In the chill room 20% of the male Ss and more than 50% of the female Ss had negative sensations. The ratings for the nose ranged between "chilly" and "very cold". The fingers were described as "chilly" to even "unbearably cold" by 28% of the females and 44% of the males at the end of their work phase. More male than female Ss indicated negative sensations also at the hands, feet and toes.

176

As expected, more frequent and also stronger cold sensations which had not subsided completely by the end of the warming-up break appeared in the cold store (cp. Fig. 5). On average, the male employees indicated cold sensations more frequently than the females. Otherwise however, female Ss experienced the specified areas as cooler.

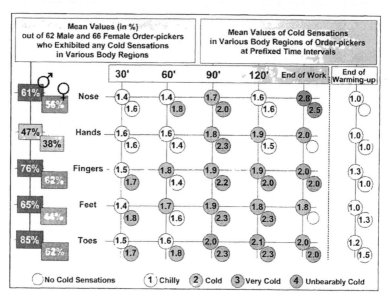

FIGURE 5. Course of cold sensations in various body regions over a work period at -24°C and successive warming-up phases. Shown are mean values of male order-pickers (upper numbers) and female order-pickers (lower numbers) who exhibited any cold sensations on a scale from "0" (no negative sensation) to "4" (unbearably cold)

The anatomical and physiological differences between males and females explain the need for the examination of possible gender-related effects of working in the cold, because the human being is subject to the laws of thermo-dynamics when under thermal strain; here, males seem to have better anatomical and physiological requirements to protect them against the cold. In order to keep the body core temperature at a normal level in case of increasing temperature gradients in relation to the environment, the energy expenditure in the organism has to be greatly intensified. The energy expenditure of females is lower than that for the males, which leads to a significantly lower production of warmth on the skin surface per m^2 and hour.

An important source of warmth is – beside the liver – the physically active musculature. So, when considering that females only have 70 % of the muscular strength compared to males and that their muscle mass is only 36 % of their overall body weight compared to 42 % for males, it is to be expected that the females' ability to develop heat is lower. It is true that the female body has 10 % more fat, which is of advantage when trying to retain warmth; however, the female skin is mostly 15 % thinner than the male one, which means insufficient insulation against loss of warmth. The ratio between body surface and body volume was expected to determine the degree of thermal storage. When compared to females, males often have a substantially larger volume with respect to their skin surface and therefore can produce and accumulate more vital body warmth.

This gives rise to the question whether females are necessarily disadvantaged when exposed to short-term or medium-term cold stress due to differences in their morphology and metabolism or whether they might be able to resist the cold because of a gender-specific, vegetative and hormonal control system, which increases the metabolic productivity in the musculature leading to an increasing production of warmth.

Although it has to be assumed that there are certain physical requirements which are beneficial to working in the cold such as thicker subcutaneous adipose tissue or more muscle mass, no tendency for a special somatotype ("strong", "normal", "athletic" or "gaunt") could be identified in the end.

ACKNOWLEDGEMENT

This research project was financially supported by the "German Research Foundation" (Grant no. STR 392/5-1 and KL 2067/1-2).

REFERENCES

DIN 33 403, Part 5. (1997). Climate at the workplace and its environments (in German). Beuth Verlag, Berlin.

Forsthoff, A. (1983). Arbeit in -28°C, Arbeitsphysiologische Untersuchung zur klimatischen Belastung bei Körperarbeit in extrem tiefen Umgebungstemperaturen unter besonderer Berücksichtigung der Kühlhausarbeit. Dokumentation Arbeitswissenschaft, Band 9.

Imamura, R., Rissanen, S., Kinnunen, M. and Rintamäki, H. (1998). Manual performance in cold conditions while wearing NBC clothing. ERGONOMICS Volume 41 No. 10. pp. 1421-1432

Kluth, K.; Penzkofer, M. and Strasser, H. (2008/2009). "Physiological responses of core and skin temperature of two age groups to working in the cold at +3°C and -24°C". OCCUPATIONAL ERGONOMICS Volume 8 No. 4. pp. 147-157

Kluth, K.; Penzkofer, M. and Strasser, H. (2010). "Age-related physiological strain of male order-pickers in deep cold". Proceedings of the 3rd International Conference of Applied Human Factors and Ergonomics (AHFE 2010), Miami, USA, 2010

Penzkofer, M.; Kluth, K. and Strasser, H. (2008/2009). "Heart rate and work pulses of two age groups associated with working in the cold at +3°C and -24°C". OCCUPATIONAL ERGONOMICS Volume 8 No. 4. pp. 135-145

Penzkofer, M.; Kluth. K. und Strasser, H. (2009). "Physiologische Auswirkungen von Kältearbeit bei +3°C bzw. -24°C auf den Blutdruck". In: Arbeit, Beschäftigungsfähigkeit und Produktivität im 21. Jahrhundert. Proceedings des 55. Kongresses der Gesellschaft für Arbeitswissenschaft, 437-441, GfA-Press, Dortmund

Chapter 20

Age-Related Physiological Strain of Male Order-Pickers in Deep Cold

Karsten Kluth, Mario Penzkofer, Helmut Strasser

Ergonomics Division
University of Siegen
D-57068 Siegen, Germany

ABSTRACT

The deep cold-storage depot – as a workplace for storing, picking and dispatching of frozen groceries – leads to the necessity that employees must remain working at temperatures of around -24° C for a considerable length of time. But it has not yet been established whether an age-related organization of the work and rest times is necessary or not. In order to assess the physiological effects, 30 male subjects (Ss) were classified into two age groups (20- to 35-year-olds and 40- to 65-year-olds). In whole working-day tests, possible age-dependent effects on the strain have been measured, to guarantee the preservation of work-ability in the long run. For the objectification of the physiological strain, "heart rate" and "skin surface temperature" were measured continuously, and, every 15 min, "blood pressure" and "body core temperature" were recorded discretely. Systematic differences of the blood pressure could not be found during cold exposures of 80, 100 and 120 min and the 20 min warming-up breaks. The heart rate values indicate a high physiological strain for both younger and older Ss, with rises above the resting level of 30 bpm and more. Due to increases over time, the endurance level sometimes was exceeded. The ability of heat generation deteriorated with advancing age, which is only shown at the results of the body core temperature taken at the tympanum. Regarding physiological strain brought about by maximum heart rate decreasing with age and declined heat generation, correspondingly adapted working-time break regimes have to be provided for older employees to ensure their work-ability in the long run.

Keywords: Physiological Field Study, Age-adjusted Strain, Working in Severe Cold, Manual Material Handling

INTRODUCTION

An effective protection against perishable influences on refrigerant or frozen food is only provided if a continuous cold chain exists from the manufacturer, through intermediate storage and the trade to the consumer. Within this cold chain, approximately one million jobs exist in Germany, of which about 70% are outdoor jobs and 30% are workplaces in technically cooled buildings. According to estimations, at present in Germany there are, however, only a few thousand jobs with environmental temperatures below -10°C. Due to a strong increase in consumption of frozen food with improved quality and product variety, also employment for working in deep cold probably will rise substantially in the future.

Despite rather successful automation in logistics, order-picking frozen food characterises the main work in the cold and has largely remained "manual work". This leads to the necessity that employees must stay at temperatures of around -24°C for a considerable length of working time, which is interrupted by several warming-up breaks.

Working in the extreme cold at temperatures below -20°C demands considerable physiological responses in thermoregulation to maintain a constant body core temperature of approximately +37°C, even when wearing protective clothing. The thermo-regulation occurs autonomously via physiological mechanisms, i.e., more or less pronounced changes of heart rate, skin surface temperature, energy conversion, and blood pressure. In spite of protective clothing and peripheral vasoconstriction, certain heat losses in the body always occur which have to be compensated for (cp. Strasser and Kluth, 2006). This can only be managed by considerably increased energy expenditure. Therefore, due to its "heating effect", the physically strenuous work of manual handling of deep-frozen groceries is beneficial and necessary rather than disadvantageous. Since the physical capacity and the energy metabolism, however, decrease with advancing age, it has to be assumed that the ability of employees to protect themselves from undercooling must inevitably grow worse with advanced age (cp. Boothby et al., 1936; Hettinger, 1968).

To monitor the comfortable thermic zone, the parameters "core temperature" and "skin temperature" as shown in the following can be used. According to a German standard DIN 33403 (2000), the core temperature can drop to 35°C without any health risk. But any further cooling down would be life-threatening. Yet, a sufficient number of scientific study results (cp., e.g., [Boennemark, 1969; Forsthoff, 1983; Veghte and Clogston, 1961) show that substantial drops of the core temperature – even at extremely low temperatures – can be prevented through an increase of the work performance and/or through a sufficiently adapted increase of the warming insulation of the cold-protective clothes.

The skin temperature, as expected, shows significantly higher variations under different climatic conditions than the core temperature. The blood circulation near the body surface is of decisive importance because the heat transmission via the skin surface is mostly controlled through the variation of the blood circulation (vasoconstriction and vascular dilatation). Skin temperatures are additionally of special importance for the heat regulation since up to 95 % of the entire heat exchange with the surroundings is carried out via the skin.

In addition to the above-mentioned parameters (cp. Strasser and Kluth, 2006), the effects on the heart rate, which allows a rating of the strain for the whole body, and the blood pressure in response to the cold exposure were of special interest in this field study. According to the laboratory study of Forsthoff (1983), there is a pronounced increase in blood pressure relative to comfortable climatic conditions as a response to cold exposure due to peripheral vasoconstriction.

Aside from some older (cp. Adolph and Molnar, 1946; Budd and Warhaft, 1966; Hellstrom et al., 1970) experiments that were typically carried out in the laboratory – e.g., Forsthoff (1983), Griefahn (1995), Häcker (1989), and Kleinöder (1988) – there are rarely any studies that examine the effects of working in the cold on the human body during real tasks (cp. Toshihara et al., 1995). Thus, neither suitable and humane regimens of work and rest periods nor instructions with respect to the ergonomic design of workplaces in cold-storage depots in accordance with work-physiological, organizational, and legal requirements are based on established work-scientific knowledge. Therefore, this work-physiological field study is an attempt to objectify the physiological strain of work in deep cold environment.

METHODS

To comply with absolutely real working conditions, the investigations of this field study were all carried out in a cold store of an industrial company with temperatures around -24°C. With the aim of assessing whether order-picking in deep cold is sustainable at any age in the long term, and of objectifying protection against the cold, an age-related analysis of the effects of working in the cold was carried out. Fifteen male subjects, each in two age groups (20- to 35- and 40- to 65-year-olds; cp. Table 1 and 2), had to carry out whole working-day tests under predetermined, realistic working conditions. In order to make the stay in the cold store bearable, the employees had to wear cold-protective clothing. Thermo underwear, a pullover, a pair of trousers and more importantly a cold-protective suit consisting of a thick jacket and long trousers, a thick hat, special thermo gloves – normally made of leather – and cold-insulating boots were worn.

The test variables at a temperature in the cold store of -24°C were the cold exposure duration of 80, 100, and 120 min per shift. Identical warming-up phases of 20 min, each at about +20°C and a work load adapted to the real job guaranteed comparable results of the skin temperature taken at the nose and 6 additional measuring points on the right part of the body: tip of the index finger, proximal phalanx of the ring finger, dorsal part of the big toe, sole of foot, kidney area, and shoulder. The core temperature was taken at the tympanum. Additionally, parameters of the blood circulation (heart rate and blood pressure) and the energy metabolism were measured. During the three working phases, the Ss were required order-picking of prepared pallets with a defined total weight of up to 3.2 t. On average, the order-pickers moved 227 items with a mean total weight of approximately 1.6 t/h (cp. Kluth and Strasser, 2006).

Prior to each cold exposure, a standard PWC_{130} test (cp. HVBG, 2002) was carried out to be able to make statements about all individuals' physical capacities of the cardiovascular system in the sub-maximal area (cp. Table 1 and 2). For the objectification of the physiological strain associated with working in the cold, the physiologically important parameters "heart rate" and "skin temperature" were measured continuously, and, every 15 min, "blood pressure" and "body core temperature" were recorded discretely. Moreover, energy expenditure was quantified for the identification of the intensity of the physical work and heat production with a mobile spirometry system.

Heart rate via "beat-to-beat" measurements was determined via the Polar® S810i system which is optimized for scientific investigations and performance diagnostic examinations. Temperature sensors that were attached to the body for the recording of the skin temperature and data loggers were used, which allowed the employees to move freely without being hooked up to a stationary measuring device. Furthermore, the systolic and diastolic blood pressure was discontinuously measured every 15 min via a commercially available full-automatic blood pressure monitor that uses the oscillometric method and is certified according to 93/42/EWG (1993). In the same time interval the tympanum temperature was measured by an infrared ear thermometer.

Table 1 Characteristics of 15 "younger" male subjects

Younger Ss	Age [Years]	Height [m]	Weight [kg]	$rPWC_{130}$ [W/kg]
01	24	1.75	53.9	1.45
02	26	1.82	86.3	1.70
03	28	1.85	80.0	1.41
04	28	1.95	99.2	1.97
05	21	1.80	77.8	1.86
06	20	1.75	64.6	1.72
07	29	1.82	90.3	1.26
08	21	1.78	84.4	1.74
09	31	1.79	75.0	1.04
10	29	2.00	75.8	1.40
11	29	1.80	75.4	1.18
12	25	1.85	76.8	1.02
13	29	1.86	95.3	1.21
14	24	1.80	61.5	2.03

15	24	1.88	88.6	1.92
$\bar{x}\pm s_d$	25.9±3.5	1.83±0.1	79.0±12.4	1.53±0.3

Table 2 Characteristics of 15 "older" male subjects

Older Ss	Age [Years]	Height [m]	Weight [kg]	rPWC$_{130}$ [W/kg]
01	43	1.85	73.9	1.47
02	47	1.87	100.7	0.85
03	46	1.76	65.1	1.87
04	59	1.84	125.4	1.37
05	62	1.78	116.8	1.25
06	57	1.83	89.7	1.34
07	63	1.78	99.6	1.05
08	53	1.76	81.2	1.52
09	62	1.75	73.6	1.66
10	63	1.79	90.0	1.25
11	57	1.86	94.1	1.86
12	56	1.87	105.8	1.98
13	55	1.71	82.0	1.46
14	64	1.72	63.5	1.83
15	48	1.80	89.2	1.48
$\bar{x}\pm s_d$	55.7±6.9	180±0.1	90.0±17.8	1.48±0.3

ASSESSMENT OF WORKING IN DEEP COLD FROM A PHYSIOLOGICAL POINT OF VIEW

HEART RATE

Heart rate responses – indicating physiological whole body strain – for the young Ss mark a high physiological strain. Rises of 35 beats per minute (bpm) and more above the resting level with an average of 70 bpm, were no rarity during the phases of work. Due to increases over time, the endurance level sometimes was exceeded. Individual rises of up to 60 bpm were partly reached. For approximately half of the Ss the base resting level of heart rate in the following warming-up period was not reached. The work pulses registered for the 40-to 65-year-olds were also on the range of 30-35 bpm above the resting level with an average of 77 bpm. In comparison there were just small differences measured in each of the three working phases, which are nevertheless highly significant. Work-related rises of the heart rate of approximately 30-35 bpm for the older Ss nearly match the average for the group of younger Ss. In the working phases with a duration of 80 and 100 min the heart rate values of the 40-to 65-year-olds differed highly significant from those of the 20-to 35-year-olds. In the longest phase with a duration of 120 min there are no significant age-related differences between the two age groups.

Based on established knowledge of physiological mechanisms, a useful vasoconstriction in the body's periphery, and a possibly resulting specific blood-pressure increasing effect, as a response to working in the cold can hypothetically be expected. Such an effect may also cause the heart rate to increase. The effects of the protective clothing may be involved, too. During working in a cold store, on the one hand, however, almost all body areas are covered by the protective clothing so that thermoregulatory responses to the cold may only partly take effect. On the other hand, the additional weight of the protective clothing of approx. 5 kg can also have an impact on activation of the cardiovascular system.

Nevertheless, it has to be considered that heart rate responses to a predefined load are dependent on several influencing variables, such as age and individual fitness, whereby strict limitations in heart rate with advanced age exist. Attention should be paid to the fact that, for every human, the maximum heart rate is lowered per year of life according to the following formula: $HR_{max} = 208 - (0.7 \cdot age)$ (cp. Tanaka, 2001; Whaley et al., 1992). So, the capacity utilization of heart rate increases is reduced with advancing age. Therefore, a work pulse profile with nearly identical increases above the resting level – and considering the higher resting levels of the older age group – is more problematic for older than for younger employees (cp. Figure 1).

Regarding maximum heart rate decreasing with age and reduced capacity utilization, correspondingly adapted working-time break regimes have to be provided for older employees. Despite its substantial physical stress, order-picking work itself is – due to the heat production – rather beneficial than an additional stress as long as the work takes place within the order-pickers physiological performance capacity and thus within the endurance limit. Therefore, the intensity of the physical work in the cold store cannot be recommended to be reduced.

184

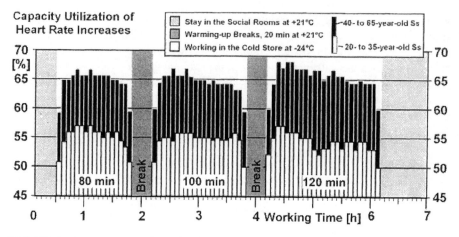

FIGURE 1. Average capacity utilization of heart rate increases of 15 male Ss at the age of 20-35 years (height of white columns) and 15 male Ss at the age of 40-65 years (height of black columns). 5-min-mean values during three working periods over 80 min, 100 min and 120 min in the cold store at -24°C

BLOOD PRESSURE

According to classification criteria of the WHO (World Health Organization), blood pressure values of up to 140 mmHg (systolic) over 90 mmHg (diastolic) are still referred to as normotonia while values of 140 to 160 mmHg (systolic) and 90 to 95 mmHg (diastolic) are called borderline hypertension. Systolic values in excess of 160 mmHg and diastolic values above 95 mmHg are considered as pathological hypertension.

The blood pressure values of the younger and older Ss were still within the upper range of what is considered normal or, in some cases, borderline hypertension (140-160 mmHg (systolic) and 90-95 mmHg (diastolic)). Averaged over the subjects of the respective age-group, maximum values of 146±20 mmHg and 83±13 mmHg were measured for the younger Ss as well as 149±20 mmHg and 85±14 mmHg for the older Ss.

Yet, it is not possible to completely rule out pathological hypertension as a potential long-term effect of repetitive cold-related peripheral vasoconstriction without further studies. Consistent with the results of other studies, it was found that fluctuations of ±10 mmHg (and more) that are unrelated to working in the cold are not uncommon. Examples for possible reasons include inter- and intradian physiological "noise," slightly varying measuring conditions, and differences in the static/ dynamic stress components related to posture. Thus, the recorded differences in the intermittently recorded blood pressure both between and within individuals can still be considered physiologically inconspicuous.

Undoubted, there is a pronounced increase in blood pressure relative to comfortable climatic conditions as a response to cold exposure due to peripheral vasoconstriction. Comparison measurements in resting as well as working test subjects showed increases in blood pressure of up to 20 mmHg (systolic) and 10 mmHg (diastolic) (cp. Forsthoff, 1983). Thus, local and sudden cold stimuli cause a pronounced increase in blood pressure, albeit of considerably smaller magnitude in persons who have been acclimatized to the cold.

SKIN AND TYMPANUM TEMPERATURE

As to be expected, the temperature sensors for the skin temperature located under the thick cold protective clothing for the cold store exhibited no substantial temperature changes in the area of the shoulder and the kidney. The skin surface temperature only dropped to an insignificant extent. The warming-up process in the breaks was finished after a very short time.

However, a considerable temperature decrease was recorded at the nose, however, which was exposed to the cold without any protection. But as quickly as the nose got cold, it was also warmed up again in the breaks – partly due to a more intensive blood circulation of the face after leaving the cold area. On average the surface temperature dropped to +15°C, causing a distinct feeling of cold for the working persons. The same was true for the fingers, whose surface temperature also reached just +16°C, on average. But also in this case, the fingers got warmed up during the breaks. A somewhat smaller but continuous decrease of temperature during an increasing working time was recorded at the toes. They only completely warmed up again within a minimum of 20 min after leaving the cold store, and only if, additionally, the boots were taken off during the break. The temperature curve of the measuring position under the sole of the foot was inconspicuous. The temperature of the foot, during continuously walking when working in the cold, was even a bit higher than during resting in the break. The mean temperature at the finger tips was as low as in the case of the nose.

It was nearly impossible to differentiate skin temperatures between the two age groups, yet for the ring finger, a slightly age-related difference could be observed. Otherwise, the results for the younger test persons without restrictions were comparable to the middle-aged group. For the other 6 measuring positions no age-related differences in skin temperature were recorded.

Figure 2 shows that the core temperature taken with an ear thermometer at the tympanum every 15 min, differed by 1.3K at a surrounding temperature of -24°C (after the 80 min exposure to the cold) to 1.5K (after the 120 min exposure) in the age group of the 20- to 35-year-olds compared to 36.6°C at the outset. Therefore, the temperature taken at the tympanum, which on principle is lower by about 2K than the temperature taken in the rectum, shows a substantial decrease. In the age group of the 40- to 65-year-olds, after 80 and 120 min working in the cold at -24°C, the core temperature dropped by 2.0K to 2.2K. After a 20-min warming-up phase, the temperature at the outset was often not quite fully reached in both groups.

186

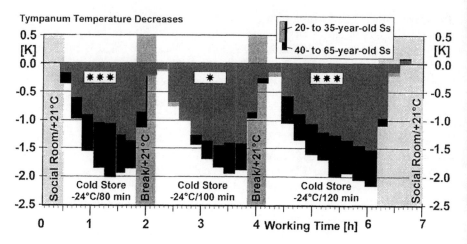

FIGURE 2. Tympanum (core) temperature decreases [K] of male order-pickers prior to, during and after cold exposures of 80, 100, and 120 min at approx. -24°C in the cold store. Means of 15 Ss in two age groups, each, with symbolic labeling of statistically significant differences of the two-sided t-test (–: p≥0.05, *: p<0.05, **: p<0.01, ***: p<0.001)

The combination of the cold protection clothing – in general satisfactory – and the manual material handling of goods showed a physiological strain, increasing with age, which leads to an improved blood circulation due to the many changes of body posture and the movement of the arms and legs, the positive effect of which can also be felt in the toes. However, for a compensation for the heat lost during working in the extreme cold while picking up and carrying goods weighing up to 15 kg, this is not sufficient.

It has to be concluded that a 20 min break is not sufficiently long enough for a complete re-warming of the body for all Ss when working in the cold store. The investigations showed that independent of age the warming-up break should at least be 30 min. No age-related regulation of working time and warming-up breaks is advisable. The breaks in the cold store, however, need to be sufficiently long. The cold protective clothing is largely adequate, but despite heat production during high demanding physical work individual body parts are highly affected by the cold. So, the clothing still needs to be improved.

According to prognoses on a national and international level, the demographic development and the resulting changes in society will lead to a clear age-increase of the working population as well as to a continuous decrease in the number of the younger work force. This means that the topic discussed here remains relevant, since the physiological strain as well as the metabolism decreases with an advanced age, just as the ability of self-protection against hypothermia.

ACKNOWLEDGEMENT

Supported by the German Research Foundation (Grant no. STR 392/5-1).

REFERENCES

93/42/EWG (1993), European Directive for Medical Products. *Amtsblatt der EG* Nr. L 169, 12.07.1993, 0001-0046.

Adolph, E.P., and Molnar, G.W. (1946), Exchanges in heat and tolerances to cold in men exposed outdoor weather. *Amer.J.Physiol.* 146, 507-537.

Boennemark, B., Johansson, L., Keiding, J., Magnusson, H., Palmgren, B., and Söderqvist, A. (1969), *Arbete i fryshus.* Rapport A-141.69, Arbetsmedicinska Institutet, Stockholm.

Boothby, W.M., Berkson J., and Dunn, H.L. (1936), Studies of the energy of metabolism of normal individuals, a standard for basal metabolism, with a nomogram for clinical application. *Amer.J.Physiol.* 116, 468.

Budd G.M., and Warhaft N. (1966), Body temperature, shivering, blood pressure and heart rate during standard cold stress in Australia and Antarctica. *J.Physiol.* 186, 216-232.

DIN 33403 (2000) *Climate at the workplace and in its environments – Part 2: Effect of the climate on the heat balance of human beings.* Beuth Verlag, Berlin.

Forsthoff, A. (1983), *Arbeit in -28°C, Arbeitsphysiologische Untersuchung zur klimatischen Belastung bei Körperarbeit in extrem tiefen Umgebungstemperaturen unter besonderer Berücksichtigung der Kühlhausarbeit.* Dokumentation Arbeitswissenschaft, Band 9, Otto Schmidt Verlag, Köln.

Griefahn, B. (1995), *Arbeit in mäßiger Kälte.* Schriftenreihe der Bundesanstalt für Arbeitsschutz -Forschung- FB 716. Wirtschaftsverlag NW, Bremerhaven.

Häcker, H. (1989), *Psychologische Determinanten von Kältearbeit bei -30°C.* Schriftenreihe der Bundesanstalt für Arbeitsschutz -Forschung- FB 563. Dortmund.

Hellstrom, B., Berg, K., and Vogt Lorentzen, F. (1970), Human peripheral rewarming during exercise in cold. *J.Appl.Physiol.* 29, 191-199.

Hettinger, T. (1968), *Isometrisches Muskeltraining,* 3. Auflage, Thieme-Verlag, Stuttgart.

HVBG (2002), *Leitfaden für die Ergometrie bei Arbeitsmedizinischen Vorsorgeuntersuchungen nach berufsgenossenschaftlichen Grundsätzen (Anhang 2).* In: Berufsgenossenschaftliche Grundsätze für arbeitsmedizinische Vorsorgeuntersuchungen. Gentner Verlag, Stuttgart.

Kleinöder, R. (1988), *Ergonomische Gestaltung von Kältearbeit bei -30°C in Kühl- und Gefrierhäusern.* Schriftenreihe der Bundesanstalt für Arbeitsschutz -Forschung- FB 562. Dortmund.

Kluth, K., and Strasser, H. (2006), *Heart rate and blood pressure responses to working in the cold.* In: Proceedings of the 16th Triennial Congress of the

International Ergonomics Association, Maastricht, The Netherlands.

Strasser, H., and Kluth, K. (2006), *Sensation of cold and physiological responses to groceries handling in cold-storage depots.* In: Proceedings of the 16th Triennial Congress of the International Ergonomics Association, Maastricht, The Netherlands.

Tanaka, H., Monahan, K., and Seals, D. (2001), Age-predicted maximal heart rate revisited. *J. Am. Coll. Cardiol.* 37, 153-159.

Toshihara, Y., Ohkubo, C., Uchiyama, I., and Komine, H. (1995), Physiological reaction and manual performance during work in cold storages. *Applied Human Science* 14 (2) 73-77.

Veghte, J.H., and Clogston, J.I. (1961), *A New Heavy Winter Flying Clothing Assembly.* Alaskan Air Command, Technical Note AAL-Tn-61-4, Archiv Aeromedical Laboratory, Fort Wainwright.

Whaley, M., Kaminsky, L., Dwyer, G., Getchell, L., and Norton, J. (1992), Predictors of over- and under-achievement of age-predicted maximal heart rate. *Med Sci. Sports Exerc.* 24, 1173-1182

CHAPTER 21

Ergonomic Assessment: Risk Exposure Among Construction Workers

Joshi Pratibha[1], Sharma Promila[2],
Sharma Avinash[3]

[1]&[2] Dept of Family Resource Management
College of Home Science, G.B.P.U.A.& T., Pantnagar
Uttarakhand,-263145, India

[3] Dept of Family Resource Management
College of Home Science
Palampur Agricultural University
Palampur, H.P., India

ABSTRACT

Construction is one of the largest industries in India, employing 7.6 million workers, or about 5% of the Indian work force. Construction workers face some of the most dangerous working conditions in the country on a daily basis. Ergonomic problem is common hazard and is one of the most common causes of injury at work. The accurate measurement of workers' exposure to the risk assessment of WMSDs has been of vital importance. Musculoskeletal system primarily concerns with dimensions, compositions & mass properties of body segments, and work related musculoskeletal disorder caused due to over exertion, adoption of asymmetric & awkward postures and unsupported positions used in task completion. Many MSDs are due to operations performed for a long period, can cause a lifetime of pain and disability. Therefore present investigation was planned to provide an objective measure of the MSD risk caused by construction work and to evaluate a job's level of risk for developing a musculoskeletal disorder in 120 construction workers with the application of REBA (Rapid Entire Body Assessment) postural assessment technique. It was found from observation that approximately 6 % of workers were in the category of AL0 indicating 'negligible' risk level means no action is necessary indicating acceptable posture, 12.7% were in AL1(low risk, further action may be needed) and 63% of workers in AL2 (medium risk, action necessary).

However, 11 & 7.3 percent was found to be in AL3 (high risk, action necessary soon), AL4 (very high risk, action necessary now) respectively.

Keywords: Awkward posture, Construction worker, MSDs, REBA, Risk Level.

INTRODUCTION

Construction is one of the largest industries in India, employing 7.6 million workers, or about 5% of the Indian work force. Construction workers face some of the most dangerous working conditions in the country on a daily basis. Ergonomic problem is common hazard and is one of the most common causes of injury at work. Since, construction workers face some of the most dangerous working conditions on a daily basis. The accurate measurement of workers exposure to the risk assessment as WMSDs has been vital importance. Musculoskeletal system primarily concerns with the dimensions, compositions and the mass properties of body segments and the work related musculoskeletal disorders caused due to frequent bending, pulling or pushing of muscles, over exertion, adoption of asymmetric and awkward postures and unsupported positions used for task completion. Posture means maintaining the normal curve of body while performing various activities. Posture in work means the whole center of gravity remains as close as possible to that of normal standing / sitting erect condition. Such postural hazards are due to operations performed mainly for a long period, can cause a pain for lifetime and disability. Therefore present investigation is intent to provide an objective measure of the postural risk of construction workers and to evaluate a job's level of risk with the application of REBA (Rapid Entire Body Assessment) technique.

MATERIALS AND METHODS

For the present study, a sample of 120 construction workers performing group of activities as basic construction activities (loading/ de-loading of materials, working with mortar, cement, crushing stones, transportation of water, bricks, etc.), welding / soldering, gas-cutting / sawing, using power tools, installation, management and supervision work were selected from Delhi/NCR construction sites. A multistage purposive cum random sampling technique were used to select the sample. The investigation adopted REBA Employee Assessment Worksheet developed by Hignett & Mc. Atamney, 2000 for postural analysis of respondents. REBA is a whole body assessment tool which was initially designed to provide a pen and paper postural analysis tool to be used in the field by direct observations (Hignett & Mc. Atamney 2000).REBA quantifies musculo-skeletal risks by coding individual body segments and providing a scoring system for muscular activity. The result is an action level with an indication of urgency. In the present study, it was used to assess the type of unpredictable working postures found in construction industries. Data was collected about the body postures, force used, type of movement or action, repletion and coupling. A final REBA score is generated giving an indication the level of risk & urgency with which action should be taken on a five-point action category scale of 0 to 4, from no action required through to action necessary now.

RESULTS AND DISCUSSION

The results regarding subjective assessment of postural risk with the help of REBA and observation technique furnished information of risk factors & work postures were assessed and enclosed in Table 1 and Fig.1.

Table 1: REBA analysis for postural assessment of construction workers

Activity Profile	REBA Action Category				
	AL_0	AL_1	AL_2	AL_3	AL_4
Basic Construction Worker (n = 20)	2 (10)	2 (10)	6 (30)	8 (40)	2 (10)
Welding/soldering (n = 20)	6 (30)	3 (15)	6 (30)	4 (20)	1 (5)
Cutting/Sawing (n = 20)	2 (10)	4 (20)	7 (35)	7 (35)	0 (0)
Installation (n = 20)	4 (20)	8 (40)	6 (30)	2 (10)	0 (0)
Handling Power Tools (n = 20)	1 (5)	2 (10)	8 (40)	7 (35)	2 (10)
Supervision/ Management (n = 20)	2 (10)	7 (35)	8 (40)	3 (15)	0 (0)
Total (N=120)	17 (14.16)	26 (21.66)	41 (34.16)	31 (25.83)	5 (4.17)

Note: Figures in parentheses indicate the percentage values

REBA Employee Assessment Worksheet

based on Technical note: Rapid Entire Body Assessment (REBA), Hignett, McAtamney, Applied Ergonomics 31 (2000) 201-205

A. Neck, Trunk and Leg Analysis

Step 1: Locate Neck Position:

Step 1a: Adjust...
If neck is twisted: +1
If neck is side bending: +1

Neck Score

Step 2: Locate Trunk Position:

Step 2a: Adjust...
If trunk is twisted: +1
If trunk is side bending: +1

Trunk Score

Step 3: Legs:

Adjust: +1, +2
Add +1
Add +2

Leg Score

Step 4: Look-up Posture Score in Table A

Using values from steps 1-3 above locate score in Table A

Step 5: Add Force/Load Score

If load < 11 lbs.: 0
If load 11 to 22 lbs.: +1
If load > 22 lbs.: +2
Adjust: If shock or rapid build up of force add +1

Step 6: Score A, Find Row in Table C

Add values from steps 4 & 5 to obtain Score A
Find Row in Table C

Posture Score A + Force/Load Score = Score A

B. Arm and Wrist Analysis

Step 7: Locate Upper Arm Position:

Step 7a: Adjust...
If shoulder is raised: +1
If upper arm is abducted: +1
If arm is supported or person is leaning: -1

Upper Arm Score

Step 8: Locate Lower Arm Position:

Lower Arm Score

Step 9: Locate Wrist Position:

Step 9a: Adjust
If wrist is bent from midline or twisted: Add +1

Wrist Score

Step 10: Look-up Posture Score in Table B

Using values from steps 7-9 above, locate score in Table B

Step 11: Add Coupling Score

Well fitting Handle and mid range power grip, good: 0
Acceptable but not ideal hand hold or coupling acceptable with another body part: fair +1
Hand hold not acceptable our possible: poor +2
No handles, awkward, unsafe with any body part Unacceptable: +3

Coupling Score

Posture Score B + Coupling Score = Score B

Step 12: Score B, Find Column in Table C

Add values from steps 10 & 11 to obtain Score B. Find column in Table C and match with Score A in row from step 6 to obtain Table C Score

Step 13: Activity Score

+1 or more body parts are held for longer than 1 minute (static)
+1 Repeated small range actions (more than 4x per minute)
+1 Action causes rapid large range changes in posture or unstable base

SCORES

Table A

Trunk Posture Score		Neck								
		1			2			3		
	Legs	1	2	3	4	1	2	3	4	
1		1	2	3	4	1	2	3	4	
2		2	3	4	5	3	4	5	6	
3		2	4	5	6	4	5	6	7	
4		3	5	6	7	5	6	7	8	
5		4	6	7	8	6	7	8	9	

Table B

		Lower Arm					
		1			2		
Upper Arm Score	Wrist	1	2	3	1	2	3
1		1	2	2	1	2	3
2		1	2	3	2	3	4
3		3	4	5	4	5	5
4		4	5	5	5	6	7
5		6	7	8	7	8	8
6		7	8	8	8	9	9

Table C

Score A (Score from Table A + Force/Load score)	Score B (Table B value + coupling score)											
	1	2	3	4	5	6	7	8	9	10	11	12
1	1	1	1	2	3	3	4	5	6	7	7	7
2	1	2	2	3	4	4	5	6	6	7	7	8
3	2	3	3	3	4	5	6	7	7	8	8	8
4	3	4	4	4	5	6	7	8	8	9	9	9
5	4	4	4	5	6	7	8	8	9	9	9	9
6	6	6	6	7	8	8	9	9	10	10	10	10
7	7	7	7	8	9	9	9	10	10	11	11	11
8	8	8	8	9	10	10	10	10	10	11	11	11
9	9	9	9	10	10	10	11	11	11	12	12	12
10	10	10	10	11	11	11	11	12	12	12	12	12
11	11	11	11	11	12	12	12	12	12	12	12	12
12	12	12	12	12	12	12	12	12	12	12	12	12

Table C Score + Activity Score = Final REBA Score

Scoring:

1 = negligible risk
2 or 3 = low risk, change may be needed
4 to 7 = medium risk, further investigation, change soon
8 to 10 = high risk, investigate and implement change
11+ = very high risk, implement change

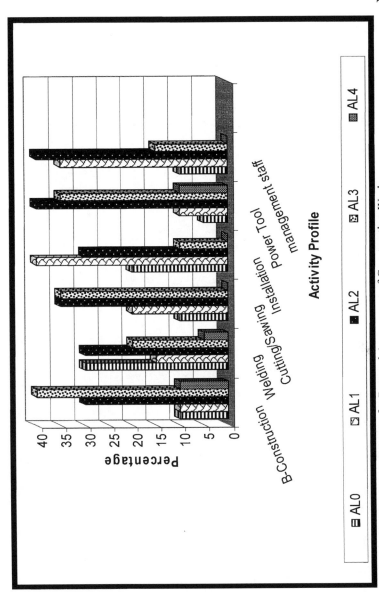

N=120

Fig: 1 Rapid Entire Body Assessment score for Postural Assessment of Construction Workers

As various postures of neck, trunk and leg in group A (plus load/force), upper arms, lower arms and wrists in group B (plus coupling) along with activity score were observed and analyzed. It was found that a sum total of 31.16 % construction workers were in AL_2 which interprets medium risk level and change is required soon followed by 25.83 % workers in AL_3 (high risk & action necessary now), 21.66 % in AL_1(low risk, further action may be needed). However, only 14.16 % of workers were found to be in AL_0 indicating 'negligible' risk level means no action is necessary i.e., acceptable posture. When the activities were analyzed individually it was observed that majority of workers who involved in basic construction activities (40%) were in the zone of AL_3 (high risk & action necessary now) followed by cutting (35%) , handling power tool (35%), welding (20%), management (15%) and installation (10%).

According to Larson and Hannihen (1995), "excessive musculo-skeletal stress at work, especially with static load, is believed to play a major role in low back pain, neck and shoulder disorders". It was further observed that in the periphery of AL_2 major proportion of power tool users (40%), management staff (40%), respondents involved in cutting (35%), basic construction activities (30%), laser operations (30%), & installation workers (30%) were found indicating medium risk level. However, in AL_4 (very high risk, action necessary now) there were mainly basic construction workers (10%), power tool users (10%), and 5% welders.

As an interesting issue of very few respondents in AL_0 corresponding to the fact that few subjects worked in construction sites in an acceptable posture are really a matter of concern.

According to Phesant (1991), "Postures which are initially adopted for occupational reasons may become habitual outside the working context and finally become irreversible owing to the shortening and fibrous contraction of muscles and soft tissues".

Thus, asymmetric postures and unsupported positions can increase the load on spine. Postural stress can increase the physiological cost and fatigue while performing the task and may also lead to pain and injuries to vertebral column in the long run.

REFERENCES

ERGOWEB. Internet address: http://www.ergoweb.com e-reference retrieved on 22/06/2009.

Hignett, McAtamney 2000, Rapid Entire Body Assessment (REBA), Applied Ergonomics 31 pp 201-205. htpp://ergo.human.cornell.edu/ahREBA.html, retrieved on 23/07/2009.

Larson D and Hannihen (1995) Surface electromyography: research clinical and sports applications. Indian J Physio Allied Sci 49: 5-9.

Pheasant S (1991) Ergonomics, Work and Health. pp 98-107. Macmillan Press, Hampshire. www.humanics-es.com/bernard/REBA_M11.pdf, retrieved on 08/07/2009.

Chapter 22

Glovebox Gloves: Effects on Forearm Muscle Activity for Gripping with Maximal and Sub-maximal Efforts

Peng-Cheng Sung, Po-Jung Cheng, Cheng-Lung Lee

Department of Industrial Engineering and Management
Chaoyang University of Technology
Wufong, Taichung County 41349, Taiwan

ABSTRACT

This study evaluates the effects of glovebox glove, layer, and thickness on maximum voluntary contraction grip strength and forearm muscle activity for gripping exertion with maximal and sub-maximal efforts. Six volunteered male comprised the subject pool. Butyl, hypalon, and neoprene gloves in 0.015" and 0.03" thicknesses were selected for evaluation. Grip strength and muscles activity were measured using a transducer equipped hand dynamometer and an electromyography system. Based upon the observations made in this experiment, 0.015" inch butyl glove in single gloving setting should be used as selection considerations for glovebox works to minimize effects on maximum grip strength. In addition, 0.015" inch butyl glove in single or triple gloving setting should be used as selection considerations for glovebox works to minimize the forearm muscle activity. This study recommends 0.015" inch butyl glove in single gloving setting to provide the desired hand protection against ergonomic-related disorders in the glovebox environment involving gripping with maximal and sub-maximal efforts.

198

Keywords: Glovebox, Glove, Muscle activity, Grip

INTRODUCTION

Work-related musculoskeletal disorders (WRMSDs) of upper extremities are among the most prevalent lost-time injuries and illnesses in almost every industry (BLS, 2007). Forearm muscles are common sites and sources for the development of upper extremity disorders even though the normal muscles function is essential in protecting ligaments and other tissues against injury or disorders (Nicolay et al., 2007). Reduced strength (weakness) of the musculature surrounding a joint also contributes to increased risk of injuries or disorders (Aagaard et al., 1998). In addition to muscle weakness, hand force has also been acknowledged by epidemiologic evidence as a mechanical risk factor for the development of upper extremity disorders (NIOSH, 1997). Gripping is one of the most commonly applied hand exertion to accomplish complex tasks for industrial jobs and has been implicated as a risk factor for several WRMSDs (Bao and Silverstein, 2005). The association between gripping and occurrence of upper extremity disorders indicates the need to explore risk factors that may affect upper extremity disorders when performing gripping tasks. Then, the desired hand protection could be accomplished against ergonomic-related disorders in the workplace involving gripping exertions.

Grip strength is the muscular power and force of the hand when an object is held in a clamp formed by the partially flexed fingers on one side and the thumb on the other (Napier, 1956). Published literature on grip strength frequently assessed the effects of different variables (e.g. age, gender, glove, posture, physical activity, etc. on maximum voluntary contraction (MVC) grip strength (Nicolay and Walker, 2005). However, the workplace activities may regularly require gripping (e.g. assembly line worker) with sub-maximal efforts. In addition, few of these grip strength studies have simultaneously investigated muscular response using electromyography (EMG) to detect muscle activity of forearm muscles (Mogk and Keir, 2003). The muscular endurance and strength are important factors to be included to product design process and workplace safety consideration (Nicolay and Walker, 2005).

Gloves are the most used protective device for human hand in many industrial tasks to protect the hand from injury. Published literature has indicated that the effect of gloves on grip strength capabilities are consistent in the sense that gloves decrease strength compared to the bare hand condition (Sung, 2006). The reduction compared to bare hand ranged from 3.7% of synthetic rubber to 50.0% of extra vehicle glove. For bare handed conditions, females produce significantly less (50-65%) grip strength than males (Nicolay and Walker, 2005; Mogk and Keir, 2003; Hallbeck and Mcmullin, 2003), but, females generally have greater muscular endurance than males when worked at sub-maximal level (Hicks et al., 2001). Hallbeck and Mcmullin (2003) found that when over several glove conditions, females averaged 74% the grip strength of males.

In terms of glove usage on muscle activity involving gripping tasks, four studies (Sudhakar et al., 1988; Fleming et al., 1997; Kovacs et al., 2002, Larivière et al., 2004) had used EMG measures (activity and/or fatigue) to assess gloves effects with MVC or sustained sub-maximal gripping exertions. No articles were found evaluating the effects of glove thickness and gloving condition on muscle activity using EMG measures for gripping tasks with maximal and sub-maximal efforts. There is a paucity of data on muscle activity due to different risk factors (e.g. glove properties, gloving conditions, etc.) for gripping tasks that are responsible for the genesis of upper extremity disorders and injuries. In addition, Larivière et al. (2004) stated that if the glove was too thin, it may not be able to change the biomechanical behavior of forearm muscles. Therefore, thicker glovebox gloves with 0.015" (0.38 mm) and 0.03" (0.76 mm) thickness that may have effects on muscle activity and fatigue will be included for evaluation in this study. Therefore, the objective of this study is to assess the effects of glovebox glove usage on MVC grip strength and forearm muscles activity for gripping exertion with maximal and sub-maximal efforts.

METHODS

Subjects

Six volunteered male free of musculoskeletal disorders/injuries (MSDs) in the upper extremities comprised the subject pool. The subject's free of MSDs status in the upper extremities was identified through interviewing during the recruiting process. All these subjects are right handed. The mean values of age, height, hand length and maximum breadth of hand are 23.7 ± 0.5 years, 172.1 ± 3.6 cm, 172.0 ± 37.0 mm, and 101.5 ± 4.2 mm, respectively.

Apparatus

Gloves and Gloving Conditions

Butyl, hypalon and neoprene gloves in two different thicknesses (0.015" and 0.03") that are commercially available for glovebox use were selected for evaluation. The cuff diameter and length of the gloves are 8" and 32". Two hand sizes, 8.5"and 9.75" were provided and the subject wore both of the gloves to pick the best fit glove.

For the single gloving condition, the subject donned only a pair of glovebox gloves. The double gloving condition includes a pair of glovebox glove as outer glove and a pair of natural rubber (Trionic© size 8, 0.02" thickness, 43.5 gm) gloves as the inner glove which was used in glovebox work to facilitate donning the outer glove. The triple gloving condition adds another pair of cotton gloves (median size, 0.008" thickness, 9.3 gm) as the innermost glove which is used in glovebox work for perspiration absorption purpose.

Electromyography (EMG)

The BIOPAC MP150 EMG System which includes amplifier & A/D converter (BIOPAC Systems UIM100C) and data acquisition software (Acq Knowledge 3.7.2) installed in a laptop were used to collect and process the EMG signals. After shaving and scrubbing the recording sites with alcohol, surface EMG electrodes were positioned over the following forearm muscles of dominant arm: (1) flexor digitorum superficialis (FDS), (2) flexor carpi radialis (FCR), (3) extensor carpi radialis longus (ECR), and (4) extensor digitorum (ED) of the subjects' dominant arm involved with gripping tasks (Larivière et al., 2004). A ground electrode was also placed at the lateral epicondile of the subjects' dominant arm. The EMG signals were bandpass filtered (20-450 Hz) and preamplified (gain: 1000) with 1k Hz sampling rate.

The root mean square (RMS) amplitude which is proportional to the energy content of the muscle activity was used in this experiment for analysis of the EMG signals. The RMS value increases indicating an increase in muscle activity in gripping exertions. The RMS amplitude for the jth gripping exertion (RMS_{ij}) for ith session was normalized for each subject using equation 1. The RMSmax,i determined from three replicates of barehanded MVC gripping and RMSbase,i determined when the forearm is in rest were measured 30 minutes before the starting of ith session. The normalized RMS (nRMS) values of EMG represent the relative level of muscle activation for the corresponding muscles.

$$\text{Normalized RMS}_{ij} = \frac{RMS_{ij} - RMS_{base,i}}{RMS_{max,i} - RMS_{base,i}} \tag{1}$$

Procedures

Upon the starting of this experiment, MVC grip strength and corresponding EMG for each subject were recorded for bare hand condition. The MVC grip strength for bare and gloved hands were measured according to Caldwell's regimen (Caldwell et al., 1974) using a Jamar hand dynamometer with Transducer (Lafayette Instrument Company). The subject was instructed to increase to maximum exertion (without jerk) in about one second and maintain this effort during a four second count. The strength datum is the mean score recorded during the first three seconds of the steady exertions. The rest period between consecutive MVC gripping exertions is two minutes. The subject was seated with their shoulder adducted and neutrally rotated, elbow flexed at 90^0, forearm in neutral position, and wrist between 0^0 and

30^0 extended and between 0^0 and 15^0 ulnar deviations to maintain neutral posture (Mathiowetz et al. 1985). A chair with adjustable arm support was used to position the forearm in neutral position. The hand dynamometer was fixed on an adjustable post made in the laboratory to place the forearm at 90^0 to the torso. The mean of three replicates of the bare hand MVC grip strength which are within $\pm 10\%$ were used as the MVC grip strength for that subject. Three levels of grip exertion, namely MVC for each bare and gloved hand condition, and 50% and 25% of bare hand MVC grip strength, were included in this experiment. The MVC grip strength data recorded are also used as one of the performance measured to be tested. To maintain target grip force, the subject viewed on the PC screen directly in front of them showing the grip dynamometer output. The display was arranged so that a major line on the screen represented the target grip force. Two other lines were also displayed to represent the $\pm 10\%$ of the target grip force. Participant was told to reach the target force level in a second and maintain the force level at the target line as steadily as possible for four seconds.

The independent variables with their respective levels tested in this experiment are summarized as below:

(1)   Glove type: butyl (B), hypalon (H), neoprene (N), and bare hand (BH);
(2)   Gloving condition: single (S), double (D), and triple (T);
(3)   Glove thickness: 0.015" (Thin) and 0.03" (Thick);
(4)   Level of exertion (%MVC): MVC and 50% and 25% of the bare hand MVC grip strength.

There are 57 treatment conditions (gloved hand: 3*3*2*3=54; bare hand: 3 levels of exertion) with two replicates of grip exertion to be performed for each treatment. In total, 114 trials will be administered to each subject in this experiment. The order of presentation of the treatment conditions were randomized for each subject. Following the two minutes rest between exertions instruction according to Caldwell's regimen (1974), 4.5 hours minimum (23~25 exertions/hour) in three two hours sessions were required for each subject to complete this experiment. A minimum of 24 hours rest period was given to each subject between sampling sessions (Bishu and Kim, 1995).

Experimental Design and Statistical Analysis

The independent variables in this experiment are glove type, gloving condition, glove thickness, and level of exertion. The respective levels of the independent variables can be found in the procedures section of this experiment. The performance measures are the MVC grip strength and mean normalized RMS values of four forearm muscles for each treatment condition representing the muscle activity corresponding to each grip exertion.

Separate analyses of variance (ANOVAs) with repeated measures were used to evaluate the effects of glove type, gloving condition, glove thickness, and level of exertion on the MVC grip strength (no level of exertion effect) and mean normalized RMS values of four forearm muscles. All data were analyzed for statistical significance at p ≤ 0.05 using the SPSS 12 (SPSS Inc, Chicago, Illinois) statistical software.

RESULTS

Effects on Grip Strength

The results show that the mean MVC grip strength for all gloved hands is significantly lower ($p<0.05$) than that of bare hand and the amount of reduction for gloved hands compared to bare hand (36.9±3.3 Kg) ranges from 13.1% to 31.8% in this study.

The results of the repeated-measured ANOVA (Table 1) show that there are significant differences between glove material (F=350.33, $p<0.001$), layer (F=510.23, $p<0.001$) and thickness condition (F=2753.31, $p<0.001$) on MVC grip strength. In addition, there are statistically significant interaction effects between glove material and layer, glove material and thickness, and among all three factors on the MVC grip strength. Bonferroni pairwise comparisons in glove material effect show that there were statistically significant differences ($p<0.001$) for all three glove materials. The butyl glove can retain significantly higher maximum grip strength followed by hypalon and neoprene gloves. In layer effect, statistically significant differences ($p<0.001$) were found for single vs. double, single vs. triple, and double vs. triple. The results show that more the layers of gloves are worn, the greater the grip strength reduction. In addition, the thinner (0.015") gloves retain significantly higher maximum grip strength (2.8 kg in mean differences) than the thicker (0.03") glove.

Table 1 ANOVA with repeated measures results for effects of glove material, thickness, layer, and their interactions on MVC grip strength

Source	SS	Df	MS	F	Sig.
Glove material (GM)	421.93	2.00	210.97	350.33	0.000
Layer (L)	660.60	2.00	330.30	510.23	0.000
Thickness (TH)	652.79	1.00	652.79	2753.31	0.000
GM * L	33.73	4.00	8.43	12.70	0.000
GM * TH	164.96	2.00	82.48	198.60	0.000
L * TH	1.40	2.00	0.70	1.61	0.221
GM * L * TH	45.35	4.00	11.34	12.54	0.000

Effects on Muscle Activity

Table 2 shows the summary of significance for the factors and their interactions on muscle activity (nRMS) for flexor digitorum superficialis (FDS), flexor carpi radialis (FCR), extensor carpi radialis longus (ECR), and extensor digitorum (ED) muscles. Level of exertion (%MVC) effect was observed for all muscles at $p < 0.001$ level. Glove material effect is observed on all four muscles at $p < 0.05$ level. Layer effect is observed on FDS ($p < 0.001$) and ECR ($p < 0.01$). In addition, Thickness effect is observed on FDS at $p < 0.01$ level.

Table 2 ANOVA with repeated measures results for effects of % MVC, glove material, thickness, layer, and their interactions on muscle activity

Source	FDS	FCR	ECR	ED
% MVC	0.000^c	0.000^c	0.000^c	0.000^c
Glove material (GM)	0.037^a	0.001^b	0.016^a	0.000^c
Layer (L)	0.000^c	0.869	0.004^b	0.080
Thickness (TH)	0.007^b	0.084	0.822	0.139
%MVC * GM	0.259	0.037^a	0.173	0.257
%MVC * L	0.000^c	0.008^b	0.001^b	0.100
% MVC * TH	0.021^a	0.000^c	0.014^a	0.053
GM * L	0.000^c	0.027^b	0.001^b	0.112
GM * TH	0.007^b	0.005^b	0.033^a	0.937
L * TH	0.013^a	0.733	0.010^a	0.229
%MVC * GM * L	0.021^a	0.960	0.060	0.004^b
%MVC * GM * TH	0.247	0.274	0.131	0.023^a
%MVC * L * TH	0.112	0.947	0.057	0.004^b
GM * L * TH	0.000^c	0.003^a	0.000^c	0.098
%MVC * GM * L * TH	0.017^a	0.249	0.005^b	0.444

[a] significance at $p < 0.05$ [b] significance at $p < 0.01$
[c] significance at $p < 0.001$

For level of exertion effect, Bonferroni pairwise comparisons revealed that the muscle activity is highest for MVC exertions followed by 50% MVC exertions and then 25% MVC exertions. Bonferroni pairwise comparisons for glove material effect shows that the muscle activity for hypalon gloves are significantly higher than butyl gloves on FDS, FCR, and ECR and significantly higher than neoprene glove on FCR. Neoprene glove also show significantly higher muscle activity than butyl glove on ED. For layer effect, the post hoc analysis indicates that the muscle activity for double gloves is significantly higher than single and triple gloves on FDS and than triple gloves on ED. Thicker gloves also show significant higher muscle activity than thinner gloves.

DISCUSSION AND CONCLUSION

The means and standard deviations of the MVC grip strength of the bare hand measured in this study are consistent with the normative data for male adults ranging from 20 to 30 years old reported by Mathiowetz et al (1985). All gloved hands impaired MVC grip strength significantly compared to that of bare hand and the observed finding are as expected and consistent based upon a review of the published literature of industrial gloves.

For the effects of glove material, layer, and thickness on MVC grip strength, the findings of the present investigation are:

(1) The butyl gloves facilitated the highest grip strength values followed by the hypalon and neoprene gloves.
(2) The single gloving allowed the highest grip strength values followed by double and triple gloving conditions.
(3) The thinner (0.015") gloves performed better than thicker (0.03") gloves.

To minimize effects of glovebox gloves on MVC grip strength based upon the observations made in this experiment, thinner gloves and single gloving setting should be used as selection considerations for glovebox works. In addition, butyl gloves should be picked over hypalon and neoprene gloves to minimize the reduction of grip strength if other criteria (chemical, physical, biological etc) were met.

For the effects of glove material, layer, and thickness on muscle activity for four forearm muscles, the findings of the present investigation are:

(1) The muscle activity is significantly higher when donned hypalon gloves than butyl gloves on FDS and ECR muscles. Hypalon gloves also generated significantly higher muscle activity than butyl and neoprene gloves on FCR. In addition, neoprene gloves generated significantly higher muscle activity than butyl gloves on ED.
(2) The double gloving allowed significantly higher muscle activity than single and triple gloving on FDS and than triple gloving on ECR.
(3) The thicker (0.03") gloves generated significantly higher muscle activity than thinner (0.015") gloves.

To minimize effects of glovebox gloves on muscle activity based upon the observations made in this experiment, thinner butyl glove in single or triple gloving setting should be used as selection considerations for glovebox works.

Kovacs et al. (2002) found no significant effect of glove type on EMG activity of FDS and ED when exerting MVC grip strength. This study also shows similar results for the MVC grip exertions. Since the glovebox activities may regularly require gripping with sub-maximal efforts in terms of maximal efforts. The results of this study recommend thin (0.015") butyl glove in single gloving setting to provide the desired hand protection against ergonomic-related disorders in the glovebox environment involving gripping exertions.

ACKNOWLEDGEMENTS

I would like to acknowledge grant NSC 98-2221-E-324-010 from the National Science Council of Taiwan for financially supporting this research.

REFERENCES

Aagaard, P., Simonsen, E., Magnusson, P., Larsson, B., Dyhre-Poulsen, P. (1998), "A new concept for isokenetic hamstring: quadriceps muscle strength ration." *American Journal of Sports Medicine*, 26, 231-237.

Bao, S. and Silverstein, B. (2005), "Estimation of hand force in ergonomic job evaluations." *Ergonomics*, 48, 288-301.

Bishu, R.R. and Kim, B. (1995), "Force-endurance relationship: does it matter if gloves are donned?" *Applied Ergonomics*, 26 (3), 179-185.

Bureau of Labor Statistics (2007), Nonfatal occupational injuries and illnesses requiring days away from work 2007, Table 5 and Table 6, United States Department of Labor, Washington, DC, 12-15.

Caldwell, L.S., Chaffin, D.B., Dukes-Dobos, F.N., Kroemer, K.H.E., Laubach, L.L., Snook, S.H.L. and Wasserman, D.E. (1974), "A proposed standard procedure for static muscle strength testing." *American Industrial Hygiene Association Journal*, 35, 201-206.

Fleming, S.L., Jansen, C.W., and Hasson, S.M. (1997), "Effect of work glove and type of muscle action on grip fatigue." *Ergonomics*, 40 (6), 601-612.

Hallbeck, M.S. and Mcmullin, D.L. (2003), "Maximal power grasp and three-jaw chuck pinch force as a function of wrist position, age, and glove type." *International Journal of Industrial Ergonomics*, 11, 195-206.

Hicks, A.L., Kent,-Braun, J., and Ditor, D.S. (2001), "Sex differences in human skeletal muscle fatigue", *Exercise and Sport Science Reviews*, 29, 109-112.

Kovacs, K, Splittstoesser, R., Maronitis, A., and Marras, W.S. (2002), "Grip force and muscle activity differences due to glove type." *American Industrial Hygiene Association Journal*, 63, 269-274.

Larivière, C, Plamondon, A., Lara, J., Tellier, C., Boutin, J., and Dagenais, A. (2004), "Biomechanical assessment of gloves. A study of the sensitivity and reliability of electromyographic parameters used to measure the activation and fatigue of different forearm muscles." *International Journal of Industrial Ergonomics*, 34, 101-116.

Mathiowetz, V., Kashman, N., Volland, G., Weber, K., Dowe, M. (1985), "Grip and pinch strength: normative data for adults." *Arch Phys Med Rehabil*, 66, 69-74.

Mogk, J.P.M. and Keir, P.J. (2003), "The effects of posture on forearm muscle loading during gripping." *Ergonomics*, 46 (9), 956-975.

Napier, J.R. (1956), "The prehensile movements of the human hand." *The journal of bone and joint surgery*, 38B, 902-913.

Nicolay, C.W., Kenney, J.L., and Lucki, N.C. (2007), "Grip strength and endurance throughout the menstrual cycle in eumenorrheic and woman using oral contraceptives." *International Journal of Industrial Ergonomics*, 37, 291-301.

Nicolay, C.W. and Walker, A.L. (2005), "Grip strength and endurance: Influences of anthropometric variation, hand dominance, and gender." *International Journal of Industrial Ergonomics*, 35, 605-618.

NIOSH (1997), "Musculoskeletal disorders and workplace factors - A critical review of epidemiologic evidence for work-related musculoskeletal disorders of the neck, upper extremity, and low back." Bernard, B. P. edited, *NIOSH publication 97-141*, National Institute of Occupational Safety and Health, Cincinnati, OH.

Sudhakar, L.R., Schoenmarklin, R.W., Lavender, S.A., and Marras, W.S. (1988), "The effect of gloves on grip strength and muscle activity." *Proceedings of the Human Factors Society*, 32nd Annual Meeting, 647-650.

Sung, P.C. (2006), "Glovebox gloves: ergonomics guidelines for the prevention of musculoskeletal disorders." Doctoral Dissertation, Industrial Hygiene Division, Department of Environmental Health Science, School of Public Health, University of California, Los Angeles, U.S.A.

Chapter 23

Factors Affecting Success and Failure in a Return-to-Work Program Established for a Department of Defense Aircraft Maintenance Facility

Edwin Irwin, MS, Kristin Streilein MS CPE PE, Richard Slife MS*

Mercer Engineering Research Center
* Warner Robins Air Logistics Center

INTRODUCTION

The strategies influencing successful return-to-work (RTW) for injured/ill employees in the civilian manufacturing sector are well characterized in the scientific literature (Pransky and Shaw, 2007). Successful strategies include: preemptive surveillance of ergonomic risks, early identification and treatment of occupational illness, early return-to-work, institution-wide support for and involvement in injury/illness reduction, and use of participatory ergonomic intervention processes. These have all been shown to reduce costs and lost-work time associated with occupational injury and illness (Talmadge and Melhorn, 2005).

Despite the documented benefits to both workers and employers, implementing a return-to-work program remains a challenge, requiring cultural adaptation as well as economic evidence to convince leaders at all levels of an organization to support it (Di Guida, 1995; D'Amato and Zijlstra, 2010). The complexities of implementation proliferate in large organizations due to the corporate environment, management structures, policies, and procedures, especially those present in many large government agencies.

The 402d Maintenance Wing at Warner Robins Air Logistics Center (WR-ALC) in Warner Robins, Georgia identified a need to develop a RTW program to address its burgeoning worker's compensation and lost-work-day expenses. WR-ALC is the largest employer in Middle Georgia, with almost 13,000 civilian employees. It also has one of the highest incidence rates of lost-work days in the Department of Defense (Bryan, 2008), a plurality of which are due to musculoskeletal disorders related to work. The 402d Maintenance Wing (MXW) on base, which employs almost 8,000 of the total WR-ALC employees, is charged with providing repair, modification, and overhaul of the: F-15 Eagle; C-130 Hercules; C-17 Globemaster; C-5 Galaxy; and all special operations aircraft and their avionics systems. Approximately 150 MXW employees are listed as having physical limitations that have resulted in them being placed in permanent limited-duty status. At any given time, there are numerous other employees with temporary restrictions or on medical leave for work-related injuries.

Employees in the limited-duty pool continue to receive their salaries, even though they are not working in their normal capacities due to their physical limitations. To maintain production capacity MXW is required to over-hire or increase overtime, leading to greatly expanded costs above the direct expenditures for employees with job restrictions. In addition, many employees in the limited-duty pool are covered under worker's compensation claims, which accrue extra costs to MXW. Expenses related to worker's compensation and lost-work days are estimated to be in excess of $10M per year (Bryan, 2008).

The MXW commander recognized that the previous process of managing MXW employees with physical limitations was ineffective in addressing the costs to the Wing. The process had developed organically in response to statutory requirements and higher-level directives, which led to non-systemic goals and diffuse management responsibilities distributed across multiple organizational entities within MXW. Not only did the process not reduce costs to the Wing, it created a significant long-term population of skilled employees placed into low-level, unskilled jobs without regard to work capacity or commitment. The result was high financial and productivity costs to the Wing, loss of wage enhancement to the employees, and a contentious relationship on both sides.

For the past 3 years, MXW has contracted Mercer Engineering Research Center (MERC) to establish and maintain a program for helping employees with permanent physical limitations return to work. The program, known as the Physician-Directed Functional Job Analysis (FJA) Program, is intended to: systematize the RTW process; inject rehabilitation engineering, ergonomics, and job analysis skills into the process; and provide occupational medicine supervision to augment existing occupational medicine resources that are overtaxed by other responsibilities.

METHODOLOGY

The FJA program was developed based on research-supported best practices that have been implemented in consonance with the bureaucratic structures and practices of the WR-ALC. Key characteristics designed into the program include: differential identification of medical and non-medical factors influencing RTW (Durand et al, 2003; Devereux et al, 2002); application of skill sets in rehabilitation engineering, ergonomics, and industrial engineering to increase options for RTW (Lincoln et al, 2002; Helm et al, 1999); and integration of occupational medicine to provide medical oversight and improve communication and coordination with employees' private physicians (Talmage and Melhorn, 2005).

The integration of multiple skills into the RTW process was considered a first priority during planning for the project. Two biomedical engineers were assigned to the project who together represent more than 35 years of experience in rehabilitation, ergonomics, human factors, biomechanics, and industrial process analysis. An orthopedic surgeon who has extensive experience in the practice of occupational medicine was subcontracted by MERC to provide medical supervision and on-demand support to the team. In order to integrate the program into the existing ALC processes, the team also required the participation of staff from the WR-ALC who are expert in ALC-specific areas of: worker's compensation; personnel; management structures, policies, and practices; program management/development; ergonomics; and occupational medicine. The FJA team consists of 8 members, including three from MERC and five from WR-ALC.

The FJA program was established with a restricted scope of practice, based on needs identified by the Brigadier General commanding MXW and based on the memorandum of agreement with the local chapter of the American Federation of Government Employees, which represents wage-grade employees at WR-ALC. First, the program is limited to working only with employees who have permanent work restrictions, though the etiology of the restrictions does not have to be work-related. Thus, employees who develop a permanent, non-work-related illness or injury are eligible for services through the FJA program.

Another restriction on the scope of the program is the prioritization of the approach. The first priority is to return employees whose labor hours earn money for the WR-ALC back to their original job in their original Flight wherever possible. If this is not feasible, an alternate job is sought within the same job series and the same Group (Aircraft, Commodities, Electronics, Software, or Maintenance Support Group). If this fails, an alternate job within the same job series, but in another Group is sought. Finally, alternate job series options are investigated.

Participation in the FJA program is strictly voluntary. Employees must sign a consent form which indicates they have been informed of the purpose of the program, and which gives MERC access to all medical records related to their work restriction. The Occupational Medicine Service Flight (OMS) at WR-ALC then provides MERC with copies of the relevant medical records. MERC conforms to all HIPAA regulations with respect to these records.

The current process of serving participants in the FJA program has evolved to include the following steps and milestones:

1. Signed consent is obtained
2. Medical records are provided by OMS and reviewed by MERC
3. MERC engineers interview the participant using validated data collection instruments to create a profile for work capacity, work

readiness, and risk. This information is briefed for the MERC physician.

4. The MERC physician evaluates the participant's physical condition to validate the engineers' findings, the medical record, and the documented work restrictions. Preliminary updates to the work restrictions, recommendations for further data collection, and a description of the participant's prognosis for improvement are reported

5. MERC engineers analyze the participant's original job with respect to general ergonomic risk, risks specific to the participant's limitations, productivity and management impacts, and potential accommodations.

6. A Phase I medical report is written detailing the issues along with the potential for placing the participant back into his/her original job and recommendations for work restrictions, accommodations, and job modification where appropriate.

7. The MERC physician reviews and approves the report.

8. OMS reviews and approves the report.

9. Once all medical approvals have been obtained, a management version is created, with all protected medical information expurgated. This version is sent to the WR-ALC team members for management review and further processing.

10. If returning to the original job is not recommended, the case is reviewed at the Group management level. MERC briefs all of the Flight managers within the Group on the characteristics of an appropriate job placement for the participant, and discusses a list of potentially appropriate jobs. The outcome of the meeting is a list of jobs MERC engineers must analyze with the same level of detail as the original job.

11. A Phase II report is prepared documenting the results of analysis.

12. MXW is responsible for ensuring that the proposed placements meet production needs and for making the official job offers. The process of job analysis and management oversight is continued until a satisfactory placement is found and management has approved it.

13. MERC engineers follow up with each participant who has been placed and his/her supervisor for a minimum of 6 months after placement in order to identify and address any problems early. A structured interview is used to collect data that can show trends in overall program outcomes over time. The interview covers areas of work health, overall health, work value (employee perceptions), and supervisor perceptions.

A simplified flow chart illustrating the evolved process can be seen in Figure 1.

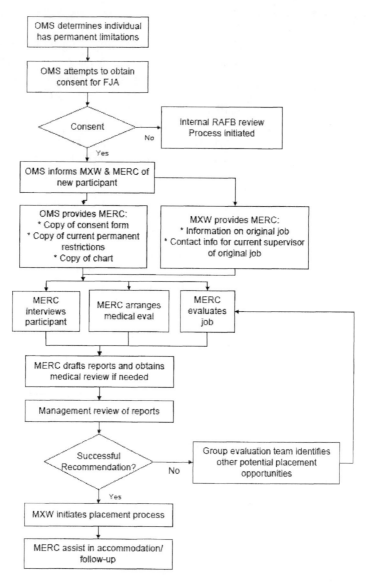

Figure 1. Illustrating the current RTW process.

The FJA team meets biweekly to discuss progress on open cases, assign action items for resolving issues that arise, discuss overall trends in the program, and determine ways to improve the program. A summary of program outcomes is presented every month to the MXW commander. Unresolved issues that require intervention at his level are also raised at that time.

RESULTS

The research-supported best practices discussed earlier formed the foundation of the FJA program, though the evolution and outcomes of the program were driven by the pragmatic requirements of the existing bureaucratic structure and culture. First, the ergonomists' skill sets in biomechanics, the etiology and physiology of musculoskeletal disorders, and process improvement practices allow them to effectively interface with injured employees, physicians, and production managers. Through structured interviews based on validated instruments the ergonomists were able to differentiate between supervision problems, production problems, and physical limitation problems, which could then be validated or refined by the physician through his interviews and discussions with participants' personal physicians. The use of scientifically validated interview tools to help differentially identify issues of symptom magnification, work orientation, self-efficacy, and physiological/biomechanical risk related to specific work factors was crucial to establishing an approach to returning an employee to work safely.

The ergonomists were further able to do detailed task analyses of jobs relative to participants' limitations. The supervisors generally expressed appreciation for having concrete guidance in task assignment based on the results of the analyses. These detailed analyses also serve as a tool for the supervisor to use to make appropriate task assignments for other employees who develop physical limitations (permanent or temporary). In addition, they identify ergonomic risks faced by non-disabled employees which the MXW ergonomics program uses to identify opportunities for proactive intervention.

In addition to the ergonomists' skills, having a physician expert in occupation medicine provided several essential services for the RTW program. These included: validating ergonomist findings during the initial employee interviews; validating ergonomist recommendations related to accommodation; validating other medical findings; and interfacing with other physicians to achieve consensus in prognosis and recovery trending. In a number of cases where work restrictions seemed excessively conservative, the MERC physician was key to working with participants' personal physicians to develop more appropriate work restrictions. When given a detailed technical analysis of work requirements, the personal physicians were better able to provide restrictions that allowed safe and productive return-to-work.

Some scientifically validated program efforts had to be abandoned, at least for the initial stages of development. Early identification and treatment of occupational illness and early return-to-work were infeasible based on the scope established for the program, though other initiatives are addressing these issues.

On the other hand, one of the characteristics most clearly correlated in the scientific literature with successful outcomes was strongly implemented in the FJA program. Institution-wide support for and involvement in injury/illness reduction was established through the active involvement of the Commander of the Maintenance Wing. He developed explicit directives for Group Commanders to foster participation in the FJA processes, and established measureable goals for the program that he reviewed on a monthly basis. This level of support was responsible for accelerating progress in placing participants in appropriate jobs, as well as in institutionalizing the priorities and processes of the program.

To-date, the FJA program has had 67 participants. Table 1 shows a breakdown of the status of these participants.

Category	Current Total
Completed	44
In Progress	19
Awaiting Medical Improvement	4
Total	67

Table 1. Participant status summary.

The completed cases break down as follows: 30 have been placed in jobs, 8 have withdrawn from the program, and 6 have resigned or retired from civilian service with the ALC. Table 2 shows a breakdown of job placements with respect to the prioritization.

Employee Type	Priority	Current Total
Direct Charging	Original Job	6
	Similar Job/Same Group	12
	Similar Job/Different Group	2
	Non-Direct Job Series	7
Non-Direct Charging		4
Total		30

Table 2. Job placements with respect to prioritization.

A key factor for ensuring continued program success involves collecting follow-up data on participants who have been placed. Structured individual interviews are performed with each participant and their supervisors. The interview questions cover areas of work value, work health, overall health, and supervisor perception. As can be seen in Figure 2, the outcomes indicate continuing high percentages of positive outcomes in every category.

The return on investment from the FJA program was calculated by the Financial Management Flight at WR-ALC in March, 2008. Based on 19 completions at that time, the return on investment was found to be 126:1. This astounding figure was verified by the Financial Management division of the Air Force Materiel Command at the Pentagon. The Department of Defense financial management figures provided to calculate ROI were significantly higher than figures estimated in concert with the Bureau of Labor Statistics due to indirect cost avoidance.

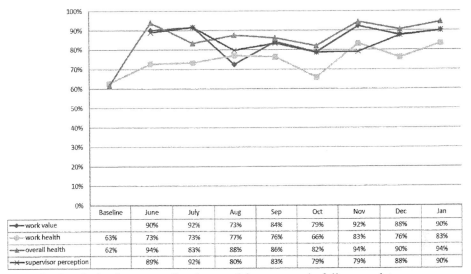

	Baseline	June	July	Aug	Sep	Oct	Nov	Dec	Jan
work value		90%	92%	73%	84%	79%	92%	88%	90%
work health	63%	73%	73%	77%	76%	66%	83%	76%	83%
overall health	62%	94%	83%	88%	86%	82%	94%	90%	94%
supervisor perception		89%	92%	80%	83%	79%	79%	88%	90%

Figure 2. Trends for the current fiscal year of the program in follow-up data.

Marketing the return-to-work program to both the employees and managers is essential to maintaining a base of participants. While initial rates of participation were fairly high, the rate at which new participants volunteer for the program has fallen dramatically in the third year. This has been partially due to a lack of continued exposure of potential participants to the benefits of the program. The lack of regular briefing of new supervisors regarding the existence and benefits of the program also decreases the amount of encouragement employees get to participate when they develop permanent limitations. Continued marketing and reinsertion of the RTW program must occur until it is fully integrated as part of the overall culture. MXW management has begun more intensive marketing to help resolve this issue.

DISCUSSION

This presentation has identified factors influencing the successes and failures we have experienced, as well as the potential solutions we have identified over the past three years of a return-to-work program established for the Maintenance Wing at Warner Robins Air Logistics Center. The key factors we have identified fall into several categories: organizational, bureaucratic, employee relations, knowledge and skills, and outcome measurement.

Organizational factors –
1. The director of a large organization having several independent units must be a strong, results-oriented leader who is completely invested in the need for positive outcomes from the program. The level of investment from top

management must be reflected in the demand for results and oversight transmitted to the unit management level.

2. Implementation of a RTW metric at the unit level can be an effective way to develop more cooperation in RTW processes. This can elevate the RTW process to the level of other bottom-line priorities.

Bureaucratic factors –

1. Communication at all levels of a bureaucratic organization is essential, but inefficient. The return-to-work program must facilitate communication at all levels in order to succeed.

2. Bureaucratic policies and procedures typically develop over time. Generally the return-to-work program must accommodate existing policies and procedures in order to succeed, since changing procedures and policies in a large organization requires extensive time and support at many levels of management. It is often easier to design the RTW program to fit within existing bureaucratic structures where possible.

3. Misunderstandings between organizational support units, such as a personnel department, and the return-to-work program often arise due to differences in both language and policy. Recognizing and reconciling these differences is crucial to achieving success, but requires extensive time and resource allocation.

Employee relations factors –

1. Establishing a posture of neutral support for both production and employee welfare is essential for the return-to-work program to garner cooperation from management and technical staff.

2. Repeated exposure to the existence and purpose of the RTW program is essential to keep a steady stream of participants.

Outcome measurement factors –

1. The return on investment provided by the return-to-work program is the most important outcome measure. The calculation of ROI must include direct and indirect cost avoidance, as well as production savings.

2. Trend analysis of employee follow-up data provides important information for early identification of problems, though the statistical validity of the data is limited due to sample frame and multiple confound issues.

CONCLUSIONS

The FJA program has had remarkably good success to-date. Return-on-investment is high, participant and supervision satisfaction is elevated and stable, and the programmatic structures are integrating well with existing bureaucratic structures.

Despite these successes, the program still faces challenges related to corporate culture and competing management priorities. High level support of the Wing Commander is extremely important to the longevity of the FJA program due to the disarticulation between shop area budgets and the costs for lost work and worker's compensation. High-level reporting and concern for RTW metrics from different managers helps maintain RTW as an important priority.

REFERENCES

Bryan J (2008), "Leadership is Big Business in 402d Maintenance Wing,"
Air Force Print News, Internet Resource:
http://www.robins.af.mil/news/story_print.asp?id=123083809 .

D'Amato A and Zijlstra F (2010), "Toward a Climate for Work Resumption: The Nonmedical Determinants of Return to Work." Journal of Occupational and Environmental Medicine, 52(1), 67-80.

Department of Defense Ergonomics Working Group (2008), "Integrated Team-Based Medical Case Management." Internet Resource:
http://www.ergoworkinggroup.org/ewgweb/IndexFrames/index3.htm.

Devereux JJ, Vlachonikaolis G, Buckle PW (2002), "Epidemiological Study to Investigate Potential Interaction Between Physical and Psychosocial Factors at Work That May Increase the Risk of Symptoms of Musculoskeletal Disorder of the Neck and Upper Limb," Journal of Occupational and Environmental Medicine, vol. 59: pp 269-277.

Di Guida AW (1995), "Negotiating a Successful Return to Work Program," Journal of the American Occupational Health Nurses, 43(2), 101-106.

Durand M, Vachon B, Loisel P, and Berthelette D (2003), "Constructing the Program Impact Theory for an Evidence-Based Work Rehabilitation Program for Workers with Low Back Pain," Work, vol. 21: pp 233 - 242.

Helm R, Powell N, Nieuwenhuijsen E (1999), "A Return to Work Program for Injured Workers: A Reassignment Model," Work, vol. 12: pp 123-131.

Lincoln A, Feuerstein M, Shaw W, Virginia I (2002), "Impact of Case Manager Training on Worksite Accommodations in Workers' Compensation Claimants with Upper Extremity Disorders," Journal of Occupational and Environmental Medicine, vol. 44(3): pp 237-245.

Pransky G and Shaw W (2007), "Return to Work - From Research to Practice." Journal of Occupational and Environmental Medicine, 49(3), 249-250.

Talmadge JB and Melhorn JM, eds. (2005), A Physician's Guide To Return To Work. American Medical Association, Chicago.

CHAPTER 24

Innovative Ways of Working: A Comparison of a Traditional and a Flexible Office Environment on Office Space Use, Performance, Privacy and Satisfaction

Merle Blok[1], Liesbeth Groenesteijn[1], Margriet Formanoy[1], Elsbeth de Korte[1], Peter Vink[2]

[1]TNO P.O.Box 718, 2130AS Hoofddorp, The Netherlands

[2]Delft University of Technology Faculty of Industrial Design Engineering, The Netherlands

ABSTRACT

The pace in which new and reliable ICT technologies are introduced is rising and the economy is changing towards an information and knowledge driven economy. This opens possibilities to collaborate and communicate with colleagues from any remote location as well as the ability to access business information. The ability of office workers to work from any remote location also implies that the way work is performed and organized will change and demands for different work facilitations. The interest for this new way of organizing work is rising, and a number of organizations has already adapted this new way of working. This includes flexible work arrangements (i.e. home working) together with innovative office concepts that support communication and collaboration. The expectations on the benefits of this new way of working are high, and it is supposed to increase productivity, satisfaction and optimal use of office space. However, the scientific proof for the benefits and effects are still lacking and there is a raising interest for scientific proof in order to optimize and found business investments. In the present study a comparison was made between a traditional and a flexible office environment and the effects on office space use, performance, privacy and satisfaction were investigated.

Keywords: New Ways of Working, Alternative Work Arrangements, Work at home, Distant work, Flexi place, Office space use.

INTRODUCTION

The present economy is structurally changing from an industrial-based nation to an information and knowledge-based global economy causing a new revolution in the nature of work (Hill et al. 2003). Companies' possibilities for flexible and innovative reaction to turbulent business environments are increased by replacing the hierarchical structures for more flexible work models. Physical labor is gradually declining and mentally complex tasks are growing. Rapid improvements and technological innovations and the power and function of new information- and communication tools could make the confidence in the reliability and user-friendliness of the systems grow. This results in an increase of new ICT technologies in many organizations. These new technologies, including video conferencing, collaborative software, internet/intranet systems converge to forge a foundation of the new workplace (Lee and Brand 2005). It causes an increase in possibilities of more flexible arrangements providing employees with new alternatives for where, when and how to accomplish work and reduces the necessity of coming to the office. In many companies acceptability of arrangements enabling flexibility in work, such as flexible working hours, teleworking (working from home or another remote locations), telecommuting and teleconferencing has gained interest (Hill et al. 2003).

Researchers suggest that the office of the future will not longer be an individually working place, but will change towards a more collaborating, meeting and communicating oriented space in which spontaneous encounter and knowledge sharing is fostered (Van der Voordt 2004; Blok et al. 2009). A few business offices have already innovated their office and created so called 'task facilitating offices'. A task facilitating office is a flexible office, designed to offer a wide range of different workspaces, such as open offices, concentration arias, discussion zones, coffee corners and is expected to encourage employees in choosing a workspace most suitable for the task to perform (Van der Voordt and Meel 2002; Blok et al. 2009; Penn et al. 1999). The task facilitating office often implies a transition towards desk sharing and is frequently arranged in combination with ICT-facilities to support teleworking. Since this reduces the required number of square meters office space, it contributes to energy savings for light and heating and diminishes building materials and maintenance (Vink 2009). Therefore, it is often seen as a considerable cost beneficial business strategy in combination with sustainability. On the other hand there are some aspects that might counter act positive findings. The privacy of flexible offices for example is of much debate, since open workspaces are known to increase visual and auditive distraction (Bauman & Arens, 1996; Van der Voordt, 2004) which can have a negative effect on productivity (Lee and Brand 2005; Banbury & Berry, 2005; Korte et al. 2007).

Most companies refer to this new organizational concept as "The New Way of Working". This way of working will be unrestrained by geography, time and organizational boundaries and employees are managed based on trust and results, causing a result-oriented culture instead of the face-time culture where hours on the job are most important (Hill & Weiner, 2003). This new way of working not only seems to meet business objectives, such as proposed productivity growth caused by the expected increase of work satisfaction, motivation and collaboration and a reduction of square meter office space; it also provides greater opportunities for workers to effectively integrate the demands of work and personal life (Hill et al. 2003), reduce or avoid unnecessary travel time and will increase attractiveness to work for the organization.

Due to great positive exposure on the proposed advantages of the new way of working, some reasonable negative (side) effects are almost neglected. The possibility of working from home for example might stimulate overtime work due to feelings of guild or mental inability to stop working, or might stimulate continuing of work in times of sickness. This might cause serious disadvantages for workers health (Hengst et al. 2008).

Despite the growing interest, there is a lack of scientific proof for the effects of this new work concept. This is partly explained by the difficulty to measure the effects (Van der Voordt 2004; De Croon et al. 2005). Additionally, the key indicators for performance will be changing. Old indicators such as hours on the job, or percentage of sick leave are not longer of any use, since they do not longer indicate performance of the new work environment (Banett & Hall 2001).

In order to gather effects of the new way of working on office space use, performance, privacy and satisfaction, a case where a group of workers changes to the new way of working was evaluated. The traditional office environment consisted of owned workstations in one or two person rooms. In the flexible office the new way of working consisted of less square meters office environment that provided a wide variety of flexible workspaces (such as lounge workspaces, small meeting rooms, concentration areas, and standing tables) and an wireless internet equipping, making it possible to work everywhere throughout the department. Additional facilities to work at home were also provided. The research question is: "What are the effects of flexible working in a new task facilitating office on office space use, performance, privacy and work satisfaction?".

METHOD

In order to study the effects of a change to a new way of working, a field study was conducted at a large telecom company. A controlled repeated measure was carried out among three departments consisting of a total of 126 employees. Two of the departments moved from a traditional office environment to a flexible office environment within the same office building (intervention group, 71 subjects). One department remained at the traditional office environment (control group, 55 subjects). The measurement consisted of a web based internet questionnaire with questions on office space use, performance, privacy and work satisfaction. The measurement was done twice. The first questionnaire (M1) was conducted while all three departments were working in a traditional office. The second questionnaire (M2) was conducted six weeks after the movement of the intervention group from the traditional office to the flexible office. Question items from several validated questionnaires were combined and used in the questionnaire.

Office space use was measured with questions on estimated percentages of time spend at the different type of workplaces.

Performance was measured using question sets on the quantity, quality, efficiency and creativity of the performed work, measured on a 7 point scale (1 = extremely low, 7 = extremely high). Specific questions were used to investigate the supported for concentration, communication, collaboration and creativity of the office environment (scale 1 to 7, 1= extremely bad, 7 = extremely good).

In order to investigate the difference in privacy experiences between the traditional and flexible office, questions on distraction of sound and vision, the experienced sound level and the privacy for private conversations and ability to work without being seen or being listenable were used (rated on a 5 point scale 1= good, 5=bad).

Work satisfaction and satisfaction with the workplace are indicated to relate well with employees individual performance (Harter et al. 2002).Therefore satisfaction with the physical office arrangement and the comfort and the satisfaction over the dimension of the provided office space were included as well as a question on overall satisfaction. Since satisfaction over the traceability of colleagues is also mentioned as a disadvantage of flexible offices (Veitch et al. 2002), a question on this topic was specifically added. Satisfaction was measured on a scale from 1 to 7 (1= extremely low, 7 = extremely high).

Although the same individuals were approached in both measures, the total response of subjects completing both questionnaires was expected to be low. In order to include all questionnaire data in the statistically analysis and compare data from the traditional and the flexible office, an independent t-test was performed using SPSS 17.0. This was done for the intervention group as well as for the control group (alpha = 5%, two tailed).

Results and discussion

A total of 150 filled out questionnaires were collected, of which 84 subjects filled out the first questionnaire (67% response) and 66 filled out the second questionnaire (52% response). The job functions of the subjects existed of managers (18%), secretary and administrative personal (14%), and office staff (68%). Sixty percent of the subjects were female; subjects had an average age of 46 years, and 14,5 years of average job experience.

OFFICE SPACE USE

Data on office space use show a significant shift in percentage of work time spend at group workplaces from 18% at the traditional office to 41% at the flexible office (see figure 1). Data on the percentage working time spend in meeting rooms, concentration rooms, library or pantry show no significant changes between the traditional and flexible office.

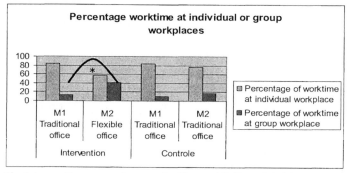

FIGURE 1. The intervention group shows a significant shift (* p< 0.05) in percentage of time spend in individual- and group workplaces.

PERFORMANCE

No significant differences in performance were found between the traditional and the flexible office regarding creativity, the quality of the work, the quantity of the work and the perceived efficiency of the work (see figure 2). Although self assessed performance was not affected by change in office, the results do show support for the hypothesis that a physical environment can affect performance. Table 1 shows significant differences between the traditional and the flexible office on the supposed effects of the physical environment on work performance. The flexible office supports less for concentration in comparison to the traditional office.

However the flexible office does better support for communication and collaboration with colleagues and supports better for creativity.

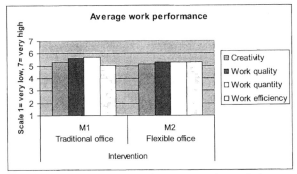

FIIGURE 2. Average work performance ratings of the questionnaire data. No significant differences were found between traditional and flexible office.

Table 1 Average score on support for work performance; ▲=significant increase, ▼=significant decrease (P< 0.05).

The physical office environment supports for: (Scale 1-7, where 1 = very bad support, 7 = very good support).	Intervention group		Control group	
	M1 Tradition al office	M2 Flexible office	M1 Tradition al office	M2 Tradition al office
Concentration	4,88 ▲	4,12 ▼	4,34	4,5
Communication with colleagues	4,95 ▼	5,59 ▲	4,97	5,32
Communication with clients / visitors	4,58	4,76	4,72	5,05
Collaboration with colleagues	4,93 ▼	5,74 ▲	4,94	5,36
Creativity	4,53 ▼	5,21 ▲	4,53	4,5

PRIVACY

A major concern in the new way of working is the lack of privacy due to the introduction of flexible open offices. This was not supported by the results, since no significant differences in scores were found between the traditional and the flexible office on privacy experience (see table 2).

Table 2 Average score on privacy aspects of the office environment;
▲ =significant increase, ▼ =significant decrease (P< 0.05).

Privacy at the office Scale 1-5, where 1 = good, 5 = bad.	Intervention group		Control group	
	M1 Traditional office	M2 Flexible office	M1 Traditional office	M2 Traditional office
Auditive privacy (distraction by sound)	3,44	3,15	2,69	3,1
Conversation privacy (listen in on)	3,47	3,06	2,72	2,9
Visual privacy (can not be seen)	3,6	3,24	3,19	3,43
Visual privacy (not being visually distracted)	3,56	3,29	3,09	3,14
Sound level	3,56	3,38	3,25	3,48

WORK SATISFACTION

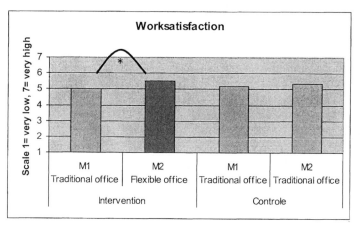

FIGURE 3. Average work satisfaction. Significant difference (* p<0.05) between the traditional and flexible office was found for the intervention group; for the control group no significant difference was found.

The average work satisfaction is significantly higher in the flexible office compared to the traditional office (see figure 3). This could partly be explained by the higher satisfaction with the office arrangement and comfort of the workplace in the flexible office (see table 3). Although it was expected that a flexible office has disadvantages for the traceability of colleagues, and reducing the square meters office space might cause feelings of overcrowding (Veitch et al. 2002), none of these topics were rated different compared to the traditional office.

Table 3 Average score on aspects of work satisfaction; ▲=significant increase, ▼=significant decrease (P< 0.05).

Aspects of work satisfaction	Intervention group		Control group	
	M1	M2	M1	M2
Scale 1-5, where 1 = good, 5 = bad.	Traditional office	Flexible office	Traditional office	Traditional office
Arrangement of the office environment	3,43▼	4,32▲	2,78	3,1
Comfort of the workplace	3,93▼	4,35▲	3,47	3,67
Dimensions of the office space	4,26	4,06	3,94	3,86
Traceability of colleagues	3,98	3,91	4	4,1

Reflection

The results confirm the hypothesis that switching from a traditional to a flexible office has some advantages. It should however not be interpreted that the found results are generalizable to all business organizations changing to a flexible office, since the form of implementation and starting points might differ largely. The second measure occurred six weeks after moving to the flexible office, and subjects did have time to get accustomed to the new work environment. It is of great interest to investigate whether same results can be found after a longer period of habituation, in order to investigate the stability of the found effects on the long-term. In the present study the subjects were positively involved in the process of moving to a flexible office, but they had not received training on office space use. Researchers suggest that training can have beneficial effects on work satisfaction and might improve individual productivity (Robertson and Huang 2006) and is therefore recommended.

OFFICE SPACE USE

Although it might be obvious to find a difference in office space use between the traditional and the flexible office, it is possible for subjects to work for some time still in a traditional way in a flexible office by for instance choosing a concentration room, which is similar to the individual workplaces of the traditional office and ignoring the new provided workspaces. Results show an increase in percentage of work time at group workplaces, indicating a change of work space use.

PERFORMANCE

It is often mentioned that work performance will increase by introducing the new way of working (Hengst et al. 2008). Results on the support for performance indicators such as communication, collaboration and creativity showed significant more positive scores for the flexible office then the traditional office, except for scores on support for concentration. This is in contrary with the results on assessed performance, such as quality, quantity and efficiency of the work performed, that did not show any significant difference between the traditional and the flexible office. It is possible that the flexible office did in fact increase communication, collaboration and creativity, but that the higher level of performance was not experienced.

PRIVACY

Lack in privacy, concentration and an increase in distraction are often mentioned disadvantage of flexible offices (Bauman & Arens 1996). This study shows that it is possible to design the office environment in such a way that privacy is not minimized. Another case study on the effects of workspace design confirms these results and found an increase of sufficient privacy for the task facilitating office in comparison to a traditional office (Blok et al. 2009).

WORK SATISFACTION

Work satisfaction and satisfaction regarding the workspace is related to individual work performance (Robertson and Huang 2006). Although the assessed work performance was not significant different between the traditional and the flexible office, scores on work satisfaction do show a significant higher score for the flexible office.

Conclusion

In conclusion, this case shows that a reduction of office space and the introduction of a variety of workspaces caused a different way of working, resulting in a better support for collaboration and communication with colleagues. There was no significant increased performance, but workers satisfaction did increase. Although the support for concentration was less for the flexible office, the experienced privacy did not decrease, indicating that it is possible to create ideal workspaces. Further research on the effects of new ways of working should be investigated in order to optimize implementations of new work environments.

REFERENCES

Banbury, S.P,, Berry, D.C. (2005), Office noise and employee concentration: identifying causes of disruption and potential improvements. Ergonomics, 48, 25–37.

Bauman, F.S., Arens, E. (1996), Task/Ambient Conditioning Systems: Engineering and Application Guidelines. Center for Environmental Design research. UC Berkerly, publication number: CEDR-13-96.

Blok M., Korte E., de, Groenesteijn, L., Formanoy, M., Vink, P. (2009) The effects of a task facilitating working environment on office space use, communication, concentration, collaboration, privacy and distraction. Proceedings of the 17th World Congress on Ergonomics, Bejing.

Croon, E.M., de, Sluiter, J.K., Kuijper, P.P.F.M., Frings-Dresen, M.H.W. (2005), The effect of office concepts on worker health and performance: A systematic review of the literature. Ergonomics, 48, 119-134.

Harter, J.K., Schmidt, F.L., Keyes, C.L. (2002), Well-being in the Workplace and its Relationship to Business Outcomes: A Review of the Gallup Studies. In C.L. Keyes & J. Haidt (Eds.), Flourishing: The Positive Person and the Good Life (pp 205-224). Washington D.C.: American Psychological Association.

Hengst, M., den, Leede, J., de, Looze, M.P., de, Krause, F., Kraan, K. (2008), Distant work, virtual teams and mobile work (in Dutch: Werken op afstand, virtuele teams en mobiel werken), TNO, Hoofddorp.

Hill, E.J., Ferris M., Märtinson V. (2003), Does it matter where you work? A comparison of how three work venues (traditional office, virtual office, and home office) influence aspects of work and personal/family life. Journal of Vocational Behavior 63, 220-241.

Hill, E.J., Weiner, S. (2003). Work/life balance policies and programs. In J. E. Edwards, J. C. Scott, N. S. Raju (Eds.), The human resources program-evaluation handbook. Newbury Park: Sage.

Korte, E., de, Kuijt-Evers, L., Vink, P. (2007). Effects of the Office Environment on Health and Productivity 1: Auditory and Visual Distraction. Ergonomics and Health Aspects, HCII 2007, LNCS 4566, 26-33.

Lee, S.Y., Brand, J.L. (2005), Effects of control over office workspace on perceptions of the work environment and work outcomes, Journal of Environmental Psychology, 25, 323–333.

Penn, A., Desyllas J., Vaughan, L. (1999), The space of innovation: interaction and communication in the work environment. Environment and planning B: Planning and Design, 26, 193-218.

Robertson, M.M., Huang, Y.H. (2006), Effect of a workplace design and training intervention on individual performance, group effectiveness and collaboration: the role of environmental control. Work, 27(1), 3–12.

Van der Voordt, T,J.M., Meel, J.J., van. (2002), Psychologische aspecten van kantoorinnovatie (Psychological aspects of office innovation). Delft/Amsterdam: BMVB & ABN AMRO.

Van der Voordt, T,J.M. (2004), Productivity and satisfaction in non-territorial offices. Journal of Corporate Real Estate, 6(2) 133-148.

Veitch, J.A., Farley, K.M.J., Newsham, G.R. (2002), Environmental satisfaction in open-plan environments: 1. Scale validation and method (IRC-IR-844). Ottawa, ON: National Research Council Canada, Institute for Research in Construction.

Vink, P. (2009), "Aangetoonde effecten van het kantoorinterieur: Naar comfortabele, innovatieve, productieve en duurzame kantoren" (Proofed effects of office environment). Alphen aan den Rijn: Kluwer.

CHAPTER 25

The Effects of Correlated Color Temperature and Illuminance on Color Discrimination

Cheng Wenting, Lin Yandan, Sun Yaojie

Dept. of Illuminating Engineering & Light Sources
Fudan University
Shanghai 200433, China

ABSTRACT

The effects of both illuminance and correlated color temperature (CCT) on people's (aged 20-25) performance of color discrimination were investigated. FM 100-hue test was used as a task under the combined conditions with 3 CCTs(3000K, 6500K and 8000K) and 3 illuminance levels (10lx, 100lx, 1000lx). The results showed that people's performance of color discrimination improved with increasing illuminance, as well as increasing CCT. It was also proved that color rendering index (CRI) of the light sources cannot reflect the effect of spectral power distribution (SPD) on color discrimination. At low illuminance(10lx), CCT had significant effects on people's performance and preference, that is, the performance increased with increased CCT, though preference decreased. The study appeals for a complete consideration in lighting design for the effects of both illuminance and CCT, and people's need for performance and preference, especially meaningful for a field with low light levels, such as cockpit, military working places, etc.

Keywords: color discrimination, CCT, illuminance, FM 100-hue test

INTRODUCTION

As early as 1970s, Boyce et al.[1] firstly used FM 100-Hue Test to study the hue discrimination of light sources. The influence of lamp types (FL, HID), illuminance(300lx, 1000lx) and subjects' age($\leq$30, 31-54, $\geq$55) were studied. Mean error scores, as well as the range in each condition were recorded. Statistical analysis showed that compared with illuminance, the effect of lamp type (spectrum) was more important, and the influence of illuminance was only important for the older age group. Knoblauch[2] et al. examined the effect of age and illuminance on the FM-100 hue test using the MacBeth Executive Daylight source. The result showed that the number of errors committed on the test increased both with increasing age and with decreasing illuminance level.

The method was still used in recent years. In Boyce's later study3 , FM-100 test was used again to study the effect of CCT on color perception and preference. Four different fluorescent sources were examined with CCTs of 2500K, 2700K, 4200K and 6300K, all of which had high CRIs. Performance significantly increased with increasing illuminance under the 2700K and 2500K lamps. Darcie. A.O'Connor and Robert G. Davis[4] used a similar color discrimination test as a visual task to study the effect of light source type and illuminance of lighting. The test was L'Anthony Dichotomous 15-Hue test[5], which was similar to FM test, but with only 15 caps to order. Three light sources were used- a halogen lamp (CRI 95), a 3000K compact fluorescent lamp (CRI 82) and a 5000K compact fluorescent lamp (CRI 84) at three illuminances(approximately 9 lx,300 lx, 700 lx). Recent study of Mark S. Rea et al[6], Ferenc Szabo and Elodie Mahler et al[7] also used color discrimination as a method to evaluate quality of LED lighting.

Most previous studies mainly focused on the effect of illuminance. Studies in recent years tend to consider the effect of spectrum (CCT) on the performance of color discrimination by using a simplified color test. However, studies on the effects differed and even conflicted. Boyce's study indicated that above about 300lx, the choice of light source (different spectral power distribution) is a much more important factor than is illuminance, but the experiments did not cover the lighting condition in low illuminance(<300lx). Darcie's study covered a wide range of light sources and illuminance levels, but used 15-hue test instead of FM 100-hue test, and got the conclusion that changes in illuminance produced much larger changes in subjects' performance than changes in light source. Vienot Francoise's experiment focused on LED light sources, but fluorescent lamps were not included. This study used fluorescent lamps as light sources, which is widely used in indoor lighting(e.g. For reading).The objective of this study was to test the performance of people on a color discrimination task under different lighting condition with the original color test, FM 100-hue test, so as to find out the effects of both illuminance and CCT.

METHOD

SUBJECTS

A total of 10 subjects participated in the experiments, aged 20 to 25，5 were female and 5 were male. They all had normal color vision, and were corrected to normal visual acuity.

Ishihara-color-test plates are used in the experiment to screen the subjects, to ensure that the test result will not influenced by people's deficit in color vision.

EXPERIMENT SET-UPS

Experiments were conducted in a dark room (FIGURE 1), with dimmable fluorescent lamps（CRI=80）on the ceiling. Spectrum and lighting parameters of each condition are as follow（FIGURE 2）.

FIGURE 1 experiment set-up

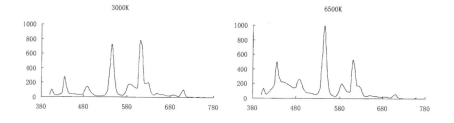

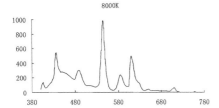

FIGURE 2 Relative spectral power distribution of the lamps used

Table 1 Values of the parameters used in experiments

	Parameters	Values
Fixed parameters	Adaptation	15 mins
	Viewing Status	Binocular
	Light Source	linear fluorescent lamp
	CRI	80
Variable parameters	Mean Illuminance on Task(lx)	10; 100; 1000
	CCT of Lamp(K)	3000; 6500; 8000

COLOR DISCRIMINATION TEST

The Farnsworth-Munsell 100-Hue Test was used in the experiments. It consists of 85 color discs arranged in four series and is used widely as a test of color vision. It has also been used to evaluate the color quality of different light sources.

Each of the four series of discs is arranged in a tray to which two anchor discs are permanently mounted at either extreme. The two anchor discs show the starting and ending points of each series. The task of the observers is to order the discs in each series according to their hue between the two anchor discs. The test was originally designed to be administered under daylight or a comparable electric light source, such as a CIE Illuminant C type lamp. There is no time limit to perform the task.

A test score is based on the number of transpositions, that is, the number of incorrectly placed color discs within each series. The error score can be calculated according to the formula:

$$ErrorScore = \sum_n (|X_{n+1} - X_n| + |X_n - X_{n-1}| - 2)$$

Where X_n is the number on the bottom of the nth cap.

An error score of zero means perfect performance, and an increasing error score means worsening performance of color discrimination. The error score gives an overall rating of the performance without regard for the directions of the errors.

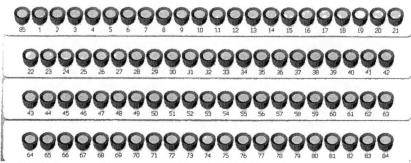

FIGURE 3 caps of Farnsworth-Munsell 100-Hue Test

PROCEDURE

Subjects were instructed as to the purpose of the study at the beginning of each experiment. A few trial presentations were conducted to aid in clarifying the procedure, until the subjects were familiar with the test, to avoid the effect of experience.

After trial experiment, subjects were asked to do the FM 100 hue test under several lighting condition one by one (FIGURE 4). Half of the subjects did the tests with one order; the others did them with the reverse order. There was no time limitation for the test, but the duration of each test was recorded.

Before each test; 15 minutes were given for the subjects to adapt to the lighting condition; enough relaxation time was also given between the tests to avoid eye strain.

FIGURE 4 One of the subjects doing FM-100 hue test

RESULT

FIGURE 5 and FIGURE 6 show the mean error scores categorized by CCT and illuminance respectively. The results showed the decrease of error scores with increasing illuminance, and also with increasing CCT.

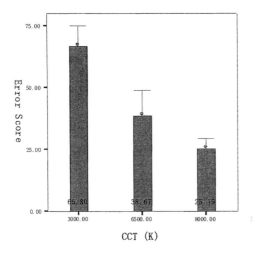

FIGURE 5 Mean error scores categorized by CCT

234

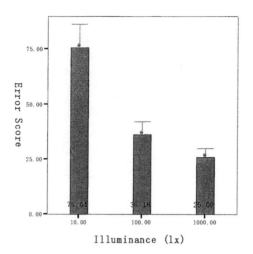

FIGURE 6 Mean error scores categorized by illuminance

In FIGURE 7, data of different illuminance was split. It showed that in each illuminance level, higher CCT lead to lower error scores.

FIGURE 8 showed the mean error scores under different illuminances with different CCTs split. A decrease of error score could also been seen under every CCT.

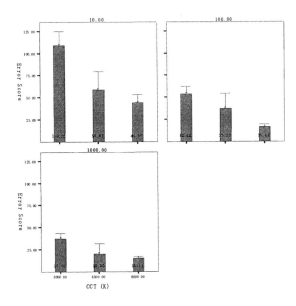

FIGURE 7 Mean error scores under different CCTs (illuminances split)

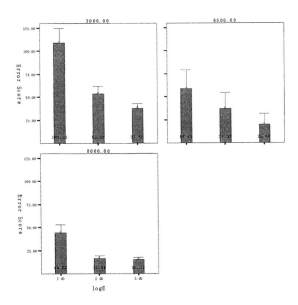

FIGURE 8 Mean error scores under different illuminances(CCTs split)

DISCUSSION AND CONCLUSION

FIGURE 9 and FIGURE 10 showed the tendency of error score categorized by illuminance and CCT. Under different illuminance, people all tend to have lower error scores, i.e. better performance under higher CCT, but the effects of CCT was much more obvious in a lower illuminance(10lx). On the other side, for whatever CCT, people's performance of color discrimination improved with higher illuminance, but the slopes changed for different CCTs.

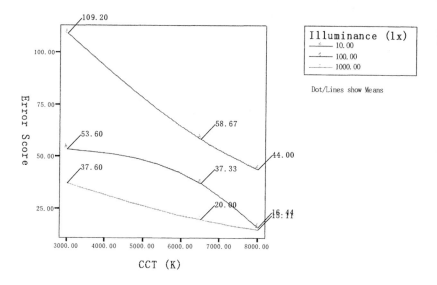

FIGURE 9 Line chart of mean error scores categorized by illuminances

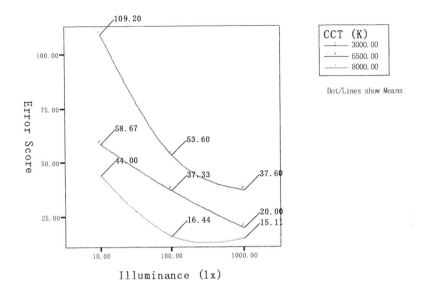

FIGURE 10 Line chart of mean error scores categorized by CCTs

The results showed that illuminance in a low level (10lx) was significantly different from higher one, and the effect became less important with increasing illuminance. The 100lx level and 1000lx one were not significantly different, that was in accord with Boyce's study, which considered illuminance level a much less important factor than light sources above 300lx. But for a wide range of illuminance, not only SPD but also illuminance influences the performance of people's color discrimination, which in accord with the results of Vienot Francoise.

With the same CRI, 3000K cases are significantly different from 6500K and 8000K ones, the results indicated that SPD of fluorescent lamps have effect on this performance, and CRI is not enough to describe the effect. The same conclusion was got by Darcie et al. and Mark S.Rea in the case of LED lighting. So the conclusion can be more generalized.

As for the low illuminance(10lx), CCT have an even more important effect. But according to Kruithof curves[8], people prefer lower CCT under low illuminance. That is to say, people's preference of lighting doesn't always accord with their performance. For a indoor lighting design, it's better to take both sides into consideration. An extreme condition is that, people don't like a 10lx 8000K lighting condition, though their performance is not bad, so it's still not suitable for their daily life.

To sum up, people's performance of color discrimination improved with increasing illuminance, as well as increasing CCT; CRI of the light sources could not reflect the effect of spectral power distribution (SPD) on color discrimination. At low illuminance(10lx), CCT had significant effects on people's performance and preference, that is, the performance increased with increased CCT, though preference decreased. The study appeals for a complete consideration in lighting design for the effects of both illuminance and CCT, and people's need for performance and preference.

ACKNOWLEDGEMENT

This project is supported by the National Basic Research Program of China (973 Program No.2010CB734102) and OSRAM.

REFERENCES

1 Boyce P R, Simons R H. Hue Discrimination and Light Sources[J]. Light Res. & Technol. 1977, 9(3): 125-140
2 Kenneth Knoblauch et al. Age and illuminance effects in the Farnsworth-Munsell 100-hue test Applied Optics 1987,vol.26, No.8 p1441-1448
3 Boyce P R, Cuttle C. Effect of Correlated Colour Temperature on the Perception of Interiors and Colour Discrimination Performance[J]. Light Research and Technology. 1990, 22(1): 19-36
4 Darcie. A.O'Connor and Robert G. Davis Lighting for the Elderly: The Effects of Light Sources Spectrum and Illumiance on Color Discrimination and Preference[J]. LEUKOS. 2005, 2(2): 123-132.
5 Lanthony P. The Desaturated Panel D-15[J]. Documenta Ophthalmologica. 1978, 46(1): 185-189
6 Mark S. Rea, Jean P. Freyssinier-Nova Color rendering: a tale of two metrics. Color research and application vol.33, no.3 Junie 2008 p192-202
7 Elodie Mahler et al. Testing LED lighting for colour discrimination and colour rendering, Color research and application vol.34, no.1 Feb.2009 p8-17
8 Kruithof A A Tubular luminescence lamps for general illumination Phips Tech. Rev. 6 65 (1941)

CHAPTER 26

Investigating Manual Work Using Human Factors Approaches and Behavioral Analysis

Glyn Lawson[1], Guido Tosolin[2], Chiara Leva[3], Sarah Sharples[1], Alyson Langley[1], Shakil Shaikh[1], Paul Liston[3]

[1]Human Factors Research Group, Faculty of Engineering
University of Nottingham, NG7 2RD, UK

[2]AARBA, Association for the Advancement of Radical Behavior Analysis
Corso Sempione 52, 20154, Milan, Italy

[3]APRG Trinity College, University of Dublin, Ireland

ABSTRACT

This paper reports an investigation into manual work using a combination of Human Factors (HF) approaches, including task analysis, ethnography and behavior analysis. The work arose from the ManuVAR (CP-IP 211548) EU-funded research project, which aims to support manual work operations through the use of Virtual and Augmented Reality (VR/AR) technologies. The investigation involved analysis of current manual working practices in the industrial partners' companies. The research aimed to analyze physical, cognitive, organizational/social ergonomics and behavioral aspects of the tasks to realize the scientific foundations of the project.

Keywords: Human Factors, Task Analysis, Ethnography, Functional Behavior Assessment, Manual Work.

INTRODUCTION

Human factors (HF) best-practice advocates applying more than one method to any given problem (Wilson, 2005). Based on this principle, and benefiting from the variety of skills and research specialties of the consortium partners, a combination of methods was applied to investigate HF aspects of manual work. These included the deployment of an innovative tool for task analysis used according to the long established method of hierarchical task analysis, but also ethnography, an approach used to investigate the work in context. The investigation also benefitted from expertise in behavior analysis, the measurement of behaviors in a particular environment, and the analysis of their causes.

Task analysis is commonly referred to as the process of collecting information about the activities that an operator or a group of operators need to perform in order to achieve a certain goal. Several authors describe different techniques for each of the above phases (Kirwan and Ainsworth, 1992; Diaper and Stanton, 2004). One of the most established methods is Hierarchical Task Analysis (HTA), which is the process of decomposing the overall goal of the worker (reason for the work) into a series of subtasks which must first be achieved. The decomposition continues until tasks are sufficiently small for analysis. The hierarchies also contain "plans" which govern the completion of sub-tasks, for example: "complete sequentially"; or "if X then do Y" (Kirwan and Ainsworth, 1992). The hierarchy is only the first stage of the analysis and can be considered primarily as task representation. The analysis is conducted afterwards; different methods can be used for this, depending on the purpose of the investigation.

Ethnography is an established, qualitative method in the social sciences and has its origins in methodological shifts in anthropological research which emphasized a move towards collecting data first-hand. It contains a strong emphasis on exploring the nature of particular social phenomena, rather than setting out to test hypotheses about them. Ethnographic studies demonstrate a tendency to work primarily with "unstructured" data, which have not been coded at the point of collection in terms of a closed set of analytic categories. Ethnography typically involves investigation of a small number of cases, perhaps just one case, in detail. The data analysis involves explicit interpretation of the meanings and functions of human actions, mainly taking the form of verbal descriptions and explanations, with statistical analysis playing a subordinate role at most (Atkinson and Hammersley, 1994).

Behavior Analysis is the science that investigates the prediction and control of behaviors through the analysis of the variables of which it is a function. These variables are called environmental contingencies and are divided in antecedents and consequences. Functional Behavior Assessment (FBA) (Daniels, 2004; Skinner, 1938) is the measurement of the behaviors and the analysis of their causes. Manual work is a matter of behavior; it is the sum of behaviors performed by workers using a number of tools. Results are defined as the product of behaviors; the difference between high and low-performance manual work is how they are performed. The main aims of FBA are to:

- pinpoint what behaviors we want to be performed, and when
- measure their actual state
- determine actual contingencies that cause actual behaviors
- build the right contingencies in the working environment to obtain them

Previous work within the project had included identification of the requirements of the industrial partners (ManuVAR D1.1, 2009) and definition of the industrial use cases (ManuVAR D2, 2010). This early work had involved interviewing the industrial partners to investigate the problems faced in their work, and the scope of the work to be investigated within the ManuVAR project. The outcome of these investigations included a list of seven "gaps", which can be considered high-level requirements categorizing the problems experienced by the ManuVAR industrial partners. The gaps, which focused the HF analysis, are as follows:

Gap 1: Problems with communication throughout the lifecycle
Gap 2: Poor interfaces
Gap 3: Inflexible design process
Gap 4: Inefficient knowledge management
Gap 5: Low productivity
Gap 6: Lack of technology acceptance
Gap 7: Physical and cognitive stresses (ManuVAR D1.1, 2009)

The focus of this paper is on a reflection of the value of the methods used in the HF investigation. The implementation of the methods is described, and samples of the outcome are provided. The relative gain from application of each of the individual methods is considered, and the importance of applying several methods to any given problem for a comprehensive analysis is highlighted.

The examples shown are taken from one of five industrial use cases: remote maintenance support in the railway sector. The aim of this use case is to allow maintenance experts to guide operators from a remote location, helping them to focus on the most important areas and to easily exchange useful information. Augmented reality (AR) is being considered as a possible technology. While the scenario involves manual work, the issues concern broader factors influencing the human interaction with the system, rather than focusing solely on posture, manual handling etc. This finding was typical of all the use cases.

METHOD

TASK ANALYSIS

A task analysis was generated for each of the ManuVAR use-cases using a tool called the Task Modeller (Leva et al., 2009). The tool supports data collection and provides a structure for the interviews and for the simultaneous graphical representation of the task. The method according to which the tool was actually deployed is HTA. The process involved decomposing the overall task goal into the constituent sub-tasks, and any plans governing their execution. The HTAs were generated based on several sources of information, including:

- the scenario definitions (ManuVAR D2, 2010)
- discussions with industrial participants
- site visits to the industrial partners

After generating the hierarchies the industrial partners were given the opportunity to review them for accuracy. Amendments were made where appropriate. A typical high-level hierarchy is shown in Figure 1. This is the first of five similarly detailed hierarchies which extend from the sub-tasks, as indicated through the link nodes. The notation and semantics used for the graphical representation of the task within the Task Modeller is that of Business Process Modeling Notation (BPMN) (White & Miers 2008). The star symbols represent tasks which parent multiple sub-tasks; diamond shapes are the gateways or plans which govern completion of the sub-tasks. The task analysis represents the current methods of work, rather than attempting to show working practices in the future following the anticipated ManuVAR intervention. This was to ensure that development follows a user-centered approach which will address the end-users' requirements, rather than focusing on available technologies.

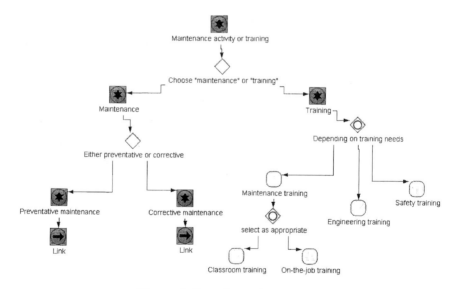

Figure 1. Sample top-level hierarchy

As mentioned, the hierarchy is primarily a representation of the tasks; analysis is conducted afterwards. In this instance, the goal was to identify the HF influences on the seven gaps. To do this, the hierarchies and interview data (ManuVAR D1, 2009) were analyzed to investigate the possible physical, cognitive, or organization ergonomics influences on the gaps. This process revealed many possible influences, often common to several industrial partners. Thereafter, the influences were scrutinized to identify the factors which could be measured using HF methods and be used to inform the evaluation work packages later in the ManuVAR project. Recommendations were made to address each of the influences on the gaps.

ETHNOGRAPHY

Ethnographic research is a tool that has the potential to give a 'thick' description of the HF, organizational, and social issues that are a feature of the tasks chosen by the ManuVAR clusters. Ethnographic research can uncover deep insights through observation and interaction with the end-user in their natural environment. Ethnography's claim to facilitate a unique humanistic, interpretative approach to studying social life owes much to its reliance on participant observation methods. Immersing a researcher in the daily lives of the participants leads to a unique insight into the social processes involved in the reality being documented by the researcher. Observations were conducted at the industrial partners' sites. There were seven key stages:

1. Access to the workers was negotiated through a 'gatekeeper' – in this instance the ManuVAR contact-person for each end-user organization.
2. Participants were briefed on the purpose of the observation.
3. Consent was obtained
4. A generic protocol was used to guide the observation based on the same one to determine the end-user requirements (ManuVAR D1.1, 2009) covering physical, organizational/social, systems ergonomics and cognitive.
5. Workers were observed in normal operational circumstances by researchers experienced in ethnographic research.
6. Researchers took notes using the observation framework as the basis for their notes. This was supported by other documentation.
7. The notes taken were analyzed resulting in a composite review of the task from a HF perspective.

FUNCTIONAL BEHAVIOR ASSESSMENT

The FBA in ManuVAR aimed to determine the actual state of behaviors and contingencies in order to provide guidelines to build a human-machine interface that is able to provide the best antecedents and consequences to maximize performance. The FBA was composed of three main steps:

1. Pinpointing (identification of behaviors of interest, target behavior)
2. On-the-job Observation or Assessment (parametric measure of behaviors)
3. Functional Analysis (Identification/measure of variables the behavior is

function of: antecedent stimuli and consequent stimuli)

Pinpointing provided a specific description of the manual task in terms of behaviors that have to be performed by workers in order to get the related results. After this process it was necessary to conduct on-the-job observations, conducted in parallel with the ethnography, thus minimizing logistical preparation and disruption to the workers. The behaviors were measured according to frequency, duration, latency, and intensity. The observation was made measuring one or more of these parameters for each behavioral item. Contingencies related to each behavioral item were also observed and divided into Antecedents and Consequences referring to the A-B-C Paradigm (B.F Skinner, 1938, 1953). The aim of data analysis was to identify of the variables for which the behaviors are a function.

RESULTS

TASK ANALYSIS

A sample from the analysis of the task hierarchies is shown in Table 1. This relates to Gap 1: problems with communication throughout the lifecycle. More specifically this gap applies to the situations when two or more stages of the lifecycle are involved: feedback, transformation of models, out-of-date documentation, duplication of information in various databases, and globalization requirements.

Table 1. Sample outcome from the Task Analysis

Gap01: problems with communication throughout the lifecycle		
Cognitive/ physical/ organizational influence on the gap	Measurable factor	Recommendation
Communication of problems identified by the manual worker to the designer is critical to the success of the product	Communication (spatial concepts/ team/ procedures)	Provide timely & easily captured communication (spatial, verbal) of problems from worker to designer.
High mental workload results from insufficient access to procedural information in the field	Communication (procedures) Workload	Provide ubiquitous procedural information (text, spatial concepts) to field operators
Communication of advice and warning is safety-critical during tasks	Communication (procedures)	Support verbal communication of advice and critical warnings
High situational awareness of engineering status is important when starting shift	Situational awareness	Support communication (verbal, spatial, procedures) of engineering status during shift handover

The first column in Table 1 shows the physical, cognitive and organizational influences on the gap. It should be noted that many of the HF influences are common to more than one use-case. The next column shows the measurable factor, which aims to identify the HF components which can be evaluated. Finally, recommendations were made to address the HF influences, as shown in the right hand column of the table. These aim to be as generic as possible such that a solution addressing the recommendations would be applicable to manual work in other companies/scenarios. However, they were written to be specific enough to be useful during the development phases of ManuVAR.

ETHNOGRAPHY

The observation protocols provided the basis for analysis and recommendations for the task. The results are categorized by areas of interest (sample reported in Table 2).

Table 2. Sample results from the ethnography, describing issues identified which relate to manuals/documentation and parts.

Manuals/documentation (In what way do they affect the task?)
The workers have been observed to carry very little documentation with them, only the checklist for preventive maintenance intervention and the work order for corrective maintenance tasks. Procedures are available in both paper and electronic format at the office of the shift supervisor where the workers meet before staring their shift, however experienced worker are thought to be able to perform their task without the use of any support manuals, while people still on training may be required to carry the work instruction around with them *Suggestion: The work instructions for the task to be performed can actually support the worker especially if troubleshooting instructions are included. To be more effective they can be made available in an electronic format.*
Parts (How do parts affect the task?)
As for the tools, only in this case for corrective maintenance and preventive maintenance tasks reported in the work instructions the materials required for the tasks are often indicated. However those procedures are not always carried by the workers and therefore task interruption occurs looking for parts. *Suggestion: As above the work instructions for the task to be performed should be made available on each trolley either in electronic format or in paper format. This could also enable a better planning system for the materials required for tasks. The stores department could then produce a list of consumable materials per train check and ensure that the materials are stocked on the trolley at the train location for the engineers to access.*

FUNCTIONAL BEHAVIOR ASSESSMENT

The results of the behavior analysis are divided into three parts, *Workflow*, *Pinpointing*, and *Functional Analysis*. The results are described below.

Workflow: This is mainly descriptive and constitutes a report of the visit. The sequences of actions performed by workers that lead to the desired result are identified. For example: *the shift supervisor receives the to-do jobs of the day, which are then distributed among the workers.*

    NOTE 1 - activities can be preventive or corrective. The complexity varies.
    NOTE 2 - The supervisor adopts three main criteria to distribute the work:
    (1) worker qualifications, (2) worker expertise and (3) worker reliability
Workers are orally informed about their tasks by the shift supervisor.

Pinpointing: This phase pinpoints the behaviors that are directly related to the desired results. This process is summarised in the following table.

Table 3. Pinpointed behaviors, which relate to the desired results. Required tools and skills/knowledge were also identified.

Desired Results	Related behaviors	Required tools	Required skills/knowledge
Reduce the time spent because of interruptions to get information from the supervisor	Take Work Instruction from office	Work Instructions	*Which* information *Where* is information stored
	Take Technical Norm of Maintenance from the office	Technical Norm of Maintenance	*Where* is the information stored
	Read work instructions and/or Technical Norm of Maintenance	Work Instructions Technical Norm of Maintenance	*Knowledge* about how to read Work Instructions and Technical Norm
	Study/read to fill individual gaps of knowledge with training	Technical Norm of Maintenance	*Knowledge* about how to read Work Instructions and Technical Norm

Functional analysis: The probability of having or not having a certain behavior is a function of its contingencies. A functional analysis was performed of the consequences of behavior in current working practices. The PIC/NIC analysis (Daniels and Daniels, 2004) classifies every consequence in the following parameters as either positive (P) or Negative (N); Immediate (I) or Future (F); and Certain (C) or Uncertain (U).

Behavioural studies demonstrate that only Positive Immediate Certain (PIC) consequences are able to effectively increase the probability for behaviour to occur again. Negative Immediate Certain (NIC) consequences are responsible for inhibiting the behaviour. PFC, NFC, PIU and NIU consequences have a smaller effect. The effect of PFU and NFU consequences is practically equal to zero. Pinpointed behaviours are shown in Table 4 and "ruling" consequences are highlighted for each.

The Functional Behavior Assessment concludes with recommended tools and solutions based on the analyses presented above. Space does not permit a description here; for more detail refer to Lawson et al. (2010).

Table 4. Sample of the PIC/NIC analysis

Behaviors	Antecedents	Consequences	P/N	I/F	C/U
Take Work Instruction from the office	Supervisor assigned me a task	I will not have to stop and walk to the office during the task	P	F	U
	I have the checklist in my hands	I will not have to ask for help from supervisor and/or colleagues	P	F	U
	It is the 1st time I have to do this assignment	I'll spend time to find them in the library	N	I	C

DISCUSSION AND CONCLUSIONS

The combination of methods provided a rich analysis of manual work in the industrial use-case. In addition to facilitating sub-level analysis and identification of the human factors influences on the gaps, the HTAs show comprehendible representations of the work. This promoted good understanding of the tasks by the analysts and clarified the scope of the industrial cases within ManuVAR. The ethnography was used to gather data on the human-centred issues which are a feature of this task. It helped understand in a deep and fundamental way the 'sharp end' of the operation – with evidence of normal, everyday operations to explain how workers compensate for deficiencies with systems and processes – something which HF as a discipline, with its hindsight bias, frequently neglects. The FBA in ManuVAR determined the actual state of behaviors and contingencies to provide guidelines to build a human-machine interface that is able to provide the best antecedents and consequences to maximize performance.

The combination of methods ensures that the situation is analyzed at different levels of granularity, traceability, subjectivity and quantification. Their application at the requirements stage allows a structured analysis of the current work situation to be obtained – this can then be used as a benchmark to understand the impact of any human factors improvements on the work situation that may occur as a result of implementation of VR/AR technologies, as in the ManuVAR project.

Wilson (2005) provides criteria for judging the quality of a method. While this case study does not allow for detailed evaluation against all the criteria, it is possible to comment on several. Concerning *validity*, it is interesting that despite the different approaches the methods revealed several similar problem areas, for example access to work instructions. The methods all resulted in useful recommendations which are suitable for guiding development of the ManuVAR platform, indicating appropriate *sensitivity*. *Acceptability and ethics* posed no particular problems for this scenario, although some of the other industrial partners had no relevant work scheduled within the period; others were concerned about commercial sensitivity. The *resources* were reasonable, with the data collection and analysis taking approximately 1-2 weeks in total for each method. It is difficult to judge the *non-reactivity* of the approach – participants may have behaved differently under observation. The *feasibility of use* was acceptable for each of the methods, although behavior analysis arguably required more specialist knowledge.

In conclusion, this paper has described the methods used to investigate manual handling in the ManuVAR (CP-IP-211548) project, including task analysis, ethnography and behavior analysis. It has given examples of the outcome of each, and provided a summary analysis against criteria for judging the quality of a method. The work highlighted the importance of applying several methods for a comprehensive analysis of work activities.

ACKNOWLEDGEMENT

The research leading to these results has received funding from the European Community's Seventh Framework Programme FP7/2007-2013 under grant agreement no. 211548 "ManuVAR".

REFERENCES

Atkinson, P., & Hammersley, M. (1994). Ethnography and Participant Observation. In N. Denzin & Y. Lincoln (Eds.), Handbook of Qualitative Research (pp. 248-261). Thousand Oaks, California: Sage.

Daniels A.C. & Daniels J.E. (2004). Performance Management, Changing Behavior That Drives Organizational Effectiveness (4th Edition). Performance Management Publications: Atlanta GA.

Diaper D., & Stanton N.A. (2004) The Handbook of Task Analysis for Human-Computer Interaction Lawrence Erlbaum As-sociates Inc., Publishers Mahwah, New Jersey

Kirwan, B. & Ainsworth, L. K. 1992. *A Guide to Task Analysis* (Taylor and Francis, London)

Lawson, G., D'Cruz, M., & Krassi, B. (eds.) (2010) High value high knowledge manual work in the European industries: identification of problem areas and their technical, business and ergonomics analysis. (working title) VTT working papers (in publication)

Leva M. C Kay A., Mattei F. Kontoggiannis, T., Deambroggi M., & Cromie S. (2009) A Dynamic Task Representation Method for a Virtual Reality Application. In Don Harris (Ed.): Engineering Psychology and Cognitive Ergonomics, HCI International 2009, San Diego, CA, USA, July 2009. Lecture Notes in Computer Science 5639 Springer

ManuVAR D1.1 (2009). Definition of requirements, analysis of standards and tools. ManuVAR CP-IP-211548. Internal Deliverable: ManuVAR 2010 www.manuvar.eu

ManuVAR D2 (2010). Definition of industrial cases. ManuVAR CP-IP-211548. Internal

Deliverable: ManuVAR 2010 www.manuvar.eu

Skinner, B.F. (1938). The Behaviour Of Organisms. Appleton-Century-Crofts: New York.

Skinner, B.F. (1953). Science and Human Behaviour. Macmillan: New York.

Wilson, J. R. (2005) Methods in the Understanding of Human Factors. In Wilson, J. R. & Corlett, N. (Eds.) Evaluation of Human Work. 3rd ed. London, Taylor & Francis.

White, S. A, & Miers, D. (2008) BPMN Modeling and Refer-ence Guide. Future Strategies Inc.

CHAPTER 27

Perceived Time, Temporal Order and Control in Boundaryless Work

Henrik Lund, Helge Hvid, Annette Kamp

Center for Occupational Health and Safety and Work Life
Roskilde University, Denmark

ABSTRACT

Contemporary working conditions are very different from just 30 years back. Many changes are characterized as new opportunities for personal development and autonomy for the individual employee. However work life researchers report of increased psychosocial strain and dissemination of work related psychological illnesses. This paradoxical tendency questions our basic knowledge about well being at work. For decades employee control has been seen as universal solution to work related psychosocial hazards, but this is now questioned. We find that control is still as important but needs to be studied in new ways. The way work has developed, the concept of time becomes a key to understand the actual degree of employee control. We cannot study control without taking the qualities of time into consideration. Hence to understand control we must study the qualities of time by looking into common routines, habits, breaks, norms and meetings that occur through the inter-personal relationships. Therefore work life research can make use of time sociology concepts to understand and study how control is gained and lost in contemporary work. The results of our study show that individualized time conflicts leave self-managing and empowered workers with feelings of being unable to get control of their work life. Time conflicts due to an increasing amount of social interactions and complexity at work have become a major psychosocial working environment problem because the sense of control erodes.

INTRODUCTION

It is becoming increasingly difficult to distinguish between positive and negative changes and tendencies for the employees in contemporary work life. Especially when work becomes more boundaryless it is difficult to study and understand the nature of the forces leading to psychosocial strain or wellbeing (Allvin 2008). Boundaryless work is assumed to have a psychosocial working environment with a high degree of individual freedom and control. Control is a core concept for psychosocial work environment. Its pivotal role can be seen in the widespread use of Karaseks D/C model (Karasek 1979). In traditional work Taylorism and bureaucracy is the enemy of employee control. However it has been questioned whether this control model provides us with a satisfactory understanding of the psychosocial work environment (PSWE) among workers in the era of post-bureaucracy, where work becomes more and more boundaryless. Team organization is widespread. The involvement of employees in the development of work organization and technology is promoted by popular management concepts. Maybe too much control is the problem when workers are empowered by self-management, team organization, self-chosen work hours etc. However there are also reports that show that employees' sense of control are decreasing despite the disseminations of these management techniques and the number of workers working under boundaryless conditions are growing (Hvid et al 2009).

'Temporal Order in Boundaryless Work' is the title of a three years research program where the authors study the interaction between tendencies toward boundaryless working conditions, time and control. We assume the transformation towards boundaryless work contains fundamental changes to the dimensions of time and control in work. This paper reports the findings from two case studies of time and control in the work of teachers at two Danish elementary schools where traditional boundaries in the teachers' work have been broken down in order to establish and maintain the best school for the pupils. A survey, mapping the PSWE among the teachers, has been carried out. Observations and interviews are used. We observed work for a week at each school and made 20 interviews with teachers, managerial personals, shop stewards and health and safety representatives. Additional the teachers have made log books mapping their activities around the clock for a week. The schools have committed themselves to improve the temporal structures of work, inspired by our baseline report. After one year a follow-up study is conducted, however this has not yet taken place. We start this paper by introducing our conceptual point of departure: Boundaryless work, time and control.

BOUNDARYLESS WORK

'Boundaryless work' is an increasingly used term for the kind of work where there is no fixed pattern of where, when, how and how much work is to be conducted. Boundaryless work is the result of the many changes that have blurred the various types of boundaries we know from work in the industrial era. E.g. boundaries between work and leisure, professions and tasks, management and employees, personal development and personal integrity etc. (Hirschhorn and Gilmore 1993, Allvin 2008, Kamp & Lund 2008). Hence boundaryless work isn't something you have or do not have but something your job can be more or less characterized by. There are many societal forces and changes making work more boundaryless: Globalization, dissemination of new ICT technology, management's pursuit of flexible organizations and employees' professional orientations and work ethics.

In a world of uncertainty the organizational solution is to be flexible according to the management concepts of the present time. It is a common understanding that it is a prerequisite for business success and an efficient public sector that traditional bureaucratic structures, hierarchies and other barriers to benefit of employees' responsibility and creativity should be broken down. By providing the employees with autonomy, management can be rewarded with the desired flexibility and innovative capacity among the employees. Pursuit of flexible organizations leads to organizational boundarylessness, where employees are working in accordance to objectives, in project groups and in an organization, which is changing all the time. These flexible organizations are associated with an order of time with few repetitions and routines. This is in contrast to the monotony of taylorism, but also an order of time with low predictability, many interruptions and few natural breaks. Statements about the impact of these changes for working life are contradictory. Some researchers emphasize a new emancipation of the workers (Piore & Sabel 1984, Kern & Schumann 1985). Others emphasize social and cultural erosion and decrease of coherence and solidarity (Sennett 1998, Bech 2002).

In an organization without boundaries, the ideal type of employee is self-managing, competent, passionate and dignified. She must be given total rewards including autonomy in return for her devotion to the firm. This ideal type of new employee can hardly be compared to a traditional subordinated, exploited and underpaid industrial worker. It is difficult to even imagine they could experience the same types of problems. This new individualised and more boundaryless work identity, could probably lead to more time spent at work. The reason for the longer working hours could be rooted in fear to fail, or it may be because the work is so compelling that it is not possible to leave it. So when using more time and energy at work, it may also be because it provides something that is difficult to obtain elsewhere. Modern management try to facilitate these work settings with a high degree of team spirit, passion and creativity. This can make it harder for the employee to set boundaries between work and leisure, and may intensify the employee's passion for work and the organization.

It is known that the temporal and spatial dissolving of boundaries may have some negative effects on the PSWE: balance between work and leisure may be disturbed to disadvantage for family life (Albertsen 2007). However the dissolution of the fixed time schedules may also contain some possible advantages for work life balance. However the traditional clock time understanding of time is insufficient to quality of time in the new world of work. We need a broader concept of time to grasp the conditions of the new forms of boundaryless work.

TIME

Since the beginning of the industrial era clock time has been the dominant time regime in the world of work. Pay is determined by how many hours the employer has bought the employee's labour. This fact makes time a key resource. As time is a limited resource it has to be optimised from a management perspective as Taylor set out to do. But even though the linear clock time regime wins hegemonic status in the industrial era the workers also structure their work time by the use of roles, rhythms and routines. In industrial work organized by the principals of taylorism and fordism, and public sector work, which is bureaucratically organised very few events stand out as extraordinary. At best the worker was bored or at worst alienated from his own work. Roles, rhythms and routines made work very predictable. Globalization, individualization and increased use of ICT have introduced new time regimes. Several researchers such as Manual Castells, John Urry, Helga Nowotny and Thomas Hylland Eriksen emphasise the dramatic change of time seen as a social institution. They give different contributions to describe the time-regime in the late modern society. However, all of them identify a new time regime characterized by compressed and accelerated time, many things must take place simultaneously and asynchronous – there are many individual, not coordinated time rhythms. However these sociologists study time in society at large and not specifically work life. Therefore we try to relate the concepts to the tendencies regarding the dissemination of boundaryless work.

Working independent of time and space is everyday practise for many workers with ICT tools. ICT makes it possible to exceed the boundaries in time and space and challenge what presumably were more natural rhythms. The technology also allows people to pack more and more different activities together in the present; multitasking becomes a characteristic feature. Urry (2000) develops the concept of 'Instantaneous Time' to capture this new time regime. Instantaneous hours are a time regime in which space and time are effectively disconnected. The space will be virtualised. This makes it possible to intensify work, speed up work processes and overcome boundaries related to presence of colleagues, office hours etc. The independence from time and space makes social acceleration possible and thereby new ways to optimise work time occur or merely expanding work hours and places. The increased speed means that time horizons are becoming shorter, and sometimes become fragmented. Finally, human activities - work, leisure, meals, etc. - are increasingly following individual patterns. People follow different time patterns and are infrequent present simultaneously in the same room. One can therefore say that there is a de-synchronization of time-space-activity relationships.

Many workers experience contemporary work as a constant demand for instant performance. Former successes and future plans seem irreverent. Many studies have confirmed this increased pace at work (ref). Helga Nowotny's concept of 'The Extended Present' is relevant to understand this phenomenon (Nowotny 1994). With this concept she wants to capture the fact that people live in a constant state of haste and do not have time to make plans for the future. We live from project to project, while the daily routines of daily life drift into the background. We are always ready to respond to new stimuli and information. If not, you are stigmatized as inflexible or as being resistance to change. Unpredictability has become an unavoidable condition in contemporary work. Workers experience this as two workdays never being alike and that one's carrier can take odd and unpredictable turns. This means that you can't plan your carrier but you have to be opportunistic. The same goes for social relations. Hylland Eriksens (2007) explanations to this is that the time in modern society is increasingly made up of short and intense moments - quick time - and this rapid time strangles the slow time: the time for immersion, time for building relations, and time for the long-term horizons. He also points out that the breaks, that previously were 'natural' now are filled out with quick time. For example, waiting at the bus stop or train station is now filled out with text messages, checking emails etc. Eriksen believes such breaks play an important role in human life for reflection, and creative distraction.

When studying contemporary work life you often find that changes is a mix of new trends and old practises. Jensen and Westenholz (2004) also observe that a new time regime develops in the world of work, but it exists together with the old time regimes. This is confirmed by recent studies of time and work, which suggests that the time conflicts - conflicts between different time regimes - are characteristic of modern work and plays an important role for psychosocial stress. An example is Ylijoki & Mätylä's (2003) study of the collective 'orders of time' in academic work. They want to understand the experience of lack of time and pace that characterizes the work. The aim is to capture the collective common time perspectives that people use when they present their work. Ylijoki and Mätylä identified, based on their interviews of more time in academic work, orders of time that are in conflict with each other.

Perlow (1999) has studied the new time regime within a group of IT engineers, who produces programs. She shows how the fragmentation of time and conflicts between individualization of time and the need for synchronization can explain their 'long hour's culture'. The main conflict in the work time is the conflict between time to do what is perceived as the core task and 'interruptions'.

These studies are all based on an understanding of time, where structures in the form of institutionalized time systems play a role, but isn't determining. Time order is not the same as what we previously called time regimes. Time orders are institutions that are shaped in local contexts, and which expresses certain standards in relation to the quality and length of time, tempo, requirements for synchronization and sequence. With emphasis on the conflicting orders of time and the actors' active negotiations of time orders. The actors are seen as co-producers of time orders. Co-producers who reproduce and transform time orders.

In most studies of working time, time is studied as something objectively measurable. In our view, time in the boundaryless work cannot only be studied with these methods, since time cannot be captured without the involvement of the employees' subjective experiences and perceptions. With inspiration from the above time sociology authors we apply aspects, such as orientation towards the presence, intensity and acceleration. These aspects need to be highlighted because they are closely related to the PSWE and may be significant factors, which can contribute both positively and negatively to the quality of work. In our empirical study of the time, control and the boundarylessness in primary schools we will look for how the quality of time depends on the experiences and perceptions of for example: Being present versus wanting to be in several places at the same time, stress and busy versus hectic but exciting activity, interruptions versus flow and common performance versus individual performance.

CONTROL

Karaseks development of the D/C model in the 70ties lead to the concept of control having a central position in research into the psychological working environment over the last 30 years. Because of this model control has been understood as individual autonomy and individual opportunities for development. The D/C-model consists of two dimensions: demands and control. The degree of job control is expressed as decision latitude, which contains two sub concepts: skill discretion and task authority (Karasek, 1979). The point was that the level of strain deriving from a given work situation is not simply dependent on the demands on workers, but also on how much control employees have over their work activities. This claim has been clearly verified through epidemiological studies. The model made it possible to recognize that a solution to working environment problems was not simply a matter of protecting the workers. Sustainable solutions would include changes at work in order to create development opportunities, and influence for employees. The most basic assumption behind the D/C model is that people are able to manage their own lives and their communal lives, and preventing them from doing so is to abuse their rights – even when people are satisfied with working conditions that do not allow this. Such abuse may ultimately lead to illness and, in some cases, death. There is overwhelming epidemiological documentation suggesting that high psychological work demands, combined with a low degree of control, increases the risk of stress symptoms, cardiovascular diseases (Belkic et al., 2004), and mental disorders (Doef & Maes 1999).

When it comes to control in contemporary working life it is difficult to say whether control is increasing or declining. The movement towards more boundaryless work is asserted with more control to the employees however some studies conclude that influence have neither improved nor declined during the last 10-15 years (e.g. Burr et al. 2003). Some Danish and European studies even state that control has been falling significantly (Pejtersen & Kristensen 2009, Birindelli et al. 2007). Seen from the perspective of the D/C model, the development of work in general seems to be going in the wrong direction, which makes the current debate about the possibility of too much control paradoxical. Grönlund (2007) examined whether employee control can be too high for groups having a high degree of autonomy in work. On the contrary she finds that in this kind of work employees are suffering from lack of control, because it is difficult for them to control the workload and working hours. This control debate is a challenge for basic assumptions related to quality of work. It seems that involvement of the employees and increased bureaucratic control can be combined (Hvid, Lund & Pejtersen, 2008).

The D/C-model has been a pivotal point for work life researchers and unions trying to overcome Taylorism and bureaucratic control. However the nature of work has changed significantly since the 1970s. The changes in work have, however, made it necessary to study control in working life. Our basic idea is that when work becomes more boundaryless there is a huge difference between 'having control' and 'being in control'. Hence we find it necessary to take the concept of time into consideration when we study control in boundaryless work. We find that the different concepts, time and control, supplement each other and offer conceptualisations of what is at stake with regards to the PSWE in boundaryless work. Therefore we have tested these concepts empirically in two case studies.

FINDINGS

First some comments to the contextualization of the concepts in our cases. In many ways teachers' work has always been more or less boundaryless. However as we shall see it is so in new forms especially when it comes to new and old orders of time. Especially because of the use of self-managing teams. It is in the teams that time is negotiated, allocated and used and by the employees.

The extend of the boundarylessness is interesting in light of the normally heavy regulated Danish labour market. This industrial relation model have for many years seen many of the boundaries that are now broken down as basic working conditions and therefore needs to be regulated. This deregulation is in many ways linked to the Danish New Public Management policies. These policies challenge teachers' autonomy in use of their working time by forcing them to do a lot of documentation activities. This means that output control and management by objective is implemented. At the same time the regulation of time in the collective agreements is changed from clock time to norms. Different tasks are given an amount of hours and this leave it op to the teachers themselves to manage their own time from their own professional perspective.

At first glance it seems the teachers have a lot of control over their work hours. They have an amount of lessons with the pupils and then have some additional tasks, which are time-regulated by norms e.g. contact with parents, meetings, preparation etc. They can choose to spend a lot of time on one task and less on others. Likewise they have a lot of control from the D/C-model perspective. They have opportunities to learn new things and develop skills. E.g. by the development of own teaching materials. They get to use all their professional skills in the job, and use their creativity to make the teaching interesting for the pupils. When it comes to control over work they also have substantial influence. They have the freedom to make work-related decisions e.g. regarding the schedules and teaching methods. These opportunities to make decisions about how to carry out the work are something the teachers use on a daily basis. In the team they have influence over how the work is organised e.g. three teachers can put their classes together, make an excursion, focus on reading etc.

From the teachers' perspective there is no such ting as too high demands, only too little time.

The teachers' express too high demands as lack of time because demands are viewed as something negative and they see their work as something positive and personal giving. When asked individually, they all express a wish or a need for learning how to manage their time better. They talk about widespread and various forms of control individually. Hence there are many stories about having control in work. But at the same time there are just as many personal stories of being out of control with their work as a hole. Likewise there are many stories about feeling drained or exhausted after work. They find that meetings, documentations and coordination with colleagues are too time consuming. At the same time they find it difficult to find the time to communicate and meet up with colleagues and management, which is required for facilitating their work. The temporal analysis of the teachers' work and their mastering of time contain many paradoxes related to control. It is difficult to capture who and what is in control of working hours and the temporal structure of work. The teachers' experience a lack of time on a daily basis.

The school bell has synchronized traditionally teaching leading to predictable rhythms and routines. However this is abandoned. Now there is a widespread de-synchronization of work. Many teachers start and end their workday at different hours. Pauses, coffee breaks and lunch follow individual patterns as the teacher just grab a bite in the classroom or at the workstation. Collective rhythms are diminishing and this makes self-management of time important. At the same time the teachers can't manage their own time because they are dependent on peers work. ICT set the teachers free of time and space but without synchronization with colleagues they can't make the necessary things happening for the pupils' learning process. This creates the need for longer working hours and flexibility. The teachers can only coordinate their work by setting up meetings in the late afternoons, evenings and weekends. Likewise this need for coordination creates a culture where it is an ideal to be accessible around the clock and everything is very hectic because of the huge amount of social interactions. This ideal about being accessible is not limited to colleagues but also to parents and pupils. It seems like a paradoxical work organization that work is increasingly individualized with self-management of time when the teachers are becoming more and more dependent on each other. The teacher's individual time strategies seem to fail all the time either because peers interrupt them or because they cannot access peers or management.

The teachers experience that different time systems are conflicting on a daily basis. Lack of control does not manifest as low decision latitude but as competing and conflicting time orders. It is a constant struggle to find time and space for the needed immersion regarding preparations and other tasks that demands a high level of concentration over a longer period. This struggle is most often moved from the school to the family. Hence the teachers report of massive work-family balance problems.

With regards to the teachers' time with the pupils, the quality of the time is in focus. It is not primarily the length of the workday, which is the focus, although extending the working day is one of many strategies to live with time conflicts. Time orders and quality time is closely connected to what is needed to facilitate successful lessons or projects with the pupils. Here are the collective standards of professionalism important. According to these standards all the norms leave insufficient time to solve the tasks. Additionally there are always some unpredicted events that need immediate attention e.g. conflicts with and between pupils, parents and sometimes even with management and peers.

DISCUSSION

Despite new management's enhanced efforts to provide individual wellbeing to employees, paradoxically many knowledge workers experience work related psychosocial illnesses. Since work is characterized by a high degree of autonomy, mainstream interpretations of causes for psychosocial illness is either employees' lack of mental capacity to cope with modern working life or lack of management quality. Hence new PSWE efforts and approaches strive towards the individual coaching of employees about how to develop coping strategies and methods and/or management development. Thereby the concept of control is excluded from the PSWE agenda. This is problematic since our study shows that the temporal orders of contemporary work erode employees' control.

The widespread use of autonomous teams and self-Management means that people must solve the dilemmas that arise when different time systems are conflicting. However, these strategies often contribute to an acceleration of pace and intensity as we have seen in our cases. This social acceleration and individualization of time management wear down the collective capacity of the workers. The PSWE in boundaryless work calls for interventions in the temporal orders of work moving towards creating collective rhythms in organisational time as a way to reach a more constructive collective capacity. The basic idea is that we should be able to control the temporalities of our productive relations. This can mean that the individual worker has to give up some aspects of freedom in work e.g. when and where colleagues and management are and can be available. In this way workers might be able to create new rhythms of work where there is a designated time for every purpose.

According to the D/C model control relates to the work tasks of the individual. However, based on the findings in our study, this understanding of control needs nuances, because actually control is primarily related to the association of different tasks and different persons in the sense that control is created in cooperation with others. That is very visible in boundaryless work, where the individuals can decide themselves when, where and to a certain extend how to do their work and solve their tasks. Lack of control within boundaryless work then emerges as a consequence of lack of time. Everybody are free to do their work as they prefer themselves, but they do not control the demands, and they do not control where their co-workers and managers are and what they are doing. That makes the workers experience, that they do not have time enough to do what they are expected to do. So therefore we have added 'Associational Control' on top of the D/C-model's individual control for the model to be applicable to study PSWE in boundaryless work.

CONCLUSION

We have studied boundaryless work as it emerges in teachers' work. The study shows, that the temporalities of work are possibly the main enemy of job control in boundaryless work. Lack of collective temporal order and rhythms erode workers' control in much of contemporary work. Studying PSWE in such work we have to be aware of the fact that a high degree of autonomy in use of time doesn't necessarily lead to employee control. In fact the degree of control is quite low in many of these so-called empowered jobs. Therefore the D/C-model is still very valuable for our understanding of the PSWE however the amount of social interactions in contemporary work control cannot be defined and studied as each individual's autonomy at work and opportunities for skills development. We must have a new understanding of control that transgress each individual's work, and look into associational aspects of the work.

By taking the concept time into our study we saw that the quality of time and the teachers' perception of time makes it possible to distinct between 'having control' and 'being in control'. The PSWE at the schools are marked by conflicts and confrontations between different time orders. As we saw in the schools the complexity of work, professional norms and many social interactions created an environment where there is never enough time. Where there is never time for breaks. Where time for immersion is threatened constantly. Where working hours are eating into leisure and family time. And where the solution of conflicts between the different time orders is left to be handled by the individual - even though individual solutions always will be inadequate. It is for example up to the individuals to find time for breaks. However, it is impossible for individuals to find the time because the collective norms say that each and everyone must be available all the time for coordination and planning. There are probably no perfect solutions to the various PSWE dilemmas but the need for creating a rhythmic relationship between work and regeneration and creating rhythms of work, which included formal and informal breaks, is obvious.

REFERENCES

Albertsen, K. A. Grimsmo, B. A. Sørensen, G. L. Rafnsdottir og K. Tómasson (2007). *Working time arrangements and social consequences – what do we know.* Temanord 2007:607. Copenhagen, Nordisk Ministerråd.

Alvin, M. (2008): New rules of work: exploring the boundaryless job. In Näswall, K., J. Hellgren & M. Sverke: *The Individual in the Changing Working life.* Cambridge.

Beck., U. (2000): *The brave new world of work.* translated by Patrick Camiller. Cambridge Polity Press, 2000

Belkic, K. L., P. A. Landsbergis, P. L. Schnall, & D. Baker (2004) Is job strain a major source of cardiovascular disease risk? *Scandinavian Journal of Work Environment & Health.* Vol. 30 (2): 85-128.

Birindelli, Lorenzo et. al. (2007) *The transformation of work? A quantitative evaluation of the shape of employment in Europe. Work organisation and restructuring in the knowledge society – WORKS project.* Project number: CIT13-CT-2005-006193

Burr, H., J. B. Bjørner, T. S. Kristensen, F. Tüchsen, E. Bach (2003) Trends in the Danish work environment in 1990-2000 and their associations with labor- force changes. *Scandinavian Journal of Work Environment and Health.* 29:270-9.

260

Castells, Manual (1996): The *Informative Age: Economy, Society and Culture Vol. 1 – The Rise of the Network Society*. Blackwell Publishers Ltd. Massachustts, USA.

Doef, V. D. M. & S. Maes (1999) The Job Demand-Control (-Support) Model and psychological well-being: a review of 20 years of empirical research. *Work & Stress*, Vol. 13 (2): 87-114.

Elgaard Jensen, T & Westenholz, A. (ed.) (2004) *Identity in the age of the new economy*. Edward Elgar Publishing.

Grönlund, A. (2007): Employee control in the era of flexibility - a stress buffer or a stress amplifier? *European Societies*, 9(3), 409-428.

Hirschhorn, Larry & Thomas Gilmore (1993): The New Boundaries of the 'Boundaryless Company'. *Harvard Business Reviews*.

Hvid, H. S. (2004) The Conductivity Model and Pragmatic Work-Change Programs. *Bulletin of Science, Technology & Society*, 24(5), 480-483.

Hvid, H., H. L. Lund & J. Pejtersen (2008) Control, flexibility and rhythms. *Scandinavian journal of working environment*. Suppl. 2008; (5).

Hvid, H., H. L. Lund, S. Grosen & H. Holt (2009): Associational control: Between self-management and standardisation in the financial sector. Forthcoming. *Economic and Industrial democracy*. Sage.

Hylland Eriksen, T. ([2001] 2007). *Øyeblikkets tyranny. Rask og langsom tid i informasjonsalderen*. Oslo, Aschehough pockets.

Kamp, A. og Lund, H. (2008). Time in Boundaryless Work. In: L.I. Sznelwar, F.L. Mascia & U.B. Montedo (eds.). *Human Factors in Organizational Design and Management IX*.Santa Monica, California, IEA Press.

Karasek, R. (1979) Job demands, job decision latitude, and mental strain: Implications for job redesign. *Administrative Science Quarterly*, 24, 285-307.

Kern, H. und M. Schumann (1985): *Das Ende der Arbeitsteilung? – Rationalisierung in der industriellen Produktion*, Bestandsaufnahme, Trendbestimmung. München, Beck

New York : Norton,

Nowotny, H. (1994). *Time. The modern and postmodern experience*. Cambridge, Polity Press.

Pejtersen, Jan & Tage S. Kristensen (2009) The development of the psychosocial work environment in Denmark from 1997 to 2005. *Scandinavian Journal of Work Environment and Health*. Vol 35 no 4.

Perlow, L. (1999). The time famine. *Administrative Science Quarterly*, 44, 57-81.

Piore, M. J. & C. F. Sabel (1984): *The second industrial divide - possibilities for prosperity*. New York. Basic Books, cop

Sandberg, Å., m.fl. (1992): *Technological Change and Co-Determination in Sweden*, Temple University Press

Sennett, R. (1998): *The corrosion of character - the personal consequences of work in the new capitalism*.

Urry, J. (2000). *Sociology beyond societies- mobilities for the twenty-first century*. London and New York, Routledge.

Ylijoki, O. & Mätylä, H. (2003): Conflicting Time Perspectives in Academic Work. *Time & Society* 12, (1), 55-78.

CHAPTER 28

Shaping Usability Engineering Methods to Fit the IT Applications Development Process

David C. Dunkle

Lockheed Martin Enterprise Business Services
Orlando, FL 32825-5002, USA

ABSTRACT

Within corporate Information Technology (IT) organizations, a wide gamut of user-centered design practices are routinely employed to help create usable software interfaces. However, just as often, valuable usability engineering methods are ignored during fast-paced application development cycles, often resulting in non-optimized or unsuitable user experiences following application deployment. This problem occurs even when time, funding and personnel are available to support application development.

Why are proven methods that save money, reduce development time, and improve software quality often ignored or discarded? Moreover, are there social engineering techniques that can be used to effect an increase in incorporation of user-centered design practices?

First, we identify a set of perspectives that routinely drive the mindset of rapid application development, and explain how these perspectives often lead to the devaluing of user-centered design activities and deliverables. We then identify deployable countermeasures designed to increase the visibility, validity and value of the activities and deliverables.

Pulling from principles of social ergonomics and relating these principles to real-life situations, we propose a set of interpersonal communication and presentation methods that have proven valuable in influencing the level of usability engineering activities included in application development schedules. These include methods of customer goal identification, project leadership control, procedural authority, and deliverable product designs that are effective in producing the critical support factors that lead to user-centered design incorporation.

Steps for blending methods of social influence into standard or existing user-centered design products are suggested. The inherent need for crafting user-centered design activities for optimal project development acceptance is explained, and the pitfalls that lead to common activity exclusion, as well as opportunities for activity inclusion, are discussed.

Keywords: IT organization, user-centered design, application development, social ergonomics, social engineering, customer goal identification, project leadership control, procedural authority.

INTRODUCTION

The "birth" of human factors as an engineering consideration can be traced to pilot errors and cockpit designs during World War II (Weiner & Nagel, 1988). From there, the practice of Human Factors Engineering has grown into a formalized discipline that is considered integral to most system development lifecycles. Usability engineering, on the other hand, is a more recent development, and while there are now a vast number of practitioners, methods, and deliverables, the incorporation of usability engineering into application development is not so established. One of the major achievements of early usability practitioners was simply the promotion of usability as a concern, and there is now a great deal of awareness and recognition of the term. However, a quick review of mainstream commercial applications and present-day business websites will confirm that usability has often taken a back seat in the development process to other concerns, such as novelty, personal preferences and time-to-market.

This is not to say that usability engineering activities are being prohibited during software application development. With few exceptions, nearly all corporate IT development organizations employ engineers for the sole purpose of focusing on application usability. A wide variety of usability engineering practices is routinely employed by these engineers, with the intent to help create usable software interfaces. However, just as often, valuable usability engineering results and recommendations are ignored during fast-paced application development cycles. This problem occurs even when time, funding and personnel are available to support usability engineering during application development.

It would seem counter-intuitive that the people entrusted with the development and deployment of an application would dismiss proven methods and recommendations that save money, reduce development time, and improve software quality (Constantine & Lockwood, 1999). Many usability engineers, in fact, will fall into the trap of regarding the dismissal of usability engineering results as illogical, and leave it at that. This does not change the fact that many applications continue to be deployed with substandard levels of usability, even while increasing amounts of usability engineering activities are performed upon them. Part of the reason for this paradox lies in the environment of modern IT organizations.

IT ORGANIZATIONAL ENVIRONMENT

Two critical environmental factors have arisen to shape perspectives and scenarios in software application development. The first is the interpersonal independence of the applications developer. Capable of working remotely as a telecommuter, and individually equipped with virtually every tool required, the developer is no longer bound by the traditional approaches of project or team management. Oversight and direction, including any sense of mentoring, has been replaced by intellectual and technological self-training by developers on their own. In short, a developer rarely needs to be told what to do anymore. Consequently, such an independent developer rarely *wants* to be told what to do. This desire is especially strong with regard to an applications user interface, arguably the most artistically creative portion of application development.

The second factor to arise is the changing processes, timelines and roles within applications development. The sequential structure and strict role/responsibilities of the "waterfall" development process have been replaced by modern processes that are more rapid and agile. Processes have been reordered into new sequences with the intent to reduce the time it takes to fully analyze or document a set of system requirements or build the system. Often, the applications developer is responsible for both the software development and the end-state design. Sometimes the design itself is documented after being developed, or a rough sense of what a stakeholder would like (identified through a series of daily communications) is used to guide development. Usability engineering and user-centered design activities are often replaced by iterative development, wish lists, and a fuzzy, abstract (if any) model of the application. The traditional role and placement of usability engineering can seem out of place in new application development processes.

CUSTOMER GOAL IDENTIFICATION

The good news is that the core of early usability engineering is still intact in agile and distributed development projects. That core is, identifying the stakeholder goals as well as the optimum environment for achieving those goals. This can be thought of as crafting the User Experience, or Customer Collaboration; in either case, this serves as the vehicle for performing, in typical usability engineering terms, a User and Task analysis. Note that even in traditional application development, these types of analyses have often become iterative, and for modern processes, they are now planned to be iterative.

One of the benefits of providing user characteristics as early design guidance stems from the fact that it is a collected, rather than a created source of information. Developers are typically more accepting of design guidance that comes from a stakeholder (which, after all, is the original source of this guidance) than from the heuristic analysis of a usability engineer. Therefore, user-centered design guidance, distilled via Customer Goal Identification, should always be the purview of the usability engineer.

PROJECT LEADERSHIP CONTROL

Applications development lifecycles have evolved over the years; however, some aspects of project management remain the same over time. Specifically, in most application development teams, even for Agile, SCRUM and/or distributed teams, there stands a team leader or project manager. That person is the one most responsible for the outcome of the development process. Responsibility, in this instance, does not suggest control. The leader or manager simply is the one who oversees the program, and represents the program outside of the development team. Since many (most?) leaders wish to share this responsibility, they will assign secondary leadership roles to other team members; in effect, establishing a cabinet-like set of leaders on which they rely and to which they assign responsibility.

Often, for a usability engineer to have real impact on an application, he or she must adopt a leadership role on the project. Ideally, this can be the System Engineer or Business Analyst role- in both cases, the engineer can focus on the user goals, requirements and business needs. Both roles are critical reports to the team leader, and both can serve as an open avenue to apply valuable usability engineering throughout development.

Note that one possible pitfall of this concept is for the usability engineer to be adopted as the "Lead Usability Engineer" of the project. This role is sometimes given a nebulous set of responsibilities, and limited authority. The best way out of this pitfall is to have a documented set of responsibilities and authority for the role, as described next.

PROCEDURAL AUTHORITY

Another factor relating to team leadership responsibility is the ongoing concern of process and procedural compliance. No serious development business survives long without a formal development process, built upon a documented set of development procedures. Most governmental applications are required to be developed according to a validated and proven process, and many businesses strive for certification of their capability to consistently follow a consistent process.

The usability engineer should then strive, as early as possible, to establish a set of usability procedures within the process. This task is not as hard as it may seem, for the documentation of process is typically not a sought-after task. Anyone who offers to perform it is usually encouraged to do so.

Documenting usability engineering methods within the development process will not hold the weight of law to force inclusion in user experience design, however, at a minimum, it easily answers the question, "Is this activity necessary?" and provides legitimacy across teams and projects for the usability engineering discipline.

PRODUCT DESIGN

One of the most distressing situations into which a usability engineer can place oneself, is a confusing presentation of valuable data due to substandard media. In essence, this is similar to the theme described by Alan Cooper in his book, *The Inmates are Running the Asylum* (2004). Cooper states that developers are the ones who have created the user interfaces for applications, but have done so through their developer eyes, building applications that only work for themselves. In other words, the inmate developers are running the application's user interface asylum.

Likewise, a usability engineer can be so entrenched in usability test data and interface layout analysis, that the resulting presentation makes sense only to them. The tragedy of this event cannot be overstated- it results in (a) valuable time being wasted, (b) good advice being wasted, (c) an engineer's reputation being tarnished, and (d) usability engineering taking a hit as a discipline. It is, however a familiar situation where organizational ergonomic factors conspire to discount the significance of usability engineering deliverables.

A usability engineer must put as much effort into the effective conveyance of user interface design concepts as is put into the development of user interfaces themselves. Thankfully, there is no better engineer for doing this. The principles of usability apply just as effectively to presentations and recommendations as they do to application interfaces.

Presentations of usability recommendations can follow a similar development lifecycle to applications. The usability engineer puts together a "prototype" presentation, building the style upon usability heuristics and graphic design principles. Once this is prototyped, the engineer then collects data on user perceptions and acceptance, via a dry-run presentation to fellow usability engineers, perhaps including a developer for a technical perspective. After refining the content, the engineer can iterate the presentation again, or move to the intended audience with the improved content. Following a productive briefing, effective presentations should be catalogued and archived to serve as models for later use.

PLAN FOR ACCEPTANCE

It's not as if usability engineers are working in the ether, attempting to describe or capture a nebulous or disembodied force or principle. User interfaces exist. If one is developed for an application, a process of user interface design does, by definition, take place. It is correct to say that user-centered design takes place as well; however, the design is often centered on the wrong user. The usability engineer is there to correct this.

Explaining this to a development team is difficult, but must be done to effect real improvements to application design. And it can go wrong for all the wrong reasons. A developer may feel protective about the user interface upon which countless hours have been spent. A Project Lead may not have a clear understanding of the role for a usability engineer. The user interface recommendations presented to the development team may be hard to decipher, and thus rejected.

None of these reasons have anything to do with the value of the usability engineer's contribution. Each of these situations exists outside the typical scenario laid out in application development lifecycles. But any experienced usability engineer has a similar story; for some reason not related to their recommendations themselves, their contribution was not accepted. We hope the methods described here can mitigate some of those situations in the future.

CONCLUSION

In is important to note that the methods and philosophies described herein do not address the actual quality (outside of the look-and-feel) of various usability engineering deliverables. The focus is solely on achieving a level of acceptance, given that the usability report has something to offer. Usability engineering methods are only as valuable as their acceptance and application to design. Highlighting the relevant content from a thorough usability evaluation after an application has been fielded, or following product deployment, brings little satisfaction to the usability engineer, and offers nothing but an "I told you so" to a development team that is, in all likelihood, already disassembled and facing a new challenging deployment schedule. Thus, the time for valuable application of usability engineering is throughout the development lifecycle.

Here we have described some methods for promoting value within the development lifecycle, in manners typically not considered by the usability engineer. In fact, there may be hesitation to employ these methods, by some who feel that they amount to a kind of dilution to the rigor of experimental research or scientific engineering that usability engineers seek.

The reality of situational and environmental factors opposing usability incorporation reveals that often, classical research and textbook engineering are not enough to achieve the final goal of the usability engineer; that of application usability. Using these methods can offer usability engineering the voice needed as a means toward that end.

REFERENCES

Weiner, Earl and Nagel, David eds. (1988). *Human Factors in Aviation*. California: Academic Press.

Constantine, L. and Lockwood, L. (1999). *Software for Use: A Practical Guide to the Models and Methods of Usage-Centered Design.* New York: ACM Press

Cooper, A. (2004). *The Inmates Are Running the Asylum: Why High-Tech Products Drive Us Crazy and How to Restore the Sanity.* Indianapolis: SAMS Publishing.

CHAPTER 29

Safety Climate in Truck Transportation: Field Studies of Their Levels Correlated with Safety Outcomes

Ya Li, Kenji Itoh

Department of Industrial Engineering and Management
Graduate School of Decision Science and Technology
Tokyo Institute of Technology
2-12-1 Oh-okayama Meguro-ku, Tokyo 152-8552, Japan

ABSTRACT

The present paper reports results of a questionnaire-based survey on safety climate in truck transportation organizations. Responses were collected from approximately 1000 drivers of 49 truck transportation companies in Japan. To elicit dimensions of safety climate, principal component analysis was applied to the entire sample of the questionnaire responses. The analysis yielded nine safety climate factors with 46% of cumulative variance accounted for. There were significant differences in several dimensions of safety climate between employment conditions, e.g., regular employees vs. non-regular drivers, and between working types such as differences in driving areas or distances.

We made correlation analysis of each dimension of safety climate with safety outcome by use of actual data of annual incident statistics collected from ten companies which participated in the survey. Significant correlations were observed for several dimensions with the rate of injured incidents. The results may suggest that the more positive overall safety climate an organization has the lower an accident risk becomes on the road. It is also implied that drivers' risk awareness and team-oriented views positively contribute to safety level in truck transportation.

Keywords: Safety climate, Safety outcome, Truck transportation, Traffic safety, Risk prevention.

INTRODUCTION

The number of fatalities by traffic accidents has been decreased about half during the last two decades in Japan, i.e., 9,211 and 5,155 persons in 1988 and 2008, respectively (ITARDA, 2008). In terms of the fatal accidents per 10,000 motor vehicles, commercial vehicles have caused accidents three times higher than private automobiles for the last ten years (IATSS, 2010). It is also suggested that ordinary drivers (non-professional drivers) seem to view truck presence on the road as a source of potential danger and it causes other drivers to have a more responsible attitude toward driving (Rosenbloom, 2009). Therefore, it is of critical importance to tackle with uncovering safety climate as well as safety attitudes of professional truck drivers to achieve higher level of traffic safety. There has been a great number of studies to investigate or measure safety culture or climate in various domains of human-machine operation domains such as aviation (e.g., Helmreich and Merritt, 1998) ship handling (Itoh and Andersen, 1999) and railways (Itoh et al., 2003), and health care (Itoh et al., 2006). However, only a few studies have systematically explored safety culture/climate in truck transportation.

With the growing importance of safety climate for traffic safety, as mentioned above, we conducted a questionnaire-based survey on this issue to explore its characteristics in Japanese truck transportation organizations. We elicited dimensions of safety climate for measuring and diagnosing its level in a specific company. To emphasize the proactive use of safety climate measurement, we examine a linkage of specific dimensions of safety climate with safety outcomes by use of actual safety records which were collected from ten companies surveyed. We also discuss some implications for improving traffic safety on Japanese roads based on the survey results.

QUESTIONNAIRE AND INCIDENT STATISTICS

QUESTIONNAIRE AND RESPONDENTS

The questionnaire used in this study was developed in Japanese with nearly half of the question items adapted from the questionnaire surveyed for train operators' attitudes to safety culture related issues (Itoh et al., 2003; 2004). This questionnaire, i.e., TMAQ, was originally based on Helmreich's FMAQ (Helmreich and Merritt, 1998) and its derivative, the SMAQ (Flight [Ship, Train] Management Attitudes Questionnaire; Itoh and Andersen, 1999). It was then translated into English. The truck transportation version of questionnaire comprised two main sections, one for driver satisfaction and the other for safety culture, as well as a demographic section and three open-ended questions relating to traffic safety. In the present paper, we focus on Section 2 which was designed to elicit dimensions of safety climate, having fifty-three statements of respondent attitudes to and recognition of general issues concerning safety such as training, fatigue, work pressure, organization, work and management. Respondents were asked to rate their agreement or disagreement with each statement on a five-point Likert scale, ranging from "strongly disagree" to "strongly agree".

The survey was made confidentially between Dec. 2008 and May 2009. Forty-nine transportation companies in Japan – mostly located in Tokyo and its near regions – accepted the invitation to participate in the survey. A total of 1901 questionnaires were distributed to these companies, and we asked a contact person in each company to send an envelope with a return stamp including a questionnaire to every driver. After respondents filled out in the questionnaire, responses were posted and returned directly to the researchers. The number of responses returned from drivers is shown in Table 1 based on age groups and employment conditions. We collected a total of 1028 responses from truck drivers (yielding an overall response rate of 54%). 78% of drivers were regular employees who were employed by the own company, while only 22% of the drivers were contracted employees or other types of non-regular employees. In our sample collected, the majority of respondents were 30's and 40's drivers, and a mean working duration for the current company was 7.8 years with averaged total experience of 14.1 years as a professional driver.

Table 1: A sample collected from 49 truck transportation companies.

Age class	Regular employees	Contracted employees	Employed by temporary staff agency	Part time workers	Other type of drivers	Total	
20's	92	13	1	1	0	107	10%
30's	296	55	3	4	2	360	35%
40's	266	47	0	8	1	322	31%
50's	122	22	2	3	3	152	15%
60's	23	19	0	8	2	52	5%
NA	2	0	0	0	33	35	3%
Total	801	156	6	24	41	1028	54%
	78%	15%	1%	2%	4%		

INCIDENT DATA

Annual incident statistics over recent five years were also obtained by 21 of the companies surveyed. We collected the incident data from 2004 to 2008 in terms of the number of injured incidents, that of non-injured incidents, and total damage (money lost calculated by Japanese yen). None of these companies made a fatal accident during these five years. We also received the annual data on the number of drivers and total driving distance from each company to calculate an incident rate of each event type per driver or million km driving distance. Table 2 summarizes mean incident rate of each event type and lost money as damage per driver in each year and its mean rate over the five years.

Table 2: Averaged annual incident rates and total damage over 21 companies.

Incident indicator	2004	2005	2006	2007	2008	Mean
Injured incident	0.030	0.021	0.021	0.026	0.016	0.022
Non-injured incident	0.130	0.190	0.234	0.223	0.197	0.201
Total incident	0.160	0.211	0.254	0.250	0.213	0.224
Total damage (JPY)	87,381	60,585	68,550	77,017	84,232	74,885

(Number of incidents per driver per year)

Mean: Mean rate across the five years.

SAFETY CLIMATE IN TRUCK TRANSPORTATION

Nine dimensions of safety climate were elicited by applying principal component analysis with the Varimax rotation to the entire sample of Section 2 responses with 46% of the cumulative variance accounted for. The analysis results are summarized in Tables 3 and 4. Eigenvalues for all the nine components were higher than 1.0. Cronbach's alpha were ranged between 0.38 and 0.92 for the nine dimensions, seven of which were higher than 0.5. As for the first component, which explained 15% of variance, factor loadings of the following items, which are related to driver attitudes to and perceptions of safety issues and organizational management in general, were particularly high: "my suggestions about safety will be acted upon if I express them to my seniors", "the drivers in this company are well informed and understood the company's policies of safety management", "leaders in the company or the distribution center listen to drivers and cares about our concerns", "the company deals constructively with problem drivers" and "leaders well understand relevant operational intentions and actions relating to driving tasks" (cf. Table 2). Accordingly, we interpreted the first principal component as a dimension of "overall climate of safety". In this manner, all nine dimensions derived by the principal component analysis were labeled as follows: (1) overall safety climate, (2) communication and coordination, (3) competence-based view, (4) recognition of human factors, (5) team-oriented view, (6) safety attitudes, (7) recognition of personal problems to safety, (8) risk awareness, and (9) motivation.

We classified driver responses to each safety climate dimension into five groups, from "very negative" to "very positive" based on a mean score of its component items. Percentage of driver respondents who perceived as each of five "negative"-"positive" classes for each dimension of safety climate is depicted in Figure 1. As can be seen in this figure, there were not many drivers who perceived overall safety climate, i.e., Dimension 1, positively, nor those who did negatively, whereas about a half of Japanese drivers viewed it rather as neutral. Regarding other dimensions of safety climate, greater than 80% of respondents well recognized importance of communication and coordination during work, which has also been acknowledged positively in other domains, e.g., healthcare (Itoh and Andersen, 2008) and railway (Itoh et al., 2004). Their safety attitudes and risk awareness were also reasonably high – 63% and 80% of respondents exhibited strong or slight agreement with these factors, respectively. There were not many drivers who had positive agreement with competence-based view, but a greater number of drivers expressed disagreement.

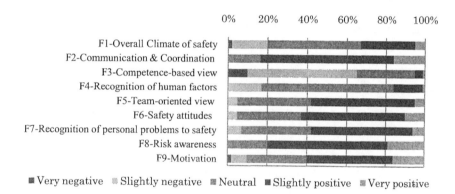

Figure 1. Truck drivers' responses to safety climate dimensions.

Table 3: Safety climate dimensions elicited by principal component analysis (1/2).

SC Factors [Var.(Cumul.)] Cronbach's α	Items highly loaded	Loading
I. Overall Climate of safety 15.2% (15.2%) α = 0.916	• My suggestions about safety will be acted upon if I express them to my seniors.	0.742
	• The drivers in this company are well informed and understand company's policies of safety management.	0.731
	• Leaders in the company or the distribution centre listen to drivers and cares about our concerns.	0.730
	• The company deals constructively with problem drivers.	0.699
	• Leaders well understand relevant operational intentions and actions relating to driving tasks.	0.697
	• Morale in the company is good.	0.695
	• Mistakes are handled well in the distribution centre	0.683
	• Colleagues are well trained in emergency procedures.	0.679
	• Drivers are well trained to cope with fatigue.	0.676
	• The distribution center provides adequate, timely information about events that might affect my work.	0.675
	• When I start an everyday driving task, I always receive a proper hand-over if necessary.	0.669
	• The concept of all personnel working as a team does not work in our distribution center. (-)	-0.668
	• This company practices the highest maintenance standards of vehicles.	0.595
	• I am encouraged by my seniors to report any unsafe conditions I may observe.	0.516
II. Communication & Coordination 6.1% (21.3%) α = 0.716	• I am normally consulted on matters that affect the performance of my duties.	0.683
	• A debriefing and critique of procedures and decisions after critical situations is an important part of safety.	0.645
	• Open, frank discussion is helpful for maintaining safety.	0.579
	• Effective job performance requires employees to take into account the personalities of other drivers and staff.	0.549
	• We should be aware of and sensitive to the personal problems of other drivers.	0.451
	• Drivers should report their own psychological stress or physical problem to the leaders before or during a job.	0.409
	• It is important that my competence be acknowledged by others in the distribution center.	0.374
III. Competence-based view (-) 5.3% (26.6%) α = 0.581	• Errors are a sign of incompetence.	0.639
	• Asking for assistance makes one appear incompetent.	0.595
	• Junior drivers should not question decisions made by leaders in the distribution center.	0.512
	• Leaders and senior staff who encourage suggestions from drivers are weak.	0.506
	• It makes no difference to me which company I work for.	0.455
	• As long as work gets done, I don't care what others think of me.	0.396
IV. Recognition of human factors 3.8% (30.4%) α = 0.513	• I am more likely to make errors in an emergency.	0.648
	• My decision making ability is as good in emergencies as in routine conditions. (-)	-0.632
	• I am less effective when stressed or fatigued.	0.609
	• Even when fatigued, I perform effectively while driving. (-)	-0.408
	• Personal problems can adversely affect my performance.	0.378

Table 4: Safety climate factors elicited by principal component analysis (2/2).

SC Factors [Var.(Cumul.)] Cronbach's α	Items highly loaded	Loading
V. Team-oriented view 3.6% (34.0%) α = 0.526	• I value compliments about my work. • As long as work gets done, I don't care what others think of me. (-) • Senior drivers deserve extra benefits and privileges. • I try to be a person that others will enjoy working with. • It is important that my competence be acknowledged by others in the distribution center. • Socializing among the drivers can help reduce stress.	0.574 -0.503 0.500 0.458 0.348 0.312
VI. Safety attitudes 3.2% (37.2%) α = 0.468	• I stop driving and take a rest when my workload is becoming (or about to become) excessive. • I always ask when I feel something I don't understand. • If I perceive someone's error, I will report it to a leader or senior staff, regardless of who might be affected. • Drivers should report their own psychological stress or physical problem to the leaders before or during a job.	0.616 0.612 0.536 0.266
VII. Recognition of personal problems to safety 3.0% (40.2%) α = 0.377	• Colleagues sometimes use mobile phone while driving. • Personal problems can adversely affect my performance. • A truly professional driver can leave personal problems behind when performing a driving activity. (-) • More attention should be paid to sleeping possibilities in a track during break or waiting for next task.	0.596 0.449 -0.425 0.302
VIII. Risk awareness 2.9% (43.1%) α = 0.511	• Alcohol presents safety problem in my company. • Accidents and near-misses are always reported according to company orders. • I stop driving and take a rest when my workload is becoming (or about to become) excessive. • This company practices the highest maintenance standards of vehicles. • Drivers should report their own psychological stress or physical problem to the leaders before or during a job.	0.644 0.440 0.398 0.281 0.244
IX. Motivation 2.8% (45.8%) α = 0.742	• I like my job. • Working for this company is like part of large family. • I am proud to work for this company.	0.566 0.459 0.423

This may be a positive sign of safety climate since too strong atmosphere of the competence-based are likely to lead to blame culture or unrealistic recognition of safety-related issues such as strong claim of "errors are a sign of incompetence" and "junior drivers should not question decisions made by leaders". Truck drivers' motivation was also high, e.g., 77% of truck respondents reported "I like my job".

In contrast, effects of human factors (including stress and workload) on their own performance were not recognized enough by truck drivers, although this trend was shared by other sectors such as railway and healthcare in Japan (e.g., Itoh and Andersen, 2008). Only 15% of truck drivers recognized human factors effects in slightly realistic on their own performance.

Regarding differences by respondent attributes in their perceptions of or attitudes to dimensions of safety climate we applied statistical tests, i.e., Mann-Whitney or Kruskal-Wallis test to examine significant differences between groups of each attribute. There were significant differences between regular employees and the other employment types, i.e., non-regular employees, for several dimensions such as communication and coordination (Dimension II; $p<0.05$), competence-based view (III; $p<0.01$), recognition of human factors (IV; $p<0.01$), Team oriented view (V; $p<0.05$) and recognition of personal problems (VII; $p<0.001$). As we expected, regular employees' views were more positive to these dimensions of safety climate than those of non-regular drivers.

Significant differences were also observed between groups of any other attributes for one or more dimensions, e.g., Dimensions III, IV, VII and IX between under and over 40 year old groups – in general the younger driver group had h more positive views and attitudes than the elder, except for motivation – and Dimensions VI through IX between respondents driving within 200km area and those driving longer than 200km distance – the long distance driver group was likely to view safety climate more negative than the short range group. As for an effect of working experience, drivers who have worked for the current company for shorter than five years exhibited significantly more positive views and attitudes to overall climate of safety than those the longer experience group ($p<0.05$). As shown in Figure 2, drivers who started to work for the current company within 12 months had the most positive attitudes to overall climate of safety. With increasing drivers' working experience, their views of overall safety climate may be perceived more negatively.

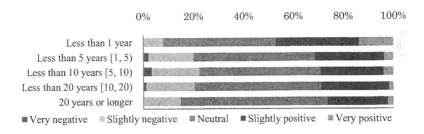

Figure 2. Overall climate of safety based on working experience for the current company.

EFFECTS ON SAFETY OUTCOME

In this section, we discuss correlations between each dimension of safety climate and safety outcomes, using the actual incident statistics which was mentioned in the previous section. To examine the relationships, we made a rank-based correlation analysis by applying Spearman's rho to the sample of ten companies, which were selected for a statistical reason, i.e., collecting ten or more questionnaire responses. Spearman's rho is shown with each dimension of safety climate in Table 5 when applying to the mean rate of injured incidents over the last five years. A significant, negative correlation was also observed with overall safety climate when using the mean total damage (lost money) as a safety outcome measure. This indicates that the more positive an overall safety climate becomes in a company, the less amount of money can be incurred by the damage of incidents or accidents.

As can be seen in Table 5, the following three dimensions of safety climate yielded significantly negative correlations with the mean rate of injured incidents: overall safety climate (Spearman's $\rho = -0.709$; $p<0.01$), team-oriented view ($\rho = -0.830$; $p<0.01$) and risk awareness ($\rho = -0.806$; $p<0.01$).

Table 5: Results of correlation analysis between safety climate dimensions and mean rate of injured incidents per driver per year (figures: Spearman's rho).

Safety Climate dimensions	Injured Inci./driver/year
I. Overall Climate of safety	-0.709*
II. Communication & Coordination	-0.394
III. Competence-based view	0.024
IV. Recognition of human factors	0.042
V. Team-oriented view	-0.830**
VI. Safety attitudes	-0.200
VII. Recognition of personal problems to safety	0.292
VIII. Risk awareness	-0.806**
IX. Motivation	-0.552

*: $p<0.05$; **: $p<0.01$; ***: $p<0.001$

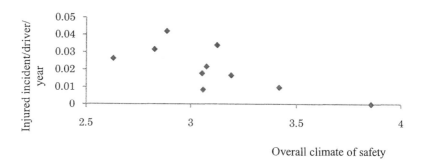

Figure 3. Correlation of overall climate of safety with rate of injured incidents.

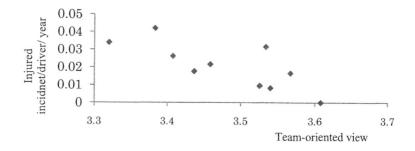

Figure 4. Correlation of team-oriented view with rate of injured incidents.

The correlations with other two dimensions are also depicted in Figures 3 and 4, plotting each company's data on the geometric plane of mean score of items relevant to a particular dimension and the mean rate of injured incidents per driver per year (average over the five years). From these results, it may be suggested that more positive trend of overall safety climate, team-oriented views and higher level of risk awareness contribute to reduction of incident rate on road traffic, and it turns to lower risk of traffic accident.

CONCLUSION

This paper primarily reported characteristics of safety climate in Japanese truck transportation organizations from results of the questionnaire-based survey and the correlation analysis with the incident statistics. The major outcomes of this investigation were, first, that we elicited a construct of safety climate which can be applied to measuring and diagnosing its present level by nine dimensions: (1) overall safety climate, (2) communication and coordination, (3) competence-based view, (4) recognition of human factors, (5) team-oriented view, (6) safety attitudes, (7) recognition of personal problems to safety, (8) risk awareness, and (9) motivation.

Secondly, we identified correlations of safety climate with safety outcome, using the actual incident data for the last five years. In particular, the following dimensions of safety climate were observed correlations in significantly negative with the rate of injured incidents: overall safety climate, team-oriented views, and risk awareness. Based on these correlations, continuous measurement of safety climate and its proactive use can be suggested as a promising management tool for risk prevention in truck transportation organizations.

In the questionnaire used in this study, fifteen items about driver satisfaction were included in Section 1. Applying similar analysis to these responses, we also elicited a four dimension construct about driver satisfaction, i.e., overall satisfaction with work and organization, satisfaction with own competence, satisfaction with training and satisfaction with vehicle. In addition, like several dimensions of safety climate, it was also found that truck drivers' satisfaction with training and with vehicle were negatively correlated with the rate of injured incidents. Also, there were significantly positive correlations between several satisfaction and safety climate dimensions, e.g., between overall satisfaction with work and organization and overall safety climate. Based on these results, we would suggest that the truck transportation's management should carefully consider training courses to improve the driver motivation and satisfaction – which also contribute to enhancement of their skills – as well as maintaining vehicles which they operate in daily work. These are expected not only to establish a good safety structure, but also to lead to positive safety climate in the organization for traffic safety, and it in turn to actual outcome of traffic safety.

REFERENCES

Helmreich, R. L. and Merritt, A. C., ed. (1998), *Culture at Work in Aviation and Medicine: National, Organizational and Professional Influencesseries.* Aldershot, England: Ashgate Publishing Limited.

IATSS (February 24, 2010), *Statistics Road Accidents Japan* (1997-2007). Tokyo, Japan: International Association of Traffic and Safety Sciences Website: http://iatss.or.jp/english/statistics/statistics.html

ITARDA (2008), *Koutsuu Toukei (Transportation statistics).* Tokyo: Institute for Traffic Accident Research and Data Analysis.

Itoh, K. and Andersen, H. B. (1999). "Motivation and morale of night train drivers correlated with accident rates." *Proceedings of the International Conference on computer-Aided Ergonomics and Safety,* Barcelona, Spain, May 1999 (CD ROM).

Itoh, K. and Andersen, H. B. (2008). "A national survey on healthcare safety culture in Japan: Analysis of 20,000 staff responses from 84 hospitals." *Proceedings of the International Conference on Healthcare Systems Ergonomics and Patient Safety, HEPS 2008,* Strasbourg, France, June 2008 (CD ROM).

Itoh, K., Andersen, H. B. and Madsen, M. D. (2006), "Safety culture in healthcare." In P. Carayon (ed.), *Handbook of Human Factors and Ergonomics in Healthcare and Patient Safety,* pp. 199-216, Mahwah, NJ: Lawrence Erlbaum Associates.

Itoh, K., Andersen, H. B. and Seki, M. (2004), "Track maintenance train operators' attitudes to job, organisation and management, and their correlation with accident/incident rate." *Cognition, Technology & Work,* 6(2), 63-78.

Itoh, K., Seki, M. and Andersen, H. B. (2003), "Approaches to transportation safety: Methods and case studies applying to track maintenance train operations." In E. Hollnagel (ed.), *Handbook of Cognitive Task Design,* pp. 603-632, Mahwah, NJ: Lawrence Erlbaum Associates.

Rosenbloom, T., Eldror, E. and Shahar, A. (2009), "Approaches of truck drivers and non-truck drivers toward reckless on-road behavior." *Accident Analysis and Prevention,* 41(4), 723-728.

Chapter 30

Usability Group Management Methods: How to Create a Cohesive, Innovative, and High-Performing Team

Leslie Tudor, Rajiv Ramarajan, Huifang Wang

SAS Institute
Cary, NC 27513, USA

ABSTRACT

Countless numbers of books have been written on the topic of management, but none have specifically addressed the challenge of managing a team of usability analysts. As one usability director and two usability managers of a single group of 16 analysts, we have iteratively refined our approach to management which has resulted in a team that is innovative, cohesive and high-performing. This poster will illustrate the various methods we employ. Specifically, our approach has focused on four initiatives: building and growing a unified team, encouraging customer outreach, fostering camaraderie and communication, and creating opportunities for innovation.

Keywords: Usability, Management, Usability Management, User Experience, Innovation. Communication

INTRODUCTION

The topic of managing for high performance is a popular one, and books and articles describing methods for creating successful teams abound. Many of these methods can be applied to the management of usability groups. However, due to the unique challenges that usability analysts face, the generic management literature is often lacking as a comprehensive approach for usability managers. Therefore, the three authors (one usability director at SAS Institute and the two usability managers who report to her) were motivated to create and refine methods that addressed needs unique to usability staff. Specifically, the challenges that staff faced were many and varied: increasing workloads, little time for exploring cutting edge technologies and design, lack of both funding and time for visiting customers, a tendency to work in silos, a demand for an ever wider expanding set of skills, and the expectation that designs would be creative and compelling but within the constraint of adhering to consistency guidelines. Though our analysts were going above and beyond - valiantly meeting deadlines, copiously producing designs with some even managing to patent and publish their work – we were concerned that this behavior would not be sustainable in the long run and that morale would suffer. Consequently, we decided to methodically tackle each issue.

ISSUES

The three authors meet with each other on a biweekly basis. During these meetings, we keep each other apprised of usability analyst activities, brainstorm solutions to issues, and exchange important company information relevant to our usability group. It was during these meetings that we decided to come up with a plan to address the aforementioned challenges that our usability analysts were facing. Over the course of several months, we had discussed issues as they came up but at some point we decided to address them in a more formal way by putting together a multi-pronged plan. When we looked across these issues, we were able to assign them to the categories of Building and Growing a Unified Team, Customer Outreach, Camaraderie/Communication, and Innovation. We focused on creating methods that would address each of these.

METHODS

1. Building and Growing a Unified Team

Temporary Assignments and Interns

As staff numbers have stayed relatively constant but project work has increased we have had to find ways to support additional products without placing undue burden on our analysts. One approach we have taken is to negotiate the temporary assignment of employees outside of our division. This works particularly well when the required support significantly impacts the perception or implementation of the UI – such as graphic arts or prototyping. Since temporary support staff are expected to work closely with UI analysts, we seek out employees who are not only available to do the work but who respect the role of usability and look forward to working closely with our staff. Most recently, to temporarily fill a graphic support role, we met with a number of graphic artists over lunch to inquire about their availability and their interest. When we found an excellent candidate, we contacted her management to convey our needs. The result was that she was able to provide us with support for several months.

We also employ summer and year round interns to fill resource needs. Each summer, we have had the opportunity to fill one or more internship slots with graduate students in Psychology, Computer Science, and Graphic Arts. Some of these students are able to stay on for the year if they attend schools in the local area. We always assign the student to a project as well as a usability mentor to work with. The support that these students have provided us with has been significant. In one case, a summer student, working in conjunction with one of our analysts, redesigned the interface for our internal defects application. A year-round student was able to create an online usability forum for one of our top business intelligence products. Other students have worked on designing slices of functionality for various products, or have worked on projects in our usability lab. We always treat interns as full members of our team, inviting them to all group meetings and group events, doing our best to make sure they feel fully integrated. When we are able to, we extend the offer of a full time position to our interns at the end of their internships. This has resulted in some excellent hires.

Project Collaboration

Due to the rapid pace of development and project deadlines, there has sometimes been an understandable tendency for analysts to be heads down, focusing exclusively on the needs of their product. To address this tendency, we set up regular meetings between groups of analysts whose products required a certain level of integration. We also hired an analyst to act as both project manager and designer in order to increase collaboration amongst analysts. The result has been a significant increase in consistency and intuitive UI integration across products.

Setting High Level Goals

As previously mentioned, there has sometimes been a natural tendency for analysts to focus on the needs of their products without working closely with analysts supporting products that are relevant to theirs. There have also been occasions were some product needs appeared to trump the needs of others from a UI perspective. We have found that setting high-level usability goals that impact all analysts has resulted in significantly more project collaboration. A good example of this is our consistency/reuse effort. When the effort was born, as it evolved, analysts began to work closely together due to the fact that they were co-designing components that they would all have to use.

Biweekly 1:1 Meetings

Biweekly 1:1 meetings are an important part of our process. One of the authors is a usability director who the other two authors, both usability managers, report to. Each manager has an individual biweekly meeting with the director. During these meetings, the director is apprised of the work that individual usability analysts are performing for their assigned project. Prior to the meeting, a document summarizing this information is sent to the director. It often contains links to prototypes and documents produced and authored by individual usability analysts. The director and her two managers all meet together on a biweekly basis as well. During these meetings, they discuss goals for the group, share important company information, and brainstorm appropriate group initiatives. These meetings are invaluable.

Balance of Skill Sets

When interviewing job candidates, we keep the balance of skill sets within our group in mind: research, UI design, aesthetics, and prototyping. It is always our desire to find a candidate who possesses skills in each of these areas and we often do. However, it is the rare candidate who is equally accomplished in all of these areas. Therefore, when we interview a candidate, we are always cognizant of the current deficiency in the group we need to address and seek to hire someone who possesses the greatest strength in that area. It is important to note that the balance of skill sets often changes over time as current analysts become more proficient in certain areas due to the needs of a project or to their own interests in growing their skills. Consequently, we are always reassessing the balance of skill sets in our group. Due to this approach, we have made what some might consider to be somewhat unorthodox hiring decisions. For example, we have hired candidates for UI positions whose strongest strengths were graphic arts and software programming. However, these decisions were good ones - not only have we filled a need in our group but we find that the mutual education that takes place between employees possessing different strengths has contributed to our overall success.

2. Customer Outreach

Our customer outreach initiative has grown significantly over the past few years. To grow this effort, we reached out to our sales teams. We explained that designing a product without meeting with end-users, observing them use our legacy products, brainstorming design ideas with them for both current and new products, and truly understanding their needs from a usability perspective is like designing in a black box. We emphasized that customer contact is part of a user-centered design (UCD) process and that designing products with user needs in mind would ultimately lead to higher sales and customer retention. To our delight, we found that many of our sales staff were very excited about our proposal for two reasons: they wanted to give us the opportunity to collect the data that we needed to design more intuitive product interfaces and they agreed that visiting with users would generate good will. The combination of better design and increased good will would presumably lead to more retention and new sales. With their help, we started small, visiting with more local customers. Sales then started funding trips to domestic customer sites that were further afield. As word of our success grew, country managers began to fund international site visits. We began to build a network of contacts amongst our sales staff and other field personnel to connect us to customers. When we presented customer contact data to development teams and management, colleagues saw the value of our efforts, and we saw acceptance and appreciation from them. Consequently, we now have our own budget for site visits.

The logistics of our current process are as follows: To initiate visits, the field will contact us to request that we visit with a specific customer, or the authors will contact any number of field personnel to inquire about accounts that use the products in our division. Contacting the field may also be initiated by analysts who need customer feedback before they can progress with work. Once we locate an appropriate customer, the authors will work with the field and the customer to review the purpose of a visit. After laying the groundwork and discussing mutual expectations, we will put the relevant usability analyst in contact with both the field employee and the customer. The authors will take care of funding logistics and often meet with analysts to discuss their pending trips, brainstorming ways to conduct the site visit so that both the customer and analyst get the most bang for the buck. During the course of the visit, the analysts often blog about their findings in order to keep their development teams informed. On the final day of the visit, analysts will often present their findings to the customers in order to validate their findings. Upon their return, analysts will present their findings to their development teams, write up a summary of their findings, and send this summary to the customer and field support. They will also post their findings to our group website where all trip reports and other artifacts collected during customer visits are archived.

3. Camaraderie and Communication

Group Events and Celebrations

We believe that camaraderie is an important component of group success and that employees who enjoy being together will communicate and work together more effectively. This is especially true for a team of UI analysts who each independently support their own products and projects. To give analysts the opportunity to spend time with one another, we have organized a variety of group trips that have included sporting events, restaurants, art museums, and local city tours. We also always make the time to celebrate personal events and have thrown numerous baby and wedding showers. When an analyst has a patent approved, a paper accepted for presentation or publication, or is promoted, we celebrate that as well. Group events are always well attended and analysts always look forward to attending future events.

Group SharePoint Site

We have always had a group SharePoint site but it was not until two years ago that we began to seriously attend to its layout and purpose. We initially conceived of it as one more way to establish our identity within the company and to educate others on usability. More recently, we decided to repurpose the site as a place where, as a group, we could post all of our work, information about our site visits, innovation efforts, and group trips. Our main site includes project relevant documents, and information about customer site visits (customers we visited, and resulting documents and blogs for each visit), usability research efforts, usability related events within and outside the company, intellectual property created by our team such as publications and patents, experimental and forward looking projects focused on innovative use of cutting edge technology, and usability related links and references. Sub-sites include those for our book club (which includes information on the books we are reading, chapter summaries, and blogs), Behavior and Interaction Seminar series (which includes links to webcasts, sites, demos, and notes from those meetings), the Usability Lab (which includes a calendar of usability tests past and future, test reports, survey findings, and related user research), our group blog, product consistency sites, product usability sites (independently maintained by the analysts who support them), and pictures from our various group activities. As a result of our efforts, the site is now seen as SAS Usability Communications Central and those outside of our group visit the site often to obtain information on various products and UI efforts.

Blogging

About 18 months ago, we began to encourage our analysts to create and post their blogs to the SAS internal blogging site. Blogging about usability issues is a great way to educate non-UI staff on important UI issues and to do so in a fun, informal way. The authors posted the first few blogs as a way to encourage analysts to do the same and also to provide them with some example blogs. After these initial entries were posted, SAS staff who supported the company's main intranet portal responded very positively, allowing all our future blogs to be posted to the main site, along with other top rated blogs. Some of our blogs have received a large number of comments and readership has been so high that in 2009 we were voted 6th in the Top 10 most popular group blogs in the company.

We also encourage our analysts to blog when they attend customer sites and participate in book club meetings. Blogging while visiting with customers serves several purposes: First, it keeps the analysts' development team in the loop so that they receive important information in a timely manner. Second, by blogging every day or sometimes several times a day, analysts make sure that they are capturing their insights and observations while they are still fresh. This also provides an opportunity for the development team to participate in interactions with users. Finally, the information that is blogged can be used as part of a larger customer site visit summary document that the analyst will create for both the development team and the customer/field personnel. As for book club blogging, the analyst leading the discussion often blogs about book chapter information and insights, and chapter-specific discussion points for the meeting.

Demo Days

About one year ago, we had our first usability day demo event. We conceptualized its purpose as an opportunity for usability to demo the products they had provided usability support to. With support from senior management, the event was held in the atrium of the building that houses much of the R&D staff. There were approximately 30 demo stations and attendance was very high throughout the 6 hour duration of the event. About 7 months later, we repeated this event, in honor of World Usability Day, this time with additional and more refined UI's. These two events were so successful that we plan to do at least one per year. It gives us a chance to educate employees outside of our projects about what we do, work on a large initiative together, and also gives project development staff a forum in which they can view each other's work and discover synergies that might lead them to collaborate in the future. These events are also high energy ones and result in a collective feeling of pride amongst all employees.

4. Innovation

New Technology and Design Exploration

It is important to keep current with trends in UI design but finding the time to simply explore and learn is challenging. To address this concern, we run an "Interaction and Behaviors" seminar series for the entire SAS usability community. The purpose of this hour long monthly seminar is to view examples of new cutting edge technologies so that we can observe and discuss their dominant design and interaction patterns. In the last year, we have met 14 times and have reviewed the user interfaces of Windows 7, Apple OS X Leopard, Office 2010, Tweetdeck, Augmented Reality concepts, Google Wave, and VoiceThread, to name a few. These meetings are often facilitated by the authors, who often select the topic and examples to be discussed, though anyone can suggest a topic and facilitate the discussion. These meetings are well attended and a fun way to learn. We intentionally keep the meetings informal and, aside from the facilitator, there is no pressure to prepare before the meeting. For those who do want to familiarize themselves with the topic ahead of time, the facilitator will send out the relevant materials. In addition to learning about the latest UI trends, usability analysts get to hear what their colleagues think of these trends and their relevance to SAS products. After each meeting, we create meeting notes, capture screenshots, and post these and other related materials to our group SharePoint site for those who missed the meeting or want to review the materials later.

Book Club

As part of our innovation effort, we run a book club that focuses on the topics of innovation, social media, and usability. The book club meets approximately once every 6-8 weeks and is open to usability analysts and those outside of our field who are interested in these topics. In the last year, we have met seven times. One of the authors will typically generate a list of relevant books and club members will vote on the book to be discussed at the subsequent meeting. Someone will invariably volunteer to lead the discussion or the authors will volunteer to fill this role. Club members are expected to read the selected book prior to the meeting though those who have not finished or even begun the book are encouraged to attend as well so that they may learn from listening to the discussion. Prior to the meeting, the person facilitating the discussion will often post chapter summaries and questions to our book club SharePoint blogging site. As for implementing the ideas that come out of the discussions, several initiatives were motivated by the knowledge acquired during these sessions including a forward looking social media work project, a new method for group UI design, and new approaches to organizing work tasks.

Forward Looking Project

In most companies, including ours, project management/marketing often generates ideas and requirements for new products. It is rare for a usability group to spearhead a new project and follow it through to fruition. As part of our innovation effort, we were inspired to do just this: We created a small forward looking project for social media. The idea was initiated by one of the authors and was refined by the knowledge acquired during the aforementioned book club. This author then met with product management to discuss the idea and received positive feedback. A small team of usability analysts with the appropriate skill sets was then recruited to further refine the concept generation and to build a prototype. After creating the prototype, the team presented it to a variety of executive employees, including R&D leadership, as well as our social media division and actual prospective end-users. We are now exploring the technical aspects of implementation and talking with development teams.

Design Competitions

Earlier in the year, we decided to encourage analysts to submit their work to internationally acclaimed design competitions. We even aligned this effort with our department goals for the year. This allowed us to address a number of issues. First, it would give analysts whose primary skills lie in aesthetics and design a well defined goal to work towards. Second, it would provide a platform for the analysts to participate with other highly skilled designers and so encourage excellence in their work. Finally, we expect that any recognition resulting from this effort will benefit the development teams and the company as a whole. We have currently outlined a few concepts which we are scheduled to enter into two major competitions relating to design and communication.

Growing Skills

We have extended our company's bi-yearly performance management process by implementing a series of stretch goals that are tailored to the dominant skill sets and/or interests of our analysts. These stretch goals are closely aligned with the broader goals that were defined for our entire department. Initially, we used a uniform set of stretch goals for all of our analysts which included tasks that pertained to the UCD process, as well as other tasks such as publishing and patenting research. We did this primarily to ensure that our analysts were including end-users in their design efforts, and that they were also growing their skill sets. We offer these goals as an a la carte menu, with both required items as well as those that can be selected from each of 3 groups of related goals. Over time, we expanded this menu to include some items that would appeal to analysts with excellent graphic design and prototyping skills. Our goal is to make the stretch goals fun and rewarding, providing analysts with a sense of accomplishment at the end of the year.

CONCLUSIONS

Managing a usability group can be a challenging yet extremely rewarding and enjoyable effort. The methods we have used in our approach to management have resulted in a group that is successful with regard to its output, and the quality of work. There is also a sense of unity and good will in our group that, we feel, has a positive impact on the daily interactions between team members and their respective successes. Further, information sharing and collaboration across usability teams has improved and so has the communications with development teams and product management. We are currently brainstorming the generation of new methods as well as ways in which to refine current ones. We have included our analysts in these discussions and have found their input invaluable.

Chapter 31

Evolving Problem-Solving Performance by Mixed-Culture Dyadic Teams

Rik Warren

Air Force Research Laboratory
711 Human Performance Wing / RHXB
Wright-Patterson AFB OH 45433-7022, USA

ABSTRACT

Disaster relief teams may be formed quickly with previously unacquainted members coming from diverse cultures. The effectiveness of such newly-formed mixed-culture teams in solving complex problems depends, in part, on how well its members communicate, cooperate, and share information. These are all factors which can be affected by cultural differences, but which also depend on the evolving history of the team and the accumulation of effects. The purpose of the study described here was to assess the effects of cultural composition on evolving problem-solving effectiveness using a role-play game in which two teammates are asked to solve a series of puzzles necessitating communication and cooperation for success. The results for 27 mixed- and same-nationality dyads are not monotonically increasing or decreasing functions of time, but rather show more complex patterns. Although the magnitudes of the correlations of nationality-mixture type with performance increase with team-experience, the directions of the correlations are not always in the expected directions. Considerations for using role-play games in research on cultural effects on teamwork are discussed.

Keywords: Culture, Teams, Cooperative Problem Solving, Role-Play Games

INTRODUCTION

In the aftermath of natural disasters, relief teams may be formed quickly with previously unacquainted members coming from diverse cultures. The effectiveness of such newly-formed mixed-culture teams in solving complex problems depends, in part, on how well its members communicate, cooperate, and share information. These are all factors which can be affected by cultural differences, but which also depend on the evolving history of the team. That is, the level of cooperation and communication within a diverse team is not static, but changes over time. There might be some initial level of good-will or caution based on cultural stereotypes held by the members, but these can be expected to change, for better or worse, as the members gain experience with each other.

Not all same-culture teams will work smoothly, nor will all different-culture teams have difficulty cooperating, of course. But communication and other differences between mixed-culture teams can be expected to have cumulative effects. At one extreme, early conflicts due to differences in interaction style and difficulties in communication might lead to a progressive deterioration in cooperation and thus to less and less effectiveness as time goes on.

The purpose of the study described here was to assess the effects of cultural composition on evolving problem-solving effectiveness using a role-play game in which two teammates are asked to solve a series of puzzles. The tasks of the puzzles are such that they are impossible to accomplish by one person, but which are easy for two people who communicate and cooperate well.

Role-play games are particularly well-suited for team research due to their immersive quality: the players become so absorbed in the game that their "true" selves can emerge. However, the analysis of the complex data-sets is not always straightforward.

For example, just what does "cumulative effects" or "progressive deterioration" mean in terms of puzzle solution times? Assume there is a way to equate puzzle difficulty or to compare performance across puzzles: Does a cumulative effect mean that a dyad's progressively cooperative performance corresponds to the upper curve in Figure 1 and progressively uncooperative performance to the lower curve? This was my original expectation. The data, however, appear to be saying that a better story is told by (a curve of) the successive correlations between a dyad's composition and each puzzle in a series. That is, the correlation between dyadic characterization and performance on the first puzzle should be zero, and the *magnitude* of the correlation for successive puzzles should increase. Further, the curve need not be smooth, only monotonic: the change in effectiveness after a particularly satisfactory or frustrating episode could be considerable.

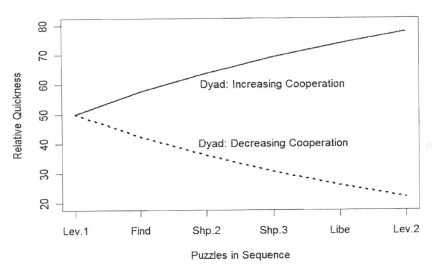

Figure 1. Predicted puzzle-solving effectiveness of an increasingly more cooperative team (upper curve) and uncooperative team (lower curve).

METHOD

THE ROLE-PLAY GAME & PUZZLES

The role-play game *Neverwinter Nights*[TM] (Bioware, 2004) provides a scripting language for scenario development and affords access to almost all game parameters thus enabling extensive data-collection for post-processing. In general, each player (here, two) has an avatar in the game space. The avatar can interact with other human-controlled avatars or scripted non-player characters (NPC's). The NPC's, for example, can be instructors during a training phase or characters in the game-space. Human-to-human communication can be either by voice or by keyboard chatting. Human-to-NPC communication is by keyboard or menu selection. All custom scenarios and game-play and data-recording functions in this study were developed by BBN Technologies using the Situation Authorable Behavior Research Environment (SABRE) (Warren, Sutton, Diller, Ferguson, & Leung, 2004; Leung, Diller, & Ferguson, 2005; Warren, Diller, Leung, Ferguson, &

Sutton, 2005).

Scenario target completion times varied from 10 to 20 minutes each with an extra 5 to 10 minutes of overtime play permitted. All tasks required extensive communication of context information and cooperative action for success. There were three training and six puzzle scenarios:

- An initial individual training scenario familiarized the players with the skills they would need to, for example, move in the game-space, open doors, pick up objects, and interact with NPC's.

- Two puzzles (lev.1 & lev.2) required the avatars to traverse a maze of rooms to reach a goal. The (often multiple) doors in the rooms were locked such that a player could not pass unless the other player pulled a lever. This task required taking turns helping each other.

- A scavenger hunt (find): one player has the list, each is charged with maximizing their own points, and neither can carry all the items due to weight restrictions. Success entails both cooperation and competition.

- In four puzzles, players balanced a ship by moving a puddle of water to the middle of a room by distributing weights around a large area. The first two puzzles (shp.0 and .1) were mid-game training. In the third puzzle (shp.2), the balance-indicator was in a separate room with an open door and was accessible by both players. In the fourth puzzle (shp.3), one player was locked in the puddle room and the other player free to roam the rest of the deck. Neither could see the other. They had to devise ways to communicate about the balance indicator and how to move weights.

- A "library" puzzle (lib) in which the players must exit an elaborate mansion. One player is locked in a room with books containing information about where things can be found and how to get past obstacles. The other player is free to roam about the the mansion.

Scenarios were always presented in the identical order for all dyads: pre-puzzle training, lev.1, find, shp.0, shp.1, shp.2, shp.3, lib, and lev.2.

OVERALL PROCEDURE

A session lasted two to three hours. In the beginning, the dyad members were briefed on the study and signed informed consent forms. Each player sat at a computer workstation such that they could not see the other player nor the other player's monitor. Using their computers, the players filled out several pre-play questionnaires. They then were introduced to the game and received individual training, at their own pace, guided by an in-game NPC. After both finished training, the dyad worked through all the puzzles in the same sequence. All communication between the players was verbal with the players wearing a microphone and headset. Keyboards were used to control avatar movement and to communicate with the NPC's. A short post-game questionnaire followed the puzzles.

PARTICIPANTS & DYADS

Fifty-four participants were recruited at a university and paid $8 per hour for a two to three hour session. Based on order of availability, 27 dyads were formed and run one dyad at a time. Cultural and other demographic information was not known prior to participants appearance at the test-site, the experiment explained, and informed consent obtained. In fact, no demographic data were processed until several weeks after a dyad's data were collected. This procedure resulted in an unbalanced distribution of cultural and demographic characteristics within and between dyads. In 17 dyads, the members came from the same nation (although the nations could differ); in 5 dyads, the members were from different nations; and in 5 dyads, the members were from the same nation, but one or two identified themselves with a hyphenated-nationality. This permitted scoring a 'dyad.mixture' variable with a '1' for different, '0' for same, and '.5' for partially-mixed nationality. In the 17 same-nation dyads, 5 nations were represented. Altogether, 11 single nations and 7 "hyphenated" nations were represented.

QUESTIONNAIRES, PRE- & POST-GAME

Pre-game questionnaires were demographics questions, cultural scales, and questions about English proficiency and computer game experience. Post-game questions were self-assessments of team and individual effectiveness, and evaluations of the teammate.

RESULTS

All puzzles were solved. The main raw performance score was the time that a dyad took to do each puzzle. But raw puzzle time is influenced by three factors that are not relevant to the goal of the study which was to ascertain how performance varies as a function of a dyad's history of interaction. Since the game is played in English, English proficiency can affect performance as a whole. Game-playing skill can also influence performance as a whole. Since role-play gaming is complex, training does not necessarily make all players equally proficient. Furthermore, individual puzzles are not equally difficult, so simply comparing performance across different puzzles does not fairly indicate the temporal evolution of relative performance (see Figure 2). Hence, the raw times needed to be transformed prior to analysis.

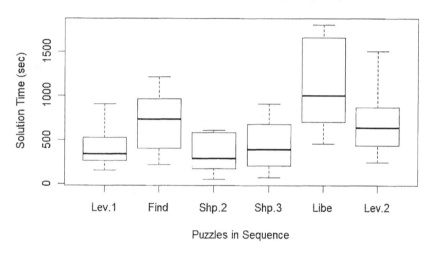

Figure 2. Raw puzzle completion times.

DATA PREPARATION: COVARIATES

In a similar fashion as Warren (2008), team English proficiency and game experience were treated as covariates and the effects were removed from the raw times using linear regression. (The small and unequal n's do not permit use of ANACOVA.) Pre-game English proficiency answers were pooled into a single score per player. The mean of the two dyad members' scores then served as the dyad's English proficiency score. Per puzzle, English proficiency accounted for between 3% and 23% of the variance in raw solution times with a median of 19%.

The per player gaming skill score was a composite of a player's pre-game self-reports about gaming experience (e.g., hours per week spent on computer games), the player's pre-puzzle game-play training time (longer time to "graduate" indicating poorer gaming skill), and the time taken to complete the ship.0 and ship.1 mid-game training scenarios (again, longer times indicating poorer gaming skill).

The mean of the two members' scores served as the dyad's gaming skill score. Per puzzle, gaming skill accounted for between 1% and 23% the variance in raw times with a median of 18.5%.

The 27 English proficiency scores, the 27 gaming skill scores, and their interaction were used as predictors of a puzzle's solve-times using linear regression. A covariate-free score per dyad per puzzle was then formed as the sum of a puzzle's mean raw time plus the *residuals* of the predicted scores (Warren, 2008).

DATA PREPARATION: UNEQUAL DIFFICULTY OF PUZZLES

The resulting adjusted puzzle solution times have a zero correlation with English proficiency and composite gaming skill (by design). However, the adjusted times have a similar distribution to that in Figure 2 since the unequal puzzle times are not due to the covariates but rather to the unequal difficulty of the puzzles.

One way to enable comparisons across puzzles is to normalize the (adjusted) times using z scores. High values of time or z scores indicate relatively poorer puzzle-solution performance or slowness of solution. By reversing the sign of the z scores, high values indicate better performance or "quickness." Multiplying the reversed z scores by 10 and adding 50 yields "relative adjusted quickness" (RAQ) scores which may be meaningfully compared among puzzles.

SEQUENTIAL RELATIVE ADJUSTED QUICKNESS (RAQ) PER DYAD

The sequential relative adjusted quickness (RAQ) curves shown in Figure 3 do not resemble the notional curves depicted in Figure 1. The curves in Figure 3 are composites formed by taking the mean of the RAQ scores of the dyads of a particular type of nationality mix (17 same-nationality dyads, 5 mixed-nationality, and 5 with at least on hyphenated-nationality). Within the mix-types, there is considerable variation in the curves for individual dyads (not shown here).

The deterioration in solution performance in the last two puzzles by the mixed-nationality dyads was not unexpected. However, the initial rise in performance was not at all expected. Also puzzling, is the curve for dyads which were "intermediate" in nationality mixture by virtue of having one member with a hyphenated-nationality. It shows the worst initial performance and the best final performance when its "intermediate" character "should" place it between the two other curves.

Puzzle Relative Adjusted Quickness (Quadratic)

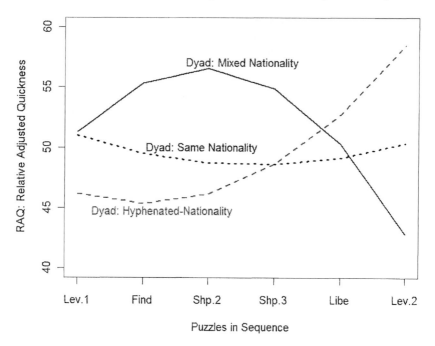

Figure 3. Sequential Relative Adjusted Quickness as a function of nationality mixture of dyads. Curves are composites.

SEQUENTIAL CORRELATION OF RELATIVE ADJUSTED QUICKNESS (RAQ) AND DYAD MIXTURE TYPE

As noted in the Introduction, the sequential relative adjusted quickness (RAQ) curves are only one way to view progressive team performance. Under the alternate view of progressive effects, the correlation between dyadic characterization and performance on the first puzzle should be zero. The correlation magnitude should increase, gradually or with jumps, as the dyad accumulates mutual experiences.

Table 1 shows the correlation between dyad nationality-mixture type and the time taken to solve a puzzle. Since, the $p=.05$ critical value of r with 25 degrees of freedom is +/- .38, the correlations for the first three puzzles are nil, but the correlations for puzzles 4 and 5 jump in magnitude. Positive correlations indicate that the greater the nationality mixture of a dyad, the faster it solves a puzzle, but this is contrary to the hypothesis. The magnitude of the correlation for the last puzzle does increase from the previous puzzles and is in the hypothesized direction. But the reversal of the direction from the previous two correlations is unexpected.

Table 1: Sequential Correlation between Mix-Type and Performance

Order	Puzzle	r	r-square
1	Levers.1	.07	.00
2	Find	.09	.01
3	Ship.2	-.04	.00
4	Ship.3	.32	.10
5	Library	.34	.11
6	Levels.2	-.39	.15

DISCUSSION

Cultural differences affect team performance, but which differences matter and to what degree are not well-understood. Considerable distance on many cultural dimensions can exist between strong friends and allies. Also, some of the worst conflicts occur in civil wars and internecine fighting in which the combatants are culturally and otherwise almost identical.

Furthermore, team performance is not static but depends on the cumulative experiences of the team as its goes about solving problems. Contrary to the tone of the arguments in the introduction, the history of experiences of a team and the time-history of its effectiveness are not inevitably monotonically increasing or decreasing, but rather can have considerable "ups and downs." The ups-and-downs are evident in Figure 3 which is based on quadratic regression of the actual relative adjusted quickness scores of each dyad (which acts to smooth the trends).

Although the results presented here are not definitive, the study does illustrate the usefulness of role-play computer games for research on cultural effects in teamwork. As with any technique, there are caveats and limitations. One such is the fact that a multi-player game generally uses one language and not all the players have native proficiency. This, of course, reflects real life in which multi-cultural teams require the use of a common language and not all members are native speakers. Another problem for research is that not all players have equal computer game skills, so certain measures such as task-time do not truly reflect abilities. This study also confronted the fact that not all tasks are equal in difficulty, but measures of across-task performance are needed for tracking the evolution of the effectiveness of a team over its history.

REFERENCES

Bioware Corp. (2004). *Neverwinter Nights*^{TM} - Platinum Edition (Computer Game Software). New York: Atari Interactive.

Leung, A., Diller, D., and Ferguson, W. (2005). ``SABRE: A game-based testbed for studying team behavior," *Proceedings of the Fall Simulation Interoperability Workshop (SISO)*. Orlando, FL.

Warren, R. (2008). ``Mixed- & homogeneous-culture military team performance on a simulated mission: Effects of age, game-experience, & English proficiency," *Proceedings of the NATO RTO HFM-142 Symposium on Adaptability in Coalition Teamwork*, held in Copenhagen, Denmark. Brussels: NATO.

Warren, R., Diller, D.E., Leung, A., Ferguson, W., and Sutton, J.L. (2005). ``Simulating scenarios for research on culture & cognition using a commercial role-play game," in M.E. Kohl, N.M. Steiger, F.B. Armstrong, and J.A. Jones, (Eds.), *Proceedings of the 2005 Winter Simulation Conference*. Orlando, FL.

Warren, R., Sutton, J., Diller, D., Ferguson, W., and Leung, A. (2004). ``A game-based testbed for culture & personality research," *Proceedings of the NATO Modeling and Simulation Group -- 037 Workshop: Exploiting Commercial Games for Military Use*. The Hague, The Netherlands.

CHAPTER 32

Measuring Performance of Office Work – An Overview of Test-Procedures

Hansjuergen Gebhardt[1], Inna Levchuk[1],
Christoph Muehlemeyer[1], Kersten Bux[2]

[1]Institute of occupational health, safety and ergonomics (ASER)
Corneliusstr. 31, D-42329 Wuppertal, Germany

[2]Federal Institute for Occupational Safety and Health (BAuA)
Unit 2.4: Workplaces, Safety of Machinery, Operational Safety
Proschhübelstr. 8, D-01099 Dresden, Germany
Email for contact: h.gebhardt@institut-aser.de

ABSTRACT

The aim of this presentation is to give an overview of test-procedures concerning measuring performance of office work. Several examinations estimating the influence of raised thermal conditions on the performance of office work have been executed recently. However in standards and directives the parameters "well-being" and "comfort" are considered for the evaluation of the climate in offices. It is important to pay attention to the fact that it is much easier to describe the correlation between physical work and performance of work (e.g. changes in heart rate is a valid indicator) than to describe the correlation between cognitive work demands and physiological parameters. The parameters "well-being" and "comfort" correlate with the satisfaction of the workers but not necessarily with the performance at work as well. For practical use it seems to be necessary to pay more attention on performance losses due to sub-optimal climate conditions.

Keywords: performance, office work, test procedures

INTRODUCTION

Several examinations estimating the influence of raised thermal conditions and air quality on the performance of office work have been executed previously (see references). However in standards and directives the parameters "well-being" and "comfort" are considered for the evaluation of the climate in offices. The most important and well known international standard is the ISO 7730: "Ergonomics of the thermal environment - Analytical determination and interpretation of thermal comfort using calculation of the PMV and PPD indices and local thermal comfort criteria". This document already exists for a long time and has been actualized in May 2006 for the last time, where it was extended on aspects of "local thermal comfort".

It is important to pay attention to the fact that it is much easier to describe a correlation between physical work and performance of work (e.g. changes in heart rate is a valid indicator) than to describe the correlation between cognitive work demands and physiological parameters. The parameters "well-being" and "comfort" correlate with the satisfaction of the workers but not necessarily with the performance at work as well. For practical use it seems to be necessary to pay more attention on performance losses due to sub-optimal climate conditions.

CHALLENGE

Although it is important to know more about performance losses due to sub-optimal conditions (e.g. climate) the selection of tests is rather difficult as this also influences the results. On the one hand it should be very close to normal office work, on the other hand the tests should result in numbers, which are more or less independent e.g. on repetitions. This is important especially under laboratory conditions, where the conditions can be controlled, so that the effect of sub-optimum conditions can be measured inter-individually.

Within the framework of a literature study, several methods measuring kinds of performance indicators such as concentration, attention, creativity, thinking etc., relevant for tasks in office work, have been examined. Criteria to be taken into account are validity of the methods as well as and transferability of the results to office work. The target is on the one hand to get as close as possible to office work and on the other to have a tool box, that can be applied in laboratory studies as well as at working places.

DIMENSIONS

To get reliable results several dimensions, which are important due to office work, should be taken into account. The following list gives some of these dimensions:

- Selective and permanent attention
- Concentration
- Memory
- Data processing
- Logical and creative thinking

The following sections give some examples for test-procedures. One may assume that there is no one single test-procedure, which gives the one number corresponding to office work. This would be office-work itself, which is hard to standardize. Also comparable numbers are hard to define when looking at office work. Therefore test-procedures are necessary, which result in numbers, but more or less focus on one or two of these dimensions.

TEST-PROCEDURES

Concerning concentration and attention the d2-test is very common and sufficiently validated. It was developed already in 1962 by Rolf Brickenkamp. A sequence of d- and p-characters is provided, where each character is shown with one to four dots (see figure 1). There is a paper-pencil as well as computer-version available. The task is to select those d-characters with 2 dots as fast and as correct as possible.

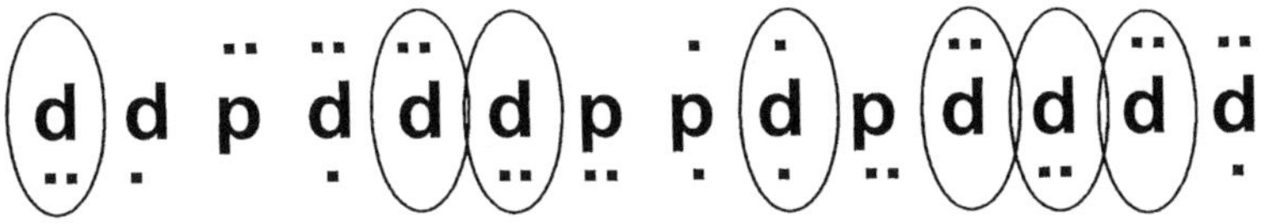

Figure 1. Example for a d2-test, the correct characters (d with two dots) to be noted are marked

The results of this test are the number of processed characters and the number of mismatches. Both numbers are noted. High concentration and high attention will result in a high number of processed characters and a low number of mismatches.

Sub-optimal conditions will result in a loss of performance due to concentration and attention, which can be characterized individually by the change in processed characters and mismatches under given conditions.

Memory-Tasks may be covered by tasks similar to the well-known memory-game (see figure 2). In this case several pairs of identical pictures are provided. Also here computer-based versions are available. While the pictures cannot be seen, the task is to find these pairs by selecting two pictures at a time and collect them in a mimimum time with a minimum number of selected pictures.

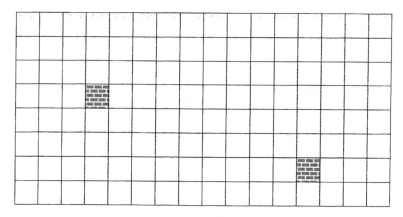

Figure 2. Example for a memory-testing, during the test only two of pictures can be seen at the same time, in this example it forms a pair

Here the time spent for solving the task and the number of steps to solve the task are indicators to be noted. Good memory means solving the task in a short time with a low number of selected pairs of pictures.

To come closer to office work word processing as well as calculations are often used for simulating data-processing. This normally means text-typing monitoring the numbers of characters and typing mismatches per time-intervall. Another form of word-processing is reading a text and searching for mismatches in a given time-intervall.

Concerning calculations it has to be considered, that normally the calculations should not exceed two digits. We suggest calculations in a form given in figure 3. Again the time total number of calculations and the mismatches are noted as indicators.

39 +	86 -	9 =
25 +	103 -	9 =
63 +	22 -	4 =

Figure 3. Example for an excerpt of calculations

Concerning logical and creative thinking there are tests available based on wordings as well as on symbols. For example the task can be to form words beginning with given two characters. This is not so applicable for laboratory studies with the same persons performing the task under several conditions as on the one hand the learning effect is dominating when always taken the same two characters. On the other hand, if the two characters are matched, the two characters should be carefully choosen, so that the noted number of tasks performed are not dominated by the number of words which are available beginning with these two characters.

Another form of tasks is giving an adjective and the subject should write down some words associated with these adjective. Again the words must be carefully choosen, so that the number of wordings, that could be found is a reliable indicator for associated as a part of creative thinking.

FOLLOWING STEPS

The selected test procedures will be performed in pre-tests under thermal neutral conditions. These pre-tests will provide more information on performing the tasks as well as effects on repetition and inter-individual differences. This is important for interpretation of the resulting numbers before going into the main testing under several defined conditions.

CONCLUSIONS

There are a brought number of more or less validated testing procedures covering aspects concerning office work which sum up in numbers quantifying the performance under given conditions.

It can be assumed, that in case of laboratory studies where one person performs the tasks under various conditions, deviations from the optimum point should be calculated. Instead of the absolute mean values, these deviations would separate inter-individual changes

ACKNOWLEDGEMENTS

This work is supported by the German Federal Institute for Occupational Safety and Health (BAuA) in co-operation with the Ziel2-project "Design4All" supported by the European Commission and the Ministry of Economic Affairs and Energy of the State of North Rhine-Westphalia.

304

REFERENCES

Haneda M., Tanabe S., Nishihara N., et al. (2007): Development of Survey Tools for Indoor Environmental Quality and Productivity. CLIMA Conference Proceedings.

Kawamura, A, Tanabe, S, Nishihara, N et al.(2007): Evaluation Method for Effects of Improvement of Indoor Environmental Quality on Productivity, CLIMA Conference Proceedings

Seppanen, O., Fisk, W.J., & Lei, Q.H. (2006): Effect of temperature on task performance in office environment. Report LBNL-60946. Helsinki University of Technology, 1-9

Seppänen O, Fisk W.J., Faulkner D. (2005): Control of temperature for health and productivity in offices. ASHRAE Transactions 111(2): 680-686. LBNL-55448

Seppänen O. and Fisk W.J. (2005): Some quantitative relations between indoor environment, performance and health. Plenary presentation and publication in Proceedings of Indoor Air 2005, pp. 40- 53. September 4-9, Beijing

Tanabe, S., Haneda M., Nishihara N. (2007): Indoor environmental Quality and Productivity. Rehva Journal – p. 26-31.

Tanabe S., N. Nishihara, M., Haneda M. (2007): Performance Evaluation measures for Workplace productivity, CLIMA Conference Proceedings,

Wargocki, P., Wyon D.P., Sundell J., Clausen G. and Fanger P.O. (2000): The effects of outdoor air supply rate in an office on perceived air quality, sick building syndrome (SBS) symptoms and productivity. Indoor Air, 10(4): p. 222-36.

Wargocki, P. (1998): Human Perception, Productivity and Symptoms Related to Indoor Air Quality, Ph.D. thesis, International Centre for Indoor Environment and Energy, Technical University of Denmark, Copenhagen, Denmark, p. 115-125.

Wargocki, P., Wyon, D. & Fanger, P. O.(2000) Productivity is affected by the air quality in offices Proceedings of Healthy Buildings 2000, Vol. 1, Helsinki, Finland, 2000: p. 635-640.

Wargocki, P. (2007): Improving indoor air quality improves the performance of office work and schoolwork. ASHRAE IAQ.

CHAPTER 33

Prevention of Work-Related Stress in Practice – A Participatory Approach

Christine Ipsen, Per Langaa Jensen, Vibeke Andersen

Department of Management Engineering
Technical University of Denmark
DK-2800 Kgs. Lyngby, Denmark

ABSTRACT

Knowledge work represents a job with high level of influence, freedom and autonomy. Earlier studies indicate that such working conditions reflect a good psychosocial environment. Although knowledge workers describe their job as being both sovereign and motivating, several studies have pointed at work-related stress and strain being an increasing problem in knowledge-intensive companies. Behaviour modification is the dominating approach to stress management. Contrary to the typical stress management practice and understanding of knowledge intensive work two qualitative studies point to potential options for preventing work-related stress based on changes in the daily activities and management.

Keywords: knowledge work, stress management, prevention, organizational level interventions, process, implications.

INTRODUCTION

The human resource strategy in knowledge intensive companies is typically to develop and sustain the expert workforce by recruiting highly educated and competent people and providing them with challenging tasks and projects. The companies can in general be characterized as decentralized embedded in a matrix organization, emphasizing management by objectives and networking in order to facilitate an efficient knowledge production. Employees are highly-qualified, competent, autonomous and cooperating in order to solve specific tasks. Furthermore there is continuous interchange with others in order to develop new knowledge. It is mainly the cultural and the mental incentives in terms of social exchange of knowledge and the unofficial recognition and trust that is gives which is of great importance. Employees seek the challenge from difficult tasks and the creation of new knowledge plus the possibility to contribute and make a difference to the customers (Alvesson 1995;Drucker 1969;Drucker 1988;Fleming & Spicer 2004;Ipsen 2006;Newell et al. 2002;Pedersen 2009)

Knowledge work is typically perceived to be a good and developing job with working conditions that are characterized by a high level of influence, control, flexibility and autonomy. These self-managed jobs are typically understood both personally and professionally as good and stimulating jobs. The high degree of freedom, influence, rewards, support etc. are valued as key factors in jobs characterized by a good psychosocial working environment for example in models such as Karaseks and Theorell's "Job Demand Control (JD-C) Model". (Karasek 1979;Karasek & Theorell 1990)

However, several studies have pointed at work-related stress and strain being an increasing problem in knowledge-intensive companies (Buch & Andersen 2008;Ipsen 2007;Ipsen & Jensen 2009;Kalimo 1999;Mogensen, Andersen, & Ipsen 2008;Stavroula, Griffiths, & Cox 2003;WHO 1999).

In two studies one of the conclusions is that the ambiguity which knowledge workers experience as part of their work can be a potential strain *and* a potential source of enthusiasm and self-fulfillment (Buch & Andersen 2009). When the balance tips in favour of strain, management of the problem is typically individual.(Ipsen 2007)

A TYPICAL STRESS MANAGEMENT PRACTICE

A study (Ipsen 2007) has shown that strain and stress related problems are often handled in an unstructured, accidental and informal way. This results in a stress management practice focusing either on the individual's capability to cope with the situation (secondary prevention) or treatment activities (tertiary) instead of organizational preventive action (primary prevention). Typical stress management practices are for example stress policies, free and anonymous counseling and awareness training to be able to spot employees with symptoms of stress.

Within the stress literature one distinguishes between 3 different interventions and prevention levels related to stress management. The levels have different targets and purposes depending on the goal of the intervention. The emphasis on second and tertiary intervention has limitations as it does not affect the root-causes of the stress and strain i.e. the organizational and managerial circumstances. The long-term primary prevention activities have this in focus. (Kompier & Cooper 1999;Murphy 1988) (See figure 1)

Intervention level	Intervention Targets	Examples
1. — Primary Goal: Reducing potential risk factors or altering the nature of the stressor before workers experience stress-related symptoms or disease	Stressors at their source; organization of the work; working conditions	Job redesign, load reduction, reorganizing the authority lines and restructuring construction of a supporting climate, Re-design or establishing a reward system
2. — Secondary Goal: To help equip workers with knowledge, skills, and resources to cope with stressful conditions.	Employee responses to stressors	Cognitive behavioral therapy, coping classes
3. — Tertiary Goal: To treat, compensate and rehabilitate workers enduring stress-related symptoms or disease.	Short term and enduring adverse health effects of job	Return-to-work programs, occupational therapy, medical intervention stress

Figure 1. Different intervention levels

The majority of the interviewees told that they had all experienced being unable to meet the demands. In those situations most of them had reacted by trying to cope with the situation either by solving the problem causing the feelings or by adjusting themselves to the situation. This was in line with general signals form the managers In severe situations different actions were initiated and some of the participating companies had besides that also introduced various initiatives which should support and help the employees to meet the challenges in their work. These practices can be categorised as secondary or tertiary interventions.

A final conclusion of the study shows that tacit knowledge of problems and possible solutions exist within the organization. But it is typically not made explicit and turned into stress-preventing activities. However, if the collective tacit knowledge is made explicit through a formalized process of collective reflections it can form the basis for the formulation of sustainable preventive changes.(Bovbjerg 2010;Ipsen 2007;Mogensen, Andersen, & Ipsen 2008)

In a subsequent research project this conclusion was further investigated and tested through a larger invention in several knowledge intensive companies. In the following the results of the two studies will be presented.

OBJECTIVE

Both studies had the overall objective of *identifying which stress management intervention options both managers and employees have and use in order to prevent work-related problems and stress in knowledge work.* They were conducted in immediate continuation of each other.

In order to change the current practice and formulate sustainable preventive changes it is necessary to understand the causes for work-related problems as well as the current management practice of these problems. (Ipsen 2007;Ipsen & Jensen 2009;Kompier & Cooper 1999;Singer et al. 1986) This paper will briefly address how shared mental models were established and how preventive actions were implemented. The main focus in this paper is on the organizational level interventions; the processes and the results based on a participatory approach.

METHODOLOGY

This paper is based on two studies carried out in cooperation with several knowledge-intensive companies in Denmark. The first study was based on a qualitative research methodology whereas the most recent was based on a mixed methods methodology. One of the basic premises in both studies has been to understand and examine the possibilities to prevent or reduce strains in the job.

In the first study (a Ph.D.-study) multiple case studies were conducted in five knowledge intensive companies with the aim to identify the causes of these problems. The succeeding study was based on a participatory action research approach where one of the premises was to explicate the corporate tacit knowledge on problems, their causes and possible solutions. It is the results of the participatory action research in both studies which is emphasised in this paper.

In the first study it was the aim to identify a series of possible ideas to primary stress management interventions based on a group reflection. The assumption behind this research activity is that organisation-directed strategies are more preventive as they focus on the sources, in contrast to the individual-directed strategies which aim at modifying the risk factors (DeFrank & Cooper 1987;Kompier & Cooper 1999;Murphy 1988;Newell 2002). A participatory workshop would thus form an adequate framework for a double-loop reflection (Argyris & Schön 1996) if the mismatch between the actual practice and the espoused theory was to be highlighted. At the same time a space would be created for development of new alternative solutions. The workshop would also at the same time be a tangible example of a collective "Ba" (Nonaka, Toyama, & Konno 2002) creating a room for developing alternative practices in order to solve the recognised problems.

A reflective workshop was thus conducted where several managers not associated with one another were asked to respond to and discuss a narrative which had been drawn up based on the preceding analysis of more knowledge intensive companies (Ipsen & Jensen, 2009). The outcome of the workshop was a list of plausible preventive organisational actions which could be taken.

In the second study the research activities had a longer time-range. For nine months five companies tried to implement a set of organizational changes which were expected to change the daily practice and thus provide the basis for a reduction of the level of work-related stress.

The changes were based on local reflections in several departments and settings where both employees and managers participated, both separately and together.

Firstly, Fish-bone charts (Ishikawa 1968) were used to map the experiences of the employees in terms of enthusiasm and strains. A corresponding workshop was carried out among the managers where the focus was still on the employees and their working life. The outcome of the workshops was a set of organisational conditions, either stated to cause strain in the daily work or to create enthusiasm.

Secondly, following the Fish-bone workshops a Priority Board Game was played by the whole department to prioritise the suggested problems and improvements. This game had been developed specifically for this project by the researchers. Based on the collective reflections and prioritising several organisational level interventions were initiated. The intervention process and results were studied and evaluated during the whole project period.

RESULTS

In order to formulate preventive changes it is necessary to understand the causes for the work-related problems as well as the current management practice of these problems. The first study revealed some of the factors which influence the psychosocial working environment. Questioned about causes and remedies to problems the employees found that it was "their own fault" that these problems occurred, as one employee expressed it. This understanding was supported by the managers by stating for example that they were too ambitious.

While acknowledging the personal influence on the problems mentioned, the focus in the study was however on the organizational factors. Using Galbraith's Star Model (2002), it is clear that the various organizational components are thus not aligned and consequently the organizational design does not support the performance and behavior which is desired.

One example of this is knowledge sharing not being rewarded even though it is a crucial part of knowledge work. Consequently, employees spend precious time searching for the knowledge they need and sometimes even insufficient solutions are developed which is experienced to be professionally unsatisfactory. Other examples are lack of support, insufficient knowledge management systems, ambiguous reward systems and conflicting demands etc.

All in all this first study concludes that individualised knowledge work in combination with individualised management systems create problems for the individual employee and supports an individual management of problems inducing secondary and tertiary interventions.

THE FIRST STUDY: COLLECTIVE WORKSHOP WITH MANAGERS

Based on these conclusions a collective workshop with managers from different companies was initiated. The aim of the workshop was to create a room for reflection where the tacit knowledge on problems and causes could be made explicit followed by an identification of potential primary preventive actions. A narrative formed the basis for the reflections and discussions of the major problems as well as the plausible preventive solutions in terms organizational level interventions. The narrative was about a successful employee who considered accepting an offer for a new job in another company as the current company provided him with a lot of frustration and no real support.

The outcome of the workshop was twofold. First of all, the statements expressed demonstrated difficulties in focusing on the organisational causes and solutions rather than the individual. Though it seemed to be difficult to address the organizational level stress interventions, the workshop made it evident that a collective reflexive setting can provide a suitable frame for addressing tacit knowledge on problems, their causes and solutions. So the second outcome of the workshop was that the participants own reflections provided the basis for alternative ideas for stress management and preventive organisational changes.

In practice the workshop was thus both a means and an objective in relation to organisational level stress interventions.

THE SECOND STUDY: A PRIMARY STRESS MANAGEMENT CHANGE - FEEDBACK FROM THE MANAGER

In the subsequent study these findings were pursued in combination with other premises and thus integrated in the research design, in practice resembled by the Fish-bone workshop and the Priority Board Game.

The intervention process

In one of the participating companies an engineering consulting company several organisational issues were identified as either occupational risks or insufficient. Four of these were prioritised and were changed during the project period. One of these issues was feedback from managers. The employees experienced that they were left very much to themselves and would like to get more feedback from their managers and colleagues. This statement was a surprise to the management team. Their general view was that people were satisfied with the feedback that they got and that the employees as knowledge workers appreciated their autonomous and individualised work. However, this change of the manager's role was prioritized as a result of the Priority Board Game and subsequently an intervention was initiated.

Initially a mapping was conducted of the persons that had gotten feedback, on what and how often. It became evident that some of the employees got more feedback than others. During the project period the change of the manager's role was addressed at the department meetings every fourth week. Then there was a status and evaluation of the activities. In this process the local co-ordinators powering the change process played a crucial role. Their job was to ensure a consistent focus on the process, give feedback and encourage their colleagues to at continuous effort. During the project period the co-ordinators got support and feed-back from the HR-department. Besides the co-ordinators there has been monitoring and visualisation of the progress in the issues chosen to be changed.

The outcome

The redesign of the manager's role had the following impact:
- A larger "we-feeling" in the department
- Less fear regarding addressing a problem.
- Less crisis management
- Increased and improved feedback to the managers

Additionally the internal dialogue has been strengthened. Problems are now being addressed with other colleagues and there is an understanding that this is a mutual problem. The dimensions for evaluating the outcome - defined before the project - show that the organisational intervention has had a significant influence on several factors for example social support, trust, credibility, clarity of roles, involvement and rewards. The changes have reduced the level of ordinary sick leave as the strains have been reduced. Finally employee turnover has dropped during the project which also has a positive effect on the bottom line.

DISCUSSION

It is often questioned whether it is possible to initiate and implement primary stress intervention. In the presented studies this question has been addressed as the research design and the participating companies have focused on the development of activities based on a participatory approach focusing on primary intervention of work -related stress. The results imply that by explicating tacit knowledge on problems, causes and possible solutions to daily problems and strains it is possible to identify primary activities to prevent work-related stress and carry out organizational level interventions. Thus, in the second study the outcome of the reflexive workshops was subsequently followed by several organizational changes here exemplified by improved feedback from managers. These changes were selected based on a collective prioritisation of the various proposals identified. The changes had a beneficial influence on the level of stress and enthusiasm besides improved bottom line and productivity.

POTENTIAL OBSTACLES AND IMPLICATIONS

As Cox et al.(2007) suggest it is necessary to pay attention to the intervention process as well as the outcome of an organisational level intervention in order to understand why an intervention becomes a success or failure.

First, both studies show the difficulties in addressing the issue of work-related stress and primary preventive actions as there is a heavy tendency to focus on the individual instead of on organisation and management. This "mental model" provides a barrier for implementing organisational-level stress interventions.

Secondly, the first study it is also evident that the strong focus on production and performance creates a barrier for carrying out preventive activities. The employees are thus named "Holy" and "Kings" by two HR-managers, meaning that the employees are not to be disturbed as it would damage the productivity. Within this thinking an organisational level intervention is thus regarded to be a disturbance.

Thirdly, in cases where interventions are approved of, it is crucial that the interventions are given same priority as the other tasks and assignments. In practice it is symbolised by an account number and a corresponding budget. If not, the intervention will run parallel to the daily activities (i.e. skunk work) and no real changes will be initiated and carried out.

Fourthly, it is of key importance that there is a persistent focus from the daily management and the co-ordinators as well as the top management.

Then it is vital that the key actors continuously follow up on the intervention and the initiated activities. In the beginning, the intervention will get a lot of attention. However, it is important that the involved department and actors become "masters in their own house". This can be achieved by a continuous follow-up and questions to the participants like "what do you think needs to be changed?", "how would you like this work place to function" etc. and present the questions and answers on the next department meeting.

Finally, one should be aware that a variation in the outcome can be due to complex social systems in which the intervention takes place. Besides, the implementation and the context in which it takes place can modify the intervention and the outcome (Randall, Cox, & Griffiths 2007).

Being aware of these potential obstacles and implications it is also important to address which circumstances would necessitate organizational-level interventions in knowledge intensive work.

The studies provide a clear answer to this. Focusing on organizational level intervention provides a workplace with an identification of aspects of work giving rise both to enthusiasm and to strains. Furthermore, organisational changes with the aim to reduce the strains at work leads to increased efficiency, improved knowledge production and processes and integrate the working life perspective in the daily activities and management. Organisational level interventions can thus be motivated by the fact that improved working processes induce an improved working life with more enthusiasm and less strain.

CONCLUSION

Overall the studies show that contrary to the typical stress management practice and understanding of knowledge intensive work, it is possible to identify primary preventive changes and carry some of them out in practice.

The studies also show that knowledge on these issues do exist within the organisation and a set of simple well-known tools within organisational development can make this knowledge explicit opening for a collective discussion of important aspects of daily work, the causal net behind these problems and proposals for improving the situations at an organisational and management level. The two studies also show that there might be a joint optimisation of both performance and working condition when proposals are implemented.

However, handling these types of risk demands a management effort corresponding to the effort needed to establish successful projects within an organisation of highly skilled, motivated, autonomous employees. A coalition has to be established between central actors, and this coalition shall over time have a dedication to this project handling the obstacles which with guarantee will show.

But the case presented from the second study shows that it is beneficial for all to handle these types of risks.

REFERENCES

Alvesson, M. 1995, Management of Knowledge-Intensive Companies Walter de Gruyter, Berlin.

Argyris, C. & Schön, D. 1996, Organizational Learning II. Theory, Method and Practice. Addison-Wesley Publishing Company.

Bovbjerg, K. M. 2010, "Yoga, meditation og mindfullness som stresshåndtering," in Ledelse og Spiritualitet, pp. 117-132.

Buch, A. & Andersen, V. 8 A.D., "Knowledge work and stress - beyond the job-strain model", ODAM IX, International Symposium on Human Factors in Organizational Design and Management.

Buch, A. & Andersen, V. 9 A.D., "Knowledge work and stress - between strain and enthusiasm".

DeFrank, R. S. & Cooper, C. 1987, "Worksite stress management interventions: Their effectiveness and conceptualization", Journal of Managerial Psychology, vol. 2, pp. 4-10.

Drucker, P. F. 1969, The Age of Discontinuity, 2 edn, William Heinemann Ltd, London.

Drucker, P. F. 1988, "The coming of the new organization", Harvard Business Review no. January-February 1988, pp. 45-53.

Fleming, P. & Spicer, A. 2004, ""You can checkout anytime, but you can never leave" Spatial boundaries in high commitment organization", Human Relations, vol. 57, pp. 75-94.

Ipsen, C. 2006, "Knowledge work and work-related stress", Congress of the International Ergonomics Association.

Ipsen, C. 2007, Vidensarbejderens særlige arbejdssituation og muligheder for forebyggelse af arbejdsrelateret stress i vidensarbejdet - Et kvalitativt studie af arbejdsrelateret stress i vidensarbejdet., Department of Management Engineering, DTU.

Ipsen, C. & Jensen, P. L. Causes of work-related stress and individual strategies in knowledge work. Human Factors and Ergonomics in Manufacturing . 2009. Ref Type: Unpublished Work

Ishikawa, K. 1968, Guide to Quality Control, Productivity Organization Tokyo, Japan.

Kalimo, R. 1999, "Knowledge jobs - how to manage without burnout?", Scandinavian Journal of Work Environment Health, vol. 25, no. 6, pp. 605-609.

Karasek, R. 1979, "Job Demands, Job Decision Latitude, and Mental Strain: Implications for Job Redesign", Adminstrative Science Quarterly, vol. 24, no. June, pp. 285-308.

Karasek, R. & Theorell, T. 1990, "The Psychosocial Work Environment," in Healthy Work. Stress, Productivity and the Reconstruction of Working Life, Basic Books, Inc., New York, pp. 31-83.

Kompier, M. & Cooper, C. 1999, Preventing Stress, Improving Productivity. European case studies in the workplace Routledge, London.

Mogensen, M., Andersen, V., & Ipsen, C. 2008, "Ambiguity, identity construction and stress amongst knowledge workers: developing collective coping strategies through negotiations of meaning".

Murphy, L. R. 1988, "Workplace Interventions for Stress Reduction and Prevention," in Causes, Coping & Consequences of Stress at Work, C. Cooper & R. Payne, eds., John Wiley & Sons Ltd, pp. 301-339.

Newell, S. 2002, Creating the Healthy Organization. Well-being, diversity & ethics at work, 1. edn, Thomson, Cornwall.

Newell, S., Robertson, M., Scarbrough, H., & Swan, J. 2002, Managing Knowledge Work Palgrave.

Nonaka, I., Toyama, R., & Konno, N. 2002, "SECI, Ba, and Leadership," in Managing Industrial Knowledge - Creation, transfer and utilization, I. Nonaka & D. Teece, eds., pp. 13-43.

Pedersen, M. 2009, Tune in, Breakdown, and Reboot : On the Production of the Stress-Fit Self-Managing Employee, Department of Management, Politics and Philosophy, CBS.

Randall, R., Cox, T., & Griffiths, A. 2007, "Participants' accounts of a stress management intervention", Human Relations, vol. 60, no. 8, pp. 1181-1209.

Singer, J. A., Neale, M. S., Schwartz, G. E., & Schwartz.J. 1986, "Conflicting Perspectives on Stress Reduction in Occupational Settings: A Systems Approach to Their Resolution," in Health And Industry. A Behavioral Medicine Perspective, 1 edn, M. F. Cataldo & T. J. Coates, eds., John Wiley & Sons, pp. 162-193.

Stavroula, L., Griffiths, A., & Cox, T. Work Organisation & Stress. Protecting Workers' Health Series 3. 2003. Switzerland, WHO. WHO. The burden of occupational illness. 8-6-1999. WHO.

CHAPTER 34

The Intermittent Duty of Information and Communication Technology Among Immediate Superiors of Social and Health Care Services - A Research and Development Project

Seppo Tuomivaara, Kaisa Eskola

Finnish Institute of Occupational Health
Helsinki, Finland

ABSTRACT

Information and communication technology (ICT) is seen as a prominent bridge for the gap between the needs of social and health care services and their diminishing resources. Middle management and immediate superiors are involved in the responsibility for improving the use, development and implementation of ICT systems. The aim of this study is to identify how the middle management and immediate superiors in social and health care use ICT in their supervisory duties, and what their experiences are of the use and development of ICT based services. The preliminary results draw a draft picture of ICT use and core task fit. They also

316

demonstrate the situation of developmental orientation of leadership in everyday work and the possibilities for service innovations in ICT-mediated tasks.

Keywords: Immediate Superior, Leadership, Information and Communication Technology, Well-being, Social and Health Care

INTRODUCTION

In Nordic welfare states, demand for social and health care services has changed greatly in recent years. The availability, quality, and cost efficiency of public services face a difficult challenge due to the ageing population, the diminishing availability of labour, and a troubled economy. It has been claimed that the solution to these problems is in the innovativeness of employees. Service innovations can be found in organizations with experienced, motivated and proficient employees working in a positive atmosphere under a supportive management. It has also been emphasized that leadership is a key factor in stimulating workers towards more innovative work behaviour.

Along with innovative employees information and communication technology (ICT) is seen as a prominent bridge for the gap between the needs of social and health care services and their diminishing resources. For example, a large project has been set up in Finland to develop national infrastructure and standardized practices for enabling so-called unbounded services. This standardization and other similar changes has also emphasized new innovative practices and the innovative use of ICT. The state of know-how of ICT use will be challenged on every level, from managers to employees. Middle management and immediate superiors are involved in the responsibility for improving the use, development and implementation of ICT systems. From the workload viewpoint, the situation can be experienced problematic if immediate superiors' core task of leadership is disturbed. The main challenge in implementing new systems is therefore how well these systems fit the practical orientation of immediate superiors' leadership activity and how the supervisory duties are experienced as meaningful at a subjective level. The aim of this study is to identify how the middle management and immediate superiors in social and health care use ICT in their supervisory duties, and what their experiences are of the use and development of ICT based services.

USE OF ICT IN LEADERSHIP ACTIVITY

Social and health care organizations are special service organizations with multiprofessional personnel. Public organizations are built up to provide services for people without making profit. This kind of service organization is traditionally seen as a hierarchical system where management is divided into strategic management, middle management, and immediate superiors, based on education-generated professional competencies.

Nikkilä and Paasivaara (2008) stated that leaders are more able than other employees to solve the changing and complex problems and situations in everyday work. Their work consists of meetings, discussions in groups and in person, administration, and substance-directed tasks. In administration duties, the core tasks are decisions about organization, resources and evaluation of work. Other tasks are documentation, calculations, and compilation of statistics. Substance-directed duties are advising and directing personnel, solving professional problems, and taking care of the occupational competency of personnel.

ICT is a useful tool for carrying out these duties. It makes systematic evaluations and quick evidence-based assessments possible at every level of the organization in the production of services. However, this is only the case if the ICT systems are well designed, they work properly, include adequate and correct information for the task at hand, and personnel and superiors are able to use them confidently and innovatively when needed.

According to Miettinen (2005), ICT and the competencies to use and develop it are crucial in the tasks of immediate superiors. They need those competencies when leading distributed teams, evaluating how short- and long-term goals are reached, for calculating cost efficiency, and working out how to improve customer-centered orientation and develop work processes. Thus it is important to realize what factors may promote or hamper the use and adoption of ICT among leaders in the social and health care sector. The barriers are negative attitudes, lack of knowledge, role adjustment related to the disruption of traditional work habits, and changes in established work roles. The expected benefits are increased quality and efficacy of patient care, information and knowledge for patient care, less paperwork, and reduced errors (Andre et al., 2008).

Earlier research has shown that lack of knowledge, strong habits, and negative attitudes are still common barriers to the development of computerized tools among social and health care workers (Andre et al., 2008; Sinervo & von Fieandt, 2005; Spenceley, O'Leary, Chizawsky, Ross, & Estabrooks, 2008; Willmer, 2007). This is a real challenge for supervisors who are supposed to be responsible for developing services, consulting and supporting their subordinates, generating a positive working culture for learning and growth, and ICT use. The challenges are even tougher if a supervisor him/herself is experiencing uncertainty when using ICT and learning new systems.

RELATION OF CORE TASK AND ICT USE

The challenges mentioned here and faced by supervisors are, from one viewpoint, personal representations from the situation in which supervisors apply ICT to their duties. In this study, we are interested in these subjective representations and experiences. One way to describe and analyze these experiences is to use the activity model (Engeström 1987) as heuristics. With this model, we can concentrate on dimensions of experience that describe a personal system of ICT use. It also makes us analyse the construction of the meaning of ICT use linked with subjectively construed understanding of the core tasks of care and nursing.

According to Benson et al. (2008), using the activity model as an analytical tool for studying ICT use means we must see ICT not as the object of activity, but as a constitutive part of the infrastructure, which impacts on all mediators of the activity simultaneously. We have to see the way in which the subjects perceive matters. They see the ICT system as impacting on at least three mediators in the model. They analyzed the influence of tools, rules and divisions of labour on activity. ICT as a tool provides possibilities on a psychological level. Rules are cultural norms and regulations that govern the performance of action, and ICT can be seen as an administrative imposition or something that must interface with these. ICT also reflects the division of labour and stakeholder groups have different roles to play out with it. These roles are changeable and negotiable.

When studying the interaction between tool and object, we focus on how, through this interaction, the object is created, and how the action produces the desired outcome. When adopting ICT at work, the perception of usefulness of the system tells us how desired goals are met and how well this system helps the performance of the work activity at hand. Especially in technological changes, when employees are studying new skills to use tools, new tasks, and rules, the subject can lose his/her focus on the object of the work activity.

In ICT-mediated work, the perceived results rely on the experienced competence of ICT use, the ICT culture of the organization, and the rules drawn from the control system. The outcome of the action is nevertheless at the top of the activity hierarchy. Here we consider it is an organization's core task, adopted by supervisors as a perceived core task. The perceived core task is the ultimate meaning of the employee's work activity. Because of new IC technology and service process development, management and leadership in social and health care are changing, which reflects on divisions of labour, communities, work rules, and in some cases on the outcome of work and the object-outcome relation.

INNOVATION AND LEADERSHIP

According to some studies, the main tendency goes to professional leaders (Viitanen et al., 2007). Leaders are not substance specialists, but experts in leader's profession. In the leader's profession, there is need for specialised knowledge about and competence in leadership and management. The place of a leader is no longer so dependent on substance knowledge and seniority.

Leadership orientation can be divided into two components: transactional and transformational (Bass 1985). Transactional leaders concentrate on an exchange process where emphasis is placed on behaviours and attitudes which mediate the quality of exchange between the leader and followers. Transformational leadership is defined according to how the leader affects followers. It is supposed that the followers feel admiration, loyalty, trust, and respect, and that the leader influences followers by connecting their self-concept to the mission of the organization or group. In this kind of situation, followers' act self-expressively and are willing to contribute to group objectives. Bass' theory (1985) suggests that transformational leaders can enhance innovative work behaviour, because a leader motivates followers to do more than they are originally expected to do. An opinion also exists that innovation arises from the contributions of flexible and open-minded individuals (Reuvers, van Engen, Vinkenburg, & Wilson-Evered, 2008) who flourish under the guidance of a transformational leader.

WELL-BEING AT WORK AND ICT USE

It has been shown that the implementation of new information technology generally raises employees' levels of stress (Aronsson, 1989; Korunka, Zauchner, & Weiss, 1997; Smith & Conway, 1997). The workload of the individual employee or leader, caused by the implementation of a new system, varies according to the size of the change, the employee's place in the organization, the employee's personal orientation and competencies, and the management of the implementation (Korunka et al., 1997). Stress usually arises from problems that originate either from the usability of the system or the implementation process (Burke & Peppard, 1995; Nah, Zuckweiler, & Lau, 2003; Ristimäki, Leino, & Huuhtanen, 2003). The implementation of a new system, along with major changes in organization structure, work processes, and work tasks, is challenging, and requires different competencies from employees.

RESEARCH QUESTIONS

This study focused on two areas. Its first aim is to research the immediate superiors' experiences of the use, implementation and changes of ICT systems. The second is to improve the immediate superiors' opportunities to use, develop, and implement new ICT in tune with their core task orientation.

Information exists on the quantity of ICT systems in social and health care and also on the amount of their users in practices as a whole. The purposes of systems are also well known. In the case of singular systems it is understood how they fit in with the work and service processes. However, there is a very limited amount of knowledge on how superiors themselves experience ICT as a tool in their duties, and also on how they use ICT in their everyday work. The first research question is: 1. How is ICT used in the leadership activities of social and health care personnel's middle management and immediate superiors?

It is known that in most ICT system changes, work processes also transform, and the division of labour is rethought. When thinking well-being at work it is a salient point to get those changes under the feeling of control. The changes also become more understandable and acceptable when they reach the expected goals set by the improvement of the management of service processes. The second and third research questions are: 2. What experiences do superiors have of the changes of supervisory duties caused by ICT and how do these experiences associate to work resources? 3. How does ICT support the core task of leadership activity from the standpoint of immediate superiors?

As stated earlier, the requirements for developing work can be seen to be behind barriers from the viewpoint of superiors if their orientation and the organization's history emphasizes the active and passive management by exception. 4. How is the superiors' leadership orientation associated with the use and development of ICT in practice?

REALIZATION OF STUDY

This research and development project focuses on three municipal organizations in social and health care services in Finland. The services selected for the study are home services for elderly people, family services, psychiatric outpatient services for young people, and welfare services for children. The home services for elderly people were studied in two organizations in which new mobile information systems for home aid personnel were under implementation. In one organization, the case under investigation and elaboration was the integrated information system that constitutes knowledge for superiors to develop client services.

From each organization we selected approximately 20 superiors to interview and about 10 to 20 superiors to fill in a questionnaire. Some of the subjects in each organization were also chosen as participants in an observation process in which their ICT use is followed for one working day. After the data acquisition period, three workshops will be held for the interviewed superiors of each organization. These workshops are to handle topics from the interviews, surveys, and observation sessions of the first stage (see figure 1). Data will be collected during the workshops, through recoding and collecting documents. After analysis of the data, network meetings will be organized for all the participants. The project is planned to last two years, and began in the autumn of 2009.

HOME SERVICES FOR ELDERLY PEOPLE

This paper concentrates on home services for elderly people in one municipal organization in Finland. The study is currently in the interview phase. In preliminary analysis, one of the organizations is introduced, and the interview results presented.

The municipality at issue has 241 000 inhabitants. In 2006, municipality comprised 22 000 inhabitants at the age of 65 or over. In 2015, the estimated number is 36 000. The corresponding number of 75 year-olds and older inhabitants is 9000 (2006) and 13 000 (2010), respectively.

The goal of home care services is to help elderly people, disabled persons over 18 years, convalescents and those with chronic diseases, to live at home safely even when their ability to function weakens. More specifically, customers are in the process of resettling at home after a stay in hospital, their ability to manage daily basic functions has weakened, they are taken care of by their relatives (who also need help), or for some reason it would be inappropriate to care for their nursing needs somewhere other than at home. The evaluation of service needs is carried out together with the customer and his/her relatives.

The tasks of immediate superiors in home services for elderly people are officially defined as management of activities, personnel management of a multiprofessional team, client work, decision-making of customer fees, services, financial support, military injury support and outsourcing services. These also include the expert development and tasks, as well as network co-operation. In other words, the job description of immediate superiors of home services is multifaceted and demanding.

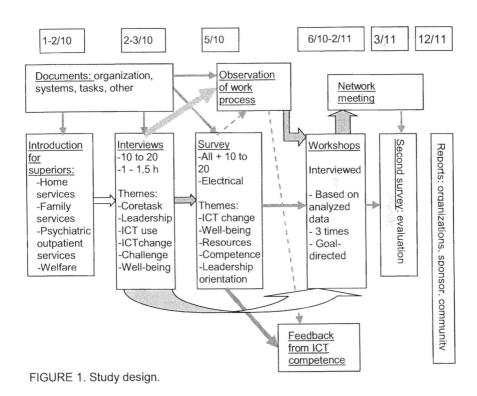

FIGURE 1. Study design.

PRELIMINARY FINDINGS

In this section, we will introduce some general preliminary interview findings. That is, how ICT is used in the leadership activities of immediate superiors of home services for elderly people, what benefits and, on the other hand disadvantages, ICT has brought into their work, and how ICT has changed their work.

We interviewed seven immediate superiors from home services for elderly people from different areas of home service. More interviews are still to come.

Immediate superiors use as many 12 different ICT programs in their management duties. Interviewees describe the biggest change as being due to the ICT customer system, which was taken into use in 2002. Customer system is an essential tool for immediate superiors for controlling work processes and monitoring the quality and quantity of services given to clients. For example the customer's service and care plan is drawn up electronically by customer system. The service and care plan is a tool for planning the given service to a particular customer and gives information which immediate superiors can use to direct and coach home care workers. It is now possible to make decisions more quickly, thanks to easy access to comprehensive information on client situations. This also improves the quality of service. Although interviewees saw customer system as an important and useful tool in their superior work, there have been problems in the inner logics of the system as well as in the attitudes and know-how of some home care workers. Some workers have been excused from using of ICT systems because they consider their own skills insufficient. This problem relates mainly to older workers. However this was not the case in all teams in home services for elderly people, and many workers possessed basic computer skills. In the interviews, superiors talked more about subordinates' computer skills than their own skills.

The main management tasks are divided into four categories, which are personnel administration/personnel management, client work, development at own team's level and at the level of the organization, and co- operation with interest groups. The immediate superiors told that they use ICT systems mainly in the first two tasks mentioned, i.e. in personnel administration/ personnel management and client work. Among these ICT programs are programs that interviewees describe as easy and functional, but also programs that they feel are heavy to use and lack the required logics. Although tasks on a computer take a lot of time in immediate superiors' jobs, they describe face-to-face personnel management as their most challenging job. Keeping up the motivation and well-being of subordinates requires the physical presence of a superior and cannot be replaced by computers.

The organization will implement mobile devices in their home services for elderly people. We asked how immediate superiors foresee the implementation process and what they expect of mobile devices. The devices were seen as a positive resource, providing that they work well. Previously (when mobile phone devices were connected to the internet) there were connection problems, and being frequently cut off naturally frustrated the users. The idea of home care workers doing their necessary office work at customers' homes received support from immediate superiors. This reduces peaks in the office, where there are not enough computers. The mobile device would enable the immediate superiors to follow the work situation in real time and ease planning in acute situations. It would help if replacement was needed, because the system would show which of the home care workers are free at a particular moment. One suspected disadvantage was that the mobile devices reduces face-to-face contact at the workplace and therefore reduces the support of the working team and immediate superior. The know-how of subordinates was also brought up as a possible problem. Some customers have also felt suspicions of home care workers using computers in their homes. On the other hand, it was mentioned that much of the computer work can be carried out in co-operation with the customer; the service and care plan, and different measures of the customers' condition. The mobile device could also quicken the service given to the customer, because it makes it possible to check many things right away.

ICT has changed the job of immediate superiors by giving them more tools for monitoring and planning their work. It has saved time by providing easy, quick access to information and reduced the amount of paperwork. Immediate superiors see that ICT systems have provided them with more resources to manage client work. They do not want to go back in time and give up ICT. Of seven interviewees only one was very critical towards the ICT policies of the organization. Another six interviewees also mentioned problems, but these attitudes are more like neutral or slightly positive. The rapid changes in the organization were brought up in many interviews, including changes in ICT systems. It was also wondered, why the pilot projects are not carried out properly, and non-functioning systems are taken into use before the problems are solved. Concern was also voiced as to why immediate superiors' opinions are not taken advantage of in the ICT planning process.

CONCLUSIONS

These results draw a draft picture of ICT use and core task fit to us. They also demonstrate the situation of developmental orientation of leadership in everyday work and the possibilities for service innovations in ICT-mediated tasks.

First of all, the usefulness and ease of use aspect of the ICT systems was highlighted in two ways. They are considered useful in everyday work and mainly promote core task execution. However, the usability of the systems was also seen as problematic. Some systems were well-functioning and definite, but others had bad logics and therefore did not fit the personal work process representation. Memory load was one main problem when using diverse systems with poor action logic. Despite this, and the time-consuming character of the systems, the use of ICT was not seen to increase mental loading in supervisory duties. The most prominent and important dimension of core task in this specific work was leading people in actual interaction, which was also seen as the most loading factor in supervisory duties.

The development aspect of leadership can be seen as very challenging in this context; also when the focus is on service restructuring in line with new ICT. From the standpoint of superiors there was no problem in their attitudes towards ICT and its use. The problems stemmed from the restrictions of the context. The main restrictions seem to be resource shortage. The most obvious shortage for superiors was the state of know-how of ICT use in the organization. It was possible that the developmental aspect of the leadership duty interrelated with ICT was quite narrow, and that, for example, the internet was not exploited in the leading process. In a way it can be argued that ICT was more a like static tool than a flexible platform for

324

developing services.

Because of the developmental approach, this project will be instantly useful to the organizations and superiors involved. On a general level, we hope to attain a more rigorous understanding of ICT in leadership activity from the viewpoint of social and health care superiors.

REFERENCES

Andre, B., Ringdal, G. I., Loge, J. H., Rannestad, T., Laerum, H., & Kaasa, S. (2008). Experiences with the implementation of computerized tools in health care units: A review article. *International Journal of Human-Computer Interaction, 24*(8), pp.

Aronsson, G. (1989). Changed qualification demands in computer-mediated work. *Applied Psychology: An International Review, 38*(1), 57-71.

Benson, A., Lawler, C., & Whitworth, A. (2008). Rules, roles and tools: Activity theory and the comparative study of e-learning. *British Journal of Educational Technology, 39*(3), pp.

Burke, G., & Peppard, J. (Eds.). (1995). *Examining business process re-engineering. Current perpectives and research directions.* London: Kogan Page Limited.

Korunka, C., Zauchner, S., & Weiss, A. (1997). New information technologies, job profiles, and external workload as predictors of subjectively experienced stress and dissatisfaction at work. *International Journal of Human Computer Interaction, 9*(4), 407-424.

Miettinen, M. (2005). Terveydenhuollon innovatiivisuuden esteitä, kannusteita ja mahdollisuuksia. In J. Vuori (Ed.), *Terveys ja johtaminen. Terveyshallintotiede terveydenhuollon työyhteisöissä.* (pp. 260-277). Helsinki: WSOY.

Nah, F. F.-H., Zuckweiler, K., & Lau, J. L.-S. (2003). ERP Implementation: Chief information officers' perceptions of critical success factors. *International Journal of Human Computer Interaction, 16*(1), 5-22.

Nikkilä, J., & Paasivaara, L. (2008). *Arjen johtajuus, rutiinijohtamisesta tulkintataitoon*: Suomen sairaanhoitajaliitto

Reuvers, M., van Engen, M. L., Vinkenburg, C. J., & Wilson-Evered, E. (2008). Transformational leadership and innovative work behaviour: Exploring the relevance of gender differences. *Creativity and Innovation Management, 17*(3), pp.

Ristimäki, T., Leino, T., & Huuhtanen, P. (2003). Call center work: customer service work with information technology. *Psykologia, 38*(5), 319-328.

Sinervo, L., & von Fieandt, N. (2005). *Tietotekniikka sosiaali- ja terveysalan osaamisen kehittämisessä.* Helsinki: Stakes.

Smith, M. J., & Conway, F. T. (1997). Psychosocial aspects of computerized office work. In M. Helander, T. K. Landauer & P. Prabhu (Eds.), *Handbook of human-computer interaction* (pp. 1497-1517): Elsevier Science B.V.

Spenceley, S. M., O'Leary, K. A., Chizawsky, L. L., Ross, A. J., & Estabrooks, C. A. (2008). Sources of information used by nurses to inform practice: An integrative review. *International Journal of Nursing Studies, 45*(6), pp.

Viitanen, E., Kokkinen, L., Konu, A., Simonen, O., Virtanen, J. V., & Lehto, J. (2007). *Johtajana sosiaali- ja terveydenhuollossa*. Vammala: Kunnallisalan kehittämissäätiö.

Willmer, M. (2007). How nursing leadership and management interventions could facilitate the effective use of ICT by student nurses. *Journal of Nursing Management, 15*(2), pp.

CHAPTER 35

Ergonomic Conditions for Social Integration of the Disabled

Jerzy Olszewski

Poznan University of Economics
Poland

ABSTRACT

The article is divided into four parts: an introduction; a presentation of research methods; a description of the empirical stage of research and concluding remarks. In the introductory part, it has been shown that the idea of integration of the disabled has been implemented for a few decades. This idea is supported by ergonomics, which contributes to the prevention of segregation and isolative tendencies, as well as the stigmatisation, intolerance and discrimination of the disabled. In part two of the study, research methods have been described, such as: a critical analysis of specialist literature, a comparative analysis and an interview questionnaire. In the empirical part, certain ergonomic solutions to reduce architectural and spatial barriers have been exemplified as a significant factor conductive to the integration of the disabled. The article ends with a summary, which includes suggestions for future ergonomic undertakings for the disabled.

INTRODUCTION

The issues of social integration of the disabled are currently receiving recognition. The problem has recently been of growing concern to economists, social activists, the media and public opinion. As follows from both theory and experience, the matter of our greatest interest should be how the process develops within the family. It is the family circle that a successful social integration of the disabled depends on.

Beyond all doubts, a family which shares their living with a disabled person ought not to be left to their own resources. Systemic solutions need to be introduced to support both the disabled person as well as the whole family. Such solutions will create favourable conditions for the process of the person's social integration.

The efforts, made in recent years, to create a uniform and complete support system have not proved entirely successful.

The economic transformation (the creation and development of a free market economy – the deregulated labour market in particular; changes in the country's social and welfare systems; changes in the system of public administration–decentralization) has played its role here and still has an effect on the socio-economic situation of the disabled.

At the current level of the socioeconomic development of our country, the issues should not be ignored. The elderly and the disabled make for a substantial part of the society. A recent analysis of demographic tendencies over the past decade shows a progressive senescence of our society and, its consequence, an increase in the number of those who become disabled with age. In such circumstances, the government needs to aim its policy at introduction of institutional, legal and ergonomic solutions in support of the disabled. Politicians could borrow a leaf from ergonomic textbooks. They can base their decisions on recent ergonomic advances, which are already being employed to remove architectural barriers and make urban areas more accessible to the disabled, and in this way make a significant contribution to their social integration.

To solve the problem, however, we need to implement a model of social integration. The model should be based on the principle of equal opportunities, i.e. equal start-out, availability of the potential for development and the lack of limitations as regards social status.

RESEARCH STRATEGIES

A survey has been carried out to check the accessibility of public transport and the design of buildings (private dwellings, public buildings, higher education institutions, and trade and gastronomy premises). The researchers employed several strategies, including analyses of films and photographs of given objects, and very often participant observation (with the researcher acting a disabled person). The research also required critical analysis of literature on various subjects, such as physiology, psychology, anthropometry and ergonomics.

The survey was carried out in Poznań, between May and November 2009, and included residential buildings, public buildings, institutions of higher education, trade and gastronomy buildings as well as city transport.

THE ACCESSIBILITY OF POZNAŃ TO THE DISABLED – A PROFILE

ARCHITECTURAL BARRIERS

Private Dwellings

To the disabled, private dwellings are minefields of challenges. Narrow doorways, for instance, often turn out impassable to individuals in wheelchairs. Under applicable Building Code, the exterior door should not be narrower than 90 cm. The recommended width is, however, 100 cm in door frame. A door wider than 100 cm, due to its overall dimensions, may be too heavy and difficult to open. The doorstep is an indispensable element in the door construction, but frequently a trouble spot, too. It often reaches up to 6-8 cm in height. Under the amended applicable Building Code, with respect to blocks of flats, the front door as well as the exterior and interior doors inside the building, the height of the doorsteps must not exceed 2 cm.

When choosing a window for a disabled person, it is very important to pay a good deal of attention to the manner of its opening. There are tilt windows, horizontal sliding windows and turn windows. Apart from that, vertical sliding windows with crank mechanisms are available. They allow for easy opening of the window and the ability to block it at any level.

Windows are usually placed 80–85 cm above the floor level. In the room of a bedridden or seriously ill person, the lower edge of the window can be placed 60 cm above the floor. A remote control can be of use to those with limited manual dexterity. It combines two methods of opening: automatic and manual, which is the best possible solution. The switches, door phones, doorbells, spy holes, grab bars and handles should be situated at the optimal height of 70-130 cm above the floor level [1].

The kitchen is for most people the centre of the house. It is, however, its location in relation to the other rooms that determines whether it is going to be functional or not. Irrespective of its type and functions (be it an open kitchen with a dining room, a purely functional closed kitchen, or one with a pantry) it should meet the measurement standards. The shape of the kitchen depends on individual preferences. It is nevertheless essential to provide the minimum 150 cm turning circle of the manoeuvre space and a free access to cupboards and kitchen appliances. It is of vital importance to anticipate the user's needs and to make the most of the working area (preparing food, cooking, etc.).

The kitchen should be situated within a comfortable proximity to the exterior door. Due to the limited manoeuvre space of the wheelchair, a minimum 150 cm turning circle should be saved. Most frequently, the kitchen is square or rectangular in shape. Depending on the length of the walls, it allows to arrange the working space in L, I, II, U or C-shaped area.

The required minimum width of the kitchen door should be 90 cm in door frame; however 100 cm is more advisable. To save space, movable or folding doors can be used.

Several factors affect the self-reliance and quality of life of the disabled. Such factors include consolidated support systems in flats and houses. One of the most important elements of the fixtures and fittings, which is indispensable, is the wiring. Recent technological advances allow for cordless systems. Their advantages are obvious: they provide flexibility in the connection; facilitate device-to-device communication and eliminate cable installations.

Alarm systems, though not entirely reliable, not only alert us to trespassers, but also function as emergency signal systems. In case of an injury or a fall they alert others to the need for help from outside.

Public Buildings

The fundamental regulation regarding the accessibility of buildings to the disabled is Article. 5, paragraph. 1 pt. 4 of the Construction Law, according to which a building together with the associated equipment should be, taking into account the expected service life, designed and built in the manner specified in the legislation, including the technical and construction aspects, in accordance with the principles of technology, providing the necessary conditions for use of facilities, utilities and multi-family residential housing by disabled persons, especially wheelchair users.

Stairs are inherent in architecture. They determine the esthetics of a building. Unfortunately, they are also among the biggest architectural hindrances.

People in wheelchairs need alternative solutions which would enable them to overcome hilly or uneven terrain. If technical conditions allow, the solutions will include systems of ramps or alternatively support devices, such as outdoor lifts as well as all types of elevators and platforms. As concerns over- and underpasses, a lift or an elevator is the only solution which allows getting form one level to another.

A ramp is an element in the construction of a building which makes public and residential buildings accessible to the disabled. Unfortunately, a correctly constructed ramp takes up much space. For this reason, due to limited space outside buildings, many a ramp is steeper than stipulated. In such cases, the translocation between levels is passive (requiring the assistance of others).

Sloping platforms and incline lifts will make it possible for individuals in wheelchairs to move up and downstairs without assistance. If, due to a lack of space in a building, an elevator cannot be installed, they remain the only alternative. Depending on the needs and the actual model, the lifts and platforms may run along single-flight stairs, double-flight stairs with a landing and winding stairs. The lifting height should be limited to two storeys. If the staircase is too narrow to install a lift, a vertical ceiling-attached platform is possible as an alternative solution.

Poznan is gradually becoming more and more disabled-friendly. Most of its valuable historic monuments are practically fully accessible, including Ostrów

Tumski with its beautiful cathedral (the pavements and square outside have recently been renovated), the beautiful baroque *Fara* (Parish Church), where organ concerts take place in summer (has its entrance at the same level as the pavement) or The Old Market, bordered by beautiful tenement houses, with its freshly renovated townhall in the centre.

Theatres and cinemas play an important role in every person's life. *Teatr Nowy* and *Teatr Wielki* in Poznan offer wheelchair seatings. Some of Poznan's cinemas are also prepared to host the disabled. Apart from the new, large, fully accessible buildings, such as *Multikino* or *Kinepolis*, smaller cinemas also show concern for all the spectators who come to visit them. *Muza* has 10 wheelchair seatings and *Rialto* has an inducting loop for individuals with hearing aids.

Several modern sport centres in Poznan are disabled-friendly, too. Keen swimmers can visit the pool in Wroniecka Street. In spite of a small flight of three stairs at the entrance, it is the only swimming pool with special wheelchair lifts which enable a safe immersion in the water. A new swimming pool on Droga Dębińska has a similar solution.

As a fair-host city, Poznan has a large accommodation base. Some of its hotels and hostels, e.g. *The Mercure, The Novotel, The Ibis, The Park and The Kemping Malta*, offer rooms for the disabled .

Likewise, the City Hall has taken measures to make certain buildings and spots disabled-friendly. At the turn of 2004 and 2005, a new elevator was put into service in a building on Libelta Street. This allows individuals in wheelchairs to get to the higher floors. Morevoer, there are outer and inner visualisations of the object, which makes it easier for people in wheelchairs to find departments and move inside the building. The customer service points in the waiting rooms are furnished with special tables and seats, which can be used by disabled persons when filling in forms and documents or when queuing up.

Higher Education Institutions

One of the main problems the disabled have to wrestle with is the fact that they have difficulty obtaining proper education and finding a job. *Poznan School of Banking* (*Wyższa Szkoła Bankowa*, abbr. *WSB*), being one of the first higher educational institutions in Poland, has prepared a comprehensive support programme to help the disabled gain higher education and be able to find satisfying jobs. The programme is called "Open University".

Within its framework, *WSB*, employing its own financial resources, got rid of architectural barriers and in this way enabled people in wheelchairs to study. In *WSB* headquarter, 5/7 Ratajczaka Street, ramps, grabs bars, toilets and wide doorways were installed. Some 50 m away from the building, parking spaces are especially allotted at the disposal of the disabled. There is also a lift, which makes it possible for disabled people to move between floors [3].

Disabled people, like all applicants, must undergo qualification examination in relevant subjects. They are offered customized forms for the entrance exam.

Institutions of higher education adjust the form of the exam so that it matches the needs of their disabled applicants, but without any allowances or exemptions from the exams.

During the entrance exams, disabled applicants are offered assistance. For applicants in wheelchairs and those who walk with difficulty, the exams take place in rooms where all architectural barriers have been removed. If they need to get to rooms with a limited access, they are provided with support. Applicants with manual disabilities can have their writing exam prolonged by 50%. People with serious visual impairments and the blind can ask for special examination papers, whose print is enlarged to an optimal size so that it suits individual needs of the applicant. It is possible to prolong the exam by 50% of the time. Applicants can also count on an examiner's help at question reading. In the case of applicants who are not able to take oral exams (the deaf), it is possible to change the examination format (oral to written) or to employ a sign language interpreter.

It is only a matter of time that the institutions of higher education will all become customised to the needs of the disabled. Advocates for the disabled and students are sparing no efforts to eliminate all barriers. The institutions of higher education in Poznań are gradually taking care of their students so that every single person is given a chance to study.

Trade and Gastronomy

Stairs, narrow passages, lack of functional washbasins and toilets, table arrangement with no access aisles – they are all factors which create barriers. They are equally problematic to those in wheelchairs and on crutches as well as to those whose physical mobility is unimpaired. People whose facial features mark their condition (e.g. those with Down syndrome or the blind) are often ignored by waiters and bartenders. Orders are taken from their caregivers. Such barriers are also encountered in Poznań. These situations are problematic to about 37 disabled people who live in the city.

An action taken by a united effort of the *Association of Friends of the Blind and Visually Impaired* from the *pion.pl* portal of *Gazeta Wyborcza* and *Radio Merkury* encourages the owners of restaurants, cafes and bars to recognise the needs of their disabled customers. In the "Tutaj jest OK" action, a commission consisting of both unimpaired and disabled people visited Poznan restaurants, cafes and bars. They were to watch and decide whether the conditions are favourable enough for a disabled person to get inside, use all the facilities (e.g. toilets, bars) and also – how the staff were tending to them. As a prize, the commission would award spoke wheels, just like those in wheelchairs. The maximum prize was 15 wheels (5 in each category). There was also a special ranking. Those restaurants, bars and cafes which ranked at the top, received prestigious badges which certify that the premises are disabled-friendly: "Tutaj jest OK" (c.f. fig. 1) [8].

FIGURE. 1. „Tutaj jest OK" badge, source: www.tutajjestok.pl

Thanks to this action, several owners decided to introduce modifications in their restaurants, bars and cafes to adapt them to the needs of the disabled.

In the course of the action, 220 restaurants, bars and cafes were visited. 61 of them entered the list of the disabled-friendly. An individual in a wheelchair can enter them, move easily inside, and use the washbasin and the toilet. The staff attends to a disabled person as to any other customer.

TRANSPORT

There are 297 overall customized stops in Poznan: 112 tram stops and 185 bus stops. By the end of our survey, the number of trams had increased by 14 low-floor Combino trams and 50 low-floor buses, thanks to which, during the day Poznan citizens can use only low-floor buses (a total of 205) [5].

Each low-floor bus is wheelchair accessible via a ramp on the rear door. It has part of the rolling stock controlled form the driver's cabin. Apart from that, low-floor buses have the tilt function, thanks to which the driver can bend the right side of the bus by a few centimetres in order to reduce the gap between the curb and the threshold.

In tram cars, the ramps are installed on the rear doors, too. The need to make use of the platform is signalled in the same way as on the bus.

Nearly 15% of tram cars are equipped with on-board information systems. The announcements are broadcast at the moment of door opening and inform about the name of the current tram stop. At the tram's departure from a stop, the name of the next stop is displayed. It is at the convenience of the disabled, especially those with visual impairments, who thanks to this system know their whereabouts. It is also at a great convenience of those in wheelchairs as windows are often shut out by crowds of commuters or simply placed too high. The announcements can also be made audible outside the vehicle, and inform about the line's number and direction.

Poznan engineers have patented a device which is an aid to the blind and those with visual impairments. PIP or Personalny Identyfikator Pojazdu (Personal Vehicle ID) looks like a cell phone. It connects with the oncoming bus with the help of GPRS (General Packet Radio Service). The sound effects and information about where the bus is heading for and what line it actually is, can help the blind move easily around the city.

The owner of such an aid chooses a bus number. When the bus arrives, PIP informs the user about the facts and gives the direction of the route. On pressing an adequate button the driver is provided with information that a disabled person wants to board.

Produced by a Poznan company in relatively small series of about 100 items yearly, the device has so far been used with bus lines only. The production cost is about 1000 zl for 100 items. However, the producer, in cooperation with MPK (the city public transportation system), wants to extend the application of the device by tram lines and zebra crossings. PKP (Polish National Railways) and PKS (National Coach Transport Company) [6] are also included in the producer's plans.

FINAL REMARKS

The research shows that Poznan is to a large extent accessible to people with disabilities.

Poznan is gradually becoming more and more disabled-friendly. Individuals in wheelchairs can easily use all means of city transport. All buses and most trams are low-floor. There are also customized taxi cabs, available 7 days a week, 24 hours a day.

The most important monuments, such as the baroque *Fara* (Parish Church) or the Old Market with its newly renovated Town Hall and bordering tenement houses, are almost freely accessible. A disabled person is also free to visit some restaurants or pubs. Besides, there are shops and supermarkets where disabled people can do shopping without assistance.

Poznan institutions of higher education are gradually becoming disabled-friendly, too. They are trying to help disabled people and encourage them to study. Even if their premises do not meet all the required standards of accessiblity, volunteers and educational aids will well make up for the shortcomings.

Poznan has been honored for its innovative approach towards the organisation of social life. Especially its involvement and enthusiasm in making the urban environment disabled-friendly has received recognition. As a prize, Poznan was offered the chance to show a promotional film to a 3000 audience at the Integration *'Gala'* Meeting, which took place in the 'Kongresowa' Hall of *The Palace of Culture and Science* in Warsaw.

However, not everything is running as smoothly as it should be. Underpasses, for instance, are still most problematic to all wheelchair users. Neglected trouble spots, they are a real nuisance to individuals in wheelchairs and create obstacles which disabled individuals have to overcome by diversion routes.

Hopefully, social integration, as a social and educational movement, is going to act counter to segregation and isolative tendencies, as well as to the stigmatisation, intolerance and discrimination of the disabled. We can expect that it will enable individuals with any impairments to participate in everyday life and make all the institutions and services accessible to them. Nevertheless, in spite of all the initiatives which have been pursued in recent years, there are still several social barriers which exclude the disabled from active social life.

334

REFERENCES

[1] ABC... dla architekta, Stowarzyszenie Przyjaciół Integracji, Warszawa 2005
[2] Balcerzak-Paradowska B.: Sytuacja osób niepełnosprawnych w UE, Wydawnictwo IPiPS, Warszawa 2002.
[3] Gonciarz B. Ostrowska A., Pańków W.: Integracja społeczna i aktywizacja zawodowa osób niepełnosprawnych zamieszkałych w małych miastach i na terenach wiejskich (uwarunkowania sukcesów i niepowodzeń).. Raport z badań, IPiS, PAN, Warszawa 2008.
[4] Informator Wyższej Szkoły Bankowej
[5] Kirenko J.: Wsparcie społeczne osób z niepełnosprawnością, wyd. Wyższa Szkoła Umiejętności pedagogicznych i Zarządzania w Rykach, Ryki 2002.
[6] Paszkowicz M.A.: Ekonomiczno-społeczny system integracji osób niepełnosprawnych w społeczeństwie polskim, AE, Poznań 2002, pp. 215 & 216.
[7] Raport za 2006 rok Powiatowego Urzędu Pracy w Poznaniu
[8] http://www.city.poznan.pl (2009-12-01)
[9] http://www.e-poznan.net (2009-12-01)
[10] http://www.niepełnosprawni.pl (2009-12-01)
[11] http://www.pion.pl (2009-12-01)
[12] http://www.ueniepelnosprawni.p (2009-12-01)

CHAPTER 36

The Paradox of Lean in Healthcare: Stable Processes in a Reactive Environment

Ander Paarup Nielsen[1], *Kasper Edwards[2]*

[1]Aalborg University, Denmark

[2]Technical University of Denmark

ABSTRACT

The principles of lean are widely being adopted in the healthcare sector. Interestingly the realized benefits appear not to warrant the interest from managers and policy makers.

This paper presents an analysis of 3 Danish healthcare organizations which all introduced lean initiatives. However, only a limited set of tools has been used and the productivity gains are limited focusing on peripheral activities and not the core medical activities.

This apparent problem with lean in health care is hypothesized to be caused by 1) the nature of healthcare work, 2) the rationality and notion of validity among different groups of healthcare professionals and 3) different rationalities in lean and professionals in healthcare.

Through analysis of three cases it is concluded that the nature of work is significantly different from manufacturing primarily because of the reactive nature of work. Finally, different rationalities are observed between different groups of healthcare professionals leading to problems employing the lean tool-box.

Keywords: Lean management, healthcare, reactive work processes

INTRODUCTION

The Danish healthcare system is a public healthcare system. Like its siblings in Europe it's always under pressure to deliver more and better healthcare services. In recent years this pressure has been increasingly accentuated.

There are several approaches for the healthcare system to deliver more care the most common is to add resources. However as scholars know, adding resources to a dysfunctional system does not provide proportionally more capacity and waiting times are not shortened. Another approach to becoming more efficient is better performing technology, which is often used but equally often the results do not match investment as the processes are not changed. Examples of CT-scanners being bought with no staff to man them are all too frequent in the press.

Recently fast-track systems as found in the UK and package systems such as in Vejle hospital, Denmark (Jakobsen, 2007) has surfaced as examples of organizing principles yielding high quality and low waiting times. Diagnostic packages are somewhat similar to fast-track systems with the noteworthy difference that all patients are offered a package – not just those suspected of a particular illness. Vejle hospital's success lie in their ability to diagnose all referred patients within a matter of days thus minimizing delay in treatment. As such the healthcare systems are beginning to use different modes of operation to obtain better performance and quality – as in production the amount of rework must be reduced to a minimum.

In this paper we focus on the use of lean in healthcare and observe that the ideas and principles from lean management are now widely being adopted within the health care sector both in Denmark but also internationally. The interest in lean from managers and policy makers, however, appear to contrast the realized benefits. This paper presents an analysis of cases reported in literature and three Danish healthcare cases, which show that organizations within health care most often only implement a limited set of tools and methods from the lean tool-box, leading to limited productivity gains.

We hypothesize that the poor results from lean is caused by 1) the nature of healthcare work, 2) the rationality and notion of validity among different groups of healthcare professionals and 3) different rationalities in lean and professionals in healthcare.

LEAN IN HEALTHCARE

Lean has been around for many years in the form of the Toyota Production System (TPS). But the concept and term surfaced as "lean" following a study of the Japanese car industry that tried to explain its high level of success (Womack et al., 1991; Liker, 2004). During the past decade we have seen lean being applied in other sectors of the economy, e.g., service industries (George, 2003), and administration (Tapping & Shuker, 2003). Lean is now being transferred to the healthcare sector. The goals are the same as in industry i.e. to increase quality of the care given to patients as well as to increase efficiency. This section will highlight some of the findings from a literature review concerning lean implementation within the healthcare sector (Edwards, Nielsen & Jacobsen, 2010).

In the process of writing this paper a thorough literature review concerning the application of lean in health care was carried out. Focus in the literature review has been on papers reporting findings from concrete cases concerning lean implementation in different types of health care organizations. Excluded from the literature review were conceptual papers and papers discussing lean implementation in very general terms, i.e. focusing on key success factors for the successful implementation of lean – for example (Spear, 2005). 17 papers concerning lean and health care were identified and examined. The papers were classified using the following two dimensions. The first dimension is concerned with the type of activity (or activities) that is being improved using lean. Here is has been decided to distinguish between core activities, i.e. treatment of and care for patients, and support activities, i.e. laboratory analysis, administrative processes, and the management of consumables. The second dimension is concerned with the approach to lean in the organization. The levels within this dimension are inspired by the prestigious Shingo Prize and the lean approach is thus split into three levels. The first level is "tool driven" which is characterized by an ad hoc use of a limited number of lean tools. The second level is "system driven" which is focused on a systematic use of a number of lean tools and principles. The third level is "principle driven" this approach is characterized by a widespread use of lean in the organization and an attempt to integrate lean principles into the organizational culture. This results in a matrix with 6 different lean implementation patterns in health care. Table 1 below reports the classification of the different papers in the literature review.

		Type of activity	
		Core	*Support*
Lean approach	*Tool driven*	King et al, 2006 Kelly et al, 2007 Ben-Tovin et al, 2007 Fairbanks, 2007 ***Count: 4***	Khandelwal & Lunch, 1999 Bushell & Shelest, 2002 Panchak, 2003 Leslie et al, 2006 Towne, 2006 Anonymous, 2007 Ballé & Régnier, 2007 ***Count: 7***
	System driven	None	Tragardh & Lindberg, 2004 Lummus et al, 2006 Weber, 2006 Fillingham, 2007 Ng et al, 2010 ***Count: 5***
	Principle driven	None	Stuenkel & Faulkner, 2009 ***Count: 1***

Table1: Lean implementation in health care.

The literature review illustrate that most of the lean activities within health care are focused on support or peripheral activities in the different organizations. Four papers have been classified as focusing on the core activities. However, all four papers focused on improving the flow of patients through an emergency ward. In all of these four cases the actual treatment procedures were not changed as a result. Improving the flow and thus reducing the waiting time can be seen as an attempt to improve a core activity directly focused on improving the care of patients by lowering their waiting times. Only one of the 17 papers reports an approach which can be characterized as principle driven lean and still the primary emphasis is on the support activities (Stuenkel & Faulkner, 2009).

The majority of the lean projects that has been analyzed are focused on the peripheral and support activities within the healthcare system such as reducing the turnover time in an operating room, improving the logistics of consumables, or improving the planning process within the wards. The administrative processes have also been analyzed, but a significant difference exists between public and private healthcare systems: billing. Public healthcare, unlike private providers, is not concerned with the front-office process of billing the patient.

Most of the papers with a focus on elements of patient care deals with implementing lean in organizational units where the patients undergo some kind of clearly identifiable medical procedure. Such procedures have distinct mechanical elements such as moving patients, drawing a blood sample and analyzing it, performing an x-ray etc. There are no studies of nursing wards where the patients are receiving care or treatment of patients with multiple, complex, and competing diagnoses. The reviewed lean implementations within hospitals are focused on activities and processes which to a large extent have the same characteristics as activities and processes in industrial manufacturing.

Also emerging from the literature review is the fact that implementation of lean is limited to the implementation of a small number of lean related tools, typically tools for process redesign or value stream mapping. For example, only one of the examined papers has an intentional focus on waste (Panchak 2003) – a key issue in the lean implementation processes and only one of the papers enters into a discussion concerning the role of lean philosophy and mindset in the organization (Stuenkel & Faulkner, 2009). Finally, many of the papers focus almost exclusively on productivity improvements and only one paper directly addresses the effects of lean implementation on patient or customer satisfaction.

Based on this literature review the paper will now go on by examining the three different explanations for the poor results of lean implementation outlined in the introduction of this paper. This analysis is based on 3 cases from Danish hospitals.

CASES

The empirical evidence is based on three different cases all from hospital wards. In the sections below the key findings from these three different cases will be presented briefly.

Case 1 is a surgery ward at a major hospital just outside Copenhagen. The ward has about 200 employees, is open around the clock and covers both acute and elective patients. It has 10 operating rooms whereas 6 rooms are for orthopaedic surgeries. The staffs were overburdened and often subject to overtime, which was a major point of complaint. Due to absenteeism and non-attendance from patients, the department was cancelling 6-7% of the planned surgeries. An analysis of the ward showed:

- Each of the 10 operating rooms showed more than 2 hours non productive time a day
- Lack of procedures surrounding surgery leading to idiosyncrasies
- Every surgeon and anaesthesia doctor had formed their own routines during operating procedures making cooperation difficult
- Clear sense of a lack of planning
- Rigid organizational structure
- Informal leaders especially among surgeons, leading nurses and anaesthesia doctors.

A lean programme was initiated to create more effective working procedures, and ensure a total continuity of care to the benefit of both staff and patients. The basic idea was to create operating rooms where surgeons do not have to leave the sterile area. In a similar manner teams were formed so that in-operating room team members did not have to leave the sterile area, thus relying on team members in the non-sterile area. With the complete operating team present in the operating room no-one is waiting and procedures can be completed without waiting.

This resulted in the establishment of two so-called "Turbo rooms" that only performs elective surgery on less complicated patients. The turbo rooms do not perform any education and is manned by the senior staff i.e. the most skilled doctors and nurses. The turbo rooms have a fixed team structure which has allowed a deeper analysis of cooperation during the actual operating procedure. The result has been overwhelming. What was previously done in three operating rooms can now be done in two and the teams are finished within their shift. This has a significant effect on morale as the teams experience a sense of accomplishment – they make a difference. While the number of patients treated by the teams are 33% higher the activity level is reportedly not higher, but the waiting times have been eliminated and absenteeism has decreased by 33%. However, the turbo rooms require more support staff which must be ready to act when needed.

Case 2 is a regional hospital and in general the hospital has very positive experiences with lean and has achieved significant positive results within, i.e., laboratory analysis and other non core activities. However, lean has primarily been implemented in support activities. The hospital now has an ambition to implement lean as part of the patient process. One example of this is a lean project in an out-patient surgical unit which tried to capture the process from patient arrival to discharge. The staff did a value stream mapping of all the activities the staff thought was involved. In the early phase of the value stream mapping there was widespread agreement on content of the different activities as well as the prioritization of the task. However, in the later phases of the project when the lean staff at the hospital began to test the value stream in real life by making test runs through ward the value stream maps did not match the real world.

The different groups involved in the process had very different interpretations concerning the value stream and actual content of activities. Differences arose over a number of issues, for example, over the necessary preparations for patients with different diagnoses, what equipment should be ready, and which specialists should be on call, where the patient should wait and how the patient should navigate. Interestingly, these different perceptions generally emerged between professions, i.e. doctors, nurses, secretaries etc.

Part of the explanation can be found in the different rationalities inherent in the professional groups. This gives rise to (mis)interpretation of the activities performed by the other professions. An internal lean consultant at the hospital stated that "doctors and nurses focus on different values and metrics. Nurses tend to focus on the quality of the care they can offer whereas doctors focus on the success rate of the treatments".

Case 3 is an oncology ward at a Danish university hospital. The ward is primarily focused on out-patient treatment of cancer patients, i.e. chemo therapy or radiation treatment. The lean activities of the cancer ward were part of a major lean initiative at the hospital. The lean activities were supported by a central lean task force and had significant top management support. The ward carried out a number of lean activities with good and positive results.

An element in the lean project in case 3 was the development of a standard operating procedure (SOP) concerning the booking of couches for chemo therapy patients. Besides initial technical problems this element of the lean project created resistance especially amongst the nurses who normally had the discretion to plan a series of sessions for the patients and book couches accordingly. Some nurses saw this new booking system as an attack on their professional judgement and discretion. They therefore resisted this standardization and argued that lean would limit their ability to provide the best possible treatment of and care of the patients.

CASE ANALYSIS AND DISCUSSION

The surgery ward (case 1) highlights the importance of stability and leveling for implementing lean. Of the 10 possible rooms only 2 was converted into turbo rooms. This was due to the mix of patients which only allowed the required stability and leveling for two operating rooms. Interestingly, emergency departments are well suited for lean despite the variation in patient mix. This observation is also supported by the classification of lean papers in table 1. The reason for this must be found in the nature of emergency departments where patients are treated with little or no preparation and a goal of being able to process patients quickly in order to either finalize the treatment and release the patient or to diagnose the patient in order to be able to transfer the patient to the relevant wards in the hospital. Unlike emergency departments, elective surgery requires extensive preparation and unforeseen switching of patients will ruin preparation and reduce productivity. All down stream activities such as wake-up and mobilization of the patients will be affected as they cannot be precisely planned. The ward receives all sorts of patients including emergencies it is not possible to turn all operating rooms into turbo rooms.

The out-patient surgical ward (case 2) illustrates another aspect in connection with challenges of implementing lean in health care. Here is found that is it very difficult to establish a true value stream due to different perceptions of the patient process. A key prerequisite in lean is that there is a shared understanding of the value the organization is providing to its customers. An example of this is found in (Stuenkel & Faulkner, 2009) where it is stated "administrators do not speak the language of nurses". Another part of the explanation can be found in the professions perception of the patient process as a series of discrete events. Each profession does his job without consideration for the prior or next activity in the process. This makes the value stream tool difficult to use as it result in what participants believe to be the process and not the actual process.

The final case from the oncology ward (Case 3) illustrates that there is a potential conflict between the development of standardized routines or SOPs inherent in lean and the professional judgment of the doctors and nurses. The health care professionals will claim standardization to be a significant problem as it limits the scope of their professional judgment, to the detriment of care quality. The implementation of lean in health care is not just limited by the different rationalities between the professions, but also by the potential conflict between the logics of standardization and need for individualized treatment and care.

THE CHALLENGES OF IMPLEMENTING LEAN IN HEALTH CARE

A challenge concerning the implementation of lean in the healthcare sector deals with hypothesis 1: the nature of the work. In industrial production the tasks are predictable, routine and they can thus be standardized. Many activities in a hospital are not to the same extent as industry predictable and routine – they are reactive. Performing surgery on – or caring for critically ill patients in a medical ward will require non-standard and specifically orchestrated care. Even though hospitals and doctors to a large extent uses standardized procedures there is always a level of variability involved as patients are different and complications might emerge during a procedure. It is therefore difficult to employ the same high level of standardization known from industrial production. Furthermore, healthcare is a service which means that it is difficult to apply methods from industrial production directly. Services are for example characterized by high levels of user/customer involvement, demand fluctuations and labor intensity (Miles, 2000). These characteristics again imply that healthcare in some areas differs significantly from industrial production.

It can therefore be concluded that it is not possible to use lean methods and principles in healthcare in general. There is a need to adapt the lean methods and principles to the special characteristics within the healthcare sector. This leads to the conclusion that lean is not the universal tool for hospital process improvement and lean cannot be applied in every activity and process in a hospital. Furthermore, in general leveling and stable processes are a prerequisite to using lean in healthcare.

Hypothesis 2 and 3 which addresses the challenges of lean implementation and the different rationalities involved. The findings from the cases confirms that core medical work is more complex than industrial work not just of because reactive nature of the work but also because it is difficult to establish a common perception of the patient process. There is no single process but many and doctors and nurses apply different criteria when analyzing the process. Furthermore, lean also put the discretion and professional judgment of the employees in the healthcare sector under pressure as a core element in lean is the standardization of work processes. This development towards standardization will in some case create a strong barrier against the implementation of lean in the core activities of health care organizations.

Lean is implemented in isolated organizational units, only few tools are being applied and the goal is to increase efficiency. Lean is implemented in those activities and organizational units that share close resemblance to industrial production. Implementation that makes changes to the core medical work and organization remains to be seen. Finally, care should therefore be taken not to implement lean in an un-reflected manner, where advocates of lean just attempt to implement lean without taking the special characteristics of health into consideration when planning the lean project. The findings in this paper calls for a more "transformative approach" to lean, where the special characteristics of health care organizations and the nature of their activities are taken into consideration.

CONCLUSION

Lean manufacturing is hailed as a universal solution to many a productivity and quality problem. Lean has proven itself as a valuable methodology in production and administration. In short "Lean works!" But lean only work in some hospital settings. The prerequisites of leveling and stability must be observed if success is to be achieved. Image diagnostics and planned surgery of select patients are examples of successful lean implantations. However, we can also observe that lean in some situations reaches its "natural limit". The meaning of this is that lean is not a universal solution which can be applied successfully in every activity in a hospital.

The reactive or responsive nature of many of the core activities in hospital limits the application of lean. The objective of stable and predictable processes inherent in lean cannot be met in many activities at a hospital. Furthermore, the standardization inherent in lean also creates challenges for the implementation of lean as some employees see the establishment of standardized processes as a threat to their professional judgment. The different rationalities in the different staff groups in a hospital also create barriers for the successful implementation of lean in the healthcare sector. These different rationalities make it difficult to develop a common understanding of value and high-quality care. The cases we report in this paper illustrate these points.

Does lean live up to its promise in health care? The answer to this question is mixed. Lean can be used to create significant productivity improvement in healthcare organizations. However, there need to fit between the nature of the activities being improved by lean and the basic assumptions within lean. In activities where fit is non-existent care should be taken not to implement "standard lean", i.e. lean as it is practiced in industrial manufacturing. This paper therefore calls for a more transformative approach to lean where the nature of work within in healthcare and the potential clashes between different rationalities are taken into account.

Is there a paradox in healthcare? Yes it is. Most healthcare work is reactive and yet the processes are very stable. The stability is tied to the different roles e.g. doctors, nurses etc. which revert to basic behaviors when having to react to an unforeseen event. The nurse and the doctor know what to do in the event of cardiac arrest. Although these processes are stable they are not part of a larger planned process – it's reactive. While lean require stable processes, lean also require larger processes that span a chain of activities. Healthcare is troubled by not having such tight integrated chains of activities, which of course is why lean does not in general apply to healthcare.

ACKNOWLEDGEMENTS

This paper has been written with funding from The Working Environment Research Fund, Denmark. The authors wish to acknowledge the help and input from our fellow researchers in the "Lean without Stress" project: From the Technical University of Denmark, Ass. Prof. Neils Møller, Ass. Prof. Peter Jacobsen. From Aalborg University, Professor John Johansen, PhD. Student Rikke V. Matthiesen, Ass. Prof. Jacob S. Nielsen. From the National. From the National Research Centre for the Working Environmen, Ass. Prof. Peter Hasle, Ass. Prof Jan H. Pejtersen and Ass. Prof. Pia Bramming.

REFERENCES

Anonymous (2007). OR Department uses LEAN to cut unneeded inventory and saves money. *Healthcare Purchasing News, 31*, 68-69.
Ballé, M. & Régnier, A. (2007). Lean as a learning system in a hospital ward. *Leadership in Health Services, 20*, 33-41.
Ben-Tovim, D. I., Bassham, J. E., & Bolch, D. (2007). Lean thinking across a hospital: redesigning care at the Flinders Medical Centre. *Australian Health Review, 31*, 10-15.
Bushell, S. & Shelest, B. (2002). Discovering lean thinking at Progressive Healthcare. *The*

344

Journal for Quality and Participation.Cincinnati, 25, 20-25.

Edwards, K., Nielsen, A.P. & Jacobsen, P. (2010), Implementing Lean in surgery - Lessons and Implications, International Journal of Technology Management (forthcoming)

Fairbanks, C. B. (2007). Using Six Sigma and Lean Methodologies to Improve OR Throughput. *Association of Operating Room Nurses, AORN Journal, 86,* 73-82.

Fillingham, D. (2007). Can lean save lives? *Leadership in Health Services, 20,* 231-241.

George, M. L. (2003). *Lean Six Sigma for service: How to use Lean speed and Six Sigma quality to improve services and transactions.* New York: McGraw-Hill Education - Europe.

Jakobsen, M. (1-12-2007). Vejle-modellen: Gavner det patienten? (in Danish). tæt på kræft [4], 4-7. Kræftens Bekæmpelse.

Kelly, A., Bryant, M., Cox, L., & Jolley, D. (2007). Improving emergency department efficiency by patient streaming to outcomes-based teams. *Australian Health Review, 31,* 16-21.

Khandelwal, V. K. & Lynch, T. (1999). Reengineering of the patient flow process at the Western Sydney Area Health Service. *Proceedings of the Hawaii International Conference on System Sciences, 145.*

King, D. L., Ben-Tovim, D. I., & Bassham, J. (2006). Redesigning emergency Department Patient Flows: Application of Lean Thinking to Health Care. *Emergency Medicine Australasia, 18,* 391-397.

Leslie, M., Hagood, C., Royer, A., Reece, C. P., & Maloney, S. (2006). Using lean methods to improve OR turnover times. *Association of Operating Room Nurses AORN Journal, 84,* 849-855.

Liker, J. K. (2004). *The Toyota Way.* New York: McGraw-Hill.

Lummus, R. R., Vokurka, R. J., & Rodeghiero, B. (2006). Improving Quality through Value Stream Mapping: A Case Study of a Physicians Clinic. *Total Quality Management & Business Excellence, 17,* 1063-1975.

Miles, I. (2000). Services Innovation: Coming of Age in the Knowledge-Based Economy. *International Journal of Innovation Management, 4,* 371.

Ng, D., Vail, G., Thomas, S. & Schmidt, N. (2010), Applying the lean principles of the Toyota production system to reduce wait times in the emergency department, CJEM Journal of the Canadian Association of Emergency Physicians, Jan 2010. 12:1; 50-57

Panchak, P. (2003). Lean health care? It works! *Industry Week, 252,* 34-40.

Spear, S. (2005). Fixing Health Care from the Inside, Today. *Harvard Business Review, 83,* 162.

Steunkel, K. & Faulkner, T. (2009). A community hospital's journey into Lean Six Sigma, *Frontiers of Health Services Management,* 26:1, 5-13

Tapping, D. & Shuker, T. (2003). *Value stream management for the lean office: Eight steps to planning, mapping, and sustaining lean improvements in administrative areas.* New York: Productivity Press.

Towne, J. (2006). GOING 'LEAN' STREAMLINES PROCESSES, EMPOWERS STAFF AND ENHANCES CARE. *H&HN: Hospitals & Health Networks, 80,* 34.

Tragardh, B. & Lindberg, K. (2004a). Curing a meager health care system by lean methods-translating 'chains of care' in the Swedish health care sector. *International Journal of Health Planning and Management, 19,* 383.

Weber, D. O. (2006). Toyota-style Management Drives Virginia Mason. *Physician Executive, 32,* 12.

Womack, J. P., Jones, D. T., & Roos, D. (1991). *The Machine That Changed the World : The Story of Lean Production.* (First HarperPerennial ed.) New York: HarperCollins.

Chapter 37

Professional Assessment and Collective Standards – Lean in a Cancer Department

Peter Hasle

Research Centre for the Working Environment
Lersoe Parkallé 105
2100 Copenhagen, Denmark

ABSTRACT

Hospitals are faced with heavy pressure from increasing demands from both patients and society, as well as a more constrained budget and still more medical opportunities. Lean has been suggested as a possible means to meet this challenge. However, lean derives from manufacturing industry, so the question is how to apply it in a hospital setting and what the consequences might be for employees. This question was studied in a cancer department in a university hospital. The psychosocial working environment was measured before and after lean implementation, and qualitative data was collected from the implementation process. The two groups most affected – lab technicians and nurses in the chemotherapy outpatient clinic – showed quite different results. The lab technicians experienced an improvement in the working environment, whereas the opposite was the case for the nurses. One possible explanation is that lean gave a greater degree of control to lab technicians, whose work had originally been rather standardized, whereas nurses, with more complex and non-standardized work, experience lean as limiting their degree of control and as questioning their professional assessment with regard to the care of individual patients.

Keywords: Lean, psychosocial working environment, hospitals, standards, professional assessment

INTRODUCTION

In Denmark as well as in other industrialized countries the health care systems – especially hospitals – face numerous challenges, including a rapid growth in demands in terms of the number of patients, quality expectations, and the availability of new expensive treatments. At the same time, budgets are restricted and there is a lack of qualified staff. Hospitals are therefore searching for possible solutions and in recent years lean has received growing popularity as one of the most important answers to the challenges in the hospital sector. It is not obvious though that lean can serve as a relevant solution. Lean manufacturing was developed in the motor industry and is based on the standardization and optimization of work flows. This paper discusses the adaptation of lean to the hospital setting and what consequences lean would have for the organization of work and for the employees.

The concept of Lean production was first formulated in the late 1980s and early 1990s (Womack & Jones 1996; Womack et al. 1990) in an attempt to make the American auto industry more competitive compared to the Japanese. To a large extent, the ideas of lean are based on Japanese production principles, with special inspiration from the Toyota Production System (TPS) (Liker 2004; Spear & Bowen 1999). The fathers of lean suggested that lean would benefit employees due to its promotion of a higher degree of participation and the opportunity to get rid of strained working conditions. This view, however, was not based on any empirical findings, and Womack (1996) soon pointed out that lean's focus on the reduction of waste implied that employees could be made redundant and that it could not be expected that the employees would be motivated to be involved in such a process. Lean's ambiguous nature with regard to the employees was thus evident from the beginning. With lean's growing popularity in the business sector, several studies have examined the propagation of lean and discussed its possible effect on the working environment. Some researchers (Anderson-Connolly et al. 2002; Babson 1993; Harrison 1994) focused on the possible negative sides and suggested that "lean is mean" and that like other ideas for workplace restructuring might lead to severe health problems. At this point, light and various shades of grey seem to have entered the debate, and positive outcomes for the employees have also been described (de Treville & Antonakis 2006; Delbridge 2005; Hampson 1999; Hasle 2009; Seppälä & Klemola 2004) and it is now a more open question as to how lean and its consequences for employees should be understood.

But almost all the research so far has been in industrial settings, and it has not been studied how lean affects employees in other sectors such as in hospitals. At the same time, lean is receiving tremendous attention in this sector. Most Danish hospitals are implementing lean projects, and the number of scientific papers and books about lean in hospitals in various countries is also growing. Most of them paint a very positive picture of the application of lean in hospitals (Ballé & Régnier 2007; Ben-Tovim et al. 2007; Dickson et al. 2009; Graban 2008; Jones & Mitchell 2006; Kim et al. 2006; Nelson-Peterson & Leppa 2007). These authors describe the introduction of lean in hospitals as straightforward and some even advocate following the original approach, as developed in Toyota, as closely as possible. This literature does not reflect on any possible difficulties in applying an auto-manufacturing concept in hospitals. On the other hand, the many case stories about positive achievements could also be taken as an indication of possible benefits for hospitals from lean. One review of the literature has been carried out (Vest & Gamm 2009), and after using more rigorous research criteria in systematic reviews the authors conclude that the evidence for the benefits of lean and other similar effectiveness transformation strategies is weak, although it points in a positive direction.

However, a few of the most recent authors take a more reflective stance towards lean – not particularly negative, but emphasizing the need to consider the special context of hospitals (Joosten et al. 2009; Winch & Henderson 2009; Young & McClean 2008). Young and McClean in particular point to a specific challenge. Lean is based on the concept of producing value for the customer. That concept is reasonably clear in many industrial settings, but this is not so in hospitals (Muir 2007). Young and McClean argue that it is not so simple to identify the customers in hospitals and likewise that value is also open to several interpretations. Based on an analysis of the value concept in hospitals, they suggest that there are at least three dimensions to value: clinical (the best patient outcome), operational (the greatest cost-effectiveness), and experiential (patients' experience of care). They furthermore indicate that most of the emphasis in the lean projects described in the literature so far has focused on the operational value. It is evident that lean improvements on one value, especially the operational one of cost-effectiveness, might conflict with one or both of the other values.

Based on the above points, the introduction of lean into hospitals has the potential to create at least two types of problems. The first one is a possible intensification of workload due an isolated emphasis on cost reduction, as has been found in a number of studies in the industrial sector (Conti et al. 2006; Landsbergis et al.1999; Parker 2003). It is possible that lean just makes the staff run faster in hospitals, too. This has not so far been reported in the literature on hospitals, and it is probably not a big risk for the core groups such as doctors and nurses, because their work is varied and complex. There are also some indications from industry that skilled workers with relatively varied work do not experience intensification (Seppälä & Klemola 2004).

The other potential problem is conflict between values. It is possible that benefits on the cost-effectiveness side may have detrimental effects on patient outcome or patient experience. In both cases, it may be experienced as a possible cause of stress for the staff (Semmer et al. 2007). Staff may experience the changes caused by lean as offensive to their professional values and as questioning their competence.

The purpose of the research presented in this paper was to investigate the possible consequences for hospital employees exposed to the implementation of lean. I present the results from an in depth case study of a cancer department – especially looking at the employees' involvement and reactions to lean.

METHODOLOGY

The case study was carried out in a cancer department at a large university hospital and lasted for a period of almost two years. The aim of implementing lean was to increase efficiency and improve the psychosocial working environment. It was thus an explicit part of the project to improve conditions for the employees.

The implementation of lean was studied with the help of both quantitative and qualitative methods. The psychosocial working environment was measured before and after the core lean implementation, using the COPSOQ questionnaire (Kristensen et al. 2005). Qualitative data was collected about lean activities and employee reactions. The data collected included observation of key activities, interviews with key actors, and individual and group interviews with employees involved. Altogether 21 interviews of 30 to 90 minutes duration were carried out. All interviews were transcribed. Moreover, an introductory workshop about the history of the department, two meetings with presentation of questionnaire results, and a feed-back seminar about the lean experience were organized, as well as participation in key events such as introductory lean training, kaizen meetings, and steering committee meetings. Meeting summaries and observation notes were made from all meetings and encounters. Written materials such as minutes, information material, lean analysis, and training material were collected.

The data material was used to make a summary of the implementation process and it was analysed for employee and management responses to lean especially with regard to psychosocial working environment, patient value, and professional integrity. It turned out that the two units with the greatest involvement in lean were the chemotherapy outpatient clinic and the local laboratory, and these two units also experienced the biggest changes in their psychosocial working environment. The analysis is therefore focused on these two units.

THE LEAN IMPLEMENTATION PROCES

The university hospital has integrated lean in its global strategy and considerable resources have been put into lean implementation. An internal lean consultant unit organizes lean implementation in waves of 5–6 departments – each lasting a year. Each wave starts with an intensive training course of 8 x 3 days for the department lean implementation groups. During and after the course, a consultant from the unit is attached to each department for 6–12 months.

The cancer department has 350 employees and includes wards, out-patient chemotherapy, radiation therapy, a laboratory, and a palliative section. A steering committee was established with the head nurse, the head consulting doctor, the leader of the lean project group, and the consultant from the lean unit. A lean project group was organized with a nurse as project leader and in addition a consulting doctor, two nurses, a secretary, a lab technician, and a radiologist.

A relatively large number of activities were initiated. The main ones concerned delivery of chemotherapy medicine, collaboration between lab technicians and the chemotherapy outpatient clinic, handling of blood samples, handling of case records, establishment of kaizen boards, and the reorganization of patient booking in the chemotherapy outpatient clinic. It was also decided to reorganize the ward rounds, but that project never got off the ground.

The change strategy was based on extensive involvement of the staff concerned, who were then supported by members of the lean project group and, if needed, the lean consultant. Some smaller changes were initiated by a kaizen workshop over one or two half–days, when representatives of the staff analysed the problems using value stream mapping and came up with solutions. In other cases, working groups were established to analyse a particular problem and come up with a solution.

Several changes which the actors describe as successful were implemented, including higher quality medicine delivery, better track of blood samples and case records, the use of kaizen boards with many implemented suggestions, and reorganization of the work of the lab technicians. But the reorganization of the patient booking in the chemotherapy outpatient clinic in particular turned out to be very problematic, and the department was still fighting to get the patient booking back on track when the project ended.

OUTCOMES FOR THE EMPLOYEES

The questionnaire results indicate that the psychosocial working environment has been considerably improved for lab technicians, but has deteriorated for nurses at the central outpatient clinic for chemotherapy. Table 1 shows the results. It is based on the COPSOC scale with baseline results, the difference after one year, and the P-value.

In the lab, all the changes are positive and there are 16 changes >7.5, which is considered to be clinically remarkable for participants (Pejtersen et al. 2010), although only two are statistically significant due to the limited number of employees in the lab. In the chemotherapy outpatient clinic, the situation is the opposite. All dimensions have changed in a negative direction with 17 significant changes, nine of which are clinically remarkable.

The question is therefore why the situation has changed so dramatically for the two groups, and I will try to identify possible explanations in the differences in the change process in the two units.

Starting with the laboratory, they initiated two major changes. The first one was regarding blood sampling and intubation of intravenous lines. Previously, the lab technicians were called to the outpatient clinic after arrival of the patient. They then had to identify the patient, search for a vacant couch, do the job and walk back to the lab. This task constituted a large proportion of the technicians' work and they spent considerable time walking from one place to another, and finding and sometimes waiting for vacant couches. They organized a kaizen workshop together with nurses from the outpatient clinic and identified possible solutions which ended in a decision to reorganise that particular task. In the new procedure the nurses ask the patients to walk to the lab where there will be one or two lab technicians on duty each in a room with a couch. The intubation and blood sampling take only a couple of minutes and the patients used time they would previously have just spent waiting in the clinic.

Table 1. Psychosocial working environment for chemotherapy outpatient clinic and laboratory*

| (scale 0–100) | Chemo outpatient clinic | | | Laboratory | | |
	Baseline	Change after one year	P	Baseline	Change after one year	P
Quantitative demands	48.2	7.6	0.01*	39.6	−2.1	0.65
Work pace	63.4	3.7	0.05*	61.1	−3.7	0.54
Emotional demands	63.1	1.8	0.40*	59.0	−2.2	0.76
Influence	52.3	−8.8	0.00*	41.0	8.5	0.29
Possibilities for development	71.2	−3.4	0.03*	60.4	5.8	0.26
Meaning of work	75.8	−3.8	0.07	74.1	7.8	0.16
Commitment to the workplace	70.6	−11.2	0.00*	60.4	13.2	0.16
Predictability	59.0	−11.5	0.00*	72.2	0.2	0.96
Rewards (recognition)	66.7	−7.0	0.01*	68.5	12.4	0.13
Role clarity	69.2	−6.2	0.00*	70.4	10.1	0.11
Role conflicts	41.7	8.0	0.00*	48.6	−10.3	0.26
Quality of leadership	57.1	−4.0	0.18	72.2	12.5	0.06
Social support from supervisor	56.2	0.2	0.95	73.1	9.6	0.04*
Social support from colleagues	62.0	−2.6	0.20	57.4	12.1	0.10
Social community at work	82.0	−6.9	0.00*	67.6	19.8	0.02*
Job satisfaction	67.2	−8.5	0.00*	63.9	13.0	0.15
Work–family conflict	40.8	3.0	0.38	27.8	−13.5	0.14
Trust regarding management	73.7	−9.2	0.00*	74.8	5.6	0.27
Mutual trust employees	77.1	−5.9	0.00*	67.6	6.5	0.43
Justice and respect	60.5	−9.2	0.00*	63.9	6.4	0.27
Self rated health	61.4	−2.1	0.40	63.9	11.1	0.24
Burnout	33.0	8.4	0.01*	38.2	−16.0	0.09
Stress	27.0	5.6	0.02*	31.9	−14.6	0.15
Sleeping troubles	21.9	1.2	0.56	29.9	−16.5	0.19

*Outpatient clinic: baseline N=39, follow up N=37. Lab: baseline N=9, follow up N=10. Significance was tested using a mixed model (PROC MIXED in SAS) of linear regression.

The other major change was the introduction of kaizen meetings. Once a week the lab technicians organize a standing meeting around a kaizen board where they suggest ways of improving everything related to the lab and the technicians' work. By the time the project ended, they had made 84 suggestions, 68 of which had been implemented. Among others things, the suggestions have resulted in more space in the quite congested lab, a more secure supply of material, and higher safety in handling of blood samples. In the interviews, both the head lab technician and the technicians said that they experienced lean as successful and also beneficial for the working environment.

The outpatient clinic shows a quite different picture. They started with an initial success with reorganization of the communication lines between the pharmacy and the clinic, which ensured that they would have medicine delivered in time. But after that effort, things got more complicated. It was decided to introduce a new IT-program for patient admissions to achieve more systematic planning of patient flow, which would reduce waiting time and utilize treatment facilities more efficiently. Among other things, the new system would make it possible to plan patient booking to fit the expected duration for the chemotherapy, which can vary quite a lot. It turned out that the technical side of the IT-implementation was much more complicated than expected, and for quite some time there were technical problems with the bookings. But the new booking system also changed the work of the nurses. Previously, they were used to individually booking their own patients, whereas now they were supposed to ask a secretary to do the booking. They explained that the new system created more work, partly because of the technical problems and partly because they had to get in touch with a secretary rather than doing the booking themselves. Parallel to the new booking system, attempts were also made to introduce kaizen meetings in the outpatient clinic, but with little success.

The nurses had quite differentiated views on the lean changes, but most of them had a negative opinion and felt that lean had created more work and their working environment had deteriorated. A large group of the nurses tended to interpret the IT-program as an attack on their professionalism, and they expressed a serious concern about the effect on their care of the individual patient. They felt that their flexibility with regard to professional care is seriously weakened by the new standardized booking system, in which they cannot personally ensure that what they see as their patients' special needs will be accommodated.

DISCUSSION

Why this difference in the reactions of the employees, in this case that of the technicians on the one hand, and the nurses on the other? One explanation for the lab technicians seeing lean as an opportunity for positive reorganization has to do with the content of their work. It consists of reasonably standardized tasks with a strong emphasis on safety matters, such as avoiding any chance of mixing or delaying blood samples. Their involvement in reorganization of work, such as the change of intubation from outpatient clinic to the lab, and the kaizen meetings actually gives the lab technicians more control of their own work and more variation. And they apparently appreciated that.

Whereas the nurses in the outpatient clinic already had a very varied job with a large degree of control and heavy demands on making decisions, so there was not the same opportunity, or need, for them to expand control. Instead they experienced the opposite situation. With its standardized booking system, lean was experienced as a reduction in their degree of control, and they experienced it as an encroachment on their professionalism and therefore a stressor. And they ended up reacting negatively, as expressed in both the questionnaire and the interviews.

The results from this study suggest that lean may challenge the traditional understanding of professionalism in hospitals – at least for nurses. Traditionally, treatment and care have had a strong element of trial-and-error. A certain treatment of care is given and the result is monitored and adjusted according to the patient's behaviour and needs. Collective standards seem obviously to conflict with the possibility for the individual nurse to take her own decisions based on her own professional judgement. In the example given here, the idea was to make patient booking more efficient, in adherence to the operational value (Young & McClean 2008), whereas the nurses were afraid that the standardized booking would have negative consequences for the experiential value as the patients may feel a lack of concern for their personal priorities. There seems to be a contradiction between values, and the new system seems to limit part of the nurses' professional judgement.

But the question is whether it needs to be so. There is little doubt the hospitals need to give priority to operational value with the growing pressure on cost efficiency, but does this need to conflict with the other core values: the clinical and the experiential? In this case, the outpatient clinic was, at the end of the project, working on the communication between the nurses and secretaries in order to ensure the same experiential value as before, and if that effort turns out to work, the nurses may save time which could be spent on direct contact with the patients. In this case it is the collaboration between different professional groups which is at stake (Gittell 2009). The same was the case for the lab. The re-organization which meant that the nurses have to ask the patients to walk to the lab rather than call a lab technician also required collaboration between professional groups, but it did not question the nurses' professional assessment.

Perhaps the increasing demands and the growing complexity of hospitals create a need to develop a new balance between the traditional individual professionalism and collective standards, especially the ones which have to do with collaboration with other professions. The lean approach can more easily help improve the standardized work tasks mostly carried out by support staff such as lab technicians, whereas it will challenge the often complex tasks of nurses and doctors, where clinical, operational, and experiential values will all be at stake. This is a real dilemma because standards may be necessary both to make work effective and to strengthen focus on the most important tasks, but at the same time standards also limit the individual assessments which are a crucial part of nurses' and doctors' professionalism, and can mean that some tasks are transferred to other professions.

REFERENCES

Anderson-Connolly, R., Grunberg, L., Greenberg, E. S., & Moore, S. (2002), "Is lean mean? Workplace transformation and employee well-being." *Work Employment and Society*, 16(3), 389-413.

Babson, S. (1993), "Lean or Mean: The MIT model and lean production at Mazda." *Labour Studies Journal*, 18, 3-24.

Ballé, M. & Régnier, A. (2007), "Lean as a learning system in a hospital ward." *Leadership in Health Services*, 20(1), 33-41.

Ben-Tovim, D. I., Bassham, J. E., & et al. (2007), "Lean thinking across a hospital: redesigning care at the: Flinders Medical Centre." *Australian Health Review*, 31(1), 10-15.

Conti, R., Angelis, J., Cooper, C., Faragher, B., & Gill, C. (2006), "The effects of lean production on worker job stress." *International Journal of Operations & Production Management*, 26(9), 1013-1038.

de Treville, S. & Antonakis, J. (2006), "Could lean production job design be intrinsically motivating? Contextual, configurational, and levels-of-analysis issues." *Journal of Operations Management*, 24(2), 99-123.

Delbridge, R. (2005), "Workers under lean manufacturing," in *The essentials of the new workplace - A guide to the human impact of modern working practices*, D. Holman et al., eds., John Wiley & Sons, Chichester, pp. 15-32.

Dickson, E. W., Anguelov, Z., Vetterick, D., Eller, A., & Singh, S. (2009), "Use of Lean in the Emergency Department: A Case Series of 4 Hospitals." *Annals of Emergency Medicine*, 54(4), 504-510.

Gittell, J. H. (2009), *High performance health care - using the power of relationship to achieve quality, efficiency and resilience*. McGraw-Hill, New York.

Graban, M. (2008), *Lean hospitals: improving quality, patient safety, and employee satisfaction*. CRC Press.

Hampson, I. (1999), "Lean Production and the Toyota Production System - Or, the Case of the Forgotten Production Concepts." *Economic and Industrial Democracy*, 20(3), 369-391.

Harrison, B. (1994), *Lean and Mean. The Changing Landscape of Corporate Power in the Age of Flexibility*. The Guildford Press, New York.

Hasle, P. (2009), "Lean and the psychosocial work environment," in *Prerequisites for healthy organizational change*, P. O. Saksvik, ed., Bentham Science Publishers, pp. 1-9.

Jones, D. & Mitchell, A. (2006), *Lean thinking for the NHS*, NHS Confederation, London.

Joosten, T., Bongers, I., & Janssen, R. (2009), "Application of lean thinking to health care: issues and observations." *International Journal for Quality in Health Care*, mzp036.

Kim, C. S., Spahlinger, D. A., Kin, J. M., & Billi, J. E. (2006), "Lean Health Care: What Can Hospitals Learn from a World-Class Automaker?" *Journal of Hos-*

pital Medicine, 1(3), 191-199.

Kristensen, T. S., Hannerz, H., Høgh, A., & Borg, V. (2005), "The Copenhagen Psychosocial Questionnaire - a tool for the assessment and improvement of the psychosocial work environment." *Scandinavian Journal of Work Environment & Health*, 31(6), 438-449.

Landsbergis, P. A., Cahill, J., & Schnall, P. (1999), "The impact of lean production and related new systems of work organization on worker health." *Journal of Occupational Health Psychology*, 4(2), 108-130.

Liker, J. K. (2004), *The Toyota way - 14 management principles from the world's greatest manufacturer*. McGraw Hill, New York.

Muir, G. J. (2007), *How to get better value health care*. Offox Press, Oxford.

Nelson-Peterson, D. L. & Leppa, C. J. (2007), "Creating an environment for caring using lean principles of the Virginia Mason Production System." *Journal of Nursing Administration*, 37(6), 287-294.

Parker, S. K. (2003), "Longitudinal effects of lean production on employee outcomes and the mediating role of work characteristics." *Journal of Applied Psychology*, 88(4), 620-634.

Pejtersen, J. H., Bjorner, J. B., & Hasle, P. (2010), "Determining minimally important score differences in scales of the Copenhagen Psychosocial Questionnaire." *Scandinavian Journal of Public Health*, 38(3_suppl), 33-41.

Semmer, N. K., Jacobshagen, N., Meier, L. L., & Elfering, A. (2007), "Occupational stress research: The "stress-as-offence-to-self" perspective," in *Occupational health psychology: European perspective on research, education and practice*, vol. 2 S. Mcintyre & J. Houdmont, eds., Ismai, Castelo da Mala.

Seppälä, P. & Klemola, S. (2004), "How do employees perceive their organization and job when companies adopt principles of lean production." *Human Factors and Ergonomics in Manufacturing*, 14(2), 157-180.

Spear, S. & Bowen, H. K. (1999), "Decoding the DNA of the Toyota production system." *Harvard Business Review*, 95-106.

Vest, J. R. & Gamm, L. D. (2009), "A critical review of the research literature on Six Sigma, Lean and StuderGroup's Hardwiring Excellence in the United States: the need to demonstrate and communicate the effectiveness of transformation strategies in healthcare." *Implementation Science*, 4.

Winch, S. & Henderson, A. J. (2009), "Making cars and making health care: a critical review." *Medical Journal of Australia*, 191(1), 28-29.

Womack, J. P. (1996), "The psychology of lean production." *Applied Psychology-An International Review-Psychologie Appliquee-Revue Internationale*, 45(2), 119-122.

Womack, J. P. & Jones, D. T. (1996), *Lean thinking*. Simon & Schuster, New York.

Womack, J. P., Jones, D. T., & Roos, D. (1990), *The machine that changed the world*. Rawson Associates, New York.

Young, T. P. & McClean, S. I. (2008), "A critical look at Lean Thinking in healthcare." *Quality and Safety in Health Care*, 17(5), 382.

Chapter 38

Status of Ward Nurses in Health Care Industry of Uttarakhand State of India: An Approach Through Hospital Ergonomics

Karki Indu[1], Sharma Promila[2]

[1]&[2] Dept of Family Resource Management
College of Home Science, G.B.P.U.A.& T., Pantnagar
Uttarakhand,-263145, India

ABSTRACT

Various researches has focused on sleep loss and fatigue related risks in health care professionals such as doctors and nurses. Nursing, the profession of caring for the sick and the convalescent, the disabled and the helpless, is one of the noblest professions throughout the world. From the general ward to the operation theatre in the hospital, nursing is the most important component of patient care. Keeping all these points in mind, the study has being planned with the objectives like studying the medical history and hazards associated with nursing job. So, for this purpose 120 nurses from government and private hospitals of districts U. S. Nagar and Nainital in the Uttarakhand state were purposively selected. The results indicated that maximum of 40 per cent nurses were suffering from headache, whereas minimum of 17.50 per cent nurses were suffering from back pain. Regarding occurrence of chronic illness among nurses, it was revealed that anemia was prevalent among 19.17 per cent nurses and least common was thyroid problem (2.50 per cent). The results regarding hazards as experienced by nurses revealed that majority (94.17 per cent each) of the nurses were encountering problem with repetitive motion and awkward posture during work schedule. Biological hazards like skin infection were reported by about 34.17 per cent. Most of the nurses (40.83 per cent each) got injured with solvents and dust and about. 60 per cent felt noise hazards, radiation, heat and cold (30.83 per cent each).

Keywords: Hazards, Hospital ergonomics, Medical history

INTRODUCTION

Nursing, the profession of caring for the sick, the disabled and the helpless, is one of the noblest professions throughout the world. The process of restoring patients to normal health depends as much, if not more, on nurses as on doctors or surgeons. In fact, the availability of effective nursing services is an indicator of the health of a country's medicare system. The common aspect of all jobs in nursing is that there are always a lot of people; a nurse has to interact with. Nurses are always expected to display a calm, perceptive approach during crises that are only to be expected in hospitals. Studies have shown that certain unsatisfactory features of work such as poor opportunities to exert influence and insufficient social support networks increase the risk of careworkers' physical condition and functioning capability deteriorating (Brown et al., 2006).

On the whole, not much attention is being paid to conduct researches on sleep problems of the workers, working in 24 hours work industry. But recently due to the introduction of multinational companies in India, this area of research has become an important area to be exercised. Therefore the time has come now where we need to give deep thinking about humanizing the work of nurses. In order to provide safe working environment and making them dedicated group of workers i.e. nurses needs several approaches to protect them from any ill effect. Keeping all these points in mind, the study was planned with the objectives like studying illness suffered by nurses during last one year and hazards associated with nursing job.

MATERIALS AND METHODS

In order to achieve the objectives of the study, descriptive-cum-experimental research design was selected. For the present study, Uttarakhand State was selected in which districts like U.S. Nagar and Nainital were selected purposively. For collecting data a pre-coded structured interview schedule was constructed. A sample of 120 nurses ware nurses (70 from government and 50 from private hospitals) was selected. Random-cum-purposive sampling technique was used to select the sample. Percentages and mean score were used to analyze the data.

RESULTS AND DISCUSSION

Table 1 shows the details of illness suffered by nurses since last one year. It was crucial to diagnose the occurrence of illness among nurses as it has an impact on health status of themselves as well as their patients. It also affects the sickness absenteeism. On the whole, it was observed that maximum of 40 per cent nurses were suffering from headache followed by body-ache (34.17 per cent), whereas minimum of 17.50 per cent nurses were suffering from back pain. In case of nurses from government hospitals, again temporary headache due to job was reported by 27.14 per cent nurses, and for private, it was temporary back-pain (42 per cent) as a result of job performance. Thus, most common illness among nurses was observed to be headache. Headache, muscular pain and general malaise from shift work were also stated by Lipkin et al. (1998). Regarding occurrence of chronic illness among nurses, it was revealed from the table that most prevalent chronic illness was anemia among 19.17 per cent nurses and least common was thyroid problem (2.50 per cent). Based on the observation of chronic illness among government hospitals, most frequent was blood pressure (17.14 per cent) which was continuous due to other sources rather than job and the least common was continuous problem of gastrointestinal disorders (2.86 per cent) due to job related factors. Observation of private hospitals reveals that again temporary anaemia due to other sources was prevalent among 12 per cent nurses. A few proportion (4 per cent) of nurses reported problem of thyroid. As far as symptoms of illness were concerned, it was revealed that most common symptom among maximum of 44.17 per cent nurses was tiredness and very few (3.33 per cent) nurses felt excessive perspiration. As category wise data was analyzed, it was observed that 37.14 per cent nurses from government hospitals reported symptom of temporary tiredness from job, whereas, very few (2.86 per cent) felt temporary muscle tightness due to other sources. Regarding nurses from private hospitals, it was revealed from the data that most common symptom was tiredness (22 per cent).

Nursing job allows a lot of psychological problems like depression, arrogance, anger, emotional problem etc. Among all the psychological problems, fatigue in patient care was felt by about 41.67 per cent nurses and the least problem was reported to be frequent absenteeism (4.17 per cent). As data were analyzed for particular category, it was inferred from the Table 1 that among nurses from government hospitals again temporary fatigue in patient care due to job was reported by 38.57 per cent nurses and only 7.14 per cent nurses each had problem of memory loss and poor communication due to job. In case of nurses from private hospitals, it was found that maximum of 18 per cent nurses were having feeling of increased negativity and irritation from job which was temporary in nature and the least common temporary psychological problem was poor communication (2 per cent) due to their long hour of job.

HAZARD IDENTIFICATION AMONG NURSES

Along with chronic illnesses, nursing job has much of hazards also which can be categorized as mechanical, biological, physical and psycho-social. Hazards, accidents, risks or injuries are the common incidence at the hospital and the research reports have revealed that rate of accidents is relatively more in the morning shift as compared to afternoon shift. The results regarding mechanical hazards as experienced by nurses (on an average) revealed that majority (94.17 per cent each) of the nurses were encountering problem with repetitive motion and awkward posture during work schedule followed by 71.67 per cent and 67.5 per cent nurses who reported injuries with sharp edges and accidents with manual handling respectively. As far as data with regard to nurses from government hospitals was concerned, it was found that all the nurses were encountered with reported repetitive motion and awkward posture followed by manual handling (84.29 per cent) whereas, very few (5.7 per cent) had injuries with lifting the material. In case of private hospitals further repetitive motion and awkward posture were experienced by majority (86 per cent) of nurses.

Hazards may occur through contact with biological waste during pathological waste etc which may be infectious to the nurses who have to perform these tests. In the summative responses, biological hazards like skin infection were reported by about 34.17 per cent followed by rashes/allergy (13.33 per cent). When data was analyzed with regard to nurses from government hospitals particularly, it was observed that rashes/allergy and skin infection were common biological hazards among nurses (17.14 per cent each). In case of nurses from private hospitals about 58 per cent reported skin infections and only 8 per cent were suffering from skin rashes/allergy. As compared to nurses from private hospitals, nurses of government hospitals are more prone to biological hazards which showed that private hospitals do not maintain the health and hygiene practices.

Chemical hazards may happen while using different chemicals (acids or alkalis), cleaning compounds, exhaustion of waste anesthetic gases through –ray machines, ultrasound machines etc. It can be envisaged from Table 2 that most of the nurses (40.83 per cent each) got injured with solvents and dust followed by 33.33 per cent who were having infections with cleaning compounds. It was further found that in relation to nurses from government hospitals, 38.57 per cent nurses had injury through contact with solvents and cleaning compounds whereas only 7.14 per cent reported problems with waste anesthetic gases. In case of nurses from private hospitals, majority (64 per cent) of nurses felt allergy with waste anesthetic gases and a few of 12 per cent complained of skin burns from acid. Physical hazards are a combination of hazards from noise, radiation, heat and cold. It is evident from the Table 2 that majority of nurses i.e. 60 per cent felt noise hazards, radiation, heat and cold (30.83 per cent each). Data pertaining to nurses from government hospitals revealed that maximum (42.86 per cent) of nurses reported hazards with noise and very few were in problem with radiation hazards (4.29 per cent). In case of nurses from private hospitals, 84 per cent had hazards of noise followed by radiation i.e. 68 per cent. The more number of hazards like noise is not a serious problem especially in private hospitals and even in government hospitals and it could be assumed that it reflected both the organizational factor, the presence of more people like relatives of the patients and management personnel.

REFERENCES

Brown D, James G, Mills P. 2006. Occupational differences in job strain and physiological stress: female nurses and school teachers in Hawaii. Psychosom Med 68:524-530. (cited in Healthy working hours: report of the research and development project) 19th International Symposium on Shiftwork and Working Time "Health and Well-being in the 24-h Society" - San Servolo Island, Venezia, Italy, 2-6 August 2009.

www.shiftwork2009.it (Retrieved on 27/Aug/2009)

Lipkin, J.; Papernik, D.; Plioplys, S. and Plioplys, A. (1998). Chronic fatigue syndrome. The American Journal of Medicine, 105 (3) : 91-93.

Rogers,E, Wei-Ting Hwang, Linda D. Scott, Linda H. Aiken and David F. Dinge (2004). The Working Hours of Hospital Staff Nurses and Patient Safety. Journal of Health Affairs, 23 (4) : 202-212.

Sahu S. and Basu K. (2008). Individual differences in adaptation to night work rotating at different speed, Proc. of international conference of HWWE, pp-39-42, 2008.

Table 1 Illness suffered by nurses during the last one year- N=120

S. No	Illness	Occurrence	Government (n=70)				Private (n=50)			
			Temporary		Continuous		Temporary		Continuous	
			From job	Other sources	From job	Other sources	From job	Other sources	From job	Other sources
	Physiological problems									
	1. Common illness									
a)	Cough, cold, fever	38 (31.67)	-	12 (17.14)	-	-	14 (28)	12 (24)	-	-
b)	Head ache	48 (40)	19 (27.14)	4 (5.71)	-	-	17 (34)	8 (16)	-	-
c)	Body-ache	41 (34.17)	14 (20)	7 (10)	-	-	20 (40)	-	-	-
d)	Back-pain	21 (17.50)	-	-	-	-	21 (42)	-	-	-
e)	Skin-rashes, Allergy	26 (21.67)	12 (17.14)	6 (8.57)	-	-	4 (8)	4 (8)	-	-
	2. Chronic illness									
a)	Diabetes Mellitus	15 (12.5)	-	-	-	11 (15.71)	-	-	-	4 (8)

b)	Respiratory diseases	16 (13.33)	6 (8.57)	3 (4.29)	-	-	-	2 (4)	3 (6)	2 (4)
c)	BP problem	20 (16.67)	-	-	-	12 (17.14)	4 (8)	-	-	4 (8)
d)	Anaemia	23 (19.17)	3 (4.29)	11 (15.71)	-	-	3 (6)	6 (12)	-	-
e)	Gastrointestinal Disorder	20 (16.67)	-	5 (7.14)	2 (2.86)	7 (10)	-	6 (12)	-	-
f)	Thyroid	3 (2.50)	-	-	-	-	-	-	-	3 (6)

2. Symptoms

a)	Muscle tightness	14 (11.67)	7 (10)	2 (2.86)	-	2 (2.86)	3 (6)	-	-	-
b)	Excessive perspiration	4 (3.33)	-	-	;	-	2 (4)	2 (4)	-	-
c)	Tiredness	53 (44.17)	26 (37.14)	14 (20)	-	-	11 (22)	2 (4)	-	-

B. Psychological problems

a)	Increased negativity, irritation	29 (24.17)	10 (14.29)	7 (100)	-	-	9 (18)	3 (6)	-	-

b)	Greater physical demand	33 (27.50)	16 (22.86)	-	8 (11.43)	-	9 (18)	-	-	-
c)	More relationship problem	36 (30)	25 (35.71)	-	3 (4.29)	-	8 (16)	-	-	-
d)	Depression	17 (14.17)	6 (8.57)	-	6 (8.57)	-	2 (4)	3 (6)	-	-
e)	Frequent bad mood	36 (30)	11 (15.71)	-	14 (20)	-	6 (12)	5 (10)	-	-
f)	Unable to concentrate	25 (20.83)	11 (15.71)	-	7 (10)	-	2 (4)	5 (10)	-	-
g)	Lack of energy	29 (24.17)	14 (20)	-	8 (11.43)	-	3 (6)	4 (8)	-	-
h)	Memory loss	18 (15)	5 (7.14)	-	7 (10)	-	5 (10)	1 (2)	-	-
i)	Poor communication	16 (13.33)	5 (7.14)	-	7 (10)	-	1 (2)	3 (6)	-	-
j)	Arrogant	23 (19.17)	9 (12.86)	-	8 (11.43)	-	5 (10)	1 (2)	-	-
k)	Postural Fatigue	45 (37.50)	26 (37.14)	-	10 (14.29)	-	7 (14)	2 (4)	-	-
l)	Fatigue in patient care	50 (41.67)	27 (38.57)	-	11 (15.71)	-	9 (18)	3 (6)	-	-
m)	Sleep deprivation	22 (18.33)	17 (24.29)	-	-	-	5 (10)	-	-	-

n)	Performance anxiety	24 (20)	9 (12.86)	4 (5.71)	-	2 (2.86)	6 (12)	3 (6)	-	-
o)	Fear of crowd	20 (16.67)	6 (8.57)	5 (7.14)	-	-	5 (10)	4 (8)	-	-
p)	Emotional outbursts	23 (19.17)	7 (10)	8 (11.43)	-	-	4 (8)	4 (8)	-	-
q)	Anger outbursts	23 (19.17)	6 (8.57)	8 (11.43)	-	-	4 (8)	5 (10)	-	-
r)	Work related stress	35 (29.17)	7 (10)	11 (15.71)	8 (11.43)	-	5 (10)	4 (8)	-	-
s)	Low self-confidence	22 (18.33)	8 (11.43)	6 (8.57)	-	-	6 (12)	2 (4)	-	-
t)	Frequent absenteeism	5 (4.17)	-	-	-	-	3 (6)	2 (4)	-	-
u)	Low motivation	32 (26.67)	15 (21.43)	10 (14.29)	-	-	4 (8)	3 (6)	-	-

Figures in parentheses indicate the percentage values

Table 2 Occupational hazards, accidents, risks, injuries at workplace (since last 1 year)

S. No.	Type of hazards	Sources of hazard	Frequency of incidence		
			Government (n=70)	Private (n=50)	Total (N=120)
1.	Mechanical/Ergonomics/ Repetitive Strain Injuries (RSI)	Lifting	4 (5.71)	8 (16)	12 (10)
		Sharp edges	57 (81.43)	29 (58)	86 (71.67)
		Repetitive motion	70 (100)	43 (86)	113 (94.17)
		Awkward posture	70 (100)	43 (86)	113 (94.17)
		Forceful motion	25 (35.71)	21 (42)	46 (38.33)
		Manual handling	59 (84.29)	22 (44)	81 (67.50)
2.	Biological	Rashes / Allergy	12 (17.14)	4 (8)	16 (13.33)
		Skin infection	12 (17.14)	29 (58)	41 (34.17)
3.	Chemical	Solvents	27 (38.57)	22 (44)	49 (40.83)
		Chlorine	24 (34.29)	15 (30)	39 (32.50)
		Cleaning compounds	27 (38.57)	13 (26)	40 (33.33)
		Waste anaesthetic gases	5 (7.14)	32 (64)	37 (30.83)
		Dust	20 (28.57)	29 (58)	49 (40.83)
		Skin burn from acid	(27.14)	6 (12)	25 (20.83)
4.	Physical	Noise	30 (42.86)	42 (84)	72 (60)
		Radiation	3 (4.29)	34 (68)	37 (30.83)

5.	Psycho-social hazards	Heat	9 (12.86)	28 (56)	37 (30.83)
		Cold	9 (12.86)	28 (56)	37 (30.83)
		Lack of Control	26 (37.14)	20 (40)	46 (38.33)
		Fatigue from shift-work	64 (91.43)	50 (100)	114 (95)
		Over work	17 (24.29)	9 (18)	26 (21.67)
		Job pressure	19 (27.14)	8 (16)	27 (22.50)
		Job stress	15 (21.43)	5 (10)	20 (16.67)
		Sleep disorders	17 (24.29)	5 (10)	22 (18.33)
		Long working hours	46 (65.71)	44 (88)	90 (75)
		Boring work	37 (52.86)	22 (44)	59 (49.17)
		Threats to physical security	12 (17.14)	6 (12)	18 (15)

Figures in parentheses indicate the percentage values

Chapter 39

Survey of Health Problems for Long-Haul Flight Travelers

Bor-Shong Liu, Chia-Chen Wu

Department of Industrial Engineering and Management
St. John's University
Taipei, 25135, TAIWAN

ABSTRACT

The purpose of present study was to examine the health problems for long-haul flight travelers. A total of 242 travelers who have been overseas traveling were interviewed. The questionnaire were designed to determine the traveler's background information, trips, cabin seats, preferable seat position, physical activities in flight, eating diary (foods or drink), and discomfort rating for body areas during the flight period. In addition, a validated Verran and Snyder-Halpern Sleep Scale was used to measure sleep disturbance, sleep effectiveness, and sleep supplementation. Results of the analysis showed that there were less than 20% of the travelers understood economy-class syndrome or Deep Vein Thrombosis. About 67.8% of the travelers have preferable seat positions, and there were about 78%, 18% and 4% of them preferring for window side, aisle side and near the emergency exit, respectively. Only 30% of the travelers could take enough rest and less than 30% of the travelers would do physical activities (e.g. in-seat exercises, upper body and breathing exercises, short walks etc.) more than three times during flight. About 50% of travelers drink less than two cups of water. However, more than 60% of the travelers want to drink some cups of coffee or tea. In addition, 29% of the travelers would drink alcoholic beverages (beer, wine, whiskey etc.). For uncomfortable rating by body map, the most musculoskeletal problems during flight period were reported with neck (48%), low back area (46%), knees (42%), shoulders area (40%), and thigh (36%). For the overall rating, about 33% of the travelers feel uncomfortable after long-haul flight. For sleep problems, the mean scores of sleep disturbance, sleep effectiveness, and sleep supplementation were 346.4, 120.3, and 197.3 respectively. Furthermore, an ANOVA was conducted to examine the effects of trips and cabin class on sleep quality. Results of analysis showed that there were significant differences in sleep quality between trips and cabin class. Travelers have higher sleep disturbance, and lower sleep effectiveness after long-haul trips (e.g. from Taiwan to USA or European area). Of course, higher sleep disturbance, and lower sleep effectiveness could be reported by travelers in economic-class. Results of present study could provide the information for promoting health of long-haul flight travelers and airline companies have the responsibility to assist and educate in preventing health problems for long-haul flight.

Keywords: Long-haul flight, Sleep Disorders, Food Intake, Deep Vein Thrombosis

INTRODUCTION

With the prospect of ever-larger commercial aircraft (e.g. Airbus 380) and increasing population, passenger-miles will continue to increase – likely doubling by year 2020 (Bergau, 1996). A long-haul flight is a flight which is over seven hours in length and often involves intercontinental travel, for example, flights from Asia to the Americas. They are associated with negative feelings after arrivals that constitute travel fatigue (Reilly, 1997). Symington and Stack (1977) have reported cases of multi-vehicle–induced PE after travel by automobile (all were drivers), rail alone, rail/ship and aircraft. They referred to the 'economy-class syndrome' as it applied to venous coagulation problems with air travel because it is much more common in economy class (than in business or first class) because of the greater number of passengers there; they emphasized that the syndrome should not be applied only to modern aircraft. More recently this syndrome has been referred to as 'coach-class thrombosis' and 'traveler's thrombosis' (Arfvidsson et al., 2000; Eklof et al., 1996). A New Zealand study of 878 passengers who had undertaken a 10-hour flight reported a 1% incidence of symptomatic venous thromboembolism (Hughes et al., 2003).

In many cases, a long-haul flight is associated with jet lag at the other end, because the plane has crossed several time zones. A variety of techniques can be used to cope with jet lag; regular travelers often develop their own. Many people recommend setting your watch to the time at your destination when you board the plane, and trying to sleep or rest during the "night," regardless as to what time it is on the plane. The effects are due to time spent in an environment that is cramped and offers little opportunity for exercise, a restricted choice of food, dehydration due to dry cabin air, and cabin hypoxia, which increases fatigue and changes the daily profiles of some variables (Brown et al., 2001; Coste et al., 2005).

In recent years, amount of long-haul travelers grew year by year. Airplanes were the main transportation for connecting the globe. Flight time of aircraft has increased from ten minutes to more than ten hours. However, there are not enough data on the effects of prolonged sitting and flights of long duration on health on travelers. There were physical factors (atmospheric pressure, SpO2, temperature and humidity, noise, radiation, G force etc.), chemical factors (concentration of oxygen, CO2), biological factors (eating and infection), ergonomic factors (limited area, Deep Vein Thrombosis DVT, musculoskeletal injuries, and psychological factors (jet lag, sleep problems). Thus, the purpose of present study was to examine the health problems for long-haul flight travelers.

METHODS

A total of 242 travelers who have been overseas traveling were interviewed and they were drawn accidentally from travel agency in Taipei. Their ages were from 13 to 80 years old. Flight route could be divided into two categories (i.e. short and long route). The long flight route defined the nonstop flight duration more seven hours in present study, for example, arrived in Brisbane (8 hours 45 minutes), Los Angeles (11 hours 25 minutes), Frankfurt (13 hours 45 minutes), Vienna (14 hours 20 minutes) from Taiwan. Furthermore, the cases of present study were 91 and 151 cases in long haul flight route and short fight route, respectively.

SURVEY OF QUESTIONNAIRE

The questionnaire was designed to determine background information, and travel behaviors in flight route, cabin seats, preferable seat position (windows, aisle side, near emergency exit), frequencies of physical activity in flight, eating foods and drinking beverages. In addition, ratings of musculoskeletal uncomfortable during flight were evaluated by body map. The areas of neck, shoulders, upper back, lower back, elbows, hands, buttocks and thighs, knees, ankles and feet were evaluated on five levels (very uncomfortable, uncomfortable, normal, comfortable, very comfortable).

VERRAN AND SNYDER-HALPERN SLEEP SCALE

A validated Chinese version questionnaire of Verran and Snyder-Halpern Sleep Scale (VSH Sleep Scale) was applied to assessment the sleep quality (Lin and Tsai, 2003). In addition, the VSH Sleep Scale utilizes three scales (sleep disturbance, sleep effectiveness, and sleep supplementation) to characterize overall sleep quality (Snyder-Halpern and Verran, 1987). Each characteristic is measured using a 100 mm visual analogue scale and the total score for the primary outcome of sleep disturbance is a sum of the scores from each scale (total score maximum 800). A lower total score on this scale indicates a lower degree of sleep disturbance. The secondary outcome parameters for this study included the degree of sleep effectiveness and sleep supplementation (and their determinants) as measured by the VSH Sleep Scale. The sleep effectiveness scale measures both quality and length of sleep as perceived by the patient using the following four characteristics: rest upon awakening, wake after final arousal, subjective quality of sleep, sleep sufficiency evaluation. A visual analogue scale is used to measure each of the four items and these scores are summed to represent a total score. The maximum possible total score is 400 with a higher score representing greater sleep effectiveness. The sleep supplementation scale measures the degree to which the bulk sleep period is augmented with additional sleep time. The three characteristics measured are daytime sleep, morning sleep, and afternoon sleep. The scores from each of these scales are summed to obtain a total score (total score maximum 300). A higher total score on this scale represents a worse outcome, as more supplemental sleep was needed (Snyder-Halpern and Verran, 1987).

DATA ANALYSIS

Data from questionnaires were coded for SPSS format. Further, descriptive procedure displays statistics for variables and ANOVA was conducted to examine the effects of trips and cabin class on sleep quality. Statistical significance was set at a probability level of 0.05.

RESULTS

BEHAVIORS OF TRAVELERS

A summary of the background data is presented in Table 1. A total of 242 travelers who have been overseas traveling were interviewed. Table 2 showed the data of travelers' behaviors in flight. Results of descriptive statistics reported that there were less than 20% of the travelers understood economy-class syndrome or Deep Vein Thrombosis. About 67.8% of the travelers have preferable seat positions, and there were about 78%, 18% and 4% of them preferring for window side, aisle side and near the emergency exit, respectively. Less than 30% of the travelers would do physical activities in seat or aisle (e.g. in-seat exercises, upper body and breathing exercises, short walks etc.) more than three times during flight. About 50% of travelers drink less than two cups of water. However, more than 60% of the travelers want to have some cups of coffee or tea. In addition, 29% of the travelers would drink alcoholic beverages (wine, beer etc.).

Table 1 Demographic characteristics of the participants

Variables	n	Percent
Gender		
Male	133	55 %
Female	109	45 %
Age (years old)		
under 29	152	62.8 %
30~39	38	15.7 %
40~49	35	14.5 %
50 and above	17	7.0 %
Marital status		
Single	174	71.9 %
Married	68	28.1 %
Drinking alcohol alcoholic beverages per week		
Never	97	40.1 %
Once	74	30.6 %
2-4 times	26	10.7 %
More than 5 times	45	18.6 %
Smoking habit		
Never	177	73 %
Yes	65	27 %
Flight route		
Long	91	37.6 %
Short	151	62.4 %

Table 2: Traveler behaviors

Cabin Classes	n	Percent
Business-class	21	8.7 %
Economy-class	221	91.3 %
Preferable seat position	**n**	**Percent**
Yes	164	67.8 %
Windows	128	78 %
Aisle	30	18.3 %
Near emergency exit	6	3.7%
No	78	32.2 %
Physical activities in seat or aisle during fight		
Frequencies	**n**	**Percent**
Non	26	10.8%
1-2 times	160	66.1%
More than 3 times	56	23.1%
Drinking water during flight		
Cups (about 300 ml)	**n**	**Percent**
1 cup	91	37.6 %
2 cups	42	17.3 %
3 cups	43	17.8 %
4 cups	28	11.6 %
5 cups and above	38	15.7 %
Drinking Caffeine beverages during flight		
Drinking	**n**	**Percent**
Yes	161	66.5%
1 cup	61	37.9 %
2 cups	54	33.5 %
3 cups	26	16.1 %
4 cups	12	7.5 %
5 cups and more	8	5 %
Non	81	33.5%
Drinking alcohol beverages during flight		
Drinking	**n**	**Percent**
Yes	71	29 .4 %
Red wine	38	53.5 %
White wine	12	16.9 %
Beer	18	25.3 %
Other	3	4.3 %
Non	171	70.6

DISCOMFORT RATING BY BODY MAP

Only 30% of the travelers could take enough rest. The uncomfortable rating of neck, shoulders, upper back, lower back, elbows, hands, buttocks and thighs, knees, ankles and feet areas were evaluated by body map. Figure 1 indicated that travelers had very uncomfortable and uncomfortable on body areas. The most musculoskeletal problems during flight period were reported with neck (48%), low back area (46%), knees (42%), shoulders area (40%), and thigh (36%). For the overall rating, about 33% of the travelers feel uncomfortable after long-haul flight. Prolonged periods of sitting caused these results. In addition, Abramowitz and Gertz 2007) suggests that the leg movement and calf compression associated with the modified airline seat may decrease the probability of DVT due to prolonged periods of sitting by reducing venous stasis. Both seat redesign and in-flight physical activities of travelers could decrease these discomfort.

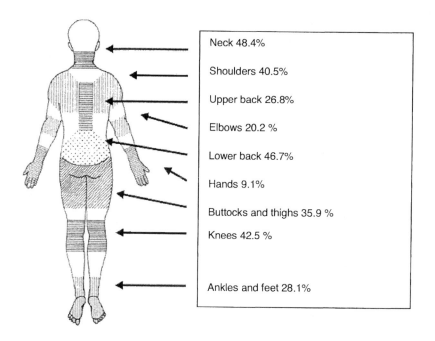

Neck 48.4%

Shoulders 40.5%

Upper back 26.8%

Elbows 20.2 %

Lower back 46.7%

Hands 9.1%

Buttocks and thighs 35.9 %

Knees 42.5 %

Ankles and feet 28.1%

FIGURE 1. Uncomfortable rating by body map during flight

SLEEP AND MEAL PROBLEMS

For sleep problems, the mean scores of sleep disturbance, sleep effectiveness, and sleep supplementation were 346.4, 120.3, and 197.3 respectively. Furthermore, an ANOVA was conducted to examine the effects of trips and cabin class on sleep quality. Results of analysis showed that there were significant differences in sleep between trips. Travelers have higher sleep disturbance, and lower sleep effectiveness after long-haul trips (e.g. from Taiwan to USA or European area). Of course, higher sleep disturbance, and lower sleep effectiveness could be reported by travelers in economic class.

For eating diary, flight attendant serves dishes depending with the flight route from once to trice. Flight attendant serves first meal on half hour or one hour after take-off. For example, lunch/dinner meal include the appetizer (roll with butter, salad), main dish (rice, noodles or spaghetti), and dessert (cake, fruit, or ice cream), and beverages (wine, tea or coffee). Results indicate that the meal service in flight do not closely match the amount of jet lag that is perceived. Chung and Kao (2008) also reported that the body clock (endogenous component) is unadjusted to the new local time after transmeridian travel immediately, jet lag occurs, as part of which the enjoyment of food decreases, indigestion is more prevalent and it is common to feel bloated after a meal. Thus, providing meal should take jet lag factor into consideration.

DISCUSSION AND CONCLUSIONS

Since economy-class seats are designed to accommodate the 'average' individual, those who are taller or obese will be more cramped, and shorter people with shorter legs will have more seat edge pressure behind their knees which may inhibit venous return. Those with marked vericosities already have some blood sludging which could be aggravated with prolonged sitting, and some birth control medicine can enhance blood coagulation with leg pooling (Sahiar and Mohler, 1994; Paganin et al., 2003). Deep Vein Thrombosis more frequently affects non-aisles passengers, emphasizing once again the role of prolonged sitting during long-haul flights. Lack of passengers' awareness of the danger contributes to the potential devastating outcome. In addition, present study reported that less than 20% of the travelers understood economy-class syndrome or deep vein thrombosis. Further, dehydration is common, due to cabin microclimate conditions, lack of sufficient water intake and increased alcohol consumption.

In overall combined factors, prolonged air travel at least doubles the risk of thrombosis in a passenger without predisposing factors. Thus, preventing deep vein thrombosis in long-haul travelers could do physical activity in flight (Rugman, 2004), for example, (1) in-seat leg exercises: flexing the calf muscles; ankle rotations; bending and straightening legs, feet and toes every half-hour; intermittent pressing of feet hard against floor/foot-rest; (2) upper body and breathing exercises; (3) short walks, at least hourly; (4) while seated, keep the thigh clear of the edge of the seat and avoid crossing legs or prolonged awkward hip or knee positions;(5) when travelling Business/First class, elevate both feet on foot or leg rests. Other strategies include adequate hydration at least one liter of water every five hours; avoid alcohol, coffee and sedatives and wear loose clothing. Results of present study could provide the information for promoting health of long-haul flight travelers and airline companies have the responsibility to assist and educate in preventing health problems for long-haul flight.

ACKNOWLEDGEMENTS

This study is supported by a grant from the National Science Council, Taiwan, and Project No. NSC 98-2221-E-129-003.

REFERENCES

Abramowitz, H.B., and Gertz, S.D. (2007), "Venous stasis, deep venous thrombois and airline flight: can the seat be fixed?" *Annals of Vascular Surgery,* 21(3), 267-271.

Arfvidsson, B., Eklof, B., Kistner, R.L., Masuda E.M., Sato, D.T. (2000), "Risk factors for venous thromboembolism following prolonged air travel Coach class thrombosis." *Hematology/Oncology Clinics of North America,* 14, 391-400.

Bergau, L. (1996), "Medical guidelines for air travel." *Aviation, Space, and Environmental Medicine,* 67, B1-B16.

Brown, T., Shuker, L., Rushton, L., Warren, F., Stevens, J. (2001), "The possible effects on health, comfort and safety of aircraft cabin environments." *The* Journal of the Royal Society for the Promotion of Health, 121(3), 177-84.

Chung, N.H., and Kao, S.Y. (2008), "The Effect of Simulated 8 Time Zone Difference Upon Food Intake." *Journal of Hospitality and Home Economics,* 5(4), 409-424.

Coste, O., Van Beers, P., Bogdan, A., Charbuy, H., Touitou, Y. (2005), "Hypoxic alterations of cortisol circadian rhythm in man after simulation of a long duration flight." *Steroids,* 70, 803-810.

Eklof, B., Kistner, R.L., Masuda, E.M., Sonntag, B.V., Wong, H.P. (1996), "Venous thromboembolism in association with prolonged air travel." *Dermatologic surgery,* 22, 637-641.

Hughes, R.J., Hopkins, R.J., Hill, S., Weatherall, M., Van de Water, N., Nowitz, M., Milne, D., Ayling, J., Wilsher, M., Beasley, R. (2003), "Frequency of venous thromboembolism in low- to moderate-risk long-distance air travellers: the New Zealand Air Traveller's Thrombosis study." *The Lancet,* 362, 2039-2044.

Lin, S. L., and Tsai, S. L. (2003), "The reliability and validity of the VSH sleep scale Chinese version." *VGH Nursing,* 20(1), 105-106.

Paganin, F, Bourdé, A, Yvin, J-L, Génin, R, Guijarro, J-L, Bourdin, A, Lassalle, C. (2003), "Venous thromboembolism in passengers following a 12-h flight: a case-control study." *Aviation, Space, and Environmental Medicine,* 74(12), 1277-1280.

Reilly T, Atkinson G, Waterhouse J. (1997), Biological rhythms and exercise. Oxford: Oxford University Press, 75-95.

Rugman, F. (2004), "Preventing deep vein thrombosis in long haul travellers" Pulse, April, 51-54.

Sahiar, F., and Mohler, S.R. (1994), "Economy class syndrome." *Aviation, Space, and Environmental Medicine,* 65, 957-960.

Snyder-Halpern, R., and Verran, J. A. (1987), "Instrumentation to describe subjective sleep characteristics in health subjects." *Research in Nursing & Health,* 10,155-163.

Symington, I.S., and Stack, B.H.R. (1977), "Pulmonary thromboembolism after travel." *British Journal of Diseases of the Chest,* 71, 138-140.

Chapter 40

An Approach to Study the Body Back Discomfort of Economy Class Aircraft Passenger

CheeFai Tan, Wei Chen, Matthias Rauterberg

Department of Industrial Design
Eindhoven University of Technology
5612AZ Eindhoven, The Netherlands

ABSTRACT

The paper describes the questionnaire study that was designed to unfold the relationship between different body back discomfort levels over time. One hundred and four anonymous self administered questionnaires were completed at Schipol International Airport, the Netherlands from October through November 2008. Long haul economy class aircraft passengers are at risk of uncomfortable for long hour sitting and experience significant uncomfortable at different body back parts such as arm, neck and lower leg.

Keywords: Body Discomfort, Economy Class Aircraft Passenger

INTRODUCTION

Air travel is becoming increasingly more accessible to people both through the availability of cheap flights and because the airlines are now able to cater for individuals of all ages and disabilities. Health problems may arise due to anxiety and unfamiliarity with airport departure procedures prior to flying, whilst during the flight, problems may arise as a result of the food served on board, differences in the environmental conditions inside the cabin (pressure, ventilation, relative humidity, noise and vibration), the risk of cross-infection from fellow passengers, seat position, posture adopted and duration of the flight. These can be further compounded by changes in time zones and meal times, which may continue to affect an individual's health long after arrival at the final destination (Brundrett, 2001). Travel by air, especially long distance, is not a natural activity for human. Many people experience some degree of physiological and psychological discomfort and even stress during flying. Excessive stress may cause passenger to become aggressive, over-reaction, and even endanger the passenger's health (Kalogeropoulos, 1998; World Health Organization, 2007). A number of health problems can affect flying passengers.

Comfort is an attribute that today's passenger demand more and more. The aircraft passenger's comfort depends on different features and the environment during air travel. Seat discomfort is a subjective issue because it is the customer who makes the final determination and customer evaluations are based on their opinions having experienced the seat (Runkle, 1994). The aircraft passenger seat has an important role to play in fulfilling the passenger comfort expectations. The seat is one of the important features of the vehicle and is the place where the passenger spends most of time during air travel. The aviation industry is highly competitive and therefore airlines try to maximize the number of seats (Quigley et al., 2001). Often this results in a very limited amount of seating space for passengers, especially in economy class (Hinninghofen and Enck, 2006).

Long haul economy class aircraft passengers are at risk uncomfortable for long hour sitting and experience significant uncomfortable at different body back parts such as neck and lower leg. Further studies concerning how to best provide comfort to long haul economy class aircraft passenger are needed. This questionnaire study was set out to examine the different body back parts discomfort of economy class aircraft passenger to help prioritize action aimed at discomfort reduction. One hundred and four anonymous self administered surveys were completed at Schipol International Airport, the Netherlands from October through November 2008.

METHODS

QUESTIONNAIRE DEVELOPMENT

The questionnaire consisted of three sections: (1) questions about the respondents' travel frequency by aircraft per year, common flight duration and the class; (2) question about their uncomfortable level of each part of their body backside after one hour and five hours flight; (3) question about demographic background.

The questionnaire begins with a short, self-explanatory introduction in which the purpose and background of the survey were explained; it was also emphasizes that data would be treated with confidentiality and analyzed in an anonymous manner. An example on how to answer the question correctly is shown.

378

The primary means of investigation is to identify the body discomfort level with regards on time during air travel. This was devised to identify the body part discomfort, to indicate the discomfort level for each defined body part for after one hour and after five hours of flight. In order to identify the body part discomfort level, a body mapping method is used. In this method, the perception of discomfort is referred to a defined part of the body. The subject is asked for the discomfort experiences during flight for each defined body part. The subject is asked to assess the discomfort level using a five point Likert scale. The scales are graded from 'extremely discomfort' to 'normal'. Figure 1 shows the body map and scales for discomfort assessment.

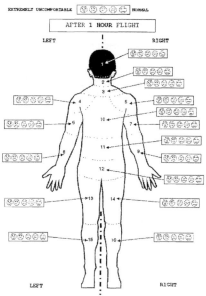

FIGURE 1 Body map and scales for body discomfort evaluation.

QUESTIONNAIRE SAMPLING

The questionnaire was completed by 104 aircraft passengers who were randomly sampled. The investigator was present on each occasion, during which aircraft passengers were approached and the aims of the investigation were briefly outlined. The questionnaire took between 3-5 min for self-completion.

RESULTS

50 females and 54 males completed the questionnaire. A wide range of ages was represented (17 to 75 years). The mean BMI of the respondents were 24.09 kg/m^2

and it is generally accepted as being in the 'overweight' category (M = 24.09, SD = 4.93).

We conducted a factor analysis on body part discomfort level after one hour flight and after five hours flight, to identify the underlying dimensions of the body part discomfort of aircraft passenger. Scores on the sixteen statements were submitted to principal components factor analysis with varimax rotation. A scree-plot indicated that the eigenvalues started to level off after three factors. Thus, a three factor solution yielded the best solution.

BODY DISCOMFORT LEVEL AFTER ONE HOUR FLIGHT

For 'after 1 hour flight', the three factors explained 72.75% of the variance in the data. Table 1 provides an overview of the composition of the three factors for body part discomfort after 1 hour flight. The first factor included five items that described the body discomfort at buttock, upper leg (left and right) and lower leg (left and right). This factor appeared to reflect the lower body of the respondent. Therefore, it was labeled as "Lower body". The second factor included four items. All four items described the body part, which are upper arm (left and right) and lower arm (left and right). The second factor was labeled as "Arm". The third factor included seven items, namely, head, neck, shoulder, left shoulder, right shoulder, upper back and lower back. The third factor was labeled as "Upper body".

Table 1 Results of Factor Analysis of Body Part Discomfort after One Hour Flight.

	Factor		
	Lower body	Arm	Upper body
Right lower leg	.910		
Right upper leg	.902		
Left upper leg	.896		
Left lower leg	.890		
Buttock	.716		405
Right upper arm		.902	
Left upper arm		.880	
Right lower arm		.763	
Left lower arm		.739	
Neck			.831
Shoulder			.772
Lower back			.683
Upper back		428	.659
Right shoulder		568	.612
Left shoulder		568	.612
Head			.588
Explained variance	46.76%	14.41%	11.58%
Cronbach's Alpha	.95	.89	.86

Note: Only factor loadings > 0.58 are selected

BODY DISCOMFORT LEVEL AFTER FIVE HOURS FLIGHT

For 'after 5 hours flight', the three factors explained 74.04% of the variance in the data. Table 2 provides an overview of the composition of the three factors for body part discomfort after five hours flight. There were four items labeled as "Arm" in first factor. The first factor included left lower arm, left upper arm, right lower arm and right upper arm. The second factor included six items that described the body discomfort at neck, shoulder, left shoulder, right shoulder, upper back and lower back. This factor appeared to reflect the upper body of the respondent. The second factor labeled as "Upper body". The third factor included five items, namely, buttock, right lower leg, right upper leg, left lower leg and left upper leg. The third factor was labeled as "Lower body".

Table 2 Results of Factor Analysis of Body Part Discomfort after Five Hours Flight.

	Factor		
	Arm	Upper body	Lower body
Left lower arm	.904		
Left upper arm	.881		
Right lower arm	.869		
Right upper arm	.829		
Head		489	
Shoulder		.866	
Neck		.843	
Lower back		.800	
Upper back		.671	
Left shoulder	550	.648	
Right shoulder	585	.603	
Right lower leg			.904
Left lower leg			.879
Right upper leg			.838
Left upper leg			.805
Buttock		428	.593
Explained variance	47.10%	15.48%	11.46%
Cronbach's Alpha	.94	.89	.90

Note: Only factor loadings > 0.59 are selected

Univariate analysis of variance was conducted to find the differences of body discomfort level between after one hour flight and after five hours flight. Figure 2 showed the comparison of body discomfort level for different body part after one hour and after five hours flight. The results showed the body discomfort level after five hours flight was more discomfort than body discomfort level after one hour flight.

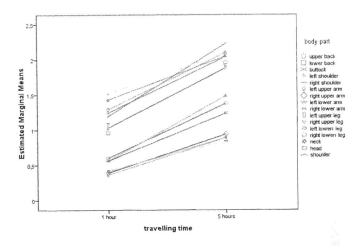

FIGURE 2 Univariate analysis result for different body discomfort level.

DISCUSSION

With respect to travel duration, passengers who travelled with 6 to 10 hours reported highest body discomfort level. The result showed that longer flight duration was causing higher discomfort at arm section. Male respondents felt higher body discomfort level at lower body section than female respondents. In the other hand, older aircraft passengers felt that their arm was more discomfort after one hour flight. The passengers with higher BMI reported that their lower body section is more discomfort after one hour flight.

Through one-way ANOVA analysis at body discomfort level after five hours flight, the aircraft passengers who travelled 11 times or more in a year have reported highest body discomfort level at upper body section. Subsequently, passengers with longer flight duration perceived higher discomfort level at arm section after five hours flight. The gender of respondents affected the body discomfort level at upper body and lower body section. Female respondents found to be more discomfort than male respondents for after five hours flight. Female respondents reported that they have highest body discomfort level at upper body section. Male respondents reported the lower body section as the most discomfort section. Neck and shoulder showed the similarity of body discomfort level for after one hour and after five hours flight. The body discomfort level for buttock was the highest in rank after five hours flight.

Based on the ranking of the body discomfort level, the main areas of body discomfort after one hour flight were shoulder, neck and right lower leg. Subsequently, the main areas of body discomfort after five hours flight were buttock, shoulder and neck. The body discomfort after five hours flight of buttock and neck were similar to the study by Quigley et al. (2001). The study found that the main areas of complaint during the flight were lower back, buttocks and neck.

The lower back was ranked after neck for the body discomfort level after five hours flight.

CONCLUSION

In the present research, we sought to gain more insights into aircraft passenger body discomfort level between after one hour flight and after five hours flight, especially with regards to flight frequency, flight duration, and gender. There are 104 respondents filled up the questionnaire about body discomfort level after one hour and after five hours flight. In line with the survey hypothesis, findings confirmed that aircraft passenger who travelled after five hours are more discomfort than passenger who travelled after one hour. The finding showed that buttock, shoulder and neck were rank as main body discomfort level after five hours flight. Economy class aircraft seat discomfort was associated with flight duration. Interventions aimed at improving the aircraft seat comfort should be prioritized when devising a discomfort reduction strategy for economy class aircraft passengers.

REFERENCES

Brundrett, G. (2001), "Comfort and health in commercial aircraft: a literature review." *The Journal of The Royal Society for the Promotion of Health*, 121(1), 29-37.

Kalogeropoulos, S. (1998), *Sky rage*. Flight Safety Australia, 36-37.

World Health Organization, (2007), "Travel by air: health considerations," http://whqlibdoc.who.int/publications/2005 /9241580364_chap2.pdf. Accessed on 3rd March 2007.

Runkle, V.A. (1994), "Benchmarking seat comfort." *SAE Technical Paper*, 940217.

Quigley, C., Southall, D., Freer, M., Moody, A. and Porter, M. (2001), *Anthropometric study to update minimum aircraft seating standards*. EC1270, ICE Ergonomics Ltd., 2001.

Hinninghofen, H. and Enck, P. (2006), "Passenger well-being in airplanes." Auton Neurosci, 129(1-2), 80-85.

CHAPTER 41

Ergonomics in the Context of Prevention, Educational Proposal and Current Situation in Colombia

Emilio Cadavid, Luz Mercedes Sáenz

Empresas Públicas de Medellín
Cra.58 #42-125, Floor 4
Universidad Pontificia Bolivariana
Circ. 1 #70-01 Bloque 10, Medellín - Colombia

ABSTRACT

This article describes the context of an ergonomic vision in Colombia. Education is considered to be a fundamental factor in the development of the subject and for this purpose educational, promotional, and applicable experiences are being developed. These experiences extend to diverse forms of human activity.

Keywords: Ergonomics, prevention, education in Ergonomics.

INTRODUCTION

In the context of prevention in the workplace, ergonomics is essential. The application of its principles to the workplace and the activity being executed procure increases not only in productivity but also in an employee's "quality of life". This suggests that ergonomics is relevant to diverse aspects of human life.

In Colombia, ergonomics has been developed mostly in an educational context. The furthering of this discipline has been achieved by implementing application strategies such as the grouping of interdisciplinary academic-scientific societies, the development of guidelines for integral attention based on evidence, the development of courses in ergonomics to undergraduate and postgraduate programs, and the creation of research groups and self care programs, among others.

Education in Ergonomics has established bases for the development of prevention criteria and has helped further the development of this discipline.

PRECEDENT OF ERGONOMICS IN COLOMBIA

Based on the historical recollection of Ergonomics in Colombia by Engineer Jairo Estrada Muñoz (Estrada, 2005), this discipline first began to appear in the country in the 1950's when the first engineers with foreign specializations in Industrial Hygiene began to arrive. From this point, ergonomic conditions in the workplace were considered in order to increase employee efficiency.

Also, between the 1960's and 1970's the first approaches to the subject were made through conferences, small trials and ergonomics courses in Industrial Engineering faculties given by professor Jorge Forcadas. These courses became an important tool which was offered to the Instituto de Seguros Sociales (government run organization in Colombia which administrates health plans) and the Seguro Social de México, among others. These projects brought academic life to Ergonomics in some universities around the country.

By the 1980's it was possible to clearly organize the influence of ergonomics in Colombia into three different levels: research or studies for specific intervention, publications, education. Today, the discipline is present in specific actions included in Professional Risk System as a strategy for prevention, health and safety in the workplace.

The 1980's was an important decade because it was then that the first Colombian professional specializations began. The Colombian Ergonomic Society (ESC) was created in 1996 with the criteria of a scientific society and oriented towards becoming a part of the International Ergonomics Association. It is also during this decade that specific ergonomic activities became present in the work environment in the form of evaluations of workspace, evaluations of work processes, improvements to some lines of production and the incorporation of techniques to completely redesign some activities.

Presently, the presence of ergonomics has been strengthened in Industrial design

Faculties and postgraduate studies as well as other continuous education courses (even at the level of specialization), which support the interdisciplinary vision of ergonomics. For this same purpose, different experiences are developed in education, promotion and application of ergonomics; these extend to diverse aspects of human activities and which are expressed by the following actions:

ERGONOMICS AND PREVENTION

Prevention is considered to be the collection of activities which allows a person to have a healthy lifestyle and enables communities to create and consolidate environments where health is promoted and risk of disease is reduced, thus, prevention implies developing anticipatory measures. In the field of health it implies a scientific conception of work, not only as a way of performing but also as a way of thinking (Zas Ros, 2001).

Ergonomics carries out analysis and diagnostics programs which observe, evaluate and carry out applications in the occupational context and other forms of human activity. The objectives of this discipline are oriented towards the reduction of errors, increase in safety and performance of the person – machine - environment system, reduction of fatigue and physical stress, making products more easily accepted and easier to use, improving the environment of use, reducing time and equipment loss and increasing productivity, among others (Chapanis, 1995).

Not only are these applications presented as solutions to specific needs but also the motives for the analysis and application of ergonomics lead to prevention strategies. By doing so, situations which may affect a person's health, welfare or integrity can be anticipated and it becomes an alternative intended to reduce risks and encourage proper practice in the occupational context, and/or the use of objects/elements in other context and generally all activities which are part of a person's everyday life.

Figure 1 shows a proposal in an occupational context with levels for prevention and intervention within the ergonomics framework. The reference point for these actions are public health diagnostics, prevention, healing and rehabilitation which are to be carried out for the benefit of the community (World health Organization, 1978) and expressed within the framework of this discipline, oriented towards the creation of a culture on the subject, the evaluation of specific conditions in the person- machine – environment relationship and the design of a work situation (the activity: objects, environment and methods of execution) which are coherent to the specific circumstances of a person's work. These conditions are considered based on health conditions which help keep an adequate physical- corporal condition, including rehabilitation and reintroduction considerations if applicable.

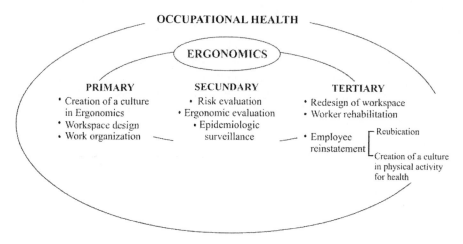

Figure 1. Levels for prevention and intervention of ergonomics based on public health actions. Proposal by E. Cadavid (2003)

INTERDISCIPLINARITY AS A STRATEGY FOR PREVENTION

This experience is developed and applied through the Corporación de Salud Ocupacional y Ambiental, which is a non profit organization that integrates the Sociedad Colombiana de Medicina del Trabajo, Antioquia chapter (Colombian Occupation Medicine Society), the Asociación Colombiana de Higiene Ocupacional, Antioquia chapter (Colombian Occupational Hygiene Association), the Sociedad Colombiana de Ergonomía, Antioquia chapter (Colombian Ergonomics Society) and the Asociación Antioqueña de Seguridad Integral (Integral Health Association of Antioquia). See figure 2. Integration of Societies for health and safety at work.

This group of associations integrates four fields of knowledge for training and development of academic activities relevant to the improvement of environmental and work conditions; welfare of the working population; the quality of work processes; productivity and competitiveness in the countries economic sector, both public and private. It also allows for interactions between different government, professional, academic and business actors in benefit of health, welfare and safety conditions of a persons occupational context and its implications on his or her family life.

Figure 2. Integration of societies for health and safety at work.

While this Corporation is an initiative originated in a specific region, it participates in the yearly Occupational Health week which is a nationwide event where recognized international professionals are invited to speak and a general view of the research and applications which are being developed in Colombia is provided. This event also constitutes a framework in which The Congreso Colombiano de Ergonomía (Colombian Ergonomics Congress) takes place and is construed as a space where education strategies and subjects which are currently researched and applied in ergonomics in Colombia are presented.

In the year 2009, the fifteenth Occupational Health week took place, here, the 8th ergonomics Congress was held, titled "Legal Framework and Normativity" (a subject which is still in an incipient stage in Colombia). In previous years, the Colombian Ergonomics Congress had presented subjects such as: Ergonomics and Design, in 2008; Ergonomics: Products, spaces and services, in 2007; Ergonomics and posture, in 2006; Ergonomics and education, in 2005; Improving work conditions and productivity, in 2004 and Ergonomics and accident occurrence in 2003, among others.

In these events proposals have been presented on the following subjects: Analysis of work conditions in diverse economic sectors, interventions on the work load based on different methodologies, specific user analysis (elderly persons, rural population, etc) and applications of ergonomics and design in diverse contexts, and methodological proposals for diagnostics of ergonomic conditions, among others.

The subjects and activities of the Corporación de Salud Ocupacional y Ambiental, related to the disciplines of the societies which conform it, Ergonomics, Occupational Medicine and Occupational Hygiene and safety constitute an interdisciplinary proposal for prevention and application of proper practice in the work context and how it translates to the wellbeing and safety of a person's personal and family life.

THE GOVERNMENT INICIATIVE USING INTEGRAL ATTENTION GUIDELINES BASED ON EVIDENCE

The ministry of Social Protection through the General Professional Risk system (Ministerio de la Protección Social, 2010) carried out a follow up to professional illnesses which had been diagnosed between the years 2001 and 2005. The conclusions pointed towards muscleskeletal disorders as the leading cause of professional morbidity in workers of the economies formal sector.

The first identified cause was carpal tunnel syndrome followed by lumbar pain and intervertebral disk disorders. The third cause of morbidity was found to be neurosensory deafness (between 2001 and 2003), in 2004 however it was bumped down to 4th place by intervertebral disk disorder which tripled from 3% in 2002 to 9% during 2004. Also, epicondylitis and tenosynovitis of the short extensor and abductor of the thumb (Quervain's tenosynovitis) are worth mentioning due to a continuous increasing tendency from 2002 to 2004.

These facts make it evident that muscleskeletal disorders (MSD) are the leading cause of professional morbidity in the contributory regime of the Sistema General de Seguridad Social en Salud (General Social Security Health System). Also a continuously increasing tendency can be observed, going from 65% during 2001 to 82% in 2004 of all diagnostics carried out. These MSD′s are affecting two body segments: upper body and spinal column.

For this reason and by means of an integral focus, the Integral attention guidelines for Occupational health –GATISO- were created. These were based on evidence which involve different disciplines that support occupational health, among them ergonomics, and from which interventions for controlling occupational risk factors are established.

Its objectives have been to constitute an important tool for unifying and approaching integral attention of professional morbidity thus generating a positive impact both on an employees health and the health personnel which prevent and diagnose said causes, as established by the Plan Nacional de Salud Ocupacional (National Occupational Health plan) for the period 2003 to 2007.

By means of resolution 2844 of August 16th, 2007 the ministry adopted the Integral Attention guidelines for Occupational Health, GATISO, as an obligatory reference and in 2005 the first 5 subjects were created:

1) GATISO for Muscleskeletal disorders related to repetitive upper body

movements.

2) GATISO for neurosensory hypoacusia induced by noise in the workplace.

3) GATISO for unspecified lumbar pain and vertebral disk illness related to manipulation of manual loads and other risk factors in the workplace.

4) GATISO for shoulder pains related to risk factors in the workplace.

5) GATISO for pneumoconiosis.

The GATISO can be consulted in the Ministry of Social Protection's website: http://www.minproteccionsocial.gov.co/vbecontent/NewsDetail.asp?ID=16481&ID Company=3 . It has encouraged dynamic interaction between diverse disciplines which have contributed to occupational health through participative meetings where experiences in which GATISO has been used are shared,

The methodology was based on the one used for the creation of practical, clinical guides which include the scientific strictness of Evidence - Based Medicine (EBM). EBM is a discipline which provides tools used to improve decision making in regards to health, concerning individual patient care, clinical service, or in a hospital or network of health providers. In the beginning, EMB was oriented towards individual patient management on behalf of clinical treatment, however, later on, its use has extended to decision making on all levels of the health systems including public and occupational health.

Subsequently, in 2008 proposals were circulated for other GATISO on the following subjects: Contact Dermatitis , Lung Cancer, Occupational Asthma, Benzene and its derivatives, Cholinesterase Inhibiting pesticides (Ministry of Social Protection, 2010).

EDUCATION AND RESEARCH IN ERGONOMICS IN COLOMBIA

In Colombia, education in ergonomics began in the mid 1970's and was specifically intended for engineering programs. It consisted of courses pertaining to the study program but not as a specific program in Ergonomics education.

Presently, education and research proposals are also included in faculties of design and engineering as a way of complementing the vision and the application processes for products which were truly conceived for users and which consider diverse contexts of use. These proposals not only take advantage of not only the subjects from the scientific disciplines which support ergonomics, such as Anthropometrics and Biomechanics, but also present courses which express the possibility of application.

Some examples of this are shown in courses such as: Ergonomics for Industrial Design, Ergonomics and information, Physical Ergonomics for Design, Ergonomics Analysis, Cognitive Ergonomics, Transergonomics, from the Universidad Pontificia Javeriana in Bogotá; Physical Ergonomics, Cognitive and Environmental Ergonomics, Product Ergonomics, Usability from the Universidad Industrial de Santander, in Bucaramanga; and Human Factors I, II, and III from the Universidad Nacional De Colombia, in Bogotá.

There are also courses in Ergonomics in undergraduate and technological study programs in Occupational Health and Occupational Therapy. On a specialization level, the offering is still incipient since there are presently only two existing in the country: in the Universidad de Antioquia, in Medellín and the Universidad del Bosque in Bogotá.

Also in process of consolidation are some laboratories dedicated to Ergonomics, Human Factors, Usability and like subjects in Universities from different regions of the country. These have fostered educational spaces for motivation, transmission and application of knowledge as well as offering services to the community.

Within the country there are also groups or lines of research articulated to design and engineering faculties whose objective is oriented towards thematic exploration (research), application and diffusion of Ergonomics and Human Factors.

An example of this is the Ergonomics research line of the Universidad Pontificia Bolivariana in Medellín, which as well as providing teaching support to undergraduate courses in the School of Architecture and Design, also provide support for postgraduate Ergonomics courses, Interior Architecture and Biomedical Engineering projects, develop research projects with and for companies of different economic sectors and encourage continuous education activities (short courses). Its goal is to a create a culture in Ergonomics, not only within the University but also projecting to the community by developing projects which contribute through the principles of participation, production and protection (Saenz and Cadavid, 2006).

PHYSICAL ACTIVITY PROMOTION PROGRAMS

Also within the context of education in Colombia, the subject of physical activity as a strategy for prevention and promotion of health is worth noting, which, in keeping with the objectives of ergonomics, encourages healthy life habits in the occupational context (and it extends to a persons everyday life), increases self esteem, teaches how to overcome obstacles and increases motivation to achieve a good academic and social performance

The National Physical activity Program: Colombia Healthy and active, constitutes a tool for the creation and implementation of public policies related to the reduction of sedentarism throughout the Colombian territory (Instituto Colombiano del Deporte, 2010). It also defines scenarios, among them the workspace, in which companies and government organizations are increasingly incorporating activities related to physical activities and healthy lifestyle habits which increase not only personal welfare (flexibility, strength, better state of health, corporal consciousness, etc) but also transcend and generate positive work dynamics and group habits.

The program "A Physical Flexibility training Program for a Group of Workers in an Utilities Company in Colombia" is a clear example of this strategy where a physical flexibility program is successfully executed with muscleskeletal risk groups focusing on knee and lumbar column (Cadavid and Sáenz, 2009).

CONCLUSIONS

In Colombia, the development of ergonomics has always focused on education as a fundamental factor for presence and growth in diverse fields of analysis and education.

Ergonomics is a part of governmental initiatives by executing, socializing and applying the Guías de Atención Integral Basadas en Evidencia (Integral Attention Guides Based on Evidence), which in the subjects of GATISO for unspecified lumbar pain and disk illness related to manual manipulation of loads and other risk factors in the workplace and GATISO for shoulder pain related to work risk factors, consider the importance of risk factors associated to physical load and muscleskeletal disorders (MSD) as the leading cause of morbidity in employees and intervention through ergonomics as an alternative to improve the situation.

The courses in the design and engineering study plan, as well as the creation of laboratories for application and transfer of knowledge, have favored the diffusion and application of ergonomics and constitute a potential for future interventions and proposals of this discipline in the occupational context as well as other human fields of activity.

The programs for physical activities and sports present an alternative based on Ergonomic prevention in order to encourage self care, improve personal welfare and achieve a good work environment.

REFERENCES

Cadavid, Emilio. and Sáenz Luz M. 2009. "Physical Flexibility Training Program Macroergonomics Proposal for a Group of Workers in a Public Services Company in Colombia" " (paper presented at the 17th Congress of the International Ergonomics Association, Beijing: Chinese Ergonomics Society, Pekín.

Chapanis, A. 1995. In "The discipline of ergonomics and human factors" in Handbook of Human Factors and Ergonomics (Third Edition), 1-5. Waldemar Karwowski. Ed. John Wiley & Sons, Inc., on line version in Willy Interscience http://www3.interscience.wiley.com/cgi-bin/bookhome/112467581/?CRETRY=1&SRETRY=0 (accessed February 5, 2010).

Estrada, Jairo. 2005 "La ergonomía laboral en Colombia, Una aproximación a la identificación de perspectivas" proceedings of Primer Congreso Internacional de Ergonomía, Cali, Colombia.

Instituto Colombiano del Deporte, Coldeportes, 2010. Colombia activa y saludable. http://www.coldeportes.gov.co/coldeportes/index.php?idcategoria=3308. Accessed February, 22, 2010.

Ministerio de la Protección Social. 2010. Por primera vez sistema general de riesgos profesionales cuenta con herramientas de prevención. Boletín de Prensa No 091 de 2007 de 29/08/2007. http://www.minproteccionsocial.gov.co/VBeContent/NewsDetail.asp?ID=16509&IDCompany=3 accessed February 15, 2010.

Sáenz, Luz M. and Cadavid E. 2006. "Creating a Culture in Ergonomics" (paper presented at the IEA2006 Congress, Meeting Diversity in Ergonomics, Maastricht: Elsevier ltd.)

World Health Organization. 1978. Declaración de Alma Ata, Atención primaria de salud. On line version www.inclusion-ia.org/espa%F1ol/Norm/AlmaAta-02-1.pdf (accessed February 24, 2010).

Zas Ros, Bárbara. 2001. La prevención en salud. Algunos referentes conceptuales. Psicología On Line, Artículos de la Gestión. Accessed January 5, 2010.

Chapter 42

The Evaluation of an Assistive Chair Design for the Elderly

Chiwu Huang

Department of Industrial Design/ Graduate Institute of Innovation and Design
National Taipei University of Technology
1 Sec. 3, Chung-Hsiao East Road,
Taipei 10608, Taiwan
E-mail: chiwu@ntut.edu.tw

ABSTRACT

Researchers have noticed that elderly people experienced difficulties in sit-to-stand and stand-to-sit maneuvers (Kao and Huang, 2008). These difficulties may be caused by the decreasing performance of their lower limb muscle or knees. In this paper a chair design with an assistive device underneath the seat was proposed to resolve this problem. The effect of the assistive device was studied via an experiment in which two identical prototypes, chairs A and B, were made; chair A was equipped with an assistive device under the seat while chair B was not. Twenty-two elderly subjects are asked to test the chairs, and their usage patterns were analyzed to determine the effectiveness of the assistive device. Results show that subjectively over 80% of subjects responded that chair A provided them with better mobility. The objective metrics show that the average time needed for getting up and sitting down on chair A is significantly longer than chair B, and this slower pace may reduce the risk for subjects to fall off a chair. The researcher also finds the average leaning angle on the subjects' trunk when using chair A much smaller than chair B. This may suggest chair A is easier to use than chair B. The experiment shows that assistive chair design enables elderly to sit down and get up more easily.

Keywords: Elderly, chair, assistive device, sit-to-stand, stand-to-sit

INTRODUCTION

Sit-to-stand and stand-to-sit maneuvers have been considered to be one of the most challenging movements elderly people encounter in their daily life (Manchoundia et al. 2006), and many fall accidents occur during these movements. Lehtola et al. (2006) reported, after a two-year observation of 555 elders (the average age is 85), that 12% of falls occurred during the daily routines of standing up and sitting down. It has been observed that elderly subjects experienced difficulties moving in and out of a chair (Kao and Huang, 2008), and therefore it is crucial to design a chair that can assist elders to safely stand up and sit down from a chair. Kao and Huang (2008) also proposed a chair design that uses an assistive device installed underneath the seat to support part of the user's weight and to minimize the pressure on his/her lower body. This study aims to test the effectiveness of that chair design.

THE ASSISTIVE CHAIR DESIGN

This design concept is to support the user through an air pump installed underneath the seat. This air pump is controlled by a handle installed on the right arm of the chair. When the user sits down or rises up, s/he can naturally press the handle to activate the air pump. Different from other conventional designs in the market that uses a slippery sloped seat, this proposed design (see Figure 1) divides the seat into two pieces that are connected by a hinge. In the rising position, the seat maintains a rather flat surface and offers users better support. The seat is 50 cm in width, 44cm in depth, and 42 cm in height. The armrest is placed 21 cm above the seat, and the back is located 34.5 cm above the seat. The seat and back are padded with 3 cm of thick sponge and fabric. After a 2-cm sponge compression, the actual sitting height is approximately 40 cm.

METHODS

To test the feasibility of the design, two similar prototypes were made with a minor difference: one has the assistive device installed under the chair (chair A) while the other one (chair B) does not. Subjects were asked to test both models in random order.

SUBJECT

Twenty-two volunteers were selected from the researcher's neighborhood at a community park near Minquan West Road MRT (Mass Rapid Transit) Station in Taipei, Taiwan. Two age groups were recruited into the subject poll to test the age difference. The selection criteria were: (1) people over 65 years old who could handle daily routines independently, and (2) people between 55 and 64 years old who had experienced discomfort in their waist or knees.

APPARATUS

As mentioned earlier, two prototype chairs were made. Chair A had an assistive device installed underneath the seat (Figure 1) while chair B did not (Figure 2). The assistive device on chair A was consisted of an air pump (lift force in 350 Newton) and a bicycle brake handle, with the brake handle wired to the device for controlling the pump in raising and lowering down the seat. A six-megapixel camera was used to record the test process, and still images were extracted from the videos for later analysis. AutoCAD 2006, a drafting software, was used to measure the angles.

FIGURE 1 Chair A, with an assistive device installed

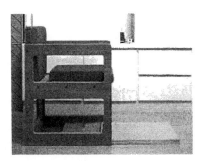

FIGURE 2 Chair B, with no assistive device installed

PROCEDURE

The experiment was carried out in 5 steps: (1) The subject was asked to fill out a questionnaire to collect his or her basic data such as gender, age, and any physical discomfort he or she may have; (2) Researcher measured the subject's weight and height; (3) The subject tested the chairs, and the process was timed; (4) Researcher recorded the subject's individual preferences of chairs A and B; and finally, (5) a semi-structured interview was conducted to collect the subject's opinion on the chairs.

During the third step, chairs A and B were tested in random order by the subjects to avoid the learning effect. The subjects were asked to sit down and then stand up from the tested chairs. The process was measured by a stopwatch and then recorded. At the same time, the subjects' movements were videotaped by a digital camera for later analysis. The test adopted the definition for sit-to-stand movement stated by Hong (1994). The start of sit-to-stand movement was defined as the moment subject's bottom moved away from the chair as he or she leaned forward. The end of the movement was defined as the moment the subject's body stretched upright during the stretching movement. Prior to testing, the subjects could practice as many times as they wanted until they felt comfortable to do the test.

After the test, subjects were asked to pick a chair they felt to be more comfortable and explain the reasons behind their choice. The tests were observed carefully and videotaped for further analysis.

OUTCOME MEASURES AND DATA ANALYSIS

Still images of standing-up and sitting-down movements were extracted from the videos for analysis. AutoCAD 2006 was used to measure the leaning-forward angles on the pictures. The leaning angle was defined as the angle between the trunk and vertical plane (see Figure 3). SPSS 12 was used to analyze the data. The independent variables include: the existence of assistive mechanism, subjects' age, stature, and weight; dependent variables include: subjective preference, time required, and the leaning angle during standing up and sitting down.

FIGURE 3 the measurement of trunk leaning forward angle

RESULTS AND DISCUSSIONS

BASIC DATA

Twenty-two subjects aged between 55 and 84 were recruited for the test. The test was conducted during October 20 – 29, 2008, and the basic data is summarized in Table 1.

Table 1 Summary of Basic Data of the Subjects

BASIC DATA		NUMBER	PERCENTAGE
Gender	male	10	45.5%
	female	12	54.5%
	total	22	100.%
Age	55-64	6	27.3%
	65-74	6	27.3%
	75-84	10	45.5%
	total	22	100%
Weight	Under 50 kg	1	4.5%
	51-60 kg	8	36.5%
	61-70 kg	12	54.5%
	Over 71 kg	1	4.5%
	total	22	100%
Height	150-159 cm	11	50.0%
	160-169 cm	7	31.8%
	170-179 cm	4	18.2%
	total	22	100.0%

In terms of gender, these 20 subjects were almost evenly divided, with 10 males (45.5%), and 12 females (54.5%). In terms of age, 6 were 55-64 years old (27.3%), 6 were 65-74 years old (27.3%), and 10 were 75-84 years old (45.5%). These subjects can be grouped into 16 over 65, and 6 under 65 (termed mid-old age).

Most subjects weighed between 51 to 70 kilograms (n=20, 91%), except two outliers (1 over 70 kilograms and 1 under 50 kilograms). In terms of height, 11 (50%) subjects were in the 150-159 cm category, 7 (31.8%) were 160-169 cm, and 4 were in 170-179 cm category.

The subjects' reported physical discomforts are listed in Table 2. Almost 70% of subjects have experienced discomforts on their waist and knees while very few subjects have experienced discomforts on their wrist.

Table 2 Reported Physical Discomforts on the Subjects

DISCOMFORTS ON		N.	PERCENTAGE
waist	yes	15	68.2%
	no	7	31.8%
	subtotal	22	100%
knees	yes	17	77.3%
	no	5	22.7%
	subtotal	22	100%
wrist	yes	3	13.6%
	no	19	86.4%
	subtotal	22	100%

RESULTS ON RISING UP

Subjective Preference

Eighteen out of 22 subjects (more than 80%) thought chair A helped them get up easier than chair B (Table 3). Four subjects thought chair A and chair B made no difference in helping them getting up from the chair. Among these four, three subjects expressed that they felt no difference because their good health and agility allowed them to get up easily; the other subject, a 78-year-old man weighing 78 kilograms, however, said that the pushing force from the assistive device was inadequate in supporting his weight and therefore he felt no difference between the two chairs. This may lead to a hypothesis that the pushing force of the assistive device needs to be calibrated according to the user's weight. This needs to be investigated further.

Table 3 Subjective Preference to Chairs for Rising up

CHAIR	N.	PERCENTAGE
A	18	81.8%
B	0	0%
No difference	4	18.2%
Total	22	100%

Time Used for Rising up

The time needed for the respondents to rise up from the tested chairs A and B are summarized in Table 4.

Table 4 Time Used for Rising up from Chair A and B (sec.)

CHAIR	N	MIN.	MAX.	MEAN	S.D.
A	22	2.91	4.16	3.44	.29
B	22	1.40	3.25	2.45	.50

The average time used for the respondents to rise from chair A was 3.44 seconds while from chair B was 2.45 seconds – a significant difference of almost 1 second (p=0.000). The standard deviation for chair A (0.29) is lower than that of chair B (0.50). This more evenly distribution of time may be caused by the assistive device on chair A provided subjects an extra force to support themselves from rising up from the chair. However, the assistive device might have slowed down the sit-to-stand process.

Some literatures suggested that elders may adopt the speed strategy to enable themselves to stand up easily from the chair. They may lean their trunks forward to gain speed without sacrificing safety while getting up from the chair. Pai (1991) pointed out that increasing speed in rising from a chair may increase the torque on hips, knees and ankles. In addition, because their declined lower limb strength, elderly people may have a greater chance to fall while adopting this strategy.

Therefore, the assistive device can help users control their bodies by slowing the movement during the sit-to-stand process.

Trunk Angle for Rising up

Still images were extracted from video by video editing software. The forward leaning angle of the body was measured using AutoCAD 2006. The result is listed in Table 5.

Table 5 Trunk Angles for Rising up on Chair A and B (in °)

CHAIR	N.	MIN.	MAX.	MEAN	S.D.
A	22	19.00	50.00	32.68	6.96
B	22	28.00	57.00	40.55	8.48

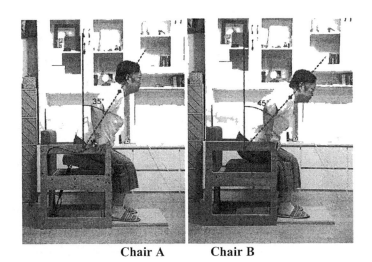

Chair A   Chair B

FIGURE 4 the comparison of the trunk angle for rising between chair A and B

Some researches have pointed out that elderly people flex their trunks to make standing up easier. As a consequence, the hip, knees and ankles will have to be stretched out and torque increased (Rodosky et al., 1989). In the test, the average leaning angle of 36.68° for chair A was significantly (p=0.000) far less than the 40.55° of chair B (see Table 5). The assistive device helped relieve the burden on the subject's knees, and consequently the subject doesn't have to flex their trunks to increase their momentum to help them stand up (Figure 5).

RESULTS ON SITTING DOWN

Subjective Preference

Nineteen of the 22 subjects said that chair A was easier to sit on than chair B (Table 6). Only 3 subjects preferred chair B to chair A. Among them, two were in their mid-old age (55-64) and thought the assistive device in chair A was unnecessary to them because they were healthy enough to manage sitting down. The third subject failed to sit down on chair A because she was too light (36 kilograms) and failed to activate the assistive device. The air pump in the assistive device needed at least 50 kilograms of weight to activate.

Table 6 Subjective Preference for Chair A and B in Sitting down

PREFERENCE	N.	PERCENTAGE
A	19	86.4%
B	3	13.6%
No difference	0	0%
total	22	100%

Trunk Angle for Sitting Down

The trunk angles for sitting down are summarized in Table 7. To avoid falling back into the chair, the subject had to lean his/her trunk further ahead to maintain stability. The smaller the trunk angle the more beneficial it is for the body. The average trunk angle for chair A ($33.31°$) is significantly ($p=0.000<0.05$) smaller than that of chair B ($38.90°$). The lifted seat in chair A supports the subject's bottom so that s/he doesn't have to lean forward as much (Figure 5). Compared to standing up from a chair, sitting down on a chair is a much more complicated and unstable movement for the elderly because lower limbs are usually weaker (Dubost et al, 2005).

Table 7 Trunk Angles for Sitting down in chair A and B (in $°$)

CHAIR	N.	MIN.	MAX.	MEAN	S.D.
A	22	20.00	46.00	33.31	6.24
B	22	29.00	56.00	38.90	7.96

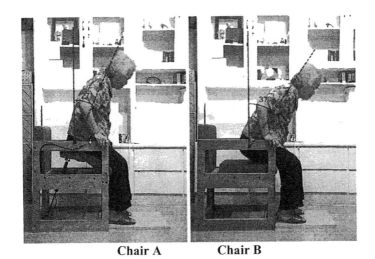

Chair A   **Chair B**

FIGURE 5 the leaning trunk angle for sitting down

As a result, the assistive device in chair A may appeared to have helped the users sit down more easily. With the support of the air pump, the subject could first sit on the raised seat and then fall down slowly to the sitting position. With the trunk angle for chair A smaller than chair B, the torque is lessened on the subject's joints and the risk of a fall is decreased.

CONCLUSION

To help elderly people rise up and sit down more easily, a chair design has been proposed to include an air pump installed underneath the seat to support much of the user weight. This paper tests the feasibility of this design by comparing two prototype chairs – one with the device and the other without. Twenty-two subjects aged 55 to 84 were recruited to test the prototypes. The result shows:

1. Eighty percent of the subjects believed that the assistive chair design (i.e., chair A) was better than the ordinary chair (i.e., chair B) in helping them get up and sit down more easily. Most subjects felt the pressure on their waist and knees were lessened when using the assistive chair design.
2. With the help of the assistive device, though taking an average of 0.99 seconds longer to complete the sit-to-stand maneuver, users considered the design to provide better stability and safety.
3. The design decreased the leaning angle and consequently lessened the pressure on users' lower limbs and decreased the risk of falls.

In summary, the proposed chair design can help elderly people in rising up and sitting down. However, the following limitations should also be considered in further studies.

1. A larger sample size and methods that would take more factors into consideration would be helpful. Other factors, such as the health condition and agility of the users could be included to pinpoint the target market of this assistive device.
2. The measuring method. A more precise device, e.g. motion capture system, can be used for capturing and measuring subjects' movements.
3. The pushing force of the air pump can be considered as a factor in the study.

ACKNOWLEDGEMENT

This study is sponsored by the National Science Council of Taiwan (NSC98-2221-E-027-044) and the Ministry of Education of Taiwan (Industry-Academy Cooperation Project M-02-106, 2007).

REFERENCES

Dubost, V., Beauchet, O., Manckoundia, P., Herrmann, F. and Mourey, F. (2005), "Decreased trunk angular displacement during sitting down; an early feature of ageing." *Physical Therapy*, 85, 804-812.

Hong, W. H. (1994), *Chair-rise in Patients after Total Knee Arthroplasty*, Master thesis, National Cheng Kung University, Tainan, Taiwan.

Kao, L. and Huang C. (2008), *A study on the rising and sitting chair design for the elderly*, (in chinese) Proceeding, Technology and Education 2008, Mingchi University of Technology, Taipei, Taiwan.

Lehtola, S., Koistinen, P. and Luukinen, H. (2006), "Falls and injurious falls late in home-dwelling life", *Archives of Gerotology and Geriatrics*, 42, 217-224.

Manckoundia, P., Mourey, F., Pfitzenmeyer, P. and Papaxanthis, C. (2006), "Comparison of moto strategies in sit-to-stand and back-to-sit motions between healthy and alzheimer's disease elderly subjects", *Neuroscience*, 137, 385-392.

Pai, Y. C. and Rogers, M. W. (1991), "Speed variation and resultant joint torques during sit-to-stand", *Archives of Physical Medicine and Rehabilitation*, 72, 881-885.

Rodosky, M. W. and Andriacchi, (1989), "The influence of chair height on lower limb mechanics during rising", *J. Orthopaedic Research*, 7(2), 266-271.

CHAPTER 43

Psychological Aspects of Residential Bathroom Design

Anna Jaglarz

Department of Architecture
Wroclaw University of Technology
St. Prusa 53/55, 50-317 Wroclaw, Poland

ABSTRACT

The necessity of doing the same basic physiological activities and personal hygiene practices causes that all users have similar needs connected with hygiene and sanitary facilities. But the final appearance and organization of theirs are diverse and to a large extent depend on many social considerations, including first of all considerations connected with family structure, social and economical status, household organization and life style. It would appear that they are the most obvious determinants of needs connected with residential bathroom and bathroom equipment. But also important in that regard is the psychological aspect refer to human needs hierarchy that need of respect and self-respect is placed on the top and take into consideration varied users behaviours connected with body treatment and perception, different attitude to hygienic activities and individual attitude toward sex, privacy and modesty. On this basis need of privacy, intimacy and seclusion connected with range and character of physiological and personal hygiene activities turns out to be the most important determinant of residential bathroom use. Possibility of necessary and required extent of privacy in home hygienic and sanitary area is the basic condition of comfortable and relaxed bathroom use and next is the principle of correct bathroom design.

Keywords: residential bathroom, bathroom design, psychological aspects of bathroom formation

INTRODUCTION

The crucial determinant of the appearance and organization of the residential bathroom and at the same time the primary determinant of the kind and extent of hygiene facilities is psychological, in terms of various individual attitudes toward the body, bodily functions, hygiene and elimination activities, sex and then privacy and modesty.

Range and character of hygiene activities and specific body attitudes determine *requirements for privacy in the bathroom*, first of all privacy which is required especially for intimate functions.

Additionally problem of privacy in the bathroom is connected with a variety of other essentially unrelated motivations, from aesthetic considerations, through need to be alone, possibility of meditations, tasks planning, until to questions concerning the importance of social structure and many others. Broad range of privacy needs may also reflect themselves in different behavior examples.

Privacy requirements emerge as the important determinant of bathroom usage.

These demands induce user to individual use of the bathroom with regard to time as well as space. If it were not for these privacy demands, the average family could probably manage with only one although spacious bathroom. Instead, in most cases, household members use "in turn" room commonly considered as a single room. When conditions permit, solution in the form of personal bathrooms for each member of the household seems to be the most comfortable. Situations, when use of the bathroom happens "in turn", force on family members severe discipline that must be scrupulously observed to avoid confusion and conflicts. This is true in most families, especially where there are school-age children whose timetables coincide with parents schedule of working.

EXTENT OF PRIVACY IN THE BATHROOM

Although the state of privacy is relative and depends on individual needs, there are a *number of degrees of privacy* that are obtainable or may be desired. It is possible to establish two major categories:
- privacy of being heard but not seen,
- privacy of not being heard and seen – other people should not even be aware of one's whereabouts or actions.

These categories generally represent degrees of tolerable (bearable) privacy rather than degrees of desired privacy. In case of possibility of total choice, most people would tend to pick *maximum privacy*. The degree of *tolerable* or *desired privacy* obviously varies considerably, depending on the individual activities and the personal preferences of user. The degree of privacy influences the various ways of physically creating the state of privacy in bathroom: the location of the bathroom relative to other areas of the house, the specific location of the entrance, the acoustics of interior space, the location and size of the bathroom windows, the functional and special arrangement, that is the extent to which facilities are intended to be for the exclusive use of one person, and the extent to which facilities may be shared and are intended to be for the jointly use - with or without compartmentalization of some sort.

The division of bathroom by means of partitions is often regarded as a method of securing greater privacy. Additionally, we should take many different factors connected with creation of proper, functional compartmentalization into consideration, including: the location of the entrance to the compartment, the kind of partitions – whether the partitions are floor to ceiling or whether they are in the form of low walls, whether the partitions are solid or open-work or transparent, whether there is a full door.

Figure 1. Examples of partial compartmentalization in residential bathroom (by means of low walls), (author's drawings)

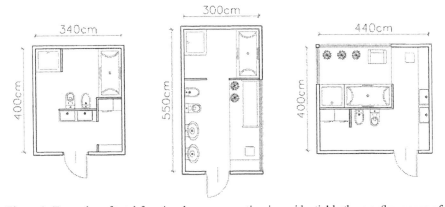

Figure 2. Examples of total functional areas separation in residential bathroom (by means of high walls), (author's drawings)

In terms of the way of partial division and the kind of partitions in this case, it offers many possibilities of securing greater privacy in those circumstances where none was had before, for example in minimal bath that was shared in use. However if a bathroom was not shared, then a greater degree of privacy was available before the partial compartmentalization (unless partitions cause total separate areas), because such division assumes that the bathroom previously unshared may now be shared and suggests possibility of jointly use. The usefulness of this way of compartmentalization obviously depends on the degree of privacy one is concerned about, and it is important, therefore, that this be clearly and precisely established at the outset.

In some instances compartmentalization is also required in order to satisfy *individual needs connected with personal feelings about space and security.* For the sake of habits or increased feelings of uncertainty and insecurity some people have come to think that certain activities can be appropriately and comfortably performed only in specific settings and conditions. This appears to be particularly true for activities involving partial or total undressing, that is bathing and elimination. Some people prefer the shower to the tub because pulling the curtain or closed the door give them a grater sense of privacy and security. A grate many people close or lock the bathroom door even when they are alone in the house because the other household members are out and there is no real danger of anyone's walking into the bathroom or of anyone's being aware of their activities. Sometimes a snug, covered space may operate as a clothing substitute.

Professor Descamps at the Sorbonne distinguishes between persons he terms "nudists", those who can stay without clothing, at last in some circumstances, and persons he classify as "textiles", those who can't be nudist under any circumstances. Some such situations pose a serious problem for them, especially in a large spaces.

Degrees of privacy, either possible or desired, are also dependent on *socioeconomic status.* In the lowest social group, for example, where living conditions in crowded urban areas force a lack of privacy and where privacy has probably never been experienced privacy need are significantly decreased. In this case social privacy norms are much less severe than they are in situations where they are social requirement.

NEED OF PRIVACY AND THE INDIVIDUAL IDENTITY

Privacy and privateness influence our sense of *individual identity* to a large extent, that their deficiency or removal has unquestionable effects on the personality that in some cases may be serious in consequences and irreparable. Direct relationship between privacy and privateness and the opportunity for self-expression and personal sense of identity and individuality is the object of observations, especially in many institutions that restrict privacy and minimize individual identity of people.

Examples of this behavior may be found among institutions that are authoritarian and have highly complex structure of hierarchy such as the military, prisons, and some educational and religious institutions.

This is also characteristic of many institutional settings such as schools, hospitals, dorms. This institutions remove individual privacy and privateness through imposition of common principles on some aspects of life that emphasize individuality, privacy during some activities among other things. The common sanitary and hygienic rooms decrease individual sense of importance and autonomy. In many cases, necessity of the adjustment to common bathrooms can also result in danger of psychopathologies (psychosis). This is also characteristic of sudden and forced privacy deprivation, which commonly occurs during catastrophes or wars.

But privacy deprivation is not a phenomenon limited to institutions. It is common distinguishing feature and at the same time a major problem of the slum housing around the world.

However even average contemporary houses offer examples. Modern "open" planning, universal multipurpose rooms, combined areas in the house intended to consolidating of unity and family bond, can evoke users irritation and completely contrary than intended result in the form of users antisocial behaviors. According to F. S. Chapin research:

The sentiment of self-respect, the respect for self as an individual with status, can hardly thrive when the person is continuously open to pressures of the presence of many others in the household. Privacy is needed for thinking, reflection, reading and study, and for aesthetic enjoyment and contemplation. Intrusions on the fulfillment of personal desires need to be shut off in order to avoid the internal tensions that are built up from the frustrations, resentments and irritations of continual multiple contacts with others.

PRIVILEGED CHARACTER OF THE BATHROOM

For the sake of strong social sanction for obtaining privacy for personal hygiene activities the bathroom assumed a special, *privileged character*. This is partly a result of present, more "open" house-planning practices. Consequently privileged character of bathroom, as the only such space in the house which makes possible of staying along with self for ever user, is used to obtain privacy for variety of purposes that have nothing to do with personal hygiene, for example experiencing and expressing own emotional conditions, such as sulking, crying. Not infrequently this specific character of bathroom is also exploited for other activities that have nothing to do with hygiene, but require for concentration, such as reading, learning.

Additionally desire of privacy in bathroom is connected with individual feelings about embarrassment or shame regarding personal hygiene activities. This is problem resulted from a fear of what others think or feel, depending on situation and circumstances. User is embarrassed when the wrong person walks into the bathroom by mistake, when his body or beauty image is faulty or when he is caught doing intimate something. In this respect strong privacy need occur among teenagers during adolescence, when bodily changes are reason for embarrassment. Continuous experiments with grooming and the time spent in front of mirror also contribute to privacy needs.

Problem of the Obsessive Compulsive Disorder is individually case and often is connected with territory of bathroom and hygienic activities. Mental compulsion of continuous hands washing, often a few dozen times a day, is one of the most common forms of this neurosis. In this case privacy is needed for hiding of own obsessions.

Specific behavior is a result of unknown reactions of other persons to personal hygiene functions. Users resort to all possible ways to avoid being heard in the bathroom simply because they don't know whether or not they might embarrass other people.

CULTURAL AND SOCIAL MEANING OF PRIVACY

Apart from purely personal and intimate aspects, privacy is *a value in a cultural and social sense*. To a large extent is a response to particular social situations. Privacy in this meaning is a necessary condition for acceptable social behavior. The most obvious social determinant of privacy is individual role in society and relationships with others. Necessity of privacy in various situations is connected with sanctity of norms specifying that people should perform certain activities in private and solitude. These norms can vary widely, depending on culture, age and sex. What is an acceptable behavior and relationship between a child and a child may not be between a child and a parent, or between children of different sexes. With respect to personal hygiene activities, age and sex are subjected to the obvious limits. Additional barriers are result of other social differences, such as social status, position, occupation. There are obviously so-called "privileged" professions, for example the physician, the nurse, the masseur, the beautician. They are excused from observing the general social rules respect to privacy of body.

Demand for privacy in bathroom may vary considerably with the specific hygiene activities. The strictest requirements apply to elimination functions. This convention concerns even persons who are close with respect to things connected with the body. On the other hand, the differences in what is deemed to be acceptable behavior are enormous. People in the some cultures, for example, who have custom to place the water closet in a separate room, are disgusted by the bathroom in which the water closet and washing or bathing facilities are all together in the same space. This is evidence of the clear discrepancy in attitude toward bathing activities and elimination activities.

Privacy requirement is also closely related to negativism degree of the bathroom activities. From the most negative like elimination functions through activities connected with cleanness of body and hair, hygiene of oral cavity, until activities inclusive of taking care about beauty, health, fitness and well-being. However there are some exceptions or even paradoxes. Among functions bound up with shaving, that are more negative rather than neutral, regular ritual of male shaving is probably the only "elimination" activity with a quite positive image. Thus paradoxically, male shaving doesn't require so much privacy in comparison with female shaving or depilation of legs, which is regarded as a very private activity.

Amid all personal hygiene activities, that are not quite positive, but at least neutral, are so-called "final corrections", for example making up, hair styling. However in many circles they are not regarded as polite and proper when they are done in public. Other functions connected with looks and grooming, such as use of the beauty cream and mask or hair curlers, are regarded as a semi-private activities, though not by everyone. Specific attitude depends on individual feelings of person who performs activity and also feelings of observer. Taking into account these occupations or make up and hair style activities, privacy requirement is not connected rather with process of this action, but is apply to relation between "before" and "after" and exposure of metamorphosis.

Regardless of all variations of motivations and attitudes, *the issues of practical hygienic and sanitary area formation* can be reduced directly to acceptance or rejection of the human body and its functions and products. Different attitudes are expressed in different views of the bathroom. Persons, who identify the bathroom with sensuality, even eroticism, and who wish to exercise, sunbathe or watch TV in the bathroom, have obviously different demands in comparison to persons who are disgusted by the notion of spending any more time in the bathroom than absolutely necessary.

Professor Alexander Kira in *The Bathroom* quotes two descriptions coming from articles of prominent architects illustrated different feelings about this place:

The proper character of bath rooms, like all the other rooms in the house, must be a function of the peculiar combination of activities which the client elects to combine in the space. Obviously the sybaritic atmosphere, designed for pleasure in warmth, steam, nudity, sex, flowers and alcohol, cannot appropriately combine with the bathinette, the toidy seat, and the diaper pail. On the other hand, the well-designed family bath, large, commodious, practical, somewhat ascetic, can be a magnificent thing, not only in terms of its expressive message. Ideally perhaps, every family should have one of each. But where for practical reasons this is impossible, a little of each, a compromise, is certainly possible, and a great challenge to the designer.

The most frequent mistake , to take a more Modest but more specific example, is the luxurious bathroom. A bathroom is a bathroom, and its equipment is still its equipment even if as beautiful as that of the Italian firm, Ideal-Standard. What takes place there is not an adventure or a luxury but a necessity. The real refinement of the bathroom consist in its not being too refined. Everything in it must be functionally perfect, orderly, sanitary, and easily maintained. this much is in good taste and nothing more.

CONCLUSION

Apart from attitudes, owing to functional and practical point of view, the residential bathroom is undeniably one of the more important and the most often visited, especially in the morning and evening, room in the house. Pleasant or unpleasant impressions connected with the bathroom can influence users mood by day and by night. In spite of different opinions and assessments of the bathroom, considerably more often it appears as a friendly and health directed environment that gives the possibilities of fitness and relaxing exercises and pleasures. Modern bathroom should be by far larger than traditional in order that it can hold additional equipment and space. Modern bathroom should be "a bathroom for living". Until recently, property developers, architects and construction engineers considered the bathroom as a smallest and least significant room in the house. They ignored the bathroom. This fact is changing gradually for satisfaction of users, who desire to have bigger and bigger bathrooms in their new or modernized houses.

REFERENCES

Kira, A., (1976) *The Bathroom*, The Viking Press, New York
Kurylowicz, E., (2007) *Higiena i jej oprawa przestrzenna w XXI wieku*, http://www.kolo.com.pl

CHAPTER 44

The Effects of Repetitive Lifting on Lumbar Spinal Motion in a Young and Middle-Aged Population

Mark G. Boocock, Grant A. Mawston, Ross Smith

Health and Rehabilitation Research Institute
Auckland University of Technology
Auckland 1142, New Zealand

ABSTRACT

High incidences of lifting-related injuries have been identified in older adults when compared to a younger population (Mackey et al., 2007). Such differences may stem from a reduction in the physical capacity of older workers when compared to their younger counterparts (Shephard, 1999; Mackey et al., 2007). During manual handling activities, lumbosacral (LS) kinematics is considered an important determinant of the risk of spinal injury (McGill, 2007). This study investigated changes in lumbar spinal motion in a group of young and older male adults when performing a repetitive lifting task.

Two groups of adult males (16 participants aged 20-36 years and 13 participants aged 43-54 years) participated in the study. Participants were required to repetitively lift and lower a box weighing 13 kg at a lifting rate of 10 lifts per minute. All participants carried out the task until they were unable to continue lifting due to discomfort or fatigue, or had completed 20 minutes. LS kinematics (maximum flexion and angular velocity) were measured at one minute intervals over the course of the lifting task using an optoelectronic motion analysis system and pairs of reflective markers attached to the upper lumbar spine and sacrum. LS

412

flexion during the lifting task was expressed as a percentage of maximum flexion prior to lifting. Generalised Estimating Equations (GEE) were used to compare temporal changes in LS kinematics between the two age groups.

Younger adults increased LS angle over the course of the lifting task, reaching approximately 100% of maximal flexion towards the end of the task. In contrast, the older adult group showed only a small increase in lumbar flexion over time, with lumbar flexion reaching approximately 80% of their maximum flexion. The slopes of the linear regression models for the younger and older participants were significantly different (p=0.012). During lumbar extension (lifting), younger adults displayed a significantly greater rate of change in maximum angular velocity over the duration of the task than the older adults (p=0.013).

The increased range of lumbar flexion observed in younger adults compared to older adults suggests that contrary to expectation younger adults are likely to be at greater risk of spinal injury during repetitive lifting. These findings have implications for the assessment and teaching of spinal mechanics, suggesting that greater emphasis should be placed on lower spine posture rather than overall lifting technique if the incidence of handling related back injuries is to be reduced.

Keywords: Low back injury, Lumbar Spine, Lifting, Lumbar Kinematics, Older adults

INTRODUCTION

Low back pain is one of the most common and disabling musculoskeletal injuries in the workplace, with a high incidence of low back injuries associated with manual handling (Dempsey, 1998). Whilst a number of manual handling risk factors (e.g. age, posture, weight of load, and experience) are known to be associated with low back disorders, their importance in the aetiology of injury is not well understood.

Mackey et al. (2007) identified that older adults have a higher rate of lifting related injuries than younger adults in the work environment. It has been suggested that the physical capacity of older adults may be lower than that of younger adults and this may increase the risk of fatigue and the potential for injury (Shephard, 1999; Mackey et al., 2007).

The characteristics of lifting technique (e.g. lumbar posture) are considered important determinants of the extent and type of tissue damage (McGill, 1997). Consequently, a large proportion of the research aimed at preventative strategies for low back injury have focused on minimising load on the lumbar spine by optimising lifting technique. A popular focus of this research has been to investigate the benefits of squat versus stoop lifting (van Dieen et al., 1999). However, there is also evidence to suggest that lifting kinematics, in particular lumbar posture, may change over the course of a repetitive lifting task due to localised muscle fatigue (Dolan and Adams, 1998; Bonato et al., 2003). The aim of this study was to investigate the effects of repetitive lifting on lumbar spinal kinematics in a group of young and older adults who self-selected their lifting posture.

METHODS

PARTICIPANTS

Twenty nine adult males were recruited into the study and were grouped according to age. Sixteen participants aged between 20 and 36 years (mean age = 26.4 years (yr) (standard deviation (SD) = ±5.8 yr), mean stature = 1.81 m (SD = ±0.09 m) and mean weight = 82.6 kg (SD = ±15.1)) made up the group of 'younger' adults. The 'older' group consisted of 13 participants aged 43-54 years (mean age = 47.1 yr (SD = ±3.1 yr), stature = 1.75 m (SD = ±0.05 m) and weight = 80.7 kg (SD = ±10.2 kg)). Participants were excluded from the study if they had: a back complaint within the last six months; undergone any previous spinal surgery; any cardiovascular or neurological condition; a musculoskeletal injury at the time of the study. Prior to the study, participants completed a health screening (Physical Activity Readiness) and Habitual Physical Activity Questionnaire. None of the participants were considered to be experienced in manual handling, i.e. undertook manual handling as part of a regular job. The study was approved by the University ethics committee and participants were required to complete a consent form.

LIFTING TASK

Participants were required to lift and lower a box weighing 13 kg at a frequency of ten times per minute. The box (30 x 25 x 25.5 cm) was held by two handles extending 6 cm from either side, at a height of 17 cm above its base. The box was lifted from a shelf 15 cm above the floor to an upright standing position, with the box held at hip height. The box was then lowered back onto the shelf. A computer generated metronome operating at a frequency of 20 times per minute provided an audible cue of when to lift and lower the box, with participants commencing each lift and lower at the sound of the metronome. Participants continued lifting either until they became fatigued and were unable to continue lifting, or had completed 20 minutes. Participants were verbally encouraged to continue and maintain the required rate of lifting throughout the task. A modified version of the progressive iso-inertial lifting evaluation criteria (Mayer et al., 1988) was used to determine whether participants had reached a state of fatigue. A person was considered to be fatigued if they were unable to keep pace with the metronome or had a heart rate greater than that recorded during a VO^2 max test, or that the participant chose to stop lifting because of subjective fatigue or excessive discomfort. All participants underwent an incremental cycle ergometer VO^2 max test (25W/minute) on a separate day prior to the lifting task.

When lifting and lowering the box, participants were required to maintain a fixed foot position as close to the shelf as possible, but not touching it. Participants were instructed to maintain their hold on the box at all times and ensure that they lifted and lowered the box to the same position each time. However, no instructions were given to participants about the lifting technique they should adopt while lifting or lowering.

KINEMATIC ANALYSIS

A nine camera motion analysis system (Qualysis Medical AB, Sweden) sampling at 60Hz was used to record 3-dimensional (3D) kinematics during the lifting task. Measures of lumbosacral (LS) angle were recorded using two pairs of reflective markers superficial to the spinous process of L1 and S1 (Figure 1). Each pair of markers was mounted on rods fixed to two small base plates that were securely attached to the skin. LS angle was defined as the angle between two lines joining the centre of each pair of reflective markers (Mawston et al., 2007). The 3D co-ordinates of each spinal marker were tracked using Qualysis Track Manager (Qualysis Medical AB, Sweden). All subsequent kinematic analysis was undertaken using Visual 3D biomechanical analysis software package (C-Motion Inc, USA). Prior to kinematic analysis, markers displacements were smoothed using a Butterworth lowpass filter with a cutoff frequency of 6 Hz.

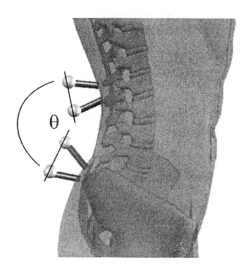

Figure 1. Lumbosacral angle measured from the two pairs of markers superficial to the spinous process of L1 and S1

Two complete lifting and lowering cycles were sampled every minute over the duration of lifting task. Measures of LS kinematics included maximum LS flexion and the maximum angular velocity of the lumbar spine when extending (lifting) and flexing (lowering) the lumbar spine. Angular displacements were expressed as a percentage of maximum LS flexion recorded from the upright standing position prior to the lifting task (Adams, 1986) (Figure 2).

Figure 2. Maximum lumbar flexion measured in the standing position

STATISTICAL ANALYSIS

Descriptive statistics were generated for each outcome measure: 1) maximum lumbar flexion during the lifting task expressed as a percentage of the total range of lumbar flexion measured prior to the task (%ROF); 2) maximum angular velocity of the lumbar spine when extending the spine (lifting); and 3) maximum angular velocity of the lumbar spine when flexing the spine (lowering). Mean values of each of these measures were calculated for the two lifting and lowering cycles recorded and at every one minute interval over the duration of the lifting task. Linear and non-linear models were fitted to the data to explore changes in outcome measures over the course of the repetitive lifting task. Generalised Estimating Equations (GEEs) with separate intercepts and slopes were fitted to the time series of observations for the younger and older adults, which allowed for correlations within data clusters (participants). A Wald test was used to test for a significant difference between the slopes of the fitted models obtained for the two groups of participants.

Data analysis was conducted using SPSS v16 (SPSS Inc., Chicago) and R statistical software package (Version 2.10.1, R Foundation). Statistical significance was set at 0.05.

RESULTS

TASK COMPLETION

Five participants from the 16 younger adults (30%) failed to complete the 20 minutes of repetitive lifting, whereas 8 older participants (61%) did not complete 20 minutes. Of those participants who were unable to continue lifting, on average younger adults stopped within 15 minutes and older adults within 11 minutes. The median time sustained by the younger group was 20 minutes, whereas the older group lifted for a median of 17 minutes. Most participants stop lifting due to excessive discomfort in the lower back.

CHANGES IN LUMBAR FLEXION

Figure 3 shows temporal changes in %ROF of the lumbar spine for a 20 minute task duration. Linear regression models provided the best fit to the data and showed that both the younger and older group of participants progressively increased lumbar flexion over the duration of the task. However, percentage increases in lumbar flexion were more pronounced in the younger adults, with %ROF approaching 100% of maximum lumbar flexion towards the end of the task. In contrast, the older adult group approached approximately 80% of maximum lumbar flexion. A test of differences between the slopes of the two regression models showed a significant interaction for time by age group (p=0.012), with main effects of age group and time being non-significant.

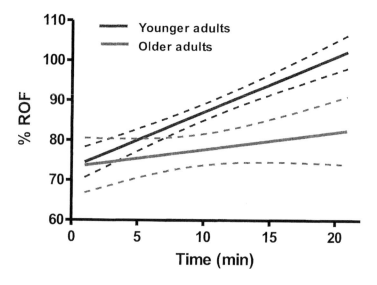

Figure 3. Linear regression lines showing changes in maximum lumbar flexion over the duration of the lifting task expressed as a percentage of the total range of lumbar flexion prior to lifting (%ROF) for the younger and older adults. Dotted lines show 95% confidence intervals

CHANGES IN ANGULAR VELOCITY OF THE LUMBAR SPINE

Figure 4 shows temporal changes in maximum angular velocity of the lumbar spine when extending the spine (lifting the box) over a 20 minute task duration. Similarly, Figure 5 shows changes in maximum angular velocity of the lumbar spine when participants flexed their spine (lowering of the box). Linear regression models provided the best fit to the data and showed that whilst the younger adult group exhibited lower lumbar angular velocities (extension and flexion) towards the beginning of the task, angular velocities approached similar values, or exceeded (extension) those displayed by the older group towards the end of the lifting task. Tests of differences between the slopes of the linear regression models for changes in angular velocities when extending the lumbar spine showed a significant interaction for age group by time (p=0.013). Main effects for time and age group were non-significant during lumbar extension. When comparing the slopes for angular velocity of the lumbar spine when flexing, there was no significant difference between the two groups.

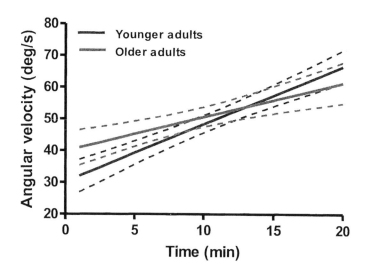

Figure 4. Linear regression models showing changes in maximum angular velocity of the lumbar spine when lifting (extending) over the duration of the handling task for the younger and older adult groups. Dotted lines show 95% confidence intervals

418

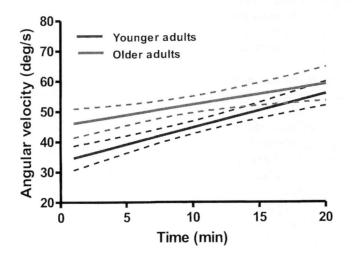

Figure 5. Linear regression models showing changes in maximum angular velocities of the lumbar spine when lowering (flexing) over the duration of the handling task for the younger and older adult groups. Dotted lines show 95% confidence intervals.

DISCUSSION

Repetitive lifting was found to increase %ROF of the lumbar spine over the duration of the lifting task. However, temporal changes in %ROF were greater amongst the younger participants than the older participants. Previous studies involving repetitive lifting have also reported increased lumbar flexion over time, although these have often involved lifting over short durations and been restricted to groups of young adults (Dolan and Adams, 1998; Sparto et al., 1997). It has been suggested that moving to a more flexed lumbar posture may be an attempt to adopt a more energy efficient stoop lifting technique (Sparto et al., 1997). However, flexion near end range has implications for possible low back injury, as at end range of lumbar flexion there is significant recruitment of the posterior ligamentous system. An increase in lumbar flexion has been shown to increase anterior shear forces acting on the lumbar spine (McGill et al., 2000; McGill and Kippers, 1994; Potvin et al., 1991) and if repeated, may lead to earlier vertebral end plate damage (Gallagher et al., 2005).

Towards the end of the lifting task, younger adults increased LS angle to approximately 100% of maximal flexion. In contrast, lumbar flexion amongst the older adults was at approximately 80% of their maximum flexion. At 80% lumbar flexion there has been shown to be minimal resistance to bending moments from the posterior ligamentous system, whereas passive forces increase significantly at 100%

flexion (Adams and Dolan, 1991; Potvin et al., 1991). It should be noted that a higher percentage of the older participants stopped lifting prior to the end of the lifting task. It may be that the older participants were more aware of their lumbar spine 'safety limit' and terminated the task prior to recruitment of the passive tissues of the spine.

Temporal changes in maximum LS angular velocity also differed between the two age groups. Increases in LS angular velocity over the duration of the task were observed during extension (lifting) and flexion (lowering). However, only the former showed significant differences between the younger and older adults. Studies investigating relationships between fatigue and changes in lumbar spine velocity are equivocal. Sparto et al. (1997) found no difference in LS extension velocity in young males when performing a high frequency, short duration lifting task to exhaustion. Similar to our findings, van Dieen et al. (1998) showed an increase in peak LS angular velocity near the end of a repetitive lifting task. Increasing trunk extension velocity has been shown to increase compressive loads on the spine (Granata and Marras, 1995). van Dieen et al., (1998) suggested that an increase in peak lumbar velocity (which would typically occur during the middle to later stages of the lift phase) may a mechanism to compensate for fatigue induced reduction in LS angular velocity at the initiation of the lift. An increase in trunk angular velocity may also be a mechanism to utilise the inertial properties of the trunk and box in order to aid lifting (Bonato et al., 2002).

CONCLUSIONS

Findings from this study suggest that fatigue and age related differences do impact on postural kinematics of the lumbar spine when performing repetitive manual handling tasks. These findings have implications for the assessment and teaching of spinal mechanics, suggesting that greater emphasis should be placed on the posture of the lumbar spine rather than overall lifting technique if manual handling related back injuries are to be reduced.

REFERENCES

Adams, M. A. and Dolan, P. (1991). "A technique for quantifying the bending moment acting on the lumbar spine in vivo." *Journal of Biomechanics*, 24(2), 117-126.

Adams, M. A., Dolan, P., Marx, C., Hutton, W., (1986), "An electronic inclinometer technique for measuring lumbar curvature." *Clinical Biomechanics*, 1, 130–134.

Bonato, P., Boissy, P., Della Croce, U., and Roy, S. H. (2002). "Changes in the surface EMG signal and the biomechanics of motion during a repetitive lifting task." *IEEE Transactions on Neural Systems and Rehabilitation Engineering*, 10(1), 38-47.

Bonato, P., Ebenbichler, G. R., Roy, S. H., Lehr, S., Posch, M., Kollmitzer, J., Della Croce, U. (2003). "Muscle fatigue and fatigue-related biomechanical changes during a cyclic lifting task." *Spine*, 28(16), 1810-1820.

420

Dempsey, P. G., Ayoub, M. M., and Westfall, P. H. (1998). "Evaluation of the ability of power to predict low frequency lifting capacity." *Ergonomics*, 41(8), 1222-1241.

Dolan P. and Adams, M. A. (1998), "Repetitive lifting tasks fatigue the back muscles and increase the bending moment acting on the lumbar spine." *Journal of Biomechanics,* 31(8), 713-721.

Gallagher, S., Marras, W. S., Litsky, A. S., and Burr, D. (2005). "Torso flexion loads and the fatigue failure of human lumbosacral motion segments." *Spine*, 30(20), 2265-2273.

Granata, K. P., and Marras, W. S. (1995). "The influence of trunk muscle coactivity on dynamic spinal loads." *Spine*, 20(8), 913-919.

Mackey, M., C. G. Maher, T. Wong and K. Collins (2007), "Study protocol: the effects of work-site exercise on the physical fitness and work-ability of older workers." *BMC Musculoskeletal Disorders*, 8(9).

Mawston, G. A., McNair, P. J. and Boocock, M. G. (2007), "The effects of prior exposure, warning, and initial standing posture on muscular and kinematic responses to sudden loading of a hand-held box." *Clinical Biomechanics*, 22(3), 275-281.

Mayer, T. G., Barnes, D., Kishino, N. D., Nichols, G., Gatchel, R. J., Mayer, H., Moony, V. (1988). "Progressive isoinertial lifting evaluation. I. A standardized protocol and normative database." *Spine*, 13(9), 993-997.

McGill, S. M. (2007). *Low back disorders: Evidence-based prevention and rehabilitation.* Champaign, IL, Human Kinetics.

McGill, S. M. (1997), "The biomechanics of low back injury: implications on current practice in industry and the clinic." *Journal of Biomechanics,* 30(5), 465-475.

McGill, S., Hughson, R. L., and Parks, K. (2000). Changes in lumbar lordosis modify the role of the extensor muscles. *Clinical Biomechanics*, 15(10), 777-780.

McGill, S., and Kippers, V. (1994). "Transfer of loads between lumbar tissues during the flexion-relaxation phenomenon." *Spine*, 19(19), 2190-2196.

Potvin, J. R., McGill, S. M., and Norman, R. W. (1991). "Trunk muscle and lumbar ligament contributions to dynamic lifts with varying degrees of trunk flexion." *Spine*, 16(9), 1099-1107.

Shephard, R. J. (1999). "Age and physical work capacity." *Experimental Aging Research* 25(4), 331-43.

Sparto, P. J., Parnianpour, M., Reinsel, T. E., and Simon, S. (1997). "The effect of fatigue on multijoint kinematics, coordination, and postural stability during a repetitive lifting test." *Journal of Orthopaedic and Sports Physical Therapy*, 25(1), 3-12.

van Dieen, J. H., Hoozemans, M. J., Toussaint, H. M. (1999), "Stoop or squat: a review of biomechanical studies on lifting technique.", *Clinical Biomechanics*, 14(10), 685-696.

van Dieen, J. H., van der Burg, P., Raaijmakers, T. A. J., and Toussaint, H. M. (1998). "Effects of repetitive lifting on kinematics: Inadequate anticipatory control or adaptive changes?" *Journal of Motor Behavior*, 30(1), 20-32.

Chapter 45

Discomfort and Other Measures for the Assessment of Postural Load

Dohyung Kee[1], Inseok Lee[2]

[1]Dept. of Industrial and Management Engineering
Keimyung University, Korea

[2]Dept. of Safety Engineering
Hankyong National University, Korea

ABSTRACT

This study is to investigate the relationships between subjective measures of discomfort and objective measures related to the assessment of postural stresses based on literature survey. Objective measures included posture holding time, maximal holding time (MHT), torque at joints, lifting index (LI) and compressive force (CF) at L5/S1. The major relationships identified in this literature survey were the following: 1) postural discomfort linearly increased with increasing holding time, and holding force, 2) whole body discomfort was inversely linearly proportional to the MHT, 3) body-part discomfort was related to objective measures such as torque at the relevant joint, 4) discomfort was strongly linearly related to LIs and CFs, and 5) the discomfort measured with the magnitude estimation was linearly related to that measured with Borg CR10. Thus, it is thought that discomfort might be used as a measure for quantifying postural stresses.

Keywords: Discomfort, Musculoskeletal Disorders, Physical Loading, Postural Stress

INTRODUCTION

Awkward, extreme or repetitive working postures have been referred as main risk factors of work-related musculoskeletal disorders (WMSDs) in various industries (Bernard, 1997; Silverstein et al., 1987). The measures to prevent WMSDs include mainly improving the working postures and consequently reduce postural stresses.

Psychophysical methods to assess postural stresses have been widely used alone or simultaneously with other objective measures like electromyography (EMG), since no objective method has been reported to be reliable to assess postural stresses (Li and Buckle, 1999; Putz-Anderson and Galinsky, 1993). Discomfort can be considered as an independent criterion for evaluating working postures (Dul et al., 1994). Although discomfort does not refer directly to the risk of WMSDs, minimizing discomfort may contribute to reduction of the risk, since discomfort and WMSDs are both known to be related to the exposure of the musculoskeletal system to biomechanical load (Miedema et al., 1997).

Many studies have evaluated postural stresses on the basis of discomfort (Putz-Anderson and Galinsky, 1993; Genaidy and Karwowski, 1993; Genaidy et al., 1995; Kee and Karwowski, 2001; Chung et al., 2003). Genaidy et al. (1995) reported a ranking system based on discomfort to determine stress induced by non-neutral static postures around the wrist, elbow, shoulder, neck and lower back. Kee and Karwowski (2001) proposed a postural classification scheme of the upper body on the basis of discomfort caused by various joint postures. Chung et al. (2003) measured the discomfort for various leg postures and proposed a scheme for evaluating stress induced in the different leg postures.

Various posture classifications have been used for quantifying postural stresses (Genaidy et al., 1994; Juul-Kristensen et al., 1997). Methods, such as OWAS and RULA, have been proven to be useful in quantifying postural stresses and contributing to the prevention of WMSDs. However, most of these methods were developed based on the rankings or subjective ratings provided by experts in ergonomics or by experienced workers, with lacking experimental results on the definite criteria of posture classification and evaluation (Juul-Kristensen et al., 1997). Some methods like LUBA and PLAS classified the postures based on discomfort for various postures to make up for the shortcomings of the previous studies (Kee and Karwowski, 2001; Chung et al., 2005).

Feelings of discomfort can reduce efficiency and job satisfaction of workers (Fellows and Freivalds, 1991), while comfort is recognized as a major selling point, as it is thought to play a significant role in product-buying decisions (Kuijt-Evers et al., 2004). The design of comfortable products like hand tools can reduce the risk of occupational injuries and result in high quality products and comfort for users (Kuijt-Evers et al., 2004). Subjective rating is most common when hand tools are evaluated with respect to comfort and discomfort. Most of these assessments are focused on the level of discomfort experience (Kuijt-Evers et al., 2007a, b).

Through the theoretical model of comfort, discomfort and its underlying factors, de Looze et al. (2003) showed that the factors underlying sitting discomfort and comfort are different from each other. The relationships among objective measures of sitting comfort and discomfort were investigated in a literature review study. The objective measures included posture, EMG, pressure distribution, spinal load and foot swelling. Of the measures, pressure distribution appeared to be the objective measure with the clearest association to the subjective ratings (de Looze et al., 2003). Kuijt-Evers et al. (2007a, b) reported that of the various objective measures of hand tool comfort, the pressure-time integral was the best predictor of comfort.

The relationships between objective and subjective measures of comfort and discomfort in sitting and hand tools were reviewed and investigated by de Looze et al. (2003) and Kuijt-Evers et al. (2007a, b), respectively. However, information about the relationship between objective and subjective measures in evaluating postural stresses is still lacking. In this study, a series of studies on assessing postures and related risk factors of WMSDs using discomfort were reviewed to show the relationships between the objective and subjective measures of postural load evaluation. In particular, the relationships between discomfort and posture holding time, maximum holding time (MHT), and biomechanical measures like compressive forces (CFs) at a lumbar disc were summarized and discussed.

RELATIONSHIPS BETWEEN DISCOMFORT AND OBJECTIVE PHYSICAL LOADING

POSTURE HOLDING TIME AND HOLDING FORCE

The relationships between holding time and discomfort were investigated by Corlett and Manencia (1980) and Manencia (1986). Based on the experiments measuring postural discomfort for pulling, torque production and tapping, these studies reported that postural discomforts developed in a similar level for the same proportion of the MHT of the postures irrespective of the extent of deviation from a neutral posture, the MHT of the posture and the imposed load. Similarly, Kirk and Sadoyama (1973) also showed that the pain levels increased linearly in proportion to the percentage of the maximum endurance time in the two different force exerting modes of pulling and torque production, which presumably require use of different muscle groups. This allowed direct comparison of different postures regardless of their absolute holding times (Manencia, 1986).

Kee (2004a) collected postural discomfort ratings at 5s, 20s, 40s and 60s for 32 one-minute holding postures from 10 male subjects using the magnitude estimation method. The postures rated in the experiment were controlled by wrist flexion/extension, elbow flexion, shoulder flexion and adduction/abduction, with four different external-load conditions. A simple regression analysis was carried out to investigate the relationship between the mean values of discomfort ratings for different postures and the posture holding time. Overall, the mean discomfort scores, which were averaged discomfort ratings over all adopted postures and subjects, linearly increased as the holding time increased, and the increase of the discomfort score as the holding time increased was statistically significant (p<0.01). Kirk and Sadoyama (1973) exhibited that the logarithmic values of posture holding times are linearly related to the percentage of the maximum holding force, which is the maximum force the subject could exert in the given mode of operation of static pull and exertion of a 2-handed static torque (Corlett and Manenica, 1980). However, the experimental conditions were limited to two postures (or tasks of 'pull' and 'torque') and six levels of external loads.

MAXIMUM POSTURE HOLDING TIME

The MHT or endurance time is a maximum duration time that a static posture can be maintained continuously from a resting state with or without external force exertion (Dul et al., 1994; Miedema et al., 1997). Moon et al. (2005) and Na (2006) measured MHTs and postural discomfort for 24 and 48 postures, respectively, which were defined by hand position, trunk rotation angle and external load. As in the previous studies (Corlett and Manencia, 1980; Manencia, 1986; Miedema et al., 1997), the hand positions were determined by two relative parameters: the percentage of the shoulder height (working height) and the percentage of the arm reach (working distance). The MHTs were significantly influenced by work height and distance, trunk rotation angle and external load. These studies revealed that the MHT decreased with the work distance and external load. Moon et al. (2005) showed that the MHT was inversely linearly related to the whole-body discomfort (R2=0.85). Na (2006) showed that the MHT is inversely proportional to the discomfort, i.e., the MHT linearly decreased as the discomfort score increased.

VERBAL CATEGORY FOR DISCOMFORT

In the studies of rating discomfort levels for varying postures of lower limbs, upper limbs and the whole body using the magnitude estimation, the subjects rated their discomfort intensity numerically using their own ratio scales without any type of anchor or standard (Chung et al., 2003; Kee and Karwowski, 2003; Kee, 2004b). At the completion of the discomfort rating for all postures, the subjects reported their numerical estimates for nine verbal categories of discomfort, which were from "extremely uncomfortable" to "extremely comfortable."

The subjects were asked to make numerical estimates for the nine verbal categories using the same scale as in the experiment measuring perceived discomforts. These values made it easy to interpret the numerical discomfort ratings using plain words, that is, the values helped relate the magnitude continua to verbal descriptors frequently used in traditional scaling techniques.

The numerical estimates for the verbal descriptors were normalized using the maximum normalization or max-min transformation methods. Representative discomfort levels for the verbal categories were obtained by averaging values of the normalized responses across the subjects (Fig. 1). Although the three studies showed slightly different values of discomfort levels for the descriptors, they all showed a consistent trend whereby the discomfort values quadratically increased as the discomfort categories moved from "extremely comfortable" to "extremely uncomfortable."

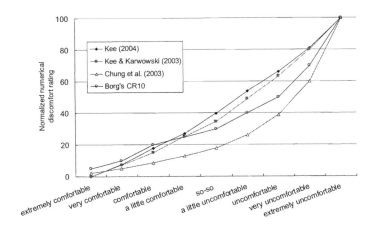

Fig. 1. Discomfort levels and Borg's CR10 corresponding to verbal categories

The correlation coefficients among discomfort levels measured with the magnitude estimation (Chung et al., 2003; Kee and Karwowski, 2003; Kee, 2004b) and Borg CR10 were larger than 0.9 with statistical significances (α=0.05), which shows that Discomfort levels were highly positively correlated irrespective of the study and measuring methods.

BORG CR10 SCALE

Chung et al. (2003) compared the Borg CR10 and the magnitude estimation method. In Fig. 1, the Borg CR10 scale for the nine verbal categories was represented with discomfort levels. Here, the Borg CR10 was multiplied by 10 for easy comparison between the two different scales of Borg CR10 and the magnitude estimation. The two different scales for discomfort or pain showed a similar trend of quadratic increases. The ratings showed slow increases in the comfort area and steep increases in the discomfort area.

JOINT TORQUES

Boussenna et al. (1982) carried out a laboratory experiment with eight male subjects to investigate the relationship between torques at the hip, knee and ankle, and perceived discomforts for four forward-bending postures with different shoulder heights. The overall and body-part discomforts were monitored every 30s, and torques at the hip, knee and ankle were calculated. The results indicated that changes in postures significantly correlated with changes in holding time and discomfort, and that discomfort levels in the body segment immediately superior to the joints under study were significantly related to torque at the joint.

Mukhopadhyay et al. (2007) showed a significant effect of torque of the forearm pronation on discomfort ratings (p=0.001) from a laboratory experiment, in which three forearm rotation angles (60% prone and supine and neutral range of motion), three elbow angles (45o, 90 o and 135 o), two exertion frequencies (10 and 20/min) and two levels of pronation torque (10% and 20% MVC) were used as independent variables. Thirty-six right-handed male university students participated in the study. In addition, Carey and Gallwey (2002) investigated the effects of exertion (10% and 20% of the MVC in the neutral wrist position), pace(10 and 20 exertions per minute) and level of simple and combined flexion/extension and radial/ulnar deviation of the wrist on discomfort for simple repetitive exertions. Eight male subjects were recruited for the study. The results revealed that exertion was the most significant factor, followed by level of deviation and then pace.

LIFTING INDEX AND COMPRESSIVE FORCE AT A LUMBAR DISC

Park (2005) conducted an experiment to measure postural stresses for 45 shoulder and elbow postures combined with external loads using Borg CR10. Twelve subjects participated in the experiment. The postures were controlled by shoulder extension and flexion (-20°, 0°, 45°, 90°, 135°), elbow flexion (0°, 45°, 120°) and external load (0.0, 1.5, 3.0 kg), which were maintained for a minute. Regression equations were used to predict discomfort levels in Borg CR10 scale for different shoulder and elbow postures with or without external load conditions. Along with the discomfort levels, lifting index (LI) values of the revised NIOSH lifting equation (Waters et al., 1993) and predicted CFs at the L5/S1 disc were additionally calculated in this study for the postures of the study of Park (2005) with the same external load conditions. The LI values were obtained under the assumption that the experimental task was to lift external loads, and that the vertical travel distance was 0.0 cm, which means no difference of LIs between starting and ending point of lifting task. The CFs at L5/S1 were assessed using the 3DSSPP (3D Static Strength Prediction Program, ver. 4.2, 1999) developed at the University of Michigan in the USA. Here, the mean anthropometry values of males in age of 20s were used.

Mean values of LIs, CFs and discomfort levels by the independent variables are illustrated in Fig. 3-5. When calculating mean values, unavailable postures due to the limitation of the NIOSH lifting equation or 3DSSPP were not included in the analysis to minimize the bias effects from unavailable data. For example, CFs for the postures with the shoulder extended by 20° were excluded since 3DSSPP does not allow shoulder extension.

CFs and discomfort levels showed the same trend for all independent variables. For example, CFs and discomfort levels increased as the shoulder was increasingly flexed from 0° (neutral position) to 90° and decreased when the shoulder flexion angle exceeded 90° (Fig. 2(a)). CFs, discomfort levels and LIs also increased with external loads (Fig. 2(c). LI values had a slightly different trend for elbow flexion. Specifically, CFs and discomfort levels were slightly larger at an elbow flexion of 45° than at other flexion angles, while LIs decreased as the elbow was flexed (Fig. 2(b)). Correlation coefficients between discomfort, LIs and CFs were larger than 0.7 with statistical significances (a=0.01), which shows a highly positive relationships among those measures.

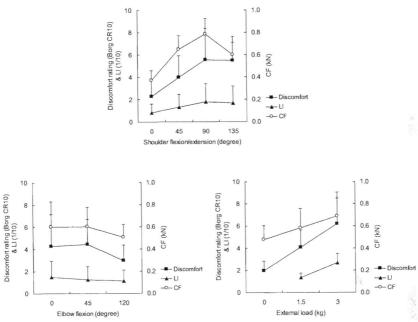

Fig. 2. Mean discomforts, lifting indices, and compressive forces at the L5/S1 disc for (a) should postures, (b) elbow postures and (c) external loads.

DISCUSSION AND CONCLUSIONS

In this literature review to investigate the relationships between subjective and objective measures of physical loading related to postural stress, discomfort was used as a representative subjective measure, and posture holding time, MHT, torque at joints, LI and CF at L5/S1 as objective measures. This study focused on surveying and analyzing the relationships between discomfort and other objective measures. The relationships identified are summarized as follows: 1) the postural discomfort measured with the Borg CR10 or the magnitude estimation linearly increased as holding time increased (Kee, 2004a; Manenica, 1986), and the logarithmic values of posture holding time are linearly related to percentage values of the maximum holding force (Corlett and Manenica, 1980), which implies that discomfort could be related to the maximum holding force; 2) the whole body discomfort was inversely linearly related to the MHT (Moon et al., 2005; Na, 2006); 3) body-part discomfort was related to objective measures such as torque at the relevant joint (Boussenna et al, 1982, Mukhopadhyay et al., 2007, Carey and Gallwey, 2002); and 4) the discomfort was strongly linearly proportional to LIs and CFs.

Relationships among subjective measures were also found. The discomfort measured with the magnitude estimation was linearly related to that measured with Borg CR10 ($r \geq 0.98$). In addition, the numerical discomfort score quadratically increased as the verbal category moved from "extremely comfortable" to "extremely uncomfortable."

It is time consuming or requires special equipment to estimate the angular deviation of a body from the neutral position to obtain objective measures of MHT, LIs and CFs. Using psychophysical scaling methods such as Borg RPE/CR10, magnitude estimation, Likert scale, etc., discomfort levels representing postural stress could be obtained easily, directly and without interrupting the worker. Of the various psychophysical scaling methods, the Borg CR10 and the magnitude estimation methods have an advantage of producing quantitative data in ratio scales, to which more various statistical analyses are applicable.

Due to the above relationships with the objective measures, ease of usage and versatility, it is thought that discomfort might represent physical loading for assessing postural stress to some extent. This conclusion is also supported by findings of previous studies, which showed 1) that the limit of posture holding time was the acceptable level of discomfort, and that discomfort was a valid measure of postural load (Corlett and Bishop, 1976; Corlett and Manenica, 1980); 2) minimization of discomfort can contribute to reduction of the risk for musculoskeletal disorders (Dul et al, 1994); and 3) discomfort can be considered as an independent evaluation criterion for static postures (Dul et al., 1994).

Although various objective measures might be replaced with discomfort, quantitative exposure effect data is not sufficiently available, which establishes ergonomic guidelines for preventing WMSDs. Therefore, more comprehensive studies collecting these data are required. It should also be noted that 1) only a study for the relations between discomfort and LIs or CFs (Park, 2005) was found in the literature review; 2) the study for torques was limited to four postures, and torque values for only lower extremities were calculated in the study (Boussenna et al, 1982), while joint torques at the wrist were used as a independent variable in some studies (Mukhopadhyay et al., 2007, Carey and Gallwey, 2002); 3) Park's study (2005) also has limitations that LIs and CFs were obtained using static biomechanical models, not dynamic approach, and that independent variable of external load was confined to 3.0kg or less. Furthermore, for certain low CFs less than 1 KN, the perceived discomfort rating was risen up to around 6 in the Borg Scale in Park (2005). This shows that the subjects might use the full range of discomfort rating within one specific experimental setup. This implies that discomfort ratings are to be interpreted with caution, especially when comparing ratings obtained from experimental setups with different postural loadings. For drawing concrete conclusion, more studies combining psychophysical scaling methods with objective measures are needed.

REFERENCES

Bernard, B.P., ed., 1997. Musculoskeletal disorders and workplace factors. Cincinnati: National Institute for Occupational Safety and Health.

Borg, G., 1998. Borg's perceived exertion and pain scales. Human kinetics, IL, USA.

Boussenna, M., Corlett, E.N., Pheasant, S.T., 1982. The relation between discomfort and postural loading at the joints. Ergonomics 25(4), 315-322.

Carey, E.J., Gallwey, T.J., 2002. Effects of wrist posture, pace and exertion on discomfort. Int. J. Ind. Ergon. 29, 85–94.

Chung, M.K., Lee, I., Kee, D., 2003. Assessment of postural load for lower limb postures based on perceived discomfort. Int. J. Ind. Ergon. 31, 17–32.

Chung, M.K., Lee, I., Kee, D., 2005. Quantitative postural load assessment for whole body manual tasks based on perceived discomfort. Ergonomics 48(5), 492-505.

Corlett, E. N., Bishop, R. P., 1976. A technique for assessing postural discomfort. Ergonomics 19(2), 175-182.

Corlett, E. N., Manenica, J., 1980. The effects and measurement of working postures. Appl. Ergon. 11(1), 7-16.

de Looze, M.P., de Kuijt-Evers, L.F.M., van Die.en, J., 2003. Sitting comfort and discomfort and the relationships with objective measures. Ergonomics 46(10), 985–997.

Dul, J., Douwes, M., Smitt, P., 1994. Ergonomic guidelines for the prevention of discomfort of static postures can be based on endurance data. Ergonomics 37, 807-815.

Fellows, G.L., Freivalds, A., 1991. Ergonomics evaluation of a foam rubber grip for tool handles. Appl. Ergon. 22(4), 225-230.

Genaidy, A. M., Karwowski, W., 1993. The effects of neutral posture deviation on perceived joint discomfort ratings in sitting and standing postures. Ergonomics 36(7), 785-792.

Genaidy, A. M., Al-shedi, A. A., Karwowski, W., 1994. Postural stress analysis in industry. Appl. Ergon. 25(2), 77-87.

Genaidy, A., Barkawi, H., Christensen, D., 1995. Ranking of static non-neutral postures around the joints of the upper extremity and the spine, Ergonomics 38(9), 1851-1858.

Juul-Kristensen, B., Fallentin, N., Ekdahl, C., 1997. Criteria for classification of posture in repetitive work by observation methods: A review, Int. J. Ind. Ergon. 19, 397-411.

Kee, D., 2004a. Investigation on perceived discomfort depending on external load, upper limb postures and their duration. J. Korean Inst. Ind. Eng. 30(2), 76-83(text in Korean).

Kee, D., 2004b. Psychophysical stress of arm motions at varying external load and repetitions. IE Interfaces 17(2), 218-225(text in Korean).

430

Kee, D., Karwowski, W., 2001. LUBA: An Assessment Technique for Postural Loading on the Upper Body Based on Joint Motion Discomfort and Maximum Holding Time. Appl. Ergon. 32(4), 357-366.

Kee, D., Karwowski, W., 2003. Ranking systems for evaluation of joint motion stressfulness based on perceived discomforts. Appl. Ergon. 34(2), 167-176.

Kirk, N.S., Sadoyama, I., 1973. A relationship between endurance and discomfort in static work. MSc report, Loughborough University of Technology.

Kuijt-Evers, L.F.M., Bosh, T., Huysmans, M.A., de Looze, M.P., Vink, P., 2007a. Association between objective and subjective measurements of comfort and discomfort in hand tools. Appl. Ergon. 38(5), 643-654.

Kuijt-Evers, L.F.M., Groenesteijn, L., de Looze, M.P., Vink, P., 2004. Identifying factors of comfort in using hand tools. Appl. Ergon. 35(5), 453-458.

Kuijt-Evers, L.F.M., Vink, P., de Looze, M.P., 2007b. Comfort predictors for different kinds of hand tools: Differences and similarities. Int. J. Ind. Ergon. 37, 73–84.

Li, G. and Buckle, P., 1999, Current techniques for assessing physical exposure to work-related musculoskeletal risks, with emphasis on posture-based methods, Ergonomics 42(5), 674-695.

Manenica, I., 1986. The Ergonomics of working postures: A technique for postural load assessment , in Corlett, E. N., Wilson, J. and Manenica, I. (Eds), The ergonomics of working posture, Taylor & Francis, London, 270-277.

Miedema, M.C., Douwes, M., Dul, J., 1997. Recommended maximum holding times for prevention of discomfort of static standing postures. Int. J. Ind. Ergon. 19, 9-18.

Moon, C., Na, S., Kee, D., Chung, M.K., 2005. Comparison of observational posture evaluation methods based on maximum holding times. J. Korean Inst. Ind. Eng. 31(4), 289-296(text in Korea).

Mukhopadhyay, P., O'Sullivan, L. Gallwey, T.J., 2007. Estimating upper limb discomfort level due to intermittent isometric pronation torque with various combinations of elbow angles, forearm rotation angles, force and frequency with upper arm at 90o abduction. Int. J. Ind. Ergon. 31, 17–32.

Na, S., 2006. An observational evaluation method for postural stress associated with an external load absed on the perceived discomfort. Unpublished doctoral dissertation, POSTECH, Pohang, Korea.

Park, G., 2005. Perceived discomfort for shoulder and elbow postures with external loads. Unpublished master thesis, POSTECH, Pohang, Korea.

Putz-Anderson, V. and Galinsky, T. L., 1993, Psychophysically determined work durations for limiting shoulder girdle fatigue from elevated manual work, Int. J. Ind. Ergon., 11, 19-28.

Silverstein, B.A., Fine, L.J., Armstrong, T.J., 1987. Occupational factors and carpal tunnel syndrome. Am. J. Ind. Med. 11, 343-358.

Waters, T.R., Putz-Anderson, V., Garg, A. Fine, L.J., 1993, Revised NIOSH equation for the design and evaluation of manual lifting tasks, Ergonomics, 36 (7), 749-776.

Chapter 46

Evaluation of Neck, Shoulders, Arms and Hands Muscles Fatigue of Sewers Using Myotonometric Method and Effectiveness of Early Multidisciplinary Rehabilitation

Zenija Roja[1], Valdis Kalkis[1], Henrijs Kalkis[2], Inara Roja[3]

[1]University of Latvia, The research centre of Ergonomics
Faculty of Chemistry

[2] University of Latvia, Faculty of Economics and Management

[3] Riga International Higher School of Practical Psychology

ABSTRACT

This research work is dedicated to occupational health problems caused by ergonomic risks of employees working in sewing industry. These problems mostly are associated with pain in neck, shoulders, arms, hands (NSAH) region. Stress can also cause excessive muscle tension and that leads to risk of NSAH complaints. Hence the early work rehabilitation is necessary in order to avoid absence of long-term illness. The aim of this study was to investigate the NSAH muscles fatigue for sewers using myotonometry and estimate the efficiency of the early multidisciplinary rehabilitation, which consisted of various physical activities and autogenic training. The investigation including rehabilitation program was done in the one year period and with focus on the sewers who suffered from chronic and repetitive pains of NSAH.

Keywords: Sewers, myotonometry, rehabilitation, muscle, fatigue, chronic pain

INTRODUCTION

Work related muscular skeletal disorders are widespread and every year results in 65% to 67% of total occupational diseases in Latvia. In general that influences employee's quality of life. Sewing is one of the oldest industries in Latvia where predominant workers are females. Injuries and muscle pain affecting the wrists, shoulders, neck and back are common problems for workers caused by ergonomic risks in the clothing industry (Gunning et al, 2001). Mostly the sewers complains about pain related to neck, shoulders, arms, hands (NSAH) and it is associated with repetitive movements working in the same position during a longer period of time. The respective muscle group tension is considered to be the main pain cause.

Stress at work can also cause excessive muscle tension and that leads to risk of NSAH complaints. It has been suggested that there is a relationship between psychosocial factors and the persistence or recurrence of shoulder pain (Kuijpers et al, 2004). Depression is common among persons with chronic pain. It is estimated that 30 – 54% of chronic pain patients suffer from severe forms of depression such as major depressive disorder (Banks and Kerns, 1996). Persons with chronic pain and depression are less active and report greater disability and interference with daily activities due to pain compared to chronic pain patients without depression (Holzberg et al, 1996; Keogh et al, 2006). In this way it creates serious social, psychological and economical consequences. To a certain extent it also influences workers work abilities and life quality in general.

The European Commission in 2007 its new strategy on health and safety at work has placed a special focus on the rehabilitation and reintegration of workers. The Member States are encouraged to incorporate into their national strategies specific measures to improve the rehabilitation and reintegration of workers excluded from the workplace for a long period of time because of an accident at work, occupational illness or disability. Such measures can include physical exercises, education, behavioral treatment and ergonomic measures. A multidisciplinary intervention involving a workplace visit or more comprehensive occupational health care intervention helps sub-acute patients to return to work faster. Moreover, it alleviates subjective disability (Karjalainen et al, 2001). Hence the early work rehabilitation is necessary in order to avoid absence of long-term illness.

Therefore, the aim of this study was to investigate the NSAH muscles fatigue for sewers and estimate the efficiency of the early multidisciplinary rehabilitation (EMDR).

METHODS

PARTICIPANTS

The investigation including multidisciplinary rehabilitation program was done in the one year period. Sewers were recruited from textile enterprise. In the inquiry participated 100 sewers (mean age 40.8 ± SD 15.9, mean duration of pain of 5.1 ± SD 4.9 years). Participants completed a questionnaire that gave information regarding which body part has pain, duration of pain, length of service in profession, age, education, about physical activities. All of them were graduated professional school, but 2 persons provided missing data. The large majority of participants had pain in the shoulders (n=52). Neck pain had 15, but pain in the hand and arms showed 22 participants. Low back pain and pain in the other body parts indicated 9 persons. Background factors are shown in Table 1.

Table 1. Background factors of the subjects, mean, standard deviation (SD), range

Variable	Mean ± SD	Range
Age (years)	40.8 ± SD 15.9	18-65
Height (cm)	165.9 ± SD 9.3	153-180
Weight (kg)	69.2 ± SD 10.9	55-90
Length of service in profession	24.6 ± SD 13.0	1-40

In the clinical measurements and in the EMDR program agreed to take part 30 sewers with chronic pain in the neck, shoulders, arms and hands of four month or more duration (mean age 37.4 ± SD 13.10). The inclusion criteria were: age; having NSAH chronic pain; full consent to participate in the study. The exclusion criteria were: acute pain in the NSAH; having not been to mandatory medical examinations.

MEASURES

Functional state of muscles group

Assessment of the functional state (tone) of muscles groups (*m. extensor digitorum, m. flexor carpi radialis* and *m. trapezius*) including determination of muscle fatigue was carried out using myotonometric (MYO) measurements with the MYOTON-3 device created in Estonia, University of Tartu (Vain, 1995). Appliance of this device is a quick and straight forward way to determine the value of the elasticity and stiffness of biological tissue, and allows estimating the cumulative trauma disorder in muscles, the stage of pathological processes, the degree of fatigue, etc. The tissue responds to mechanical impact with damp oscillations, which are registered by an acceleration sensor located on Myoton's measuring tip. A microprocessor records and analyzes signal onto the display screen, values of the tone (frequency of the oscillation Hz), and stiffness (N/m) of the muscle. It is noted that the natural oscillation frequency of muscles in their functional state of relaxation is usually 11–16 Hz (contracted 18–40 Hz), depending on the muscle. The stiffness values depend significantly on the muscle under investigation, their usual range is 150–300 N/m. For contracted of muscles the stiffness value may be higher than 1000 N/m. The procedure of muscle testing was performed in sitting position, on the chair; the muscle length was middle; for all measurements the subject takes the same position.

Pain

Pain assessment tool – Visual Analogue Scale (VAS) help persons describe their pain. The VAS provides a continuous scale for subjective magnitude estimation and consists of a straight line, the limits of which carry a verbal description of each extreme of the symptom to be evaluated (Huskinsson, 1983). Persons rated their average pain intensity in different body parts over the previous week on VAS scale: 0 = "non pain" and 10 = "worst pain possible". The effectiveness of rehabilitation after the successful program accomplishment was assessed also using the VAS.

Depression

Depression was detected using Montgomery-Åsberg Depression Rating Scale (MADRS) – 10-item checklist. Assessed were apparent sadness, reported sadness, inner tension, reduced sleep, reduced appetite, concentration difficulties, lassitude, inability to feel, pessimistic thoughts, suicidal thoughts. On the numerical rating scale, the person is asked to identify depression manifestation by choosing a number from 0 (no symptoms) to 6 (worse symptoms). MADRS designed to be used in patients with depressive disorders, both to measure the degree of severity of depressive symptoms, and particularly as a sensitive measure of change in symptom severity during the treatment of depression (Montgomery and Åsberg, 1979).

The rehabilitation programme

The EMDR endured 9 months and consisted of various physical activities and autogenic training (AT). Physical activities consisted of exercises during the work breaks (3 minutes after 60 minutes of work), exercises after the work (remedial gymnastics 45 minutes 2 times per week) or swimming (2 times per week for 45 minutes).

AT method is psychotherapeutic method, focusing on the neuromuscular system, which allows developing progressive, dynamic relaxed state of the muscles (neck, shoulders, back, arms and legs), feeling of inner peace and the stabilizing of mood. The human brain and body (central nervous system and peripheral nervous system) are involved in the production of such dynamic relaxation, and it is facilitated by feelings of heaviness in the body, sensations of pleasant warmth in the limbs, recognizing of psychotraumatic events, emotional comfort (Jacobson, 1929).

The AT method was applied in order to learn the technique of self relaxation and self positive control to workers who suffered from chronic NSAH pains. The treatment course endured 3 months twice a week, the length of the session was 45 minutes. After the session the patient's feeling is assessed with the help of a survey where the effectiveness of the applied positive self relaxation (PSR) method for each particular patient is marked. The essence of the method is as follows: the patient is sitting facing the PSR guide – the doctor hypnotherapist; the guide asks the patient to close his eyes and take a comfortable seat and then starts the relaxation training of the patient: the patient's mimicry and body muscles gradually relax, breathing and heart function rhythms even out; the imagination programme is activated in the brain, positive self-influence and good feeling arise, the patient's "ego" strengthens; the negative stress experience is processed in a healthy way; as a result changes take place in the patient's behavior, emotional reaction and life quality (Roja et al, 2006).

Life quality assessment

Life quality assessment was realized before and after the rehabilitation using Quality of Life Scale (QLS) which was offer by American Chronic Pain Association. Quality of Life Scale is a measure of function for people with chronic pain (Cowan and Kelly, 2003). Quality of Life Scale looks at ability to function, rather than at pain alone. It can help people with pain and their health care team to evaluate and communicate the impact of pain on the basic activities of daily life. The scale is meant to help individuals' measure activity levels. Quality Of Life Scale consists of 0 to 10 variables (0 = "feel hopeless and helpless about the life" and 10 = "normal daily activities, social and family life").

Statistical analysis

The results acquired were processed using statistical data processing program SPSS.11 according to popular descriptive statistical methods (Pearson's correlation coefficient r, a.o.). Reliability interval (interrater agreement) was also calculated determining Cohen's Kappa coefficient (k) (Landis and Koch, 1977). This coefficient identifies connectivity of the experimental data, the number of participants and the proportion or correlation of the participants' acceptance of the experimental data:

$$k = (P_O - P_C) / (1 - P_C), \tag{1}$$

where: P_O – correspondence proportion of objective experimental data with respondents' responses (yes or no), P_C – correspondence proportion of data with number of participants ($P_C = \Sigma p_i^2$, where p_i is acceptance of each participant expressed in percent or as fractional number).

RESULTS

VAS

As based on the Visual Analog Scale pain intensity in NSAH for all age groups (n=30) before EMDR was comparatively high (mean points 6 in scale 0-10, $r > 0.85$). After 9 month EMDR pain decreases – the level of pain (mean points 1.4 in scale 0-10). Results of investigated age groups are shown in Table 2.

Table 2. Descriptive statistics for Visual Analog Scale for age groups

Age groups	Body parts	Before the rehabilitation			After the rehabilitation		
		Mean	Standard deviation	Range	Mean	Standard deviation	Range
18-25 (n=7)	Neck	2.71	2.29	1-6	0.29	0.49	0-1
	Shoulder	8.29	0.49	8-9	1.43	0.53	1-2
	Arm/hand	6.14	2.12	4-9	2.29	1.98	0-5
26-45 (n=15)	Neck	4.67	3.18	1-8	1.00	1.13	0-3
	Shoulder	7.60	0.91	6-9	1.67	1.05	0-3
	Arm/hand	6.47	3.20	1-9	1.73	1.10	0-3
46-65 (n=8)	Neck	3.25	2.60	1-7	0.50	0.76	0-2
	Shoulder	7.88	0.64	7-9	1.63	0.52	1-2
	Arm/hand	5.50	1.69	3-8	1.75	1.91	0-4

MADRS

Light depression indications were found only in two age groups (26-45 and 46-65) before the rehabilitation. The mean MADRS score was $12.7 \pm$ SD 2.6 ($r > 0.75$) before the rehabilitation. Light depression symptoms significantly decreased and practically eliminated after the EMDR. Scores by MADRS before and after the rehabilitation are shown in Fig. 1.

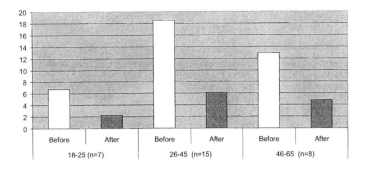

Figure 1. MADRS scores before and after the rehabilitation in age groups.

MYO

To objectively assess the muscles fatigue MYO measurements were performed. Myotonometry testing of the following muscles was done in relaxed state: *m. extensor digitorum*; *m. flexor carpi radialis*; and *m. trapezius* (upper part). According to regression analysis of MYO data, the slope of the lines (trendline) reflects the condition of the muscles after one week work cycle. It was stated that all workers can be subdivided into several categories:

Category I – subject is able to relax the muscle;

Category II – muscle is able to adapt to the work load and to relax partly;

Category III – muscle is not able to relax (muscle tone is increased which associate with muscles fatigue).

The example which demonstrates MYO data regression analysis for hand muscle (*extensor digitorum*) is shown in Fig. 2.

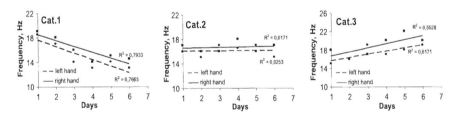

Figure 2. Results of the myotonometry testing data regression analysis of muscle frequency during consecutive 6 work days (example of MYO Categories).

MYO testing results reflected in Fig. 3 show the frequencies of different muscles in the beginning and at the end of the work week. These frequencies show changes in the muscle tone for sewers who are not adapting to the workload and whose muscle frequency exceeds the norm (11 up to 16 Hz, exist for each muscle individually) after the work week cycle. Therefore, the muscles fatigue is stated, and such workers are related to III MYO Category. These data was observed for sewers before EMDS.

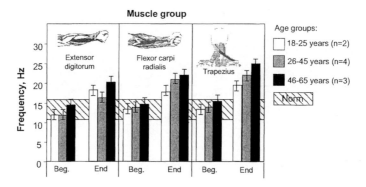

Figure 3. Illustration of frequency changing in separate muscle groups while performing the sewing at the beginning and at the end of the work week – for workers (n=9) who are not able to adapt with the workload and whose muscle frequency exceeds the norm after the work week cycle.

The percent of workers with differences in their muscle tone (MYO Categories) before and after EMDR is shown in Table 3.

Table 3. Percent of sewers (n=30) with differences in their muscle tone (before and after multidisciplinary rehabilitation, Pearson's correlation (r), and Cohen's Kappa (k)

Before rehabilitation			After rehabilitation		
Category	r	k	Category	r	k
I – 0 %	0.75	0.41	I – 33 %	0.75	0.67
II – 70 %	0.75	0.78	II – 60 %	0.75	0.86
III – 30 %	0.75	0.65	III – 7 %	0.75	0.55

To observe the changes of the muscles tone for several workers age groups in the whole EMDR period (9 month long-term active rehabilitation programme) the MYO measurements where done after each 3 month (duration of the measurements was one work week at the end of the each 3rd month). The results are shown in the Fig. 4.

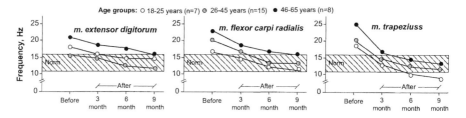

Figure 4. The development of the muscles tone among several sewers age groups before and after rehabilitation programme (data show measurements at the end of working week).

QLS

Before the EMDR based on the QLS questionnaire all sewers (n=30) with chronic pain in NSAH marked low quality of life (limited social activities, lack of plans for weekends etc.), but after the EMDR quality of life increased and sewers started to actively participate in social activities (family life, outside activities, active working hours). Descriptive statistics for quality of life scale before and after rehabilitation is shown in Table 4.

Table 4. Descriptive statistics for quality of life scale

Age groups	Before the rehabilitation			After the rehabilitation		
	Mean	Standard deviation	Range	Mean	Standard deviation	Range
18-25 (n=7)	7.00	0.82	6-8	9.43	0.53	9-10
26-45 (n=15)	6.53	0.52	6-7	9.00	0.38	8-10
46-65 (n=8)	7.00	0.93	6-8	9.25	0.46	9-10

DISCUSSION

The research work mainly focused on investigation of muscles functional state applying myotonometric method before and after EMDR in relation with muscles chronic pain, depression and quality of life. Inquiry data showed that the sewers (n=30) most frequently complain on feeling discomfort after the work, fatigue or chronic muscle pain in NSAH region and relate these problems with psychological factors like family life problems, unsuitable workplaces a.o.

All age groups indicate severe pain in shoulders and less intensive pain in neck and arm/hand before EMDR. Self-report of pain intensity using VAS presented in shoulders mean 7.9 points ($r > 0.85$), in arms/hands mean 6.0 points ($r > 0.85$), and in neck mean 3.5 points ($r > 0.85$) before the rehabilitation. The highest level of pain intensity (mean 8.29 points, $r > 0.85$) was in shoulders in the age group 18-25. After EMDR programme the chronic pain intensity in NSAH significantly decreased: shoulders mean 1.6 points ($r > 0.84$); arms/hands mean 2.0 points ($r > 0.85$); neck mean 0.6 points ($r > 0.85$).

Self-report about level of depressive mood using Montgomery-Åsberg Depression Scale was observed light depression symptoms in two investigated age groups before the EMDR. It is in accordance with the finding by Holzberg and Keogh (Holzberg et al., 1996; Keogh et al., 2006). It was shown that the main depression symptoms in age group of 26-45 years (total score 18.5, $r > 0.75$) were sleep disorders, apathy and tiredness, light appetite decrease, but in age group of 46-65 years (total score 12.9, $r > 0.75$) complaints about sadness, sleep disorders, short-term inner tension. In the youngest age group (18-25) the depression symptoms were not observed despite they marked highest pain intensity (total 8.29 of 0-10 points, $r > 0.85$). It is contradictious with a study by the World Health Organization that individuals who have persistent pain are four times more likely than those without pain to suffer from depression and more than twice as likely to have difficulty working (Gureje O et al., 1998). In our case it could be explained with short length of service in profession (2.7 ± SD 1.3 years) and lack of adaptation to work performance. At the end of EMDR light level of depressive mood eliminated for all investigated age groups: mean MADRS scores 4.4 ± SD 0.9 ($r > 0.75$).

Our main interest in this investigation was concentrated on the time period when the EMDR programme (including remedial gymnastics, swimming and AT) achieves effectiveness and allowed chosen muscle groups to relax and restore capacity to workload. The objective results (muscles possibility to adapt to the work load) showed MYO measurements.

Determination of muscle frequency (Hz) which support also muscles stiffness (N/m) is of great importance, for fatigue can be subdivided into high-frequency fatigue (HFF) and low-frequency fatigue (LFF), what differs from mechanic (also electric) features of muscles. HFF fatigue is characterized by an excessive loss of force at high frequencies of stimulation and rapid recovery when the frequency is reduced. Frequencies in excess of 50 Hz are rarely observed by voluntary activation of human muscle, and for this reason there has been some doubt as to whether high-frequency fatigue is a significant feature of normal activity. These findings are consistent with the accumulation of K^+ in the t-tubules and inter-fiber spaces of the muscle (Jones, 1996).

LFF is characterized by a relative loss of force at low frequencies of stimulation and a slow recovery over the course of hours or even days and there is evidence provided by intracellular measurements that low-frequency fatigue is a result of reduction in Ca^{2+} release (Jones, 1996). It has been observed that for several persons physiological LFF of muscles (this type of fatigue is called also – lasting fatigue) is accompanied by pain. If the work load requires too much muscle loading and stretching, when eccentric contractions are created, active muscle fibres resist the stretching and the cause of pain is the ultra-structural damage of the muscle.

Our investigation based on MYO data shows that for the sewers the greatest load was put on *m. flexor carpi radialis* and *m. trapezius* because their work usually involves fast movements of arms and hands, strain in the neck and shoulder region, as well as sitting long times in compulsory posture. It was stated there were no differences in the muscles contraction frequencies (ie. muscles tone) of both arms. More typical muscles strain which was supported also by pain in the shoulder region before rehabilitation programme was obtained for 80 per cent of sewers. This strain verifies increased *m. trapezius* contraction frequency more than 1.5-2 times at the end of the work week. Therefore, the EMDR programme was performed which allowed investigating the dynamic of muscles ability to relax and adapt to the work load. These results are better demonstrated in Table 3 and Fig. 4.

It was shown that yet after 6 month the tone of all investigated muscles groups achieve parameters nearly norm (12-17 Hz). Based on MYO measurements data muscles fatigue significantly decrease for all participants at the end of EMDR programme. For many sewers *m. trapeziuss* tone decreased usually below the norm (9-10 Hz), therefore 33 per cent of participants can be referring to MYO category I which testifies a good muscles ability to relax. For 60 per cent workers after rehabilitation muscle tone remained within the normal range, which means that, they were able to adapt to the existing work load.

In our work low muscle frequencies were not investigated because device doesn't allow measuring the deep muscles groups. Therefore, it was not possible to prove that muscles pains complained by all sewers (n=30) were caused by physical muscles fatigue (before EMDR programme MYO measurements show increased muscles fatigue only for 9 persons). Presumably, for most of sewers the intensity of pain is associated with psychosocial risks. This is in accordance with studies which shown that chronic pain causes can be linked with negative stress at workplace, emotional conflicts at family, psychosocial problems, depressive mood, because brain is low natural anti-pain and anti-anxiety capacity (Tyrer, 2006). In future research will elicit what are the risk factors are decisive in the origin of pain. In this case the electromyography measurements are necessary.

Significant focus in our EMDR program was on teaching sewers with chronic NSAH pain of self relaxation and self positive control, as well as performance of physical activities. AT as psychotherapeutic modality was effective accordingly to ability elevating natural anti-pain and anti-anxiety capacity (Kanji, 2000). As a result such combined EMDR was progressive for muscle relaxation, depression and pain intensity decrease. Recommended rehabilitation programme including education, reassurance, exercises and behavioral pain management and advice to stay active is also advised by Verbeek J.H. (Verbeek, 2001).

It is generally stated that chronic pain has a negative impact on quality of life (Kempen et al, 1997; Schlenk et al, 1998; Stewart et al, 1989). That corresponds to our investigation applying VAS and QLS. Notable results were found for all quality of life measured activity levels by individuals before and after the EMDR. If QLS indicated low quality of life before the EMDR (mean 6.8 of 10 points), sewers reported comparatively high pain intensity (mean points 6 in scale 0-10, $r > 0.85$). After 9 month EMDR quality of life improved (mean 9.2 of 10 points) and the level of pain significantly decreased (mean points 1.4 in scale 0-10). To improve good reliability, validity and responsiveness for measure of the quality of life of depressed workers the further investigations could be complemented with Quality of Life in Depression Scale (Hunt and McKenna, 1992; Tuynman-Qua et al, 1992).

CONCLUSION

Myotonometric measurements are suitable for objective determination the state before and after early multidisciplinary rehabilitation of such muscle groups: *m. extensor digitorum, m. flexor carpi radialis* and *m. trapezius*. The early multidisciplinary rehabilitation is an effective method for sewers who suffer from chronic NSAH pain. It significantly decreases chronic pain and depression symptoms. Physical activities rapidly decrease muscles tension and tone, but AT teaching of self relaxation and self positive control decreases chronic pain intensity and depression symptoms and improves quality of life.

442

REFERENCES

Banks, S.M., Kerns, R.D. (1996), "Explaining high rate of depression in chronic pain: a diathesis-stress framework". *Psychol Bull*, 119, 95–110.

Cowan P., Kelly N. (2003). *Quality Of Life Scale*. The American Chronic Pain Association. http://www.theacpa.org/documents/Quality_of_Life_Scale.pdf

Gunning, J., Eaton, J., Ferrier, S., Frumin, E., Kerr, M., King, A., Maltby, J. (2001). *Ergonomic Handbook for the Clothing Industry*. Published by the Union of Needletrades, Industrial and Textile Employees, Toronto.

Gureje, O., Von Korff, M., et al. (1998). "Persistent pain and well-being: A World Health Organization study in primary care". *JAMA*, 280, 147–151.

Holzberg, A., Robinson, M.E., Geisser, M.E., Gremillion, H.A. (1996), The effects of depression and chronic pain on psychosocial and physical functioning". *Clin. J. Pain*, 12, 118–25.

Hunt, S.M., McKenna, S.P., (1992), The QLDS: A scale for the measurement of quality of life in depression. *Health Policy*, 22, 307–319.

Huskinsson, E.C. (1983) "Visual analogue scales", in *Pain Measurement and Assessment*. R. Melzack (Ed.), New York: Raven Press, pp. 33–37.

Jacobson, E. (1929). *Progressive Relaxation*. Chicago, University of Chicago Press.

Jones, D.A. (1996) "High-and low-frequency fatigue revisited". Acta Physiol. Scand., 156(3), 265-270.

Kanji, N. (2000). "Management of pain through autogenic training". *Complementary therapies in Nursing and Midwifery"*, 6(3), 143–148.

Karjalainen, K., Malmivaara, A, van Tulder, M., Roine, R., Jauhiainen, M., Hurri, H., Koes, B. (2001), "Multidisciplinary biopsychosocial rehabilitation for neck and shoulder pain among working age adults: a systematic review within the framework of the Cochrane Collaboration Back Review Group", *Spine*, 26(2), 174–181.

Kempen, G.I., Ormel, J., Brilman, E.I., Relyveld, J. (1997), "Adaptive responses among Dutch elderly: the impact of eight chronic medical conditions on health-related quality of life". *Am. J. Public Health*, 87(1), 38–44.

Keogh, E., McCracken, L.M., Eccleston, C. (2006), "Gender moderates the association between depression and disability in chronic pain patients". *Eur. J. Pain*, 10, 413-22.

Kuijpers, T., van der Windt, D.A., van der Heijden, G.J., Bouter, L.M. (2004), "Systematic review of prognostic cohort studies on shoulder disorders", *Pain*, 109(3), 420–431.

Landis, J.R., Koch, G.G. (1977), "The Measurement of Observer Agreement for Categorical Data". *Biometrics*, 33, 59–174.

Montgomery, S.A. & Åsberg, M. (1979), "A new depression scale designed to be sensitive to change". *British Journal of Psychiatry*, 134, 382–389.

Roja, I., Bertholds, M., Zalkalns, J., Roja, Z. (2006). *Autogenic training*, Riga (in Latvian).

Schlenk, E.A., Erlen, J.A., Dunbar Jacob, J., McDowell, J., Engberg, S., Sereika, S.M. (1998), "Health-related quality of life in chronic disorders: a comparison across studies using the MOS SF-36, Quality of Life Research". *International Journal of Quality of Life Aspects of Treatment, Care and Rehabilitation*, 7(1), 57–65.

Stewart, A.L., Greenfield, S., Hays, R.D., Wells, K., Rogers, W.H., Berry, S.D. (1989), "Functional status and well-being of patients with chronic conditions. Results from the Medical Outcomes Study". *JAMA*, 262(7), 907–13.

Tuynman-Qua, H., de Jonghe, F., McKenna, S.P. (1997), "Quality of Life in Depression Scale (QLDS). Development, reliability, validity, responsiveness and application". *Eur. Psychiatry*, 12, 199–202.

Tyrer, St.(2006), "Psychosomatic pain". *The British J. of Psychiatry*, 188, 91–93.

Vain, A. (1995). "Estimation of the Functional State of Skeletal Muscle", in: *Control of ambulation using Functional Neuromuscular Stimulation*, Veltink, P.H. (Ed), Boom HBK, Enschede, University of Twente Press, pp. 51–55.

Verbeek, J.H. (2001),"Vocational rehabilitation of workers with back pain", *Scand. J. Work Environ Health*, 27(5), 346–352.

Chapter 47

Ergonomic Assessment of Farm Activities of Women Farmers of Mountain of India: Approach towards Drudgery Reduction

Kaliya Anita[1], Sharma Promila[2], Singh Sucheta[3], Karki Indu[4]

[1] Dept of Family Resource Management
Thapar Institute
Patiala, Punjab, India

[2]&[4] Dept of Family Resource Management
College of Home Science, G.B.P.U.A.& T., Pantnagar
Uttarakhand,-263145, India

[3]Directorate of Extension Education
G.B.P.U.A. & T., Pantnagar
Uttarakhand,-263145, India

[2]Department of Family Resource Management
College of Home Science, G.B.P.U.A. & T., Pantnagar
Uttarakhand,-263145, India

ABSTRACT

The farm and home are inseparable in India. An overwhelming majority of women in rural India are associated directly or indirectly with agricultural production, processing and distribution. They substantially contribute towards the labour force required in farm. Poor women in developing countries continue to be responsible for the time and labour intensive tasks of farm. Therefore, the study was designed with objectives to identify the extent of drudgery producing activities performed by women in various farm, household, and livestock activities through physiological response and identification of improved drudgery reducing implements and tools. The locale of the study was Nainital district of Uttarakhand state. Multistage purposive-cum-random sampling technique was used. For descriptive data 114 women and for experiments 14 women were selected. The results indicated that the activity, digging of land (4.76), harvesting (4.7), hoeing (4.65), and fetching of fuel (4.57) were the activities produced maximum drudgery. The extremely heavy activities were those where HR>175, RR>120, heavy activities HR=125-150, RR=107-114 and moderately heavy activities with HR=100-125, RR=100-107. Digging of land was perceived as extremely heavy activity. Mopping, fetching of water and fuel, weeding and harvesting were perceived as very heavy activities by majority of the women farmers. Milking activity followed by fetching of and ground leveling were the activities where much grip strength decreased of the women farmers. While studying physiological cost of the selected activities, it was found that blood pressure, heart rate, respiration rate and pulse rate of the women farmers increased in activities. But with use of improved implements the energy consumption decreased.

Keywords: Drudgery, Energy Expenditure rate (EER), Farm Women, Heart Rate, Respiration Rate, Awkward posture, Construction worker, MSDs, REBA, Risk Level.

INTRODUCTION

The farm and home are inseparable in India. An overwhelming majority of women in rural India are associated directly or indirectly with agricultural production, processing and distribution. They substantially contribute towards the labour force required in farm. Poor women in developing countries continue to be responsible for the time and labour intensive tasks of farm.

MATERIALS AND METHODS

The study was designed with objectives to identify the extent of drudgery producing activities performed by women in various farm, household, and livestock activities through physiological response and identification of improved drudgery reducing implements and tools. The locale of the study was Nainital district of Uttarakhand state. Multistage purposive-cum-random sampling technique was used. For descriptive data 114 women and for experiments 14 women were selected.

RESULTS AND DISCUSSION

Drudgery Index – The activity digging of land had highest drudgery score i.e. 4.76 followed by harvesting (4.70), hoeing 4.25, fetching of fuel 4.55, thus producing maximum drudgery to women farmers, fetching of fuel (4.53), fetching of water. (4.37) weeding (4.30) were some of the activities which produced maximum drudgery in work.

Table 1: Drudgery index (n=114)

Sl.No.	Activities	Mean score	Ranking
Household			
1	Fetching of water	4.37	5
2.	Fetching of fuel	4.53	4
Farm			
1	Digging of land	4.76	1
2.	Levelling	3.64	9
3.	Sweing	2.54	14
4.	Hoeing	4.25	7
5.	Application of fertilizer	3.29	12
6.	Weeding	4.30	6
7	Transplanting	3.37	10
8.	Harvesting	4.70	2
9.	Transportation	3.98	8
10.	Threshing	2.35	15
11.	Drying of grains	2.88	13
12.	Husking	2.22	17
13.	Storage of grains	3.20	18
Livestock			
1	Cleaning of cowshed	2.16	19
2.	Bathing animals	2.25	16
3.	Fetching fodder	4.55	3
4.	Cutting fodder	3.42	11
5.	Milking	2.05	20

The stress on muscles for activates was measured in terms of the stress of grip muscle strength (table 2). The grip strength reduced to 60% in case of milking indicating that muscular stress was maximum in this case. The percentage decrease in muscular case of fetching of fuel, and 54.5% increase of leveling of land. The percentage increase over rest was an indication of stress.

Table 2: Muscular stresses felt in work (n=114)

Sl.No.	Activities	Decrease in strength	% increase over rest
Household			
1	Fetching of water	3	16..70
2.	Fetching of fuel	9	52.90
Farm			
1	Digging of land	5	45.5
2.	Leveling	6	54.5
3.	Sowing	1	5.6
4.	Threshing	8	8.6
5.	Weeding	5	33.3
6.	Hoeing	7	38.9
7	Application of fertilizer	4	8.7
8.	Harvesting	4	22.2
9.	Transportation		
10.	Threshing	1	4.50
11	Drying of grains	3	27.3
12.	Husking	3	20.0
13.	Storage of grains	1	5.6
Livestock			
1	Cleaning of cowshed	1	5.0
2.	Batching animals	1	6.25
3.	Fetching fertilizer	10	55.5
4.	Cutting fodder	8	35.7
5.	Milking	12	60.0

Table 3: Physiological costs of activities performed with traditional implements

Activities	Mean pressure/min.			Heart Rate/min.			Respiration rate/min.		
	BA	AA	PI	BA	AA	PI	BA	AA	PI
Digging	89.17	94.50	6.0	92	214	132.6	73	110	50.7
Leveling	86.6	95.6	10.3	80	123	153.7	88	120	36.4
Sowing	89.3	93.3	4.5	80	97	21.25	75	115	53.3
Transplanting	88.3	93.3	5.7	80	110	37.50	92	150	63.0
Weeding	90.42	95.25	5.3	73	128	75.3	80	106	32.5
Hoeing	91.17	95.75	5.0	88	188	113.6	80	83	3.8
Application of fertilizers	94.4	104.0	9.2	96	128	33.30	80	107	33.6
Threshing/ Harvesting	97.87	100.4	2.6	75	163	117.3	80	108	34.2
Transportation of fertilizer	93.3	104.7	12.1	75	105	40.0	75	104	38.7

Table 4: Physiological cost of activities with traditional and improved implements

Activities	% increase in physiological cost in working with traditional implements				% increase in physiological cost in working with improved implements			
	AMP	HR	R	P	AMP	HR	R	P
Digging of land	3.84	25	29.17	16.24	2.31	12.5	16.67	3.7
Weeding	11.53	38.57	50.00	14.87	3.84	31.25	4.17	11.11
Hoeing	2.31	40.00	4.14	13.67	2.31	37.6	2.78	3.26
Leveling of land	3.16	21	27.11	14.25	2.1	16.1	17.37	4.75
Sowing/ transplanting	2.3	24.8	61	23	2.0	21.7	41.0	18
Collection of cow dung/agriculture/ home wastes	3.23	24.0	27.17	17.24	2.38	13.5	17.67	3.8
Transportation of fertilizer	2.71	22.15	26.03	16.66	2.12	14.17	16.66	3.5
Preparation & processing of fertilizer	2.5	26.8	61	23	2.1	21.3	38	15
Application of fertilizers	4.11	23	25.17	12.25	2.2	11.3	12.5	3.70
Harvesting/ threshing	2.4	27.8	60	24	2.0	21.3	40	17

AMP = Average Mean Pressure   HR = Heart Rates (beats/min)

R = Respiration (rate/min)   P = Pulse (rate/min)

Table 5: Physiological cost of activities with traditional and improved implements

Sl.No.	Activities	% increase in physiological cost of work			
1	Digging of land	Ti	16.24	Ii	3.7
2.	Hoeing	Ti	13.67	Ii	3.26
3.	Weeding	Ti	14.87	Ii	9.05
4.	Fetching of fodder/fuel	Ti	18.67	Ii	3.26
5.	Fetching of water	Ti	20.16	Ii	11.75

Ti = Traditional Implement   Ii = Improved Implement

There was sharp decrease in physiological cost of work i.e. heart rate per minutes. The recorded data in table showed that while digging land decrease in heart rate per minutes goes down to from 16.24 to 3.7 per cent beat per minute, in hoeing from 13.67 percent to 3.26, in weeding 14.87 to 9.05 from 14.87 to 9.05 percent in weeding in fetching of fodder/fuel and water it was noticed 18.67 to 3.26 and 20.16 to 11.75 percent decrease in heart beat per minutes.

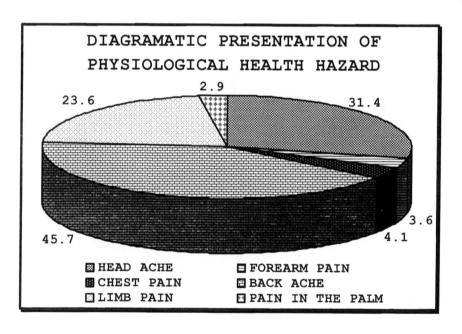

FIGURE 1 Diagrammatic presentation of physiological health hazards

Besides this the ventilation depth respiration rate and pulse rate per minutes was also found to be decreased while performing above activities with traditional implements.

Thus when subjectively, it was asked by the women farmers about impact of technology on them with statement i.e. increased/decreased and no change, it was reported by all that it has reduced the negative impact to positive i.e. they found work comfortable. The data of point gradation scale for posture analysis of selected activities also reveated that improved implement lead to sharp decline in awkward bent of body in performing the activities.

Table 6: Difference between mean pressure, heart rate and respiration rate before and after the activities

Parameters	T (values)
Blood pressure	-1.165419
Heart rate	1.95322*
Respiration rate	1.918338*

* Significant at 5 percent level

To test some of the personal/family situational variables with physiological parameters it was revealed that heart rate, respiration rate was significantly effected while at rest and after the activity is completed where tralve was 1.95322* and 1.918338, whereas it was negative in case of blood pressure.

REFERENCES

ERGOWEB. Internet address: http://www.ergoweb.com e-reference retrieved on 22/06/2009.

Hignett, McAtamney 2000, Rapid Entire Body Assessment (REBA), Applied Ergonomics 31 pp 201-205. htpp://ergo.human.cornell.edu/ahREBA.html, retrieved on 23/07/2009.

Larson D and Hannihen (1995) Surface electromyography: research clinical and sports applications. Indian J Physio Allied Sci 49: 5-9.

Pheasant S (1991) Ergonomics, Work and Health. pp 98-107. Macmillan Press, Hampshire. www.humanics-es.com/bernard/REBA_M11.pdf, retrieved on 08/07/2009.

Report of ICAR Adhoc Project (1996-98) G. B. Pant University of Agriculture & Technology, Pantnagar, India

Report of AICRP, ICAR Project, (2000) India

CHAPTER 48

Influences of Office Tasks on Body Dynamics using Dynamic Office Chairs

Liesbeth Groenesteijn[1,2], Rolf Ellegast[3],
Kathrin Keller[3], Helmut Berger[4], Peter Vink[1,2]

[1] TNO Quality of Life, P.O. Box 718, 2130 AS Hoofddorp
The Netherlands

[2] Delft University of Technology, Industrial Design Engineering
The Netherlands

[3] BGIA – Institute for Occupational Health and Safety of the
German Accident Insurance, Germany

[4] VBG – Accident Prevention & Insurance Association for
Administration, Germany

ABSTRACT

In this study the influence of office tasks on posture, movement, muscular activity and chair position is investigated in a laboratory study. The cross-sectional laboratory research was performed in a simulated office workplace involving seven office tasks and five chairs. Body part postures, muscle activation, body part movements and chair part positions were gathered. This paper reports the findings from 10 subjects, both male and female.

Findings from the experiment demonstrated that the tasks performed exerted strong significant effects on subjects' muscular activity, postures and movement.. The sorting file task was associated with the highest muscle activity, while mouse use was associated with the lowest activity. Error correcting tasks were associated with the most pronounced forward bended posture of the spine. Sorting tasks on the other side showed an upright trunk with substantially flexed head positions.

Keywords: Office Seating, Tasks, EMG, Posture, Physical Activity, (Dis)Comfort

INTRODUCTION

Forty-seven percent of employees in the EU perform white-collar (predominantly office) work, predominantly in offices. As the proportion of white-collar work continues to increase (Parent-Thirion et al., 2007), so does the importance of ensuring that this population can be productive, comfortable and free of symptoms.

The functionality of office chairs is important to accommodate variations in user postures and movements as several studies showed these variations with performing different tasks (Adams et al., 1986, Van Dieën et al., 2001, Dowell et al., 2001, Commissaris & Reijneveld, 2005, Babski-Reeves et al., 2005, Ellegast et al., 2007). Also, users' chair preferences in relation to function type are different with diverge functions (Legg et al., 2002). Therefore it seems that different types of tasks need different chair characteristics. An experiment with a focus on task support of the office chair showed that office workers performing a reading task required a larger back rest inclination range compared with a VDU task (Groenesteijn et al., 2009).

Task specific chair requirements are poorly understood. To improve the match between the characteristics of the task and features of the chair we need to define requirements based on theory, user experiences and physical parameters in the interaction with the task and the office chair. A few experiments are performed where task type is distinguished and physical parameters are specified. According to the study of Luttmann et al. (2003) muscular activity and fatigue, differ per task during sitting. The highest muscular activity in the shoulder region was shown for paper work whereas mouse application showed the highest activity in the lower arm. Physical activity in terms of postural change showed that spending more time in telephoning leads to an increase of extreme spine postures (Benninghoven et al. 2005). This shows that besides the type of task, duration is also an important issue to consider.

The paper of Ellegast et al. (2009) shows that the body posture and the muscle activity of the m. erector spinae and m. trapezius depend more on the tasks performed than on the use of a particular type of office chair. This paper focuses on what the task effects of the specified tasks are on physical parameters using five chairs with different dynamic systems. The selection of tasks performed when using office chairs involves error correcting on paper, typing text data, intensive mouse use, sorting paper files and telephoning. The research question of this study is:

What are effects of different office work tasks like error correcting on paper, typing, mouse use, sorting files and telephoning on postures and movements of body parts and positions of parts of the chair?

METHOD

The study consists of a laboratory research, carried out in a simulated computer office workplace (see figure 1) and set up as a cross sectional study with seven office tasks and five chairs. Body postures, muscular activity by electromyography (EMG), body movement and chair movement data were gathered. This laboratory test is part of an extended study with a laboratory experiment and a broader field study where Ellegast et al, (2008) report about. Ten healthy subjects (5 men and 5 women) volunteered to participate in the laboratory study. The mean ages were 35.2 years (SD 12.3 years) for the men and 34.8 (SD 12.7) for the women. Body heights ranged from 1.75 to 1.86 m (mean: 1.82 m) for the men and from 1.62 to 1.68 m (mean: 1.65 m) for the women. Body weights varied from 76 to 100 kg for the men and from 47 to 78 kg for the women. All subjects had performed the majority of their duties VDU workplaces for several years.

Figure 1 The simulated computer office workplace

Four chairs selected for their specific dynamic characteristics, labeled A, B, C and E and one reference chair labeled D were used in this study (see figure 2). They were all covered with dark blue textile. The manufacturers and types were blinded. All specific dynamic chairs have features of a conventional dynamic office chair and in addition to that they all come with specific dynamic features that are supposed to facilitate a range of postures and prevent statically or passive behavior of users.

Figure 2 The experimental chairs: special dynamic chairs A,B,C,E and reference chair D

During the lab test the subjects performed the following standardized and precisely defined tasks (duration of activity in brackets):

1. Reading and correcting text data on a printed standard text on paper that contains textual faults (10 minutes)
2. Typing words in a Word document with keyboard and mouse. The text (A4 format paper) was presented in a paper stand on the left side of the screen (20 min)
3. An intensive mouse task in a game following and hitting a target on the screen (20 min)
4. Reading and correcting text data continuing of the first task (10 minutes)
5. Typing words continuing the second task (20 min)
6. Sorting paper files on the desk in various document files (10 min)
7. Telephoning performing one call (10 min)

The tasks were for all ten subjects offered in the same order.

Joint body angles and movements of body parts and chair parts were measured with the CUELA system (Ellegast & Kupfer, 2000). Surface electromyography (EMG) was used for measuring the muscle activity of the m. trapezius (right/left) and m. erector spinae (right/left) with the CUELA EMG signal processor for long-term analysis (Glitsch et al. 2006).

From the measured signals, the following body/joint angles were calculated: Head inclination, cervical spine inclination, flexion/extension and lateral flexion of the spine in the thoracic (Th3) and lumbar spinal regions (L1 and transition to L5), trunk inclination and the spatial position of the upper and lower legs (right and left). From the EMG signals percentage of activation was expressed in relation to the Reference Voluntary Contraction (RVC). From the kinematic measurements of all sensors physical activity intensities (PAI) were determined by calculating a sliding standard deviation of the high-passed filtered vector magnitude of the 3D acceleration signals. From the chair signals the angles of seat inclination (in for / backward and sideward directions) and backrest inclination was calculated. For extended analysis descriptions see also Ellegast et al. (2008).

For statistical analysis of the lab study ANOVA for repeated measures was used for comparisons of the, EMG, PAI and chair data. Post hoc LSD was used to compare task by task. The significance level for all statistics was determined at a 0.05 level.

RESULTS

MUSCLE ACTIVATION

There is a main effect of task on all four measured muscles for the medium level (50 percentile). For the peak level (95 percentile) of the left and right trapezius also a main effect of task was found. The most significant differences in EMG between tasks are seen in the right and left trapezius muscles, but also in the erector spinae activity are some significant differences found. The sorting task shows over all muscles the highest muscles activity which is significantly different for both left and right trapezius from all other tasks. The mouse task shows the lowest muscles activity which is significantly different from all other tasks in the right trapezius muscle. The first correcting task is for the EMG of the right trapezius different from the mouse task. The first correcting task shows lower muscle activity compared to the sorting task in all measured muscles. The first typing task shows higher muscles activity compared to the mouse task in the peak levels (95 percentile) of the trapezius muscles and in the medium level (50 percentile) of the left erector spinae. The first typing task shows lower muscle activity compared to the sorting task in left and right trapezius muscle. The first correcting task shows also lower muscle activity compared to the telephoning task in the peak level of the right erector spinae.The mouse task shows lower muscle activity compared to all other task in the right trapezius muscles. Lower muscle activity is also seen in the left trapezius muscle compared to the second typing task and sorting. The second correcting and typing tasks show also lower activity compared sorting in both left and right trapezius and at the peak level of the left erector spinae. The sorting task shows higher muscle activity in all muscles compared to telephoning.

PHYSICAL ACTIVITY OF BODY PARTS

A main effect of task was also found for physical activity intensity of head, thoracic spine, lumbar spine L1 and L5, and left and right thigh and lower leg. There are many significant differences in physical activity intensity between tasks. Among the many differences some are interesting to report to demonstrate the relative differences between tasks in physical activity intensity. Sorting files showed the highest physical activity in the head followed by telephoning. Sorting files showed also the highest activity of thoracic and lumbar spine again followed by telephoning. The correcting and mouse task showed the lowest activity of the head. The correcting and mouse task showed the lowest activity of thoracic and lumbar spine.Telephoning showed the highest physical activity in both upper and lower legs where correcting the first time showed the lowest leg activity. Typing showed also low activity in the legs. There are also significant differences between the first and the second time correcting and typing.

JOINT BODY ANGLES

For the joint body angles also main effects of task were found for al measured angles. And between tasks significant differences in joint body angles are also found in many comparisons. Between correcting the first time and correcting the second time performed there are also significant differences which is the same task. These p-values of the comparison of the first and second time correcting are between 0.05 and 0.01 where the values between different tasks are <0.01. Among the many differences some are interesting to report to demonstrate the relative differences between tasks. Telephoning shows the highest L5 inclination which is significantly different of all other tasks. The mouse task shows the second high L5 inclination also significantly of the other tasks. Sorting files shows the least L5 inclination. Both the correcting tasks show the highest cervical spine flexion and head inclination. The sorting task shows also cervical spine flexion and considerable head inclination. The other tasks show cervical spine extension with the mouse tasks with the highest extension and the lowest head inclination. The second time correcting shows the highest lumbar spine flexion closely followed by telephoning and first time correcting. Sorting files shows the lowest lumbar spine flexion. The subjects are sitting pretty upright possibly to overview their paper documents. Correcting, mouse task and telephoning show all comparable high trunk flexion. Sorting files shows the lowest significantly different trunk flexion. All tasks show a little trunk lateral flexion to the left except for some peak values. Correcting shows the highest lateral flexion. The second time correcting shows the highest kyphosis together with the mouse and the telephoning task. Sorting files shows the lowest kyphosis.

CHAIR PART POSITIONS

Main effects of task were found for seat pan inclination in forward/backward direction and in sideward directions Telephoning shows the highest backward inclination of the seat pan and is significantly different from the other tasks. Sorting files shows a forward inclination which is significantly different from the other tasks except for typing. Sorting files shows the most neutral sideward seat pan inclination which is significantly different from the other tasks with more inclination to the left. Typing shows, together with correcting and the mouse task the most sideward inclination to the left. Telephoning shows the highest backward inclination of the backrest and is significantly different from the other tasks. Sorting files shows the lowest backward backrest inclination and is significantly different from telephoning, correcting and mouse task.

DISCUSSION

Different office tasks were investigated in a laboratory setting with five different dynamic chairs. The tasks showed many differences in body dynamics and postures between tasks. A summary of the results per task is shown in table 1. The first and second time typing and correcting are despite some differences presented as one as these differences were smaller than differences between tasks.

Table 1 Summary of results per laboratory task with relative comparison between tasks

Task	Influence			
	muscle activation	physical activity	postures	chair positions
Correcting	- 'medium' activity	- lowest head activity - low leg activity	- highest kyphosis and high trunk flexion - highest head inclination and cervical spine flexion	- close to neutral seat pan inclination - little back rest inclination
Typing	- 'medium' activity	- low leg activity - 'medium' trunk and head actvity	- high trunk flexion and kyphosis - fairly upright position of cervical spine with some head inclination	- close to neutral seat pan inclination - little back rest inclination
Mouse use	- lowest actvity	- lowest head activity - lowest trunk activity	- highest trunk flexion and high kyphosis - highest lumbar spine flexion and high L5 inclination - highest cervical spine extension and upright head position	- high backrest inclination with some backward seat pan inclination
Sorting	- highest activity	- highest trunk and head activity	- least L5 and trunk inclination - high cervical spine flexion and head inclination	- most forward seat pan inclination and side ward inclination to the right
Telephoning	- 'medium' activity	- highest leg activity - high trunk and head activity	- high L5 and trunk inclination - fairly upright position of head and cervical spine	- highest backrest inclination

This study showed that posture variation as an effect of tasks is large. This is in line with the study Van Dieën (Van Dieën et al. , 2001).

The high L5 inclination together with high backrest inclination with telephoning is supported by Groenesteijn et al. (2009) where subjects preferred a large backrest inclination with a telephoning task.

The significant differences in the same tasks, correcting and typing, for the first and the second time performed in physical activity and joint body angles in the lab study showed that there is also within these tasks variation. These differences are smaller than between different tasks and significant values are less significant than between tasks, but still below the 0.05 significance level. Because of the systematic order this can be caused by either increased learning or fatigue where the higher physical activity the second time is due to more degrees freedom or a way to compensate fatigued muscles. But this is speculative and further investigation is needed with varied order for founding these speculations.

The position of head and cervical spine is for all tasks highly determined by the target location of the view of the eyes. Tasks with the computer like typing and mouse use have cervical spine extension together with a little inclination of the head. This is in contrast with a target location at the desk which showed cervical spine flexion and head inclination. Telephoning has no direct target location for the eyes view and leaves the most independent posture of the head. Observational studies have shown that office workers usually perform their tasks in upright or forward leaning postures (Dowell et al. 2001). The visual demands of the task and the reach distances can play a role in leaning forward (Lueder, 2004). Reclined postures with substantial engagement between the sitter's torso and the chair backrest account for only about 15 percent of work postures for workers performing a range of office tasks. In this study is specified that reclined postures are related to telephoning. The more reclined postures of conversation and telephoning seem more preferred as Gescheidle, Miller & Reed (2004) found that preferred postures are substantially reclined. Because the task of telephoning is less restricted in posture by input devices, screen and desk work the subjects are more able to have a preferred posture.

CONCLUSION

Considerable effects of tasks on postures and movements of body parts and chair part positions are found with many differences between the task types.

The correction task showed 'medium' muscle activity and the lowest physical activity of the head in comparison. The body posture in this task is a forward flexed spine with the highest kyphosis, head inclination and cervical spine flexion and high trunk flexion. The chair has a little backward backrest inclination and a nearly horizontal seat pan.

The typing task showed 'medium' muscle and physical activity of trunk and head. The body posture with typing is a fairly upright cervical spine position together with a little inclination of the head. The chair positions have a little backward backrest inclination and a nearly horizontal seat pan.

Mouse use showed the lowest muscle and physical activity of the head and trunk. It is a very static task with a body posture that showed the highest trunk flexion, the highest lumbar spine flexion and the highest cervical spine extension head position.

The sorting files task showed the highest muscle activity with high physical activity in the trunk and head and is therefore the most dynamic task. Sorting files showed the least L5 and trunk inclination with cervical spine flexion and high head inclination, and the smallest backrest inclination of the chair.

Telephoning was in between the muscle activation extremes, but still different from other tasks. Telephoning showed the second highest physical activity in the trunk and head and the highest physical activity in the lower legs. The body posture with telephoning showed the most backward position of L5 and trunk together with the highest trunk flexion and kyphosis and a fairly upright position of head and cervical spine. This is in line with the found highest backrest inclination of the chair. This is also a forward flexed spine as in the correction tasks, but with further inclined lumbar spine and assumed different body load by leaning more backward to the backrest.

ACKNOWLEDGEMENTS

This project was initiated by the VBG German Accident Insurance for Administration and funded by the German Social Accident Insurance (DGUV). The authors are grateful to Rene Hamburger, Ingo Hermanns and Merle Blok for their support. Special thanks also to Beverly Schlenther, Melanie Niessen and Daniel Annemaier for their help with data collection and analysis.

REFERENCES

Adams, M., Dolan, P., Marx, C., Hutton, W., 1986. An electronic inclinometer technique for measuring lumbar curvature. Clin. Biomech. 1, 130-134.

Babski-Reeves, K., Stanfield, J., Hughes, L., 2005. Assessment of video display workstation set up on risk factors associated with the development of low back and neck discomfort. International Journal of Industrial Ergonomics 35, 593-604.

Benninghoven A.,Bindzius F., Braun D., Cramer J., Ellegast R., 2005. CCall - Healthy and Successful Work in Call Centres. International Journal of Occupational Safety and Ergonomics (JOSE), Vo. 4, 409 – 421

Commissaris, D.C.A.M., Reijneveld, K., 2005. Posture and movements during seated office work; results of a field study. In: Veiersted, B., Fostervold, K.I., Gould, K.S. (Ed.), Ergonomics as a tool in future development and value creation; Proceedings of the 37th Annual Conference of the Nordic Ergonomics SocietyNES and NEF, Oslo (Norway), 10-12 October, p.p 58-61.

Dowell, W.R., Yuan, F., Green, B.H., 2001. Office seating behaviors: An investigation of posture, task and job type. Proceedings of the Human Factors and Ergonomics Society, pp. 1245-1248.

Ellegast, R.P., Kupfer, J., 2000. Portable posture and motion measuring system for use in ergonomic field analysis. In: K. Landau (ed.) Ergonomic software tools in product and workplace design. Verlag ERGON GmbH, Stuttgart 47-54

Ellegast, R., Hamburger, R., Keller, K., Krause, F., Groenesteijn, L.,Vink, P., Berger, H., 2007. Effects of using dynamic office chairs on posture and EMG in standardized office tasks. Springer, Berlin, pp. 34-42.

Ellegast, R.P., Keller, K., Hamburger, R., Berger, H., Krause, F., Groenesteijn, L., Blok, M., Vink, P., 2008. Ergonomische Untersuchung besonderer Büroarbeitsstühle. BGIA-Report 5/2008, ed.: Deutsche Gesetzliche Unfallversicherung (DGUV), Sankt Augustin 2008.

Gscheidle G. M., Miller H.and Reed M. P., 2004. Sitter-selected postures in an office chair with minimal task constraints. Proceedings of the Human Factors and Ergonomics Society 48th Annual Meeting

Glitsch, U., Keller, S., Kusserow, H., Hermanns, I., Ellegast, R.P., Hüdepohl, J.: Physical and physiological workload profiles of overhead line service technicians. In: Pikaar, R.N., Koningsveld, E.A.P., Settels, P.J.M (eds.) Proceedings IEA 2006. Elsevier Ltd., Oxford (2006)

Groenesteijn, L., Vink, P., De Looze, M., Krause, F., 2009. Effects of differences in office chair controls, seat and backrest angle design in relation to tasks. Applied Ergonomics 40 (3), 362-370.

Legg, S.J., Mackie, H.W., 2002. Milicich W. Evaluation of a Prototype Multi-Posture Office Chair. Ergonomics 45, 2, 153-63.

Lueder, R.K., 2004. Ergonomics of seated movement: A review of the scientific literature. Humanics Ergosystem Inc.

Luttmann, A.; Kylian, H.; Schmidt, K.-H.; Jäger, M., 2003 Long-term EMG study on muscular strain and fatigue at office work. In: Quality of work and products in enterprises of the future: proceedings of the Annual Spring Conference of the GfA on the Occasion of the 50th Anniversary of the Foundation of the Gesellschaft für Arbeitswissenschaft e.V. (GfA) and the XVII Annual Conference of the International Society for Occupational Ergonomics & Safety (ISOES), Munich, Germany, May 07th - 09th.

Parent-Thirion A, Fernández ME, Hurley J, Vermeylen G., 2007 Fourth European Working Condi¬tions Survey. Dublin: European Foundation for the Improvement of Living and Working Conditions.

Van Dieën, J.H., De Looze M.P., Hermans, V., 2001. Effects of dynamic office chairs on trunk kinematics, trunk extensor EMG and spinal shrinkage. Ergonomics 44, 7, 739-750.

CHAPTER 49

Whole-Body Vibration Exposure and Musculoskeletal Complaints in Urban Mail Carriers

Hsieh-Ching Chen, Yung-Ping Liu, Cheng-Tsung Chiang

Department of Industrial Engineering and Management
Chaoyang University of Technology
Wufong, Taichung County 41349, TAIWAN

ABSTRACT

This study explores the dose response between whole-body vibration (WBV) exposure and musculoskeletal complaints among mail carriers riding motorcycles for postal delivery. A questionnaire survey was performed and, then, daily exposure was determined. Questionnaire results indicate that male mail carriers (n=103, average 39.9±7.8 years) have a higher subjective pain than male back-office staff (n=34, average 49.5±5.1 years) in the post office at all investigated body regions. Twenty-five mail carriers (average 39.1±6.5 years) were recruited for determining the daily WBV exposures from carriers participating in the questionnaire survey. Only one mail carrier had a vibration dose value (VDV) exceeding the 17 m/s^2 boundary, indicating a high probability of adverse health, as specified by ISO 2631-1 (1997) guidelines. However, 13 of the 25 evaluated mail carriers had S_{ed} exceeding the 0.8 MPa boundary, indicating a high probability of adverse health, as specified by ISO 2631-5 (2004) guidelines. Analytical results demonstrate that these mail carriers have experienced WBV containing high shocks. By integrating analytical results derived from a questionnaire survey and daily exposure measurements, we believe that a dose-response relationship may exist between

WBV exposure and musculoskeletal complaints of postal workers.

Keywords: Motorbike, Postman, Musculoskeletal Complaint, ISO 2631

INTRODUCTION

Previous studies indicate that driver vibration exposure depends on road or traffic conditions, vehicle characteristics such as speed, type, weight, seat, maintenance and engine size, and driver characteristics such as age, characteristics, experience, sitting posture, and body weight (Peitte and Malchaire, 1992; Ozkaya *et al.*,1994; Malchaire *et al.*, 1996; Donati, 1998; Chen *et al.*, 2003; Mansfield and Griffin, 2002). Commonly reported health effects caused by whole-body vibration (WBV) exposure include discomfort, musculoskeletal problems, muscular fatigue, reduced stability, and altered vestibular function (Seidel, 1993; Wasserman *et al.*, 1997; Bongers *et al.*, 1988; Griffin, 1998). Several studies have also indicated that long-term WBV exposure is associated with early spinal degeneration (Frymoyer *et al.* 1984), low back pain, and herniated lumbar disc (Bovenzi and Zadini 1992, Boshuizen *et al.* 1992).

Motorcycles are a common transportation mode in Asia. An estimated 33 million motorcycles are used in mainland China, 18 million in Vietnam, 13.8 million in Taiwan, 13 million in Japan, 5.8 million in Malaysia, 2.7 million in Korea, and 1.8 million in the Philippines (IRF, 2006). Most motorcycles in Taiwan can be categorized as scooters (*i.e.,* no clutch, seated riding position) or motorbikes (equipped with clutch, straddled riding position). These motorcycles generally have 125cc engines or smaller, are ridden on shoulders or in reserved lanes, and are convenient for accessing driving lanes. Although motorcycles are typically used only for short-distance transport, they are the main transportation mode for workers such as postal workers, delivery workers, and urban couriers. Consequently, these occupations are likely associated with high WBV exposure.

Health problems related to WBV exposure in motorcycle riders are often neglected despite the large population size potentially affected. Chen *et al.* (2009) described high WBV exposure in twelve male motorcycle riders traveling on a 20.6 km rural-urban paved road based on ISO 2631-1 (1997) and ISO 2631-5 (2004) standards. However, the possible adverse health effect and dose-response relation among motorcycle riders remain unclear.

This study explores the WBV exposure and subjective feelings of regional pains in mail carriers riding motorbikes for postal delivery. A questionnaire survey is performed and, then, daily exposure is measured. The WBV exposure in mail carriers was compared with the boundary values of health guidance caution zone (HGCZ), as specified by ISO 2631-1 (1997), and with the limit value associated with a high probability of adverse health effects based on ISO 2631-5 (2004) guidelines.

METHOD

A questionnaire survey was performed to collect characteristic data from a group of target mail carriers. Collected data with respect to age, gender and work experience were later used to stratify and sample mail carriers from those participating in the questionnaire investigation in order to determine daily WBV exposure.

QUESTIONNAIRE SURVEY

One hundred and ten mail carriers and sixty-six office duty personnel from two major post offices in downtown Taichung City, Taiwan, participated in the questionnaire survey (Table 1). The office duty personnel were recruited to serve as the control group in this study. All participants filled a questionnaire distributed by the Office of Safety and Health, Chunghwa Post. Questions on personal characteristics, occupation history, WBV exposures conditions, and subjective feelings of regional pain over the previous year were modified from Pope *et al.* (2002). Pain in the neck, shoulders, arms, wrists and hands, as well as upper and lower back areas were recorded. Response categories were "no pain," "mild pain," "discomforting," "distressing," and "intense", as scored on a scale from 0 to 4.

The same gender groups did not significantly differ in stature, whereas mail carriers and office duty personnel significantly differed in gender constitution ($\chi^2 = 42.4$, $p < 0.001$). Additionally, the male to female ratio was approximately 1:1 among the office duty personnel and 15:1 among the mail carriers (Table 1). The average age and years of experience among male mail carriers was 10-year and 6-year younger than that among the male office duty personnel ($p < 0.001$), respectively.

Table 1: Characteristics of participants in the questionnaire survey (mean ± s.d.)

Group	Mail carriers¶		Office duty personnel	
Gender	Male	Female	Male	Female
N	103	7	34	32
Age (yr.)	39.9±7.8**	45.9±9.2	49.5±5.1	48.8±6.0
Experience (yr.)	13.0±8.8**	19.4±10.0	19.4±9.5	22.6±9.8
Height (cm)	170.2±5.8	155.6±6.3	167.9±6.2	152.3±8.4
Weight (kg)	71.1±9.1	59.4±7.1	66.6±8.6	53.7±16.8

¶$p < 0.001$, gender constitution difference (χ^2 test); **$p < 0.001$, group difference (t-test, Male)

DAILY EXPOSURE MEASUREMENT

Participants and vehicles

Twenty-five mail carriers (twenty-four males and one female) were sampled and recruited for motorcycle riding tests. Their age (39.1±6.5 years), body height (169.4±7.1 cm), body weight (68.6±11.3 kg), years of experience (14.4±8.0 years) and gender constitution did not significantly differ from those mail carriers participating in the questionnaire survey. All subjects rode standard motorbikes (Sanyang Corp., 125 c.c., straddled riding) and performed daily postal delivery. Each participant described riding a motorbike in an upright sitting position without leaning forward or backward by more than 10°.

Equipment

Vibrations transmitted to the seated human body as a whole through the supporting surface of the buttock were measured using a triaxial ICP seat pad accelerometer (model 356B40, Larson Davis Inc., USA). The accelerometer had a frequency sensitivity range of 0.5–5000Hz, and was pre-calibrated for excitation of 1g RMS/159.2Hz with a hand-held calibrator (model 394C06, PCB Piezotronics Inc., USA). Seat pad outputs were connected to a 3-channel amplifier (model 480B21, PCB Piezotronics Inc., USA) with a signal conditioning gain of 10. Outputs of the amplified signals were recorded on a portable data logger at a rate of 5000 samples/s per channel. A pressure switch, constructed by a 0.8-cm-thick hollow sponge sandwiched by two thin copper foils, was affixed on the seat pad surface with tape to detect ingress and egress status of the seat pad. The switch status was digitized by a voltage divider and was recorded by the portable data logger. The logger can continuously store collected data on a 4GB compact flash (CF) memory card up to 8 hours. The logged data were downloaded onto a personal computer using a card reader for further data processing.

Location and riding speed of the participant were recorded every 5-sec during the task by using a GPS data logger (DG-100, GlobalSat, Taiwan). A portable media recorder (CNF-200, Carry Media Electronic Ltd., Taiwan) and a mini-camera recorded the riding status of a motorcycle rider. The recorded video was synchronized with the logged acceleration data using a remote transmitter, which sent radio frequency signals at the beginning and end of the riding task.

Field testing procedure

The standard testing procedure for the riding test was explained, with detailed instructions given to all participants before the test. Each participant performed daily postal delivery tasks. Each task lasted 2-8 hours. Duration of each task was determined from the time the participant exited the post office parking lot to the time of their return. Each participant was allowed to leave the seat during the riding task.

For each riding task, a seat pad accelerometer was placed on the seat beneath the buttocks of each subject in compliance with ISO 2631-1 (1997) standards. The positive x and z directions were anterior and upward, respectively. Each test motorbike had a mini-camera and portable media recorder affixed to its rear frame and a GPS data logger taped beside its speedometer. The signal amplifier, data logger, remote transmitter, and a rechargeable battery set were placed in a plastic box that was put in the mail basket near the rear seat (Fig. 1).

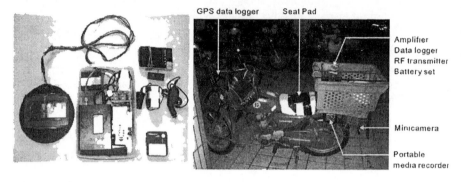

Figure 1. Field measurement equipment and experimental setup

Data processing procedure

Artifact signals caused by ingress or egress were removed from the downloaded data according to the seat pad status by using a pre-processing program. If the participant did not sit on the pad longer than 1 sec, acceleration signals were suppressed to zero from 0.5-sec before the instance of egress until 1-sec after the ingress. The data were combined with the taped video by using 'Viewlog' software, as described elsewhere (Chen *et al.*, 2006). Its vibration analysis module, specifically designed to perform batch computing, evaluates WBV exposure and transmits the results to a user-defined MS Excel template (Taiwan IOSH, 2007). For each participant, root mean square (*RMS*) acceleration and daily vibration dose (*VDV$_d$*) were computed according to ISO 2631-1 (1997) standards. Additionally, the daily equivalent static compression dose (*S$_{ed}$*) was calculated based on ISO 2631-5 (2004) standards. The detailed data processing procedure and computation of risk factor (*R*) were the same as those described elsewhere (Chen *et al.*, 2009).

DATA ANALYSIS

Statistical analysis was performed for questionnaire data with SPSS 10 for Windows. Group differences in participant characteristics were assessed by an independent *t*-test for all scale variables, by χ^2 test for gender constitution, as well as by the Mann-Whitney test for subjective regional pains. A p value less than 0.005 was considered statistically significant. Due to the significantly higher number of male than female mail carriers, the questionnaire survey focused on summarizing the results of male participants. The WBV exposures obtained in the motorcycle ride tests were compared with the boundary values of HGCZ recommended by ISO 2631-1 (1997) and ISO 2631-5 (2004) guidelines, as well as with the test results of 20-km motorcycle rides described elsewhere (Taiwan IOSH, 2008).

RESULTS

Except for gender differences in body height and weight, the two male groups did not significantly differ in exercise, smoking, and alcoholic beverage drinking, as well as in motorbike use rate. Nevertheless, the mail carriers had a significantly greater annual motorbike riding and automobile driving distance (ranging from 8000–12000 km/year), and a greater annual exposure duration than the office duty personnel (ranging from 4000–8000 km/year). The male mail carriers had a significantly higher subjective feeling of regional pain in the upper back ($p<0.003$), bilateral forearms ($p<0.001$), left hand ($p<0.001$), hip ($p<0.002$), and right foot ($p<0.001$) than the male office duty personnel did (Table 2). Moreover, the groups did not significantly differ in the neck and lower back regions; even these regions had relatively high subjective pain.

Table 2. Subjective regional pain in the male mail carriers (N=93) and the office duty personnel (N=32). Group differences are denoted by the p-value for Mann-Whitney test.

	Neck	Back		Shoulder		Forearm	
		Upper*	Lower	Right	Left	Right*	Left*
Carrier	1.09±1.01	0.87±0.92	1.49±1.07	1.01±1.01	0.92±0.95	0.63±0.82	0.62±0.77
Officer	0.66±0.97	0.34±0.60	0.94±0.98	0.47±0.72	0.47±0.84	0.16±0.45	0.13±0.42
p–val.	n.s.	<0.003	n.s.	n.s.	n.s.	<0.001	<0.001

	Hip*	Hand		Knee		Foot	
		Right	Left*	Right	Left	Right*	Left
Carrier	0.51±0.79	0.91±0.99	0.94±0.97	0.82±0.93	0.88±0.93	0.80±0.88	0.86±0.89
Officer	0.09±0.39	0.44±0.67	0.19±0.54	0.34±0.70	0.38±0.71	0.22±0.71	0.47±1.02
p–val.	<0.002	n.s.	<0.001	n.s.	0.005	<0.001	0.005

Table 3 lists the mean and range of riding distance, duration, peak and 50th, 75th, 95th percentile riding speeds measured by the GPS data logger. The average riding speed of mail carriers was relatively lower than the regular vehicles running on urban routes. Figure 2 shows 25 WBV exposures (RMS, VDV_d, and S_{ed}) in this study and 90 WBV exposures of 20-km motorcycle ride tests reported previously by the Taiwan IOSH (2008). All VDV_d values, except for one, fell in the HGCZ (8.5–17 m/s^2), as specified by ISO 2631-1 (1997) guidelines. Nevertheless, more than 50% of all mail carriers had a S_{ed} value exceeding the upper boundary value of HGCZ (0.8 MPa), indicating a high probability of adverse health, as specified by ISO 2631-5 (2004) guidelines. Moreover, two 51-year old carriers had risk factor (R), as calculated by the individual participant's age, years of experience, and S_{ed}, according to ISO 2631-5 (2004) guidelines, over the value of 1.2, indicating a high probability of adverse health. These two mail carriers both scored "distressing" at a lower back region in the questionnaire survey. In contrast with the health risk predicted by S_{ed}, 14/25 RMS values were less than the lower boundary value of HGCZ (0.5 m/s^2), indicating a low probability of adverse health, recommended by ISO 2631-5 (2004) guidelines.

Table 3. Mean±s.d. and range of riding distance, duration, peak and 50th, 75th, 95th percentile speeds

(N=25)	Distance (km)	Duration (hour)	Speed (km/hr)			
			P50	P75	P95	Peak
Mean±s.d.	24.3±22.3	4.9±2.2	3.9±3.2	14.7±9.9	33.3±9.0	55.8±9.7
Range	6.9–83.8	1.3–8.2	0.9–12.0	3.3–29.3	19.7–46.7	36.0–84.8

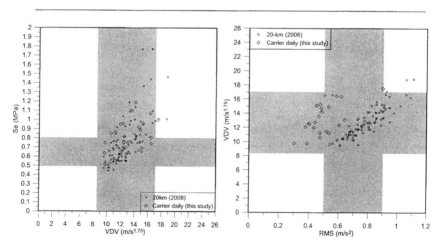

Figure 2. Daily WBV exposures (RMS, VDV_d and S_{ed}) of 25 mail carriers (this study) and 90 WBV exposures of 20-km motorcycle ridings (Taiwan IOSH, 2008). HGCZ: grey bands.

DISCUSSION

Questionnaire survey results have indicated a possible relationship between work nature and subjective regional pains. Although the mail carriers were on average 10 years younger than the office duty personnel, the mail carriers experience significantly higher regional pains than the office duty personnel do. According to onsite observation of the investigator, subjectively greater pain in the hand-arm and right foot region of the carriers are likely associated with the repetitive task of sorting mail, using a motorbike hand brake, and operating the motorbike foot clutch. Whereas a significantly greater pain in mail carriers' hip and upper back and the highest pain in carriers' lower back than that in the office duty personnel are likely attributed to WBV exposure. Unfortunately, the measured WBV exposure parameters (RMS, VDV_d, and S_{ed}) were not strongly correlated to any subjective regional pains, possibly because this study is a cross-sectional study using a small sample size (N=25). Future longitudinal research should investigate how these WBV exposure parameters and musculoskeletal difficulties are related by using a considerably larger sample size.

According to ISO 2631-5 (2004) guidelines, more than 50% of the tested mail carriers had a high possibility of adverse health risk associated with WBV exposure containing multiple shocks. Computed R estimated that more than 13 out of 25 mail carriers would reach a value of 1.2 before their retirement (65 years old). This observation suggests that mail carriers with many years of experience may face health hazards caused by WBV exposure. Follow-up studies are warranted to investigate whether carriers changed jobs or job tasks prematurely due to WBV induced disability. We also recommend conducting health checks on senior mail carriers (aged 45 and above) to determine whether early degeneration of the spine occurs in this occupational group.

Approximately half of the test cases had a health risk categorized differently when using ISO 2631-5 (2004) and by ISO 2631-5 (2004) guidelines. When more than half of the tests had a RMS lower than the lower boundary value of HGCZ, as specified by ISO 2631-1 (1997), more than half of the tests had a S_{ed} value greater than the upper boundary value of HGCZ, as specified by ISO 2631-5 (2004). This contradiction was largely owing to the motorbike ride incurring shock vibrations, which is often underestimated by RMS index (Chen et al., 2009). Especially due to long pauses and egress periods during mail delivery tasks, the RMS value computed in an average sense were significantly smaller for mail carriers than for continual 20-km motorcycle rides described by Taiwan IOSH ($p<0.001$, t-test).

The daily WBV exposure, VDV_d and S_{ed}, obtained in this study had a distribution range similar to that as VDV and S_e measured in 20-km motorcycle ride tests reported by Taiwan IOSH (Fig 2), despite the mean duration of daily WBV exposure (4.9 hours) in the mail carriers was approximately 4 folds higher than that for a continuous 20-km ride. This observation suggests the daily VDV and S_e for the mail carriers cannot be accurately estimated by "extrapolating" WBV data acquired in regular motorcycle rides. In contrast with normal motorcycle riders, the relatively slow riding speed of mail carriers and a greater weight bearing of postal motorbikes may introduce a possible estimation error.

CONCLUSIONS

Questionnaire survey results indicated a highly subjective feeling of regional pain in the mail carriers possibly attributed to WBV exposure. This study determined the WBV exposures of 25 mail carriers delivering post in an urban area. Analytical results based on ISO 2531-5 (2004) suggest a high possibility of adverse health effect among certain senior mail carriers. Experimental results further demonstrate that using the *RMS* value alone can underestimate the health risk of mail carriers exposed to shocks. This finding also suggests that the ISO 2631-1 (1997) standard is less stringent than the ISO 2631-5 (2004) standard in terms of assessing the health risk of mail carriers who use motorbikes as major transportation mode.

ACKNOWLEDGEMENTS

The authors wish to thank the National Science Council and the Institute of Occupational Safety and Health, Taiwan for financially supporting this research study (NSC96-2213-E-324-025- MY3, IOSH98-H317). Ted Knoy is appreciated for his editorial assistance.

REFERENCES

Bongers, P.M., Boshuizen, H.C., Hulshof, T.J., and Koemeester, A.P. (1988), "Back disorders in crane operators exposed to whole-body vibration." *International Archives on Occupational Environmental Health*, 60(2), 129–137.

Boshuizen, H.C., Bongers, P.M., and Hulshof, C.T., (1992), "Self-reported back pain in fork-lift truck and freight-container tractor drivers exposed to whole-body vibration." *Spine*, 17, 59–65.

Bovenzi, M., and Zadini, A. (1992), "Self-reported low back symptoms in urban bus drivers exposed to whole-body vibration." *Spine*, 17, 1048–1059.

Chen, J.C., Chang, W.R., Shih, T.S., Chen, C.J., Chang, W.P., Dennerlein, J.T., Ryan, L.M., and Christiani, D.C. (2003), "Predictors of whole-body vibration levels among urban taxi drivers." *Ergonomics*, 46, 1075-1090.

Chen, H.C., Chen, C.Y., Lee, C.L., Wu, H.C., and Lou, S.Z. (2006), "Data logging and analysis tools for worksite measurement of physical workload", proceedings of the Sixteenth World Congress of the IEA, Maastricht.

Chen, H.C., Chen, W.C., Liu, Y.P., Chen, C.Y., and Pan, Y.T. (2009), "Whole-body vibration exposure experienced by motorcycle riders: An evaluation according to ISO 2631-1 and ISO 2631-5 standards." *International Journal of Industrial Ergonomics*, 39(5), 708–718.

Donati, P. (1998), "A procedure for developing a vibration test method for specific categories of industrial trucks." *Journal of Sound and Vibration*, 215, 947–57.

Frymoyer, J.W., Newberg, A., Pope, M.H., Wilder, D.G., Clements, J., and MacPherson, B., (1984), "Spine radiographs in patients with low-back pain: An epidemiology study in men." *Journal of Bone & Joint Surgery- American volume*, 66, 1048–55.

Griffin, M.J. (1998), *General hazards: vibration, Encyclopedia of Occupational Health and Safety*. International Labour Organization Geneva, 50.2–50.15.

International Organization for Standardization (1997), *ISO2631-1. Mechanical vibration and shock– evaluation of human exposure to whole-body vibration. Part 1: General requirements*. Geneva: ISO.

International Organization for Standardization (2004), *ISO2631-5. Mechanical vibration and shock– evaluation of human exposure to whole-body vibration. Part 5: Method for evaluation of vibration containing multiple shocks*. Geneva: ISO.

IRF (2006), *World road statistics – data 1999 to 2004*. International Road Federation (IRF), Geneva, Switzerland.

Malchaire, J., Piette, A., and Mullier, I. (1996), "Vibration exposure on fork-lift trucks." *Annals of Occupational Hygiene*, 40, 79–91.

Mansfield, N. J., and Griffin, M. J. (2002), "Effects of posture and vibration magnitude on apparent mass and pelvis rotation during exposure to whole-body vertical vibration." *Journal of Sound and Vibration*, 253, 93–107.

Pirtte, A. and Malchaire, J. (1992), "Technical characteristics of overhead cranes influencing the vibration exposure of operators." *Applied Ergonomics*, 22, 121–127.

Pope, M., Magnusson, M., Lundström, R., Hulshof, C., Verbeek, J., and Bovenzi, M. (2002), "Guidelines for whole-body vibration health surveillance." *Journal of Sound and Vibration*, 253(1), 131–167.

Seidel H. (1993), "Selected health risks caused by long-term whole-body vibration." *American Journal of Industrial Medicine*, 23, 589–604.

Taiwan IOSH (2007), *Development of software and field testing tools for assessing whole-body vibration with shocks*. Research Report No. IOSH 96–H318, Taiwan Institute of Occupational Safety and Health (IOSH), Council of Labor Affairs, Taiwan. (in Chinese)

Taiwan IOSH (2008), *Onsite assessment and personnel training program of whole-body vibration exposure*. Research Report No. IOSH 97–H317, Taiwan. Institute of Occupational Safety and Health (IOSH), Council of Labor Affairs, Taiwan. (in Chinese)

Wasserman, D.E., Wilder, D.G., Pope, M.H., Magnusson, M., Aleksiev, A.R., and Wasserman, J.F. (1997), "Whole-body vibration exposure and occupational work-hardening." *Journal of Occupational and Environmental Medicine*, 39, 403–407.

Chapter 50

The Relationship Between The Subjective Perception Of Physical Capability And Muscular Strength During Repetitive Lifting Tasks

Ren-Liu Jang

Department of Industrial Engineering and Management
Mingchi University of Technology
Taiwan, R.O.C.

ABSTRACT

The main objective of this study was to investigate the relationship between the subjective perception of physical capability and muscular strength during repetitive lifting tasks. The subjective perception of physical capability was defined as the sustainability that was the duration the individual was willing to perform based on his/her muscular strength. RPE-10 scale was used to report the change of perceived level of exertion. The results found that the exertion for the lifting task from floor level was mainly from legs muscle strength and the exertion for the lifting task from 50 cm above the floor was from torso muscle strength. The sustainability for the lifting tasks was affected by stature, lifting posture, and muscular strength. The duration showed highly correlated with the RPE value.

Keywords: Physical Capability, Muscular Strength, Repetitive Lifting Tasks

INTRODUCTION

In Taiwan, there were many convenient stores most located at major intersections in the cities. To quickly fulfill the demand of goods to the stores, order selectors in the distribution centers or warehouses often worked with tight schedules. Jang and Chen evaluated manual material handling in warehouse and found that the risk factors for the order-selecting tasks from storage area to a pallet cart were the weight of cases and the working environment. Overreaching was often seen when the case on the back row was lifted (Jang, 2002). In such conditions, the order selectors might suffer low back pain in the long term and that eventually turn into low back disorders.

Low back pain has been widely recognized as the main occupational health problem in most industrialized countries. Many studies had identified loads, awkward postures and trunk motions as related risk factors for developing low-back disorders suggested reducing the physical workload to lessen the risk of low back injuries. (Ayoub, 1997; Kuiper, 1999; Marras, 1997; Marras, 1995; Neumann, 2001; Punnett, 1991).

Karwowski et al. simulated all possible lifting conditions to obtain the range of the recommendation weigh limit (RWL) of NIOSH. The results suggested that the proper weight to be lifted for 1 hour period should be less than 13 kg, 12.5 kg for 1 to 2 hours and 10.5 kg for 8 hours period (Karwowski, 1999). Wright and Haslam performed a field study in a soft drinks distribution centre to investigate the risk of manual handling in order-selected and delivery. The results suggested the weight of the case full with bottles of soft drink should be reduced according to the RWL of NIOSH (Wright, 1999).

Granata et al. studied the effect of repeated lifting exertions and found that the spinal load increasing in an upward trend from one exertion to the next identical task. Boxes' weight and personal experience influenced the trunk moments generated during the lifting phrase (Granata, 1999). Davis compared the spinal loads generated during lifting tasks with during lowering tasks under the combination of different positions, trunk velocities and weights. The results indicated that lowering strength needed was about 56% greater than lifting strength. The maximum compression forces in lowering motions were greater than that in lifting exertions. The difference was approximately 600 N (Davis, 1998).

Jorgensen et al. used the engineering intervention to influence workers to avoid using the awkward postures to lift. The results showed that workers lifted faster to compensate the walking time for the increased distance between the pallets in fear of losing the productivity (Jorgensen, 2005). Lavender et al. quantified the peak dynamic bending moments on the spine during lifting as a function of the initial height of lift, the lifting speed and the load's magnitude. The results found that the peak moments were significantly greater when lifting from lower height (7.5cm from the floor), at faster lifting speeds (lift as fast as possible), and with heavier loads (300N) (Lavender, 2003).

Khalil et al. examined the differences among four techniques for the determination of lifting abilities and proposed a new psychophysical submaximal isometric test that was based on the concept of acceptable maximum effort which was the level of static exertion the individual was willing to perform voluntarily and comfortably without over-exertion. This new approach indicated that people tended to overestimate their ability to lift an object when no motion involved such as MVC values. The results also recommended that MVC be used only to establish subjects' muscular strength. Maximum acceptable weight (MAW) yielded the best estimate of the maximum acceptable level at which the individual was willing to exert (Khalil et al., 1987).

Many studies had investigated workers' ability from biomechanical, physiological and psychophysical points of view and recommended that weight of goods to be lifted be reduced to a reasonable level and lifting tasks to be performed match the capabilities and limitations of workers. However, individual differences such as unwillingness to do labor work and weak muscle strength nowadays may psychologically influence their physical capability when performing lifting tasks. The study was to investigate the relationship between the subjective perception of physical capability and muscular strength during repetitive lifting tasks through the understanding of how workers exerted their muscular strength to the job performed and how they perceived their sustainability during the repetitive lifting task as well.

METHODS

APPROACH

Two lifting tasks were selected to be performed because they were two worse cases often seen in the distribution center. One was lifting from floor level to 75 cm above the floor at the frequency of 6 lifts per minute; the other was lifting from 50 cm above the floor (about knee height) to 75 cm above the floor at the frequency of 10 lifts per minute. The case used in this study was the case most frequent being lifted in the distribution center which contained 24 bottles of water or soft drink and weighed about 14.7kg. The case dimension was $40.5 \times 26.5 \times 23.5$ cm (length ×width× height) with no handle. To make sure the experimental setup was proper, 1991 NIOSH equation was used to calculate the Lifting Index (LI) for these two tasks. LI for the task lifting from floor level was 1.24; LI for lifting from knee height with high frequency was 1.5. Both lifting tasks met the requirement that workers need to make enough effort but won't cause them a high risk of muscular discomfort.

PARTICIPANTS

Five participants from a local university of technology (mean age 21.4 (SD 2.6) years; mean stature 168.2 (SD 5.4) cm; mean body mass 65.1 (SD 3.5) kg) voluntarily participated this study. The participants were no material handling experience and had no musculoskeletal discomfort or prior musculoskeletal disorders of the torso at the time participating in the experiment. The objective and procedures of the study were explained before the start of the experiment. They were also informed that they can stop the experiment anytime when they felt discomfort during the experiment.

PROCEDURE

Participants' maximum static lifting strengths and voluntary exertions were measured using load cell (Jackson Strength Evaluation System, JSES Model 32628, U.S.A.). The participant was asked to pull a handle against a stationary chain to the maximum level of voluntary exertion according to the testing posture and maintain their maximum voluntary exertion for a period of 3-5 sec. Therefore, the peak value obtained was recorded as the maximum voluntary contraction (MVC). Three testing postures in the experiment were arm lift, leg lift and torso lift. A week later, the participant was asked to perform a 20-min session of one of lifting tasks randomly and the other lifting task in another day. In each session, the participant continuously lifted and transferred a 14.7 kg case without handles to the table, a fixed surface (75 cm above the floor) on the right side, and a minor twisting motion of the trunk occurred during the transferring. The frequency of lifting was regulated by playing the record of the sound of metronome at 60 Hz. mixing with a sound of beeping to signal the participant when to start the lifting task. No instruction was given on how to perform the lifting task. The participant was allowed to use his preferred way to lift. The subjective perception of physical capability was defined as the level of difficulty he perceived when performing the lifting task. The participants were asked to continuously report the level in an increasing order of RPE-10 scale (0 means nothing at all; 10 means extremely strong). In the period of 20 minute, the participant once perceived the level of difficulty increasing and immediately reported 'level up' until it reached 10. The whole lifting task session was video-taped from the sagittal view. The heart rate was also recorded during the whole trial. In the end of the experiment, the participants were asked to assign a percentage for each muscle strength to indicate its contribution for the lifting tasks.

RESULTS

MUSCLE STRENGTH

The average static strength (N) for arms muscle group, legs muscle group, and torso muscle group, were 199 ± 58, 648 ± 227, and 313 ± 68, separately. The participants muscle strength had only 52%, 68%, and 57% of average value of US industrial workers, compared to average static strength values from Chaffin, et al. (1977).

THE DISTRIBUTION OF MUSCLE GROUP STRENGTH

Table 1 showed the percentage of exertion from each muscle group when performing repetitive lifting tasks. On the average, when lifting from floor level, the participants perceived using 49 percentage of leg muscle strength. When lifting from knee height, the participants perceived using 49 percentage of torso muscle strength. Such results indicated the participants were well aware of adjusting their strength usage to the different work requirement.

Table 1. Exertion from each muscle group when performing lifting tasks
(unit: percentage)

Subject	Lifting from floor level			Lifting from 50 cm above floor level		
	Arms	Legs	Torso	Arms	Legs	Torso
1	15	35	50	40	0	60
2	20	60	20	20	40	40
3	30	40	30	40	40	20
4	25	50	25	30	10	60
5	25	60	15	25	10	65

THE COMPRESSING FORCE ON LOWER BACK

The postures for each lifting task were selected at 1 picture per min for 20 minutes from the video tape. Trunk angles, limb angles and hand positions were estimated by reviewing the videotape. The University of Michigan's 3D static Strength Prediction Program (3DSSPP) was used to estimate the compression force on the low-back. Figure 1 showed the postures in two different lifting tasks. The average compressing force on lower back for lifting from the ground level was 3392 ± 310 N and for lifting from knee height was 2240 ± 244 N.

HEART RATE

The mean of heart rate lifting from ground level was 129.8 beats/min; the mean of heart rate lifting from knee height was 123.3 beats/min. There was no significant difference between two lifting tasks.

THE SUBJECTIVE PERCEPTION OF PHYSICAL CAPABILITY

The subjective perception of physical capability was defined as the sustainability that was the duration the individual was willing to perform based on his/her muscular strength. Figure 2 and 3 showed how the duration changed as a function of RPE value. The sustained time showed highly correlated with the RPE value (the correlation coefficient ranged from 0.94 to 0.99).

DISCUSSION

The study was to investigate the relationship between the subjective perception of physical capability and muscular strength during repetitive lifting tasks. The sustainability was defined how much time psychologically they were willing to sustain to continuously perform the tasks. When the participant reported level up to 10 in RPE-10 scale, it meant he encountered extremely strong feeling on the lifting task. This suggested that his ultimate physical capability had reached. Figure 2 and 3 showed, if the correlation between RPE value and time can hold, it is possible to develop a regression model to predict the limitation of sustainability which could be used as a reference point when to take a rest. For example, most participants can sustain more than 20 minute when lifting from floor level but subject 3 may not sustain for 20 minute. The results also showed the participants while lifting cases from ground level had higher heart rate to meet the physiological need and used more leg strength; the workers while lifting cases from knee height had lower heart rate and used more torso strength. However, when the workers' strength is weak on the demand, they would adjust the working posture to utilize the stronger muscle to acclimate to the lifting task.

CONCLUSIONS

Two lifting tasks were used to examine how participants exerted their muscular strength to the job performed and how they perceived their sustainability during the repetitive lifting task as well. The results showed the sustainability for the lifting tasks was affected by stature, lifting posture, and muscular strength. The duration showed highly correlated with the RPE value.

Figure 1. Lifting form floor level (left) and lifting from 50 cm above the floor (right)

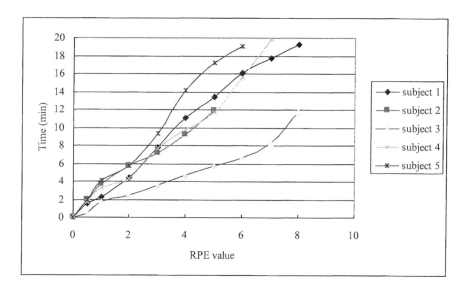

Figure 2. The relationship between RPE value and time when lifting form floor level

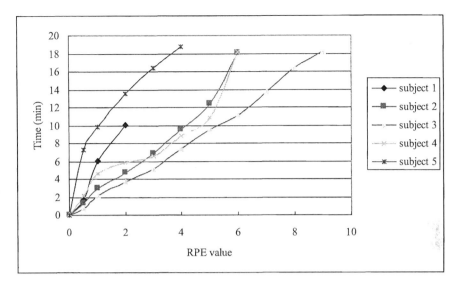

Figure 3. The relationship between RPE value and time when lifting from 50 cm above the floor

ACKNOWLEDGEMENTS

This research was supported by Mingchi University of Technology, Taiwan, R.O.C. (98-M-11).

REFERENCES

Ayoub, M.M., Dempsey, P.G. and Karwowski, W. (1997). Manual material handling, in *Handbook of Human Factors & Ergonomics*, G. Salvendy, Ed., John Wiley, New York, 1085-1123.

Chaffin, D.B. Herrin, G. Keyserling, W.M. and Foulke, J., 1977, *Pre-employment strength testing*, CDC-99-74-62. (Cincinnati: Department of Health, Education, and Welfare, NIOSH)

Davis, K. G., Marras, W. S. and Waters, T. R. (1998). "Evaluation of spinal loading during lowering and lifting." *Clinical Biomechanics*, 13, 141-152.

Granata, K. P., Marras, W. S. and Davis, K. G. (1999). "Variation in spinal load and trunk dynamics during repeated lifting exertions." *Clinical Biomechanics*, 14, 367-375.

Jang, R. L. and Chen, A. (2002). *The evaluation of manual material handling in warehouse*, Research report for Institute of Occupational Safety and Health, Taiwan, R.O.C. (In Chinese)

480

Jorgensen, M.J., Handa, A., Veluswamy, P. and Bhatt, M. (2005). "The effect of pallet distance on torso kinematics and low back disorder risk." *Ergonomics*, 48, 949-63.

Karwowski, W., Gaddie, P., Jang R. and Lee, W.G. (1999). A population-based load threshold limit (LTL) for manual lifting tasks performed by males and females, *The Occupational Ergonomics Handbook*, ed. W. Karwowski and W.S. Marras, CRC press.

Kuiper, J.I., Burdorf, A., Verbeek, J., Frings-Dresen, M., Beek, A.J., Viikari-Juntura, E. (1999). "Epidemiologic evidence on manual materials handling as a risk factor for back disorders: a systematic review." *International Journal of Industrial Ergonomics*. 24, 389-404.

Khalil, T.M., Waly, S.M., Genaidy, A.M. and Asfour, S.S. (1987). "Determination of lifting abilities: a comparative study of four techniques." *American Industrial Hygiene Association Journal*, 48, 951-956.

Lavender, S. A., Andersson, G. B. J., Schipplein, O. D. and Fuentes, H. J. (2003). "The effects of initial lifting height, load magnitude, and lifting speed on the peak dynamic L5/S1 moments." *International Journal of Industrial Ergonomics*, 31, 51-59.

Marras, W., Lavender, S., Leurgans, S., Fathallah, F, Ferguson, S., and Allread, W. (1995). "Biomechanical risk factors for occupationally related low back disorders." *Ergonomics,* 38, 2, 377-410.

Marras, W. and Granata, K. (1997). "Spine loading during trunk lateral bending motions." *Journal of Biomechanics*, 30, 697-703.

Neumann, W.P., Wells, R.P., Norman, R.W., Frank, J., Shannon, H., Kerr, M.S., the OUBPS Working Group. (2001). "A posture and load sampling approach to determining low-back pain risk in occupational settings." *International Journal of Industrial Ergonomics*. 27. 65-77.

Punnett, L., Fine, L. J., Keyserling, W. M., Herrin, G. D., and Chaffin, D. B. (1991). "Back disorders and nonneutral trunk postures of automobile assembly workers." *Scand J Work Environ Health*, 17, 5, 337-346.

Wright, E. J. and Haslam, R. A. (1999). "Manual handling risks and controls in a soft drinks distribution centre." *Applied Ergonomics*, 30, 311-318.

Chapter 51

Good Ergonomic Setup for Healthy Posture at Work

Singh Sucheta,[1] Sharma Promila[2]

[1]Directorate of Extension Education
G.B.P.U.A. & T., Pantnagar
Uttarakhand,-263145, India

[2]Departmentt of Family Resource Management
College of Home Science, G.B.P.U.A. & T., Pantnagar
Uttarakhand,-263145, India

ABSTRACT

Posture is the position in which one holds his body upright against gravity while standing, sitting or lying down. Poor posture can affect the health of a person as there exists a cramped position of the internal organs in the body. These organs consist of varied abdominal, the lungs and the heart. The body cannot function normally if the organs are pushed against each other by the presence of poor posture. The circulation of the body is slowed down and ligaments can be strained. When the body isn't functioning normally then the diminished blood flow can lead to chronic disorders. When the blood isn't allowed to flow as it properly should then other parts of the body can become congested with blood. The pelvic organs can suffer if other organs are sagging. If a person tries to read by looking down too far or by looking up too high then the eyeballs can become distorted. Poor sitting habits can lead to bad reading habits and also then to defective vision. Neuromuscular fatigue can result from poor posture and cause the body to have less energy. The joints become strained and become also painful. Often in later years a person can develop severe pain from backaches caused by many years of poor posture. A person can also become "pot bellied" with constant fatigue and stretching of nerves in the spinal cord. The abdominal organs in the body can also sag. Sometimes the result of poor posture can be a great deal of foot pain as the feet fatigue very easily. Good posture involves training body to stand, walk, sit and lie in positions where the least strain is placed on supporting muscles and ligaments during movement or weight-bearing activities. Proper posture: Keeps bones and joints in the correct alignment so that muscles are being used properly, helps decrease the abnormal wearing of joint surfaces that could result in arthritis, decreases the stress on the ligaments holding the joints of the spine together, prevents the spine from becoming fixed in abnormal positions, prevents fatigue because muscles are being used more efficiently, allowing the body to use less energy, prevents strain or overuse problems, prevents backache and muscular pain and contributes to a good appearance. Many of the agricultural domain viz., farming, industries, academics, etc. includes the posture of carrying load, static posture in standing, bending or squatting position which needs to be taken care off.

Keywords: Ergonomics, healthy posture, Back problems, agricultural domain and postural problems

INTRODUCTION

The word "ergonomics" has been derived from Greek word which means Ergon i.e. work and Nomos i.e. rule. Branch of science that deals with relationship between work, worker and environment is known as ergonomics. This relationship has to be very close for better work efficiency, maximum body comforts and good health of workers.

Workers in the agriculture are most at risk to lower back disorders, and those in agriculture, forestry, and fisheries face the greatest risk of work related upper limb disorders. Health problems range from discomfort, minor aches and pains to more serious medical conditions requiring time off work, medical and hospital treatment. In more chronic cases, treatment and recovery are often unsatisfactory, and the result can be permanent disability, with loss of job. Most common cause of neck problems is the result of static muscle work and faulty postures (Greico, 1986). Work-related musculo skeletal disorder (WMDs) accounted for 60-86 percent of all occupational illness among workers. Early symptoms of musculo skeletal disorders are referred to as unpleasant sensation or discomfort associated with fatigue, perceived exertion and poor posture along with symptoms of pain and discomfort in the shoulders, elbows, lower back, hips, knees, wrist and hands. (Bureau of labour statistics, 1990 and Pinzke and Gutsafsson, 1995). Musculo Skeletal Disorders (MSDs) can occur across all types of jobs and work sectors in agriculture, but:

- Manual workers, both skilled and unskilled, are most at risk;
- Upper limb disorders affect women workers more than men largely because of the type of work they do;
- Older workers report more MSD problems;
- Workers in precarious employment, such as those on fixed-duration or employment agency contracts, are significantly more exposed to repetitive work and working in painful or tiring positions.

GOOD POSTURE

Posture is the position in which you hold body upright against gravity while standing, sitting or lying down. Good posture involves training body to stand, walk, sit and lie in positions where the least strain is placed on supporting muscles and ligaments during movement or weight-bearing activities. Proper posture:

- Keeps bones and joints in the correct alignment so that muscles are being used properly.
- Helps decrease the abnormal wearing of joint surfaces that could result in arthritis.
- Decreases the stress on the ligaments holding the joints of the spine together.
- Prevents the spine from becoming fixed in abnormal positions.
- Prevents fatigue because muscles are being used more efficiently, allowing the body to use less energy.
- Prevents strain or overuse problems.
- Prevents backache and muscular pain.
- Contributes to a good appearance.

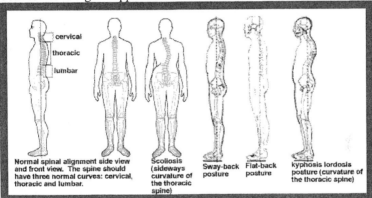

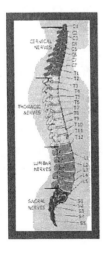

FIGURE 1 PHYSIOLOGICAL ASPECTS OF A POSTURE

REQUIREMENTS OF GOOD POSTURE

1. Correct Sitting Position

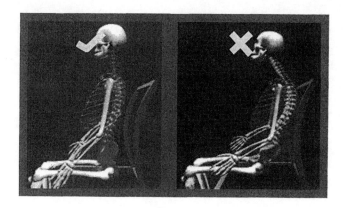

FIGURE 2 Correct and wrong sitting posture

- Sit up with back straight and shoulders back. Buttocks should touch the back of chair.
- All three normal back curves should be present while sitting. A small, rolled-up towel or a lumbar roll can be used to help you maintain the normal curves in back.

Here's how to find a good sitting position when not using a back support or lumbar roll:

- o Sit at the end of chair and slouch completely.
- o Draw body up and accentuate the curve of back as far as possible. Hold for a few seconds.
- o Release the position slightly (about 10 degrees). This is a good sitting posture.
- Distribute body weight evenly on both hips.
- Bend knees at a right angle. Keep knees even with or slightly higher than hips. (use a foot rest or stool if necessary). Legs should not be crossed.
- Keep feet flat on the floor.
- Try to avoid sitting in the same position for more than 30 minutes.
- At work, adjust chair height and work station so you can sit up close to work and tilt it up at you. Rest elbows and arms on chair or desk, keeping shoulders relaxed.
- When sitting in a chair that rolls and pivots, don't twist at the waist while sitting. Instead, turn whole body.
- When standing up from the sitting position, move to the front of the seat of chair. Stand up by straightening legs. Avoid bending forward at waist. Immediately stretch back by doing 10 standing backbends.

2. Correct Driving Position

- Use a back support (lumbar roll) at the curve of back. Knees should be at the same level or higher than hips.
- Move the seat close to the steering wheel to support the curve of back. The seat should be close enough to allow knees to bend and feet to reach the pedals.

3. Correct Lifting Position

- If you must lift objects, do not try to lift objects that are awkward or are heavier than 30 pounds.
- Before you lift a heavy object, make sure you have firm footing.
- To pick up an object that is lower than the level of waist, keep back straight and bend at knees and hips. Do not bend forward at the waist with knees straight.
- Stand with a wide stance close to the object one is trying to pick up and keep feet firm on the ground. Tighten stomach muscles and lift the object using leg muscles. Straighten knees in a steady motion. Don't jerk the object up to the body.
- Stand completely upright without twisting. Always move feet forward when lifting an object.
- If you are lifting an object from a table, slide it to the edge to the table so that you can hold it close to body. Bend knees so that body is close to the object. Use legs to lift the object and come to a standing position.
- Avoid lifting heavy objects above waist level.
- Hold packages close to body with arms bent. Keep stomach muscles tight. Take small steps and go slowly.
- To lower the object, place feet as to lift, tighten stomach muscles and bend hips and knees.

POSTURE PROBLEMS

The major causes for poor posture need to be mentioned as these can be due to many factors including inherited or organic defects, faulty habits, clothing or fatigue. There are many people who have inherited limbs of uneven length, crooked spines, defective bones, weak muscles, or deformities such as bowed legs or constitutional disorders. Sometimes even malnutrition especially during the childhood formative years can be a contributing factor in poor posture. For the majority of people having poor posture it is due most likely to faulty or bad habits in sleeping, walking or sitting. Fatigue can be a common cause of bad posture as habits of slumping can be from muscle relaxation. Having a profession where one deals with delicate work requiring very close concentration can be a factor contributing to poor posture. Sometimes a person can also have poor vision that leads to faulty posture as it would require leaning over to read.

There is a common type of poor posture caused by postural defects due to curvature of the spine. There exists in the human body four curves, two are present at birth and these are the sacral and thoracic. As a person matures and grows two others are added, the lumbar and cervical. With these four curves the balance of the body becomes maintained. If the spine if perfectly straight these curves can be seen when viewed from the side. This includes both a view of the front or the view from behind a person.

Poor posture can affect the health of a person as there exists a cramped position of the internal organs in the body. These organs consist of varied abdominal, the lungs and the heart. The body cannot function normally if the organs are pushed against each other by the presence of poor posture. The circulation of the body is slowed down and ligaments can be strained. When the body isn't functioning normally then the diminished blood flow can lead to chronic disorders. When the blood isn't allowed to flow as it properly should then other parts of the body can become congested with blood. The pelvic organs can suffer if other organs are sagging. If a person tries to read by looking down too far or by looking up too high then the eyeballs can become distorted. Poor sitting habits can lead to bad reading habits and also then to defective vision. Neuromuscular fatigue can result from poor posture and cause the body to have less energy. The joints become strained and become also painful. Often in later years a person can develop severe pain from backaches caused by many years of poor posture. A person can also become "pot bellied" with constant fatigue and stretching of nerves in the spinal cord. The abdominal organs in the body can also sag. Sometimes the result of poor posture can be a great deal of foot pain as the feet fatigue very easily.

FACTORS AFFECTING POSTURE

The manner in which you sit, stand, or walk will determine the quality of posture. The efficiency of body will depend on having good posture. The weight of body will be distributed more evenly along the body's center of gravity. Each bodily activity such as sitting, reclining, running, or walking will require a different posture. Each activity if accomplished correctly will allow freedom of movement with the least use of energy.

Many elements can effect posture such as diet, weight, feelings, sleeping habits and fitness. Often people who are tall will walk with their heads down and their body bent thinking others will see them as not being so tall. When standing, the body should be erect but not stiff, controlled and also balanced. Support should not be necessary such as standing against a wall or a chair. One might practice standing in front of a mirror to obtain better posture. The weight of the body can be shifted to one leg or the other and can cause a standing posture that is a "hunched" position leading to bad posture.

Physical causes of musculoskeletal disorders (MSD) include:

- Manual handling
- Loads
- Poor posture and awkward movements,
- highly repetitive movements,

- forceful hand applications,
- direct mechanical pressure on body tissues,
- vibrations, and
- cold work environments.

Causes due to the organization of work include:

- pace of work,
- repetitive work,
- time patterns,
- payment systems,
- monotonous work, and also
- psychosocial work factors.

Special attention must be given to the feet as they are very important to having good posture. Wearing shoes that are comfortable and fit and also being aware of these features instead of adhering to the current style will help posture. There are so many shoes to be found on the market today that are not proper footwear as they can easily cause back strain, poor posture, and also injuries to the foot. A common foot problem known as the fallen arch, or flattening of the arch of the foot can cause pain in the arch, ankle, and also in the lower leg muscles and can be a leading factor in poor posture.

EXERCISES FOR HEALTHY POSTURE

- Having regular natural exercise daily including in activities that help to maintain good posture can be very beneficial. There are a few exercises that can strengthen muscles such as the ones below.
- The body can be trained to keep parts in proper alignment by standing straight with the back against a wall letting the head, shoulders and hips touch the wall. The heels should be about four inches from the wall, no less than three. The back can slide up and down the wall as far as possible while keeping balance.
- For the spine to be exercised and to remove excess curves, lie on the back and bend the knees over the body, bringing the forearms around the knees and clasping the tips of the hands together holding the legs together and rocking the body from side to side.
- The abdominal muscles can be strengthened by sitting on a low bench with the feet under the edge of a bed, then lying back on the bench slowly lift the trunk of the body with the abdominal muscles. This will also help an over bulging abdomen.
- Lying on the back with the lower legs resting on a chair with the arms folded across the chest. Lifting the hips a few inches off the floor first letting the weight rest on the shoulder blades, returning the hips to the floor, taking deep breaths. Then by slowly exhaling the air the spinal curvatures will be strengthened. Another exercise in this same position can be achieved by

extending the legs on the floor, taking a deep breath, and then raising the chest high and contracting the abdomen.

- Bicycle exercises can be done with the legs in the air, using first one leg and then the other to pedal in large circles. The legs must be kept in the same plane and the knees must not touch as they pass from one to another. The legs should not swing sideways in their movement.
- To exercise the spine and to remove curves, one can lie on the back and bend the knees over the body, bringing the forearms up around the knees and clasping the tips of the hands together to hold the legs together then rocking the body from side to side like a cradle.
- Excellent posture can be obtained by regular exercise as stated above and by just being aware of the need for practicing good posture habits. The quality of a person's life especially in later years can be determined by a lifetime of good posture.

HOW TO CORRECT POSTURE AT WORK

Step 1: Sit nice and tall in desk chair. It may seem obvious, but it is important think about posture at work. Posture problems develop from bad habits. Sit tall to correct posture problems at desk before they start.

Step 2: Tighten core to control posture. Bracing midsection helps take pressure off of the lower back. It helps to take a deep breathe in and brace abdominal muscles.

Step 3: Keep chest up to improve posture. Visualize a string attached to chest and the ceiling. This helps you have better posture at computer.

Step 4: Pull shoulders back and down. Slouching the shoulders deteriorates the supporting muscles of the spine. This is a where a posture correction brace can really help.

Step 5: Consider a posture corrective brace to promote proper alignment of the back and shoulders.

If you work at a desk, consider stretching throughout the day to keep muscles flexible and avoid injury. In this way, minimizing the work related discomfort may improve workers performance, increase output, decrease production costs and increase are workers satisfaction Sehoenmarklin et. al. (1989).

REFERENCES

Bureau of labour statistics, (1990), "A report on the survey of occupational injuries and illness". U. S. Department of labour. 92,731-732

Greico, A., (1986), "Sitting posture: An old problem and a new one". *Ergonomics*, 29, pp. 345-362.

Pinzke, S. and Gutsafsson, B., (1995), "Working environment in dairy borns". Part 3. Musculoskeletal problems in Swedish dairy farmers, Report: Department of Agriculture bio-systems and technology. Swedish University of Agriculture Sciences, pp. 95-96.

Sehoenmarklin, R. W. and Marrac, W. S., (1989), "Effect of handle work orientation on hammering". *Human factors*. 31,4, pp. 413-420.

Chapter 52

Ergonomic Assessment of Dairy Activities Performed by Rural Women of Punjab (India)

Krishna Oberoi, Shivani Sharma, Poonam Kataria

Punjab Agricultural University
Ludhiana 141004, Punjab, India

ABSTRACT

Participation of women in dairy sector was found as high as 75 million women against 15 million men. In spite of the fact that women participation in dairy activities is very high; their role in livestock production has often been under estimated or ignored. Tools to perform dairy activities are rarely designed as per the needs of women and they often work with faulty tools which are responsible for higher physiological and muscular stress. Keeping these problems in mind, the present study was undertaken to know the percentage of women involved in dairy sector, time spent and the difficulties faced by them. Most time consuming, frequently performed and drudgery prone activities performed by women were identified on the basis of collected data and two improved tools viz, sickle for cutting fodder and low height revolving woven stool for sitting while milking animals were designed as per the ergonomic needs of women. The ergonomic assessment of improved tools designed for the women involved in drudgery prone activities was done by using suitable ergonomic tools and techniques. The impact of these two improved tools designed for women were assessed on the basis of reduction in ergonomic and muscular stresses. The results of the study showed that out of all the dairy activities performed by the women, cutting & bringing fodder and milking animals were found as most time consuming and difficult tasks and these were performed by the maximum number of women in Punjab. The impact of improved tools was found positive and significant. Rural women are therefore motivated to use the improved tools to reduce their ergonomic and muscular stresses.

Key words: Physiological and Muscular stresses, Drudgery

INTRODUCTION

Dairy sector had played a significant role in generating gainful employment for women particularly for the landless, small and marginal land holding women. Despite the significant involvement and contributions of women in dairy activities, no attention is paid to design the suitable tools for them. They perform most of the dairy activities with the faulty tools in unnatural postures like stooping or squatting and these faulty tools create several musculo-skeletal problems to them, such as pain in back, knee, cervical region etc. Frequent stooping and lifting heavy load in bending posture wear out the spinal column and induce hazards in the supporting system of the spine. The adoption of long static postures like bending and squatting for milking animal, collecting dung, and cleaning animal sheds lead to high ergonomic and physiological costs to the work. Higher ergonomic and physiological costs may lead to irreparable damage to the body of worker as observed in some other related activities (Oberoi *et. al* 1999). It was therefore felt imperative to study the profile of dairy activities performed by the women and to design few ergonomically sound and drudgery reducing tools for women workers and test these tools with the suitable ergonomic techniques in terms of reduced ergonomic and physiological stress. The present study is therefore undertaken with the following objectives:

- To find out the percentage of women involved in the dairy activities and the frequency of performing these activities.
- To assess the time spent and difficulty score in the performance of these activities.
- To find out the drudgery index of performing the said activities.
- To design few ergonomically sound women friendly tools for performing the drudgery dense dairy activities.
- To study the impact of newly designed improved tools in terms of reduced ergonomic and muscular stress.

METHODOLOGY

Methodology for the present study was under taken in two phases. In phase 1 the analysis of the dairy activities was done on the basis of involvement of women in dairy activities and the frequency of performance. Analysis of the activities in terms of time spent and the drudgery caused to the women was also done.

In phase 2, two improved tools; one for cutting fodder and second for sitting while milking animals, were designed on the basis of women's ergonomic parameters and the ergonomic assessment of these tools was done with the suitable ergonomic techniques to know there suitability to women in terms of reduced ergonomic and muscular stress.

Data for the phase 1 of methodology was collected from three agro-climatic zones representative of Punjab state. The data was collected from 500 farm women, selected from different land holding categories, with the help of well-structured, pre-tested interview schedule.

The data were analyzed and assessment of drudgery prone activities was done by consolidating the different parameters viz., time expended per year on each activity, frequency of activity performance and degree of difficulty experienced by the women in performing the activity as explained below in different steps.

STEP I

Time spent in doing each of these activities per year was calculated on the basis of time spent per day and number of days a particular activity is being performed.

STEP II

Frequency of performance of each activity was obtained on five point scale viz. daily (5), alternate day (4), weekly (3), fortnightly (2) and seasonal (1) and calculated the number of days per/month each activity is performed.

STEP III

Difficulty score of each activity performed was obtained on five point scale viz. very easy (1), easy (2), neutral (3), difficult (4) and very difficult (5).

DRUDGERY INDEX

The three parameters viz., time spent (min.), frequency of performance (obtained on five point scale) and difficulty score were compiled by linear combination method as given below after drawing coefficients of each activity and the drudgery Index was calculated.

Co-efficient for Time (x_i)
= Time spent per day for the ith activity/ Total time spent
Co-efficient for Frequency of Performance (y_i)
= Frequency score for the ith activity/ Total weight of frequency scores
Co-efficient for Difficulty score (z_i)
= Difficulty score obtained for the activity/ Total weight of difficulty score

Drudgery Index = $[(x_i+y_i+z_i) / 3]$ x 100

In phase 2 of the methodology, two ergonomically sound women friendly tools given below were designed. The pictures/ design of these tools and important features which were useful in reducing the ergonomic and muscular stress of the women workers are given below.

Figure1. Improved sickles for fodder cutting and fodder cutting activity being performed

The improved sickle has long serrated blade, which works on the friction and sheer principle to cut the plant with less force. Sharpness of the blade is long lasting. Handles of the improved sickles are specially designed to reduce the stress of grip muscles while cutting fodder and also to ensure safety against arms and hand injuries.

Figure 2. Revolving woven stool and the milking activity being performed

The revolving stool for sitting while milking animals was designed as per the sitting height of women. The use of this stool for sitting while milking animals improved the posture of women. The provision of wheels made the movement of the women easy without getting up time and again.

Impact of two drudgery reducing tools viz., improved sickle for cutting fodder manufactured by a local firm of Ludhiana (Falcon Industries) and revolving stool designed by the Scientists of Family Resource Management, PAU for sitting while milking animals was assessed with the suitable ergonomic techniques. Detail of these techniques is given as below

SELECTION OF EXPERIMENTAL SAMPLE

For each selected activity 30 rural women, who were in good state of health, were selected. Suitability of the experimental women was ascertained by measuring the following physiological parameters before the start of the activity.

Body temperature recorded for 3 minutes	: Not above 99°F
Blood Pressure	: 120/80 ± 10
Heart Rate	: 70-90 bpm

Recording of Ergonomic Parameters viz. Heart Rate and Energy Expenditure

The heart rate was recorded with the help of heart rate monitor and the energy expenditure was worked out with the help of following formula

Energy expenditure $=0.159$ x heart rate (bpm) $- 8.72$

Recording of Muscular Stress

Muscular stress of selected drudgery prone activities viz., milking of animals and cutting fodder was measured in terms of four parameters i.e. Grip fatigue, frequency of postural change, angle of deviation of backbone and the incidence of musculo skeletal problems while performing the selected activity. Grip strength was measured by using the Grip Dynamometer, frequency by simple counting the number of time the women changed the posture, angle of deviation by Flexi curve and Body map (Corlette and Bishop, 1976) was used for recording musculo skeletal problems.

Grip Strength

Grip strength was measured separately both for the right hand and left hand and the Grip strength was determined with the use of following formula

Grip strength in $\% = [(Sr - Sw)/ Sr]$ x 100

Where, Sr means strength of muscles in rest and
Sw means strength of muscles in work
The decreased grip strength with the performance of the activity was interpreted as grip fatigue of the muscles.

Postural Analysis

Postural analysis of the lumbo-sacral region during the Performance of the activity was done by subtracting the angle of bend in the normal sitting posture from the angle of bend during the performance of the activity.

Musculo-Skeletal Problems

Incidence of musculo-skeletal problems were identified using the Body Map indicating pain in different parts of the body after the completion of the activity on five-point scale viz., 5, 4, 3, 2 & 1 as very severe, severe, moderate, mild and very mild respectively.

RESULTS

Results for the present study have been discussed phase wise as per the methodology. Tables 1-4 are based on the results of phase 1

Results given in Table 1 show that out of all the dairy activities milk processing was found as the maximum performed dairy activity by all the categories of women viz. landless, marginal & small land holders, medium and large land holder (94.3 %, 94.8 %, 89.8 % and 97.2 5) respectively followed by milking of animals (37.1 %, 61.1 %, 72.9 %, 75 % respectively). Though activities of collecting and bringing fodder for dairy animals were not performed by the women of medium and large land holding but these activities were performed by the majority of landless, marginal and small land holding women. Jain and Verma (1995) also revealed that in Hisar district too, out of the seven operations of dairy farming two operations namely, cleaning of animal sheds and milking of animal (87.5%) and other three operations viz. preparing feed for animals, protection of animals from ticks and lice and offering water to animals (above 80 %) were predominantly performed by rural women.

Table1. Number of women performers and their performance frequency score (PFS)

Dairy activities	Land holding category							
	Landless		Marginal & small		Medium		Large	
	No. (%)	PFS	No. (%)	PFS	No. (%)	PFS	No. (%)	PFS
Collection of fodder	10 (5.1)	4.10	10 (4.7)	3.6	-	-	-	-
Bringing fodder	13 (6.7)	4.61	12 (5.7)	3.91	-	-	-	-
Feeding animals	50 (25.8)	4.96	96 (45.5)	4.96	28 (47.4)	5	13 (36.1)	5
Washing animals	57 (29.4)	4.68	79 (37.4)	4.83	19 (32.2)	4.95	7 (19.4)	5
Grazing animals	17 (8.8)	4.70	12 (5.7)	4.25	4 (6.8)	4.25	-	-
Collecting dung	64 (33)	4.91	69 (32.7)	4.87	8 (13.5)	5	-	-
Cleaning shed	62 (31.9)	4.81	88 (41.7)	4.87	12 (20.3)	5	1 (2.8)	5
Milking animals	72 (37.1)	5.0	130 (61.1)	5	43 (72.9)	5	27 (75)	5
Processing milk	183 (94.3)	4.78	200 (94.8)	4.92	53 (89.8)	5	35 (97.2)	5
Delivering milk	-	-	8 (3.8)	3.5	2 (3.4)	5	1 (2.8)	5

Data pertaining to frequency of performance of dairy activities showed that most of the dairy activities mentioned above were performed daily by all the women of all categories except collecting and bringing fodder and delivering milk which were performed on alternative days by the women of small and marginal land holders.

The data of time spent on each activity has been presented in Table 2 and it is observed that women from landless category spend more time on milking of animals (263.66 hr/yr) followed by bringing fodder (210.08 hr/yr) and spent least time on grazing animals. Milking of animals took more time out of all the other land holding categories too viz. marginal & small (264.3 hr/yr), medium (272.21) and large (353.43 hrs/yr) land holders. Grazing animals was found to be least time consuming dairy activity for landless (94.33), marginal & small (74.03) and medium (78) land holding women whereas large land holding category of women spent least time on feeding animals (142.95 hr/yr).

Table 2. Time spent (hr/yr) by women on dairy activities

Dairy activities	Landless	Marginal & small	Medium	Large
Collection of fodder	158.76	89.7	-	-
Bringing fodder	210.08	84.48	-	-
Feeding animals	187.83	160.86	174.58	142.95
Washing animals	135.81	155.03	194.48	178.45
Grazing animals	94.33	74.03	78	-
Collecting dung	134.86	129.76	167.28	-
Cleaning of animal shed	122.20	128.53	155.75	182.5
Milking of animal	263.66	264.3	272.21	353.43
Processing of milk	117.53	187.53	227.93	295.65
Delivering milk	-	252.85	136.86	243.33

Table 3 reveals the difficulty score of dairy activities perceived by women of different landholdings and it is observed that activity of collecting fodder was perceived as the most tiresome activity by the women of landless category because the difficulty score of the said activity was found 3.4 as well as 3.3 in case of women of marginal and small land holding category where as cleaning animal shed and delivering milk were considered as most difficult tasks by the women of medium land holding (3.42) and large landholding (5) categories.

Other activities, which were considered difficult to perform, were washing animals by the women of landless category (3.24), medium (3.26) and large (3.57) land holding categories whereas cleaning animal shed (3.19) was considered difficult by the women of marginal and small land holding categories. On the whole it was observed that majority of the activities performed by the women were perceived moderately tiring having difficulty score less than 4 for all the activities

Table 3. Difficulty score of dairy activities perceived by women

Dairy activities	Landless	Marginal & small	Medium	Large
Collection of fodder	3.4	3.3	-	-
Bringing fodder	3.08	3.18	-	-
Feeding animals	3.06	2.62	2.71	3.46
Washing animals	3.24	3.10	3.26	3.57
Grazing animals	3	2.5	1.75	-
Collecting dung	3.06	3.13	3	-
Cleaning shed	3.18	3.19	3.42	3
Milking	3.03	2.73	3	3.04
Processing milk	1.82	1.76	1.81	2.03
Delivering milk	-	3.12	2	5

Data given in table 4 indicate that milking of animals was perceived as the most drudgery prone task by all categories of women viz; landless, marginal, small and medium land holding category. (68.7, 84.9 and 86.7) whereas delivering milk was perceived as most drudgery prone task by the women of large land holding category. Feeding animals (64.3), delivering milk (76), washing animals (78.5) and milking animals (86.9) were perceived as the second most drudgery prone tasks by the women of landless, marginal & small, medium and large land holders.

Table 4. Drudgery index of different activities performed by women

Dairy activities	Landless	Marginal & small	Medium	Large
Collection of fodder	59.1	57.3	-	-
Bringing fodder	63.3	57.9	-	-
Feeding animals	64.3	70.8	72.8	69.9
Washing animals	60.6	72.4	78.5	74
Grazing animals	56.8	54.3	49.5	-
Collecting dung	60.6	69.7	73.8	-
Cleaning shed	60.3	69.9	75.2	70.5
Milking	68.7	84.9	86.7	86.9
Processing milk	50.8	68.2	73.4	74.7
Delivering milk	-	76	63.4	89.6

Data highlighted in the above four tables showed that out of all the dairy activities performed by the farm women, cutting and bringing fodder and the milking animals were found most time consuming and difficult tasks and these two tasks were performed by the maximum number of farm women. In order to reduce the stress of women for performing these above mentioned activities, two improved tools viz, sickle for cutting fodder and low height woven stool shown in the methodology were identified / devised and ergonomic assessment of these tools was done. The results of this section are presented in the tables 5-7.

Table 5. Heart rate and energy expenditure with the use of improved tools /posture

Activity	Working HR (beats/ min.)	Energy expenditure (kj/min.)
Sickle for fodder cutting		
Traditional Sickle	142.32	13.90
Improved Sickle	135.08	12.86
% Reduction	5.08	7.48
Milking of animals		
Traditional posture of squatting	140.17	13.57
Improved sitting posture on revolving stool	134.17	12.61
% Reduction	12.30	12.35

From the results of table 5, it is observed that Heart rate and energy expenditure of the women were reduced when they performed the activities of cutting fodder and milking animals sitting on revolving stool. The reduction of average heart rate and energy expenditure were however found non-significant.

Table 6. Grip fatigue, angle of deviation and postural change with the use of improved tools for performing the selected activities

Activity	Grip fatigue	Angle of deviation	Postural change
Sickle for fodder cutting			
Traditional Sickle	5.78	1°	11.50
Improved Sickle	6.46	0.21°	10.70
% Reduction	11.76	79*	8.69
Milking of animals			
Traditional posture of squatting	20.1	19.33°	13.14
Improved sitting posture on revolving stool	18.5	2.17°	10.10
% Reduction	7.7*	88.77	20.1

Data presented in table 6 shows that with the use of improved sickle and woven stool for cutting fodder and milking animals, grip fatigue and angle of deviation were reduced and reduction was found significant for cutting fodder and for milking animal. It is further observed that with the use of improved sickle and woven stool for sitting during milking, reduction in postural change was also achieved.

Table 7. Intensity of pain in different parts of the body

Body parts	Intensity of pain				
	Very Severe	**Severe**	**Moderate**	**Mild**	**Very mild**
Neck	-		20	50	-
Shoulder joint	-		30	40	10
Upper arm	-		-	20	20
Elbows	-	10	40	30	-
Lower arm	10	20	-	30	-
Low back	-		-	30	10
Wrist/hands	10		-	70	-
Buttocks	-	-	-	-	10
Upper leg/Thigh	-	-	10	10	-
Knees	10		-	50	-
Calf muscles	-		-	20	10
Ankles/feet	40		-	70	-
Upper back	-	-	10	-	-

Intensity of pain in different parts of the body while performing the selected activities is presented in table 7. It is observed that with the use of improved tools maximum of the women felt mild pain in most of the body parts. However few women experienced severe to very severe pain in elbows, lower arms, knees, and feet too, while performing the selected activities.

CONCLUSIONS

From the results of the above study it may be concluded that the ergonomic and physiological stresses of women in performing the drudgery prone dairy activities can be reduced with the use of improved tools. They should therefore be motivated to use these improved tools through different extension programmers' so that they can become comfortable in performing the selected activities and increase the work output.

REFERENCES

Corlette, E.N, and Bishop, R. P. (1976), "A technique for assessing postural discomfort." *Ergonomics*, 19, 175–182.

Oberoi, K., Sidhu, M., Gupta, R., and Varman, P. (1999), Ergonomic *study of farm women's drudgery in home, farm and allied activities.* All India Coordinated Project in Home Science, Department of Family Resource Management , Punjab Agricultural university, Ludhiana.

Jain, V., and Verma, S. K. (1995,) "Decision making pattern of farm families in animal husbandry." *Indian Journal of Social Research*, 36, 215–23.

Chapter 53

Musician Related Cumulative Muscular Disorders as Assessed Using a Taiwanese Questionnaire Survey

Chin-Lung Chen

Department of Visual Communication Design
Mingchi University of Technology
Taiwan, R.O.C.

ABSTRACT

Although work-related musculoskeletal disorders in musicians have received little attention from researchers, growing evidence suggests that musicians are also highly vulnerable to such disorders. Thus this study implemented a preliminary questionnaire to survey over 300 subjects majoring in music at universities in Taiwan to determine the prevalence of cumulative muscular disorders in musicians. The survey revealed a high prevalence of such problems among musicians in Taiwan. Unsurprisingly, symptoms were concentrated in the upper limbs. Furthermore, string and keyboard players had higher incidences of such injuries than other instrumentalists. Most musicians suffered at least one problem involving pain in the arms, fingers, wrist, neck and back. The finger, the parts most closely related to performing technique, exhibited a U shape relationship between rates of disorder and years of experience of performing. Furthermore, the arm, back, neck and wrist, all of which were used to maintain posture while performing, also exhibited disorders in some cases, the rates of disorder increased with years of experience of performance demonstrated that performance posture becomes fixed during the initial period of learning to play an instrument and later cannot easily be changed. On the other hand, the arm, back and neck, all of which contribute to maintaining performance posture, easily suffer disorders as a result of frequent practice, and moreover the increase is quite steep. The prevalence of hearing disorders was lower than others since hearing problems tend to be neglected by musician. So the incidence of hearing disorders was likely under reported.

Keywords: Musicians, Cumulative Disorders

INTRODUCTION

Work-related musculoskeletal disorders cause pain, disability and loss of employment for workers in many occupations. Musicians have rarely been a focus of research on work-related musculoskeletal disorders, but there is growing evidence that musicians are also highly vulnerable to such disorders (Lambert, 1992; Pascarelli & Hsu, 2001).

Over 120,000 children aged 3-12 years old play a musical instrument in Taiwan, but few studies have examined the health implications of practicing music. Therefore, further epidemiological research regarding cumulative muscular disorders that focuses on musicians is regarded. A preliminary questionnaire study was implemented to survey over 300 subjects majoring in music in Taiwan to determine the prevalence of cumulative muscular disorders in musicians. The survey results provide the foundation for further research and preventing strategy.

METHODS

A preliminary, anonymous, retrospective questionnaire was distributed to students and teachers in music departments in Taiwanese colleges as well as among professional musicians working in orchestras in Taiwan. The questionnaire comprised three parts and the questions focused on the present status of respondents with regard to cumulative muscular disorders. The first part dealt with the demographic characteristics of the study sample, while the second part dealt with instrumental practice and performance habits, and the last part was dealing with muscular disorders. The results were analyzed via descriptive statistics and illustrated in tabular form.

On the other hand, since the physical demands involved vary significantly among different instruments, disorder frequency is determined for specific individual instrument types and body parts. Trends in prevalence of disorders relative to practice and playing exposure were also identified for separate body parts.

RESULTS

The study respondents comprised a total of 302 musicians, and were predominantly female (87.1%) and students (88.7%). In terms of age distribution the respondents were concentrated in the 16-24 years old group (91.3%). The largest single instrument represented (in terms of major) was piano (45.7%), while the most prevalent degree of learning experience was 10 years and above (53.3%), and the most common mean frequency of weekly practice sessions was 5-9 times (67.2%).

The prevalence of muscular disorders among musicians in this investigation, as listed in Table 1, is 73.2%. The symptoms of victims were mainly concentrated in the upper limbs (with 27.5% affecting the arms, 22.8% the fingers and 17.5% the wrists), followed by the back (18.2%) and neck (17.9%). With the exception of flute performers (38.1%), over half of all performer complained of muscular disorders. Ranking disorder prevalence by instrument, they followed the order cello (100%), violin (78.6%), piano (76.1%) and clarinet (66.7%).

As listed in Table 2, the proportion of disorders among performers increases with increasing average practice sessions per week (R^2=0.998). Furthermore, the arm (R^2=0.989), back (R^2=0.968), and neck (R^2=0.924) tend to exhibit disorders with increasing practice frequency.

Overall, as listed in Table 3, after just 0-4 years of performance experience a high proportion of participants (62.3%) already suffer muscular disorders. With 5-9 years of performance experience the proportion increases further to 75.0%. However, after that point the prevalence of muscular disorders stabilizes, remaining between 75.0% and 75.8%.

Table 1: Rate of muscular disorder varied with instrument

	Instrument						Total
	Cello	Violin	Piano	Clarinet	Flute	Others	
Body parts							
Arm	58.8[a]	42.9	34.8	25.0	19.0	12.0	27.5
	(10)[b]	(6)	(48)	(3)	(4)	(12)	(83)
Finger	35.3	50.0	20.3	50.0	33.3	15.0	22.8
	(6)	(7)	(28)	(6)	(7)	(15)	(69)
Back	35.3	35.7	22.5	00.0	00.0	13.0	18.2
	(6)	(5)	(31)	(0)	(0)	(13)	(55)
Neck	17.6	35.7	23.9	8.3	9.5	10.0	17.9
	(3)	(5)	(33)	(1)	(2)	(10)	(54)
Wrist	23.5	28.6	26.1	16.7	14.3	4.0	17.5
	(4)	(4)	(36)	(2)	(3)	(4)	(53)
Waist	23.5	21.4	8.0	0.0	0.0	8.0	8.6
	(4)	(3)	(11)	(0)	(0)	(8)	(26)
Palm	41.2	21.4	9.4	0.0	0.0	2.0	8.3
	(7)	(3)	(13)	(0)	(0)	(2)	(25)
Eyes	23.5	7.1	5.8	0.0	0.0	2.0	5.0
	(4)	(1)	(8)	(0)	(0)	(2)	(15)
Ears	0.0	7.1	0.7	0.0	0.0	2.0	1.3
	(0)	(1)	(1)	(0)	(0)	(2)	(4)
Others	5.9	21.4	9.4	58.3	33.3	10.0	13.6
	(1)	(3)	(13)	(7)	(7)	(10)	(41)
Total							
Disorder rate	100.0	78.6	76.1	66.7	38.1	72.0	73.2
n of disorders	17	11	105	8	8	72	221
N of responders	17	14	138	12	21	100	302

[a] Disorder rate (%) = (n of disorders)/(N of responders)×100%.

[b] Number in parentheses is n of disorders.

Table 2: Rate of muscular disorder varied with weekly practice frequency

	Average practice sessions per week			Regression model		
	0-4	**5-9**	**≥10**	**Intercept**	**Slope**	**R^2**
Body parts						
Arm	23.0[a]	29.1	33.3	21.3	1.030	0.989
	(20)[b]	(59)	(4)			
Finger	25.3	21.7	25.0	24.2	-0.030	0.006
	(22)	(44)	(3)			
Back	12.6	19.7	33.3	7.4	2.070	0.968
	(11)	(40)	(4)			
Neck	13.8	18.7	33.3	8.3	1.950	0.924
	(12)	(38)	(4)			
Wrist	16.1	18.7	8.3	19.8	-0.780	0.519
	(14)	(38)	(1)			
Waist	8.0	8.9	8.3	8.2	0.030	0.107
	(7)	(18)	(1)			
Palm	6.9	8.9	8.3	7.1	0.140	0.465
	(6)	(18)	(1)			
Eyes	6.9	4.4	0.0	8.6	-0.690	0.975
	(6)	(9)	(0)			
Ears	2.3	1.0	0.0	2.7	-0.230	0.994
	(2)	(2)	(0)			
Others	8.0	16.3	8.3	10.7	0.030	0.001
	(7)	(33)	(1)			
Total						
Disorder rate	67.8	74.9	83.3	64.5	1.550	0.998
n of disorders	59	152	10			
N of responders	87	203	12			

[a] Disorder rate (%)= (*n* of disorders)/(*N* of responders)×100%.

[b] Number in parentheses is *n* of disorders.

Table 3: Rate of muscular disorder varied with years of playing experience

	Average years of playing experience			Regression model		
	0-4 yr	5-9 yr	≥10 yr	Intercept	Slope	R^2
Body parts						
Arm	22.6[a]	25.0	30.4	20.5	0.780	0.953
	(12)[b]	(22)	(49)			
Finger	24.5	21.6	23.0	24.1	-0.150	0.267
	(13)	(19)	(37)			
Back	9.4	18.2	21.1	8.0	1.170	0.922
	(5)	(16)	(34)			
Neck	11.3	13.6	22.4	8.0	1.110	0.897
	(6)	(12)	(36)			
Wrist	11.3	18.2	19.3	10.7	0.800	0.851
	(6)	(16)	(31)			
Waist	15.1	6.8	7.5	15.1	-0.760	0.681
	(8)	(6)	(12)			
Palm	5.7	6.8	9.9	4.5	0.420	0.930
	(3)	(6)	(16)			
Eyes	5.7	5.7	4.3	6.2	-0.140	0.750
	(3)	(5)	(7)			
Ears	1.9	1.1	1.2	1.9	-0.070	0.645
	(1)	(1)	(2)			
Others	18.9	17.0	9.9	21.6	-0.900	0.900
	(10)	(15)	(16)			
Total						
Disorder rate	62.3	75.0	75.8	61.6	1.350	0.794
n of disorders	33	66	122			
N of responders	53	88	161			

[a] Disorder rate (%)= (n of disorders)/(N of responders)×100%.

[b] Number in parentheses is n of disorders.

DISCUSSION

The prevalence of muscular disorders among musicians in this study roughly agrees with the findings of Zaza (1998) (range from 39% to 87%), and moreover indicates a high prevalence of medical problems among musicians in Taiwan. Symptoms of sufferers in this work, similar to other studies (Lederman, 2003; Hagberg et al., 2005), are concentrated in the upper limbs. The concentration in the upper limbs is no surprise given the extraordinary demands music places on these parts of the body.

Analysis by individual musical instruments, as in previous studies (Zaza & Farewell, 1997; Heming, 2004; Bragge et al., 2005), found that string (cello and violin) and keyboard (piano) players had higher incidences of injury than other instrumentalists. Since most musicians suffered at least one muscular problem involving the arms, fingers, back, neck or wrist, the contribution of body parts in playing an instrument can be dealt with in two categories, namely: (1) performance technique and (2) performance posture, according to body part function. The finger, the parts that were closely related to performance technique, exhibited a U shape relationship between disorder rates and experience of instrumental performance. This result indicated that experience were beneficial to players for a certain period of time, but become less of an advantage after a certain time. Meanwhile, for the arm, back, neck and wrist, all of which simply maintain performance posture, disorder rates increased with performance experience. This phenomenon indicated that performance posture is fixed during the initial learning period and subsequently is difficult to change. Therefore fostering correct posture, performance technique and habits are particularly important to beginners. On the other hand, pain or discomfort can accumulate over time, meaning that the prevalence of injury increases with practice time. The arm, back and neck, all of which maintain performance posture, can quite easily suffer muscular disorders as a result of frequent practice, and moreover the rate of disorder increases sharply with practice time.

The prevalence of hearing disorders is low (1.3%), corresponding with the findings of Palin (1994), but conflicting with those of Laitinen et al. (2003) and Hagberg et al. (2005). Hearing problems have frequently been neglected by musicians, and moreover are difficult to study owing to subject reluctance to admit their existence. This may be one reason for the divergent results and necessary for further improving understanding of this area.

ACKNOWLEDGEMENTS

Professor Chia-Fen Chi is appreciated for her assistance in completing this paper. Ted Knoy is appreciated for his editorial assistance.

REFERENCES

Bragge, P., Bialocerkowski, A. and McMeeken, J. (2005), "A systematic review of prevalence and risk factors associated with playing-related musculoskeletal disorders in pianists." *Occup Med*, 56, 28-38.

Hagberg, M., Thiringer, G. and Brandström, L. (2005), "Incidence of tinnitus, impaired hearing and musculoskeletal disorders among students enrolled in academic music education—a retrospective cohort study." *Int Arch Occup Environ Health*, 78, 575-583.

Heming, M.J.E. (2004), "Occupational injuries suffered by classical musicians through overuse." *Clinical Chiropractic*, 7, 55-66.

Laitinen, H.M., Toppila, E.M., Olkinuora, P.S. and Kuisma, K. (2003), "Sound exposure among the Finnish National Opera personnel." *Appl Occup Environ Hyg*, 18, 177-182.

Lambert, C.M. (1992), "Hand and upper limb problems of instrumental musicians." *Br J Rheum*, 31, 265-271.

Lederman, R.J. (2003), "Neuromuscular and musculoskeletal problems in instrumental musicians." *Muscle Nerve*, 27, 549-561.

Palin, S.L. (1994), "Does classical music damage the hearing of musicians? A review of the literature." *Occup Med*, 44, 130-136.

Pascarelli, E.F. and Hsu, Y-P (2001), "Understanding work-related upper extremity disorders: clinical findings in 485 computer users, musicians, and others." *Journal of Occupational Rehabilitation*, 11(1), 1-21.

Zaza, C. (1998), "Playing-related musculoskeletal disorders in musicians: a systematic review of incidence and prevalence." *CMAJ*, 158, 1019-1025.

Zaza, C. and Farewell, V.T. (1997), "Musicians' playing-related musculoskeletal disorders: an examination of risk factors. *American Journal of Industrial Medicine*, 32, 292-300.

Chapter 54

Designing Lifting Task in Shoe Industry Using Genetic Algorithm

Sanjay Srivastava[1], Kamal Srivastava[2], N. Swati[3], Nitin Singh[4]

[1,4]Department of Mechanical Engineering
Dayalbagh Educational Institute, Agra, India

[2,3]Department of Mathematics
Dayalbagh Educational Institute, Agra, India

ABSTRACT

In this paper we present an application of a genetic algorithm (GA) based method to the design of hide-unloading job, an asymmetric lifting task in shoe industry of Agra, India. In India, which has the second largest shoe industry in the world, it is labor intensive and concentrated in the small and cottage industry sector with Agra being a major production hub. Due to awkward postures and high load handling in hide-unloading, workers are exposed to ergonomic hazards. We focus the design of hide-unloading job in the present work with an aim to reduce the risk to back injury in the purview of revised NIOSH equation. We carried out our study in four shoe manufacturing firms in Agra. The study was conducted on a total of 20 workers, 5 from each firm. It is observed that workers assume different awkward postures mainly due to bulky and unstable load handling. The potency of the study is to present multiple optimal solutions to the design problem using GA while meeting safety and productivity requirements in a given workplace environment conditions. Multiple optimal solutions provide greater agility to ergonomist to implement the recommended design solutions.

Keywords: Hide unloading, asymmetric lifting task design, genetic algorithm.

INTRODUCTION

Low-back pain is often the result of incorrect lifting methods and posture. Statistics from around the world shows that nearly one-quarter of the working population falls victim to manual handling injuries especially low-back injuries (Graham et al., 2009). Repetitive lifting, bending, and twisting motions of the torso affect both the degree of severity and frequency of low-back pain. In addition, low-back pain may also be the result of bad lifting habits. The NIOSH lifting equation, a tool to evaluate two-handed lifting tasks, was developed by the National Institute for Occupational Safety and Health to assist in the identification solutions for reducing the physical stress associated with manual lifting. Recommended weight limit (RWL) is the principal product of the revised NIOSH lifting equation. It is the weight of the load that nearly all healthy workers could perform in a specific set of task conditions over a substantial period of time (eg. up to 8 hours) without an increased risk of developing lifting-related low back pain. The revised lifting equation for calculating the RWL is based on a multiplicative model that provides a weighting for each of six task variables. The weightings are expressed as coefficients that serve to decrease the load constant. The load constant represents the maximum recommended load weight to be lifted under ideal conditions. The RWL is defined by the following equation:

RWL = LC x HM x VM x DM x AM x FM x CM

Where, LC – Load Constant = 23 kg., HM – horizontal multiplier, VM – vertical multiplier, DM – distance multiplier, AM – asymmetric multiplier, CM – coupling multiplier, and FM – frequency multiplier. The term multiplier refers to the reduction coefficient that serves to decrease the load constant. The Lifting Index (LI) is a term that provides a relative estimate of the level of physical stress associated with a particular manual lifting task. The LI is defined by the following equation:

LI = Load Weight (L) / Recommended Weight Limit (RWL)

Lifting tasks with a Lifting Index greater than 1.0 pose an increased risk for lifting-related low back pain. If the magnitude of the LI increases: (1) the level of the risk for the worker performing the job would be increased; and (2) a greater percentage of the workforce is likely to be at risk for developing lifting- related low back pain. It is believed that nearly all workers will be at an increased risk of a work-related injury when performing highly stressful lifting tasks (ie. lifting tasks that would exceed a LI of 3.0). Therefore, the goal while designing a lifting job is to achieve a LI of 1.0 or less.

We attempt to design hide-unloading job within established working environments wherein the ability to provide flexible recommendations for ergonomist in the form of multiple design solutions becomes a very important issue for a successful intervention strategy. To adequately address this problem, the ergonomist would have to rely upon an efficient optimization technique. The difficulty encountered using conventional optimization techniques is that only one "optimal" solution is typically presented to the user. Although linear programming is capable of finding multiple optima, it can only be used in situations where the function to be optimized is a linear summation of decision variables. This additivity assumption is violated in the case of the Revised NIOSH equation where the task parameters have a multiplicative effect in determining the lifting index (LI). One alternative method for determing multiple solutions to an optimization problem is genetic algorithm (GA). Using GA's ability to process a population of solutions in a single iteration, we explore here a number of potential solutions to the design problem, each with its own degree of usefulness (Carnahan and Redfern, 1998). Genetic algorithms are search algorithms based on the mechanics of natural selection and genetics to search through decision space for optimal solutions (Goldberg, 1989). In GA, a potential solution to a problem is represented by a string. A genetic algorithm uses a population of solutions in each iteration of its search procedure instead of a single solution, and therefore the outcome of a GA is also a population of solutions. This unique feature of GA makes it a very appropriate technique for finding multiple optimal solutions and that is how GA based methods transcend classical search and optimization techniques.

In this study we included five design parameters in a typical GA solution which are horizontal distance (load to ankles), asymmetry (between load and sagittal plane), vertical distance of hands to floor (end of lift), vertical distance of hands to floor (start of lift), and duration of lifting task. The constraints imposed on each solution are the number of lifts required on the part of operator, the minimum and maximum time allotted to perform lifts, weight of the load handled (in N), quality of hand/handle interface (good, fair, poor), and the lifting index (LI) considered acceptable by ergonomist. Initial fitness of each GA solution is based on the RWL set by the revised NIOSH equation (Waters et al., 1993). To measure the closeness of the RWL to the weight limit determined by the LI and weight of load (L), we compute standard fitness (SF) of each GA solution. This value is based on the absolute difference between the RWL and the ratio (L/LI).

The purpose of this paper is to present multiple solutions to designing a hide-unloading task with a genetic algorithm. The solutions attempt to meet the productivity and safety criteria set by the ergonomist. The structure of GA is based upon the Revised NIOSH equation for Manual Lifting by Waters et al. (1993).

GENETIC ALGORITHM FOR HIDE-UNLOADING DESIGN PROBLEM

We view the task requirements as environmental conditions while solving hide-unloading design problem using GA. Potential solutions have only two goals; to survive, and reproduce offspring within these environmental constraints. Whether or not they achieve these objectives is dependent upon their physical make-up and the characteristics of the environment. The following is a brief description of the stages in the GA used to solve for the multiple solutions as applied to the lifting design problem (Davis, 1991). The environmental conditions are established by setting the constraints on the solutions. The environmental conditions are five task parameters associated with hide-unloading job:

1. weight of the load handled varied: 12 kg to 18 kg
2. quality of the hand/handle interface: poor
3. the number of lifts required on the part of the operator as set by the employer: 0.67 lifts/minute
4. the maximum time allowed to perform the lifts: 30 minutes
5. the Lifting index (LI) not to exceed 1 is considered acceptable

Parameters 1 and 2 form the product constraints, parameters 3 and 4 comprise the production demand restrictions and parameter 5 corresponds to the safety requirement as set by the ergonomist. The problem for the GA is to search for multiple design recommendations for manual lifts which meet the product and production requirements while satisfying the Lifting Index criteria.

A GA solution here is a string which represents an instance θ of the lifting task design (Fig. 1); Each parameter ti of an n-tuple string, T, can assume values as set for the lifting task. Five types of associated fitness with each individual string are determined as described in the following paragraph.

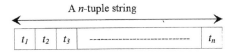

FIGURE 1. An instance of lifting task design

The initial population, consisting of 250 solutions (chosen based on computational experiments), is generated by randomly selecting 250 individual strings from the feasible search space, i.e., each t_i of a string consisting of five parameters is chosen randomly from the defined bounds shown below.

Parameter	Definition	Minimum	Maximum
H	Horizontal distance (load to ankles)	25.4 (cm)	63.5 (cm)
A	Asymmetry (between load & sagittal plane)	90 (deg)	135 (deg)
V_e	Vertical distance (hands to floor - end to lift)	0 (cm)	60 (cm)
V_s	Vertical distance (hands to floor - start of lift)	80 (cm)	80 (cm)
D	Duration of lifting task	25 (min)	30 (min)

Further, a Genetic Tag, an integer between 1 and 6, is randomly assigned to each of the 250 solutions. Each tag represents a 1 of 6 potential "species" of solution and will play a significant role in developing multiple optimal solutions to the problem. An example of a design solution developed by a GA is shown in Table 2. For this solution, the task should be designed so that the load is no more than 37.0 cm in front of the ankles and that the angle of asymmetry is no more than 110 at the start of the lift. The 12 kg loads is positioned 80 cm above the floor at the start of the lift as fixed by the truck base height with the vertical distance at the end of lift as 30 cm. Lifting 0.67 loads per minute for approximately 30 min under these conditions will enable the operator to perform the required 20 lifts within the time allotted. The "'fitness" of each solution in the population is determined as per the following procedure. In the context of the lifting design problem, fitness is a quantifiable measure of how well the solution meets the productivity and safety constraints which comprise the environment conditions. A solution's fitness is based not only on the environment that exists but also on the fitness of other solutions within the population. Each solution is comprised of five types of fitness:

1. Raw fitness: It is nothing but the recommended weight limit (RWL) as calculated by the revised NIOSH equation (Waters et al., 1993)

2. Standardized fitness (SF): It is a measure of how close the RWL is to the weight limit determined by the LI and weight of load (L), which is computed using the formula shown below.

$$SF = 1/[1 + [Sg * mod(RWL - (L/LI))]]$$

3. Species fitness: It facilitates GA to search for multiple solutions in parallel. It is computed by dividing the standardized fitness (SF) of each individual solution by the total number of solutions bearing the same Genetic Tag. The standardized fitness of each solution will decrease linearly with an increase in the number of solutions of the same species. This prevents any single type of solution from overwhelming the population and allows other solutions to co-exist and co-evolve (Spears, 1994).

4. Relative fitness: It is simply calculated by dividing the species fitness of a single solution by the sum total of species fitness within the population.

5. Cumulative fitness: The running total of the relative fitness starting with the first solution is a measure of cumulative fitness.

After fitness assessment, solutions are chosen for reproduction using stochastic selection with replacement. Under this form of selection, the contribution of an individuals solution to the next generation is proportional to its fitness relative to the fitness of the rest of the population (Fogel, 1995). In other words, superior solutions are chosen, on average, more often than relatively less desirable solutions. Stochastic selection with replacement is repeated until 250 solutions have been chosen for the reproductive stage. In this stage, the solutions are randomly chosen to "reproduce'" the next generation of design solutions. This reproductive stage is carried out in two phases: crossover and mutation. Crossover reproduction produces intermediate solutions, known as "offspring", that are a combination of the 5 design parameters. A blend crossover is used here. Thereafter mutation is carried out on the population with mutation probability as 0.02. In the final stage the selected population, altered by crossover and mutation, replace the initial population of solutions. After this stage, a single generation of solutions has evolved. The genetic algorithm then repeats the stages in a cyclic fashion to produce a number of succeeding generations. This succession of generations is known as a Run. The total number of generations, representing the length of the Run, is determined by the user.

RESULTS

Fig. 2 represents the improvements in standard fitness of evolving six subpopulations over the generations. Each subpopulation represents a potential solution to the hide-unloading problem. In observing changes in the standardized fitness, increasingly better solutions evolve with each iteration of the algorithm. The average standardized fitness of each of the six subpopulations is plotted as a function of generation within a single Run of the GA. Note that subpopulations reach their respective peak average fitness values at different rates. The algorithm's use of genetic tags successfully allows relatively less fit subpopulations to survive until better solutions emerged during the course of a Run. The number of solutions with the same genetic tag roughly mirrored the average fitness of its corresponding subpopulation across generations. In Fig. 3 the number of solutions in a subpopulation is plotted, which starts with 35 initial solutions and shows an increasing trend during the first 37 generations of the Run. Later due to co-existence of other subpopulations the number of solutions decrease and stabilizes to about 50 solutions over 100 generation.

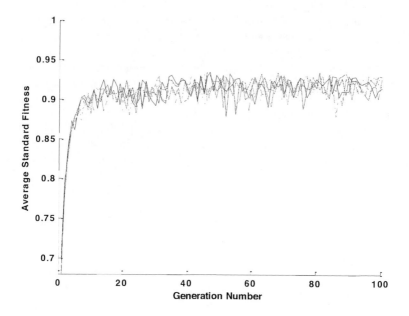

FIGURE 2 Improvements in average standard fitness over the generations for hide-unloading problem

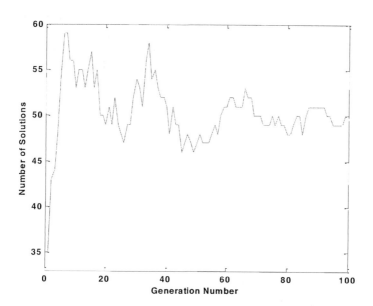

FIGURE 3 Number of hide-unloading designs solutions in a subpopulation over the generations

CONCLUSIONS

The proposed GA based method effectively solves asymmetric lifting design problem by incorporating the concept of multiple optimal solutions for a single objective problem. GA is employed here to solve hide-unloading design problem of shoe industry in Agra. It can search for many-optimal solutions after examining an extremely small fraction of possible solutions. Multiple solutions to the same problem provide greater flexibility to the ergonomist to implement the solution in the existing environment of shoe factory under consideration.

ACKNOWLEDGEMENT

This work is supported in part by UGC, New Delhi under Grant F. No. 36-65/2008 (SR), dated 24/03/2009.

REFERENCES

Carnahan, B. J. and Redfern, M. S., 1998. *Application of genetic algorithms to the design of lifting tasks*. International Journal of Industrial Ergonomics 21, pp. 145-158.

Davis, 1991. L. Davis. In: *Handbook of Genetic Algorithms*, Van Nostrand Reinhold, New York (1991), pp. 1–22.

Fogel, 1995. D.B. Fogel, *Evolutionary Computation: Towards a New Philosophy of Machine Intelligence*. In: , IEEE Press, New York (1995), pp. 90–91.

Goldberg, 1989. D.E. Goldberg, *Genetic Algorithms in Search, Optimization, and Machine Learning*. In:, Addison-Wesley, Reading, MA (1989), pp. 2–7.

Graham, R. B., Michael, J. A., and Stevenson, J. M., 2009. *Effectiveness of an on-body lifting aid at reducing low back physical demands during an automotive assembly task: Assessment of EMG response and user acceptability*. Applied Ergonomics 40(5), pp. 936-942

Spears, 1994. W.M. Spears, *Simple subpopulation schemes*. In: A. Sebald and L. Fogel, Editors, Proceedings of the 3rd Conference on Evolutionary Computation, World Scientific, New Jersey (1994), pp. 296–307.

Waters et al., 1993. P.R. Waters, V. Putz-Anderson, A. Garg and L. Fine, *Revised NIOSH equation for the design and evaluation of manual lifting tasks*. Ergonomics 36 7 (1993), pp. 749–776.

516

Prevalence of Low Back Pain in Long Distance Truck Drivers of Mountainous Terrain

Manish Goon, Sutanu Ghoshal,
Baskaran Chandrasekaran, Bidhan Chandra Sharma

Department of Physiotherapy, SMIMS
Sikkim Manipal University
Sikkim 737102, India

ABSTRACT

The incidence of back pain in the truck drivers of mountainous areas is unknown. The occupational wellbeing and the incidence of low back pain in professional Truck Drivers were investigated regarding low back symptoms, personal, occupational, environmental, social and ergonomical factors using a well validated questionnaire. Of 200 truck drivers included, the prevalence of back pain is 73.52%.

Keywords: low back pain, Truck Drivers, Questionnaire, prevalence.

INTRODUCTION

Low back pain is a growing pandemic in the Indian drivers with prevalence 40% to 69% (Kumar 1999). Back pain in truck drivers is of multifactorial such as vibrations, strained postures for long hours, etc. (Laxmaiah 2000)

Sikkim, a Himalayan state is connected to the rest of India only through the National highway 31-A which is its lifeline. Throughout the year, the truck drivers of Sikkim are encountering the rough hilly terrain, hair pin bend narrow roads, adverse climatic conditions and natural calamities. The incidence of the low back pain in these high hill truck drivers is unknown.

Though earlier literature had reviewed the epidemiology of back pain in the truck drivers of plain areas, the epidemiology of back pain in drivers of hills are not stated so far. Hence the purpose of this study is to find the prevalence of low back pain in the truck drivers of Sikkim.

PARTICIPANTS & PROCEDURE

The study was approved by Central Referral Hospital Ethical Committee, Sikkim. A cross sectional survey was conducted in 200 truck drivers at their association office with the informed consent of the office secretary. But only 34 members complied with the survey and were of mean age group of 28 years.

MODIFIED OSWESTRY MCGILL BACKPAIN QUESTIONAIRE

The tool used for interviewing the participants was a questionnaire which was compiled with reference to Oswestry's low back pain questionnaire and Mc Gill's Low Back Pain Questionnaire, with suitable modification made, considering the cultural, social, environmental, and educational and the geographical factors. The questionnaire consisted of two headings.

The first division of the questionnaire included the symptoms, predisposing factors, past history of low back pain, diurnal variations, onset, duration, severity, progression, relation of pain with rest, activities, effect of pain upon ADL's posture habits like alcoholism, smoking, diet, work hours and time spent with family.

The second division consisted of questions pertaining to road conditions, maintenance details of the vehicle (servicing, etc.), comfort issues of the sitting arrangements, Use of ergonomics aids like cushion for extra back support, footrest etc. After compilation of questionnaire, its validation was discussed with the experts and was translated into local language (Nepali).

STATISTICAL ANALYSIS

The results would be analyzed through **descriptive analysis** in **Microsoft excel 2007.**

RESULTS

Out of 200 truck drivers from the association, only 34 complied with the research and completed the questionnaire.

1. Prevalence of back pain in truck drivers

Among the thirty four (34) truck drivers participated, 25 were symptomatic for back pain. This can be clearly seen from the fig: 1.

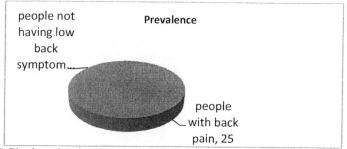

Fig. 1.1. Pie chart showing the prevalence of the low back pain in the truck drivers.

2. Inter individual differences in pain facilitation during driving

Out of 25 participants having low back pain, 9 of them complained of pain during uphill driving. 3 of them experienced pain during activity downhill and 13 participants could not relate their pain with uphill and downhill driving. fig: 2.

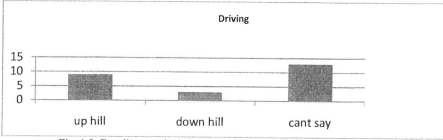

Fig. 1.2. Bar diagram depicting the pain demography during driving.

3. Age differences in back pain

4 out of 9 participant's experiences early morning pain and stiffness were above the age of 40 years. Fig 3 shows 44% of the total population examined suffers a back pain who above 40 years old.

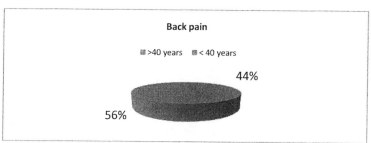

Fig.1.3. Age related variations in incidence of back pain in truck drivers

4. Psychological stress

Out of 25 participants, 4 of them complained of psychological stress (fig: 4)

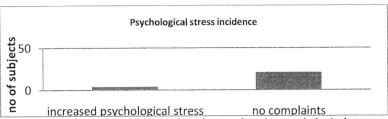

Fig.1.4: Bar diagram showing the drivers undergoing psychological stress.

5. Ergonomical dysfunctioning

As for occupational factors, Out of 34 truck drivers interviewed, 9 of the drivers complained that they experiences discomfort on their driving seat (Fig. 5)

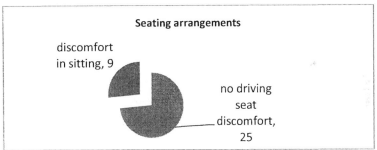

Fig. 1.5: Discomfort due to improper designing of the driving seats.

6. Bad rough terrain

19 truck drivers interviewed thought that the road conditions of the daily route (N.H.31A) were bad most of the time (fig. 6).

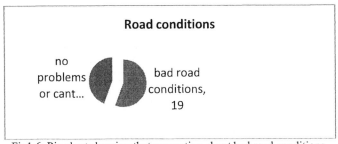

Fig1.6: Pie chart showing that perception about bad road conditions.

7. Postural variations during seated driving

19 of the 34 admitted that they assured relaxed normal sitting posture while driving, 10 of them admitted their postural dysfunction during seated driving, while 5 of them left no comments as depicted in fig. 9

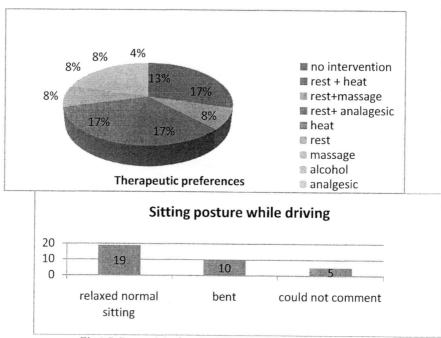

Fig.1.7: Postural dysfunction perception during seated driving.

8. Variations in therapeutic preferences among the drivers

25 participants experienced back pain. The truck drivers preferred various therapies as noted from their questionnaires as stated in Fig: 8. 25% preferred pharmacological and 50% preferred some forms of non pharmacological therapy.

Fig.1.8: Therapeutic preference variations among drivers.

9. Ergonomic aids preference
On interviewing it was evident that out of 34 drivers, only 9 used ergonomic aids like a cushion to support their lumbar spine while driving. This is depicted in fig: 9

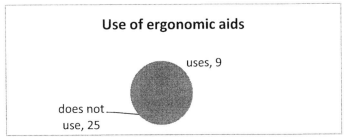

Fig1.9: Ergonomic aids adapted by the drivers for their back pain.

10. Foot rest preference

36% of the total population (9 out of 34) used foot rest ensure relaxation while driving.

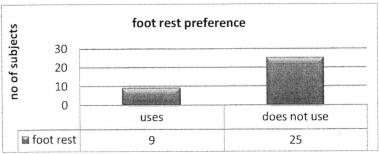

Fig. 1. 10: Foot rest preference.

DISCUSSION

From the analysis of the results presented, it can be clearly depicted that the prevalence of the low back ache in the long route truck drivers of the National highway 31-A is 74% (fig. 1). Earlier Indian literature by Kumar Adarsh (1999) et al and Laxmaiah Manchikanti (2000) concluded that the prevalence of back pain in truck drivers is 40% – 69%. Our study's prevalence rate is higher than that of the national prevalence rate. This may be due to the fact that earlier researches were conducted in the plain areas where as our study included drivers of hilly areas with rough terrains.

We have found only 36% (nine of 25 symptomatic) of the subjects had morning stiffness suggesting of some degenerative changes may be due to modifiable factors like vibrations & sudden torsional forces (Kumar Adarsh 1999), prolonged seated posture (Laxmaiah 2000), frequent smoking (Jane Lyons 2002) or due to non modifiable factors like age related degenerative changes (Jane Lyons 2002, Silvia Ferreira 2006), osteoporotic effects and malnourishment.

As we have found from our survey, 56% (fourteen out of 25 symptomatic subjects) have pain during uphill driving (fig.2). This may depict the incidence of lumbosacral radiculopathies and disc involvement in these subjects (Silvia Ferreira 2006). We did not find any remarkable changes during driving uphill or downhill. This is in accordance to earlier literature (Silvia Ferreira 2006).

As we expected, drivers who are more than 40 years of age had increased incidence of back pain than their counterparts of age less than 40 years (fig.3). Hence we confirm with the earlier literature by Jane Lyons (2002) and Silvia Ferreira (2006) that age is an important risk factor in back pain incidence in truck drivers. We suggest that the increase incidence of back pain in this aged truck drivers may due to degenerative changes, disc nutrition loss, muscle weakness, ligament laxity and osteoporotic changes as stated earlier (Silvia Ferreira 2006).

Though there is a claim that occupational psychological factors are highly associated with back pain, we have found less association between psychological factors and low back pain (fig.4). This is in contrary to the earlier established literature (Steven James Linton 2001).

Only 36% of the survey population was accessible to ergonomic aids like (foot rest and seating arrangements) as stated in fig. 9 & 10. This may depict the urgent need for the development of necessary intervention strategies for drivers by the concerned regional welfare society for the drivers and state welfare committees.

CONCLUSION

The overall prevalence of the back pain in the truck drivers is 74%. It is on the part of welfare societies to implement transport and vehicle servicing, ergonomical modifications, strategy planning in working hours and psychological counseling.

REFERENCES

Jane Lyons (2002). Factors contributing to low back pain among professional drivers: A review of current literature and possible ergonomic controls. *Work: A Journal of Prevention, Assessment and Rehabilitation*; Volume 19 (1); 95-102.

Kumar, Adarsh M.Tech; Varghese, Mathew MS; Mohan, Dinesh PhD; Mahajan, Puneet PhD; Gulati, Praveen MD; Kale, Shashank MCh (1999) : "Effect of Whole-Body Vibration on the Low Back: A Study of Tractor-Driving Farmers in North India". *Spine.* Volume (24); 2506.

Laxmaiah Manchikanti;(2000); "Epidemiology of Low Back Pain". *Pain Physician* Volume 3(2), 167-192.

Silvia Ferreira Andrusaitis; Reginaldo Perilo Oliveira; Tarcísio Eloy P. Barros Filho ;(2006); "Study of the prevalence and risk factors for low back pain in truck drivers in the state of São Paulo, Brazil". Clinics vol.61 no.6 .

Steven James Linton. Occupational Psychological Factors Increase the Risk for Back Pain (2001).A Systematic Review. *Journal of Occupational Rehabilitation*: Volume 11 (1).

CHAPTER 56

Ergonomic Aspects of CNC-Units – A Survey Among Trainees

André Klussmann, Inna Levchuk, Andreas Schäfer, Karl-Heinz Lang

Institute of Occupational Health, Safety and Ergonomics (ASER)
Corneliusstr. 31
42329 Wuppertal, Germany

ABSTRACT

During the last decades, a huge amount of heavy physical work load was increasingly replaced by Computerized-Numerical-Control (CNC) Units in manufacturing areas. Hence manual work is decreasing, while static work (e.g., standing, sitting) and cognitive demands rise in these fields of activity.

Machine operators have a high responsibility, because mistakes may lead to human injuries as well as to product losses – and in consequence may lead to high monetary losses as well. An intuitive usability and an ergonomic organization of CNC workplaces can be an essential basis to reduce the risk of failures in operation as well as physical complaints (e.g. pain or diseases because of bad body posture during work).

Within this study, a standardized questionnaire was developed and a survey was carried out among trainees learning the operation of CNC operated machines. The content of the questionnaire were sociodemographic data, questions on "weak spots" or troubles during operating the CNC machines as well as questions about general working conditions and musculoskeletal symptoms and disorders. Besides, the design of the workplaces was documented in a check list.

112 male trainees (age 17 to 56 years) from 4 training centers were included in the first evaluation. This survey discovered that there is a need of optimization of the

ergonomic design of CNC machines and the CNC control panel in particular. This regards the arrangement and appearance of operating elements and the operation of touch screens. Besides, the general organization of the workplace (height-adjustability of CNC control panels, arrangement to the working process) should be on focus.

INTRODUCTION

Since beginning of the 1970's hard manual work in the production is replaced by computer-controlled machines. Today, modern machines are equipped with complex CNC control units, which make highly complex and precise phases of operation possible. Manual work decrease but static work (e.g. long standing) and cognitive demands in these fields of activity increase. Machine operators have a high responsibility, because operating errors can lead to machine and/or product damage as well as to human injuries. High economic and/or human losses could result in consequence.

The current developments in the industrial production area go away from homogeneous mass production to a product usage production. This "lean production philosophy" leads to the fact that machines must be reequipped more frequently. From economic view, it is important that after a re-equipment, the first workpiece fits. This increases the psychosocial pressure to the persons employed at the machines to avoid errors.

An intuitive operability and an ergonomic organization of CNC units can be a substantial basis to reduce the risk of operating errors and physical complaints, too.

The research questions in this study are: How is the usability of CNC units? Which faulty operations arise and which damage occurs with it? Which ergonomic optimizations are needed?

METHODS

Within this study, a standardized questionnaire was developed and a survey was carried out among trainees learning the operation of CNC operated machines. The content of the questionnaire were socio demographic data, questions on "weak spots" or troubles during operating the CNC machines as well as questions about general working conditions and musculoskeletal symptoms and disorders. Besides, the design of the workplaces was documented in a check list. 112 male trainees (age 17 to 56 years) from 4 training centers were included in the first evaluation.

RESULTS

How is the usability of CNC units evaluated by trainees in first education and by people learning the operation of CNC machines in further education?

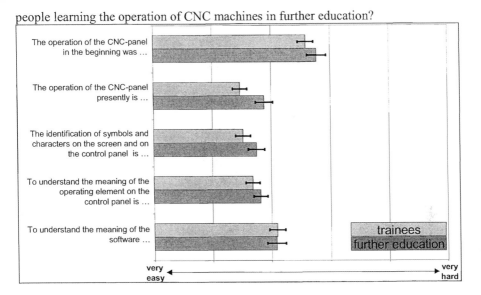

Figure 1: Estimation of the usability of CNC panels. Trainees in first education compared to people learning the operation of CNC machines in further education. Mean and standard error of the mean.

How is the usability of CNC units evaluated by trainees in first education and by people learning the operation of CNC machines in further education?

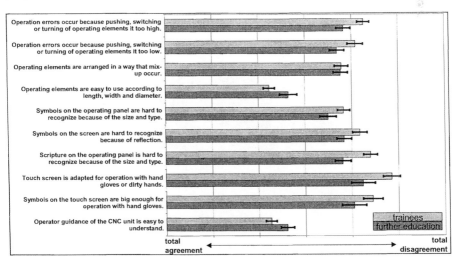

Figure 2: Statements about the usability of CNC panels. Trainees in first education compared to people learning the operation of CNC machines in further education. Mean and standard error of the mean.

Which operating errors arise and which damages occur with it?

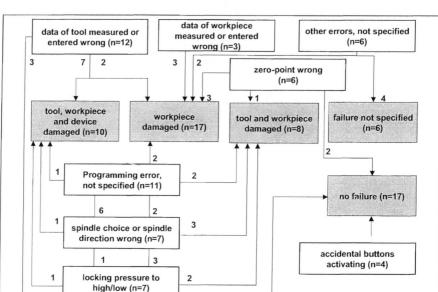

Figure 3: Operating errors and damages among 112 male trainees in first and further education within the last 4 weeks.

How is the organization of the workplace and operating surface of the CNC units?

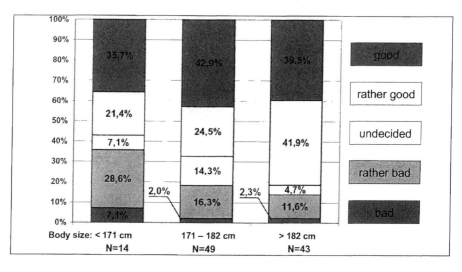

Figure 4: Visibleness of the production process from the CNC unit. People smaller

than 171 cm compared to people 171-182 cm and higher than 182 cm.

In the estimation of the usability of CNC panels among trainees in first education compared to people learning the operation of CNC machines in further education, only a few relevant factors were judged different.

Trainees in first education estimated the operation of the CNC-panel in the beginning and at present less difficult than people learning the operation of CNC machines in further education (Figure 1).

The recognition of symbols and scripture on the operation panel and on the screen were judged relatively easily. However, regarding the character size and type it was found to be estimated somewhat heavier among trainees in first education than the in further education. The operator guidance of the CNC machine seems to be predominantly understandable in both groups. However, trainees in first education reported more difficulties in understanding than the people in further education. The operating elements on the control panel were evaluated positively in both groups. However, touch screens were identified as unsuitable for the work with gloves and/or dirty hands in most cases. This is also because of the symbol size on the touch screen (Figure 2).

Wrong measurement of tools and workpieces, wrong entry of dimension data of tools and workpieces as well as mix-up of the axles were the most occurring errors, which usually led to tool, workpiece and/or device damage (Figure 3).

The trainees were asked about the organization of the workplace and operating surface of the CNC units. 33% of the interviewees rate the "space within reach" as "rather bad" or "very bad". 32% of the interviewees could not monitor the working process from the CNC control panel (Figure 4). 80% of the interviewees worked on not-height-adjustable CNC control panels. Accordingly 37% of the interviewees felt the height of the CNC control panel as not appropriate. 65% feel that their body posture during the operation of the machine is (rather) strained.

If a touch screen was available, it was described as rather ergonomic, but the unintentional activating of neighboring keys/functions was criticized. Every 10th interviewee indicated that the operating elements are confusing because of their arrangement or their appearance. Even among the persons who work in their job for more than 5 years, 14% still have difficulties with the arrangement of the operating elements on the surface of CNC units.

CONCLUSIONS

In the estimation of the usability of CNC panels among trainees in first education compared to people learning the operation of CNC machines in further education, only a few relevant factors were judged different. This survey discovered that there is a need of optimization of the ergonomic design of CNC machines and the CNC control panel in particular. This regards the arrangement and appearance of operating elements and the operation of touch screens. Besides, the general organization of the workplace (height-adjustability of CNC control panels, arrangement to the working process) should be on focus.

AKNOWLEDGEMENTS

This study is part of the "Design4All" project, which started in spring 2009 and will run until spring 2012. This project is funded by the Ministry of Economic Affairs and Energy of the State of North Rhine-Westphalia in combination with the "NRW Ziel 2 Programm" of the European Regional Development Fund (ERDF).

CHAPTER 57

Identification of Hazards at a Constant Work Place by Means of the Map of System Work

Teresa Musioł, Jarosław Grzesiek

Faculty of Organisation and Management
Institute Engineering of Production
ul. Roosevelta 26, 41-800 Zabrze
Poland

It is not about security but about safe behavior.

ABSTRACT

The purpose of the paper is analysis of identification of hazards existing at a selected constant work place. The analysis has been carried out by means of a tool of work system mapping. As opposed to the methods of forecasting consequences of faults existing during a work process, mapping should enable reduction of potential hazard. Muscle memory, which is the result of training, used in the process of production management cannot be a creator in work system, and this means response delays to unexpected appearance of hazard.

The introduction describes the rules and the essence of work system mapping on the basis of maps and special attention has been paid to work system. The paper presents mapping of particular activities resulting from the techniques of product manufacturing. This is shown by means of such organizing techniques as devising of process as a component of the system. The authors described the rule of defining a process at a constant work place by determining work methods and control rules. Possibilities of identification of key and unnecessary operations in the work process

530

were then used to identify hazard existing at the mentioned-above constant work place.

In the practical part the method of case study has been used to present the analysis results. On the basis of design works carried out by students during the course of research and work organization, hazard related to the process of assembly of car components has been identified by means of a map of system work.

Keywords: Identification of Risk, Map of System Work, Potential Hazard.

THE PURPOSE OF UNDERTAKING THE SUBJECT

After reading the book „Globalization" by Joseph E. Stiglitz and in confrontation with the professional experience of the authors (industry, didactics) some thoughts cross my mind. Many companies investing in developing countries to forget about the relationship of working conditions with the quality of work (Stiglitz, 2004).
Success of an organization is conditional not only by technical and economic or marketing factors which create favorable conditions for a sense of personal safety of a member of the organization, such as: health, an accident-free job and satisfaction from the activity. In order to make it possible, there must be the atmosphere favorable to basic activities promoting safety. It means that the climate of organizational culture must be a characteristics of a particular organization (Musioł, Ujma Wąsowicz, 2008).

The process of globalization began changes in thinking not only in macro but also in micro scale. The result of thinking in the micro scale in the scope of working conditions is the work place located in the work system, understood as the core element. The interactions between the work place and other elements of the work system are subject to quantification, all kinds of analyses and first of all diagnosis. Figure 1 presents a diagram of this system.

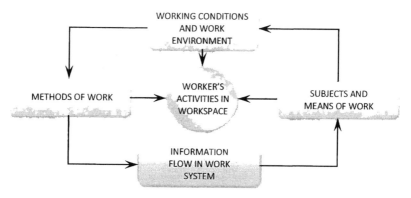

Figure 1. Ideogram of a work system. (Musioł, Grzesiek, 2008)

A balance of the work system is a resultant of statistic-dynamic parameters in the scope of measurable criteria (using the formalized language of numeric data) and non-measurable criteria (ontological reduction, e.g. intuition, emotional or intellectual effort). In the case of non-measurable criteria we deal – party of completely – with a human and a role that a human fulfils in the system and how s'he perceives it.

A work system, like every other system, is subject to hazards of endogenous and exogenous character. A hazard can be defined in many ways taking into account its character, kind, degree and results.

The term *hazard* includes all factors and phenomena accompanying the destructive energy which impacts people, property or environment. In order to fully determine any hazard the following elements should be identified (Weisman, 2007):

- hazardous factors, i.e. all media (objects, people, processes, phenomena) which potentially have so called destructive energy, i.e. the energy which exceeds immunity threshold of a victim (a human, property, environment),

- hazardous situations (possibility of destructive energy impact), i.e. circumstances in which people, property or environment are exposed to hazardous factors,

- harmful events (the form of destructive energy impact), i.e. events in which hazardous situations turn into damage.

As it has been already mentioned above, the way of perceiving reality in a work place, the way individuals perceive their work, strongly and directly influences their behavior and a way of thinking – about themselves, their job and the environment. It determines a method of operating, which in turn impacts achieved results, as Figure 2 shows.

Figure 2. The way of perceiving reality in a work place.(Authors)

The structure of perception changes when human teams in the configuration of a chain, pair, star or triangle are formed.

In order to prevent distorted perception of reality specialists recommend to evaluate own expectations using two criteria. The first of them is the criterion of truth, which consists in being guided by timeless, universal rules which govern the world. The second one is the criterion of realism, also known as the criterion of reasonable hope. It recommends to strive to achieve such goals which are feasible in a particular situation and within particular time.(Evans Ch., 2005)

That said, the hypothesis is that a map of processes of work system is a tool enabling correct perception of hazards in a work place within the criterion of realism.

THE ESSENCE OF PROCESS MAPPING

"The term *process* in organization management should be understood as a group of actions or activities and decisions which transform input data into output data. Input data in a process are usually materials, information including plans which are created on different levels of management, people, money and environment conditions indispensible to initiate the process and make it progress successfully in order to achieve the intended objective. Output data in a process can be a material or non-material product or a service. Output data is a purpose for which the process is realized. In order to initiate any process its input data has to be determined. For every process which transforms inputs into outputs (process objectives) it is possible to determine the structure, connections with other processes and to monitor/measure input data, accomplished objectives (compliance of input data with process objectives) and actions which are important from the point of view of a process supervisor. Usefulness of a map is not determined by its level of detail but its intended application. A map also allows to determine: where we are, where we are heading, how to achieve the goal and what obstacles we can encounter on the way." (Ścierski, 2010)

While drawing up a process map we must decide on a level of detail. Less detailed maps are created to present general rules of process functioning, whereas more detailed maps – when we are looking for reasons of failures in the process which consists a lot of operations. In the course of creating a process map we are able to identify a lot of hazards.

Process mapping is graphical presentation how a process or a group of processes function and what their mutual connections are. Elements of a process map are presented in a form of normalized graphic symbols. (Galloway, 1994)

In carrying out the process, the output (target) can be achieved by different paths. The choice of path, which runs the process determines a number of factors. Performing actions accompanied by decision-making. Decisions shall be accompanied by any man, but in most cases they are made to intuitively. (Ścierski, 2010)

The linear production processes employed relatively rarely is forced to make decisions. The decision is accompanied by processes of control (self-control) or tests.

Due to the crossbreeding of the production processes with the processes related to the flow of information (exploitation or the creation of documentation) during the mapping may be necessary to use graphic symbols for production processes maps and information and business processes maps.

Table 1: The symbols of process mapping. (Adapted from strefa-iso.pl, 2009)

Production symbols		Information and business symbols	
▽	Material pickup, acceptance of delivery	▱	Data, information, input or output material
△	Storing	▢	Process, operation, action
○	Operation (e.g. CNC milling)	⏢	Manual operation
▢	Control (e.g. meter reading, measurement, etc.)	⬠	Document, entry to the data base
◉	Operation + control	◇	Decision, test, inspection
◇	Decision (control, test, etc.)	⬭	Terminator (the beginning or the end of a map)
⌓	Waiting, delay	○	Connector (between pages of a process map)
⇨	Transport (internal, external, etc.)		

In the paper two groups of symbols are used (Table 1): the symbols normally used to describe production processes and the symbols used to describe business processes.

CASE STUDY

In order to start the process of identification hazard in the work system using the tool of mapping borders of the process and the system should be determined.

Every process exists in the framework of a system. According to ISO 9000:2005 *a system* is a collection of mutually connected or mutually interacting elements, which means that a system can be also understood as an arrangement of mutual relations and connections which exist between processes or actions involving resources.(Ścierski, 2010).

For purposes of the hypothesis stated in the paper the borders of the work system in a shock absorber assembly work place were limited to:
- one workstation,
- working conditions and work environment (Fig.1),
- worker's activities in workspace (Fig.1).

Table 2 presents a list of workstations which participate in the shock absorber assembly process taking into account their assignment to work places linked by a closed technological cycle.

Table 2: The list of workstations (Authors)

Process	Work place	Workstation/operation
Assembly of shock absorbers	Work place 1	Plastic injection/OP70
		Testing/OP80
		Manual assembly/OP90
		Control of force/OP100
	Work place 2	Welding/OP50
		Press assembling/OP60
		Connecting/OP110
	Work place 3	Testing/OP170
		Optical control/180

Out of workstations/operations presented in Table 2 Manual assembly/OP90 was chosen. Figure 3 shows identification of hazard by map of OP90.

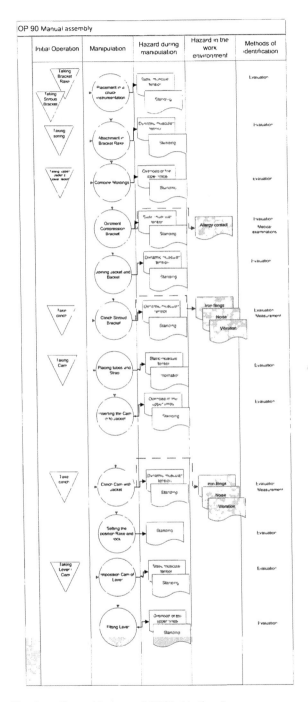

Figure 3. Identification of hazard by map of OP90. (Authors)

DISCUSSION

During manual assembling a worker should see a technological instruction in the form of a map of processes. Before starting work he should be given training sessions not only in the form of practical instructions but also an emotional and cultural training.

The frequent result of both increasing pace of movement and numerous adverse factors affecting our body, for example high temperature or other stressors, are disorders of internal environment and dysfunctions of central nervous system. The consequences are usually slowing down of conveying information from the brain and disturbances in receiving them from receptors and in motion control. It can be demonstrated by a longer time of reaction or criteria of work efficiency, which decreases from 20 to 80% during physical and intellectual activities independently of motivation and involvement of a person in intended task result. The brain encodes retained stereotypes of movement for every type of work. (Gwóźdź, 2003) This is a very individualized phenomenon because as Edyta Stein noticed whereas the law of causality applies to the world of nature, the law of motivation applies to the world of spirit and the relation of experiences of an aware subject cannot be definitively explained on the psycho-physical ground. (Dec, 2002)

A process map of work system can never be created as a result of creating Self Organization Maps (Kantola, 2005) because of the phenomenon of individualization of work process for every person. On a certain level of process detail individual differences begin. In manufacturing it can lead to creating movement dysfunctions and all kinds of disorders of sensory system.

CONCLUSION

To sum up, the hypothesis stated at the beginning of the article has been positively verified. Such a tool as a process map can be used in training sessions and workshops as a teaching material which arises awareness about hazards and self-control during work process. In the subsequent phase it can be the element that supports vigilance and attentiveness during work and eliminates routine and bad habits. It does not promote safety but safe behaviors.

REFERENCES

Dec, I., (2002) Edyta Stein (1891-1942) – Wzór poszukiwania i integrowania wartości. Edyta Stein Patronka Europy. Materiały Międzynarodowego Sympozjum „VII Dni Edyty Stein" Wrocław 12-16.09.2001 pod red. R. Zajączkowskiej. Wrocław. Wyd. Tow. im. Edyty Stein we Wrocławiu.

Evans, Ch., (2005) Zarządzanie wiedzą. Warszawa. Polskie Wydawnictwo Ekonomiczne.

Galloway, D., (1994) Mapping Work Processes. Milwaukee. American Society for Quality. Quality Press.

Gwóźdź, B., (2003) Efektywność pracy w ujęciu neuroergonomii. Praca. Zdrowie Bezpieczeństwo. Stowarzyszenie Inżynierów i Techników Przemysłu Hutniczego w Polsce. No 2 (188)

Kantola, J., Karwowski, W., Vanharanta, H., (2005) Creative tension in occupational work roles: a dualistic view of human competence management methodology based on soft computing. Ergonomia. An International Journal of Ergonomics and Human Factors. Published by the Committee on Ergonomics of the Polish Academy of Sciences in cooperation with the International Ergonomics Association. Vol.27. No.4 pp. 273-286.

Musioł, T., Grzesiek, J. ed. (2008) Podstawowa problematyka projektowania stanowisk pracy. Bytom. Wyd. WSEiA w Bytomiu.

Musioł, T., Ujma-Wąsowicz K., (2008) The Method of Interactive Reduction of Threat of Isolation in the Contemporary Human Environment, been accepted for publishing in the book "Advances in Human-Computer Interaction" Aleksandar Lazinica Scientific Manager I-Tech Education and Publishing KG Kirchengasse 43/3, A-1070 Vienna Austria, EU http://www.i-techonline.com/

Ścierski, J., (2010) Mapowanie procesów. Gliwice. Zeszyty Naukowe Politechniki Śląskiej. Seria Organizacja i Zarządzanie (in printing).

Stiglitz, J., (2004). Globalizacja. Warszawa: Wyd. Naukowe PWN pp.74

strefa-iso.pl (November 24, 2009) Mapowanie procesów, Narzędzia jakości. IPO Website: http://www.ipo.pl/zarzadzanie_w_firmie/jakosc/narzedzia_jakosci_-_mapowanie_procesow_592703.html

Weisman, A., (2007) Świat bez nas. Gliwice: Centrum Kształcenia Akademickiego CKA

Occupational Safety Management and Application of Economic Guidelines in the Shaping of Working Conditions in Supermarkets

Katarzyna Lis, Jerzy Olszewski

Poznan University of Economics
Poland

INTRODUCTION

The issues of occupational safety management in trade have always played a vital role in most literature on the subject. This has not only been determined by the concern for customer safety, but, primarily, also by the concern for the safety of the staff, who perform their tasks over many hours, often in a rigid position, in diverse physical, biological and – quite frequently – chemical environments [7]. The issues of occupational safety gain importance in view of the fact that stores, as workplaces, have diverse product lines and organisational systems. In relation to the size and variation of product lines, the literature on the subject distinguishes the following types of stores (cf. Table 1).

Hence, from Table 1, lighting as an environmental factor, complies with the standards. With microclimate, however, the situation is slightly different, especially in large stores, where crowds of customers make the microclimate vary considerably. The same applies to the sound environment and biological factors in large stores [5]. Chemical factors, in the commercial establishments surveyed, appear to a minimal extent. Static load appears to the greatest extent in large stores and to the least extent in shops with articles for daily needs or general stores.

DESCRIPTION OF OCCUPATIONAL SAFETY CONDITIONS IN LARGE STORES

In large stores, errors occur mainly from storage, maintenance of patency and orderliness along transport lines, as well as the furnishings and equipment in the sanitary-hygienic premises [1]. Lack of patency along transport lines usually results from irregular deliveries and excessive departmental overflow throughout a given commercial establishment. Storage of goods in the wrong place (e.g. along transport lines; across emergency exits) turns out to have been the main reason for blocking access to emergency exits which, in an emergency situation, would hamper prompt evacuation of employees and others (customers). Moreover, there have been cases of permanent lock up of emergency exits (the doors being, for instance, shut off with bars or locked and with limited access to the key) – [cf. Fig. 1].

Table 1. Store Categorization by Product Line and Turnover in the Context of Occupational Safety Conditions [based on [2], pp. 100-106, [3] pp. 511-518, [10] pp. 258-265]

Type of Commercial Establishment	Environmental Factors					
	Lighting	Micro-climate	Sound	Biological Factors	Chemical Factors	Static Load
Specialist stores	General	Invariable	Variable[a]	Considerable hazard of infection	Minimal	Medium
Shopping Mall (Department Store)	General and spot lighting	Invariable	Variable[a]	Hazard of infection	Minimal	Medium
Stores with articles for daily needs	General	Invariable	Minimal	Minimal	Minimal	Small
General Stores	General and spot lighting	Variable[b]	Variable[a]	Considerable hazard of infection	Minimal	Small
Discount Stores	General	Variable[b]	Variable[a]	Minimal	Minimal	Small
Off-price Stores	General	Invariable	Variable[a]	Minimal	Minimal	Small
Catalogue Showrooms	General and spot lighting	Invariable	Variable[a]	Minimal	Minimal	Small
Superstores	General and spot lighting	Variable[b]	Variable[a]	Considerable hazard of infection	Minimal	Small

| Supermarkets | General and individual[c] | Variable[b] | Variable[a] | Considerable hazard of infection | At chemistry departments | Too deep baskets – unfit for customer service |
| Hipermarkets | General and individual[c] | Variable[b] | Variable[a] | Considerable hazard of infection | At chemistry departments | Too deep baskets – unfit for customer service |

a depending on the number of customers; b depending on the number of customers and adopted solutions; c shelves illuminated with halogen bulbs and fluorescent tubes

In smaller establishments, more errors have been detected. These, most often, concern employee preparedness for work (professional training, medical check-up, and evaluation of occupational risk), their protection against electrical current, and the marking off and securing of trouble spots (cf. fig. 2).

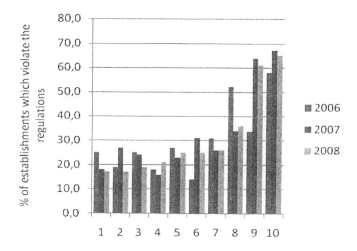

Key:
1. Provision of relaxation breaks for employees who continually perform their work in a standing position.
2. Use of electrical devices and wirings.
3. Immediate access to emergency exits.
4. Compliance with standards as regards moving heavy load by hand.
5. Technical state of workplace objects and rooms.
6. Use of engine carts.
7. The marking off and securing of trouble spots.
8. The furnishings and equipment in the sanitary-hygienic premises.
9. Storage.
10. Orderliness along transport lines.

Figure. 1. Violation of Work Hygiene and Safety Laws in large stores
[http://www.pip.gov.pl/html/pl/sprawozd/08/spraw_08.htm (2009-10-01)]

In the surveyed establishments, irrespective of size, problems have also occurred with access of daylight to workstations. The reason usually has to do with surplus of goods which, due to lack of space in the repository or warehouses, are placed in workrooms, in stacks that shut out the view from the window .

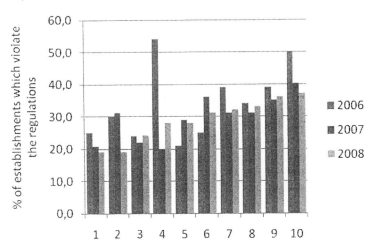

Key:
1. The furnishings and equipment in the sanitary-hygienic premises.
2. Signalling equipment: 'Man in the cold room'
3. Open access to safety regulations.
4. Traffic regulations for interior lanes.
5. Storage.
6. Orderliness along transport lines.
7. Evaluation of occupational risk.
8. Evaluation of the efficiency of fire protection.
9. The marking off and securing of trouble spots.
10. Access to storage instructions.

Figure. 2. Violation of Work Hygiene and Safety Laws in other commercial establishments of retail trade [cf. Fig. 1.]

Considering the scale and type of errors detected, the results of the inspection bear out that the observance of safety regulations in commercial establishments is rather unsatisfactory. It is worth noting that most errors relate to organisational matters [8].

EVALUATION OF OCCUPATIONAL RISK AT A GIVEN WORKPLACE

The concept of risk is used both in colloquial language as well as in the scientific discourses of economists, lawyers and technologists. It has a lot of meanings and thus can be called polysemous. "The Dictionary of the Polish Language" interprets it as follows: 'Risk is a possibility (likelihood) of a failure of an enterprise whose outcome is uncertain or open to doubt'. In economics, the concept of risk most often occurs in the context of insurance risks: it is a possibility of an occurrence provided for by the insurance and the awarding of damages that follows. In legal terms, the concept of risk is similar to its economic counterpart. It is 'a possibility of an occurrence of a damage that will charge expenses to the person who incurs it, irrespective of their being or not being guilty (unless the contract or rule of law obliges another person to compensate for the damage)'1. In so far as risk in its general or economic aspect is taken into consideration, one can talk about so-called actual risk, whose essence is the possibility of an occurrence of damage and its degree. In the case of legal risk, however, the possibility of an obligation to award damages regardless of the doer's guilt plays an essential role.

The definition of risk in its legal aspect additionally indicates, in view of the Labour Law, the risk bearer (who is charged with damages), and also specifies the negative consequences of a particular conduct in a given situation. In the literature of Labour Law, the following concepts of risk are specified:

- personal risk,
- economic risk,
- technical risk,
- social risk.

To avoid going too far into theoretical deliberations on these concepts, one can say that personal risk amounts to a situation where a given establishment takes responsibility for the consequences of an inadvertent error inflicted by an employee, which results in pecuniary prejudice.

As for occupational risk (Chapter X Article 226 of the Labour Code), the person in relation to whom an occurrence is investigated, though not only, is an employee. The negative consequences may be the likelihood of an accident at work or, broadly understood, occupational disease (related to their work, even if it is not included in the list of occupational diseases).

With regard to Work Hygiene and Safety Laws, occupational risk as an actual risk is not a legal concept and for this reason should be interpreted from the economists' point of view, with a focus on:

- the reasons why the risk arose,
- the risk factors,
- methods for its evaluation,
- precautionary measures that should be taken against it.

To sum up, occupational risk is the probability of unwelcome occurrences related to a particular work, which bring about damage, especially health problems that result from occupational risk in the working environment or the way a given job is performed.

Evaluation of occupational risk is an obligation imposed by the rule of law and ought to be observed at every workstation. It basically serves the following aims:

- to check whether all the hazards have been recognised; whether the occupational risk they relate to is known and if adequate precautionary steps to limit its occurrence are being taken,
- to demonstrate, both to the employees and the supervisory board of a given company, that an analysis of hazards has been carried out and that adequate precautionary measures have been taken to limit the risk,
- to make the right choice of furnishings, equipment and accessories as well

as other materials like chemical substances and work schedule for the workstations,

- to establish priorities in operations which aim to eliminate or limit occupational risk,
- to ensure a systematic improvement of working conditions,
- the results of the evaluation of occupational risk should be used to plan corrective action and prevention, which aim to provide for safety and health protection of the employees as well as to improve working conditions.

Evaluation of occupational risk should be treated as a continuous process, leading to a systematic improvement of working conditions. It should be repeated periodically, depending on the hazards. It should be repeated whenever changes are introduced at a given workplace or when some pieces of information (e.g. the requirements imposed by the law) used during the evaluation process have changed [4]. In the surveyed supermarket, the designer's workstation underwent evaluation.

DESCRIPTION OF THE WORKSTATION

It is a one-shift work which, as regards energy expenditure, can be classified as medium-task. The basic range of activity at this workstation includes making posters with product prices; cooperation with the management when preparing promotional campaigns of products, placing promotional posters and occasional banners in the sales gallery; and cooperation when preparing occasional decorations. When doing their work, the designers make use of a computer, a plotter, a scanner, laminators, equipment for work at high altitude, a cutting-platform, a ladder and other manual tools. The designer works in an allotted room, where the illumination is 500 lx and the temperature 180C at the minimum [4]. Special consideration should be given to the way the work is performed, most often at high altitude – by placing promotional posters. The designers may, intermittently, be faced with the necessity of doing their work in a forced or unnatural position [9].

IDENTIFICATION OF THE DESIGNER'S WORKPLACE PROBLEMS

Information on problems at the designer's workplace is included in Table 2. The table shows the results of an analysis of the problems, their sources, their possible consequences and precautionary measures.

Table 2. Trouble spots at the designer's workstation (based on my own survey in hipermarket 'Real')

Hazard	Source of Hazard (Cause)	Possible Consequences	Precautionary Measures
Hand-operated electrical devices (drills, screwdrivers)	Movable and loose parts, an insecure device in hand	Injured hands, cuts, injured eye balls	Measures for individual safety (shades, gloves), acting in accordance with safety regulations
Hitting at motionless objects	Workplace furnishing; work at the area of the supermarket	Bruises, cuts, fractures	Increased alertness, tidiness
Fall at the same level	Uneven, slippery surfaces; mess	Bruises, fractures	Increased alertness, proper footwear, tidiness
Fall down the stairs	Stair surface covered with a layer of filth, water or fat	Limb fractures, spinal injuries, bruises, general injuries	Proper maintenance of the steps, frequent cleaning, proper footwear
Fall onto a lower level	Work at high altitude (ladders, platforms, scaffoldings, lifts)	Briuses, fractures, concussions	Acting in accordance with safety regulations, preventive measures for individual safety, equipment in a good technical condition, increased alertness
Vehicles (different means of transport) in motion	Horizontal and vertical transport; getting hit by an engine cart or a pushcart	Body injuries, serious or minor, depending on the velocity	Increased alertness at crossing internal transport lanes, proper marking of transport lanes
Work in a changeable microclimate	Changeable microclimate, sudden changes of temperature	Frequent colds	Acting in accordance with safety regulations
Wiring	Electric shock when decorating the store with electric lights (230V)	Electric shock	Necessary qualifications, appropriate tools, acting in accordance with safety regulations
Lighting, light-emitting devices	Inappropriate use	Damage to eye or temporary impair-ment of vision	Acting in accordance with safety regulations
Contact with a rough surface	The tools and devices in use	Hand injuries	Increased alertness, preventive measures for individual safety
Getting grasped by moving parts of a device	Electric tools	Finger amputations, crushes, cuts	Increased alertness, preventive measures for individual safety

Work up the ladder	Work at high altitude, fall off the ladder	Limb fractures, spinal injuries, bruises, internal injuries	Medical certificate to ascertain that the employee is fit for work at high altitude, training in work safety, use of ladders in a good technical condition, use of peripheral ladders
Chemical allergens	Toxic, irritant or corrosive chemical substances	Intoxication, chemical burn, irritation	Use of appropriate protective measures as recommended by the producer
Chair, seat	Moveable parts, fall off the chair	Bruises	Acting in accordance with safety regulations
Work in front of the screen	Lack of definition, contrast and transparency of the image	Impairment of vision	Corrective glasses
Body position at work	Physical and static load, lumbar injuries and pains, wrist and elbow pains	Injured disk, degenerative diseases of the spine, deformation of body posture	Proper organization of the workplace, relaxation breaks for the whole organism and the organs of locomotion
Fire	Ignition of a dustbin or other piece of furnishing	Thermal burn, death	Acting in accordance with the safety regulations and fire instructions
Noise	Amplifiers	Impairment of hearing	Preventive measures for individual safety
Stress	Increased alertness while working at high altitude and operating on a low-speed lift platform at the same time	Psychosis, neurosis, sleep disorders, illnesses of the alimentary tract	Training in stress recovery techniques

An analysis of the information included in Table 2. reveals that the most common problems occur through the use of manual tools, when hitting at motionless objects, during a fall and while working up a ladder. Problems also arise when working with lighting and electronic devices. The important health risk factors include fire, noise and stress.

EVALUATION OF OCCUPATIONAL RISK.

The Risk Score method has been chosen to evaluate occupational risk at the workplace. The method requires three parameters: the probability of problem occurrence (P), the seriousness of the consequences (S) and exposure to risk factors (E).

Table 3. Evaluation of Occupational Risk.

Hazard	Probability of Hazard Occurrence (P)	Level of Seriousness of the Consequences (S)	Eksposure to Risk Factors (E)	Initial Risk **WR**	Risk after Correcting Action **WR**
Fall at the same level	1	3	6	18	9
Fall onto a lower level	1	3	6	18	9
Work at high altitude (platforms, scaffoldings)	0,5	15	3	22,5	7
Getting grasped by a moving part of a device	0,5	7	1	3,5	3,5
Work up the ladder	0,5	15	3	22,5	7
Wiring	1	3	2	6	3
Work in a changeable microclimate	1	3	2	6	3
Lamps, lights, light-emitting devices	1	3	2	6	3
Chair	1	3	2	6	3
Body position at work	1	7	6	42	21
Boiling water (common room, coffee, tea)	1	3	2	6	3
Fire	1	3	2	6	3
Noise	1	7	3	21	3
Stress	0,5	3	2	3	3
Horizontal and vertical transport; getting hit by an engine cart or an electric cart	3	2	1	6	3

The highest level of threat against professional risk reduction occurred among workers adopting inappropriate posture during their work and those working at high altitudes (platforms, scaffoldings and ladders). The worked out plan for the reduction of the level of occupational risk has led to the implementation of solutions for the improvement of safety at work and, hence, its consequent reduction, by as much as 50%.

As follows from the evaluation, the risk level at the designer's workplace is acceptable. Based on the results of the evaluation, it has been concluded that: the norms and rules are fulfilled and the rules for risk limits are implemented accurately. To provide for employee security, the newest available technical and organisational devices have been employed and adequate security measures for individuals and human population have been provided for. The fulfillment of the above-mentioned conditions allows to state that occupational risk at the designer's workplace is permissible.

CLOSING REMARKS

The employed Risc Score method of occupational risk evaluation has allowed to determine the actual level of occupational risk at the designer's post in hipermarket 'Real'. It has also allowed to indicate how, at little expense, the risk can be reduced. The evaluation aims to alert the employee, and the employer, to the types of hazards they can encounter when performing their work. It reveals how to reduce hazards at a given workplace or how to exclude them entirely so that the employee can safely perform their everyday tasks and the employer remains safe from damage effects.

REFERENCES

[1] Bojarski M.: *Warunki techniczne i usytuowanie budynków. Zagadnienia administracyjno-prawne*, wyd. LexisNexis, Warszawa 2009, s. 46 i n.

[2] Garbacik E.: *Ekonomika obrotu towarowego i usług*, Państwowe Wydawnictwo Naukowe, Warszawa-Łódź 1980, s. 100–106.

[3] Kotler Ph.: *Marketing, analiza, planowanie, wdrażanie i kontrola*, wyd. Felberg SJA, Warszawa 1999, s. 511–518.

[4] Lis T., Nowacki K.: *Zarządzanie bezpieczeństwem i higieną pracy w zakładzie przemysłowym*, Wydawnictwo Politechniki Śląskiej, Gliwice 2005, s. 63.

[5] Olszewski J.: *Najnowsze tendencje w ergonomii*, w. Prace z zakresu nauk rolniczych i leśnych, Wydawnictwo Polskiego Towarzystwa Przyjaciół Nauk, Poznań 2007, s. 62.

[6] Olszewski J.: *Podstawy fizjologii pracy i ergonomii*, AE, Poznań 1997, s. 5 i n.

[7] Olszewski J.: *The role of the social work inspection agency in work safety and hygiene management*, w. Education in ergonomics and occupational safety, wyd. Publishing House of Poznan University of Technology, Poznań 2009, s. 81 I n.

[8] Olszewski J.: *Wykorzystanie zaleceń ergonomii w ograniczaniu osobom niepełnosprawnym barier architektonicznych i urbanistycznych (na przykładzie miasta Poznania)*, Wydawnictwo Polskiego Towarzystwa Ergonomicznego, PTErg oraz Komitetu Ergonomii Polskiej Akademii Nauk, Wrocław 2009, s. 10.

[9] Praca zbiorowa, *Ocena ryzyka zawodowego. Wykorzystanie systemu STER*, wyd. CIOP, PIB, Warszawa 2008, s. 5 i n.

[10] Praca zbiorowa pod red. J. Altkorna, *Podstawy marketingu*, wyd. Instytut Marketingu, wyd. IV, Kraków 2004, s. 258–265.

[11] Praca zbiorowa pod red. M. Gałusza, *Bezpieczeństwo i higiena pracy w Unii Europejskiej*, wyd. Tarbonus, Kraków-Poznań-Tarnobrzeg 2007, s. 10.

[12] Ślęzak J.: *BHP. Poradnik społecznego inspektora pracy*, wyd. Sanbonus, Sandomierz 2009.

CHAPTER 59

A Challenge for the Future: Efficient, Attractive and Sustainable Factories

Stina Johansson, Åsa Wikberg Nilsson

Department of Human Work Science
Luleå University of Technology
Sweden

ABSTRACT

The concept of organizational development is of essential importance in order to maintain a sustainable industry; this can be seen in numerous articles and research projects presented in scientific magazines, books and conferences every year. In June 2008, the Future Factory project started at Luleå University of Technology, Sweden, inspired by participatory research, aiming at breaking existing patterns in organizational development by involving stakeholders that often are neglected or underrepresented, such as women and young people. The project as such is interdisciplinary; researchers with deep knowledge from different areas; industrial production, design, gender and ergonomics are all contributing with their specific expertise in order to achieve an understanding of the phenomena. This paper will present some of the result achieved so far. The aim is to develop principles that contribute to an attractive, effective and sustainable industry and also to identify or develop methods that support effective change processes within the industry. Early results points at the need of creating flexible organizations with the human factor in focus; offering the employees the possibility to plan their schedule, be part of an ongoing structured development process and stay connected to the inner and outer environment through different networks.

Keywords: future factory, macro ergonomic, life time trainee, participatory

INTRODUCTION

The purpose with this paper is to introduce the reader to the early result from the Future Factory project; a project with the aim to develop principles that support the idea of an effective, attractive and sustainable factory and to test and evaluate methods of an interactive research approach in organizational development. Identifying key factors, seen through a system perspective, will be of special interest for this paper, the result from our method evaluation is presented in another conference contribution. Organizational development is an often discussed field and in order to deepen the existing knowledge the project has taken a somewhat unusual approach, to invite and involve stakeholders that often are neglected or underrepresented in development processes in the industry, in Sweden this is often the case when it concerns women and young people. In the project, young people, labor union representatives, employers within manufacturing industry, shop floor workers and other industrial representatives are stakeholders in the project, but the main contribution comes from the work that has been performed in a design team, solely consisting of women from the industry. As we understand it, women are rather seldom given the opportunity to take an active part in the development process in the industry, this might depend on the fact that women are in a strong minority within the manufacturing industry in Sweden, only twenty percent are women, and by those are a further minority in a position where they can influence the result. At the same time, Swedish manufacturing industry are facing another dilemma, it is a low amount of women (and young people) that are positive to work within the industry or take part of a technical education. By gathering the opinions from these two groups we hope to identify key factors that, if introduced, can lead to a shift in this opinion. A result from this shift could be an increase of the industries attractiveness, lead to a wider recruitment base and thereby an increased possibility to employ high skilled workers at all levels, something that naturally would affect the industry in a positive direction.

THE FUTURE FACTORY PROJECT

In order to predict the future, the project begin from the past, drawing on the concept of 'good work'; a field that originates from the socio-technical school (Thorsrud & Emery, 1969) and became popular in Sweden in the middle of the 1980's when the Swedish Metal Workers Union (today IF Metall) adapted these thoughts (Abrahamsson & Johansson, 2009). The concept of 'good work' focus on the need of creating conditions for good work and a good work place in the industry and can be seen as a way of fulfilling goals of both efficient as well as attractive factories. The theoretical base in this project can also be found in the field of industrial work environment, gender, organization development, design and in modern production technique; this implies that we are not focusing solely on the somewhat hard aspects in a production system; social and cultural aspects such as gender identity, profession identity and workplace culture (Abrahamsson & Johansson, 2006) is of most importance as well.

To reach a understanding of which key factors that are important in creating an attractive, effective and sustainable factory, and in order to find effective ways of working with organization change processes, a number of activities have been performed, which in total have included nearly 140 participants (including 7 researchers) described in brief in table 1 below.

Table 1 Activities within the Future Factory project

ACTIVITY	WHEN	WHAT	Σ
Comprehensive survey of manufacturing industry and literature review	2008	Study and analysis of research grants and contemporary industrial organizations	7
Workshop with Young people	2008	Study and analysis of young peoples values and preferences	23
Workshop with Trade Unions	2008	Values and preferences from a Trade Union perspective	12
Workshops, focus groups- and contextual interviews	2008 – 2010	Values and preferences from stake-holders: industrial employers (CEO'S), shop floor workers ,industrial employees, Swedish Work Environment Authority etc.	83
Design of a proposal for a Future Factory	2009-2010	Workshops with the design team of women technicians, system designer, CEO's, human resource manager, engineers, students, architects and researchers	14 +4.
Interactive approach test/ development of methods	2009-2010	Test/evaluate and develop to reach a 'Future Factory Approach'	(7)
Analysis	2010	Analysis of the material from all of the above performed activities	(7)
Prototyping	2010	Developing a prototype of a future factory, based on the material gathered during the project	(7 +15)
Research team	cont	Collaborative inquiries within research team; discuss, understand and share knowledge of performed activities/studies	(7)
		Total number of participants in project (including 7 researchers)	**139**

The material from the different activities is in the form of recorded interviews, recorded and photographed workshops, notes, mind maps etc and has later been used in the analyse work. The participants have all represented different organizations and industrial sectors; that imply that the key factors were discussed on a system level, not regarding a specific industry.

OUR APPROACH

It is our belief that it is necessary to encompass multiple perspectives when trying to analyze, change or design a new organization, a belief that is shared by many other researchers (e.g. Kleiner, 2006; Haro & Kleiner, 2008; Holden, Or, Alper, Rivera & Karsh, 2008). This insight, together with our earlier knowledge and experiences has guided us through our choices of underlying theories and methods.

SOFT SYSTEMS THINKING

We believe that in order to grasp a true understanding of a complex situation, is it necessary to analyze the situation by iterating between the whole and its parts, understanding relations in the studied system as well as the whole, this will lead to a holistic picture of the studied phenomena.

Soft Systems Thinking, SST, is an interpretative framework that was developed by Peter Checkland and his colleagues at Lancaster University in the 1970's. SST does not, as one might think, consider the world as organised in systems, it stress that in order to understand it better, we can organise it in system models. In order to reach a rich understanding is it necessary to take many aspects in consideration, such as relations and interactions between the different parts, functions, contexts, patterns etc. all with the underlying understanding that the whole is always more important than the parts (Checkland, 1981, 2000; Checkland & Holwell 1998). It's interpretative approach implies that humans perceive the world in different ways, and this has to be addressed in the research or design work as well; we start from an unstructured situation, add the different participant's perspectives and then compare the wanted situation with the existing situation in order to define the changes that need to be implemented (Checkland, 2006).

Successful organizational-level change requires a holistic systems approach, it is necessary to pay attention to all levels of the system, including macro level elements such as culture, management, and the environment, as well as to the interaction-rich system as a whole (Holden et al, 2008).

MACRO ERGONOMICS

One main concern in our project has been to discuss and learn from the experience and knowledge that the participants in the different activities has shared with us in order to reach a understanding of what preferences that is essential in creating a efficient, attractive and sustainable industry. The participants in the project represent different organizations and have contributed with their own thoughts; these have been on several levels, from thoughts concerning design of their actual workplace to complex organizational design concerns. These various levels leads to the field of macro ergonomic, an approach based on the socio-technical system framework. In contrast to the more traditional ergonomic perspective, macro ergonomic is derived out of the need for a larger system perspective, all in order to raise the result of the performed activities (Kleiner, 2008). If traditional ergonomic often concerns micro work environment, how human and a workplace fit to each other, macro ergonomic deals with the overall physical work environment and how this is connected to the business objectives of an organization (O'Neill, 1998). Today it's necessary to understand the complexity of the studied system; the environment is turbulent, it exists in a challenging competitive market and there is a continually ongoing interaction within the organisation (Kleiner, 2006). In order to reach a successful large-scale change in industrial organizations is it necessary to pay attention to several important aspects. The traditional human-technology-organizational structure interface and, not to forget, the interactions that take place between different sub-systems within and outside of an organization needs to be addressed, together with an integrated micro- and macro ergonomic approach at all levels (Kleiner, 2004). Macro ergonomics stress the importance of the employees and other stakeholders participation as essential in order to reach a positive outcome, it is they that have the knowledge of *where* and *what* the problems are and with assistance of trained ergonomics they can correct this problems (Hendrick, 2003, 2008; Zink, 2006). Carayon (2006) also points at the need for researchers to address the end-users when designing a socio-technical system, otherwise we as researchers will fail in producing knowledge, concepts and methods that they can use and apply.

Finally, Zink (2006), stress that in order for macro ergonomic to reach its full power, is it necessary that it takes the step from being a specialist task to becoming a management topic. He continues, claiming that the difference between companies is mainly people, technology and organizational structures may be imitated, whereas the qualifications and motivation of a work force are difficult to imitate. Therefore, he continues, are human resources the most relevant enabler for the success of an organization.

METHODS

PARTICIPATORY APPROACH

It is our belief that involving stakeholders in a change process increases the possibility of reaching a result that will be sustainable and effective; the final solution is developed by the ones that are going to use it and is, if not accepted, at least understood by all involved. This belief has influenced us in taken a participative approach in the project. The project as such has a directly forward striving approach; we are trying to look at what might be, instead of what is, thoughts that often characterize as a design perspective (Simon, 1969). Based on our understanding this implies involvement of stakeholders in the process, to explore their understandings and knowledge's. Design is a research field that has later gained a lot of interest in many other fields where attempts has been made to implement another, more solution oriented, way of thinking and problem solving (e.g. Edeholt, 2007). Therefore, our research project is characterized by an interactive research approach, inspired by the concept of design-thinking and the field of Participatory Design. Participatory design includes methods such as future workshops (Jungk & Müllert, 1987), personas (e.g. Cooper, 2000) and scenarios, described in detail elsewhere (Wikberg Nilsson, Fältholm & Abrahamsson, 2009).

RESULTS FROM THE PROJECT - SO FAR

We are now in the final state of gathering and analyzing the material from all of the performed activities in the project, in this paper we will try to address some of the results.

HUMAN RESOURCE

One thing that has been brought up to discussion in every activity that we have performed is the statement that the human resource must be recognized as the most important aspect in order to make efficient, attractive and sustainable future industries become a reality, whether it concerns individuals or a group of people, the human resource is, the most important factor to address.

LIFETIME TRAINEE

It is our understanding that we can see a shift in attitudes and values where the employees no longer strives at having a lifetime employment at the same company, instead many, especially young people, see their work and employer as a station along their development in life. For instance, under an interview we received information that there had been an unfamiliar situation in a company, former seen as the 'optimal and must secure' employer at the location, a received employment there could be kept until retirement. One employee refused to accept an employment-offer, instead she wanted to work as long as she 'felt for it' and then continue traveling around the world or taking another work – it would be less complicated and she didn't have to take the same responsibility for the performed work with a time limited employment.

At the same time, we have been receiving signals from our participants that many people are searching for another form of lifetime employment, this time with the content of a continuous individual development, a kind of a *'lifetime trainee'*.

To ensure that every individual contributes and develops within the organization, our participants have an idea of organizations that performs individual development – and carrier plans for everybody within that organization, no matter which level or workplace you begin at. The idea is that he/she works in various parts and processes in the organization during a certain period, this way, knowledge and experience can be shared within the organization and each employee gets an overview of the activities. These plans should include continuous training and education, whether it is shorter training in for an example welding or longer education in universities.

One important focus in this approach is to reach a higher level of rotation within the organization. According to our participants a problem today is that there are very few options open for changing work or positions within a company as people often stay at the same position for a long period of time. Adopting this somewhat different strategy could lead to an organization where people are changing workplaces in a more structured way, a way of marketing the industry as an attractive workplace for the employees. Another side-effect is the fact that the industries can keep employees with proper education and an excellent knowledge of the entire organization (that can contribute to an effective and sustainable industry). We understand this as a natural consequence of the development of the knowledge society; to either consider the workforce as disposables (following the same line is to move industries to developing countries), or, which our participants prefers, follow the knowledge societies development and develop industry in the same direction. The idea would also contribute to flexibility for the organizations, since all employees have an understanding of the processes.

FLEXIBILITY

As concluded above, the concept of lifetime trainee has many advantages for an organization; one is flexibility which we will highlight as a concept of its own. This concept has been brought up in a number of discussions, one concerning the employee's wishes of having the possibility to have individual schedules in contrast to the traditional inflexible shift forms that most of the industries apply today. In the flexible schedule, each group, consisting of people from the actual department, has the responsibility and possibility to plan and schedule their presence at work for a couple of weeks ahead, planned together according to what roles that has to be manned during each hour. This idea is already successfully used within Swedish Health Care but is in minor use within Swedish industry, even though both can be seen as a kind of process industry with an input, a transformation of material or services leading to a final output. According to our participants, flexible schedules could be a way of making it easier to balance the private life and the time spent at work. Depending of where you are in your life you have different possibilities to manage shift work; children needs day care and transport to school, making it difficult to start early in the morning or working late. When people are getting older it often implies that it's harder to alternate between the different shift-forms (night, day, afternoon) with problems of getting enough sleep as a result which can result in difficulties for the employee to perform, both at work and in the private life.

NETWORKING

Another key factor, this time seen from the employers point of view, is the possibility to have networks between e.g. local companies that jointly are responsible for certain types of competencies, such as e.g. welders, sheet-metal workers etc. The same thing when it comes to skills that, at least small, businesses find difficult to have in-house, such as specialist competences in ergonomics, organizational and industrial design etc. In Sweden, we see a lot of companies outsourcing the most monotonous and health concerning operations. Our participants see this as one way for organizations to buy out of ergonomically issues. If organizations plan and answer for those issues together, the solution might be better for all concerned. Furthermore, our participants also have an idea of another form of networking, considered to contribute to both an effective and attractive organization (not to mention an innovative organization). Each employee is part of one or several networks, both within and outside the organization. The networks could be about quality, safety, ergonomics or product or production knowledge. Having groups that are responsible for a specific area is not new within an organization, but to have participants in networks from other industries and from all levels of an organization is a somewhat newer idea. According to our participants, the idea of networks would make manufacturing industry a more attractive workplace and can, when fully implemented, contribute to an innovative and effective work force.

DISCUSSION

A large number of activities have been performed in this project. We have involved nearly 140 participants, people representing different industrial sectors, professions, hierarchical levels, genders and ages. We have tried to adopt a true multi perspective throughout the entire process, from the researchers and other stakeholders that are participating in the project, over chosen methods and theories and to our participative approach. From our perspective, this is an advantage as our focus is to achieve understanding of preferences and needs of attractive, effective and sustainable factories. This multi perspective approach has a price, it has been difficult to discuss this kind of questions looking at a special type of industry or organization, so the discussions and the result has been performed on a 'system-level'.

It is our belief that neither of the concepts efficient, attractive or sustainable is working in isolation from each other. An activity that influences a factory concerning its efficiency is most likely affecting its sustainability as well; the same thing with attractiveness, or turning it the other way around.

Finally, it is our hope that this glint into the Future Factory can inspire other researchers and practitioners to reflect upon their work in organization development and maybe see the benefits that an approach as the one we have used can contribute in reaching a richer understanding.

FUTURE WORKS

As for future research, we would like to take on the challenge of realizing our ideas in a real-life situation, as inside an organization that realize the need of new ideas and concepts to meet the future challenges as suggested in the initial future scenario.

As for now, our future works also include finalizing a model or a prototype which is intended to address all the aspects we have found as a proposal for a future factory.

ACKNOWLEDGMENTS

We would like to express our gratitude to all those who has participated in the activities within the project and have contributed with their experience and expertise, without them the project could not have been conducted. The project is funded by AFA Insurance, FAS (Swedish Council for Working Life and Social Research) and the European Regional Development Fund.

REFERENCES

Abrahamsson, L., and Johansson, J. (2009), "The good work – A Swedish trade union in the shadow of lean production." *Applied Ergonomics,* 40, 775-778

Abrahamsson, L., and Johansson, J. (2006), "From grounded skills to sky qualifications – a study of workers creating and recreating qualifications, identity and gender when meeting changing technology in an underground iron ore mine in Sweden." *Journal of Industrial Relations,* 48(5), 657 – 676

Carayon, P. (2006), "Human factors of complex sociotechnical systems." *Applied Ergonomics,* 37, 525-535

Checkland, P.B. (1981), *Systems Thinking, Systems Practice.* Chichester: John Wiley

Checkland, P.B., and Holwell, S. (1998), *Information, Systems and Information Systems,* Chichester, John Wiley

Checkland, P.B. (2000), "Soft Systems Methodology: A Thirty Year Retrospective." *Systems Research and Behavioural Science* 17, 11-58.

Checkland, P.B. (2006), *Learning for action: a short definitive account of soft systems methodology and its use for practitioners, teachers and students,* Chichester, John Wiley

Cooper, A. (1999), *The Inmates Are Running the Asylum: Why High-Tech Products Drive Us Crazy and How to Restore the Sanity.* Indianapolis: SAMS

Edeholt, H. (2007), Design och innovationer. In S. Ilstedt-Hjelm (Ed) *Under Ytan: En antologi om designforskning. (In Swedish), pp. 222-235* Stockholm: Raster förlag

Haro, E. and Kleiner, B.M. (2008), "Macroergonomics as an organizing process for systems safety." *Applied Ergonomics,* 39, 450-458

Hendrick, H.W. (2003), "Determining the cost-benefits of ergonomics projects and factors that lead to their success." *Applied Ergonomics,* 34, 419-427

Holden, R.J., and Or, C.K.L., and Alper, S.J., and Rivera, A.J., and Karsh, B-T. (2008) "A change management framework for macroergonomic field research." *Applied Ergonomics,* 39, 459-274

Jungk, R. & Mullert, N.(1987), *Håndbog i fremtidsvaerksteder.* (In Danish). Kobenhavn: Politisk revy

Kleiner, B.M. (2006), Macroergonomics; Analysis and design of work systems. *Applied ergonomics,* 37, 81-89

Kleiner, B.M. (2004), "Macroergonomics as a Large Work-System Transformation Technology", *Human Factors and Ergonomics in Manufacturing,* 14(2), 99-115

Kleiner, B.M. (2008), Macroergonomics: Work System Analysis and Design, *Human factors,* 50 (3), 461-467

O'Neill, M. (1998), *Ergonomic design for Organizational Effectiveness*, CRC Press LLC, Boca Raton

Simon, H. A. (1969), *The Science of the Artificial.* Cambridge, Mass.: MIT Press

Thorsrud, E., Emery, F.E. (1969), *Medinflytande och engagemang I arbetet; norska försök med självstyrande grupper*, (In Swedish), Stockholm, Utvecklingsrådet för samarbetsfrågor

Wikberg Nilsson, Å., Fältholm, Y. and Abrahamsson, L. (2009), The Future Factory – a concept designed by women and young people. *In proceedings of 17th World Congress on Ergonomics IEA 09, Beijing China*

Zink, K.J. (2006), Human factors, management and society, *Theoretical Issues in Ergonomics Science,* 7 (4), 437-445

Chapter 60

Eco - Ergonomics in Architecture

Andrzej Skowronski

Department for Architecture
Wroclaw University of Technology
ul. B. Prusa 53/55, 50-317 Wroclaw, Poland

ABSTRACT

Global warming, along with the growing prices of traditional energy sources and the shortage of drinking water in the summertime, attracted the world attention to the problem of ecological as well as ergonomic and economical architecture, the problem which apparently has been noticed and even become trendy. In Poland people also want to live 'ecologically', which involves not only a healthy diet or jogging, but also living, working and relaxing in ecological environment, i.e. in the environment characterized by a healthy micro - climate and the space adapted to the needs and aspirations of the inhabitants. In extreme situations, one can notice the phenomenon of escaping from civilization and so called 'return to nature' which often means just moving to the countryside. In architecture, this phenomenon is observed as a desire to build a new, artificial environment for a human in a balanced way, a home which is understood as being:

- ecological, i.e. healthy and built from natural materials
- as autonomic as it is possible, independent of the water and energy supplies from the outside, and not polluting the natural environment (with sewage)
- ergonomic, i.e. adequate to meet the needs of people, according to their preferences and traditions and therefore creating the feeling of well-being
- economical, which means cheaper than the existing building techniques.

Keywords: ecology, ergonomics, economy, architecture

INTRODUCTION

The conducive atmosphere for eco-ergonomic building in Poland is the result of the country's membership in the European Union which imposes the measures to reduce the energy demand, the emissions of carbon dioxide and the preservation of natural resource, especially drinking water. Joining the EU has also resulted in a growing wealth of the Polish society.

Poland belongs to medium-sized (area of 312 685 sq km) and moderately wealthy countries, flat and level in most parts, situated between the Baltic Sea in the north and the mountain chains of the Sudety and Karpaty Mountains in the south. The climate of Poland is characterized by:

- medium insolation (approx. 1000 kWh/m2/year)
- moderate winds, with stronger ones occurring seasonally along the coastline of the Baltic Sea
- medium and small rainfall (approx. 700mm/year)
- low temperatures in winter (at night down to minus 35 degrees of Celsius) and high temperatures in summer (at daytime up to 35 degrees of Celsius).

The factors mentioned (geographical, geological, climatic) do not favor the development and preferring of one only source of renewable energy (such as solar power) and that is why in Poland more economically, ergonomically and ecologically justified are the hybrid systems adapted to the existing local geographical conditions. The geographical location and geological structure of Poland make it possible in some regions to use the underground geothermal water as an important source of so-called green energy.

The building methods in Poland ought to be adapted to the customs grown from the building tradition and adjusted to the technical and economic possibilities of the society. The eco-ergonomic architectural approach should be expressed through the rational building art which often results from the tradition and in the best possible way fulfills the needs of people, as well as:

- applies the technical powers created by the new techniques with regard to the local bio-climatic conditions
- reduces the costs of building, maintenance and recycling.

As it was mentioned above, Poland is a country situated in the zone of moderate climate. Even if the summer temperatures go up to 35 degrees of Celsius, in winter they may fall down to minus 25, or in some regions even to minus 35 degrees. The most intensive sun radiance in the year takes place from April to September, which is the summertime. However, the number of the sunny days in the year does not allow to use solar power as the main energy source for heating purposes. This method can be used only to support the main heating system and for heating water. Poland is a country of moderate insolation (exposition to sun rays) and when it comes to the number of sunny days it takes the last but one position (before Belgium) among the EU countries.

The pure, clean energy produced by water power plants covers only a slight percentage of the general energy production in Poland. Water power stations in the world cover approx.18% of the general electricity demand, whereas in Poland it is as little as 2.5%. In European countries the average resources of surface water reach approx. 4600 m3/year per one inhabitant, which is three times more than in Poland. As far as drinking water is concerned, the situation in Poland is similar to Egypt since Poland is the area where the influences of two different climates, i.e. the oceanic and continental ones often clash. In Europe, while going from the west to the east, the rainfall becomes smaller although evaporation remains stable.

As for Poland, only along the Baltic Sea coastline (length: 450 km) the conditions for situating wind power plants are favorable. Consequently, only there one should build big windmill farms, as well as smaller wind plants which could provide the neighboring settlements with so-called green energy. The amount of energy produced in this way is smaller than 1%, even though it becomes more and more popular. The growth is also visible in the energy produced from so-called biomass, which, however, does not make a significant contribution to the whole energy balance.

Poland has good geothermal conditions, since approx. 80% of the country area is covered by three so-called 'geothermal provinces' (of Central Europe, the Carpathian foreland, the Carpathians). The temperature of water in these areas falls between 30 and 130 degrees of Celsius, and the depth of water layer in sedimentary rocks is from 1 up to 10 km. Actually, a natural outflow of warm water hardly ever takes place. Up till now, in Poland eight geothermal heat stations have been working, and a few more are being built.

WHAT IS ECO - ERGONOMICS

In order to explain the idea of eco - ergonomics one should consider and discuss not only the general definitions of ergonomics, but also the ones of ecology and economy.

ERGONOMICS

The etymology of this word should be traced by the analysis of the ancient Greek words: *ergon*, which means 'work', and *nomos*, which means 'law' or 'principle'. Thus, ergonomics is a scientific discipline that concerns work and its adaptation to the psycho-physical abilities of man.

Ergonomics is the scientific discipline concerned with the understanding of interactions among humans and other elements of a system, and the profession that applies theory, principles, data and methods to design in order to optimize human well-being and overall system performance. Ergonomics is the application of scientific information concerning humans to the design of objects, systems and environment for human use.

ECOLOGY

The word "ecology" can be heard everywhere. It is used and understood in different ways, e.g. when we want to say that something is healthy, we may say: 'it is ecological'. The people who care about the nature are called 'ecologists', etc. The notions: 'ecology, ecological' are often used in colloquial language and their meaning is often wide and far from precision, not always connected with the science of ecology. Ecology is also mistaken for sozology, i.e. the science of natural environment preservation. Therefore, one may ask: what, precisely, is ecology? The word 'ecology' originated from the combination of two ancient Greek words: *oikos*, which means 'home' and *logos*, which means 'study' or 'science'. By analyzing this etymology we may conclude that ecology is a branch of science which deals with our home, especially home understood as planet Earth.

All in all, ecology is a study of the animated nature as a whole, with its structure and working, as well as the mutual relationships of individual organisms or groups of living organisms and inanimate environment. Ecology examines the concrete area, along with the natural habitat of living organisms, which is called 'the ecosystem'. Ecological architecture ought to respect the existing interdependence between man as a living organism and his home understood quite broadly (the place of living, the place of work, relax, a settlement, town, etc) as inanimate environment in which man just exists.

ECONOMY

The economy of the building process is another specific factor of eco-ergonomic architecture. The word 'economy' comes from the Greek words: 'home' (*oikos*) and 'law, principle' (*nomos*). This can be understood as the rules of housekeeping. Some people say that the second part of the word comes from the Greek word 'man, human' (*nomeus*), which – when connected with the word 'home' (*oikos*) - may be interpreted as a man who manages the house, a house keeper (in Greek: *oikonomeo*). The economical aspect should be an important component of eco-ergonomic architecture.

It is not a big achievement to build an autonomic building which would be fully independent of energy and water supplies from the outside, not producing waste and pollution at the investment cost of millions of dollars and affordable only for very few customers. A building is not economical if the cost of the investment due to the building materials and technical equipment is higher than the energy effectiveness and economical profit resulting from the maintenance and finally recycling of the building. The economy of the solution should be assessed at every stage of building life, which means calculating the whole cost of the building process, maintenance and the recycling of the materials after the technical death.

ECO - ERGONOMICS IN ARCHITECTURE

Eco-ergonomic architecture is understood by the Author as a kind of architecture in which the qualities described by the three notions: ergonomic + ecology + economy, briefly: „3 x E", are combined. From the etymology of those words the definition of eco – ergonomics can be derived as:

- ergonomics understood as work + rule / law (in Greek: *ergon* + *nomos*)
- ecology understood as home / planet Earth + science (in Greek: *oikos* + *nomos*)
- economy understood as home + law, rule (in Greek: *oikos* + *nomos*) or according to another definition: home + a man to manage the house (in Greek: *oikos* + *nomeus*).

Eco-ergonomic architecture ought to be first of all rational, harmoniously developed, which would create optimum solutions fulfilling the conditions of ergonomics, ecology and economy to guarantee the well-being of people, the future customers. Eco-ergonomic architecture is not very spectacular and does not break any records in respect of height, gigantic glazing on elevations or innovatory form of the building. Eco-ergonomic architecture is often hidden below the ground or behind the wall of green plants (so-called biological partitions) and is exposed through glazing and sun collectors to the sun. While creating eco-ergonomic architecture one should aim at applying optimum solutions of techniques and materials, adequate for the given climatic conditions and the local cultural traditions in order to organize the material, artificial surrounding for a man meeting his physiological, mental and social needs, which in consequence results in achieving the social satisfaction and the state of well-being.

Obviously, the world highest building, such as the Burj Dubai, cannot exemplify eco-ergonomic architecture, even if they were equipped with the most ecological and cost-effective devices and technical solutions, because the building like that has been constructed almost beyond the human powers as for technical capacity, rational work and economic ability. It does not mean, however, that it cannot be regarded as the world biggest engineering achievement and the symbol of prestige. We also may ponder if the Empire State in New York is more rational than the Burj Dubai, as far as the price of land in New York is considered, but in both cases the cost of building materials, construction and maintenance, as well as the cost of creating the bio-climatic conditions inside those buildings, not forgetting the dangers of keeping it in existence, are higher than the profits resulting from such an intensive, extreme way of building. The maintenance costs of the buildings such as skyscrapers, which are now fashionably glazed, in the summertime may be even higher (due to the necessary cooling) than in the winter, and this is proved by the figures illustrating the consumption of energy. Moreover, the air conditioning installations may be a good living place for bacteria which are harmful for health or even life. If we additionally calculate the costs of the future alterations and demolishing of the building, as well as the recycling of the building materials, it will turn out that the final balance looks even worse.

Similarly, the residential quarters composed of big multi-level blocks of flats cannot be treated as eco-ergonomic architecture, since they take such a large space that within the same area a complex of terraced houses, as spacious as 90m2, on a building lot of 130m2 for every inhabitant could be built. Many residential districts exist in Poland and anywhere in the world. Whether it is nice to live in them, everybody knows. There we have:

- noise and variety of smells coming from neighboring apartments
- insects and rodents living in the installations, especially riser pipes
- feeling not safe and totally anonymous, etc.

The people who live in such residential quarters, if they only can afford, usually escape to the suburbs or outskirts to be closer to nature hoping to find peace and quiet and a better life there.

Eco-ergonomic approach to architecture can be illustrated by the results of the contest held in 2009 whose aim was to reconstruct the building of the Polish Science Foundation [Fundacja na Rzecz Nauki Polskiej] in Warsaw. The winning project by the company FABB Figurski A. and Białobrzeski A. assumed that walls would be covered with green plants growing (so-called „vertical garden"), and the building itself equipped with wind turbines, heat pumps and rainwater collectors.

ECO - ERGONOMIC BUILDING METHODS

In eco-ergonomic architecture more and more popular become natural materials, such as wood, straw, clay, stone, sand metal, glass, as well as any combinations of those materials processed by means of new techniques (for instance: burnt clay – brick, wood-like boards, mineral wool, up-to-date plaster work, clay work, etc.).
The simpler and less processed the material, the better - since its price is lower, it is easier to get, and the micro-climate created inside the building is more favorable.

One of the examples of new ecological materials is a wood wool mat, the material which has been manufactured on a mass scale for several years and is becoming more and more popular as a replacement for mineral wool. It is made of wood chips and hemp, which are natural, renewable raw materials. The wood used in the production process comes exclusively from the forests grown in accordance with so-called ' good forest management rules' where in the production process there are no harmful chemical additives. Wood wool mats are perfect for thermal and acoustic insulation of outside walls, as well as inside walls, roofs and ceilings.

The fashionable trend to build houses from solid wood or wooden logs, as well as wooden framework, which are popular mainly in Canada and the United States and therefore known in Poland as 'Canadian systems', have now been extended by applying new technologies based on straw and clay. This method involves building the supporting walls from pure clay that is a natural and easily available raw material while the filler is made of straw chaff that is renewable too. The building process starts with preparing semi-liquid mixture which is then poured in the wall timbering or, alternatively, hollow blocks are first made from clay and straw and then brick walls are constructed of them on the base of clay mortar. In multi-level buildings the process is completed with a framework construction (mainly wooden pillars and beams) and clay is used as a filler for the supporting framework.

Apparently, the first Polish building made of straw and clay was constructed in 2000 by an architect Mrs. Paulina Wojciechowska in the village of Przelomki. This object could hardly be called 'a house' since its usable area was as small as 30 m2 (outside dimensions: 5 x 7 m) and, having been built without a special building license, it was notified as a farm building. To construct it local raw materials coming from the fields nearby were used. The plain technique involved making walls form clay, sand, straw and water (in the United States the system is called 'load bearing'). The outside walls here do not have any additional supporting construction from wooden beams and can themselves support the garret ceiling and the roof. Not only the wall construction is far from the building standards, but also the ground work. It consists of a thick layer of gravel on which a foundation made of cobblestones cemented with mortar is laid, which created a brickwork as high as 0.5 m. This is an important component protecting the clay and straw walls from harmful moisture. The roof was designed traditionally as a thatched roof. Additional thermal insulation is done by applying so-called eco-fiber which the product obtained by the milling of waste paper with the use of boron salt.

In Poland and in the world houses made from straw are becoming more and more popular. The method involves making a wooden framework which is then filled with blocks of pressed straw or straw bales. Individual straw bales are bound by means of semi-liquid clay mortar. Such straw walls are usually plastered with clay which is very flexible. The people who built such houses praise their values, such as a very good insulation of the outside walls, small weight of the building, a fantastic and healthy micro-climate inside the house and cost effectiveness of the investment – since the cost is more or less a half of the cost of the traditional and popularly used methods. The houses of this type are popular in the United States.

In Poland several houses built by the straw bale method have been started recently, and in the branch catalogs many architects advertise their ready-to-use projects of this type. One of the examples is a house in the village of Lubla (in Podkarpacie province), built fully legally, i.e. according to the Polish building law. This is a one-family house completed in 2008. On its roof there are two sun collectors supporting the central heating and water heating systems.

In Europe and around the world there appear straw and clay houses which are technically more and more advanced. For example, there is a residential house in Eschenz, Switzerland, designed by Felix Jerusalem in which natural materials and prefabrication are combined in an interesting way. In this solution large, specially pressed, prefabricated straw elements are shielded from the outside with a transparent siding. The whole building is situated over the ground which enabled to avoid a complicated ground job and apply only a light foundation (information and photographs can be found on the Internet website: [www.strohhaus.net]. Apart from prefabrication, more and more interesting stylistic patterns of the houses based on the straw bale method can be noticed. In higher buildings the wooden framework is replaced with steel or ferroconcrete constructions.

In mountainous regions in Poland houses are often covered with shingle roofs made of thin wooden boards cut into short sections more or less as big as typical roof tiles. A shingle roof is made of those species of wood that are moisture proof. This kind of material comes from the regional building tradition of Podhale (the Tatra mountains), but it is still trendy in that region and extensively used in newly built objects. In other regions of the country more popular for residential houses, motels and restaurants are roofs made from reed or straw mats .

Another quite popular building material is natural stone. It is very strong and durable and that is why is used to make foundations and to strengthen foundation walls. It does not have a good thermal insulation ability so it is not often used for outside walls. Yet, stone has one more important quality - a high calorific capacity resulting from its weight which makes it useful for making inside walls and floors where it can accumulate warmth.

Nevertheless, the most popular building material in Poland and in the world is undoubtedly brick. It is a natural material, although due to the burning process it is for sure less ecological and less economical than natural clay. The amount of energy necessary for manufacturing brick or hollow brick blocks, the cost of labor and transport and the costs of pulling down the former building make this method much more expensive and that is why we cannot regard it as eco-ergonomic. Anyway, it is all relative since it is hard to imagine that a block of flats as high as a dozen or so levels can be built only from straw, wood and clay. Obviously, higher buildings have to be built, too.

CONCLUSION

Some new building techniques discussed above are not new at all actually. Cheap regional houses, especially those built in the country, have existed for centuries and have always been based on similar materials and construction methods. It can be assumed that the modern building trends often refer to the old, well proved and cheaper methods and techniques, which is the result of not only fashionable ideas, but also of the fact that people wish to live in natural conditions, in harmony with nature. Eco-ergonomic architecture has developed around a great circle going back to the sources, the traditional building methods supported by the new technology enabling to gain so-called green energy.

The building methods in Poland ought to be adapted to the customs grown from the building tradition and adjusted to the technical and economic possibilities of the society. The eco-ergonomic architectural approach should be expressed through the rational building art which often results from the tradition and in the best possible way fulfills the needs of people.

Eco-ergonomic architecture first of all ought to be rational, make a balanced, harmonious progress, apply optimum ideas of ergonomics, ecology and economy to achieve the well-being of man, its user. Therefore, eco-ergonomic architecture can be shortly defined through „3 x E" formula (ergonomics + ecology + economy).

REFERENCES

Górecka, M. (2004), "Architektura energooszczędnego domu mieszkalnego polskiej wsi w aspekcie zrównoważonego rozwoju" ["Architecture of energy - saving residential house of Polish village in view of balanced development"], *Oficyna Wydawnicza Politechniki Warszawskiej, ISBN 8372074712, 9788372047713*, Warszawa, pp. 160

Vale, B., James, R., and Vale, D. (1991) " Green architecture: design for an energy-conscious future", *Little, Brown, ISBN 8389192470, 9788389192479*, pp. 192

Drapella-Hermansdorfer, A. and Cebrat, K. (2005) "Oblicza równowagi: architektura, planowanie na progu międzynarodowej dekady zrównoważonego rozwoju" ["Aspects of balance: architecture, planning on the threshold of the balanced development decade"], *Oficyna Wydawnicza Politechniki Wrocławskiej, Vol. 1*, Wroclaw, pp. 637

http://www.faab.pl/Fundacja_na_Rzecz_Nauki_Polskiej_90_html

http://strohhaus.net/

http://www.strawbale.pl/

http://www.strawbale.pl/straw-bale-w-polsce/dom-ze-slomy-i-gliny-w-przelomce

http://www.strawbale.pl/straw-bale-w-polsce/dom-w-lubli-nowe-wiesci/

Chapter 61

Building with Reclaimed Materials in View of Ergonomic Design

Maciej Skowronski, Jerzy Charytonowicz

Department of Architecture
Wroclaw University of Technology
B. Prusa 53-55, 50-317 Wroclaw, Poland

ABSTRACT

The building sector has a massive impact on the environment in terms of energy consumption, use of natural resources and waste production. The construction industry is under increasing pressure to become sustainable. Modern buildings ought to save energy, be ecological and planned according to the rules of ergonomic designing, i.e. they should protect a healthy environment and comfortable surrounding conditions for the people inhabiting them, and they should not be a nuisance to the natural environment at all stages of their existence. Nowadays re-using materials from the waste stream appears as a critical component of sustainable development. All over the world there is a huge amount of construction waste. The potential to re-use them and to reduce landfill and new materials production growth is enormous. When reclaimed materials are secured from an existing building site, the environmental impact is very low. Even when they are sourced from far away, reclamation process is still one of the most environmentally friendly options for supplying materials to the construction industry.

Keywords: Green architecture, sustainable design, ecology, energy saving, alternative technology, material re-use, deconstruction process

INTRODUCTION

The beginning of the 21st century can be characterized by a high level of urbanization, very high figures of energy consumption and much more intensified circulation of different materials. Natural resources, similarly to the situation in the 20th century, are intensively taken from the environment, processed to obtain the needed materials and used by different industries. This tendency would not awake suspicions, but for the fact that eventually most of those ready-to-use products, after the stage of exploitation, finish their way on the dumping ground. Raw materials that were originally neutral to the environment after being processed, in their lifetime contribute to the pollution and degradation of the natural environment. At the beginning, the problems mentioned, obviously resulting from the industrial revolution, were neglected. It had to take a long time to recognize the negative consequences of the civilization progress and its influence on the environment.

For the welfare of humankind the attitude to nature as an inexhaustible source of raw materials and a place for storing waste ought to be significantly changed in the near future. The growth of population and the increasing exploitation of resources make it necessary to notice and consider the new ecological strategies of engineer's designing. The building sector, as well as other industrial branches, ought to strive to perfect the methods that are currently used. One of the available solutions aimed at improving the existence of contemporary man is the re-consumption of materials. A proper selection and application of reclaimed materials will make it possible to achieve the aim, still bearing in mind a wide spectrum of ecological and ergonomic aspects. Designing and manufacturing according to the natural cycle seem to belong to the key strategies of the balanced development in the 21st century.

REASONS FOR RE-CONSUMPTION IN THE BUILDING SECTOR

Designers and their customers seem to apply or at least realize the idea of using reclaimed materials in architecture. Up till now it has occurred quite seldom and not on a big scale, being mainly applied for building one-family houses and when old monuments were restored. Nowadays, this idea is more often appreciated by leading designer's companies and gradually introduced in order to complete bigger investments. There are three main reasons why the idea of re-consumption in architectural designing is becoming more and more popular: the desire to reduce a destructive influence of the building sector on the natural environment, the opportunity to gain measurable profits due to the kind of investment performed, the chance to create a positive image of the activity of people who take part in such investment.

Anyway, the superior reason for re-consumption is a chance to reduce and finally eliminate the destructive influence of the building sector on the natural environment which involves: excessive exploitation of nearly exhausted resources, pollution of atmosphere as the consequence of new materials production and transport, degradation of the landscape. While the consumer's society in the beginning of 21st century appreciates mainly the advantages of new, energy-consuming materials and technical solutions, the engineers responsible for designing and creating buildings must realize that the influence of the building sector on the environment is the main, dominating factor defining the quality of life and existential conditions for the future generations. Nowadays, in highly developed countries the range of natural resources exploitation exceeds the acceptable limits. If all the world countries claimed as high demand for resources as the most developed countries now, it would be necessary to have the Earth three times bigger for humankind to survive.

Therefore, the re-consumption of potential waste in different branches of economy seems to be inevitable. In Europe the leading position, when it comes to introducing innovatory solutions in this area, is taken by Scandinavian countries, Germany, Austria, the Netherlands and the United Kingdom. While analyzing these examples, one may conclude that the level of re-consumption, as well as the growing automation of reclaiming and segregation of the materials gained from demolishing old objects, brigs measurable effects in the battle with the growing amount of waste. In the segregation process it is quite easy to reclaim for example different kinds of metal products. Steel and aluminum can be sorted out by means of electromagnetic methods which are very efficient. Other materials which can be selected only by human eye and hand cause more problems. However, the process of segregation is steadily perfected and different technical innovations are introduced, such as sorting bricks or hollow blocks by automatic recognition of the color. Moreover, high prices of waste dumping, as well as more precise legislation should help to reduce the large amount and scale of building waste. In this respect, the most expensive European countries are: the Netherlands and Switzerland.

Re-consumption of reclaimed materials for new buildings may often bring measurable profits - both for the investment and for the investor. The most common privileges encouraging people to use the new solutions are: the permission to take no account of the cost of the former demolition and constructing of a large part of the new object in case of an architectural adaptation of the existing object, a chance to exclude a share of costs resulting from the necessity to stores building waste on a dumping ground -- instead, the elements reclaimed from demolition are directly used on the building site. One should also mention an easier procedure for obtaining the planning permission, especially when the investment requires an opinion of the monument conservator, as well as the cost reduced due to the using of reclaimed sub-assemblies within the installations (usually cheaper than in case of new equipment). Next come additional privileges in case of a necessary assessment describing the influence of the building on the natural environment, in some situations it also means the chance to obtain financial support from the government and the opportunity to advertise both the designing team and the investor due to emphasizing their interest in the state of the environment.

The level of the natural environment degradation has become a serious problem recently, being extensively discussed in the media and scientific circles. Due to the growing consciousness of the society in respect of ecology and potential dangers, the public opinion and the governments of many countries have become less tolerant towards the people and companies that do not attempt to make the investment process clean and environmentally friendly. The ecological consciousness and the certainty that the negative influence on the environment will be significantly reduced are the factors that are becoming more and more considered while the contractor is chosen, also in the building sector.

DESIGNING IN ACCORDANCE WITH THE LOGIC OF NATURE CIRCULATION

To solve the problem of the growing amount of waste, later in the 20th century a hierarchy of solutions was established which can be defined through the 3xR rule – reduction, re-consumption, recycling. The order is not accidental since the biggest advantage comes from reduction of the excessive consumption and, consequently, the repeated using of the same products, which means that the decision to classify the product as waste can be made as late as possible. Eventually, the waste that is processed rationally relieves the burden caused by the gaining of new products from original raw materials which - due to the waste accumulation mentioned before - can be reduced to minimum. The 3xR rule ought to be completed with another „R" which symbolizes the re-orientation of views. It must be realized that the problem of waste is the consequence of the traditional, archaic approach to the ways of re-using it since waste is understood as a side effect of the production or exploitation process. The phenomenon of producing so much waste results from the consumer's society activity which does not apply the rules of reclaiming materials and by no means reflects the principles of the nature circulation. Owing to ecological education, engineers in their design work more and more often start noticing the necessity of including the life cycle of materials in their projects and can early enough predict the circulation of the materials in economy.

The concept of re-consumption can be clearly explained by the analysis of the life cycle of the materials (see Figure 1). Since the industrial revolution was begun up till now the majority of industrial production processes (which in the building industry means manufacturing and assembling the components in order to achieve the final product – a building) is performed according to the rule „from cradle to grave" which reflects the linear circulation of components. Yet, the desirable solution aimed at reducing waste would rather close the circle through the repeated usage of materials - the same as it takes place in nature („from cradle to cradle").

In the building industry, it still seems apparently impossible to achieve the closed circulation scheme for the circulation of components. This results from many factors, with the most important being the traditional method of constructing objects, neglecting the fact that the life cycle will be finished and a careful deconstruction will be necessary. It is possible, however, to introduce the idea gradually by the re-using of the materials obtained from demolition, as well as by making new objects specially designed, which will make the process of reclaiming materials easier in the future. Among the existing, accomplished projects it is not easy at all to spot many good examples in which the future deconstruction was assumed from the beginning. They appear only occasionally and the method of assembling the prefabricated elements was used rather to make the building process easier. Now the designers must realize that their work should assume the necessary demolition of the object designed in the future, which will significantly improve the situation on the market of reclaimed materials.

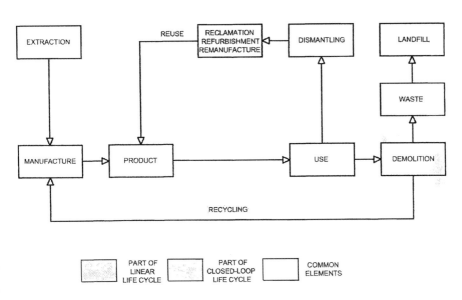

Figure 1. The life cycle of materials.

KINDS OF MATERIAL'S RE-CONSUMPTION IN ARCHITECTURE

The re-consumption of materials in architecture can be observed in three different areas, such as: architectural adaptation directly on the building site (renovation of the existing building to prolong its exploitation in the old or a new function, as well as direct using of individual elements obtained from demolition in the same place), making use of materials and components, obtained from demolition but used in a different place (re-using them and if needed repairing those elements which are out of condition), application of new products, after recycling, containing a partial or full percentage of waste materials.

The smallest intervention in respect of ecological point of view is needed when the building or its part is saved or adapted to a new function in the original place. The undertaking of this kind exerts the environmental influence which is much less negative than the process of a demolition and later reconstruction with the use of new materials only. In this case a highly energy-consuming phase of demolition and transport is eliminated and therefore there is not so much waste produced and it is not needed to bring new building materials. If the investment allows the architectural adaptation, the necessity of replacing the elements which can be easily disassembled ought to be predicted. From ecological point of view, it is essential to try to save the existing foundation, the bearing structure and if possible – the facade of the object. It is usually necessary to have a technical expert's report and, if needed, to improve the parameters of individual elements.

When the keeping of the structure or its elements is not possible in the same location, they may possibly be used in other investments. This will probably require the repair work and upkeep in order to improve the qualities lost in the process of original exploitation. The range of those modifications can be very wide and take different forms depending on the individual component. For instance a steel beam may be just cleaned and once more protected against corrosion, cut for dimension and in ending parts adapted to being joined with the new bearing construction.

Another group embraces the re-processed materials. They are obtained from the waste recycling and their new function is different than the original one. Typical examples of this kind in the building sector are: chip boards made of wood chips which are obtained in the woodworking process in sawmills, the production of cardboard 'timbering' (used for the bearing ferro-concrete elements) from the waste paper, etc. Generally, three sources of waste, which in the next stage of their life cycle become substrates for recycling, can be mentioned: side products while obtaining the destination raw materials (secondary materials), waste materials that are side product of industrial production (by-products) and worn-out elements that are at the end of their life cycle (post-consumer waste).

THE COURSE OF THE INVESTMENT PROCESS INVOLVING THE APPLICATION OF SECONDARY MATERIALS

The main barrier discouraging re-consumption in architecture is a lack of knowledge concerning the mechanisms of propagation for this kind of investment emphasizing its positive influence on the environment. Another obstacle is the still fashionable trend towards newness. Besides, the fact that recycled materials or renovated components are used does not significantly change the course of designing work and the building process. The components and equipment are chosen on the grounds of technical parameters specified. The only difference is the necessity to regard their origin and to choose a contractor who would understand how important it is to use this kind of elements. Other differences resulting from the limited availability of such materials come into view when reclaimed secondary products are used or when the adaptation of existing building structures is performed. If the investors declares his readiness to use the existing object as a whole or a part of it, as early as at the beginning of the designing work and investment planning some important things ought to be done. First of all, one should spot in the area at least a few objects that potentially could be adapted for a new function. The next step would be a full stocktaking and expert's report concerning the technical condition of the object. If the improving of the bearing construction parameters is needed, one should estimate the method and cost of such modification. Also, some architectural expertise seems reasonable to assess the usability of the object, as well as the functional layout of its rooms to know if they may perform the new functions planned, which makes it possible to avoid the expensive changes later, during the investment accomplishment.

The application of secondary building elements purchased on the secondary building market (reclaimed material centers, demolitions, etc.) makes the most important difference in comparison with the traditional investment process. During the designing process preceding the construction from traditional materials the designing team presents the building concept, prepares the building project assuming the application of a given method and makes a detailed project including specification of adequate materials. While accomplishing the investment the contractor just purchases the materials specified. When secondary materials are to be used, the investment process will be much more developed and often requires the needed elements to be purchased immediately after the architectural concept has been approved, before the stage of a detailed design is started. The whole decision-making process ought to be considered in a complex way since one individual component of the object may determine the way in which the other components are incorporated. The decision concerning the method statement, as well as the source of construction elements required, cannot be made before the set of the components needed in the building process is specified. At the very beginning, still before the stage of the investment planning, it is necessary to analyze the situation on the market in respect of secondary building materials available. After gaining the required information one should prepare the concept of the object assuming the application of those. Another phase of the designing process involves the listing of the specified materials and components necessary for accomplishing the object. What should be specified about them is: their approximate shape, size and the technical parameters required (it is worth suggesting a few alternative solutions which will make it possible to continue the investment process when some of the materials are not available, which will result in avoiding the necessity of the general concept modification). Now, the designer starts to identify the places where the specified components are to be obtained (reclaimed materials centers, stock-taking of the objects that are to be demolished, etc.). When this is done, one should assess the quality of the secondary products and next estimate the methods and costs of purchasing them (especially when they are to be obtained from a demolition of other buildings). After the expected financial conditions are set, not exceeding the investment budget planned, the secondary materials can be purchased. When this is done, it is often advisable to verify the whole concept of the designed object and making it more precise and detailed. Next comes the moment of the physical gaining, delivering the products. The deconstruction method of the purchased object is established in order to gain as many needed elements as it is possible. The way of transport and the place of storing should be prepared as well. Now the components ought to be renovated and tested in order to confirm their utility and construction values. After having it done, the materials are delivered to the building site and stored in favorable conditions. As soon as the designer is informed about all technical parameters of the building materials collected, the detailed final working project can be settled.

CONCLUSIONS

The analysis of the investment plan involving the application of secondary materials shows that it is quite complicated in comparison with the conventional building methods. Apart form a positive attitude of the investor and the proper training of the designing team, the success of the undertaking depends on additional factors, such as the more or less advanced level of the secondary building materials market, the knowledge of the building objects in the area which are to be demolished, as well as the ability to estimate the technical condition, costs and time needed to obtain the given components. This, however, does not depreciate the fact that nowadays – due to the environmental degradation growing and with the energy resources shrinking – designers and investors are made to pay attention to the great potential rooted in the materials described above which are paradoxically often omitted in any lists of materials. A proper selection of those makes it possible to achieve the feeling of – let us say generally – comfort for the surrounding of the newly built architectural objects. It also guarantees that the conditions of the human existence due to the reduced pollution of the environment in which people have to live will be improved.

REFERENCES

Addis, B. (2006), *Building with Reclaimed Components and Materials. A Design Handbook for Reuse and Recycling.* Earthscan, London

McDonough, W., and Braungart, M. (2002), *Remaking the Way We Make Things.* North Point Press, New York

Wines, J. (2000), *Green Architecture.* Benedikt Taschen Verlag GmbH, Koln

Chapter 62

Beauty and Ergonomics of Living Environment

Wojciech Bonenberg

Poznan University of Technology
Poland

ABSTRACT

Environment protection or restoring and regeneration of the environment, if it was already damaged, are a purpose of many disciplines. Here, the efficiency of these actions is measured by a set of standard environmental indicators, such as: purity of air, water, soil, and noise level, etc. Whereas aesthetical qualities of the environment, including architectural beauty, are frequently avoided, as an aspect of small value – unnecessary and impractical. It has been forgotten that the beauty visibly influences people, triggers good reactions, promotes good behaviours and attitudes. Beauty sets off behavioural response, documented by ergonomics experiments. Harmonious environment has an impact on human emotional development, activates cognitive processes and processes of identification. For a man, beautiful architecture is a source of satisfaction, a fascinating subject. On the other hand, monstrosity sets off negative behaviours in people living in devastated environment.

Keywords: beauty, living environment, emotional values, urban space

INTRODUCTION

There is a prevailing opinion that contemporary technological environment of a city is both a great achievement of humanity as well as a danger to our social and biological life. Technology is considered a valuable form of human activity within the environment but, at the same time, the consequences of technological activity are more and more frequently criticised. The critique is raised from the point of view of human needs and experience, human emotions related to beauty, truth, and happiness, which failed to be solved by the technology.

Spatial visions of industrial cities were based on the opposition of areas of life related to work (work as an unfortunate duty) and spare time (the only way to express one's individual personality).

From the historical perspective the urban development may be presented in the following way:

1850 – 1917. Industrialisation, spatial expansion of industry. Social changes, intensive growth of urban population. The construction of new technological infrastructures of cities begins.

1917 – 1970. The need for transportation appears on a large scale; rapid development of motorization. Development of large companies, integrated hierarchically - from acquiring resources, semi-finished goods manufacturing, to the production of consumer goods.

1970 – now. Cities in crisis. Deterioration of traditional industries, abandoned post industrial areas. Motorization as a main problem of cities spatial development. Extra-urban localisation strategies of contemporary industries. Ongoing deterioration of city centres, suburbanisation, city centre depopulation. Decapitalisation of city centre buildings, city centre technical infrastructure deterioration.

The crisis of cities triggers reflection on contemporary living conditions of people in urban areas, and is a source of quite common feeling of disappointment. Not only the manufacturing urban areas but also residential areas, city centres, based on automotive transport, have been criticised. The direction of urban technology development has been reflected on; negative opinions of residents have been indicated.

It might be noticed that this is a result of a method which gives the urban environment technical and functional qualities but not emotional ones. Typical division of space according to its functions such as industrial, residential, recreational and services is not balanced by a division which gives the space attributes related to beauty, happiness and satisfaction.

EMOTIONAL VALUES

It is worth mentioning that the post-industrial reality requires new perception method of the role of emotional values (including beauty) in the development of a city. The beauty of a city in functional sense becomes a product (Throsby, D., Withers, G. 1993). It acquires economic value. It is associated with material qualities, which as visual attraction, image, sign, and symbol, are aimed at consumer market (Landry, Ch. 2006). The cityscape is a typical example of such product. It constitutes a contour of the form of a city, built over centuries with the landscape as a background, creates scenic interiors which vary due to their attractiveness to the viewers (Bonenberg, A. 2009). The art of urban composition is able to emphasise qualities such as picturesqueness, uniqueness, atmosphere, for which consumers (e.g. tourists) are ready to pay certain price.

From the economic point of view, the beauty of the architecture, charm of the squares and streets, intimate scale, interesting urban composition, architectonical detail, presence of water and greenery acquire further meaning and have an impact on property prices. They constitute positive factors in the new urban economy, become attractive to these who want to bring in the creative potential of the economic growth. Economic analyses have raised the concept of emotional value of urban space as an important growth factor.

RESEARCH PRELIMINARIES

This article presents results of a study which attributes various emotional conditions to each areas of the Poznań metropolitan area. The metropolitan area covers an area of 2161 km² with a population of 870 000 people.

The study is based on an assumption that people's investment decisions and spatial behaviours depend on their feelings towards urban environment. Urban composition, beauty, harmony, and spatial order can influence behaviours in an effective way. Beautiful surroundings require beautiful behaviours and positive actions, it has an impact on human emotional development, activates cognitive and identification processes. For a man beauty is a source of satisfaction.

And contrarily: monstrosity triggers negative behaviours in people who live in a degraded environment, it discourages, repels, makes people feel unfavourably towards certain places. Between beauty which has a positive impact on residents, and monstrosity which results in negative emotional behaviours, there are a number of transitional states. We can differentiate spaces which cause: fear, anger, disgust, depression, peace, boredom or curiosity, happiness, and admiration.

It is known that the behaviours and investment decisions of space users (investors, residents, tourists) are based on emotions rather than cold calculation. This phenomenon, known well from the theories of marketing and advertising, is reflected also in the settlement areas, which may evoke negative or positive emotions. This aspect of space evaluation has become a subject of a growing number of publications on architecture, sociology, and urban ecology.

MAPS OF THE EMOTIONAL QUALITIES

The aim of these studies is to assess the Poznań metropolitan area suitability for various functional purposes from the point of view of consumers preferences. The preferences have been measured by certain emotional states of people who use the urban space.

Following spaces have been distinguished, which were connected with prevailing emotional states:

1. Space which evokes fear.
2. Space which evokes anger.
3. Space which evokes disgust, aversion.

4. Space which evokes depression.
5. Space which evokes pleasant feelings and happiness.
6. Space which evokes admiration (exciting space).
7. Space which evokes hope (optimism).
8. Space which calms down.
9. Space which evokes boredom.
10. Space which awakes curiosity.

The analyses of the emotional impact in the study which was targeted at the whole area of the Poznań metropolitan area have been based on expert method. The results of the study have been illustrated by maps showing distribution of the emotional qualities attributed to certain areas of the agglomeration (fig. 3.1, 3.2, 3.3, 3.4).

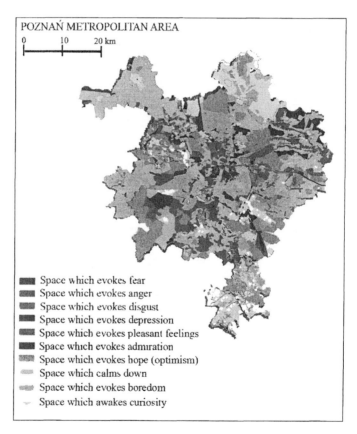

FIGURE 3.1 Distribution of emotional states in the Poznań metropolitan area.

FIGURE 3.2 Aggregation of positive feelings in the Poznań metropolitan area.

FIGURE 3.3 Aggregation of neutral feelings in the Poznań metropolitan area.

FIGURE 3.4. Aggregation of negative feelings in the Poznań metropolitan area.

CONCLUSIONS

Following conclusions have been made based on the study:

- Various areas of the agglomeration show clear spatial segmentation of emotions.

- Preparation of a map of the emotional qualities may constitute a base for locating various functions within the metropolitan area.

- This is an effective (author's) assessment method of usability of an area from the point of view of consumer preferences.

It needs to be emphasised that the study has a form of preliminary survey analyses. It was targeted at the whole metropolitan area, was based on expert method, and carried out by architecture students taking part in summer urban training in July 2009. Undoubtedly these analyses need to be supplemented by surveys carried out on representative groups of residents of communities. Therefore, the results of the study need to be approached with care; this applies especially to the detailed interpretations. General conclusions, however, achieved by the aggregation of positive emotions (happiness, admiration, hope), neutral emotions (peace, boredom, curiosity) and negative ones (fear, anger, disgust, depression) seem to be very interesting. It is worth to notice the map (fig. 4.1) showing spatial segmentation of positive emotions towards the metropolitan area. It might be noticed that the spatial distribution of such emotions is linked with areas designated for development in the city development plans.

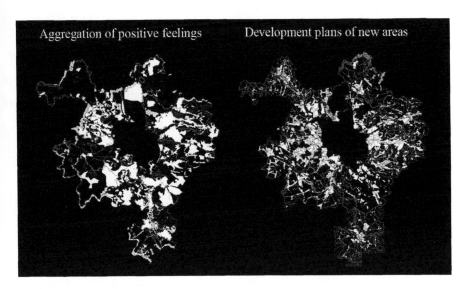

FIGURE 4.1 The comparison of the map showing positive emotions with development plans of new areas in the Poznań metropolitan area.

It means that the city authorities have ascertained correctly that the clients are interested in potential new attractive investment areas and included them in the spatial development plans. There is however the remaining issue of whether such strategy of giving up beautiful landscape, even at a good price, will bring real benefits.

REFERENCES

Landry, Ch. (2006), *The Art of City Making*. Earthscan Publications, London.
Bonenberg, A. (2009), "Media, context and architecture." *Zeszyty Naukowe Politechniki Poznańskiej. Architektura i Urbanistyka*, 17(2009), 61-66.
Throsby, D., Withers, G. (1993), *The Economics of the Performing Arts*. Edward Arnold, London.

CHAPTER 63

Temporary Architecture – Building, Exploitation, Demolition

Przemyslaw Nowakowski

Wroclaw University of Technology
Department of Architecture
Wroclaw, Poland

ABSTRACT

It is generally assumed that temporary buildings are erected to use in a relatively short period. The industrialized countries have a great technical potential for the permanent architecture with a long lifetime and period of usefulness. Also, the economic considerations encourage architecture to be used for several decades. However, both in the past and nowadays there is a strong trend in architecture connected with temporary buildings, exploited over a short period of time. They are erected mainly for special occasions to hold special events and in emergency situations. Paradoxically, despite its short lifetime, many temporary buildings have found their place in the history of architecture and building techniques, sometimes becoming a kind of model or standard for objects with a long exploitation period.
In the paper the following issues will be discussed: evolution of temporary architecture, social, political, economic, functional and technical conditions for its design, erection, exploitation and demolition, and also its influence on social changes and evolution of the traditional architecture, etc.

Keywords: temporary architecture, lifetime of buildings, investment process, ergonomics

INTRODUCTION

The temporary architecture accompanies people since the beginning of their building activity. Its impermanence and transitory nature has created a very little interest over the centuries. Unlike the traditional, permanent architecture, the general use value of the temporary architecture have been considered as less significant. This tendency got turned not until the last decades, when the temoporary architecture has become a means of testing and implementing more modern building systems, technology and proecology solutions, and what is more it has become the answer to short-term changing social requirements, in crisis situations in particular. It has turned out that technical solutions tested in practice in objects used for a short period of time are often applied later in the area of permanent architecture.

OLD AND MODERN TEMPORARY BUILDINGS

The material activity of man in the prehistoric times was related to a temporary character of the living environment. Obtaining food (hunting) forced people to migrate in quest for game. Societies then had a nomadic lifestyle. Elements essential to build a shelter against unfavourable climatic conditions (shelter, shack – household) were carried along with (leather) or were obtained on the spot when stopped (branches, reeds, moss, grass, earth, clay, etc.). Also, natural hiding places such as caves, created by geological factors were sought.

In a further period, tools were made and social division of work and goods manufactured ensued. Development of farming, agriculture, pasturing and further craft forced people to get attached to land and build more permanent and durable households. With the passing of time the first settlements emerged. Developing social norms, since the Antiquity, sanctioned importance of owning land and having own permanent household. Taking advantage of the civilizational benefits then required having a settled life style and creating a permanent urbanized space. Temporary objects lost its significance and did not play an important role any more in the civilizational and cultural development of societies. They had only a secondary supporting role (lavatories, inventory buildings, workshops, etc.). With the passing of time temporary tents and pavilions were erected only, for example, for organizing social events, balls or art performances (theatres, circus performances, etc.). Those objects were often addressed to higher social classes.

Many people, however, did not participate in the civilizational development of countries and led (and is still leading) a nomadic life. The most frequent decisive factor there was living in harsh natural conditions, in an environment unfavourable for leading a settled lifestyle (deserts, mountains, steppes, the Arctic, etc.). The

Eskimos, Indians, Beduins, Mongols, Tuaregs and others have still lived in shelters and tents and changed places of their nomadic lives.

Development of industry caused serious changes in the urban planning and design organization and social processes. Propagating technical solutions required new means of presentation of goods to a growing number of consumers. In the nineteenth century numerous exhibitions and fairs were organized (i.e. on the world scale). Temporary character of such events and moving them from one city to another – important economic centres of Europe and the USA – contributed to a more common use of temporary buildings. Joseph Paxton's Crystal Palace (1851) belonged to a pioneering realizations of that time. That very large exhibition building had a modular steel construction and a glass elevation. It was prefabricated entirely and the assembly process was carefully organized. Since that time temporary buildings have been a permanent architectural and urban element of contemporary cities, particularly during exhibitions (e.g. EXPO), outdoor events, etc.

Despite the expected short-term purpose, temporary buildings, apart from the traditionally erected, now still perform an important urban role and satisfy changing social needs. They can serve the following:

- trade and services;
- exhibitions, installations, fairs;
- mass events (concerts, religious meetings);
- the housing industry (prestigious, social, etc.);
- army and rescue teams;
- exploration and research expeditions in harsh conditions.

Development of use needs, in the scope of art and entertainment in particular, and also research and rescue activity provides explanation for building modern temporary objects.

Known and unchangeable over the centuries temporary forms such as tipis and yurtas have become a model for modern permanent objects, frequently very large and performing complex functions. The principle of a tight and stretched membrane of Beduins and Tuaregs' tents supported of a system of poles has been applied in modern construction systems, e.g. when covering stadiums (the olympic stadium in Munich in Germany by Frei Otto), and also when building suspension bridges. The principle of building tipis in the North American Indian tribes has been applied in spacious bar constructions, and its self-ventilation in designing air circulation in technologically compound buildings (double facades) has been highly valued. The principle of geodetic design of a Mongolian yurta made from elements adjusted to each other has been applied in building stable and light roof coverings with large spans (Fuller's geodetic dome). The above mentioned features, developed in temporary objects known since time immemorial, are successfully used in the contemporary permanent architecture (Kronenburg, 2008, Waterson, 1998).

STRUCTURE, ROLE AND FUNCTION
OF THE TEMPORARY ARCHITECTURE

Modern temporary buildings are usually expected to fulfill short-term users' expectations. They are often erected and disassembled in a hurry, in a short period of time and in unfavourable climate and localization conditions. A success of such an undertaking requires meeting the following demands:

- choice of rational construction system (simple, modular elements and spatial structures – cubics);
- fast and easy production of an object elements outside the place it is designed for;
- fast and easy transport of light elements and half-ready structures of an object;
- easy assembly and disassembly, replacement of damaged elements and structures of an object;
- little participation of people and equipment in the assembly and disassembly processes;
- possibility of immediate handing over and opening an object for use;
- potential integration of a transporting system with modules of an object (mobile architecture, mobile buildings);
- recycling and utilization of used elements of an object.

Quick erection of a temporary building, frequently in different places, requires designing and building a specific construction and building a system. Singular separate elements and modules (cubics) should be made in a production factory independently of the intended location of the object to be erected. They should be assembled in a complete object on the spot. Fulfilling this condition requires application of systems with a considerable level of typization and prefabrication.

Furthermore, a success of systems of multiple use requires constant and instant access to substitute elements of a system replaced in case of damage or devastation.

Not only building technologies affect a form and construction of temporary buildings (particularly the ones used many times). An important technical factor limiting freedom of design is transport possibilities of entire objects and their parts. The sea and railway transport enable to move even entire buildings or their large size parts (usually in packages). Ultimately, however, an object usually reaches its destination by car, and sometimes with the help of draught animals and power of human muscles. Specificity of vehicles, traffic capacity, carrying capacity of roads (width of a traffic lane, height under the flyovers and viaducts) determine then a size of particular building elements. Transport requirements must also be considered and included in the design of forms of buildings themselves as well as in some assembly processes. Popular mobile homes are "caravans" or "trailers" from which many housing estates and settlements in the USA have been erected.

The transport and assembly costs comprise a considerable part of investment costs and may decide about its profitability and cost-effectiveness. Threfore, they must be included and encompassed in the entire economic balance in advance, already at the desgin stage.

Choice of a specific construction system and application of particular building and finishing materials are not free and accidental since they must consider specificity of a quick assembly and disassembly at destination. What is of particular importance here is including people and some building equipment in this process. The economic calculation indicates preference for system solutions with participation of possibly few assembly workers (also not qualified) as well as some light building equipment (mainly manual mechanical tools, seldom electric ones). First of all it refers to erecting temporary buildings in places of calamities, natural disasters, conflicts and in regions with poor underdeveloped technologies (e.g. army camps, refugee camps, camps for victims of natural disasters). In these cases participation of the local people, unqualified technically, is generally essential in the assembly processes.

However, there are numerous exceptions to the rule of the economic calculation and building simplicity, in particular when the market, artistic and aesthetic requirements determine the need of erecting a prestigious temporary object. It refers mainly to various mass events such as concerts, religious meetings, sports tournaments, exhibitions, etc. Temporary objects may be then erected on a one-off basis (e.g. pavilions of countries at EXPO exhibitions, altars at the time of papal pilgrimage journeys, etc.), or the ones built and assembled repeatedly (e.g. concert stages, pavilions at outdoor exhibitions).

Considering the economic calculations and assumption of the temporary nature of objects for some reasons, should not cause (if it is possbile) functional and practical quality deterioration in comparison with permanent buildings. It refers mainly to objects intended for living and work. Safety conditions for their use should be even more strict for a potential mobility, adaptability and transformability, and for the fact, that they are often designed for untrained persons. The temporary architecture then should be assessed in the same way as permanent buildings in terms of the use and technical quality as well as changeability (adaptability for changing needs).

TEMPORARY ARCHITECTURE AS A TECHNOLOGICAL, SOCIAL AND AESTHETIC EXPERIMENT

A dynamic technical progress and social transformations affect faster and faster changes of views and tastes of technologized societies, forms and ways of living, urban structure of cities and more and more modern forms of work environment and manners of performing work. On the other hand, ecologic and environment-friendly trends propagating the "balanced development" have appeared, also renewable energy resources have been used and exploitation and natural resources waste reduced (especially water). Buildings erected conventionally turn out to be useful and suitable for current needs. Many of them, however, quickly lose their functional and practical values and are not compatible with needs changed with time. Like never before, buildings get old morally being little damaged in terms of the technical usefulness. In spite of assessed as having usually over hundred year life-span, many of the buildings are earmarked for modernization or even demolition, already after a few or dozen years. That being so, many architecture theoreticians call for a more extensive use of low processed and renewable building materials, recycling and reuse the existing ones, etc. As a consequence, technical products are required to be more flexible and mobile and are expected to be able to adjust to changing places and functional needs. It refers also to the architectural objects that are supposed to be more easily transformable in terms of form, destination (function), energy and water consumption as well as their localization in space. Sometimes, temporary and frequently mobile objects by definition intended for users mobile spatially and mentally.

Erecting buildings, objects and temporary installations is often favoured by a momentary social demand for an object, not necessarily being a building in a classic understanding of the term. In such a case an investor may commission only a specific "functional space" intended for an exhibition or a concert, etc., limited even only to a roof. A contract formulated in such a way gives a designer a creative freedom, restricted only by a short-term marketing and functional criteria as well as an object location. Technological aspects may be then of minor significance. Therefore, temporary and mobile buildings are often characterized by unrepeatable unique forms and aesthetic values.

The contemporary building industry best performs well and turns out to be good and efficient in some extreme conditions unfavourable to people. The objects in questions are i.e. buildings erected in harsh location and weather conditions (e.g. research stations in polar areas, on surfaces and in depths of oceans, in the outer space, etc.), and also used repeatedly in various places in a very short time (e.g. stages on concert tours of pop stars). Also, it plays an important social role in critical crisis situations. It concerns first of all protection of war refugees, victims of natural disasters (floods, earthquakes, forest fires, etc.), evacuation and quarantine of epidemic and contamination victims, etc. In such circumstances, the injured and harmed are resettled into isolated and controlled areas, to "flats" of poor functional and technical quality (rapidly erected tents, barracks and containers, etc.). Temporariness of such situations in each and every aspect should be regarded literally. It should be remembered that a prolonged stay in crowded objects of a poor quality in non-standard conditions may cause various frictions, tensions and social conflicts.

Two major tendencies in aesthetic shaping of temporary buildings are noticeable. Short-term objects frequently distinguish themselves unrepeatable unique aesthetic values – an individual and complex form, a selection of elegant and expensive finishing materials and a varied functional arrangement. Such features are determined by marketing aspects, a prestige, or a solemn character of an undertaking or event that the object is designed to serve. Objects intended for a repeatable use or mobile ones are characterized by a clear and simple design that is a consequence of its assumed purpose, function and construction. Aesthetic values then may be of minor importance and may result only from the above mentioned primary factors. Constructions logic and simplicity, lightness of objects, a selection of prefabricated materials, ease of assembly may carry themselves significant natural aesthetic values.

Buildings made of recyclable materials may be numbered among temporary building techniques. Living houses built from civilizational waste (recycling) in 60. and 70. of twentieth century counted as examples of an anti-establishment trend in architecture, promoted (in the USA in particular) by groups rejecting the current natural social order. Nowadays, similar objects are numbered among some prestigious trends in architecture, or even avant-gard and snobbish (Slawinska, 1995). What determines a choice of a temporary object for living (e.g. a hut on a tree, a barge or boat on a river, a container in an old factory, etc.) is factors such as: the natural environment protection, favourable location in a city, a unique (even eccentric) house character, aesthetic values, fashion, etc.

A TEMPORARY BUILDING LIFECYCLE

Temporary buildings are frequently designed by design teams attending to permanent architecture. Experiences gained with experimental temporary objects are sometimes used in permanent architecture designs, buildings commonly erected. Contemporary temporary buildings are characterized mainly by a construction lightness and an easy assembly and disassembly. Spatial and coating or layer constructions are also commonly applied. Realization of three-dimensional light constructions is possible first of all thanks to CAD techniques, which in the virtual space allow to check and test correctness of solutions. Use of materials such as polymer, polyester, glass and coal fibres, kevlar, fabrics with a special structure, etc. let create lighter and lighter and more durable constructions.

It is assumed that a period of temporary buildings exploitation is short by definition, especially in relation to the inteded period of permanent buildings exploitation. In case of temporary objects it is easy to design the entire exploitation process of a building and its shutdown and finally demolition. Production, assembly, exploitation, disassembly and shutdown of a building should conform with economic calculations and social expectations as well as they ought to be possibly neutral for the natural environment. A short life cycle of a temporary building provides then a chance to apply low processed building and finishing materials, simple building technologies, and to reduce water and energy consumption in the entire life cycle of objects. It is also possible to use recyclable materials, earmarked for utilization and recycling. When demolishing the object, a possibly full recycling of building waste should be taken into consideration, seldom energy consuming utilization. The example of such ecological building was the Japan pavilion at the EXPO exhibition in Hanover in 2000 with a construction made of cardboard pipes (by architect Shigeru Ban).

The temporary buildings are peculiar experimental objects in which new functional and technical solutions are tested (here also constructional and material ones), since it is rare that models of building designs are realized in a real scale. In the temporary building industry various interests and experiences meet – of designers, builders and industrialists – when testing new designing and technological solutions. Simultaneously, an individual character of temporary buildings makes tested solutions considerably go beyond the area of the conventional building industry. Paradoxically, the success of prototype technical solutions tested in the temporary and mobile architecture do not often find application in the permanent conventional architecture. The example can be realization only two projects *"Wichita House"* from earlier commissioned 37,000 and designed by the famous engineer Buckminster Fuller in 1946 (Kronenburg, 2008).

Apart from a short-term purpose, experiences gained at building, exploitation and demolition of such buildings may and should find a greater application in various fields of technology: the building industry, material engineering, space technology, transport, etc. Furthermore, promoting some solutions such as reducing consumption of building materials, energy and water, may fall into a trend of ecological activities

CONCLUSION

The temporary architecture has been a constant element of contemporary cities and a supplement of the permanent architecture since XIX century. It is a positive and advantageous answer to short-term and changing social needs. Ease of erection and demolition and its short-term exploitation perfectly fits momentary demands at the scale of the country (natural disasters and wars), a city or a region (artisitc or religious undertakings and events) and families (houses and flats). It has got an interventionist character and serves improving the quality of artificial environment of man and satisfying specific their needs. It should be expected that accelerating pace of life, more and more increasingly changing needs of societies and more dynamic technical progress will make the demand for temporary buildings and their role in the city structure now grow in in the future.

REFERENCES

Kronenburg, R. (2008), "Mobile Architektur: Entwurf und Technologie", Birkhuser – Verlag AG, Basel, 8 – 22.

Slawinska, J. (1995), "Ruchy protestu w architekturze wspolczesnej", Oficyna Wydawnicza Politechniki Wroclawskiej, 24 – 48.

Waterson, R. (1998), "Mobility in Traditional Architecture", Detail No. 8, Institute fuer internationale Architktur Dokumentation, Munich, 1372 – 1374.

Chapter 64

Ergonomic Harmonization of Universal Mobility Design in Urban Environment

Sadao Horino[1], Kazutaka Kogi[2]

[1]Kanagawa University, Yokohama, Japan

[2]Institute for Science of Labour, Kawasaki, Japan

ABSTRACT

Access and mobility conditions for wheelchair users in Kamakura, a historic city west of Tokyo, and user-oriented road signs in Yokohama were studied. The study aimed at developing proactive roadmaps for developing universal mobility design in urban environment. For securing mobility for wheelchair users and aged visitors, action checklists listing three criteria were applied to major tourist routes and shops: accessibility for wheelchairs, easy-to-understand signs and multi-purpose toilets. The results were compared with the questionnaire replies of visitors interviewed. This led to some improvements as to height differences, road signs and toilets. Similarly, road signs in complex crossings and intersections were assessed to improve designs of route signs. These results indicate the need for taking participatory steps applying universal design principles in improving mobility along tourist routes for wheelchair users and at complex crossings and intersections.

Keywords: Ergonomics roadmap, universal design, mobility, wheelchair users, urban environment

INTRODUCTION

Awareness is growing about the impact of ergonomics roadmaps on research and practice for improving ergonomic conditions in various environments. Recent experiences in universal design projects and participatory work improvement approaches indicate the need for developing and using action-oriented toolkits applying ergonomics and universal design principles meeting local priority needs.

In identifying and meeting priority needs in daily work and life situations, participatory steps applying locally adjusted toolkits are particularly important. The use of such toolkits in public facilities and route designs has proven useful for improving universal design aspects and mobility for people with disabilities.

A typical example is the design of routes and relevant facilities frequently visited by wheelchair users and various visitors. Recent achievements in improving universal mobility design along tourist routes and complex sections in urban environment are reported.

IMPROVING MOBILITY FOR WHEELCHAIR USERS ALONG MAJOR TOURIST ROUTES

A working group consisting of ergonomists, designers, wheelchair users, students and volunteers examined mobility conditions along major tourist routes in Kamakura, a historic city west of Tokyo. For securing mobility for wheelchair users, action checklists based on the following three criteria were applied to major tourist routes, railway stations and shops: access for wheelchairs, easy-to-understand signs and multi-purpose toilets. These criteria were found appropriate to identify actions necessary for improving the mobility of wheelchair users and aged visitors. The results were compared with the questionnaire replies of visitors interviewed and confirmed valid for pointing out priority improvements.

Figure 1 shows the rates of shops along two main routes in terms of access for wheelchair users. The shops were easily accessible for the users on in about one-third of the shops along the routes. Some of shops with poor accessibility then agreed to improve wheelchair access.

The need for improving route signs in crossings and for installing multi-purpose toilets was found acute along these major routes.

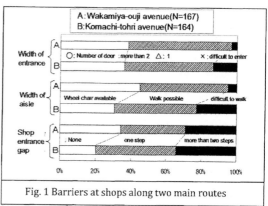

Fig. 1 Barriers at shops along two main routes

On the basis of the study results, barrier-free maps showing accessibility conditions for wheelchair users along five main tourists' routes were designed and uploaded on the internet. Further, at the suggestion of the working group, the city and the railway companies agreed to reduce inaccessible height differences, improve some of the signs and plan new multi-purpose toilets.

The action checklists applied by volunteers and wheelchair users thus proved useful. Different action checklists were used for identifying necessary actions regarding the universal design of the routes, railway stations and access to public facilities. It is suggested to develop locally adjusted toolkits for assessing mobility and identifying necessary improvement actions for use by participatory steps.

IMPROVING ROAD SIGNS AT CROSSINGS AND INTERSCTIONS IN URBAN ROUTES

Another working group comprising ergonomists and students examined the road signposts at crossings and intersections known for confusing signage in Yokohama, By analyzing the traffic conditions from universal design points of view, practical improvements could be suggested and implemented.

A typical example was the design of signposts at a complex traffic intersection where signposts for two crossings were combined. The intersection was used by a large number of pedestrians and car traffic congestion occurred frequently in rush hours. As a result, many car accidents used to be reported at the intersection. The working group recorded the movements of cars crossing the intersection by means of a video camera, and identified causes of confused car movements. Improved design of critical signposts and markings on the road surfaces were proposed and adopted. This led to the elimination of accidents due to confusing signposts.

Improved designs of confusing intersection signposts along the metropolitan expressways were further studied. By recording the car movements due to these signposts, it was found that near-accidental conditions occurred when cars hesitated which way to take, overbraked abruptly or suddenly changed their directions. These conditions resulted from the poor directional signage lacking in the contextual information essential for safe driving. Improved designs of critical signposts were proposed and implemented.

Figures 2 and 3 show examples of improved signages based on the suggestion of the working group.

Fig.2 Improved road sign to the shrine at Kamakura

Fig.3 Suggested road sign at confused intersection in Yokohama

On the basis of these experiences, new toolkits for improving road signposts have been developed. They included ergonomic design criteria for improving the designs of signposts in terms of (1) visibility, (2) legibility, (3) easiness of understanding and (4) context clues. At our suggestion, these criteria are adopted and used by Expressway Public Corporations and road administrators for improving road signposts from ergonomic points of view.

An important aspect of improving confusing signposts, particularly those at complex crossings and intersections, is to take participatory steps that apply universal design principles based on local data and user opinions.

CONCLUSIONS

These results indicate the need for applying universal design principles through appropriate participatory steps in improving mobility along tourist routes for wheelchair users and at complex crossings and intersections. Action-oriented toolkits, such as action checklists listing practical improvement actions based on these principles, are particularly useful in the improvement process. Practical hints may be drawn from these experiences in developing ergonomics roadmaps for orienting ergonomics research and practice towards human-centered design of accessible mobility and urban environment.

REFERENCES

Horino S., "Roadmaps for ergonomics practice in universal design", Proceedings of the Applied Human Factors and Ergonomics International 2nd International Conference, CD-ROM (2008-7).

Mori M., Horino S., Kitajima S., Ueyama M., Ebara T., Itani T., "Ergonomics solution for crossing collisions based on a field assessment of the visual environment at urban intersections in Japan", Applied Ergonomics 39, pp. 697-709 (2008-8).

Mori M. and Horino S., "Participatory approach to evaluation of user-friendly guide sign system at large-scale railway station complex" 41th Annual Nordic Ergonomics Society Conference, (Elsinore, Denmark, 2009-6).

Horino S., Mori M., Kogi K., "Participatory universal design practice at the historic tourist city of Kamakura - Development of barrier-free road maps for wheelchair visitors –", the 17th World Congress of the International Ergonomics Association, (Beijing, 2009-8).

Horino S. and Kogi K., " Developing ergonomics roadmaps responding to diversifying needs in East Asia" , Proceedings of the 1st East Asian Ergonomics Federation Symposium, UOEH CD-ROM (2008-11).

Kogi, K. (2008)Facilitating participatory steps for planning and implementing low-cost improvements in small workplaces. Applied Ergonomics, 39: 475-481.

Ohashi, T., Ebara, T., Shin, H., Mizuno, M., Horino, S., Kogi, K., Sakai, K. and Kishida, K. (2009) Examination of the processes of developing ergonomics roadmaps by means of participatory methods. Journal of Science of Labour, 85: 75-88.

Chapter 65

Ergonomic Assessment of Old Age Homes and Related Health Hazards on Elderly People

Kashyap, S. N.[1], Sharma Promila[2]

[1]Dept of Family Resource Management
College of Home Science, G.B.P.U.A.& T., Pantnagar
Uttarakhand,-263145,India

[2]Dept of Family Resource Management
College of Home Science
Rajendra Agricultural University
Pusa, Bihar, India

ABSTRACT

Old age homes are sharing the traditional role of Indian family due to the some factors like increased urbanization, migration of the younger adults of the family to town, in some cases to foreign countries and increased participation of women in paid employment attributing to increase old age homes in India. The design of old age home and its environment will affect more elderly people and its implications nationwide. There is an urgent need for interventions in order to cope with the consequences of ill design. The purposive random sampling was adopted to select districts, blocks. A sample of sixty elderly people from four old age homes was selected randomly for the study. The study was conducted in Kumaon and Garhwal regions of Uttarakhand. Descriptive cum experimental research design was adopted. Experimental data was recorded through various instruments whereas descriptive data was collected through interview method using interview schedule, checklist and arbitrary scale. The data were analyzed using both descriptive and the relational statistics. The findings showed that the elderly people belonged to 60-80 years age group and the mean age was 72.5 year. Unmarried female and widows dramatically outnumbered male widowers. The females had better smell & hearing organs than males, but general health status of males was better than females. Accidents due to falls/slips occurred among elderly people belonged to the age group of 70-80 years, the cause was slippery flooring. Females were more prone to accidental falls than males. The old age homes and its furniture, infrastructural facilities, fittings and fixtures and storage spaces were not constructed according to the anthropometric data and recommendations given for furniture, building codes of Indian standard and require to be building ergonomically. Majority of males were in moderate risk category, while the females were in high risk prone zones yet they affirmed to be satisfied with the available facilities and in other cases though satisfied but they perceived higher risk. Elderly people were found to be in moderate range of satisfaction level. Skid resistance "Polycemtiles" and "Ergonomically designed Chair" are recommended for the old age homes. This will reduce health hazards and ensure safety. Some of the diseases identified can be treated in the present health system through a health care programmes. The findings of the investigation brought out a number of implications for concerned users, builders, architects, manufacturers, and equipment designers for a variety of products and facility design applications.

Keywords: Ergonomics, old age people, health hazards

INTRODUCTION

Old Age Home is defined as "a residential complex comprising dwelling for one or more types established to provide accommodation for retired person or the aged and may include provisions for community facilities such a recreational or medical facilities and the life for person comprising such community, but does not include an institutional home or a hospital" (Thorne R, 1986). As per recent statistics, there are 1018 old age homes in India today. Out of these, 427 homes are free of cost while 153 old age homes are on pay and stay basis, 146 homes have both free as well as pay and stay facilities and detailed information is not available for 292 homes. A total of 371 old age homes all over the country are available for the sick and 118 homes are exclusive for women. A majority of the old age homes are concentrated in the developed states including Gujarat (Directory of Old Age Homes in India, Help India, 2002). Among the population aged 55 and above in Asia, there are about 90 men for every 100 women; for those aged 75 and above, there are only about 70 men for every 100 women (Lee and Mason, 2000). Projected increases in both the absolute and relative size of the elderly population in many third world countries is a subject of growing concern for public policy (Kinsella and Velkoff 2001; World Bank 2001; United Nations 2002; Bordia and Bhardwaj 2003; Liebig and Irudaya Rajan 2003). In India, the population of the elderly is growing rapidly and is emerging as a serious area of concern for the government and the policy planners. Consider the facts first—7 percent of India's population is elderly today. The definition of elderly as defined by WHO and other agencies is above the age of sixty years. But interestingly, now the agencies divide the elderly population into three age groups:

1. Young-old-aged---- 60 to 70 years old
2. Middle-old-aged---- 70 to 80 years
3. Old-old-aged---- above 80 years.

Research has shown that one of the serious problems in elderly care is the unsuitable residential environment for the elderly who are physically and psychosocially fragile (Choi 2000; Lee 1998; Oh 2000; Park et al., 1998). The role of the environment in the health improving process is a growing concern among architects, designers, environmental psychologies, and health care providers (Devlin, 2003). The Indian family has traditionally provided natural social security to the old people. However, in more recent times, institutions such as old age homes are sharing the traditional role of family due to some factors like increased urbanization, migration of younger adults of the family to towns and in some cases to foreign countries and increased participation of women in paid employment outside the home. As the proportion of elderly people in the Indian population continues to increase; there are demands for increase in number of old age homes in India and consequent to these there is need to ensure safety. Until recent times, ergonomics have tried to improve industrial situation by improving the condition of work but workstations and furniture in residential building particularly in old age homes have been neglected to a great extent. Certain research questions like Are the old age homes build according to the National Building Codes and with the guidelines of ergonomics principles? Are the infrastructure facilities provided appropriate? Are the furniture build according to the Indian furniture standard and by keeping the anthropometric data of elderly people? Are the fittings and fixtures suitable to them? Do they find themselves experiencing any discomfort/pain while using storage spaces and furniture? Are the environmental parameters-temperature/humidity, noise and illumination level comfortable to them? Are the constructions of flooring in habitat room and bath non slippery? and so forth intrigued the investigator. Hence, in a quest to seek answers to some of these queries the present study was envisaged on the topic entitled "ergonomic assessment of old age homes and related health hazards on elderly people" with following objectives:

1. To find out the general profile of elderly people.
2. To study the existing construction and dimensions of old age home in the light of ergonomic principles.
3. To study the infrastructure facilities available in old age home.
4. To find out the satisfaction derived by the elderly people from the old age home.
5. To study related health hazards faced by the elderly people in old age home.
6. To develop a guidelines for functional constructive unit for elderly people.

MATERIALS AND METHODS

The purposive random sampling was chosen for selection of area and samples. The present study was conducted in two regions of Uttarakhand State, India i.e. Kumaon-Nirmala and from Garhwal-Geeta Kutir, Vridh Sewa, and Prem Dham were selected randomly. Descriptive data were collected from randomly selected four old age homes. Fifteen elderly people were selected randomly thus making total sample of sixty elderly people. The stature was measured in cm with a tape measure and the rest of anthropometric measure with a standard professional anthropometric kit. Anthropometric measurements of twenty per cent out of total elderly people (N=60) for the study were randomly selected from the old age home (Geeta Kutir) of Haridwar, District located in the Garhwal region of Uttarakhand State, India The data were collected in two phases, in first phase descriptive data through interview method, whereas in second phase the experimental data recorded through various instruments.

Different tools interview schedule, checklists and experimental tables were used as instruments for collecting data. Corlett and Bishop CRS was used for perceived pain intensity level. A three-point rating satisfaction scale was developed and was judged for content validity and reliability using split-half method. All the responses received on the data sheet were categorized and analyzed using both descriptive (frequency, percentage, average and standard deviation) and the relational statistics (chi-square, r value, fisher t test). Percentile of 5th, 50th, and 95th for the various body dimensions were estimated.

RESULTS AND DISCUSSION

Majority of elderly people belonged to 60-80 years age group. The mean age was 72.5 year with 7.18 S.D. This finding is consistent with other study (NP Das, 2004) that about 61 percent were above the age of 69 years, 32 percent were between 60 and 69 years while the remaining 7 percent were less than 60 years old. As for gender, the Census indicated that the 60 + category favored the males, but in the 70+ age group, the ratio of females was higher than males, which is explained by the higher life expectancy at age 60+ for females in comparison to males. The studies show that it was the women who suffer most as larger proportion survive longer than their spouses. Widows were with sizeable proportion than the elderly. Specifically, the Indian women married to men, who were 10-15 years older than their wives. Hence, they are the have to endure longer periods of widowhood. The educational characteristics indicate that the elderly people were literates and represented different caste/religion. In old age home Ist (Vridh Seva), IIIrd (Nirmala), and IVth (Prem Dham) majority of the elderly people were independent, respectively, and they are paid for maintenance, while in IInd (Geeta Kutir) majority of the elderly people had no income and were dependent on the authority. Recent Indian Council of Medical Research (ICMR) studies in Chennai, Lucknow, Delhi and Mumbai have revealed that out of the surveyed older population, 52 per cent did not have any income. A study in Bangalore (Jai Prakash 1998) found that the fear of physical dependency, (including being sick, or disabled) rather than economic dependency was a major cause of worry for the elderly. Elderly (53.3 per cent) usually led fallen between the age group of 70-80 years. This finding is consistent with other studies (Donmez and Gokkoca, 2003) that majority of accidents involving older people, both fatal and non-fatal, are falls. Out of which 75 per cent were females and had restriction to daily activities. Falling is not only an important marker of frailty (Davis JW, et al., 1999, Walston J, et al.,1999 Nourhashemi F, et al., 2001) but frequent falls can play a role in accelerating the downward spiral in a frail older person (Fried LP, et al., 2001). Women had a higher fall rate in all age groups and the severity of their injuries appeared as their age increase. The findings on gender differences match other research showing that elderly women report more accidents than elderly men (Mary LM Gilhooly, 2007). It was proven that the occurrence of accidents depends on the age and sex in (Nirmala and Prem Dham old age homes. The findings of the study revealed that the many accidental injuries occurred in the bathroom followed by separate toilets and corridors. Causes of accidents were slip followed by fall. Tripping was not a cause of any single accidents in the studied group. Lord et al.

(2001) have summarized and classified the risk factors for accidents as psychosocial and demographic factors, postural stability factors, sensory and neuromuscular factors, medical factors, medication factors and environmental factors. Many accidental injuries occurred in the bathroom and could have been avoided with common sense and inexpensive measures. Research has shown that many home accidents are preventable (Davison et al. 2005). Studies that aim to determine the causes of home accidents in elderly are crucial for taking safety measures.

The good old age "health is wealth" is even truer in respect of old people. Ageing is a time of multiple illness and general disability. Along with the changes in the biological compositions, life style factors are also important for disorders and diseases in old age. Old age diseases are not always curable, implying a strain on financial as well as physical health infrastructure resources, both at the macro and micro levels. At this point, it is sufficient to note that those who perceived their health as "excellent" or "very good" were found to be suffering from physical disabilities such as visual or hearing problems. In both rural and urban areas, more females than males are physically disabled (Kohli, 1996). However, the females had better smell organs than males in all the four old age homes. Depression was major psychological health problem in both the sexes. Major diseases identified were hypertension, osteoporosis, heart disease, asthma in both the sexes. The role of the environment in the health improving process is a growing concern among architects, designers, environmental psychologies, and health care providers (Devlin, 2003).

A dwelling must satisfy user wants and provide user needs, the most important part is to adequately dimension and order the functional spaces of the physical environment as these are often difficult to change after construction. The findings of Fisher's t values showed that the dimensions of habitat room were found in accordance with National Building Codes, whereas the separate water closet and store room were different from National Building Codes and require to be building ergonomically.

Safety measures are also related to comfort. Storage that does not require too much twisting, bending or reaching ensures that people do not challenge their bodily limits too often, inviting trouble. Easy-to-use storage also reinforces the user's sense of capability and independence, and hence his/her psychological well-being and confidence. The existing storage facilities provided in old age homes were uncomfortable specifically the lower shelves were much more stressful and require to be building ergonomically.

The infrastructural facilities in old age home IIIrd were superior to others but still not appropriate and require to be building ergonomically. The furniture i.e. chairs, tables and beds were not constructed according to the anthropometric data and the recommendations given for furniture codes of Indian Standard.

Supportive environments, both physical and social, are not only key determinants of health but also essential conditions for healthy ageing. The average relative humidity level of four old age homes was 67.75%; with 2.62 S.D. falls in humid range but found to be within the recommended value. The average ambient temperature in all the four old age homes was 27.6°C in the month of August, 2006. The average noise level in all the old age homes was 41.95 dB; with 3.19 S.D. The average illumination level of four old age homes in habitat rooms was 35.92 Lux with 6.11 S.D. and in bathrooms 21.37 Lux with 3.44 S.D. was found to be far less than the prescribed Indian electrical standard i.e. 50 Lux for habitat room and 100 Lux for bathroom which showed that the old age homes provided insufficient illumination level which may be near fatal to them. Improved design necessitates coupling age bands with a range of environment (e.g. housing, work environments, public places) and assessing associated physical and mental demands (Neil Charness, 2008). Elderly people remain seated at old age home for a considerable amount of time. Static posture and prolonged sitting in a forward bending position, as elderly often acquire, puts an extreme physiological strain on the muscles, the ligaments and in particular on the discs (Bendix, 1987; Brunswic, 1984).

For the assessment of risk, the scores earned by elderly people were categorized in three ranges: low, moderate and high risk. Insufficient lightings in bathrooms, stairs, corridors, absence of number of luminaries for light during night hours, lack of generator facilities, lack of bell, absence of grab bars in bathrooms, slippery flooring, improper drainage during rainy season, absence of rubber mats in likely to be wet area were perceived risk by both the sexes. The elderly females perceived higher risk for features like-unreachable height of ventilators, latches of doors, windows, and storage spaces. Majorities of elderly males were found to be in moderate risk category, while the females were in high risk category. In old age homes Ist and IVth the elderly people were satisfied but they perceived higher risk. Consequent to these findings (Gill and colleagues, 2000) affirms that 44% of falls occurring in the presence of one or more environmental hazards suggests that their elimination might result in a decrease in falls and fewer significant injuries. A recently published study assessing floor slipperiness also indicates that both objective and subjective measures of slipperiness are important in field studies and that an average friction co-efficient and subjective perception may be in fair agreement and both might be good indicators of slipperiness (Chang et al, 2004). Lord et al. (2001) have summarized and classified the risk factors for accidents as psychosocial and demographic factors, postural stability factors, sensory and neuromuscular factors, medical factors, medication factors and environmental factors

Housing satisfaction is an area of research that has received considerable attention. Much of this literature is concerned with examining the types of factors that may have an influence on satisfaction. For example, among the factors found to affect housing satisfaction most strongly is inadequate space (Kinsey and Lane, 1983). Majority of elderly people were found to be in moderate range of satisfaction level Consequent to these findings Bagga (1997) mentioned that the under-engaged and idle residents of old age homes feel more depressed and listless. The elderly people belonging to the old age home IInd though were found to be high risks prone yet they affirmed to be satisfied with the available facilities.

CONCLUSION

To promote healthy ageing among the elderly with chronic illness, we need the collaborative efforts of government agencies, non-government organizations, and the community as a whole to carry out interventions that focus on meeting individual needs and enhancing the quality of life. By providing a supportive environment, we would encourage aging in place while preserving the life-styles and preferences of our senior citizens. Government intervention will be needed in formulation of standards and protocols for specialized products in order to protect older people's rights as consumers. The Eighth and Ninth Five Years Plans, however, incorporated fairly more specific and comprehensive welfare measures for the elderly such as provision of old age homes, day care centres, Medicare and no institutional services. However, the issue of older persons' learning has not been given any importance in the government policies and programs.

RECOMMENDATIONS

As an older person requires attentive care and special treatment, there is an urgent need to provide a comfortable living to elderly through old age homes by simple modification in designing the old age homes and carefully selecting the infrastructural facilities, the recommendations are as under:.

FITTINGS AND FIXTURES

Installation of door level handles, larger and well located electric switches, visual as well as auditory alarm system, sliding doors (particularly for cabinet and cupboard), pull-push handles, and handrails along both sides of building corridors are essential requirements in old age homes.

BATH FIXTURES

- A raised toilet seat with hand grips on both sides helps. Also, showers should be of non-slip material underfoot and grips at both standing and sitting level.
- The toilet should be positioned next to a sidewall to allow the installation of an assist grab bar.
- Provide a toilet with a seat height of about 18″ (the standard is 16″).
- Avoid the available high seat toilets that are 20″; the feet of a short person may not reach the floor.
- The electric vent fan either wall or ceiling mounted, gravity flow vents, and shafts that allow air to escape to the outside.

TOILET

The standard 15- to 17-inch height of toilet seats creates a problem for many people, especially those with arthritis, hip, knee or back problems. Elevating the seat from 5 to 7 inches will give better leverage in regaining a standing position. There are several types of removable and permanently fixed raised toilet seats available, two examples are:

- A molded plastic seat is the simplest way to increase seat height by about 4 inches
- An adjustable seat will add from 3 to 6 inches of height.
- For a more permanent raised toilet, a plumber can put the stool on a wooden platform made to fit the toilet bowl base.

If building a new bathroom, consider a wall-hung toilet (see Figure 4) that can be hung at any height.

SPECIAL FEATURE (PORTABLE BIDETS)

A special unit (portable bidet) for cleaning the perinea area without hands or paper may be attached to any standard toilet bowl. It is an electrically powered unit with a mechanism for spray washing with warm water and drying with a flow of warm air. This promotes independence for persons with every limited hand/ arm fuctions.

GRAB BARS

Grab bars around the toilets are for safety. Many types are available. The choice will depend on 1) available wall space near the toilet; 2) nearness to other fixtures in the room; and 3) needs of people in the household. Basic types of toilet support bars include:

- Wall-mounted on the back wall behind the toilet
- Wall/floor mounted
- Slip-over guard rails
- Other safety features

LIGHTING FIXTURES

Good lighting not only assists with reading and safety, it helps prevent depression for those confined indoors.

- A flashlight at bedside and night lights along the route to the bathroom are essential and pay for themselves in accidents prevented.
- If cost is a concern, the improved, inexpensive fluorescent light bulbs may be satisfactory.
- The stairs should be well illuminated by natural or artificial light. A light switch should be located at each end of the stairs. The fixtures should not cause glare, nor should it cause any shadows.
- Wall fixtures are preferred over ceiling fixtures for easier and safer bulb changing. Provide night lights in hallways and bathrooms.

ENVIRONMENTAL SAFETY

It can be maximized through sound insulation; improvement in residential temperature control would avert temperature extremes. Maintain a relative humidity of 30-70% at 65-75 degrees Fahrenheit to accommodate a decrease in thermal range tolerance that comes with age.
Furniture and furnishing

The types of furniture chosen for a facility can impact on a person's mobility and, therefore, their independence. Furniture is also a safety concern for the elderly because older people are often prone to falls due to general weakening as a person ages.

Here are some things to keep in mind regarding furniture in a facility: Special care should be taken in selection of furniture and furnishing at old age homes.

- The chair redesigning can be done with proper cushioning and back rest with slop of 1100-1150 angles with arm and foot rest.
- Lightweight chairs with high arms will help a person move the chair while seated in it.
- The arms of chairs need to extend to the edge of the seat to facilitate pushing off. Chair arms should have flat tops and ends designed for grasping.
- A high back on a chair will decrease the chance of falling backward for those people who rock to get out of a chair.
- A seating height of 17-18 inches and a seating depth of 18-19 inches will allow a person's feet to touch the ground without putting pressure behind their knees.
- The bottoms of chairs should be open so a person's feet can be tucked underneath as their weight is shifted forward to stand.
- Chairs with casters are unsafe and should not be used.
- A redesign of current operating room tables may be required to meet the ergonomic guidelines, table height should be 66.6 cm.
- A round table with a sturdy pedestal base will allow more seating space with the ability to adjust for wheelchairs.
- The bed should not be too low/high, it can be an appropriate height for easy transfer on and off. The metal cots can be replaced with wooden with thick mattresses to avoid back pain.
- Beds should be set up so that when seated at the side of the bed, a person's feet will rest firmly on the floor with their legs at right angles.
- Mattresses should have firm edges to make getting in and out of bed easier and decrease the occurrences of falls.
- Use an attachable handrail on beds when needed. The handrail should attach securely to the frame, be height-adjustable, and swivel and lock in place.
- Consider alternatives to standard spring and coil mattresses, such as an "opencell" foam mattress or a large-celled alternating pressure mattress, in facilities where pressure ulcers may be a concern. Sheepskin pads or rubberized sheets with a soft backing can also be used.
- The wooden walking cane should be with a crook handle and length adjustable from 74 cm to 97 cm.
- Furnishings should be stable, with rounded corners for safety.
- A variety of heights should be available for counters and shelves.

COLOUR CODES

There was lack of colour codes and visual contrast and this can be achieved either through brightness or colour difference to make the key features more perceptible.

- Paint doors and their trim a contrasting colour from the walls. Door thresholds should be flush with the floor, or painted a contrasting colour. Doors left halfway open can be dangerous. One way to solve this problem is with gentle self-closing devices.
- Interior colour schemes with good contrast help elderly people with diminishing eye sight identify doors, entrance ways and changed surface conditions.
- Switch plates, and electrical sockets, that are of contrasting colour from the wall will be easier to see. Use light shades of paint on walls. Flat finishes help eliminate glare.
- Light filtering curtains, blinds, shades and awnings cut down on glare through windows, but allow plenty of indirect sunlight into the house.
- Safety architectural symbols should be marked for electric installation like fuse board, fan, regulator, switch outlets and so forth.

DESIGN REQUIREMENTS

Home modifications can include changes or additions to the structural layout of the home, such as widening doorways; installing special equipment like grab bars, handrails and ramps; changing the layout of furniture; and adjusting the way the clients use their homes, such as moving to the bedroom to the first floor or changing the use of a room. Elimination of level differences, rounded corners and edges, easily operable facilities like accessible storage spaces (cabinet, cupboard and rack).

STAIRS AND STEPS

Stairs and steps both indoors and out are formidable obstacles to many elderly people and are the location of many seriously debilitating accidents. Stairs and steps should be avoided in environments designed for the elderly people or minimized as much as possible and these can be replaced by providing stair lift.

- Handrails should be placed on both sides of the stairs, even within the dwelling unit.
- The rail should extend beyond the steps at the top and the bottom

Design consideration should be more for female than male as the female survive and suffer more. Knowledge about the biomechanical, physiological, and psychological effects of aging is needed when designing ergonomic spaces for the elderly. A regular visit to old age homes by geriatric doctors and nursing care such as dressing, medication and injection etc. are highly needed. The old age homes cannot offer the same satisfaction as the person's own home, but the institutional environment can be enhanced through the following:

- An attractive décor.
- Inclusion of the individual's personal possessions.
- Respect for privacy and personal territory.
- Recognition of the individuality of the resident.
- Allowance of maximum control over activities and decision making
- Environmental modifications to compensate for deficits.

- Social, recreational and educational activities provided to develop their potential to promote a family atmosphere and to enhance mutual support among elderly people.

FLOORING

As we reach our later years we can start to feel more vulnerable within our homes. Often, this is simply due to us being less mobile and more at risk of suffering a fall slip or trip but there are other factors which elderly people and their families should also consider. Accidents in the home unfortunately tend to occur more frequently and become more distressing as we get older. Falls, slips and trips are the most common cause of physical injury to the elderly but the risks of them occurring can be reduced if we take some precautionary measures. The floor surface contributes too many trips and falls.

- Extra power points eliminate the tangle of loose extension leads that can cause unexpected falls
- Eliminate unnecessary changes in floor levels wherever possible
- Using more impact-absorbent floor surfaces reduces the potential level of injury caused by a fall
- Mosaic tiles are also popular but should be used on walls instead of floors as they are slippery. Changes in floor levels must be avoided.

Smooth floor surfaces should be minimized. In potentially wet areas such as kitchens, baths, entries or laundries, unglazed "Skid Resistance Polycemtiles" which was developed by Central Building Research Institute, Roorkee, (India), and this tile is recommended for the old age homes.

REFERENCES

Brunswic M. (1984), "Ergonomics of seat design". *Physiotherapy* 70 (2), 40-43.

Bendox T. (1987), "Adjustment of the seated workplace- with special reference to heights and inclinations of seat and table". Dan. Med. Bull. 34 (3), 125-139

Bagga Amrita. (1997), "A study of women in old age homes of Pune", In: Chakravarty Indrani (ed.), 1997. Life in twilight years. Calcutta: Quality Book Company, 1997, pp.171-175.

Bordia, Anand and Bhardwaj, Gautam (ed.) (2003), "Rethinking Pension Provision for India."Tata Mcgraw Hill Publishing Company Limited New Delhi.

Choi M K. (2000), "Analysis of housing environments of he elderly households and its suggestions in preparation or aging society". *Journal of Architectural Research*, Vol.16, No. 9, 29-39.

Chang W.R;Li, K.E;Yueng-Hsiag;Huang. Y H;Filiaggi, A; Courtney TK., (2004), "Assessing floor Slipperiness in fast food restaurants in Taiwan using objective and subjective measures" 35 (4) 401-408.

Devlin A. S. (2003), "Health care environments and patient 0utcomes: A review of the literature". *Environment and behavior* vol. 35 No.665-694.

Davison J, Bond J, Dawson P, Steen IN, Kenny, RA (2005). "Patients with recurrent falls attending Accident and Emergency benefit from multifactor intervention? a randomized controlled tria". *Age and Ageing* 34, 162-168.

Donmez, L and Gokkoca, Z (2003), "Accident profile of older people in Antalya City Centre, Turkey". *Archives of Gerontology and Geriatrics*, 37(2) 99-108.

Davis JW, Nevitt MC, Wasnich RD, Ross PD, (1999), "A cross-cultural comparison of neuromuscular performance, functional status, and falls between Japanese and white women". *J Gerontol Med Sci* 54A:M288-M292.

Fried LP, Tangen CM, Walston J, et al. (2001), "Frailty in older adults: evidence for a phemotype". *J Gerontol Med Sci* 56A:M146-M156.

Gill, T. Williams, C. and Tinnetti, M. (2000), "Environmental hazards and the risk of nonsyncopal fall in the homes of community-living older persons". *Medical Care.* 38:1174-1183.

Irudaya Rajan, S, Mishra, US and Sarma, PS. (2003), 'Demography of Indian Ageing, 2001-2051', *Journal of Ageing and Social Policy*, 15(2&3), 11-30.

Jai Prakash, I. (1998) "Maintenance of Competence in Daily Living and Well-being of Elderly". *Research & Development Journal.* Vol. 4: 2 & 3.

Kinsey J and Lane S. (1983), "Race, housing attributes, and satisfaction with housing". *Housing and Society* 10 (3): 98-116.

Kohli, AS. 1996. "Social Situation of the Aged in India". Anmol Publications Pvt. Ltd. New Delhi.

Kinsella, Kevin and Velkoff, Victoria A. (2001), "An Ageing World, U S Census Bureau Serie"s, P 95/01, Washington D.C.

Lee, S. and Mason, A. (2000), "Population Aging Raises Questions for Policymakers." Asia-Pacific Population and Policy. Number 53, April. And Policies. New York: The Haworth Press The difference between the heights of seat and desk must be adjustable. Suitable height for a desk is from 67 to 75cm.

Lord, S.R., Sherrington, C. Menz, H.B. (2001), "Falls in older people". Cambridge University Press, Cambridge, U.K.

Liebig, Phoebe and Irudaya Rajan, S. (ed.) (2003), "An Ageing India: Perspectives, Prospects"

Mary L. M. Gilhooly, 2007. "Home Accidents amongst Elderly People: A Locality Study in Scotland". *Journal on Social and psychological Gerontology*, Issue 01/2007.

Nourhashemi F, Andrieu S, Gillette-Guyonnet S, Vellas B, Albarede JL, Grandjean H, (2001), "Instrumental activities of daily living as a potential marker of frailty". A study of 7364 community-dwelling elderly women (the EPIDOS Study). *J Gerontol Med Sci* 56A:M448-M453.002 nited Kingdom: Department of Trade and Industry.

N. P.Das. (2004), "A Study of Old Age Homes In the Care of the Elderly in Gujarat". Project Report Population Research Centre Department of Statistics Faculty of Science M.S. University of Baroda Baroda-390

Neil Charness (2008), "Aging and Human Performance". *Human Factors*, 50: 3, 548-555.

Oh E J. (2000), "The relationship between the therapeutic qualities of architectural environment and behavior of elderly persons with dementia in long-term care settings". Doctoral dissertation, Yonsei University

Park, S. H. (1998), "A study on architectural planning for the elderly". *Journal of Architectural Research*, Vol. 18, No. 1, 89-94.

Thorne, R. (1986), "The Housing and Living Environment for Retired People In Australia", South wood Press Pty. Ltd., Mauriceville.

United Nations. (2002), "World Population Ageing", 1950-2050, Department of Economic and Social Affairs, Population Division, New York.

Walston J, Fried LP, (1999), [review]. Frailty and the older man. *Med Clin North Am* 83:1173-1194. "Assessing floor Slipperiness in fast food restaurants in Taiwan using objective and subjective measures" 35 (4) 401-408.

World Bank. (2001), "India: The Challenge of Old Age Income Security, Finance and Private Sector Development, South Asia Region", Report 22034-In, Washington.

CHAPTER 66

Visibility Profiles of Stadium Stands Derived from Profiles of Comfortable Stairs

Zdzislaw Pelczarski

Facultuy of Architecture
Bialystok University of Technology
Grunwaldzka 11/17
15-893 Bialystok, Poland

ABSTRACT

In the practice of stadium designs, the profile of the stands is usually determined taking into account parameters related to the assumed visibility standards. As a consequence, the geometry of tiers and their inclination, as well as the dimensions of steps of the stands and the parameters of radial gangways result from the visibility profile. In such conditions, directives concerning the proportions of steps of radial gangways stipulate only maximum and minimum values of tread and riser. This, in fact, precludes the use of Blondels formula *2 x riser + tread = pace length,* which is suitable for designing comfortable, safe, and ergonomic stairways.

The present article provides results of the author's research into profiles of stadium stands derived from profiles of comfortable stairs, which both ensure appropriate visibility conditions and facilitate ergonomic stairway design adjusted to human psychophysical standards.

Keywords: Visibility Profiles, Stadium Stands, Profiles of Comfortable Stairs, Stadium Design, Stadium Architecture.

INTRODUCTION

Any architect designing a stadium stand, apart from complying with the assumed visibility standards, faces the necessity of considering comfortable and safe routes of egress and evacuation. This stems from the fact that suitable conditions for movement of spectators in both lateral and radial directions have to be ensured. The former kind of movement requires appropriate seating row depth, the latter – occurring in the perpendicular direction to the edges of steps of the tiers – entails the usage of stairways. These stairways, allowing access to successive seating rows, form what is called *radial gangways*. Such an arrangement of spectators' passage over the auditorium is the most typical one, and has been in use since the times of theatres and amphitheatres of ancient Greece and Rome.

Vertical and horizontal geometry of steps of the stand results mostly from the assumed visibility standards of so-called *point of focus*. On the other hand, geometry of stairways is obtained by adjusting the stairs to the visibility profile of the stands. Accordingly, several consequences for the design arise here. As the design limitations related to good visibility are not in any kind of cause-effect relation with directives concerning comfortable stairs, the stairways obtained in the design, though usually adhering to construction standards, are far from fulfilling the requirements of comfortable stairs formula. These standards (EN 13200-1, 2003) establishes the maximum allowed riser of stairs in spectator stands at $r_{max} = 200$ mm, and minimum tread at $t_{min} = 250$ mm. At the same time, the slope of the stairs in such conditions cannot exceed 35 degrees.

The height and the depth of steps of the stand are multiple values of risers and treads of steps of the stairways. With each of the steps of the stand corresponds typically a set of two or three steps of stairs. In the case of stands with straight-line profile of tiers, geometry of individual steps of stands and stairways, and that means also their slopes, are identical for the entire stand, or at least for its large part. Relationships between the profile of the stands, the profile of stairways, and sightlines are presented in Figure 1.

THEORETICAL BASIS OF STAIRWAY DESIGN

The main challenge encountered in designing stairs involves establishing the dimensions of their treads and risers, and in consequence the slope of their pitch. Commonly used for this purpose is the computational approach relying on the so-called comfortable stairs formula $2r + t = pace\ length$, an algebraic equation where r denotes step's riser, while t denotes step's tread. The formula, dating back to ca 1672, was introduced by Francois Blondel, Director of the Royal Academy of Architecture in Paris. In the original Blondel's formula the pace length was equal to contemporaneous 24 inches (Sinnott, 1985). Stairways whose steps fulfill the equation are well adjusted to human anatomical structure and physical abilities, guaranteeing economical energy expenditure and feeling of psychophysical comfort

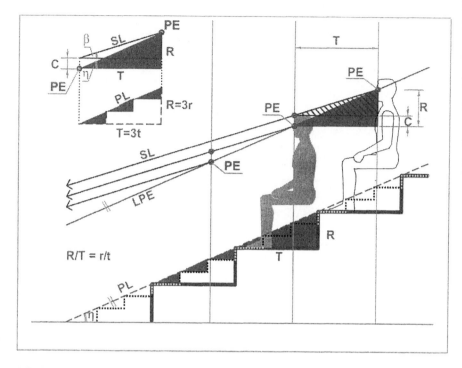

FIGURE 1. Geometrical relations between steps of the stand, stairs of radial gangway, and sightlines.

Legend: **PE** - *Point of eye,* **SL** - *Sightline,* **PL** - *Pitch line,* **LPE** - *Line of eye-points (parallel to PL),* **C** - *'C' value or the so-called 'sight line elevation',* **R** - *Riser of seated stands,* **T** - *Tread of seated stands (seating row depth),* **r** - *Risers of steps in radial gangways,* **t** - *Tread of steps in radial gangways,* η⁻ - *Slope of the stand,* β - *Slope of the sightline.*

for the people using them. The link between stairs' geometry and energy expenditure was a subject of interest for German physiologist G. Lehman. Among others, he confirmed experimentally that a stairway with the slope of 30 degrees and the stair dimensions equal 17/29 cm., satisfying the equation $2r+t=63$ cm., are optimal in respect of energy expenditure (10cal/KGM) and comfort of use, (Grandjean, 1978). In design practice, the averaged length of man's and woman's pace is commonly accepted as 630 mm (Neufert, 2005; DIN 18065, 2009 or PN-EN 13056, 2002).

Humans have an ability to move on the horizontal surface with the so-called horizontal pace (HP), in the vertical direction – e.g. on a ladder – using the so-called vertical pace (VP), and up/down a ramp or stairs with the so-called slopy pace (SP). In the first case, the movement of the whole body occurs horizontally, in the second – vertically, an in the third – both horizontally and vertically simultaneously. This movement can be represented by vectors. The vector of the slopy pace is a resultant of two components: horizontal (t) and vertical (r). It is quite obvious that when the vertical component *r=0* the pace becomes horizontal pace, whereas when the horizontal component *t=0* the pace becomes vertical pace. Research conducted within the field of anthropometry and ergonomics confirmed that the height of a human vertical pace constitutes half of the length of a horizontal pace. Hence, its averaged height can be assumed as 315 mm.

Blondel's formula can be illustrated with the comfortable stairs triangle (see Figure 2). Relationships between the riser (*r*) and the tread (*t*) of a stair are described by a linear function whose graphic representation in a two-dimensional Cartesian coordinate system is a line segment (LBE) defined by the *y=mx+b* equation. Considering the notation of comfortable stairs formula, this linear dependency takes the form: *r = -0,5t + 315*.

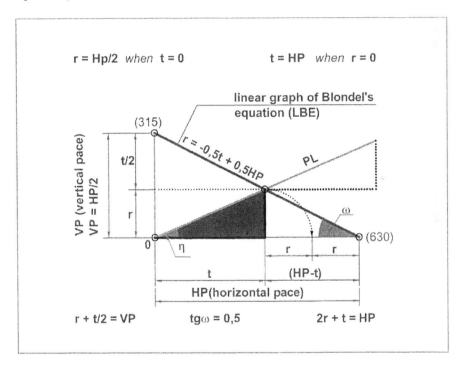

FIGURE 2. Triangle of comfortable stairs based on the Blondel's equation

Legend: **HP** - *Horizontal pace,* **VP** - *Vertical pace,* **LBE** - *Linear graph of the Blondel's equation,* ω *-Inclination of LBE. Remaining symbols as in Figure 1.*

The inclination (ω) of this line towards the horizontal axis (x) is constant for any value of horizontal pace length, and since the slope *tgω=VP/HP=0.5*, it is equal to ω=26.565051 degrees. The abovementioned line segment (LBE) is the hypotenuse of a right-angled triangle whose horizontal leg has the value of averaged horizontal pace length (630 mm.), and whose vertical leg represents the value of averaged vertical pace height (315 mm.). Any given point lying on the hypotenuse (LBE) defines the corresponding values of riser and tread lying on the axes (of the coordinate system). These parameters always satisfy the equation *2 × riser + tread = pace length*. If a straight line is drawn from the vertex of the right angle of the triangle across the discussed point on the LBE segment, it will be tangent to convex or concave edges of the first step of the stairs. The line also establishes the position of successive stairs and the slope of their pitch. Accordingly, this line was given the name of a guideline of the stairs slope (PL).

Awareness of the described geometrical and mathematical dependencies allows for the use of a simple and effective graphical method in the design practice. This method makes it possible to quickly draw graphs of stair profiles, as well as to determine optimal stair parameters (see Figure 3).

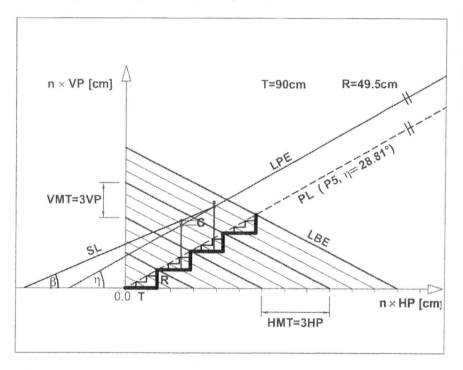

FIGURE 3. Example graph of the stadium stands visibility profile derived from the profile of comfortable stairs (P5: R/T=49.5/90 [cm], r/t=16.5/30 [cm], η=28,81°, β=22,62°)

*Legend: **HMT** - Horizontal module of tier step, **VMT** - Vertical module of tier step, **P5** - Number 5 straight-line visibility profile of the tier derived from the profile of comfortable stairs. Remaining symbols as in Figures 1 and 2.*

The approach employs a graphical template in the form of a grid of LBE lines that correspond with successive stairs. The guideline of the stairs slope (PL), intersecting individual LBE lines, fixes the position of the successive steps.

Both research and practical experience indicate that the slopes of stairways are usually contained within the range of 10 to 45 degrees: at slopes below 10 degrees or above 45 degrees, respectively ramps and ladders are used. With regard to the slope and stair geometry, stairs can be divided into three groups: flat stairs with the slopes between 10 and 21 degrees; comfortable (normal) stairs with the slopes between 21 and 35 degrees; and steep stairs characterized by the slope between 35 and 45 degrees. Classification of stairs is based on Hykš, Gáborik and Wrona (Hykš, 1984), introducing author's own modification concerning the range of normal stairs' slope (Pelczarski, 2009). Specifically, lower boundary of the normal stairs slope interval was established relying on the DIN 18065 standard assuming the minimum riser $r=14$ cm, whereas the upper boundary assuming minimum tread $t=26$ cm.

Taking into account the condition that the maximum tier inclination should not – according to standards – exceed 35 degrees, the steep stairs are practically unacceptable in the context of spectator stand design. Furthermore, typical inclinations of stand profiles that are dictated by visibility standards indicate that the main kind of applicable stairs are comfortable stairs, or, in some exceptional circumstances, flat stairs with proportions close to normal stairs.

PROFILES OF STADIUM STAND DERIVED FROM THE PROFILES OF COMFORTABLE STAIRS

The described above method for drawing up the pitch line of stairs can be also applied to designing straight-line profiles of stadium stands with geometry derived from the geometry of comfortable stairs (see Figure 3). Let us assume, for instance, that for each step of a stand there are three corresponding steps of a radial gangway. Then, in order to create a template for the steps of the stand it is enough to single out graphically every third guideline LBE of the stair step geometry. They constitute guidelines of both the stand steps geometry and the stair steps geometry. Following the rule expounded when explaining the method of drawing up the pitch line of stairs a graph of steps of the stand is obtained. A characteristic feature of these steps is that the height of each one of them is equal to the height of three risers of the corresponding stairs of a gangway, while the depth of each one of them is equal to the depth of three treads of these same stairs. All in all, the discussed approach yields a straight-line profile of a stand with a defined inclination, and – as a result – with definite sightlines, along with comfortable stairs incorporated into it. Since the slope of comfortable stairs must be contained within the 21 to 35 degree interval, there is only a limited number of corresponding steps of stairs. Figure 4 presents a selection of ten graphs of stand profiles derived from comfortable stairs (P1 –P10), drawn up with the assumption that for each step of the stand correspond three steps of comfortable stairs. In devising these graphs, the principal idea was to adopt a series of modular steps of the stairs - the steps with practically achievable geometry in modern concrete technologies. In the assumed series of dimensions, treads of successive steps of the stairs vary 10 mm in length, whereas the difference

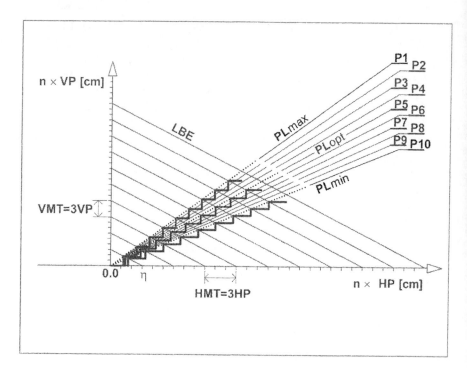

FIGURE 4. Parameters of the optimal stadium stands visibility profiles derived from profiles of comfortable stairs

*Legend: **PLmax** - Pitch line of maximal slope, **PLmin** - Pitch line of minimal slope, **PLopt** - Pitch line of optimal slope, **P1-P10** - Straight-line visibility profiles of the tiers derived from the profiles of comfortable stairs. Remaining symbols as in Figures 1, 2 and 3.*

between their risers equals 5 mm. In consequence, the dimensions of the steps of the stand change 30 mm. in terms of depth, and 15 mm. in terms of height. With these assumptions, the pitches of stairs, as well as the equivalent pitches of the steps of the stands are the end-values for the particular profiles. Slopes of these pitch lines define also the slope of the line of eye-points which running in parallel to the guideline PL, with a nominal rise of 120 cm. Additionally, for each of the profiles sightline slopes were determined, accepting for the purpose that the 'C' value equals 12cm. Practically, having gained sufficient experience in creating the discussed graphs, it is enough to use the guidelines and the sightlines only, postponing the detailed drawing of the chosen profile until the final stages of the decision-making process.

Table 1: Parameters of ten optimal stadium stands visibility profiles derived from profiles of comfortable stairs. *Symbols as for Figures 1-4.*

	Parameters			
Profiles of Tier	η	R/T	r/t	β
	[°]	[cm/cm]	[cm/cm]	[°]
P1	35,43	55,5 / 78	18,5 / 26	29,15
P2	33,69	54,0 / 81	18,0 / 27	27,41
P3	32,00	52,5 / 84	17,5 / 28	25,71
P4	30,38	51,0 / 87	17,0 / 29	24,15
P5	28,81	49,5 / 90	16,5 / 30	22,62
P6	27,30	48,0 / 93	16,0 / 31	21,16
P7	25,84	46,5 / 96	15,5 / 32	19,77
P8	24,44	45,0 / 99	15,0 / 33	18,44
P9	23,10	43,5 / 102	14,5 / 34	17,16
P10	21,80	42,0 / 105	14,0 / 35	15,95

CONCLUSIONS

The described principles of creating profiles of stands integrated with profiles of radial gangways constituted a basis for the author's development of templates of profiles slope for stadium stands derived from comfortable stairs and templates of sightline slopes for each of the profiles. Ultimately, a tabular catalogue of parameters of such profiles were compiled. As a result of this study, a method for modeling the profile of the stand with a beam of sightlines was developed. Hence the application of the template of sightline slopes using the author's *furthest eye method* for graphical determining of profiles of stands derived from profiles of

616

comfortable stairs became possible. However, considering the limited scope of the present article, it is impossible to provide further details concerning these issues. They are more extensively discussed in the author's 2009 monograph entitled: 'Spectator Stands of Contemporary Stadiums – Determinants and Design Problems' (Pelczarski, 2009).

REFERENCES

DIN 18065 (2009), The German Standard, *Trappen (Stairs in buildings - Terminology, measuring rules, main dimensions).*

EN 13200-1 (2003), The European Standard which has the status of a Polish Standard PN-EN 13200-1: 2005, *Spectator facilities - Part 1: Layout criteria for spectator viewing area – Specification.*

Grandjean, E. (1978), *Ergonomics of the Home.* Arkady, Warszawa, pp. 164-165.

Hykš, P., Gáborik, M., Wrona, O. (1984), *Stairways.* Arkady, Warszawa, p. 14.

Neufert, E. (2005), *Podręcznik projektowania architektoniczno-budowlanego (Architects' data).* Arkady, Warszawa, p. 194.

Pelczarski, Z. (2009), *Spectator Stands of Contemporary Stadiums – Determinants and Design Problems.* Oficyna Wydawnicza Politechniki Bialostockiej, Bialystok.

PN-EN 13056 (2002), The Polish – European Standard, *Ships of inland navigation. The stairs with inclination angle of 30 degrees to 45 degrees. Requirements, types.*

Sinnott, R. (1985), *Safety and Security in Building Design.* Collins, London, p. 133.

Chapter 67

Ergonomic Assessment of Interiors in Residential Dwellings and Impact on Residents

Agarwal Shalini[1], Sharma Promila[2], Gupta Luxmi[3]

[1] Dept. of Human Resource and Family Welfare
School of Home Science
B.B.A.U, Lucknow, U.P. India

[2] Dept of Family Resource Management,
College of Home Science, G.B.P.U.A.& T., Pantnagar
Uttarakhand,-263145,India

[3] Texas, USA

ABSTRACT

Construction is one of the largest industries in India, employing 7.6 million workers, or about 5% of the Indian work force. Construction workers face some of the most dangerous working conditions in the country on a daily basis. Ergonomic problem is common hazard and is one of the most common causes of injury at work. The accurate measurement of workers' exposure to the risk assessment of WMSDs has been of vital importance. Musculoskeletal system primarily concerns with dimensions, compositions & mass properties of body segments, and work related musculoskeletal disorder caused due to over exertion, adoption of asymmetric & awkward postures and unsupported positions used in task completion. Many MSDs are due to operations performed for a long period, can cause a lifetime of pain and disability. Therefore present investigation was planned to provide an objective measure of the MSD risk caused by construction work and to evaluate a job's level of risk for developing a musculoskeletal disorder in 120 construction workers with the application of REBA (Rapid Entire Body Assessment) postural assessment technique. It was found from observation that approximately 6 % of workers were in the category of AL0 indicating 'negligible' risk level means no action is necessary

618

indicating acceptable posture, 12.7% were in AL1(low risk, further action may be needed) and 63% of students in AL2 (medium risk, action necessary). However, 11 & 7.3 percent was found to be in AL3 (high risk, action necessary soon), AL4 (very high risk, action necessary now) respectively.

A residential dwelling should be planned, furnished and built in such a way that it accommodates in best manner all the activities besides being simple, comfortable and interesting rather than ornate or disruptive. Residential dwelling must fulfils the needs for the daily activities of the residents, and more precisely it must give them: satisfaction, security, comfort, and independence or at least one of them. Visual ergonomics is the investigation of human physical characteristics and limitations with respect to the environment as related to visual communications. Tints of primary and secondary colour were preferred on walls of different rooms except in store and bath room where white or its nearby colours were preferred. Most of the respondents preferred designs on two alternate walls. Majority of respondents had fair knowledge regarding elements of art. Most of the respondents had the opinion that there is high association between visual task performance and elements of art. In most of the residential dwellings, infrastructural facilities were lacking and if available they were not as per the requirement of the residents. Respondents of different residential area were found suffering from eye problem, diabetes, hyper tension and joint problems. The eye problems may be due to improper lighting facility at different work stations. Most of the colours symbolized positive characteristics except black which symbolized negative characteristics. Different types of lines (vertical, horizontal, zig-zag, diagonal, curved and circular) symbolized movement of eyes and activity. Designs on two walls especially, alternate walls was preferred the most followed by designs on ceiling only. The reason may be designs on all walls may make the room look busy and crowdy. Lighting facility available at work place affects the eye movement like contraction and expansion of pupil diameter, blinking frequency and number of blinks per minute. The visual task performance of individual is affected by the type and source of lights used. Inadequate lights used in a work place also affect the heart rate and blood pressure of the person. In most of the cases, heart rate increases in high intensity light.

Keywords: Visual ergonomics, interior decoration, lighting, illumination

INTRODUCTION

A residential dwelling should be planned, furnished and built in such a way that it accommodates in best manner all the activities besides being simple, comfortable and interesting rather than ornate or disruptive.

Residential dwelling must fulfills the needs for the daily activities of the residents, and more precisely it must give them: satisfaction, security, comfort, and independence or at least one of them.

Interior space planning together with the design of color, finish, and lighting treatments must support the day-to-day activities of building users. It leads to psychological satisfaction, which is related to interior environment and interior have impact on task performance.

Visual ergonomics is the investigation of human physical characteristics and limitations with respect to the environment as related to visual communications.

Visual ergonomics is concerned with the combination of finish color, finish texture and material, and lighting produces the visual environment within a space. It is critical that room finishes and lighting be coordinated to provide adequate seeing conditions for the activities to be conducted in the space.

Many of our oldest and wisest sayings are related to eyes. That's probably because vision is our primary connection with the world. The process of vision begins with visible light — a portion of the radiation spectrum that stimulates the nerve endings in the retina.

Illumination is important to humans because it alters stimuli to the visual system and the operating state of the visual system itself. Interior lighting is almost always sufficient for the visual system to be operating in the photonic region. Differences between persons in visual capabilities are common and are usually dealt with by providing lighting that is more than adequate for visual performance and visual comfort.

The visual sense is the most important channel of information in information- intensive work. From the point of view of seeing and eye fatigue, the ordinary visual displays are not the most optimal solutions. Interiors of rooms have various visual effects on humans. Color, form, space and light are the principal interactive ingredients of the architectural environment, but color is the one element that most affects the others. The most obvious and basic requirement in the dwelling is lighting as it permits vision to perform different activities. Home is a place to relax and to carry necessary household work and other activities. Lighting in the dwelling should provide certain emotional and aesthetic satisfaction. Poor lighting has been associated with a variety of problems including low productivity, high human error rates, eyestrain, headache, reduction in mental alertness, general malaise, and low employee morale. A line is the beginning point for designing, because it establishes shape or form and is a valuable element of composition. In interior designing, straight lines are considered intellectual rather than emotional, classic rather than romantic and sometimes severe and masculine. Curves, on the other hand, are used to achieve a more joyful, subtle and rich effect. Texture is most often used to describe the relative smoothness and roughness of a surface. Texture expresses the artist's meaning and accentuate the impact created. Designers tap the potential of the textural qualities of building materials.

As human being spends most of his time in his dwelling thus functional and comfortable dwelling along with different elements (proper entry of light, colour, line and texture) in the interiors is must for his health as well as to increase working productivity.

OBJECTIVES OF THE STUDY

1. To study the family profile of the respondents.
2. To study the existing structure and lighting system of the residential dwelling.

3. Application of Visual ergonomics to the elements of design.
   A] To study the functional characteristics of light and its impact.
   B] To study colour vision and preference.
   C] To study visual search and effect of texture, design and line in the interiors.
4. To study the physical and psychological cost of work in terms of visual performance, task performance and productivity.
5. To study the symptoms and causes of visual discomfort.
6. To develop approaches to improve performance and visual comfort.

MATERIALS AND METHODS

Exploratory cum Experimental research design was chosen for collecting data. The present study was carried out in district Udham Singh nagar. Purposive cum Random sampling design was used to select the study area and to select residential unit, systematic random sampling was adopted. Total sample size was 120 residential units, 40 each from three different selected residential colonies (partially self build, fully self build and build by builder).

Pre-coded personal interview schedule was constructed to gather general information about respondents and residential dwellings. Checklist was used to collect in-depth information about the residential dwelling. Three point scales were developed to collect knowledge and opinion of respondents on different aspects of elements of art. Test was developed to collect information on psychological effects of different elements of art. Experimental tables were constructed to record data related to eye fatigue and visual task performance in different lighting conditions. Various equipments were used for collecting the data and these included anthropometric kit, polar heart rate monitor sphygmomanometer, eye head tracker, lux meter, stopwatch and vision testing equipment for experimental data. The data was collected in two phases, in phase first, descriptive data was collected and in second experimental data was collected. In the present study frequency, percentage, mean, standard deviation were used to analyze descriptive data. Analysis of variance and chi-square were used for testing hypothesis.

RESULTS AND DISCUSSION

Maximum selected respondents both male and female were between 26-45 years age group. As far as educational qualification was concerned, it was found that most of the respondent were highly qualified, that is nearly 37.0 per cent female and 45.0 per cent male respondents were post graduate. Nuclear type family system was predominant (73.3 per cent) most of the male respondents (98.3 per cent) were employed whereas only 41.6 per cent female respondents were working.

It was found that 26.6 per cent female and 40.0 per cent male respondents were possessive private jobs. Forty per cent were having their total monthly income ranging from Rs. 25001-30000/-. Nearly 81.7 per cent respondents were having their own residence and 70.8 per cent were having single storied dwelling while 29.2 per cent had double storied dwelling.

It was found that 36.7 per cent respondents had their residence facing towards east while most of the respondents (81.7 per cent) had open area in their residential dwelling in which 51.7 per cent had open area on two sides of their residence.

Lighting facility in different rooms showed that 39.2 per cent respondents had more than two light sources in their living room. Whereas 61.7 per cent respondents were having two types of light sources in

bedroom and in dining room most of the respondents (70.0 per cent) had only one type of light source. Similarly, in kitchen and bath / w.c., most of the respondents (65.8 percent & 75.8 percent) respectively had only one type of light source. It was found that in store room all the respondents were having only one type o light source. Results also revealed that nearly, 35.8 per cent respondents were having three different light sources (fluorescent straight light, incandescent filament bulb and CFL) in their living room, where as in bed room, combination of incandescent filament bulb and CFL was used by most of respondents. In dining room most of the respondents used either CFL or fluorescent straight light while in kitchen CFL was by 29.2 per cent respondents. In store room 45.0 per cent used incandescent filament bulb. In bath / w.c. also, incandescent filament bulb were used by majority of respondents.

Data revealed that major health problems faced by the respondent were eye problem (35.8 per cent), diabetes (30.0 per cent), hypertension (28.4 per cent) and joint problem (25.0 per cent). Very few respondents complained of either pain or stiffness in any body part.

When asked about eye related discomfort, results showed that nearly 20.0 per cent respondents complained of pain in eyes and of eye fatigue respectively. About 15.8 per cent respondents complained of redness in eyes while 13.3 percent told about irritation in eyes as well as blurred vision.

Only 15.0 per cent stated heredity reasons for eye related discomfort and rest stated other reasons like working in low light, improper lighting facility at work places, insufficient natural lighting, etc.

Results revealed that 64.2 per cent respondents had fair knowledge about elements of art, where as 20.0 per cent had good knowledge and only 15.8 per cent had poor knowledge about the same.

It was also found that about 44.1 per cent respondents felt that there is high association between visual task performance and elements of art, where as 30.8 percent felt that there is fair association while 25.0 per cent felt there is poor association between visual task performance and elements of art.

When infrastructural facilities in different rooms were studied, results revealed that most of the infrastructural facilities were lacking in all the rooms and those available, they were not as per requirement of family members though height of doors and windows was suitable for family members. The width of doors and windows was sufficient for ventilation and natural lighting in most of the rooms especially in living room, bed room, dining room and kitchen. Results also depicts that the colour of wall as pleasant for all family members and the floor was non-slippery in all the rooms.

When choice of colour on walls of different rooms were studied by showing shade card of nerolac paints, results revealed that 66.6 per cent respondents preferred Lafleur, followed by 62.5 per cent respondents selecting After Glow. None of the respondents preferred white smoke colour in their living room, the reason for their preference may be to make their living room lively and bright.

In bed room most of the respondents (70.8 per cent) preferred Belinda colour where as 66.6 per cent respondents preferred Cool pool, Green Grape and Zesty Lilac. In this room also none of the respondents liked using White Smoke colour. The reason for selecting such colours may be, because, these colours are cool in nature and may be relaxing. In dining room, about 45.8 per cent respondents liked to use Green grape and Marble Mist. In kitchen, more than half of the respondents (55.0 per cent) preferred Green grape. It was also found that bright colours were avoided in kitchen. The reason may be because bright colours absorbs light and in kitchen good amount of illumination is required. In store room, majority of the respondents preferred shades of white, like Marble Mist and in bath room, most of the respondents (79.1 per cent) preferred Marble Mist where as few respondents (12.5 per cent, 11.6 percent) preferred White Smoke and Green Grape respectively.

When preference for designs on walls and ceiling were studied, results revealed that majority of respondents (70.8 per cent) preferred designs on alternate walls, whereas 53.0 per cent preferred designs on adjacent walls. It was also seen that more than fifty percent (54.0 per cent) respondents preferred designs on ceiling and only 37.5 per cent respondents liked to have designs on ceiling.

A test was developed to study the psychological effect of colour, line and texture on individuals. Results of the test showed that for 40.0 per cent respondents, yellow colour was the symbol of gaiety, while red colour was the symbol of love for nearly half of the respondents. Nearly 40.0 per cent felt that this colour is stimulating while only 10.0 per cent respondents felt that red colour shows aggression also. Nearly 20.0 per cent respondents believed that blue colour denotes honor and happiness whereas, 14.0 per cent respondents felt that this colour also depicts truth. Nearly half of the respondents believed that orange colour symbolize courage whereas 14.0 per cent respondents believed that the colour also depict intellect.

Results also showed that for nearly half of the respondents (48.0 per cent), green colour was a symbol of spring and nearly 16.0 per cent felt that the colour represents rest also. Similarly 40.0 per cent respondents felt that purple colour symbolize mobility while 14.0 per cent felt the colour reflects vagueness too.

When neutral colours were studied, results revealed that white colour depicts formal effect for 34.0 per cent respondents where as 30.0 per cent symbolize white colour related with clarity but for 10.0 per cent respondents this colour represented something horrible too. Results also depicted that for nearly 42.0 per cent respondents black colour depicted horrible things and for 34.0 per cent respondents it is a symbol of sophistication and darkness. Only 5.0 per cent respondents felt that the colour represented solidity too. Similarly, 38.0 per cent respondents felt that brown colour symbolize boredom, while 14.0 per cent felt, it represents masculine properties. Grey colour represented gentleness for nearly 36.0 per cent respondents.

When psychological properties of different lines were studied, results revealed that majority of respondents (80.0 per cent) felt that vertical lines produces illusion of height, while 90.0 per cent respondents felt that horizontal lines broaden the shape, majority of respondents (80.0 per cent) felt that zig-zag lines represents activity. About 64.0 per cent respondents felt that diagonal lines represent activeness. Majority of respondents (90.0 per cent) felt that curved lines are smooth where as 70.0 per cent respondents felt that circular lines gives the feeling of compactness.

Further the results revealed that majority of respondents (96.0 per cent) felt that smooth textures attracts attention and about 94.0 per cent respondents felt that such textured objects reflects light while, 90.0 per cent respondents felt that rough textures absorbs light and about 58.0 percent felt that they add style too.

To access visual stress while performing activities, an experiment was conducted on eye- head tracker in DRDO laboratory, New Delhi. Results of the same revealed that as the intensity of light decreased pupil diameter increased, which may be due to stress on eyes in low intensity light.

Similarly, to study relationship between visual task performances under different lighting conditions an experiment was conducted in visual ergonomics lab. in the dept of family resource management, college of home science, G.B.P.U.A.&T., Pantnagar. To study the task performance, parameters like number of mistakes made and time taken to complete the task were studied and results revealed that maximum mistakes were made under incandescent halogen bulb of 60 watt and minimum mistakes were made under incandescent halogen bulb of 100 watt while minimum time was taken under CFL of 14 watt. Physiological parameters like heart rate and blood pressure were studied under different lighting conditions and results showed that increase in heart rate was seen after the activity in each light, but maximum increase was observed under incandescent halogen bulb of 100 watt. Similarly, there was increase in systolic and diastolic blood pressure. Maximum increase in systolic blood pressure was observed under incandescent halogen bulb of 60 watt and CFL of 14 watt.

It was also found that chi-square value was found non significant between health problems and residential area as well as eye problems and residential area which depicts that health as well as eyes problems are affected by the residential area.

Similarly chi-square value was computed for visual task performance and different lighting conditions was found non significant which depicts that visual task performance was affected by the lighting system used for conducting the activity. Analysis of variance was computed to study the relation between knowledge of respondents about elements of art and residential area as well as to find out opinion of respondents regarding elements of art and visual task performance and f-value in both the cases was found partially significant which suggests that knowledge of respondents is partially affected by the residential area and respondents felt that some elements affect the visual task performance.

CONCLUSIONS

On the basis of findings of this investigation the following conclusions were drawn:

Tints of primary and secondary colour were preferred on walls of different rooms except in store and bath room where white or its nearby colours were preferred. Most of the respondents preferred designs on two alternate walls. Majority of respondents had fair knowledge regarding elements of art. Most of the respondents had the opinion that there is high association between visual task performance and elements of art. In most of the residential dwellings, infrastructural facilities were lacking and if available they were not as per the requirement of the residents. Respondents of different residential area were found suffering from eye problem, diabetes, hyper tension and joint problems. The eye problems may be due to improper lighting facility at different work stations. Most of the colours symbolized positive characteristics except black which symbolized negative characteristics. Different types of lines (vertical, horizontal, zig-zag, diagonal, curved and circular) symbolized movement of eyes and activity. Designs on two walls especially, alternate walls was preferred the most followed by designs on ceiling only. The reason may be designs on all walls may make the room look busy and crowdy. Lighting facility available at work place affects the eye movement like contraction and expansion of pupil diameter, blinking frequency and number of blinks per minute. The visual task performance of individual is affected by the type and source of lights used. Inadequate lights used in a work place also affect the heart rate and blood pressure of the person. In most of the cases, heart rate increases in high intensity light.

Hence, it is concluded that improper use of different elements of art (light, colour and texture etc.) in work place may reduce task performance and effect ones health thus in residential dwellings selection and usage of elements of art should be proper to have healthy, comfortable, cheerful and peaceful environment which is essential for better performance and healthy living.

Chapter 68

Ergonomic Design of School Furniture: Challenges for the Portuguese Schools

Ignacio Castellucci [a], Maria Antónia Gonçalves [b], Pedro M. Arezes [a]

[a] Ergonomics Laboratory, School of Engineering
University of Minho - Guimaraes, Portugal

[b] School of Managements and Industrial Studies
Porto Polytechnic Institute, Porto, Portugal

ABSTRACT

It can be observed an increased concern about the school classrooms, in particular about the study and design of school furniture fitting the students' needs and anthropometrics characteristics. The main aim of this study was to perform an anthropometric survey, considering the main anthropometric dimensions and, in accordance, to define the dimensions and characteristics of school furniture for Portuguese students of the 1st education cycle using valuable and validated ergonomic criteria. The analyzed sample includes, so far, 432 students. Obtained results seem to reinforce the need to consider the specificity of the anthropometric characteristics of the Portuguese students by adding an extra size mark for school furniture in a future revision of the corresponding standard.

Keywords: school, furniture, anthropometry, Portuguese.

INTRODUCTION

It can be observed an increased concern about the school classrooms, in particular about the study and design of school furniture suitable to the needs of the students and with appropriate dimensions according to the students' anthropometrics characteristics. An important milestone in this increasing concern is the publication of the European standard EN 1729 (CEN - European Committee for Standardization., 2006), which determines the dimensions and characteristics of different types of school furniture for the whole European population.

In addition, there is a large number of studies worldwide that shows a clear mismatch between anthropometrics characteristics and the dimensions of classroom furniture (Castellucci et al.; Chung & Wong, 2007; Gouvali & Boudolos, 2006; J. F. M. Molenbroek et al., 2003; Panagiotopoulou et al., 2004; Parcells et al., 1999; Saarni et al., 2007). This mismatch might affect the learning process, even during the most stimulating and interesting lessons (Hira, 1980) and can produce some musculoskeletal disorders, such as low back pain and neck-shoulder pain (Grimmer & Williams, 2004).

Despite that, in Portugal there is still no specific legislation or standard for the definition of the appropriate furniture characteristics to be used by schoolchildren. This situation can be a consequence of both the lack of knowledge from the governmental authorities and the lack of a representative anthropometric database of the population in concern (Molenbroek et al., 2003). Therefore, it seems that Portuguese schoolchildren are using school furniture that has been acquired without any ergonomic criteria, which most likely will result in some changes and problems in their musculoskeletal system, as well as in a possible decrease in their education performance.

The main objective of this ongoing project is to perform an anthropometric survey of the most important anthropometric dimensions regarding the use of the furniture and, in accordance, to define the dimensions and characteristics of school furniture for Portuguese students of the 1st education cycle, by using valuable and validated ergonomic criteria. Currently, the work is centered on a specific aim, which comprises the definition of the furniture dimensions for the mentioned students.

METHODS AND MATERIALS

SAMPLE

The studied sample includes, so far, 432 volunteer students (216 male and 216 females) from 9 schools belonging to the 1st cycle of the Portuguese educational system. The students aged 7 to 10 years, with an average of 8.5 (±1.2) years old. After giving written and verbal information about the study to the headmaster of the school, written authorization was obtained from the teachers, parents and students.

It should be noted that the sample was a sample of convenience and so far, the measurements were taken only in the Northern part of the country, near the city of Porto.

INSTRUMENTS

One of the specific objectives of this study was the design and validation of a new anthropometric chair. This developed tool should allowed to gather more anthropometric data than previous similar models, such as the model developed by Gouvali et al (2006).

For the validation of this new tool, 20 subjects were measured with a Holtain portable anthropometer (exception made with subjects' stature) and with a fixed (or wall) anthropometer. Afterward, these measures were compared with those obtained by using the developed anthropometric chair (Figure 1).

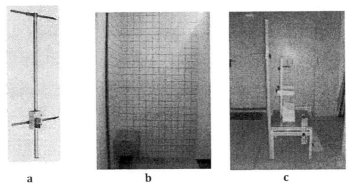

a   b   c

FIGURE 1. Instrument used during this study. (a) Portable anthropometer. (b) Fixed or wall anthropometer. (c) Developed anthropometric chair.

ANTHROPOMETRIC VARIABLES CONSIDERED

A group of 13 anthropometric variables were defined and collected using the anthropometric chair. Anthropometric variables were taken with the student wearing a t-shirt and shorts, and without shoes.

All measurements were made by the same measurer and recorded in centimeters by an assistant along nine different sessions. Accuracy and reliability of the measurements were achieved by undergoing through a specific training with a certified anthropometrics specialist and practice in performing measurements at the pilot study carried out previously.

Body anthropometric variables with the subject seated were collected while they were sitting in a relaxed and erect position on the anthropometric chair, with their upper and lower legs at a 90° angle, and with their feet flat on an adjustable footrest.

The following anthropometric variables (ISO 7250, 1996; Pheasant, 2003) (Fig. 2) were considered and collected in this study:

- Stature (S): determined as the vertical distance between the floor and the top of the head, and measured with the subject erect and looking straight ahead (Frankfort plane).
- Shoulder Height (SHH): determined as the vertical distance from the floor to the acromion.
- Eye Height (EH): Vertical distance from the floor to the inner canthus (corner) of the eye and measured with the subject erect and looking straight ahead (Frankfort plane).
- Sitting Height (SH): vertical distance between the top of the head and the subject's seated surface, and measured with the subject erect and looping straight ahead (Frankfort plane).
- Elbow Height Sitting (EHS): taken with a 90° angle elbow flexion, as the vertical distance from the bottom of the tip of the elbow (olecranon) to the subject's seated surface.
- Shoulder Breadth (bideltoid) (SHB): Maximum horizontal breadth across the shoulders, measured to the protrusions of the deltoid muscles.
- Popliteal Height (PH): measured with 90° knee flexion, as the vertical distance from the floor or footrest and the posterior surface of the knee (popliteal surface).
- Buttock-Popliteal Length (BPL): taken with a 90° angle knee flexion as the horizontal distance from the posterior surface of the buttock to the popliteal surface.
- Buttock-Knee Length (BKL): Horizontal distance from the back of the uncompressed buttock to the front of the kneecap.
- Hip Width (HW): the horizontal distance measured in the widest point of the hips in the sitting position.
- Thigh Thickness (TT): the vertical distance from the highest uncompressed point of thigh to the subject's seated surface.
- Sitting Eye Height (SEH): vertical distance from the seat surface to the inner canthus (corner) of the eye and it was determined with the following calculation: SH-(S-EH).
- Sitting Shoulder Height (SSHH): Vertical distance from the seat surface to the acromion, determined with the following calculation: SH-(S-SHH).

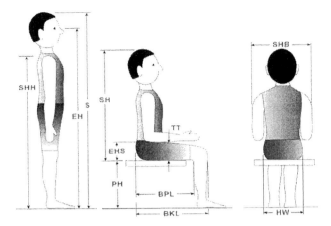

FIGURE 2. Selected anthropometric variables.

APPLICATION OF THE MEASURES

During the design of school furniture many aspects of human comfort must be taken into consideration to make it suitable for the students. Furthermore, the furniture dimensions are one of the main aspects and, amongst these, the seat height is typically the starting point for the design of this type of furniture.

In order to illustrate how the dimensions can be applied to define the furniture dimensions, it is possible to present the potential relationship between anthropometric data and furniture and defining criteria for the furniture design.

As an example, it is typically that PH should be higher than the SH (J. Molenbroek & Ramaekers, 1996; Parcells et al., 1999), but it does not have to be higher than four centimeters (Gutiérrez & Morgado, 2001) or 88% of the PH (Parcells et al., 1999). This mentioned limitation is due to the need to avoid the compression of the buttock region (García-Molina et al., 1992).

Accordingly, it is possible to define a criteria for establishing seat height, using the criteria described by Gouvali and Boudolos (2006). Besides, it also possible to include a correction related with a shoe height of 2.5 centimeters. Using this data, seat Height (SH) can be defined according to equation 1.

Eq. [1] $(PH+2.5) \cos 30° \leq SH \leq (PH+2.5) \cos 5°$

Similarly, it is possible to establish a criterion for the Seat to Desk Height (SDH). Based on available evidence, EHS is the major criterion for SDH (García-Acosta et al., 2007; Milanese et al., 2004; J. F. M. Molenbroek et al., 2003; Sanders et al., 1993). Additionally, Parcells et al. (1999) suggested that SDH may also depends on the shoulder flexion and abduction angles. Other researchers recommended that desk should be 3 to 5 cm higher than the EHS (Pheasant, 1991; Poulakakis and Marmaras, 1998). Using this data, the defined criterion for the dimension of the SDH can be obtained through a modified equation that accepts EHS as the minimum height of SDH, in order to provide a significant reduction on spinal loading (Occhipinti et al., 1985). Simultaneously, the equation considers that the maximum height of SDH should not be higher than 5 cm above the EHS, as represented in equation 2.

Eq. [2]. $EHS \leq SDH \leq EHS + 5$

RESULTS AND DISCUSSION

VALIDATION OF THE ANTHROPOMETRIC CHAIR

An Independent t-test (with 95% confidence interval) was performed to examine the differences in measurements between the different applied tools for gathering the anthropometric data. Obtained results show that no statistical significant difference (p>0.05) was identified between the three measurement methodologies/tools for all the anthropometric variables gathered.

Although it is not possible to quantify, it is important to mention the easiness of using the developed anthropometric chair. Nevertheless, there are still some problems related with the use of this tool, in particular the difficulty to carry the device from the laboratory to the different schools.

ANTHROPOMETRIC VARIABLES

According to the obtained results (Table 1), it is worth to mention that the current anthropometric variables are normally distributed. Moreover, it also possible to notice the existence of a strong Pearson correlation coefficient between stature and a group of other anthropometric variables, such as the popliteal height (r=0.90), buttock-popliteal length (r=0.84), sitting height (r=0.92), shoulder breadth (r=0.74) and buttock-knee length (r= 0.89). The correlation between variables can be an important point, as most of recommendations for furniture selection tend to use, as reference, the stature, assuming that all the other characteristics will be also appropriate, However, some authors, such as Molenbroek et al. (2003), suggests that the furniture selection can be carry out using the politeal height instead of stature.

Table 1. Anthropometric data obtained from the studied sample (cm).

Anthropometric variables	Mean	S.D	Percentile		
			5th	50th	95th
Stature (S)	131.3	8.9	117.1	131.2	145.3
Shoulder Height (SHH)	106.4	10.4	93.6	106.6	119.9
Eye Height (EH)	122.1	8.8	108.2	122.3	135.7
Sitting Height (SH)	67.0	4.2	60.1	67.0	73.7
Elbow Height Sitting (EHS)	16.5	2.0	13.4	16.3	20.0

Shoulder Breadth (SB)	31.9	2.8	27.9	31.5	36.7
Popliteal Height (PH)	33.8	3.0	29.2	33.6	38.8
Buttock-Popliteal Length (BPL)	38.2	3.3	33.2	38.3	43.7
Buttock-Knee Length (BKL)	45.9	3.9	40.0	45.9	52.6
Hip Width (HW)	28.2	3.0	24.2	27.8	33.8
Thigh Thickness (TT)	11.4	1.6	9.2	11.1	14.4
Sitting Shoulder Height (SSH)	44.0	7.6	38.8	44.5	50.6
Sitting Eye Height (SHE)	59.8	4.1	53.4	59.8	66.3

CLASSROOM FURNITURE DIMENSIONS

Considering the data of popliteal height and elbow height sitting gathered from the 432 subjects, as well as the definition of the appropriate height for furniture, both for chair and table, it is possible to compare it and establish different sets of furniture to cover the entire observed population.

According to the obtained data, it is necessary to development 4 different sets of furniture to allow students to be seated in the correct position (table 2).

Table 2. Proposed dimensions for each type of furniture (cm) and % of match.

Type of furniture	Seat height	Table height	Users (%)
Furniture #1	28	45	10.0
Furniture #2	32	51	50.0
Furniture #3	36	56	36.3
Furniture #4	40	61	3.7

From table 2, it is possible to verify that furniture #2 and #3 together can fit 86.3% of the analyzed students. Moreover, it is also possible to highlight that these two types of furniture have similar dimensions with the Size mark 2 and 3 from the EN 1729 (Table3).

According to the BS EN 1729 (British version of the standard), if it is assumed that the entire group of students is comprised between 7 and 10 years old, one third of the chairs and tables should be size mark 3 and two thirds should be size mark 4. However, if data of the obtained study is compared with the size mark scheme proposed in EN 1729, it is also possible to compute the percentage of match population for ach size mark, as presented in table 3.

Table 3. EN 1729 size mark dimensions for seat height needed to fit 100% of the studied population.

	Size mark 1	Size mark 2	Size mark 3	Size mark 4
Seat Height (cm)*	26	31	35	38
Users (%)	3.7	43.0	44.6	8.7

* Considering an angle of -5° to +5° +for the seat

Considering the recommendation of BS EN 1729 for 7 to 10 years-old students and table 3, it is possible to notice the difference between the English and the Portuguese population. It seems that the low stature of the Portuguese student population will imply the need to revise the size mark scheme presented in the EN 1729. This change in the size mark scheme was already noticed by Molenbroek et al. (2003), for the Dutch population, but in the case during the revision of the design of a standard for the dimension of school furniture.

What seems to result from this analysis is the possible inadequacy between the dimensions of desks and chairs proposed in the EN1729 and the Portuguese students population, in particular for the population younger than then studied age range, 7 to 10 years-old. Accordingly, it seems also possible to admit that this inadequacy is most notorious for younger populations, for children between 3 and 6 years old. As the standard presents only one size mark smaller than the size mark presented in the table 3, which is the size mark 0, it is possible that there will be a need to revise the size mark scheme. This revision may include the need to establish an extra size mark, below the size mark 0, as also suggested for the Dutch population (Molenbroek et al., 2003), but in that specific case, to be above the largest one.

CONCLUSION

The obtained results allowed, among other things, the identification of all the static anthropometric measures needed to develop the school furniture, which so far indicates the need of 4 different types of furniture sets to cover the observed variability within students, against the currently existing 2 types. Furthermore, and as expected, a low stature has been observed for the studied Portuguese students population, particularly when compared with other European countries. This seems to highlight the need to consider specific Portuguese anthropometric characteristics in the future adaptation of the EN 1729, which will result, most likely, in the need to define an additional size mark, below the existing ones.

REFERENCES

Castellucci, H.I., Arezes, P.M., & Viviani, C.A. (n.a.), "Mismatch between classroom furniture and anthropometric measures in Chilean schools.". *Applied Ergonomics*, In Press.

CEN - European Committee for Standardization., (2006). prEN 1729. Furniture - Chairs and tables for educational institutions Part1: Functional dimensions.

Chung, J.W.Y., and Wong, T.K.S. (2007), "Anthropometric evaluation for primary school furniture design.", *Ergonomics*, 50 (3), 323-334.

García-Molina, C., Moraga, R., Tortosa, L. , and Verde, V. (1992), *Guía de Recomendaciones para el Diseño de Mobiliario Ergonómico*. Valencia: Instituto Biomecánico de Valencia.

Gouvali, M.K., and Boudolos, K. (2006), "Match between school furniture dimensions and children's anthropometry.", *Applied Ergonomics*, 37, 765-773.

Grimmer, K., and Williams, M. (2004), "School furniture and the user population: an anthropometric perspective.", *Ergonomics*, 47, 416-426.

Gutiérrez, M., and Morgado, P., (2001), *Guía de recomendaciones para el diseño del mobiliario escolar,* Chile: Ministerio de Educación and UNESCO.

Hira, D.S. (1980), "An ergonomic appraisal of educational desks.", *Ergonomics*, 23, 213-221.

ISO 7250. (1996), *Basic human body measurements for technological design.*

Molenbroek, J., and Ramaekers, Y. (1996), *Anthropometric design of a size system for school furniture.* In S.A. Robertson (Ed.), Proceedings of the Annual Conference of the Ergonomics Society: Contemporary Ergonomics. (pp. 130-135). London: Taylor & Francis.

Molenbroek, J.F.M., Kroon-Ramaekers, Y.M.T., and Snijders, C.J. (2003), "Revision of the design of a standard for the dimensions of school furniture.", *Ergonomics*, 46, 681-694.

Panagiotopoulou, G., Christoulas, K., Papanickolaou, A., and Mandroukas, K., (2004), "Classroom furniture dimensions and anthropometric measures in primary school.", *Applied Ergonomics*, 35, 121-128.

Parcells, C., Stommel, M., and Hubbard, R.P. (1999), "Mismatch of classroom furniture and student body dimensions: empirical findings and health implications.", *J. Adolesc. Health*, 24(4), 265-273.

Pheasant, S., (2003), *Bodyspace* (Second ed.): Taylor & Francis.

Saarni, L., Nygård, C., Kaukiainen, A., and Rimpelä, A. (2007), "Are the desks and chairs at school appropriate?", *Ergonomics*, 50(10), 1561-1570.

Chapter 69

Design Development of Toilet Chair on Ergonomic Principles, Through (CAD)

Nirmal Kaur*, Promila Sharma*, Krishna Oberoi**

*G.B.P.U.A.& T., Pantnagar,
U.S.Nagar, Uttarakkhand, India

**P.A.U. Ludhiana, Punjab, India

ABSTRACT

It is well known that the world's population is ageing, with more developed regions, because of increasing survival to older ages as well as smaller numbers of births. Consequently, the support of this ever expanding elderly population has become of increasing concern. In India approximately 20 per cent of people aged 70 years or older, and 50 per cent of people aged 85 and over, report difficulties in such basic activities of daily living as bathing and toileting. The increase in the number of old people is presumably also increasing the number of old people with disabilities specially related to joints which creates hindrance or cause pain in using Indian toilets.

The middle income group families in India are unable to construct European type toilets for their older parents due to reasons like space problem in their homes, financial problem and sometimes the family doesn't bother about elders to take so much pain for them. Besides this, no consideration has been given by the manufacturers (who are involved in the production of toilet chairs) about the body dimensions of these people, for designing products for them. It has been also seen that these age group faced lots of problem like fall and slip and other kinds of accidents due to the use of these malfunctioning toilet chairs. Taking into account this study was planned to sort out the dimensions of old age people for designing of toilet chair through CAD to reduce the cases of accidents among them. For this total 200 people were taken for data collection where 50 per cent were male and 50 per cent were female. All the required dimensions were used on percentile bases and also the adjustments were used by taking 5th, 50th and 95th percentile values for this chair. Besides this some additional features were also added in it on the basis of the limitations of this age group people like, curved edges and softness etc. After designing this was converted into actual product and was also given to the third age people so that they can use it and can say about the positive as well as negative points of this toilet chair. The responses against this chair were that 60 per cent of people were highly satisfied by it on the other hand there were people (i.e. 10 per cent) who were showed low range of satisfaction and the cause behind this was the weight of toilet chair. Those people who lived alone and have no people in the family to favor them in moving this chair from one place to another find difficult to handle it themselves. But if we want to avoid accidents, weight can't be reduced but there is an alternative that locking wheels can be added in it to make it easily movable. In this way the problem among third age people can be cured in this area to make their life easy going.

INTRODUCTION

It is well known that the world's population is ageing, with more developed regions leading the process, because of increasing survival to older ages as well as smaller numbers of births **(United Nations Population Division, 1998 revision)**. Consequently, the support of this ever expanding elderly population has become of increasing concern. Approximately 20 per cent of people aged 70 years or older, and 50 per cent of people aged 85 and over, experienced difficulties in such basic activities of daily living as bathing, dressing and toileting. **Barbaccia (1995)** stated that the musculoskeletal dimensions, mechanical performance, flexibility of joints, muscle strength, gait speed, bone densities are all important factors in the physiological system of changes which occur in these with aging. Arthritis, heart disease, diabetes, difficulty with vision of hearing is common in older people. These problems with muscle of joints as hip fractures really contribute to a decrease in activities of daily living. Changes in body composition are well known to occur with old age, but there is limited knowledge of the normal values rate of changes in anthropometric indices of body composition or the distribution of these changes in the elderly. The anthropometric standards derived from adult population may not be appropriate for the elderly because of body composition changes occurring during ageing. Specific anthropometric reference data for the elderly are necessary, before designing any kind of product or furniture or similar related equipment.

Now a day's the knee problem is very common among third age people who creates hindrance in using Indian toilet seat because of their problem to sit down on the feet and folding their knees. Thus to avoid it seat height should be 20"-24" that will give better leverage in regaining a standing position and the fall and slip injuries can be checked which are associated with this knee problem as it is quoted by **Heikkinen et.al 2002**, that more than one third of adults' of ages 65 years and older fall each year.

Coleman (1997) pointed out the importance of consulting elderly people during the design process, because they claimed that elderly people are surrounded by things which do not work well for them, or that they simply cannot find the things that they want.

JUSTIFICATION OF THE STUDY

The major part of the research is based on the requirement of third age people regarding their toilet seat. Many companies are manufacturing toilet seats but there is no consideration for third age people having joint problem as a common illness in this age. There are so many factors which should be considered especially of third age people like comfort, softness and support which they need in this age while using their toilets. Many of them are unable to sit on the Indian toilets due to their joint problem and will not able to stand after using the toilet without support but these all factors are ignored by the designers. Most of the third age people are not able to move, thus they need folding type of toilet chair so that they can use it nearby their beds. Besides this there are families who are not able to construct European type toilets in their homes due to reasons like, space problem, financial problem, disliking by other family members or sometimes the family members do not bother about elders to take so much pain for them. Just to solve this problem there are locally made folding chairs available in the market but these are not up to the requirements of third age people, as these are not ergonomically designed and thus not suitable and also the comfortability is totally ignored. Their heights and widths are improper and there is no scope of support in any ways. Keeping the above issue in mind, it is proposed that the toilet chair for third age people should be ergonomically sound, comfortable and should be constructed on certain guidelines by keeping in mind the special requirements of third age people.

MATERIALS AND METHODS

Experimental design was planned to study eight anthropometric measurements, necessary for designing of toilet chair through CAD. The whole list of males and females of old age people of Pantnagar and Rudrapur block was made than out of that list, 100 old age people each from both the blocks were chosen for data collection, using simple random sampling without replacement, out of which 50% were male and 50% were female, thus altogether making total size of 200 people of third age having joints problem or other which cause hindrance in using Indian toilet, so needs a toilet chair for them. Different scientific instruments were used for collection of experimental data were anthropometric rods, weighing balance and grip dynamometer. For designing of toilet chair 5th, 50th and 95th percentile values were calculated.

I. 5th percentile

It is used to measure the minimum requirement with regard to height, length and width etc.

II. 50th percentile

It is used to measure the average requirement with regard to height, length and width etc.

III. 95th percentile

It is used to measure the maximum requirement with regard to height, length and width etc.

$$Pp = 1 + \frac{PN - F}{fp} \times i$$

Where,

Pp = required percentile rank

1 = Lowest value of the class interval where Pp falls

PN = Cumulative frequency in relation to the Pp point

F = Cumulative frequency below the lowest value of Pp class

fp = actual frequency of Pp class

i = class interval score

RESULTS AND DISCUSSION

The anthropometric measurements of third age people were taken and on the bases of that the total value was calculated as shown in table 1, for the designing of toilet chair. It was found by the correlation coefficient that there was a significant relationship among the dimensions of males and females of third age. It can be seen from table 2 that for seat height all the three values of 5th percentile, 50th percentile and 95th percentile was taken so that all kinds of people i.e. shortest, average and tallest can be easily accommodate. For seat width 95th percentile and for seat depth 5th percentile value was taken. In hand rest and back rest 95th percentile was taken as well as the provision of adjustment was given. As the old age people were having the problem of support, in the front, support handle was also given in the toilet chair. The weight of the toilet chair was made as per the 50th percentile value of the grip strength of old age people.

On the basis of these data the dimensions of toilet chair was decided for its construction and the additional features were decided to add in it on the basis of the needs of these age group people as described under justification of the study. The design of toilet chair was developed through CAD.

CONCLUSION

As the design of toilet chair was made on ergonomical principles by using anthropometric data of old age people for giving them all kinds of features which they required due to their limitations, then it is expected that the problem among these group can be solved to some extent and we will be able to avoid the cases of accidents during the use of different kinds of toilets, which are available in Indian market, by these people.

REFERENCES

Barbaccia, J.C. 1995. Changes in physical function and the ageing process : implications for facility design. In D. Driver (ed.). *Proceedings of Blueprint for Ageing*. CA, Oakland.

Coleman, R. 1997. Breaking the age barrier, speech at the Royal Society of Arts, June 97 as part of the Design Council – Design in Education Week, London: Design Age, Royal College of Art, March 12th. http://valley.interact.nl/DAN/NEWSLETTER/NEWS97/home.html, 6, 6, 1998.

U.S. Bureau of the Census (1996). Population Projections of the United State by age, Sex, Race and Hispanic Origin: 1995-2050, Current Population Reports. p 25-1130, U.S. Government Printing Office, Washington, D.C.

Heikkinen, R.L.; Berg, S.; Avlund, K. and Tormakangas, T. (2002). Depressed mood : changes during a five years follow-up in 75 years old men and women in three Nordic localities. *Ageing Clin. Exp. Res.*, 14 (3 Sup.): 16-28.

DIFFERENT VIEWS AND PARTS OF THE TOILET CHAIR

Front view of toilet chair

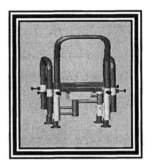

Side view of toilet chair

Back view of toilet chair

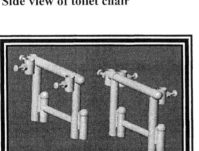
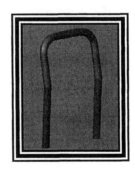

Legs of Toilet chair

Back rest rod of Toilet chair

**Hand rest rods of
toilet chair**

Assembled parts of toilet chair

Pan of Toilet chair

Seat Plate of Toilet chair

Table 2 Dimensions for designing of toilet chair for the old age people

Measurements	5th Percentile	50th Percentile	95th Percentile
Sitting Height (For extending backrest height)	-	-	86 cm
Sitting Shoulder Height (For Backrest Height)	-	-	61 cm
Buttock Popliteal Length (For Seat Depth)	38.5 cm	-	-
Popliteal Height (For Seat Height)	39 cm	43 cm	46 cm
Elbow rest Height (For Hand rest Height)	15 cm	21 cm	26 cm
Hip Breadth (For Seat Width)	-	-	44 cm
Hand Length (For Circumference)	16 cm	-	-
Grip Strength (For weight of chair)	-	10 kg	-

Table 1 Anthropometric measurements of Old Age People

Sitting Measurement	Male (cm)					Female (cm)					Total (cm)				
	M	SD	5th	50th	95th	M	SD	5th	50th	95th	M	SD	5th	50th	95th
Sitting Height*	80	4.9	72	80.5	8.6	75.5	3.0	71	76	80	78	4.9	71	77	86
Sitting Shoulder Height*	55	3.7	51	54	61	54	3.0	50	53	59	54	3.5	50	54	61
Buttock Popliteal Length	45	3.7	40	45	53	44	3.1	39.5	44	49	44	3.4	38.5	44.5	50
Popliteal Height*	44	2.6	40	44	49	41	4.8	36	42	46	42	4.1	39	43	46
Elbow Rest Height*	21	3.5	16	21	26	20	3.5	15	20	25	20	3.5	15	21	26
Hip Breadth*	38	4.0	34	37	44	38	4.4	30	37.5	43.5	38	4.2	33	37	44
HandLength*	17	1.3	15.5	17.5	19	17	1.2	15.5	17	19	17	1.2	16	17	19
Grip Strength	16	10.4	0	15	31	12	9.6	0	9	31	13	10.2	0	11	31

*Significant relationship

CHAPTER 70

Virtual and Augmented Reality Technologies for Supporting Heterogeneous Multidisciplinary End-Users Across Product Lifecycle

S. Kiviranta[1], M. Poyade[2], K. Helin[1]

[1]VTT Technical Research Centre of Finland
Tekniikankatu 1, FIN-33101 Tampere, Finland

[2]University of Malaga, ETSI de Telecomunicación
Campus de Teatinos, s/n, 29071 Malaga, Spain

ABSTRACT

This paper presents ManuVAR project plans to utilize Virtual and Augmented Reality (VR/AR) technologies for supporting high knowledge high value manual work. It is proposed to employ the VR/AR technologies as a medium for accessing and providing views of the existing product information through the end-user context-adapted virtual environments. The goal is to enable bi-directional information flows between the heterogeneous multidisciplinary end-users across the product lifecycle. This system will be tested in the laboratory environment, validated and exploited on the factory floor demonstrations in several industrial cases and these plans are presented.

Keywords: Virtual Reality, VR, Augmented Reality, AR, Manual Work, Product Lifecycle Management, PLM, Ergonomics, Virtual Model, ManuVAR

INTRODUCTION

Virtual and Augmented Reality (VR/AR) technologies and Virtual Environments (VEs) have been studied in various domains such as safety engineering (Lind et al. 2008), assembly instructions (J. Brough et al. 2007), real-time ergonomic analysis (U. Jayaram et al. 2006), collaborative teamwork (IST- IP project CoSpaces) and support material generation (J. Ritchie et al. 1999, Leino et al. 2009). Along with this ongoing research, industry has been using VR/AR in marketing, design, education and training as dominating application areas. The capabilities of VR/AR have been widely studied and successfully applied on focused problem domains since the early 90's. Still, despite the long history of development, VR/AR technologies are considered to have a low industrial and user acceptance.

ManuVAR project aims to develop technological platform and methodological framework for providing means to systematically address most influential manual work related issues at different stages of product lifecycle (LC). This will be done by the means of combining Product Lifecycle Management (PLM), for accounting various end-users; VR/AR technologies, for interfacing these users to the product data; and Virtual Model (VM) concept as link between existing industrial Product Data Management (PDM) back-end structures.

As stated above, VR technologies have been studied in various domains and are seen as suitable medium for interfacing end-users to product data. It was identified that alone this technology is not enough to address the seven fundamental problems of manual work in European industries (Krassi et al. 2010). Communication – the most prominent problem to be solved by means of VR/AR – is far more complex than just visualizing content, as the system has to be able to account feedback acquired from all sources at present, but also be capable of handling earlier product generations and giving means for storing the current information for later generations. This information needs to be further on adapted to the specific areas of expertise for heterogeneous multidisciplinary end-users, Figure 1.

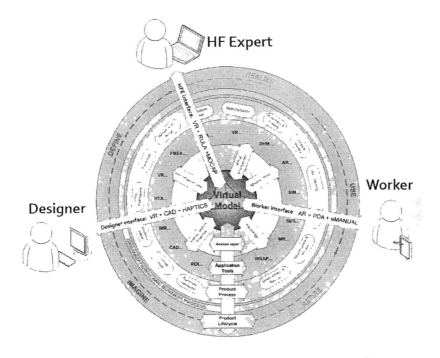

FIGURE 1. Illustrating heterogeneous multidisciplinary end-users across product lifecycle (Source: ManuVAR project)

Such structure forms set of near-unlimited relations and variables that are extremely hard to address with one specific technology, considering all the varying requirements and taking into account the human cognitive capabilities, no technology alone is sufficient to handle all information at once. For the complexity management purposes it was identified that VR/AR technology as interfacing technique needs to be combined with Virtual Modeling concept to handle the information complexity.

This setup gives opportunity to tie together the existing information within company PLM structure by VM concept, manage the complexity through adapting the available content to the user specifics through VR/AR technologies and in addition to handle PDM to end-user link. This also provides means for capturing and storing the product information back to PDM on various stages of lifecycle. So not only can the ManuVAR system provide communication link at specific lifecycle stage, but the approach channels the communication through the virtual model from where all feedback and data are distributed to other relevant end-users. Meanwhile this also enables the end-users to take information from previous product generations into account to the current stage of lifecycle while embedding the current decisions and reasoning to the product for the use of next stages of lifecycle.

MATERIALS AND METHODS

In first stages of the ManuVAR project, information from the companies was acquired by interviews and literature review, to determine the key issues on European manual work industry. Comprehensive literature study was performed to determine the current status of VR/AR technologies within the context of manual work; this was further on reflected to previous project results from the consortium. In the deductive phase this information was then generalized for defining high level requirements for the ManuVAR system, the 7 gaps of manual work industry in EU. The methods for determining the gaps, forming requirements for the ManuVAR system, were following the inductive – deductive work flow (Krassi et al. 2010). Based on these requirements, technical structure was outlined by adapting PLM and Virtual Prototyping methodology along with Virtual Modeling concept for technological basis and ergonomics as domain to support with the technology.

This "inductive – deductive" workflow allowed us to form the projects foundation by generalizing the issues identified within the 5 Clusters, from here it was possible to identify the commonalities in order to form generic solutions applicable to European industry in context of manual work. Within the scope of the ManuVAR project, companies were associated to research partners in clusters numerated from 1 to 5. Interviews were performed in these Clusters with the focus of supporting satellite assembly, manufacturing line design, remote support of train maintenance, power plant task planning and training and also heavy machinery productization and maintenance. Results formed the following plans to utilize VR/AR technologies within the 5 Clusters.

Brief description of Cluster 1 - The focus of this cluster is on developing systems and procedures using VR/AR for overcoming difficulties in the work carried out in spacecraft integration and assembly facilities. The target of Cluster 1 is to provide VR/AR support tools to plan and prepare assembly operations, train and validate test procedures as well as providing Non-conformance and feedback collection support through Augmented Reality. The aim is to increase efficiency and effectiveness of manual work on implementation and quality control activities and to utilize VE for information exchange between two shift working teams. The use cases suggested by Cluster 1 include the development of an on-line AR application to provide interactive instructions and development of an off-line VR application for integration and testing to be performed before integration phase. Human factors side of the applications presented in (Tossolin et al. 2010). This off-line application aims to validate and familiarize worker to the integration and testing by performing all procedures in a Virtual Environment before task execution through the means of VR visualization and haptic manipulation technologies as well as streamlining the clean room work procedures with help of AR.

Brief description of Cluster 2 - The focus of this cluster is on increasing sustained productivity, flexibility, ergonomic performance in small and medium sized companies in assembly industry. Cluster 2 has a special interest in developing use cases dealing with design and testing of existing or virtual production processes, and navigation and monitoring of existing production processes by the means of VR/AR and Mixed Reality (MR). Design and testing of existing or virtual production processes will be based on evaluation of combination of actual or virtual human with actual or virtual product and actual or virtual production environment. This aims to achieve several improvements in actual production process by providing navigation and monitoring of existing production processes. The setting will be based on 4 scenarios such as (1) supplying step by step work instructions at factory floor level using AR technologies; (2) providing fast and direct (sound or visual) feedback to operators on the factory floor, aiming to enhance existing tools to be more visual and more flexible; (3) Analyzing and monitoring productivity of workers, production environments and design phase; (4) monitoring ergonomics and working postures by using AR and Ergomix (Vink 2005). This provides Small and Medium-sized Enterprises (SME's) cost efficient means for detecting potential physical overloads or incorrect postures and effective ways for detecting needs for redesigning production environment.

Brief description of Cluster 3 - The focus of this cluster is on providing remote on-line maintenance support based on AR technology in the railway sector. The target of Cluster 3 is to provide expert-driven remote assistance to maintenance operator, helping end-user to focus on most important places and to easily exchange useful information in real-time with maintenance experts. Furthermore, a diagnostic system will also propose additional information as typical locations where anomalies reside. The use cases suggested by Cluster 3 are based on corrective maintenance and preventive maintenance. Corrective maintenance includes expert providing remote support and maintenance using 3D model based AR and voice transmission technologies. Also, Cluster 3 considers corrective maintenance to support workers by providing information accessibility from diagnosis system and from technical manuals, displayed using AR technologies. Cluster 3's preventive maintenance targets to training purposes by presenting typical tasks, which have already been performed to the device, by the means of AR.

Brief description of Cluster 4 - The focus of this cluster is on developing systems and procedures for training and assisting personnel in charge of power plants maintenance based on the metallographic replica technique, utilizing advanced VR technologies such as Virtual Environments and Haptic interaction. The target of Cluster 4 is Metallographic Replica, which is an on-site non-destructive sampling technique used to measure the obsolescence of an inner material located in pipe work within the nuclear power plant.

The use cases suggested to be used on the ManuVAR platform are, based on manual skills training in the performance of metallographic replica task, task planning and knowledge management as re-use of information during task planning and performance. The training use case will consist in developing a simulator allowing workers to improve their manual skills performing once and again the metallographic replica task in a computer generated environment. The training session will be previously set up by an instruction in order to adjust several environmental and ergonomics conditions. Worker position will be constantly tracked in order to satisfy postures requirements related to the performance of the task in real condition. Interaction within the virtual environment will be ensured using haptic interfaces that provide force feedback to user.

The use case for planning will provide to workers a specific VR tool to prepare, schedule and plan on-site metallographic replica, by using product data centralized through Virtual Model. Then, the use case for knowledge management will consist in the update of the information through the VM after on-site metallographic replica performance and its re-use during the planning stage of metallographic replica.

Brief description of Cluster 5 - The focus of this cluster is on improving heavy machinery product process by utilizing ergonomic methods supported by VR/AR technologies, while embedding the system to existing PLM structures. Goal of Cluster 5 is to develop methods and processes for re-using existing 3D models and product data to Virtual and Augmented Reality applications and Kiosk-PC, as well as development of PDM (product data management) and PLM (product lifecycle management) links to ensure effective distribution of product development data. In this context, new generation engine module will be virtually verified by applying participatory design method with VR/AR technologies as addition to ordinary review meetings (Leino at al. 2010).

Developed systems will be utilized in virtual prototyping, design reviewing, manual work, designing and simulating new production line and also in ergonomics, safety and risk analyses. The targets are on helping all end-users along product lifecycle to work on same up-to-date information. This connects manual workers, designers and engineers by exploiting VR/AR technologies (FIGURE: 2).

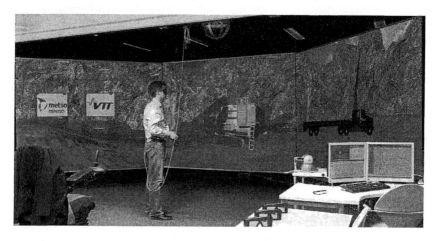

FIGURE 2. Virtual Environment created in ManuVAR predecessor VIRVO-project for maintainability review (Source: VTT)

RESULTS

Preliminary results of ManuVAR project established a list of 7 gaps, highlighting the most prominent problems of the European manual work industries in the context of high knowledge high value manual work. Based on the gaps, communication was detected to be the critical problem domain solvable by means of Virtual and Augmented Reality technologies. One of the aims of ManuVAR is to harness VR/AR to streamline communication within product process. This is done by the means of enabling all participants involved in product development to work collaboratively together through the Virtual Environment reflecting all design relations to the product developers with VM, making sure the actors are working upon up-to-date information and providing bi-directional information flows across the product LC.

ManuVAR proposes VR/AR technologies as a medium to access the VM for providing end-user context adapted view on the existing product data, meaning that depending on area of expertise, users are given the information relevant to them. The proposed VR/AR solution has to take into account various users along the product lifecycle, ranging from CAD designer to Human Factors expert and from end-user to the workers responsible of recycling the product. System needs to be capable presenting feedback from previous generations and capable of gathering feedback from the current state of design and to distribute this information to all relevant actors involved with the product. In this situation, the VE can be seen as universal translator between heterogeneous multidisciplinary actors for communicating and interacting with same virtual representation of the real product.

The ManuVAR system architecture is introduced in (Krassi et al. 2010). It is obtained as a projection of the ManuVAR PLM model, and it contains following main elements, Virtual Model, Virtual Environment, Application Tools and end-users. Here we further elaborate the technical aspect of the architecture (Figure 3).

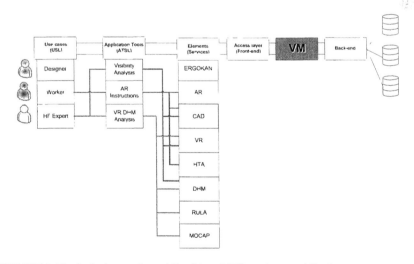

FIGURE 3. Technical overview at the ManuVAR system architecture

In order to make reusable Application Tools that can be interchangeably used in a number of industrial problems, we separate the use case specifics and the application tools specifics. User Specific Logic (USL) captures the lifecycle end-

650

users roles. The USL allows us to adapt the end-user VE view on the VM, thus visualizing only the relevant parts of the VM for the end-user.

The ManuVAR application tools contain Application Tools Specific Logic (ATSL) and technology and methods elements. The role of ATSL is to orchestrate the elements into coherent Application Tools (AT). Each AT addresses one or several identified gaps and can serve several end-users.

The elements of AT's include VR/AR software and hardware as well as ergonomic methods. Each element is wrapped into a software shell with clearly defined interface. This makes it possible to re-use the elements in variety of ATs.

The front-end Access Layer (AL) contains a set of operations that allow the element to communicate with the VM. At the back-end, the VM provides semantic references to the existing PLM repositories

Building the ManuVAR platform on top of existing PLM and PDM structures makes it possible to provide the end-users with up-to-date information. This is done by creating the VM in the way that the interactions of the user in the front-end are always reflected to the data in the PLM repositories. This provides means for the end-users for submitting their contribution to the product development through their specific VE view.

Instead of trying to build a monolithic system that can cover everything, the system will be designed upon concept of ATs, introduced earlier on this paper, for compiling interchangeable technological and methodological elements. This enables creation of generic platform and allows testing of its feasibility by the different use cases spread along the product lifecycle. The figure below details the use-case preparation per cluster on the whole product life cycle. Feasibility of the ManuVAR system will be first implemented in laboratory demonstrations and further validated and exploited in factory floor demos as follows, see Figure 4.

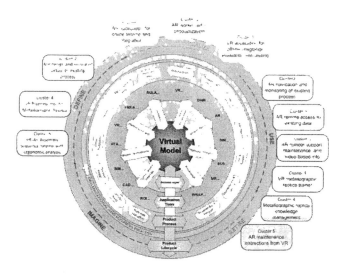

FIGURE 4. Cluster use-case distribution to ManuVAR PLM model covering the stages define, realize and use.

DISCUSSION AND CONCLUSIONS

With the preliminary results of ManuVAR, it was identified that VR/AR technologies are currently used for solving specific problems in isolated stages of system lifecycle and by providing direct non-transparent communication links between the heterogeneous multidisciplinary end-users. ManuVAR tries to tackle this problem by providing information on how VR/AR should be applied in the manual work context by utilizing existing PLM structures, end-user relations, product data relations and complexity management.

This system will provide possibility to reuse the existing and information such as 3D data at multiple stages of lifecycle, giving heterogeneous multidisciplinary end-users an easy access to the other disciplines for design decisions and provide possibility give direct feedback regarding their own discipline. Even if this system will eventually allow the companies to solve specific product related problems, the driving force is to provide systematic way for the methods, supported with technology, to blend efficiently to the companies' product process.

This setting builds up a unique perspective to the industry and research partners for understanding the value of VR/AR across the product lifecycle, giving insight on what are the most prominent problem domains, what methods are related to these domains and what technologies are needed to successfully apply the methods. Based on the preliminary results, this is one of key obstacles for the technology acceptance, that there are little to no efforts on utilizing VR technologies to cover the whole product lifecycle. This is critical for understanding the value of VR/AR, as investment-to-profit ratio over time on a single problem area may not be convincing enough for companies to invest on the technology. The platform by nature enables various LC actors to use the same VE for assessing product design through VM and distributing their development progress to all related participants.

When approaching the issues in this manner it is possible to provide systematic solution on how the VR/AR technologies should be used, for what purpose they need to be used and provides means for directly acknowledging the actual financial benefits gained from the complete product lifecycle. Hence, the ManuVAR system will not only support the companies to solve individual problems, but it will provide the foundation for the technology and methods to blend efficiently with the companies' product process ensuring that they can acknowledge the requirements of manual work in all stages of the LC.

ACKNOWLEDGEMENTS

The authors of this paper express their gratitude to all their colleagues who helped to prepare this paper, especially Dr. Boris Krassi of VTT. The results presented in this paper are the outcome of the joint efforts of the ManuVAR consortium. The authors are grateful to all contributors in the ManuVAR project who proposed ideas, gathered and analyzed data, and provided feedback and comments. The research

652

leading to these results has received funding from the European Commission's Seventh Framework Program FP7/2007-2013 under grant agreement 211548 "ManuVAR".

REFERENCES

Lind, S., Leino, S., Multanen, P., Mäkiranta, A., Heikkilä, J., (2009) "A Virtual Engineering Based Method for Maintainability Design" *Proceedings of 4th International Conference in Maintenance and Facility Management*

Brough, J.E., Schwartz, M., Gupta, S.K., Anand, D.K., Kavetsky, R., and Pettersen, R. (2007) "Towards Development of a Virtual Environment-based Training System for Mechanical Assembly Operations" *Virtual Reality, 2007*

Jayaram, U., Jayaram, S., Shaikh, I., Kim Y., Palmer, C. (2006) "Introducing quantitative analysis methods into virtual environments for real-time and continuous ergonomic evaluations" *Computers in Industry 57, 2006*

IST- IP project CoSpaces, www.cospaces.org

Ritchie, J.M., Dewar R.G., Simmons, J.E. (1999) "The Generation and Practical Use of Plans for Manual Assembly using Immersive Virtual Reality" *Journal of Engineering Manufacture, 1999*

Leino, S.-P., Lind, S., Poyade, M., Kiviranta, S., Multanen, P., Reyes-lecuona, A., Mäkiranta, A., Muhammad, Ali. (2009) "Enhanced industrial maintenance work task planning by using virtual engineering tools and haptic user interfaces" *3rd International Conference on Virtual and Mixed Reality. HCI International, 2009*

Krassi, B., Kiviranta, S., Liston, P., Leino, S.-P., Strauchmann, M., Reyes-lecuona, A., Viitaniemi, J., Sääski, J. (2010) "ManuVAR PLM model, methodology, architecture, and tools for manual work support throughout system lifecycle" *3rd International Conference on Applied Human Factors and Ergonomics, AHFE, 2010*

Lawson, G., Tossolin, G., Leva, C., Sharples, S., Langley, A., Shaikh, S. (2010) "Investigating Manual Work Using Human Factors Approaches and Behavioral Analysis" *3rd International Conference on Applied Human Factors and Ergonomics, AHFE, 2010*

Vink, P. (2005) "Comfort and design: principles and good practice" *CRC Press, 2005. (ISBN0849328306, 9780849328305)*

Leino, S.-P., Kiviranta, S., Rantanen, P., Heikkilä, J., Martikainen, T., Vehviläinen, M., Mäkiranta, A., Nuutinen, P., Hokkanen, I., Multanen, P. (2010) "Collaboration between Design and Production by exploiting VR/AR and PLM - Case Metso Minerals" *3rd International Conference on Applied Human Factors and Ergonomics, AHFE, 2010.*

Chapter 71

ManuVAR Training: Supporting the Implementation and Design of VR/AR Quality-improvement Tools for Manual Handling

Paul Liston[1], Sam Cromie[1], Maria-Chiara Leva[1], Kaj Helin[2], Mirabelle D'Cruz[3]

[1] Aerospace Psychology Research Group, University of Dublin, Trinity College, College Green Dublin 2, Ireland

[2] VTT Technical Research Centre of Finland Tekniikankatu 1, FIN-33101 Tampere, Finland

[3] Human Factors Research Group, Faculty of Engineering University of Nottingham, NG7 2RD, UK

ABSTRACT

This paper takes as its starting point the premise that implementation of the ManuVAR solutions will only be effective if the technology transfer is supported by appropriate knowledge transfer and training. The paper has four main foci. Firstly it defines the particular training challenges which the ManuVAR project presents and those challenges which the VR/AR tools must face, and situates this in the current state of the art by offering an overview of the literature and best practice guidelines in VR & AR training. Secondly it sets out the training concept which is informing the development and specification of the training activities. Thirdly it outlines the framework that will be adopted to face the challenges, meet the requirements and ensure the success of the tools and solutions developed in the project. And finally it outlines the implications for training from the preliminary results of the requirements gathering process.

Keywords: technology training, implementation support, technology design, human factors in VR/AR

INTRODUCTION

The objective of the ManuVAR project is to develop a technology platform - exploiting virtual and augmented reality - and a methodological framework to improve the efficiency and reliability of high-value, high-knowledge manual work in Europe.

The project addresses five applied case studies that all require, and benefit from, manual work. The experience, intelligence and adaptability of the manual worker all provide the flexibility necessary for organizations to navigate a productive path through the sometimes-competing forces of quality and efficiency. The project has four stages: (i) identify industrial problems - targeting real issues for real organisations, (ii) develop innovative VR and AR solutions to these issues, (iii) demonstrate value for European industry, and (iv) produce a commercial tool to assist more European industries fight outsourcing.

Success of the ManuVAR project rests on effectively meeting two challenges: (1) Effective solution design, and (2) Effective solution implementation. These challenges are clearly mutually dependent. Poorly designed solutions are difficult to implement; indeed good design will have taken the implementation process into account. Solution implementation needs to have an effective feedback loop to solution design. This paper outlines the approach the project is taking to address these two critical challenges through the development and specification of training which supports design and implementation of the ManuVAR solutions.

THE CLUSTERS & LIFE-CYCLE FOCUS

The applied focus of the ManuVAR project is centred on five clusters, each of them located in a different geographic area and developing a specific industrial case:

Cluster 1. Terrestrial spacecraft assembly
Cluster 2. Assembly line design
Cluster 3. Remote maintenance of trains
Cluster 4. Personnel training for maintenance of nuclear reactors
Cluster 5. Large machine assembly

The five cluster case-studies are also spread across the system lifecycle, in line with the project goal of optimizing the overall leanness and agility of the system as a whole – not just in manufacturing, but also design, operations, maintenance and recycling.

THE TRAINING CHALLENGE

The challenge for training centres on how to develop and specify a training concept that will be incorporated into the design of the ManuVAR technology and how to structure and deliver a training program to maximally support the implementation of the ManuVAR technology in the end-user organisations such that its utility can be demonstrated and the tools and technology can be exploited commercially for the benefit of European industry. In order for the ManuVAR project to be successful two main tasks must be achieved. They are:

1.   *Train users in how to use the VR/AR solution (one for each cluster)*

Users will need to be trained in how to harness the ManuVAR technology for practical benefit.

2.   *Implement the VR/AR solution in existing organizational systems*

Each end-user organisation will need a tailored implementation program accounting for existing organizational systems and practices.

CHALLENGES OF VR/AR TECHNOLOGIES

A literature review of existing VR applications and previous studies of training in virtual reality highlight some interesting issues that need to be considered and faced during the development of the ManuVAR training. The issues identified are:

Transfer to real world tasks. Rose et al. (2000) argue that "virtual environments (VEs) are extensively used in training but there have been few rigorous scientific investigations of whether and how skills learned in a VE are transferred to the real world" (p.494).

Realistically-priced environment. How complex should a VE be; so as to keep it easy to achieve the goals within a limited amount of time/budget; and still remain sufficiently sophisticated to be realistic? As pointed out by Foss (2009) "while virtual environments offer flexibility and location independence, they are still lacking absolute authenticity where an accurate model of a location is required" (p.560).

Interaction with other human actors. As highlighted by Benford et al. (2001) virtual worlds need a much better representation of self, and the perception and awareness of other users' presence.

The learning curve for trainers using a VE: "do we know how to use this power?" (Dorsey et al., 2009, p.197). As pointed out by Dorsey et al. (2009) "fundamental theories and findings from educational and training psychology may or may not apply in the world of virtual environments" (p.197).

CHALLENGES OF THE PROJECT

The inherent features of the ManuVAR project present further challenges. Five different clusters with five different case studies in five different industrial contexts –manual handling being the only commonality; each cluster with different system lifecycle characteristics; implementation in five different organizational contexts, and embedding the training in existing organizational systems and processes.

The project will develop VR/AR solutions to assist procedure compliance in spacecraft component assembly clean rooms (Cluster 1), to help design low-cost VR systems for improving assembly lines in SMEs (Cluster 2), to enhance remote online maintenance support in the railway sector (Cluster 3), to help improve training on nondestructive techniques (Cluster 4), and improve the assembly and maintenance of large machines (Cluster 5), all the time maintaining a lifecycle-focus.

LESSONS FROM APPLIED RESEARCH

EU funded VR/AR research and development projects such as TRAIMWE (funded under Esprit 4), AITRAM (funded under the IST program), and VIRTHUALIS (funded under the 6th Framework Program) have, respectively, looked at distributed maintenance training systems for aircraft maintenance, integrated human factors (HF) and technical training in VR environments for aircraft maintenance training, and integrated systems incorporating VR and HF methods for training, risk assessment, safety management and accident investigation. Of particular interest is the AITRAM project as it represented an attempt to integrate human factors and technical training for aviation maintenance using a VR environment. The use of a VR solution demanded that specific human factors-related actions be specified and defined (as best practice) so that they could be represented in the VE – this would help with training 'good' actions and extinguishing 'bad' actions. Ultimately this issue hindered the success of the project outcomes in that, while the project produced a demonstrator version of the VR training tool, it could not achieve its goal of integrating human factors and technical issues in this system (Liston, 2006). The task of defining human factors best practice proved to be particularly elusive and it was impossible to come up with agreed statements about best practice human factors actions – what was considered appropriate in one organisation was frowned upon in another, and regulatory requirements offered no common framework either. This presents a unique lesson-learned and an opportunity to inform the best practice guidelines about applied VR solutions – a task must be veridically represented in the VE if it is to be of any use. If operators cannot perform or behave how they normally behave then the system, and the training it produces, will have no transfer back to the job. This highlights how a *real* and *true* understanding of what a job or task involves underpins, in a fundamental way, the capability of VR/AR solutions to effect positive change. This is especially true of social tasks and activities.

MANUVAR TRAINING CONCEPT

Based on the lessons learned from previous research projects and from the particular challenges of the project a training concept was specified to help inform and situate the training framwork. The ManuVAR training concept is based on 4 central tenets:

1. Embed training in all 4 phases of the project

Typically training for technology is something that is considered long after the technology has been designed and usually just about when a prototype has been developed. That training is embedded in the whole ManuVAR project agenda from the outset is a distinct advantage and strength of the ManuVAR project design. It affords the training developers the chance to influence and guide the requirements gathering process such that subtle training issues can be explored and investigated. In addition the training development partners were involved in the requirements gathering exercises and case study definition – further ensuring that training issues were always considered from the outset.

2. Learn from past mistakes/failures/successes

Building upon the experiences of other EU-funded VR development projects and keeping abreast of research developments documented in the literature helps ensure the currency and innovative edge of the ManuVAR training tools.

3. Obtain an accurate representation of true operational practice

In order to build a training system which tackles real issues and trains employees to deal with real situations it is imperative that the training is built upon an accurate representation of true operational practice – i.e. the training must reflect the way in which tasks are done on a day-to-day basis, work-arounds and procedural deviations included. To this end the case study definition in the ManuVAR project incorporated ethnographic studies of normal operations in the analysis of tasks and processes. This ensures that the reality experienced by employees performing their work will make its way into the training curriculum – simultaneously ensuring more transfer-of-training and greater face validity.

4. Plan for implementation support

Designing a training system as a stand-alone intervention to ensure safe and efficient operations is a short-sighted approach. Training is part of a whole range of interventions that can help ensure safety and efficiency at work.

MANUVAR TRAINING FRAMEWORK

The training framework is based around four key activities: (1) Requirements gathering, (2) Case study specification, (3) Training development, and (4) Training implementation support. These four activities can be considered the operationalisation of the training ManuVAR concept and are critical to a successful outcome for ManuVAR (as measured by implementation and sustained use of ManuVAR tools in the end-user organisations). The last activity in particular, training implementation support, represents an innovative and sustained approach to ensuring successful implementation.

REQUIREMENTS GATHERING

The ManuVAR project, from inception, has deliberately involved the training partners in the requirements gathering phase of the project – the initial phase of the project which covers the high-level requirements that the industrial partners have for VR/AR solutions. The premise here is that without a fundamental grasp of the real needs and requirements of the industrial partners it would be impossible to design a training system to support the implementation of the VR/AR tool developed to target those needs and requirements. Similarly a requirements gathering process that is undertaken without taking training into consideration from the outset is one which will hinder the successful implementation of the VR/AR solution in the end-user organisation because the barriers to implementation of training will not have been considered.

CASE STUDY SPECIFICATION

A deep and fundamental understanding of the actual activities which will be modelled or improved by the VR/AR systems is a critical input to training specification and development. Experience from the AITRAM project has shown how crucial it is to be able to model all aspects of a task in a VR/AR system – especially the socio-technical dimensions. To ensure that the case study specification (an important task detailing all aspects of the task targeted by that cluster such that it can be modeled) is as detailed and valid as possible, ethnographic research methods are used. Ethnography, a qualitative social research methodology with roots in anthropology, can be used to particular advantage when investigating socio-technical and HF issues associated with tasks like manual work (Liston, 2006). If training designers are to really understand why procedures are violated, why shortcuts are performed, and how workers sometimes compensate for organisational deficiencies we must understand the task from the perspective of those who engage in these behaviours. This information ensures not only the content validity of the VR/AR representation of a task (the VR/AR model will represent the reality as experienced by workers) but also the face validity of the solution (workers will appreciate the accurate representation of complexity and the trueness of the representation of the task they perform in normal everyday circumstances).

TRAINING INFORMATION AND SUPPORT (TIS)

A key component of the ManuVAR training approach involves broadening the notion of training to include not just training development and specification, but also information support processes. The diagram below illustrates a model currently being specified by Cromie (2009) which has benefits for ManuVAR.

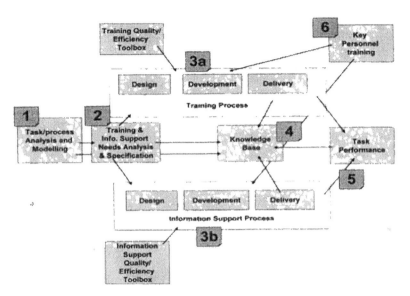

FIGURE 1: Training Information and Support (TIS) Model, Cromie (2009)

A comprehensive training/information support process should be based on a thorough analysis of the relevant task/process (conducted in the Requirements Gathering phase of the project) which should form the basis of a principled analysis and specification of requirements for training and information support, including specification of which knowledge elements should be supported by training and which by information support. This in turn should lead to parallel design-development-delivery processes for training and information support, each drawing on a dedicated toolbox of quality/efficiency tools. Then, the knowledge generated at each stage in the process should be captured and integrated in a knowledge base with user-centred designs for information and training-supply interfaces.

This TIS process has the potential to inform the training development activities in a profound way, and given that the training workpackage of the ManuVAR project (in which the training development activities will take place) has not yet begun there may be the potential for synergies with the ICMR project if the model can be demonstrated to be of use.

TRAINING DEVELOPMENT

Implementation of the ManuVAR solutions will only be effective if the technology transfer is supported by appropriate knowledge transfer and training. The target groups for the ManuVAR solutions are managers, in-house trainers (to ensure the ManuVAR knowledge is preserved in the industrial partner companies after the project is complete) and users of the technology, each of which will have different training requirements. The training development task will be split into five mini-training programs, each one adopting a multilevel instructional strategy as suggested by Dorsey et al. (2009) and incorporating not just off-line training but also the various information supports that are a feature of the TIS process. A range of methods and techniques will be explored with a view to developing a customised and ecologically powerful information system to directly support task performance. These will include: Task guidance, Embedded training, Integration in process management information and control systems, Messaging and communication capability, Performance reporting and feedback, Threat and error management, and Active knowledge capturing technology.

This would be based on a 'design-in-depth' philosophy in which the user-interface is seen as a set of nodes in a system-wide information management system. It is the deployment and management of information in the system as a whole which delivers leverage over process performance improvement, not just the design of the user interface itself, though this is an important component. This philosophy has been deployed in major task support development projects in aviation flight operations and maintenance.

RESULTS FROM REQUIREMENTS GATHERING

As outlined above Requirements Gathering and Case Study Specification are the first elements of the ManuVAR training framework and the results of these activities (reported in Krassi, et al. 2010; Kiviranta, et al., 2010 and Lawson, et al., 2010) are of particular interest.

In some of the ManuVAR case studies training is key to the envisaged benefits of the proposed VR solutions. In the Cluster 4 case study for instance, where the operator needs to be trained in performing a task whose results largely depend upon experience and whose conditions are often safety critical, the capability to train the operators in a much more effective way is one of the main desired results. For this case study the end user has the following requirements:

Training – training the operator to perform the task in three different levels (skills, procedural and managing hazardous situations).

Planning - managing critical situations, low accessibility, radiation spots and stressful situations so that workers can plan their activities to reduce task completion and radiation exposure.

Knowledge Base of previous tasks – repository where previous tasks are stored in the VR simulation and documented so that the operator can access previous tests (and any associated comments or results).

For this cluster (Cluster 4) the expected advantages from a ManuVAR solution centre around a reduction in accidents and health injuries during real world metallographic replica realization - brought about by a low-cost solution compared to real world physical training that supposes a reproduction of the whole working area which is not always possible or affordable; an unlimited possibility for retrieving previous stored training sessions; the creation of scenarios within the same virtual environment to simulate different operational situations with the aim of training workers under different conditions; and advance planning of the working area in which the task will be carried out in the field, so that radiation exposure can be minimized, so that the location can be chosen carefully in order that the task is completed more interactively, and so that costs can be reduced.

Training was implicated also in the results from the Cluster 5 case study where the issue of how to train workers to quickly learn how to assemble mass-customized products was raised. In this instance the challenges for a potential ManuVAR solution are about producing 3D models of new mass-customized products and developing AR: identifying product features to recognize possible assembly errors and to guide the reassembly. One key factor of success for training in this context is the possibility of developing a system that could facilitate learning from experienced worker (correct assembly order from the work done).

Another training issue comes from the results of the requirements gathering fieldwork in relation to Cluster 3 where as a result of the ethnographic observations performed on the shop floor it was found that the requirements expressed by management did not take into account the opinions and attitudes of workers, and that workers and operators were not familiar with the possibilities which AR/VR could offer in terms of improving problematic situations that they encountered in their working lives. This finding highlights the importance of socially-situated requirements gathering that examines normal practice. These findings are being fed into further work packages in the project so that the needs of both management and operators are met through consultation and iterative design. In this cluster the issue of task support is key – where management seek improvement to procedural compliance in a context where the culture believes that using instructions means poor preparation. Only by tackling both issues and demonstrating the tangible benefits of VR/AR to both management and operatives can the Cluster produce an effective solution.

662

SUMMARY

Set out in this paper is the training strategy for the ManuVAR project. The key innovations in this strategy relate to requirements gathering and case study specification and the way these in turn inform training development and implementation support. By following the specified framework to support the implementation process and manage worker reactions to the new initiatives it is anticipated that the implementation will be successful. The ManuVAR project is midway through it's first year (of three) and will be evaluating the progress it makes in developing training to support the implementation of VR/AR quality-improvement tools in high-value, high-knowledge manual work. The outcomes of evaluations of the training challenges and how well they are met, together with reviews on the implications for training literature will be particularly interesting and both are to be produced later in the project timeline.

ACKNOWLEDGEMENTS

The research leading to these results has received funding from the European Commission's Seventh Framework Programme FP7/2007-2013 under grant agreement 211548 "ManuVAR". www.manuvar.eu

REFERENCES

Benford, S., Greenhalgh, C., Rodden, T. & Pycock, J. (2001). "Collaborative virtual environments." *Communications of the Association for Computing Machinery*, 44, 7, 79–85.

Cromie, S. (2009). *"Training Information and Support: A model."* Unpublished manuscript, Trinity College, University of Dublin, Ireland

Dorsey, D., Campbell, G. and Russell, S. (2009). *"Adopting the instructional science paradigm to encompass training in virtual environments."* Theoretical Issues in Ergonomics Science, 10:3,197 - 215

Foss J. (2009). "Lessons from learning in virtual environments." *British Journal of Educational Technology*, Vol 40 No 3, 556–560.

Kiviranta, S., Poyade., M., Helin, K. (2010) "Virtual environments for supporting high knowledge , high value manual work design." *Applied Human Factors and Ergonomics - AHFE 2010*, USA

Krassi, B., D'Cruz, M. & Vink, P. (2010) "ManuVAR: a framework for improving manual work through virtual and augmented reality." *Applied Human Factors and Ergonomics - AHFE 2010*, USA

Lawson, G., Tosolin, G., Leva, C., Sharples, S., Langley, A., Shaikh, S. (2010) "Investigating Manual Work Using Human Factors Approaches and Behavioral Analysis." *Applied Human Factors and Ergonomics - AHFE 2010*, USA

Liston, P. (2006). *"Human Factors Competence"*. Doctoral Thesis. Trinity College, University of Dublin, Ireland.

Rose F. D., Attree E. A., Brooks B. M., Parslow D. M., Penn P. R. & Ambihaipahan N. (2000) "Training in virtual environments: transfer to real world tasks and equivalence to real task training." *Ergonomics*, Vol. 43, No. 4, 494-511

Integration of Human Factors Methods to the ManuVAR Product Lifecycle Management Model

Susanna Aromaa[1], Juhani Viitaniemi[1], Chiara Leva[2], Boris Krassi[1]

[1] VTT Technical Research Centre of Finland
Tekniikankatu 1, FIN-33101 Tampere, Finland

[2]Trinity College, University of Dublin
College Green, Dublin 2, Ireland

ABSTRACT

It is necessary to acknowledge human factors (HF) more systematically throughout the lifecycle. Human factors engineering is a significant part of the product and system design process, but now it is time to widen the scope from design phase to cover all situations during the whole lifecycle. This paper aims to describe how to integrate HF methods into the ManuVAR Product lifecycle model (PLM) and how to make HF methods more visible and usable throughout the entire lifecycle. By dividing methods and services into four groups: (1) manage requirements, (2) support design, (3) evaluate and get feedback and (4) support training and sharing of information it is possible to respond to the needs and requirements of ManuVAR goals, lifecycle, gaps, cluster cases, high value, high knowledge manual work and technology applications.

Keywords: Human factors, product and system lifecycle, human centered design, information systems

INTRODUCTION

BACKGROUND

Human factors engineering is a significant element of product design processes and also of the whole product lifecycle. Humans have an important role in work processes: they can be the most intelligent and flexible parts of the work process but in the other hand they can also be the one who are responsible that planned action or task may fail to achieve their current goals (Reason et. al., 2003). Moreover to provide safety, ergonomic and usable working conditions, it is necessary to design those elements into the products at an early stage, because it increases the real and perceived value of design (McClelland et. al., 2005). Human factors (HF) are considered more and more as a part of the lifecycle, however HF is mainly considered in the design phase in a way which may not cover the whole lifecycle concurrently or similarly.

Lifecycle concept can be approached with Product Lifecycle Management (PLM), which is an integrated, information-driven approach comprised of people, processes/practices, and technology to all aspects of a product's life, from its design through manufacture, deployment and maintenance – culminating in the product's removal from service and final disposal. By trading product information for wasted time, energy, and material across the entire organization and into the supply chain, PLM drives the next generation of lean thinking (Grieves, 2006). PLM also helps to extend beyond traditional lean in the sense of providing holistic optimization of lifecycle, communication and knowledge management. PLM is a holistic business activity addressing many components. Product data is just one component of PLM. But PLM represents the all-encompassing vision for managing all the data relating to the design, production, support and ultimately disposal of manufactured goods. Thus the product data include the products themselves, organizational structure, working methods, processes, people and information systems. Addressing them together leads to better results than addressing them separately. The whole is greater than the sum of the parts. (Stark, 2006).

HF is mainly integrated to the life cycle the way that user needs and requirements are considered in the design phase. It can also be called user-centred design (UCD), or human-centered design (HCD) and is one step towards integrating HF into the lifecycle. It makes users more motivated to learn and can also increase productivity, enhance quality of work, reduce support and training costs, and improve user satisfaction (ISO 13407, 1999). There are plenty of studies introducing ways to integration of HF and UCD/HCD to the system development processes (e.g. Helin et. al., 2007, Vilpola, 2008, Kuusinen, 2009). Participatory approach (e.g. Kuorinka, 1997, Wilson 1995, Hirschheim, 1989 and Muller et. al., 1993) is one way to obtain user needs and requirements of different actors during the lifecycle. Participatory approach is a good way to make the users more satisfied with the products/systems by giving them responsibility for influencing on decisions of the next steps in the process. It can also increase knowledge of the users, context of use and involved organizations (Vink et. al., 2006).

666

AIM ON THIS PAPER

The aim of this paper is to describe critical issues and viewpoints on how to integrate HF methods into the ManuVAR PLM and how to make HF methods more visible and usable to all throughout the lifecycle as a continuity not just as an intervention.

INTRODUCTION TO MANUVAR PLM MODEL

The ManuVAR project (Manual Work Support throughout System Lifecycle by Exploiting Virtual and Augmented Reality, CP-IP 211548) (ManuVAR, 2010), funded under the European Union 7th Framework Program, aims to develop an innovative technology platform and a framework to support high value, high knowledge manual work throughout the product lifecycle. ManuVAR's ideology bases on supporting the high value, high knowledge manual work by the concept that includes for each user an own personalized Virtual Environment (VE) view to allow immersion to the virtual environment i.e. work activity system (technical, environmental, and human subsystems) that is created and updated using the information from Virtual Model (VM) (Figure 1). VM manages all the databases (PLMs, ERPs, CCMSs, etc.) required for creation each of the personalized VE views and information (e.g. application tools, methods and services providers, etc.), that is needed to create users' own personalized VEs.

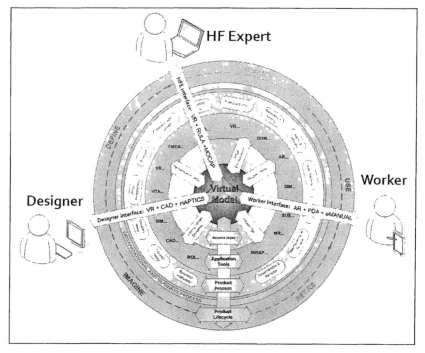

FIGURE 1. ManuVAR Virtual Model and Product Lifecycle Management approach. (ManuVAR D3, 2010)

The important issues are supports for bi-directional communication, efficient interfaces and knowledge creation and share. ManuVAR VM represents the way of storing and retrieving the information form different databases, information being essential to the case in hand to be solved or resolved. Also ManuVAR PLM is in many ways a paradigm partially seen as an enabler of efficient collaboration and communication between actors (e.g. engineers, managers, human factors experts and workers) in complex business environment and partially also as technical elements to support methods and services, and partially as technologies to allow support for the use of VR, AR, simulation, animation and interaction in VEs. An assumption is that PLM will connect the entire lifecycle of a product or system into one analytical participatory session bringing together the people, processes, business systems, and knowledge, which will higher the level of common comprehension and ability to achieve better conceptual alternatives, and solve/resolve to the problems.

The development of details into ManuVAR PLM and VM means that the previously introduced industrial based scenarios need to be elaborated and the use cases and specifications defined from them. This provides the information to define application data, tools, and knowledge flow to different users with various VE view applications, and between all the life-cycle stages. Compromises between industrial cases will be needed that the more scoped and precise definition, design and implementation for ManuVAR PLM and VM could be worked out.

REQUIREMENTS AND NEEDS FOR HUMAN FACTOR METHODS IN LIFECYCLE

There are different kind of requirements and needs to be considered when integrating and choosing HF methods to ManuVAR PLM model. In the beginning of the ManuVAR project were requirements collected from the companies and also the state-of-the-art research. Information was collected and formulated to seven gaps that relate to manual work: (1) problems with communication throughout lifecycle, (2) poor interfaces, (3) inflexible design process, (4) inefficient knowledge management, (5) low productivity, (6) lack of technology acceptance and (7) physical and cognitive stresses.

In ManuVAR there are also five company cluster cases, for which manual work-related issues needed to be solved, and in which the ManuVAR results will be tested. The five cluster cases are related to satellite assembly, assembly line design, remote maintenance of trains, training of maintenance of industrial plants and large machine assembly process design. The manual work done in these clusters is high value, high knowledge manual work that cannot be automated or outsourced.

VR/AR technologies will be also used in cluster cases to improve manual work. The purpose is to use HF methods together with VR/AR technologies as an application tool. Application tool in this instance means a combination of human factors methods and technology tools e.g. motion capture system, digital human model and posture analyses; together they will be used for supporting manual work.

Lifecycle considerations can also generate new requirements for HF methods. HF has already been implemented at least on a theoretical level in the design phase, but when managing HF in whole lifecycle, it establishes different needs and requirements to the HF methods example in managing knowledge and information. Consideration of fluent information flow will improve HF, because it will make it easier to collect information from different lifecycle phases e.g. information from the end-users, information from the assembly line and information from the maintenance workers and use that all to make requirements for the new product. This use of HF information can also support e.g. the training phase. To produce the right information, the right HF methods must be applied.

The activities of the ManuVAR network have the potential to radically transform the way manual work is supported in the future. This results in significant needs and requirements at a variety of impact levels:

a. Individual employee level – improve working practices, procedures and conditions; enhance usability of and accessibility to products for operations and maintenance

b. Organisational level – reduce costs; better knowledge and risk management; increase levels of innovation; reduce numbers of safety-related incidents; better process orientation; improve operational performance

c. Sectoral level – improve safety; lower life-cycle costs; improve productivity and competitiveness; faster system-wide diffusion of innovation

d. Wider impact – applicability of approach to other complex, safety-critical industrial environments.

Gaps, cluster cases, the high knowledge manual work, the use of VR/AR technology, lifecycle and impact levels sets certain requirements and needs for HF methods that will be used and integrated to ManuVAR PLM model. These things should be covered when choosing usable and applicable methods for ManuVAR.

Helping the manage methods in lifecycle some approaches were applied: general system theory (Bertalanffy, 1973) and cybernetics (Wiener, 1948) approach. A system of any nature can be represented as the following four elements: reference – plant – output – feedback. In our case, the plant is a workplace. There is also a similarity to the HCD standard (ISO 13407, 1999), which highlights the importance of the requirements and the feedback (sections 5.2 and 7.1 of the standard). However, the standard focuses on the design and it does not propose any clear grouping of methods.

RESULTS

As a result from mentioned approaches and requirements of ManuVAR goals, gaps, cluster cases, high knowledge manual work, lifecycle, technology applications and impact levels, four groups were established (Figure 2):

1. Manage requirements (reference)
2. Support workplace/task design (plant)

3. Evaluate actual situation and get feedback (actual output of a planned or a working system)
4. Support training and sharing of information (feedback loop).

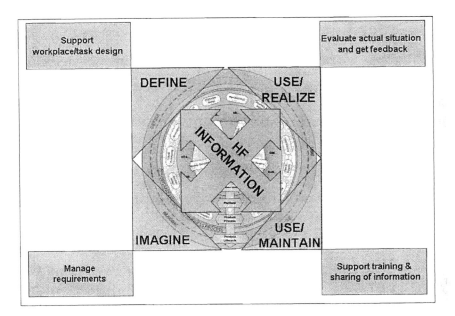

FIGURE 2. Four groups and information change of human factors methods in ManuVAR product lifecycle management model. (ManuVAR D4, 2010)

The first group, manage requirements, includes methods that establish requirements coming from human (e.g. physical, mental), tasks, tools, environment, organization, business and legislation and also from the product (e.g. from previous generations). It is good to remember that all requirements discussed here are somehow human related and other requirements e.g. technical are not included here. The second group concentrates to design support and includes methods that correspond to workplace design, task design and production design needs. The third group highlights the importance to evaluate and get feedback from the real users and actors from the whole lifecycle. It concentrates actual situations either in the real world or in Virtual Environment (VE). The fourth group supports training and sharing of information and means methods that deliver work instructions, provide training (in off-line) and support motivation. Four groups are covering the lifecycle and also the information sharing, but they are not in a certain order or time-line of use.

Examples from cluster cases how integration of HF method to ManuVAR PLM model and results of ManuVAR affects also different impact levels:

• At individual and organizational level ManuVAR has to support training for knowledge and skill acquisition that is normally acquired on the job and after many years of experience, speeding up the process. A ManuVAR tool can also provide solutions for task support that may improve the performance at individual and organizational level (easy accessibility of required documentation in a paper free and user friendly format and remote assistance).

- At a sectoral level ManuVAR technology can be used to support the design stage of a process or of an assembly line, further it can also be used to support the planning stage of safety critical tasks across multiple organizations

- At organizational level some of the possible benefits can also be transferred then to a wider audience of complex operations industry (Wider impact). For instance the solutions to be developed for support a meaningful way of storing knowledge can have a high level of transferability to other similar context. For example the availability of a 3D model for plant accessed on repeated occasion can support a clever way of storing information about places where interventions have been previously realized that can be marked in the VR simulation and documented. Operator could use to see where tests have previously been performed, any comments and the results. Or in some case ManuVAR is currently studying a system to store knowledge about successful trouble shooting performed on non routine tasks, to be made available for further consultation by the workforce whenever needed. Both approaches could easily be transferred to other industrial context than the ones they have been generated within.

DISCUSSION AND CONCLUSIONS

Four method groups present integration of HF methods to the ManuVAR PLM model and they are a good approach to control and use more efficiently the HF information throughout the lifecycle. It makes also easier to find right methods for ManuVAR-project and to technical oriented people to approach HF methods. Four groups are also giving good basis for further application designs were HF methods and technology tools will be integrated together. It also highlights the importance and management of HF information throughout the lifecycle. It is important to get all needed HF information from whole lifecycle and deliver it there where it is needed.

It is hard to make strict lines between groups and also with methods, because some methods can be used in several groups. It should be remembered that this deviation to four groups is meant to do some harmonizing and generalizing from all human factor methods from the ManuVAR point of view. So there may also be other applicable approaches when integrating HF methods to lifecycle. This deviation to four groups should also be tested and validated more, if it is desired to make it more general and applicable to other projects and lifecycles.

These four groups give new possibilities to manage the HF information and it is one step further from the integration of human centered design and product development processes (e.g. Ulrich & Eppinger, 2004), which is the approach usually today. As said, today good HF design criteria are quite well known at least in theoretical level, but in lifecycle also managing HF information and other HF needs e.g. assessing existing work places or acceptance of technical assistance tools is needed to consider.

ACKNOWLEDGEMENTS

The research leading to these results has received funding from the European Commission's Seventh Framework Programme FP7/2007-2013 under grant agreement 211548 "ManuVAR".

REFERENCES

Bertalanffy von, L. (1973) *General system theory: Foundations, development, applications.* New York, 336 p.

Helin, K., Evilä, T., Viitaniemi, J., Aromaa, S., Kujala, T., Patel, H., Pakkanen, T., Raisamo, R., Salmenperä, P., Miettinen, J., Kilpeläinen, P., Rannanjärvi, L. and Vähä, P. (2007). *HumanICT. New Human-Centred Design Method and Virtual Environments in the Design of Vehicular Working Machine Interfaces.* VTT, Espoo. 85p. + app. 1 p. VTT Working Papers: 73. http://www.vtt.fi/inf/pdf/workingpapers/2007/W73.pdf

Hirschheim, R. (1989). *User participation in practice: experiences with participative system design.* In: Knight, K. (ed.). Participation in systems development, Columbia GP Publishing. Pp. 194-204.

ISO 13407. (1999). *Human-centered design processes for interactive systems.* International Organization for Standardization, Geneva.

Kuorinka, I. (1997). "Tools and means of implementing participatory ergonomics." *International Journal of Industrial Ergonomics* 15, 365–370.

Kuusinen, K. (2009). *Integrating Human-Centered Design into Product Development Process.* Helsinki University of Technology. M.Sc. Thesis.

ManuVAR. (2010). Manual Work Support throughout System Lifecycle by Exploiting Virtual and Augmented Reality, CP-IP 211548, www.manuvar.eu

ManuVAR D3. (2010). Analysis of validity of VR/AR for manual work support and human factors aspects of manual work; conceptual specification and technical basis alternatives for VM. ManuVAR CP-IP-211548. The University of Nottingham: Nottingham, England.

ManuVAR D4. (2010). Specification of ManuVAR platform, definition of methods and tools. ManuVAR CP-IP-211548. VTT Technical Research Centre of Finland: Tampere, Finland.

McClelland, I. & Fulton Suri, J. (2005). *Involving people in design.* In: Wilson J. R. & Corlett E. N. (Eds.), Evaluation of Human Work (pp. 281-333). CRC Press

Muller, M.J. & Kuhn, S. (1993). "Participatory design." *Communications of the ACM,* 36, pp. 24-28.

Reason, J. & Hobbs, A. (2003). *Managing Maintenance Error: A Practical Guide.* Ashgate Publishing Company, Burlington.

Ulrich, K. and Eppinger, S. (2004). *Product Design and Development.* 2d ed. Boston: McGraw Hill Irwin.

Vilpola, I. (2008). *Applying User-Centred Design in ERP Implementation Requirements Analysis.* Tampere University of Technology, Publication 739

Vinka, P., Koningsveldb, E. A. P. and Molenbroeka, J. F. (2006). "Positive outcomes of participatory ergonomics in terms of greater comfort and higher productivity." *Applied Ergonomics* 37, 537–546

Wiener, N. (1948) *Cybernetics, or, control and communication in the animal and the machine.* New York, 194 p.

Wilson, J.R. (1995). "Solution ownership in participative work redesign: the case of a crane control room." *International Journal of Industrial Ergonomics* 15, 329–344.

CHAPTER 73

Supporting Workers Through the Combination of Mixed Reality And Mobile Devices in Wide Areas

Nikos Frangakis[1], Ioannis Karaseitanidis[1], Roland Blach[2], Tim Gleue[2]

[1]Institute of Communication and Computer Systems
Dep. of Electrical and Computer Engineering, 15773, Athens, Greece

[2]Fraunhofer IAO, Germany
CTVT Fraunhofer IAO, Nobelstr 12, 70569 Stuttgart, Germany

ABSTRACT

Many VR-systems lack a standardized 2D interface for access and control. Web technology provides these standards, but these are not used widely for immersive 3D virtual environments. Mobile devices have gained in computation power and are increasingly used for permanent web access. We would like to exploit these standardized and ubiquitous technologies to improve the access to and the control of VR system engines and its applications in order for remote users-workers to be able to access centralized information within Virtual Environments. We will describe some implementation examples, on how web technology can be used for accessing and controlling VR engines. These examples include a couple of interaction paradigms and an implementation example for a "Google-maps" like interface to virtual worlds. This work can be used as a proof of concept design of how annotated information on 3D models can be accessed from mobile devices in remote locations.

INTRODUCTION

Many VR-systems lack a standardized 2D interface for access and control. One of the reasons is that many of them have been developed in the mid to late nineties when web technologies were not standardized or ready to provide more than the delivery of HTML pages. Also due to real-time interaction requirements for immersive systems, web technologies have always been considered too slow. That has changed in the last decade significantly with the development of asynchronous communication between server and clients (AJAX), thereby eliminating the need of a complete web page refresh. Moreover, service oriented architectures dominate the web development and system independent protocols as SOAP have been defined and settled. These methods and protocols have obviously proven their maturity. Also many tools already exist to facilitate the access and the authoring of the necessary data structures.

Another observation is that mobile devices like mobile/smart phones, PDA's, Blackberries etc., have gained in computation power and presentation layout. Especially in the last few years great efforts have been made in enabling these devices to access the World Wide Web in a way similar to desktop computers. This has resulted in the development of very sophisticated mobile web browsers, which are capable of handling not just the traditional HTML but also the latest versions of CSS and Javascript. Moreover, the hardware interfaces that are provided nowadays, like multi touch displays, gyroscopes, Bluetooth and WLAN capabilities make them natural candidates for interaction devices for virtual environments.

To benefit from these ubiquitous technologies in hardware and software a further investigation for the integration in VR-system frameworks is necessary. Specifically two main ideas seem to be promising for a pragmatic integration:

- Lightweight Browser based access support which benefits from ubiquitous hardware. The VR-system acts as a web service and is able to provide standardized control access.
- Using a standard protocol for remote object access to abstract from a specific VR-system which allows for more generic control functionality and possibly also for simple exchange of configuration or predefined modules e.g. as navigation patterns.

In the following sections we will describe our generic integration concept and the description of an implementation example based on our VR Framework. (Blach98)

Related Work

The integration and usage of web technology for immersive VR-systems has already been extensively explored by other researchers and can be structured as follows:

- Developing specific clients to use with PDA's and Mobile devices as system control devices as (Watsen 1999, Hill00, Benini 2002).

- Building on top of Web technology and standards an authoring and runtime systems as Contigra (Dachselt 2002).
- Using XML as container for system and application configuration with appropriate templates and translators as e.g. VrJuggler (Bierbaum 20010, inTML (Figuer et al 2002) or other custom systems [Gaitatzes06] .
- Building systems based on X3D which integrates well web technology as the instantreality system [Behr04].

All these approaches have shown the feasibility and the flexibility of web technology in combination with VR-systems. Our contribution is built on top of these experiences and addresses two issues:

- Lightweight extension of existing VR-systems with web technology
- Description of system functionality in a system agnostic protocol which is well integrated in current web technology

Web technology for VR-Engines

To provide the most basic structure to access the system via web browsers, the VR-engine has to act as a web service where the client can access the states of the running application. The main advantage is that devices which are able to access the web-server with a browser can easily turn into sophisticated interaction devices for virtual environments. This can be easily achieved by integrating a web server with the application. Three different approaches can be found on this:

- Running an external webserver (e.g. Apache) and providing scripts which connect to the VR-engine
- Starting the webserver inside a scripting environment as part of the application
- Linking the webserver functionality into the runtime system itself.

The webserver acts as dispatcher for the concurrent access which not only allows for multi user support but also for novel multi interaction paradigms. An illustrating example is a navigation scheme where the user could have a PDA and a mobile phone with different interaction modes and use it simultaneously. The concurrent access can be handled seamlessly by the web server. Also all involved users can control a virtual environment with their personal devices by simply accessing the web server which mediates the concurrent access to the VR-engine by default.

To provide a system independent access to the objects which define the virtual environment we propose to adopt basically a system independent remote procedure protocol. The primary requirements to exploit such a protocol is that the VR-engines adopts the notion of objects with attributes and methods which can be accessed from outside. This is the case for most of the existing VR-engines and is specifically true for our system.

For web applications two different flavors can commonly be found: SOAP or JSON. Both provide a system independent remote procedure call framework as was provided by e.g. CORBA or DCOM. The web oriented remote procedure call mechanisms are based on text based messages which are by nature binary independent, so no modules have to be integrated at link time. The downside of this is obviously the reduced performance due to the parsing and packing of the messages. Also these protocols have been evolving alongside with web services and service discovery which made it more widely adopted as the classical frameworks. SOAP uses XML as wrapper and is used in combination with the web service discovery language WSDL. JSON is based on JavaScript and is not as generic but more lightweight in its application. For many web applications the system independence seems to outweigh the performance issues considering that network and compute bandwidth have increased and in 2D desktop like interaction the user is the slowest component compared to the rest of the system. This is not the case for the inner simulation loop of 3D real-time interactive systems as VR-engines in general, but for system control and object access very often only 2D desktop like interfaces are used. Some are part of the virtual environment itself as 3D menus or some are running on external devices with dedicated client software. For these interfaces the real-time requirements are comparable to the web interfaces we have today to access server functionality as databases or search-engines. For these cases we think interfaces based on web standards would lead to better integrated and more ubiquitous access to VR-engines.

ARCHITECTURE

To be able to use generic functions the VR-engine has to provide a translation module between the system agnostic remote procedure protocol and the actual engine. If such an access component has been established an interesting side effect will be that software interaction patterns might be described in a system independent manner such that users can bring their own patterns which then can be transferred to the VR-engine assuming that comparable modules are available. For this specific application it is desirable that the VR-engine is able to expose its functionality to the web client. That is the reason why we have chosen the SOAP protocol instead of JSON because the integration with the WSDL seems to be more mature. JSON seems to be well suited if the functionality is completely known outside the VR-engine and serves only for message exchange. The two levels described in the section above will be shown in the following figures.

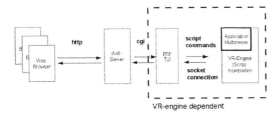

FIGURE 1 Simple Web Server approach with cgi interface

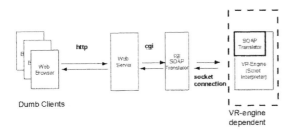

FIGURE 2 Simple Web Server approach with SOAP Translator (Dumb Client)

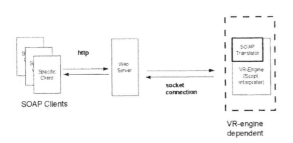

FIGURE 3 Pure SOAP Client approach - No need for Web browser besides routing

It can be seen here that if a specific client is written, which already provides the SOAP messages, the web server basically acts as proxy. If a generic web browser is used the dispatching between the interface in the browser and the SOAP messages is performed on the server site. Ideally, all the clients should be kept in sync with the server in order for the information transfer to be consistent. This implies a stateful system, where client and server have uninterrupted communication with each other in order to be kept up to date. Unfortunately, because of the stateless nature of the web protocol and the performance issues we would have if the clients were constantly polling the server for information the resulted architecture is stateless. However, this architecture does not really comply with the REST architecture, since the messages exchanged do not contain all of the information necessary for a system transition, but are dependent on the server state and evaluated according to that.

IMPLEMENTATION CONSIDERATIONS FOR THE SOAP DISPATCHER

We are describing now the integration of a SOAP to C++ translation module in our VR-engine xyz as an example as most of the runtime modules of VR-engines due to the real-time requirements are written in C/C++. We have chosen to use the gSOAP library, which does the translation handling and also the provision of the WSDL files for the webserver or other clients. The implementation of this translation module is unfortunately not straightforward because a simple wrapping of the object declarations in the header files cannot be handled properly by these template generation tools. Especially if more modern C++ language constructs are used, the mapping to the WSDL and the SOAP templates is not one to one. Instead we have created a specific interface component which can be exposed to the SOAP functions and mediates between the VR-engine core and the translation modules as can bee seen in figure 2 and 3. This also controls the exposed API which can be kept more lean and clean. We have just started to implement the SOAP based dispatcher and cannot report on specific experiences with the application of this architecture. Nevertheless the approach seems to be reasonable to be pursued.

Sample Implementation: Navigation and "Google-Maps" like interface for an arbitrary Virtual World

Based on the first approach as seen in Figure 1 of the proposed architectures, we implemented a sample web interface to navigate and control virtual environments through standard web browsers, using either desktop PC or hand-held devices like Android Phone and IPhone/IPod. This implementation serves as a proof of concept and as a testing bed for speed and usability tests. To experiment on these a virtual world two different paradigms were implemented:

- One classical paradigm using arrows to control the movement. It contains 2D controls for all 3D available movement, like pitch/raw/roll, forward/backward movement and turning around, allowing the user to fully control his motion in the virtual environment.
- The second implementation aims at touch devices and enables the user to move and turn around by sliding his finger on the surface. Moreover, devices that contain orientation sensors enable the user to control height by operating the device in landscape mode.

We did not implement a fully fledged SOAP communication protocol, but rather a phase 1 implementation, using as dumb client the web-browser and relaying the translation work to the web-server. On the VR-engine side we developed a simple application for a Virtual World with a snapshot-camera serving as a map. The interface is designed to be functional on any web browser supporting JavaScript, xmlHttpRequest API and CSS2. This way we can use AJAX technology to send HTTP POST requests to the underlying real-time VR-engine without the need to refresh the web pages, providing this way a smooth and consistent experience to the user.

IMPLEMENTATION

The system is composed from of main components:
- The HTML pages, which define the structure and appearance of the web pages and the JavaScript functions, which take care the client-server communication, by sending commands and receiving the results in asynchronous mode, using AJAX.
- The server side scripts, which are written in PHP or TCL, take care of the translation of AJAX commands to actual script language understood by the VR system, then execute it, receive the outcome and relay it back to the web pages.
- Finally the VR part includes a socket listener to receive commands. Moreover, a multiplexer has been implemented to allow dual control from "normal" input devices like keyboard and mouse, but also input from the web interface.

The HTML pages can vary from very simplistic to really complicated, accordingly to the user needs. We have implemented two different version, one using arrows for the various navigation instructions and a second one, which is completely blank, is used as touch navigation from touch enabled devices. This serves as proof of concept on how easy is to integrate different paradigms of motion control on the same interface. The obvious advantage of including both paradigms is the easy exchange between them.

The server side scripting is completely agnostic to the interface and the user. It is simply used to translate the HTTP POST requests to commands understood by the underlying VR-engine, in our case xyz. We have two different implementations, one using TCL scripting in combination with a TCL web server and another using PHP scripting in combination with an apache web server. The main functions of calculating new positions, setting/resetting flags etc., are done from the server side scripting. The VR server receives simple, straight forward commands.

The final element is a multiplexer, embedded in our virtual reality applications script, which facilitates the combination of movement between traditional input methods, like mouse and keyboard, and the input from the web navigation. This way the system achieves multi user input, from users located either in the Virtual System and/or at their web terminals. It also facilitates a smooth movement of the camera, overcoming this way the bandwidth limitation and lag of the network communication between server and client. If one attempts to move directly the camera, without the use of a multiplexer, a fast network can achieve repetition time of 30ms, ca. 30Hz, which results in choppy movement. Moreover, due to the nature of the TCP/IP communication and since AJAX will initialize each time a new connection, it is possible that a t_{n+1} command can arrive earlier than a t_n command, resulting in jumps in the movement. For these reasons, the use of intermediate translator of the movement, the multiplexer, is imperative. This way, we do not need to send continuous updates of the camera, but only the changes in the movement, which occur only when the user reacts with the interface. Finally, the necessary components for the implementation of the map have been included in the Virtual World, which include an orthographic projection camera.

VIRTUAL GOOGLE MAPS

Another interesting implementation was a Google-Map like interface to the Virtual Environment, where the user could move the map around and also zoom in and out. The interface is an exact copy of Google-Maps featuring the same navigation functions. However, the JavaScript implementation is customized, since the virtual and the real world have some significant differences in how they define coordinates and altitude, namely:

- Real world defines coordinates in longitude and latitude, while in virtual worlds we usually define Cartesian coordinates
- Real world has an absolute value of 0 altitude, while in the virtual worlds this has no meaning, since this is just the z-coordinate, where objects can well be positioned below that.
- In the virtual world it makes sense to be able to move the camera-user around, whereas in a real world it makes sense to move only the map.

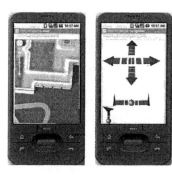

FIGURE 4 Virtual Google Maps interface and Arrow navigation running on the android emulator

Moreover, in this application we have added context menus, which allow the user to move the camera in the virtual world through the map interface, providing this way also a "World in a Miniature" [Stoakley95] feeling.

Conclusion and Future Work

STATUS OF THE CURRENT IMPLEMENTATION

In our current implementation we have explored the utility and usability of a 2D web interface for controlling Virtual Environments.

The main advantages of the HTML implementation relate to the range of the devices upon which the interface can be used. It has been tested on desktop PCs, the Android Phone, the IPhone/IPod devices, even simpler mobile phones. Most notably, on the IPhone/IPod devices, where the JavaScript implementation allows access to the touch events, a touch navigation interface was implemented, which facilitated a very natural and easy navigation. Another very important aspect is the ease and speed of the interface development, since HTML is popular and easy to use. This in combination with the new features of CCS3 and the speed of the new JavaScript engines, as demonstrated from Google, shows the future in web interfaces development.

The main disadvantages are related to the networked nature of the devices. In certain occasions network connections can lead to delays or out of sequence communication.

Secondly, although HTML and CSS are standardized, their implementation in different browsers does not follow the standards. This leads to small implementation differences between the different branches of browsers located in either in PCs or hand held devices, which in turn are responsible for implementation differences in the applications.

The client server approach applied in the communication between the devices and the Virtual Reality platform fits perfectly the ManuVAR "Virtual Model". All the core functionality is served from a central point and the mobile device access is just a view to the common "Virtual Model", where all changes are occurring there.

RELEVANCE TO MANUAL WORK AND FUTURE WORK

We are well aware that the abstraction proposed in this paper does not create a higher semantic layer which is another important issue for system independent descriptions of virtual worlds. It mainly describes a technical exchange mechanism on the message level, which is more general than X3D but also has less semantics. We believe that this more open protocol might be nevertheless another starting point for a discussion about semantic descriptions which embraces more variety of system approaches than X3D can possibly do due to some inbuilt world description paradigms, e.g. the event propagation via data flow, which does not fit for all existing system architectures.

Regarding our interface, we plan to standardize the communication between the web-server and our VR-engine, xyz, using SOAP. We will need to modify both xyz and finalize the integration of the SOAP translator, as well as our server-side communication scripts. From the interface point of view we are experimenting with different navigation techniques, best fitted for touch screen hand-held devices and we will include also some inertia navigation method as soon as the gyroscope values of the devices are accessible through their web-browser.

We do envisage a series of applications for the proposed architecture dealing with manual work. Training for wokers on the field dealing with complex maintenance, or assembly/disassembly tasks, navigation on a virtual model, sharing of information between different stakeholders are among numerous applications.,

Acknowledgements

The research leading to these results has received funding from the European Community's Seventh Framework Programme FP7/2007-2013 under grant agreement no. 211548 "ManuVAR" (www.manuvar.eu).

This project has been partially supported by the Fraunhofer IAO internal project "augmented identity".

REFERENCES

Behr, J., Dahne, P., and Roth, M. 2004. Utilizing x3d for immersive environments. In Web3D '04: Proceedings of the ninth international conference on 3D Web technology, ACM, New York, NY, USA, 71–78.

Benini L., Bonfigli M.E., C. L. F. E. R. B. 2002. Palmtop computers for managing interaction with immersive virtual heritage. 183–189.

Bierbaum, A., Just, C., Hartling, P., Meinert, K., Baker, A., and Cruz-Neira, C. 2001. Vr juggler: a virtual platform for virtual reality application development. 89–96.

Css 3d transforms module level 3. http://dev.w3.org/csswg/css3-3d-transforms/.

Dachselt, R., Hinz, M., and Mei, K. 2002. Contigra: an xml based architecture for component-oriented 3d applications. In Web3D '02: Proceedings of the seventh international conference on 3D Web technology, ACM, New York, NY, USA, 155–163.

Gudgin, M., Hadley, M., Mendelsohn, N., Moreau, J. J., and Nielsen, H. F., 2003. SOAP version 1.2 part 1: Messaging framework.

Hill L., C.-N. C. 2000. Palmtop interaction methods for immersive projection technology systems. In Fourth International Immersive Projection Technology Workshop (IPT 2000).

Jung, Y., Keil, J., Behr, J., Webel, S., Z̈O 380 LLNER, M., Engelke, T., Wuest, H., and Becker, M. 2008. Adapting x3d for multi-touch environments. In Web3D '08: Proceedings of the 13th international symposium on 3D web technology, ACM, New York, NY, USA, 27–30.

Watsen K., R. D., and Capps, M. 1999. A handheld computer as an interaction device to a virtual environment. In Third International Immersive Projection Technology Workshop (IPT 1999).

Blach R., Landauer J., A. R. A. S. 1998. A flexible prototyping tool for 3d real-time user-interaction. In Proceedings of the Eurographics Workshop on Virtual

Environments 98.

Schmidt, D. G., and Kuhns, F. 2000. An overview of the real-time corba specification. Computer 33, 6, 56–63.

Web services description language (wsdl) 1.1.http://www.w3.org/TR/wsdl.

Gaitatzes, A., Papaioannou, G., Christopoulos, D., Zyba, G. 2006 Media productions for a dome display system In VRST '06: Proceedings of the ACM symposium on Virtual reality software and technology.

Behr, J., Dahne, P., Roth, M., 2004 Utilizing X3D for immersive environments In Web3D '04: Proceedings of the ninth international conference on 3D Web technology

CHAPTER 74

A Virtual Interactive Training Application to Support Manual Work

Matthias Strauchmann[1], Tina Haase[1], Juhani Viitaniemi[2], Sauli Kiviranta[2,] PeterVink[3]

[1]Fraunhofer Institute for Factory Operation and Automation IFF
39106 Magdeburg, Sandtorstrasse 22, Germany

[2]VTT Technical Research Centre of Finland
33101 Tampere, Tekniikankatu 1, Finland

[3]TNO / Delft University of Technology
2132 JJ Hoofddorp, Polarisavenue 151, Netherlands

ABSTRACT

The demands on professional qualification and training courses have changed in recent years. The trend is toward more flexibility in terms of time, location and content.

Workers who solely learn operations or maintenance procedures in different industrial units from text manuals and 2-D graphics may misinterpret instructions and make mistakes in a real work situation. Realistic and visually intuitive representations of complex machinery and technical equipment reduce the risk of misinterpretation. Following the concept of active learning, interactive 3-D training additionally facilitates flexible, time-independent and user-adapted training irrespective of the availability of a machine or piece of technical equipment.

An interactive 3-D-training application is utilized to design interactive training lessons in different fields of application, e.g. mechanical and plant engineering, shipbuilding and power engineering. This paper presents the didactic concepts on which it is based, discusses decisions made about the user interface and design and reports on experiences acquired during its industrial use.

Keywords: 3-D training, ManuVAR, virtual reality, maintenance, manual handling

MOTIVATION

The development, design and commissioning phases in the product life cycle are increasingly characterized by shortened production and commissioning times and increased ranges of variants. The expertise at a company's disposal is fundamental to the rapid startup and reliable operation of production systems and is reflected in the qualifications of every employee involved in the startup process.
Operators must be able to immediately react to any malfunctions of a machine or plant during their operation. Inaccurate data sheets or instructions with contradictory part designations cause confusion. Therefore, qualification courses, assistance systems and knowledge transfer should be integrated in employees' work.

Spatial and temporal constraints on trainers and real machinery are difficulties faced by trainees in traditional learning environments. Moreover, processes inside machinery are often hidden by parts such as the housing and operators use complex controls (e.g. computerized interfaces), which are often spatially separated from the actual machine. In some cases, processes are too fast or too complex to be perceived by the human eye.

These are typical conditions in a number of fields of application including mechanical and plant engineering, shipbuilding and power engineering.

Industrial demands have generated a need for a flexibly configurable qualification platform that supports such work, factors in employees' different levels of knowledge and promotes cross-company information exchange. The interactive 3-D training platform presented here is intended to meet these requirements. The platform bundles knowledge on the startup and operation of production systems. Its methodological and didactic organization of tasks enables trainees to acquire the knowledge necessary to start up and operate a system and continuously supply the knowledge storage system with feedback. It reproduces complex interrelationships and trainees' interaction with the interactive system requires them to actively handle tasks and solve problems on their own. Numerous cross-industry solutions for different training and educational objectives have been developed on the basis of this learning system (Belardinelli et al 2008; Blümel, 2007; Hintze et al 2000; Schenk, 2006).

Norros (2004) applied the concept of a core task (CT) to define the content and context of studied technologically mediated work activity for training simulators. Norros also introduced the theory of affordances as an alternative approach to cognitive information processing (Gibson, 1977). Affordance defines the human subject in relation to the environment and identifies the ability to grasp objects in real-time training environments as a challenge.

Martens et al (2001) presented a tutoring system for Web-based and case-oriented training in medicine based on three different models chosen and adapted to the domain of medicine, namely a:

- tutoring process model,
- case knowledge model and
- medical knowledge model.

Although it is very domain specific, it exhibits certain parallels to the approach presented in this paper, e.g. a knowledge base that enables users to receive information on demand and a guided learning tour that supports the learning process step-by-step (by creating and regenerating real-time knowledge repositories for that purpose).

THE INTERACTIVE TRAINING PLATFORM

CONCEPT FOR LEARNING IN 3-D WORK ENVIRONMENTS

The development of a special training solution requires initially defining the educational objectives in collaboration with the client and based on Bloom's taxonomy of educational objectives (Bloom, 1976).

An interactive training application is intended to improve cognitive skills (paper focus). It does not focus on the affective and psychomotor level of Bloom's taxonomy since it is hardly affected by an interactive visualization.

Once the educational objectives have been identified, the storybook for the virtual scenario is created. It includes detailed information on the working processes visualized, the visualization techniques employed and user interactions that enable users to control the scenario.

Different training modes (see Figure 1), which require different activities, support trainee learning of knowledge about working processes. The modes and their features are:

Discovery Mode
No task or procedure is performed and complex functionalities are controlled interactively, e.g. by using the mouse to move complex processes.

Presentation Mode
A professional's approach to the completion of the task at hand is demonstrated in a presentation comparable to an interactive video in which users can pace their own progress. Users may diverge to acquire more details on specific topics, select different viewpoints or immerse to obtain a realistic sense of targeted objects, events, and related phenomena.

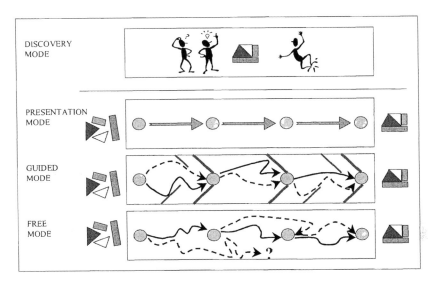

FIGURE 1: The four training modes available in the VR based training

Guided Mode
Having the knowledge of the presentation mode, the guided mode defines concrete tasks a trainee has to perform. The trainee receives instructions on and explanations of the task. The system specifies the sequence of subtasks.

Free Mode
Trainees are not provided any specified procedure to complete the task. The training objective stresses independent action and comprehension of causalities. Users may freely manipulate virtual objects. They may make errors, which the system evaluates and incorporates in its update of the knowledge database.

These modes have been applied in virtual reality training scenarios (Blümel and Jenewein. 2005) that allow trainees to specifically select the learning mode.

USER INTERFACE AND INTERACTION CONCEPTS

The interactive training system is used by trainees in advanced on-the-job training and by service technicians in their working when need to review information.
The wide application field necessitated designing an easy-to-use and intuitive user interface.

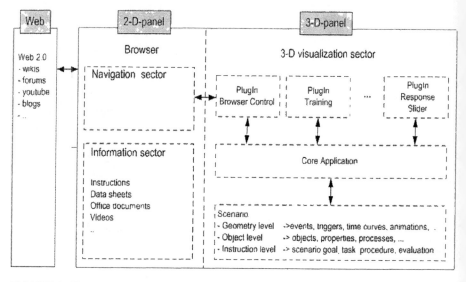

FIGURE 2: The system architecture of the interactive training platform

Therefore, the training system workspace is divided into three sections (see Figure 3):

- navigation sector,
- information sector and
- 3-D visualization sector.

Navigation sector
The navigation sector in the upper left corner of the workspace contains the table of contents from which all the scenario's functionalities may be accessed. The navigation sector has been implemented in html and flash, thus making it easy to use, even for non-computer experts.

Information sector
The information sector in the lower left corner displays additional information, e.g. videos, PDF data sheet files or 2-D animations coupled with the 3-D scene. The 2-D panel can be unlocked to allow trainees to a full screen view its contents. This is especially useful for PDF instruction or maintenance documents.

3-D visualization sector
The 3-D visualization sector to the right of the workspace contains a 3-D scene, which trainees may navigate and interact with.

Workflow
Direct communication was established between the three sectors to ensure the workflow is uninterrupted (see Figure 2). This facilitates initiating animations or other actions in the 3-D-scene directly from the 2-D-panel as well as accessing

additional information in one of the left-hand sectors by interacting with the 3-D-panel, e.g. by pushing a virtual button. The aforementioned training modes are extensively integrated in this workspace and not only allow the use of 3-D visualization but also 2-D-animations, videos and documents.

A scenario author formulates the interaction concept and determines which options are offered to trainees. The use of html and flash makes it possible to easily adapt the layout of the system to a client's cooperate identity and to dynamically handle feedback and knowledge creation and regeneration. The training platform is used as a knowledge base that compiles dispersed knowledge from highly skilled workers with valuable experience.

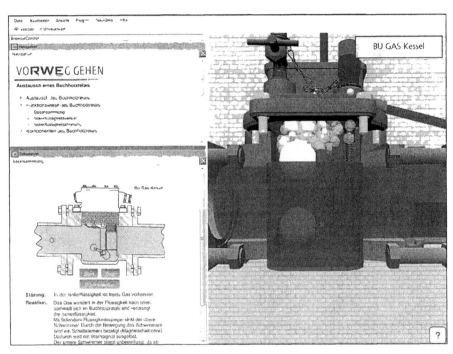

FIGURE 3: Interactive training system workspace (see best practice example)

ASSESSMENT

Assessments are an important means of providing trainees feedback and enabling them to rate their own performance. Therefore, the training system has options for integrating self-assessment tests.

Such tests are designed by the scenario author. Assessment tests distinguish between open and closed response formats (Proske and Körndle 2001).

Closed response formats such as multiple choice tests, drag and drop tasks or connection tasks (see Figure 4) are best suited for use in computerized systems

since the system itself can evaluate them immediately and automatically. This allows users to specify their learning activities based on their learning progress.
The following response formats have been implemented in the training platform:

- *Multiple choice test*
  A classic multiple choice test with at least one correct answer.

- *Drag and drop task*
  Pairs that belong together, e.g. components and their names, must be matched by dragging and dropping.

- *Arranging task*
  A number of items must be arranged to place the work steps in the correct order.

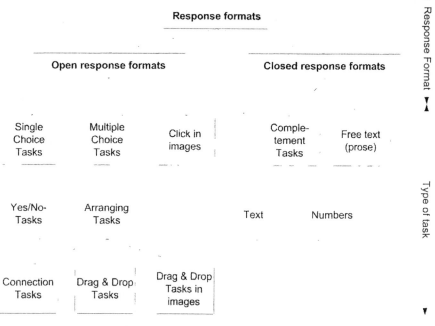

FIGURE 4: Response formats used in this system base on Weber (2006)

BEST PRACTICE EXAMPLE

The interactive 3-D training application is being used in numerous industry projects, including one that deals with high voltage equipment.

Manual operations such as the installation, commissioning and maintenance of high-voltage equipment require expert knowledge of the equipment's function, its basic data and thresholds and profound know-how to professionally assess inspection results.

The 3-D learning environment enables technicians to familiarize themselves with the working processes without the constraints or risks of a real work environment.

Trainees may interactively explore the function of a pole column and the operating mechanism by sliding its mechanical behavior (see Figure 5). The operation of real components is impossible to observe because their reaction time is within milliseconds and thus invisible for the human eye and any movements occur inside the components.

Additional parameters, e.g. electricity and gas, are visualized and may be used in a training course to explain functional dependencies.

The learning environment allows the inclusion and linking of videos, documents and other media with the 3-D-environment to optimally prepare users for their work environment. The inclusion of available media boosts acceptance of the system among users because "their" operations are recognizable and not abstractions far removed from real experience.

The solution also includes an assessment that provides learners feedback on their knowledge. One task entails matching pole column components by dragging and dropping. Another requires trainees to demonstrate their proficiency by placing work steps in the correct sequence.

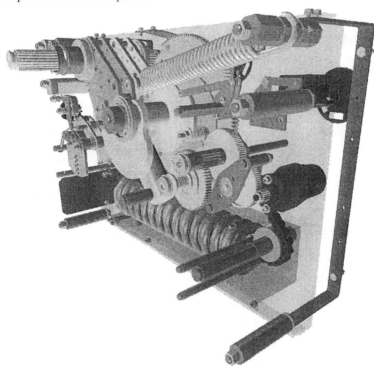

FIGURE 5: Virtual operating mechanism

CONCLUSIONS

Virtual reality technologies have experienced a sizeable leap in the development in recent years. Interactive 3-D training systems are able to reproduce extremely complex realities.

Interactive VR technologies in basic and advanced vocational training of technical specialists will be technically and economically feasible in the near future. This fact makes this technology attractive to support manual work processes e.g. in satellite assembly, in task planning and training of industrial plant maintenance, and in maintenance of heavy machinery. This should serve researchers and developers as an impetus to focus more intently on the potentials and needs of training and education in interactive 3-D environments at factory floor. The technological developments presented here support training on realistic virtual products, machinery and plants even when access to real objects, which are often not available for training at all or only to a limited extent, is limited. A challenge in daily use of this technology is the integration of virtual interactive training environments in real work processes. This topic is under research within ManuVAR e.g. the feedback capturing at factory floor to establish an improvement cycle between operation phase and design phase in the life cycle.

The theoretical construct constitutes the foundation for researching the didactic and technical potentials of visualization systems for training and education. Work is being done to formulate a conceptual theory to research learning actions in real and virtual technical systems.

ACKNOWLEDGEMENTS

The research leading to these results has received funding from the European Commission's Seventh Framework Programme FP7/2007-2013 under grant agreement 211548 "ManuVAR".

REFERENCES

Belardinelli, C., Blümel, E., Müller, G., Schenk, M. (2008) Making the virtual more real: research at the Fraunhofer IFF Virtual Development and Training Centre. Journal Cognitive Processing, Springer Verlag, p. 217 – 224, ISSN 9217-224

Bloom, B.S. Ed. (1976). Taxonomie von Lernzielen im kognitiven Bereich. Weinheim, Basel: Beltz Verlag.

Blümel, E., Jenewein, K. (2005) Kompetenzentwicklung in realen und virtuellen Arbeitsumgebungen: Eckpunkte eines Forschungsprogramms. In: Schenk, M. (Hrsg.): Virtual Reality und Augmented Reality zum Planen, Testen und Betreiben technischer Systeme. Magdeburg: IFF, 177-182

Blümel, E. (2007) Stand und Entwicklungstrends des Einsatzes von VR/AR-Techniken für Qualifizierung und Training im Arbeitsprozess. In: Gesellschaft für Arbeitswissenschaft e.V. (Ed.), Bericht zum 53. Arbeitswissenschaftlichen Kongress, Dortmund: GfA-Press, p. 241-244, ISBN 978-3-936804-04-1

Gibson, J. (1977) The Theory of Affordances. In: R. Shaw & J. Brandsford (eds.).Perceiving, Acting and Knowing. Toward and Ecological Psychology. New York: Lawrence Erlbaum Associates. Pp. 67-82.

Hintze, A., Schumann, M., Stuering, S. (2000) Employing Augmentation in Virtual Environments for Maintenance Training. Proceedings of the Industry/Interservice Training Simulation and Education Conference (I/ITSEC), November 26 – December 1, 2000, Orlando, FL/USA

ManuVAR. (2010). Manual Work Support throughout System Lifecycle by Exploiting Virtual and Augmented Reality, CP-IP 211548, www.manuvar.eu [Accessed: 23 February 2010]

Martens, A. and Bernauer, J. and Illmann, T. and Seitz, A. (2001) Docs 'n Drugs - The Virtual Polyclinic" An Intelligent Tutoring System for Web-Based and Case-Oriented Training in Medicine. In: Proc. Of the American Medical Informatics Conference, p. 433-437, Washington, USA, AMIA

Norros, L. (2004) Acting under Uncertainty. The Core-Task Analysis in Ecological Study of Work. Espoo: VTT, Available also URL: http://www.vtt.fi/publications/vtt_pdf.jsp

Proske, A., Körndle, H. (2001). Lern- und motivationspsychologische Wirkungen beim Bearbeiten von Lernaufgaben unterschiedlichen Antwortformats. Vortrag auf der 8. Fachtagung Pädagogische Psychologie, Landau

Schenk, M. (2006) Virtuelle Realität – Trends und Anwendungen für die Zukunft. In. Schenk, M. (Ed.): Wettbewerbsvorteile im Anlagenbau realisieren. Zukunftsszenarien und Erfahrungsberichte. Tagung Anlagenbau der Zukunft. Magdeburg, p. 97-103.

Weber, C. (2006) Entwicklung und Evaluation mediendidaktischer Konzepte zur Computer-basierten Lernerfolgskontrolle im Fach Molekulare Pharmazie. Inauguraldissertation, Basel.

CHAPTER 75

ManuVAR PLM Model, Methodology, Architecture, and Tools for Manual Work Support Throughout System Lifecycle

B. Krassi[1], S. Kiviranta[1], P. Liston[2], S.-P. Leino[1], M. Strauchmann[3],
A. Reyes-Lecuona[4], J. Viitaniemi[1], J. Sääski[1], S. Aromaa[1], K. Helin[1]

[1]VTT Technical Research Centre of Finland
Tekniikankatu 1, FIN-33101 Tampere, Finland

[2]Trinity College, University of Dublin
Aerospace Psychology Research Group, School of Psychology
Dublin, Ireland

[3]Fraunhofer Institute for Factory Operation and Automation
Magdeburg, Germany

[4]University of Malaga
Malaga, Spain

ABSTRACT

This paper defines the core elements of a high knowledge high value manual work support system – the ManuVAR – across the entire system lifecycle. The central result is that manual work can be efficiently supported only by combining the following three elements: (1) the lifecycle approach to ensure bi-directional communication throughout system lifecycle, (2) a virtual and augmented reality technology solution to implement the user interfaces contextualized and adapted to all actors in the lifecycle, and (3) ergonomics methods to prototype workplaces, evaluate manual work performance, gather the feedback improve delivery of instructions, training, motivation, and management of human-related requirements. In this paper, the ManuVAR product lifecycle management (PLM) model, methodology, system architecture and application tools are proposed. It is shown that the virtual model is essential for communication across the lifecycle. The main principle of the virtual modeling is formulated and an explanation of how it helps to manage change and complexity of the ManuVAR system is offered.

Keywords: Manual Work, Product Lifecycle Management, PLM, Virtual Reality, Augmented Reality, Ergonomics, Virtual Model, ManuVAR

INTRODUCTION

This work-in-progress paper summarizes the recent results of the European research project ManuVAR (211548) – Manual Work (MW) Support throughout System Lifecycle (LC) by Exploiting Virtual and augmented Reality (VR/AR), see (ManuVAR, 2010; Krassi, et al., 2010). The scope of the project is high knowledge high value MW. The work is valued high because of the relevance to the unique, complex, and precious machinery throughout its LC: from design, manufacturing to operation, maintenance and recycling. Here, the human constantly accumulates and brings the tacit knowledge and skills, which are scattered and poorly accessible. This type of MW refers to (Rasmussen & Goodstein, 1985) "knowledge-based behavior" and partially to sophisticated "skill-based behavior".

In this paper, we concentrate on the main theoretical results of the project: the PLM model, methodology, system architecture and VR/AR-based tools for supporting high knowledge high value MW. Throughout the paper, we refer to the seven most prominent problem areas that are faced by the European industry regarding high knowledge high value MW. These areas are presented in (Krassi, et. al., 2010) and briefly listed below:

1.   *Problems with communication throughout the lifecycle.*
2.   *Poor interfaces.*
3.   *Inflexible design process.*
4.   *Inefficient knowledge management.*
5.   *Low productivity.*
6.   *Lack of technology acceptance of AR/AR technology.*
7.   *Physical and cognitive stress.*

The ergonomics aspects of the methodology and the application tools (ATs) are considered in (Aromaa, et al., 2010), the training concept is presented in (Liston, et. al., 2010) and the virtual environment side is discussed in (Kiviranta, et al. 2010).

METHODS

The ManuVAR overall method of work is described in (Krassi, et al., 2010). There are five industrial clusters in the project covering support of satellite assembly, remote training for locomotive maintenance, manufacturing line design, task planning and training in industrial plant maintenance, and design, productization and maintenance of heavy machinery.

During the inductive phase of the project, the seven problem areas related to high knowledge high value MW that are relevant to a number of the European industries were identified (Krassi, et al., 2010). In this article, we represented these areas in the PLM model, which allowed us to formulate and to justify the overall ManuVAR solution, the methodology, the system architecture, and the list of the ATs that combine ergonomic methods and the VR/AR technology to support MW throughout the system LC. The architecture and methodology link the ergonomic methods and the supporting VR/AR technology to the problem areas. This makes it clear why a specific method or piece of technology is employed.

During the deductive phase, the ATs will be implemented, tested in the laboratory environment and demonstrated on the factory floor in the five industrial clusters. Not only have we to find out whether the ManuVAR system technically works, but also we have to demonstrate that it is accepted on the factory floor and it delivers a business benefit. While developing the five industrial cases we will also test all stages of the methodology. The results of the tests and demonstrations will be used for adjusting the ManuVAR methodology, architecture and tools. In ManuVAR, instead of trying to build a monolithic system, we will build architecture and a set of interchangeable ATs. This will help to make the project results usable to a large variety of industrial problems.

RESULTS

MANUVAR PLM MODEL

We adapted the conventional PLM model – see e.g. (Grieves, 2006) – for the ManuVAR context of high value high knowledge MW to represent and to explain the indentified seven problem areas, Figure 1.

The ManuVAR PLM model contains several layers:

• System or product LC stages with the five main stages: imagine, define, realize, use, and dispose. This layer also captures business and organizational processes – all in the context of MW.

• Product process with respect to MW, where we identify four main groups of areas (the grouping is inspired by the general cybernetic system (Wiener, 1948): reference-plant-output-feedback) where the human has to be accounted for along the entire LC: (1) management of requirements, (2) workplace design including production of instructions, physical and virtual prototyping, (3) MW evaluation in the real and virtual environments; handling the direct human feedback, and (4) provision of instructions, training and motivation (Aromaa, et al., 2010).

- ATs helping to improve MW and comprising ergonomics methods in the mentioned four groups and supported by the VR/AR technology.
- Virtual model (VM) in the center and its access layer. The VM provides a semantic reference to various PLM repositories and to the CAD, PDM, PLM, ERP functionalities that, in analogy with the production part of the activity theory triangle (Engeström, 1987) have to contain information on (1) the human actors, (2) the product, and (3) the human and product environment throughout the LC.

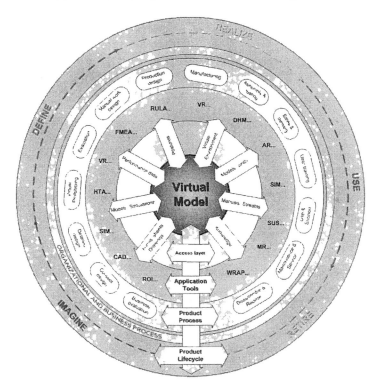

FIGURE 1. ManuVAR PLM model (Source: ManuVAR project)

The adapted PLM model is sufficiently rich to represent the seven problem areas. Thus, the communication flow in the LC is traditionally from the designer to the worker – linear and unidirectional. For the actors at later stages of the LC, it is hard to provide their feedback to the actors at the earlier stages. This situation is captured by the first and second problem areas. The VM is very complex and primarily suitable for the designers, production engineers and management. The worker needs his/her own focused and natural interface for instructions and feedback. This is captured by problem areas two, and four to seven. The reduction of physical and mental stress requires prototyping of the workplace to be done hand-in-hand with the product, production and work environment design (problem areas three and seven). However, prototyping needs the knowledge on how the MW is performed and the feedback from the workers (problem areas one, two and four).

ManuVAR Solution to the Manual Work Problems

Analysis of the identified problem areas in the context of the PLM model shows the following main ways to improve MW:

1.   Providing means for bi-directional communication throughout system LC. This can be achieved by using the VM as a communication mediator – a single access point to the variety of data, information, models on the system for all actors in the LC;

2.   Implementing the user interfaces contextualized and adapted to all actors in the LC. The role of the interfaces is to provide a natural and efficient medium between the human and the complex VM. Here the application of VR/AR in combination with ergonomics methods is very promising (Talabă & Amditis, 2008);

3.   Adopting ergonomics methods to prototype work places, to evaluate MW performance, gather the feedback for the LC actors (e.g. workers' feedback or experience to designers, designers' feedback or recommendations to the workers), and to improve delivery of instructions, training, motivation, and management of the human-related requirements, i.e. standards, legislation, organization and business requirements.

Such a combined multidisciplinary approach is essential to address all of the identified problem areas. This is especially important for knowledge management. Knowledge, unlike data or information, is very difficult to formalize and, therefore, impossible to be gathered, stored, processed, and transmitted in the same way as data or information. Modeling of knowledge is beyond the scope of the ManuVAR project. Here we assume that the human is the main source and carrier of knowledge. Therefore, we are developing the system that inherently supports knowledge communication and re-use by means of bi-directional communication (e.g. the designer-worker-designer chain), together with interfaces that enhance the provision of feedback, and training and instructions that motivate the humans to rely their knowledge, skills, and experience.

ManuVAR Methodology

The ManuVAR methodology describes the process of applying the ManuVAR system (PLM model, ATs, VM) to solving industrial problems related to high knowledge high value MW. The methodology is presented in the form of the flowchart. The elements of the flowchart correspond to the layers of the PLM model when one penetrates the model from the outer LC layer towards the center VM, thus forming a "vertical" projection of the PLM model, see Figure 2.

The main idea of the methodology is to capture the problem (industrial case) and to decompose it into the seven known problem areas, for which ManuVAR offers its ATs. This allows for the reuse of solutions – the ATs with some technological adaptation for the specific case – thus avoiding the situation when each new industrial problem has to be solved from scratch.

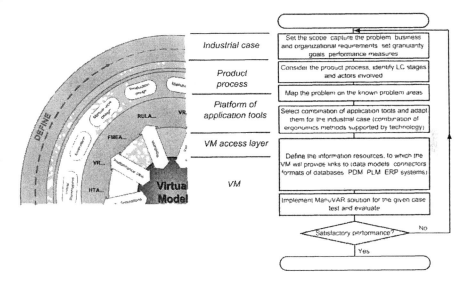

FIGURE 2. ManuVAR methodology (the flowchart) as a "vertical" projection of the PLM model (Source: ManuVAR project)

MANUVAR SYSTEM ARCHITECTURE

The ManuVAR system architecture describes the role and interaction of system elements and system implemented options. The architecture is derived as a "horizontal" projection of the PLM model. Each layer of the model has a counterpart in the architecture, Figure 3. Here we concentrate on the front-end (actors↔VM). The back-end (VM↔PDM/PLM) is the subject of further research.

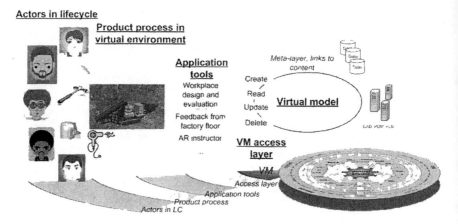

FIGURE 3. ManuVAR front-end architecture as a "horizontal" projection of the PLM model (Source: ManuVAR project)

The core of this projection is the VM that provides reference to the information on (1) the human actors, (2) the product, and (3) human and product environment throughout the LC. The VM (a digital mockup or a virtual prototype) is a computer-based model that substitutes a real system. It is a systemic, semantic aggregation of all information, models, processes, and simulations that describe the system in evolution throughout its LC. The VM has been studied in other domain e.g. in (Krassi, 2006), the semantic side of it e.g. in (Simantics, 2010), and it will be studied further in future ManuVAR publications.

The term VM has been used in other research studies in a variety of contexts, but mostly with the reference to VR/AR technologies. Here, "virtual" means "as good as real, yet not existent in the real world" so that the presentation side of the VM related to VR/AR is handled by the virtual environment (VE).

Because the VM is very complex, it has to have a set of adapted and contextualized interfaces to make the VM usable to the human actors across the LC. The complexity of the VM interior is encapsulated in the VM object, which is accessed through the VM access layer and the Virtual Environment (VE), which offer natural real-time actor-focused view of the VM. A number of ATs are built to operate in the VE, e.g. ergonomic analysis, delivery of contextual instructions, and capture of the worker feedback.

All essential properties of the VM such as its systemic and phenomenological nature, ability to be focused for different actors, its interaction and communication properties can be expressed in the following principle.

The main principle of virtual modeling: the VM has to be manipulated in the same way as the real object that it models. The properties of the VM are learned through interacting with it by means of virtual experiments.

Virtual experiments are similar to physical experiments – simulation and system identification. To obtain length of an object, one needs a ruler; to obtain mass of an object, one needs a scale; to obtain color of an object, one needs a source of light and a receiver of the reflected light; if the object becomes modified, e.g. painted, this is immediately visible to everyone who interacts with the object.

The main principle of virtual modeling has three implications in the architecture. First, the VM is kept minimal and its derivative properties (ergonomic evaluation, task instructions of MW etc) are obtained by virtual experiments in the VE (in analogy with generative design). This prevents the complexity explosion of the VM.

Second, if an actor makes a change in the VE, the change has to be fed to the VM (persistent reference) with the subsequent regeneration of the VE (volatile presentation). This helps to keep the VM and VE in synch and decoupled.

Third, the actors across the LC should communicate not directly, but through the AT, VE and VM. It is similar to the open-source software community with many contributors (actors) working on a single entity (piece of software) in a mediated environment (mediators). A dual service-broker model is also possible: the services are associated with the actors and the broker plays the role of the VM. In both cases, a kind of mediator is required; otherwise it is very difficult to ensure the synchronous operation of the actors. In summary, only the VM, the central common reference, is able to synchronize multiple actors and to handle their heterogeneity.

MANUVAR APPLICATION TOOLS

The ManuVAR application tools (ATs) are designed to provide combined methodological and technical solutions to the identified MW problems in accordance with the ManuVAR system architecture. Each AT is linked to one or more problem areas and it consists of an ergonomics method supported by VR/AR technologies; through these elements of the tools it is possible to formulate the requirements to the VM. Here are preliminary examples of the ATs that we will implement and test in the ManuVAR project:

- Workplace design and evaluation in VR combined with a digital human model and ergonomic evaluation methods. This tool can be merged with one for task design, productization, and sequencing of operations.
- AR instructor to deliver contextualized augmented instructions on the factory floor.
- Bi-directional feedback system (a) from the factory floor to evaluate the MW performance and to gather direct feedback and (b) from the experts to the factory floor to provide recommendations supporting the factory floor operations.
- Training in VR for two types of training: perceptual-motor skills and procedural. This is intended to improve the depth of training (a larger spectrum of training scenarios are possible in VR that in a physical prototype) and its dynamics to respond to fast product changes and a shorter time to market.

The work on defining the ATs and their content in terms of ergonomics methods and VR/AR technologies is ongoing and it will be reported in future publications. Here, as an example, we list the elements of the planned workplace design and evaluation tool. On the methods side, it follows the general lines of a participatory design approach, when various actors join their efforts on designing a workplace.

Since physical prototyping is expensive, a virtual prototype (VE view of the VM of the product, the humans and their environment and tools) is employed to reduce costs and to make testing of various design options faster. At this level, we exploit VR for visualization, digital human model for embedding the human into the workplace, motion capture systems for measuring real human postures and motions. These are combined with the ergonomic evaluation analytical methods such as RULA (Rapid Upper Limb Assessment) posture analysis.

DISCUSSION AND CONCLUSIONS

Traditionally the MW issues are referred to as problems seven (physical and cognitive ergonomics) and five (leanness). Coffee & Thornley (2006) point out that leanness and ergonomics have to go hand-in-hand to avoid overloading and robotizing the human. Recently, in the context of fast product changes, leanness has been considered in conjunction with agility (Narasimhan, et al., 2006), which is captured in problem area three. A great number of technical solutions employing VR/AR and aiming to solve human-machine interface and ergonomics problems have been reported in more than 40000 papers (Engineering Village search engine), about 2000 patents (MicroPat database) and studied in tens of European research projects. However, in accordance with our study, VR/AR is still considered a risky technology with low industry user acceptance (problem area six). Our interpretation of the situation, considering the VR/AR technology high maturity level and its potential as a natural medium for the human-machine interfaces, is that industries cannot justify the investment into VR/AR by a single factor such as improvement of an interface, solving an isolated productivity or ergonomics issue. In the interviews conducted in the requirements-gathering phase of the project, the highest concern and expectations were associated with problem areas one (communication in the LC) and four (knowledge management in the LC), which require a multidisciplinary approach.

By adopting the PLM model and considering the system LC, we acknowledge the importance of the LC to represent and to solve MW problems. Currently, PLM covers human aspects, but the coverage is fragmentary. For example, (Stark, 2004) includes the worker view and management of human resources and (Grieves, 2006) positions the people as one of the key elements of the PLM, but there is no reference to the comprehensive ergonomics and work support framework. In several projects (e.g. EU FP6 CoSpaces, DiFAC), the LC and ergonomics are considered together, but the link to PLM has to be strengthened. Here ManuVAR makes a significant contribution to the PLM: on the practical level it covers the entire LC of PLM from the MW perspective, which is harder to formalize than other processes or systems in PLM. It also offers new insights to the change and complexity management in the future PLM by applying the main principle of virtual modeling.

The ATs similar to those proposed in ManuVAR have existed before, e.g. OSKU (Helin, et al, 2007) or Ergomix (Vink, 2005). In ManuVAR, we provide a unified PLM-based framework for the complementary ATs that are better adaptable for the needs of industries and have a strong and dynamic connection with the VM. The novelty of the presented ManuVAR results is threefold (1) it covers the entire LC span of PLM (2) from the perspective of supporting MW, which is hard to formalize, and (3) it offers new insights to the change and complexity management in the future PLM systems. For this it combines in a systematic way three complementary areas – PLM, VR/AR technology, and ergonomics – and it is this combination that makes it possible to address the MW challenges.

ACKNOWLEDGEMENTS

The results presented in this paper are the outcome of the joint efforts of the ManuVAR consortium. The authors are grateful to all contributors in the ManuVAR project who proposed ideas, gathered and analyzed data, and provided feedback and comments. The research leading to these results has received funding from the European Commission's Seventh Framework Programme FP7/2007-2013 under grant agreement 211548 "ManuVAR".

REFERENCES

Aromaa, S., Viitaniemi, J., Leva, C., Krassi, B. (2010) "Integration of Human Factors Methods to the ManuVAR Product Lifecycle Management Model." *Applied Human Factors and Ergonomics - AHFE 2010*, USA

Coffey, D., Thornley, C. (2006) "Automation, motivation and lean production reconsidered." *Assembly Automation* 26/2, p. 98–103

Engeström, Y. (1987) *Learning by Expanding: An Activity - Theoretical Approach to Developmental Research*. Orienta-Konsultit, Helsinki

Grieves, M. (2006) *Product Lifecycle Management*. McGraw-Hill

Helin K., Viitaniemi, J., Aromaa, S., Montonen, J., Evilä, T., Leino, S.-P., Määttä, T. (2007) "OSKU Digital Human Model in the Participatory Design Approach - A New Tool to Improve Work Tasks and Workplaces." *VTT working papers 83*, ISBN 978-951-38-6634-1

Kiviranta, S., Poyade., M., Helin, K. (2010) "Virtual environments for supporting high knowledge , high value manual work design." *Applied Human Factors and Ergonomics - AHFE 2010*, USA

Krassi, B. (2006) *The Application of Dynamic Virtual Prototyping to the Development of Control Systems*. Doctoral Thesis, Helsinki University of Technology, 183 p., [http://lib.tkk.fi/Diss/2006/isbn9512281686]

Krassi, B., D'Cruz, M., and Vink, P. (2010), "ManuVAR: a Framework for Improving Manual Work through Virtual and Augmented Reality." *Applied Human Factors and Ergonomics - AHFE 2010*, USA

Lawson, G., D'Cruz, M., Krassi, B. (eds.) (2010) "High value high knowledge manual work in the European industries: identification of problem areas and their technical, business and ergonomics analysis." (working title) *VTT working papers* (in publication)

Liston, P., Cromie, S., Leva, C., Helin, K., D'Cruz, M. (2010) "ManuVAR Training: Supporting the Implementation and Design of VR/AR Quality-improvement Tools for Manual Handling" *Applied Human Factors and Ergonomics - AHFE 2010*, USA

ManuVAR project web site, [www.manuvar.eu] Referred in February 2010

Narasimhan, R., Swink, M., Kim, S.W. (2006) "Disentangling leanness and agility: An empirical investigation." *Journal of Operations Management,* p. 440-457

Rasmussen, J., Goodstein, L. (1985) "Decision support in supervisory control" *Analysis, Design and Evaluation of Man-Machine Systems, 2nd IFAC conference*, Italy, p. 33-42

Stark, J. (2004) *Product Lifecycle Management: 21st Century Paradigm for Product Realisation*. Springer

Simantics web site, [www.simantics.org] Referred in February 2010

Talabă, D., Amditis, A. (eds.) (2008) *Product Engineering: Tools and Methods Based on Virtual Reality*. Springer, 563 p.

Vink, P. (2005) *Comfort and design: principles and good practice*. CRC Press

Wiener, N. (1948) *Cybernetics, or, control and communication in the animal and the machine*. New York, 194 p.

CHAPTER 76

From Teaching Machines to the Exploitation of Virtual and Augmented Reality: Behavior Analysis Supporting Manual Workers in Aerospace Industry Within the European Project ManuVAR

Guido Tosolin[1], Fabio Tosolin[1], Alessandro Valdina[1],
Carlo Vizzi[2], Enrico Gaia[2]

[1]A.A.R.B.A. Association for the Advancement of Radical Behavior
Analysis, Milan, Italy

[2]Thales Alenia Space Italia S.p.A.
Business Line Space Infrastructures & Transportation, Turin, Italy

ABSTRACT

The ManuVAR project was launched in May 2009 under the EU 7th Framework Program (ManuVAR, 2010). Its aim is to demonstrate that high value, high knowledge manual work presents an opportunity to improve the competitiveness of EU industries. This objective will be achieved by employing virtual and augmented reality technologies (VR/AR) to improve communication between people and systems. The application dimension of the project is split into five Clusters, each of them developing a specific industrial case, based on the general ManuVAR platform and methodology thus proving the main project concepts in different industrial areas (Krassi et al., 2010).

This paper will show how ManuVAR platform can be supported by behavioral methodologies through the development of teaching machines, generally meant as machines that are able to control human behavior. In the first section the methodologies and the tools that constitute the basis for the study are briefly introduced. Then it will be investigated how methodologies and tools can be combined to develop teaching machines able to support manual work.

On a general level, two types of tools will be proposed. The first ones are the *off-line* applications, employed for training and tuning of the work instructions. The second type is represented by the *on-line* applications, to be used on-the-job.

Finally these general concepts will be exemplified describing five use-cases, that will be developed in one of the Clusters of the ManuVAR project, to support manual integration of satellites components.

INTRODUCTION

In 1958, professor B.F. Skinner of Harvard invented teaching machines by applying the laws of behavior to stimuli-providing machines. These machines combined learning methodologies derived from Behavior Analysis with the technology tools represented by electromechanical systems. Notwithstanding the great effectiveness of these instruments, their application was limited. In fact, methodologies were well consolidated, the technology of that time offered restricted possibilities to measure behaviors and to provide sensorial stimuli in a simulated (virtual) environment.

Nowadays, the development of technologies such as Virtual and Augmented Reality presents the opportunity for building new generations of teaching machines, which are based on the same methodologies, but with highly improved devices for measurement of responses and provision of stimuli (ManuVAR D1.1, ManuVAR D3). In many cases machines of this kind are already employed in training application like flight, driving and even surgery simulators (Tosolin et al. 2005, 2009, Judkins et al., 2006): this paper will show that areas of application can be further widened.

In 2009, EU project ManuVAR started with the aim to exploit Virtual and Augmented Reality to support manual work throughout system lifecycle. The project will develop a technological platform and will apply it in five different industrial cases throughout Europe. The application of VR/AR technologies to high-value/high-knowledge manual work led to the question: can we combine Behavior Analysis methodologies and VR/AR tools to build "teaching machines" (generalizing this term as much as possible) able to support industrial manual work?

Cluster 1 will try to answer to this question developing a set of tools to improve performance of manual workers in the system integration facility of Thales Alenia Space Italia (TAS-I, 2010) in Turin - one of the worldwide leaders of Space Systems design, development and integration.

The Cluster is headed by the Association for the Advancement of Radical Behavior Analysis (A.A.R.B.A., 2010), a non-profit scientific association founded in 2001, whose mission is the development, enhancement and support of Behavior Analysis initiatives including research, education and practice, in academic, governmental and social/organizational environments.

METHODOLOGIES AND TOOLS INVOLVED

The methodologies involved are based on the principles of Behavior Analysis, also known as *Learning Psychology*. The Behavior Analysis scientifically investigates *behavior*, to be intended as anything a person or animal does that is observable and measurable, including verbal and emotional behaviors (Skinner, 1938). The aim is to control and predict behavior thanks to the framework of scientific laws, whose fundamental paradigm is the "Three Contingencies Model", or the A-B-C Model (Skinner, 1938). From the strictly scientific point of view, the behavior can be perfectly predicted and controlled by its control variables, which are Antecedent (A) and Consequent (C) stimuli. The behavior (B) is always evoked by the environmental stimuli which occur before it and it's increased or decreased by what comes after it. The events produced by the behavior and perceived by the performer are called consequences (C). The events that come before the behavior and provide information about the consequences are called antecedents (A). The antecedents can only evoke the behavior providing information about which behavior has to be performed to get a certain consequence. On the other hand, the consequences are responsible for increasing/decreasing the probability that the behavior will occur again under the same circumstances (i.e. under the same antecedents).

The methodologies derived from applied science of Behavior Analysis are employed in many fields with specific methods, aimed to control the crucial aspects of human performance: *training* and *motivation*. Among methods for training are Precision Teaching, Fluency-Based Training (Lindsley, 1992, Binder, 1990, Tosolin, 2005), etc. Several methods for motivation are generally organized in a coherent framework of behavioral "tools" known as *Performance Management* (Daniels, 2004).

While Behavior Analysis represents the conceptual, scientific core of the methodology, VR and AR are the technological basis for the application tools.

Virtual Reality is the creation of a three-dimensional representation of the real world with which we can interact (ManuVAR D1.1, 2009). Augmented Reality is the superimposition of virtual elements in the real world. Nowadays, a complete set of tools is available to implement these technologies in different settings and with different levels of accuracy in reproducing reality. In the case of the A.A.R.B.A.-led Cluster of ManuVAR, the following technological devices will be employed:

- **VR**: Power Wall with 3-D Gloves and a Haptic Manipulation device;
- **AR**: Head Mounted Displays (HMD with see through lenses) supported by Ultra Mobile PCs (UMPC);
- **Optical Motion Tracking:** for provision of contest oriented data.

LINKING BEHAVIOUR ANALYSIS AND AR/VR TO SUPPORT MANUAL WORK

Teaching machines are defined as computer or mechanical devices used for programmed instruction, which provide immediate corrective feedback and, often, extra practice if needed (Webster's new world dictionary, 2009). In this article we propose to develop specific teaching machines based on VR/AR technologies to modify behavior of manual workers both on-the-job (*on-line*) and during training (*off-line*). In the first case the machine is integrated in the work instructions; in the second it is a device to be used before real working activity.

"OFF-LINE" TOOLS

The potential of teaching machines in training is well known since 1958, when B.F. Skinner developed the first electro-mechanical devices. The basic principle behind all the learning methodologies developed within Behavior Analysis is the exploitation of the ABC Model (Skinner, 1950, Lindsley, 1992, Binder et al., 2002). The added value of VR applied to the teaching machines is the way the Antecedents, the Behaviors and the Consequences are managed. VR allows to recreate extremely realistic training conditions (A); it allows the real-time measurement and capture of physical movements, emotional responses and verbal behaviors (B); and it makes it possible to give high-frequency and conveniently timed feedbacks and consequences (C). Both the VR technology (ManuVAR D1.1, ManuVAR D3) and the learning methodology (Tosolin et al., 2005, Binder & Watkins, 1990, Binder, 1988) are mature and their synergistic combination has been already exploited in many training environments (e.g. flight simulators, Tosolin et al., 2009).

However, the limitation of many systems is that the technology tools are often dedicated and costly. Therefore the challenge is to build a content-free teaching machine that is cost competitive and easy-to-program for many different training purposes. For many cases, it is not necessary to recreate the whole virtual environment, because many crucial skills can be trained at a convenient level using only a selection the relevant stimuli, activating only specific learning channels (senses).

The most powerful way to arrange the consequences under VR is pairing the correct answers (the behaviors which lead to the result, e.g. choosing a tool or screwing at the right strength) with an immediate *positive feedback* (e.g. to see the screw becoming immediately green) or, sometimes, with a *positive reinforcement* (e.g. to see on the VR glasses' screen a number representing points which will be hoarded by the worker to get reward). At the same time, in the Virtual Environment the workers will get mainly *extinction* of the undesired or wrong behavior, not *punishment*. In this way it will avoid stress, improving the motivation to be involved in the learning process. Only when needed (e.g. in learning to extinguish a fire), the ManuVAR engineers will eventually develop teaching machines able to provide *negative reinforcement* (Tosolin et al., 2005).

Sometimes it is crucial to develop creative behaviors for new unexpected situations, that were not specifically trained. The teaching machine can realize this process, called *generalization* (Tosolin et al., 2005, Sanguini et al. 2009, Martin & Pear, 2007). In this case it in necessary to set conveniently the stimuli to provide many different multiple educational situations. On the other hand, sometimes the trainee has to distinguish in great detail between different stimuli in order to perform different behaviors for imperceptibly different situations. In this case the training contingencies will be managed in order to improve *discrimination*. The same machine can apply both *generalization* and *discrimination* paradigms. This is done through a precise management of the stimuli.

"ON-LINE" TOOLS

A relatively new field is the use of Augmented Reality to support behavioral methodologies for performance enhancement (e.g. Performance Management, Daniels, 2004). In this case, the most interesting possibilities are offered by devices for on-the-job support. These are commonly intended to be used to give detailed instructions (Antecedents). From this point of view they could improve usability and accessibility to information that are already on-paper. Anyway their potential is much wider than this.

In fact, ManuVAR engineers can rely on Augmented Reality as a powerful *Antecedent discriminative stimulus* (Martin & Pear, 2007), prompting the manual action which is needed to get a successful result. The behavior/result could then be measured (by a capture system or by a real-time evaluation of outcomes) or self-measured (e.g. signing check-boxes). Then the system could provide to the worker an immediate *feedback about his/her performance*. As we said any antecedent is effective in relationship with the consequence that it forecasts (Skinner, 1953). Continuous measurement of workers allows to find out how the workers are performing. Indeed, frequent measurement is the only way to make right decisions about giving frequent consequences (Daniels, 2004); if the teaching machine cannot immediately and consistently measure the worker's behavior, than the machine cannot choose the immediate Feedback and/or Positive Reinforcement to be provided to improve motivation. AR and systems for behavior capturing allow motivating the manual workers by a high number of immediate artificial contingencies. As a matter of fact they can be far much more performing than the people working only under natural contingencies, under the three dimensions of work performance: productivity, quality and safety.

Finally, Augmented Reality can be also used to improve correct actions for less trained workers (i.e. perform on-the-job training), to get the right behaviors even if they do not know or remember which tool is to be used in a determined situation or which way the movements have to be performed. AR can be adopted to reduce the probability of making a mistake too, because a successive action is not allowed as far as the correct one has not been correctly completed (indirect negative feedback). Furthermore, a registration of the behaviors and results can be also used by supervisors to individuate possible training needs.

In Figure 1 it is reported a schematic representation of AV/AR tools, putting side by side AR/VR technologies/devices with the ABC Model.

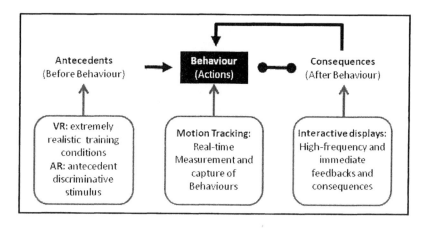

Figure 1 - Integration of VR/AR tools in the ABC Model

POTENTIAL IMPROVEMENTS INTRODUCED BY USE OF VR/AR IN SPACE INDUSTRY

During the initial phases of the ManuVAR project, the gaps about actual manual practice have been detected through interviews within the clusters (ManuVAR D1.1, 2009). Persons with different roles and positions within the organization participated to the collection of data to get a complete overview of the areas of improvement. The data were processed to find a common list of high-level gaps within the five Clusters, with peculiar specification for each industrial case. A summary of these gaps is reported in Table 1 for A.A.R.B.A.-led Cluster.

Table 1 - Common gaps and their specific declination in Cluster 1

General Gaps	Impact on TAS-I
Problems with communication throughout lifecycle	Reporting of activities and non- conformances to logbook on paper it is time consuming. Training is performed on-the-job but it would be necessary for critical activities to be performed for the first time.
Poor user interfaces	Instructions mainly on paper, difficult or tough to read and to transform in actions
Inefficient knowledge management	Difficulty in keeping reference configuration updated and getting all the teams informed. Accounting for lessons learnt takes too much time.
Physical and cognitive stresses	Some operations have to be performed under stressful contingencies, increasing probability of errors
Inflexible design process	Designers experience problems in forecasting assembly (accessibility) problems during design phases.
Low productivity	Reporting and meeting activities takes too much time. Integration for spacecrafts must be certified and errorless.

TAS-I proposes to use VR&AR to improve activities in its integration department. In these laboratories experienced technicians integrate different equipment and sub-systems of spacecrafts by hand. Manual labor is essential for tasks that are often unique. Automating these activities, apart from being almost impossible, would also mean to ignore the flexibility, adaptability and resourcefulness of the human worker that in this prototype activity can make the difference between a bad or well done job.

DESCRIPTION OF THE CASE-STUDIES

TAS-I and A.A.R.B.A. identified five possible applications of VR/AR to improve actual gaps. A brief description of each application tool is provided in this section.

1. CRITICAL INTEGRATION AND TEST PROCEDURE VALIDATION

Technologies: *Power Wall, DHI, Motion Capture, Haptic Manipulation.*
Methodologies: *Ergonomic Analysis, Performance Management (Pinpointing of crucial Behaviors, Functional Analysis, Performance Feedback).*

The aim of the validation is to prevent any potential difficulty during the actual integration process, so that avoiding or reducing the design phase produced inconsistencies. Moreover this makes it possible to tune the work instructions.

The validation will be performed when no hardware is available using integration VR simulated processes before the actual integration starts. During the simulation, ergonomics and consequences for the workers can be analyzed.

Difficulties and hazards can be rated and the designers will receive feedback about their work. Quality of design will be improved too, providing positive reinforcements to the engineers for well designed integration activities.

2. INTERACTIVE ON-THE-JOB INSTRUCTIONS

Technologies: *HMD, UMPC, Motion Capture.*
Methodologies: *Performance Management (Pinpointing of crucial behaviors, Performance Feedback, Positive Reinforcement).*

Interactive instructions will have to: (1) Provide powerful antecedent discriminative stimulus, prompting the manual action which is needed to get a successful result (Augmented Reality); (2) measure the behavior/result (motion capture system) and allow self-measure (e.g. signing check-boxes: this will also integrate checking activities and collection of Non-Conformance Reports, NCR); (3) provide feedback to the worker about his/her performance; (4) provide both continuous and interval-variable ratio positive reinforcement as soon as desired behavior is detected; (5) integrate all information and functionalities in a unique tool.

These features will be implemented using the HMD supported by a UMPC: the former will be the main instrument for visualization of information, feedbacks and consequences, while the latter is used by the worker to insert data and to navigate through different features.

In this way work instructions will provide structured antecedents & feedback to reduce or to avoid errors that generate non-conformances. Based on such instructions, the teaching machines will allow collection of a large amount of data that will feed checklists and, if needed, support NC reporting. They can also collect information about the state of the integration process.

3. CRITICAL INTEGRATION AND TEST PROCEDURE TRAINING

Technologies: *Power Wall, DHI, Motion Capture, Haptic Manipulation Tools.*
Methodologies: *Fluency-Based Training, Precision Teaching, Instructional Design.*

When complex integration activities have to be performed for the first time and when the cost of errors can be very high, virtual training will be used to increase the probability of success.

Training will take place in a VR-based teaching machine where the subject can learn by consequences, exploiting neural mechanisms activated by the "operant conditioning" paradigm. Specific skills can be trained through a driving sequence of antecedent stimuli (virtually simulated assembly situations), which evoke trainee's response, and consequent stimuli provided by the machine immediately after the occurrence of the responses. Realistic reproduction of working environment, real-time measurement of behaviors, organization of learning items respecting shaping principles (Skinner, 1953) and a high number of positive reinforcements will be the main elements of the training.

4. SUPPORT CHECKING ACTIVITY AND NC DATA COLLECTION FOR THE MANUFACTURING REVIEW BOARD (MRB)

Technologies: *HMD, UMPC.*
Methodologies: *Performance Management (Performance Feedback, Positive Reinforcement).*

Checking activities and Collection of Non Conformance Reports will be integrated in the same tool that provides work instruction. This will reduce the time needed for checking and allow easier registration of reports in electronic format.

The tool will be able to: (1) ask the worker to execute the check with perfect timing on the basis of steps just performed and registered; (2) improve checking activity, e.g. collecting videos/pictures about the work that has been done; (3) provide immediate feedback to the engineering department about Non Conformances originated from the design phase; (4) create direct communication channels between the integration and engineering department.

5. SUPPORT TO HAND-OVER MEETING BETWEEN TWO INTEGRATION TEAMS WORKING IN SHIFTS

Technologies: *Power Wall, DHI.*
Methodologies: *Performance Management (Performance Feedback, Positive Reinforcement).*

Data and information collected during integration activity will be used during hand-over meetings. On the one hand, this will support logistics and organization (e.g. providing information about advancement of the work, NC detected, etc). On the other it can support motivation because workers can have feedback about their performance and positive reinforcements can be provided.

CONCLUSIONS

The paper highlighted the potential of behavioral methodologies to develop teaching machines. In these tools, VR/AR are utilized as means to provide antecedent and consequent stimuli to the manual workers in order to get the desired behavior and to make it stronger, faster, more accurate, durable or resistant to extinction according to the needs of the company and to the requirements of the task.

The VR/AR devices will be programmed with the aim to allow the worker to apply the learned behaviors to new, quite different situations (*generalization*) or shifting to new more effective behaviors as soon as small differences in the work environment suggest to substitute the learned behavior with a different one (*discrimination*).

Specific application tools that will be developed in Cluster 1 of ManuVAR have been briefly presented. These VR/AR-Based machines will be developed in order to improve: design of manual processes; motivation and errorless integration on-the-job; training for critical activities.

Even if the scientific community agrees on the astonishing effects of positive procedures in bringing out the best in people, very few managers and foremen are skilled to effectively adopt such paradigms. AR/VR machines are the only way to do it consistently and in time, much more than humans can do.

REFERENCES

Association for the Advancement of Radical Behavior Analysis www.aarba.org

Binder, C., & Watkins, C. L. (1990). "Precision Teaching and Direct Instruction: Measurably superior instructional technology in schools". *Performance Improvement Quarterly*, 3(4), 74-96.

Binder C., Sweeney L.(2002). "Building fluent performance in a customer call center", *ISPI*.

Daniels A.C. and Daniels J.E. (2004). "Performance Management, Changing Behavior That Drives Organizational Effectiveness" (4th Edition). *Performance Management Publications*: Atlanta GA.

Judkins, T.N., Oleynikov, D., Stergiou, N. (2006), "Real-time Augmented Feedback Benefits Robotic Laparoscopic Training". *University of Nebraska*, USA.

Krassi, B., D'Cruz, M., and Vink, P. (2010), "ManuVAR: a Framework for Improving Manual Work through Virtual and Augmented Reality". *Applied Human Factors and Ergonomics - AHFE 2010*, USA

Lindsley, O.R. (1992). "Precision Teaching: Discoveries and Effects". *Journal of Applied Behavior Analysis*. 25(1), 51-57.

ManuVAR www.manuvar.eu

ManuVAR D1.1 (2009). "Definition of requirements, analysis of standards and tools". *ManuVAR CP-IP-211548*. Fraunhofer IFF: Magdeburg.

ManuVAR D3 (2010). "Analysis of validity of VR/AR for manual work support and human factors aspects of manual work; conceptual specification and technical basis alternatives for VM". *ManuVAR CP-IP-211548*. University of Nottingham: UK.

Martin G, Pear J. (2007). "Behavior Modification: What It Is And How To Do It" (8th Edition). *Prentice Hall*: New Jersey

Sanguini R., Tosolin F., Bacchetta A., Algarotti E., Gatti M. (2009). "Experimentation of Precision Teaching learning methodology in helicopter pilot training". *2009 Didamatica Conference* - Trento, Italy.

Skinner, B.F. (1938). "The Behaviour Of Organisms". *Appleton-Century-Crofts* - New York.

Skinner, B.F. (1950). "Are theories of learning necessary?" *Psychological Review*, 57(4), 193-216.

Skinner, B.F. (1953). "Science and Human Behaviour". *Macmillan*: New York.

Thales Alenia Space
    http://www.thalesgroup.com/Markets/Space/Related_Activities/Thales_Alenia_Space/

Tosolin F., Orlandi G., Truzoli R., Casarola G. (2005). "Precision Teaching as a scientific tool to obtain motivation, learning and resistance to extinction in e-learning: historical and methodological issues and applied researches in large railway companies in Italy". *WIT Transactions on Information and Communication Technologies, 2005*.

Webster's new world dictionary, 2009

Augmented Reality in Factory Environments - System Usability and Human Factors Studies

J. Sääski[1], A. Langley[2], T. Salonen[1]
[1]VTT, Technical Research Centre of Finland
Metallimiehenkuja 6, FIN-02044 Espoo, Finland

[2]Human Factors Research Group, Faculty of Engineering
University of Nottingham, NG7 2RD, UK

ABSTRACT

The aim of this paper is to assess the applicability of augmented reality (AR) technology to support different manual work activities, in order to implement improvements in industrial systems. Augmented Reality refers to software applications that integrate typically video stream with additional digital information such as 3D models, animation and text. Currently AR techniques are used in various application areas to support manual work. The literature shows that manual work processes are very diverse. Therefore, the technology needed to support these processes is varied. Today AR technology is used primarily in design and analysis. However working applications at later lifecycle stages (production, maintenance) rarely make use of it outside laboratories. This work investigates whether existing systems, tools, devices and interfaces are sufficient to support manual work processes and if it has to be improved from the technological/technical point of view in order to improve manual work processes.

INTRODUCTION

Human workers are flexible and able to solve problems. Hence, their significance in the complex high value, high knowledge work is important even today. The challenge is to deliver relevant information to the workers about each task at a given time. Augmented Reality (AR) technology provides new opportunities to enhance this manual work. AR is a field of computer research which deals with the combination of real world and computer generated data. At present, most AR research is concerned with the use of live video imagery which is digitally processed and "augmented" by the addition of computer generated graphics. AR is interesting for companies as it provides a possibility to give work instructions effectively. AR can support the worker by providing information in-situ, for example, projecting a virtual user manual onto the workplace, thus facilitating decision making and increasing safety. Moreover, The rapid development of small-sized low-cost technology such as handheld computers, head-mounted displays and camera phones with enough processing capacity and long lasting batteries has enabled light-weight mobile AR systems and thus increased the potential of this technology.

This study is part of ManuVAR project (Manual Work Support throughout System Lifecycle by Exploiting Virtual and Augmented Reality). ManuVAR is a European 7th Framework project that runs from 2009 through 2012 and comprises 18 partner organisations across 8 countries. Through the use of virtual and augmented reality technology, the ManuVAR project will demonstrate that high value, high knowledge manual work presents an opportunity to improve the competitiveness of EU industries (http://www.manuvar.eu/). The aim of this paper is to assess the applicability of augmented reality (AR) technology to support different manual work activities, in order to implement improvements in industrial systems.

The outline of the paper is as follows. First, the augmented reality technology is described. In the next section, usability studies dealing with augmented reality are described. The section that follows describes potential of augmented reality to improve human factors and the final section contains discussion and conclusions.

AUGMENTED REALITY

Augmented reality is a field of computer research which deals with the combination of real world and computer generated data. At present, most AR research is concerned with the use of live video imagery which is digitally processed and "augmented" by the addition of computer generated graphics.

The characteristic features of AR systems are the combination of real and virtual objects in a real environment; interactivity in real time; and registration (alignment) of real and virtual objects with each other. The basic components in AR applications are a display, a camera and a computer with application software (Azuma et al.

2001). Different kinds of hardware can be used to implement this, e.g. camera phones, handheld computers, laptops, head-mounted displays (HMD) etc.

The augmentation process is as follows (www.hitl.washington.edu/artoolkit/): The live video image is turned into a binary (black or white) image based on a lighting threshold value. This image is then searched for square regions. AR-software finds all the squares in the binary image. For each square, the pattern inside the square is captured and matched against some pre-trained pattern templates. AR-software then uses the known square size and pattern orientation to calculate the position of the real video camera relative to the physical marker. AR software computes the transformation matrix that describes the position, rotation, and size of an object. The virtual object is thereafter located in relation to that marker. Thus the user experiences video see-through augmented reality, seeing the real world through the real time video with virtual models. The diagram in Figure 1 shows the image processing used in AR-software in more detail and in Figure 2 shows an example how a virtual object is overlaid on the marker.

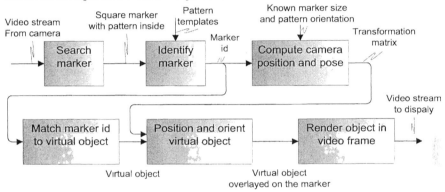

FIGURE 1. Image processing used in AR-software.

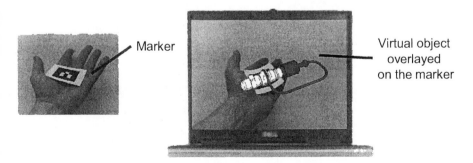

FIGURE 2. An example of the marker-based augmentation. From the video stream of the camera the marker is recognized and the corresponding virtual object is put on the palm.

A typical AR system consists of the following components:

- Computer hardware: stationary AR systems use a desktop computer and mobile AR systems use a wearable computer, e.g. laptops in a backpack configuration to process real-world images from a digital camera and to calculate positions of virtual objects for augmentation. Smart camera phones and PDAs (personal digital assistant) are also used for mobile purposes (Hakkarainen et al, 2008).

- AR displays: AR displays allow for the overlaying of virtual graphical objects in real-world images. Typical AR displays for AR systems are head mounted displays (optical see-through or video see-through), handheld displays or typical displays.

- Tracking: accurate tracking of the user's head is essential for all 3D overlays and augmentations of real-world objects. For text augmentations, the tracking accuracy is usually not so important because the text is not registered to anything in the real world. The choice of the tracking technology is dependent on the required accuracy. A typical method is to use markers (2D markers, images or beacons); in other words recognisable elements that are placed in the environment.

- Software: software is used to calculate the position of virtual 3D objects in real-world images supplied by digital cameras which react to the user's input and provide a logical application.

- Application data: AR systems need engineering/authoring tools which enable the creation of application tasks. The majority of the product data is created in design systems and stored to PDM/PLM system. One of the challenges is the retrieval and refining of this information into VR/AR system on the factory floor. For example, Figure 3 shows the content creation process from CAD systems to factory floor.

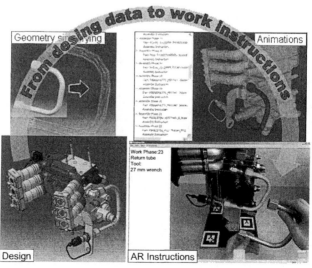

FIGURE 3. Image processing used in AR-software.

USABILITY OF AUGMENTED REALITY SYSTEMS

Augmented reality integrates video stream from a real environment with digital information. The display technology is probably the most important factor to increase usability of this technology. A head-mounted display (HMD) is a device that has video information displayed directly in front of one eye (monocular HMD) or both (binocular HMD). Most of the HMDs are binocular, but there are also some monocular versions available. Nowadays the LCD and the OLED technologies are the most used display types. For AR applications video see-through and optical see-through applications are the most used technologies. Video see-through HMD uses integrated video camera, and optical see-through HMD uses optical combiners (essentially half-silvered mirrors or transparent LCD displays) (Bimber, 2005). Recently more compact HMDs are available that look almost like common sunglasses. However, the compact size usually means lower properties. High-tech HMDs are bigger and heavier (see Inition, 2009).

For the main evaluation criteria, price can be used along with resolution and field of view (FOV) however the same resolution can look quite different based on the distance between eyes and a display. If we have a HMD with a small resolution and a small FOV, the picture can look much better than a HMD with a big resolution and a large FOV. Because of this, there is a need for normalised resolution called "angular resolution". It can be calculated as follows: angular resolution = 60 * diagonal FOV / diagonal resolution. In addition to the main evaluation criteria, there are also a number of elements to consider, e.g. stereo overlapping, human match factor and global quality factor (Bernatchez, 2007).

Jeon et al. (2006) have investigated the comparative usability among three different viewing configurations of augmented reality system that uses a desktop monitor instead of a head mounted display. Their results indicated that mounting a camera on the user's head and placing a fixed camera in the back of the user was the best option in terms of task performance, user perceived performance and easiness of setting up the environment. Jeon et al. (2006) argue that HMD configuration is bound to cause usability problems because of the mismatch in the user's proprioception, scale, hand eye coordination, and the reduced 3D depth perception.

Also a study (Sääski et al., 2008a) showed that HMD was considered unpractical. In the usability test the participants wished most improvement to the head-mounted display. The image was told to be vague and difficult to watch. Similarly, the system was too slow to react to the head movements. AR system should be interactive in real time, i.e., when the user moves his head, AR software should be able to update the view quickly. This is emphasized in accuracy demanding tasks.

Assessment of HMDs is difficult because of the variety of configurations that are available on the market. It is possible to do some comparison between them based on technical specifications. However, the implementation or the compatibility of the HMDs is still difficult to evaluate without testing a HMD with an actual computer.

POTENTIAL OF AR TO IMPROVE HF'S

The initial point of ManuVAR was requirements analysis and the interviews with industrial partners. The requirements analysis provided the gaps that are the basis to derive possible application fields for the AR technology in the ManuVAR context (ManuVAR D1.1, 2009 & ManuVAR D2, 2010). It was found that the following gap items could be enhanced with the use of AR:

- Provision of appropriate information on working floor
- Documentation, manuals, instructions, etc. on a proper medium
- Mobile trainings and feedback cell at working floor
- Support to define integration sequences for different parts (accessibility and installability)
- Riskless/safe training of skills
- Training considering different user skills (different level of training)
- Support of high customised work
- Description of clear assembly procedures
- Shortening of the learning time (training on work floor level)
- Visualisation of large models with realistic rendering
- Tracking worker actions for analysing assembly time

There are also a number of studies that highlight where augmented reality technologies have provided improvements relating to human factors (see Ong et al., 2007). The following studies relate to the requirements of the ManuVAR project.

One challenge is assembly work that requires skilled manpower to perform tasks in a specified sequence with careful attention and particular skills. Tang et al. (2003) described an experiment that tested the relative effectiveness of AR instructions in an assembly task. Three instructional media were compared with the AR system: a printed manual, computer assisted instruction using a monitor-based display, and a head-mounted display. An assembly task was based on Duplo blocks consisting of 56 procedural steps. For each step, subjects were required to acquire a part of a specific colour and size from an unsorted part-bin and insert the part into the current subassembly in a specific position and orientation according to the instruction. Results indicated that completion times with these four media were: printed manual 14 min. 24 sec., computer assisted instruction using a monitor-based display 11 min. 25 sec., computer assisted instruction using a head-mounted display 11 min. 8 sec., and augmented instructions 10 min. 39 sec. Significant difference was in error rate. Users with AR system made an average ten times less errors during the task. Also measurement of mental effort indicated decreased mental effort in the AR case, suggesting some of the mental calculation of the assembly task is offloaded to the system.

Eursch (2007) uses AR to assist workers when dealing with the critical environment by superimposing important information directly onto the participant's view of their working environment. The study showed that this function supplied the user with all the necessary information for the task, it eased decision making, reduced stress and increased comfort and therefore significantly increased the safety of the task. The study used AR in combination with an immersive camera system.

In Siltanen et al (2007) a multimodal AR system to aid assembly work which allowed for speech and gesture input is used. The system was used as a demonstrator and as a test bed to evaluate different input modalities for augmented assembly setups. The study compared printed instructions to augmented instructions. The difference was subtle but that all users shared the opinion that the AR system was useful and the multimodal input interface was favourably judged.

A study that used AR to support information delivery in high precision defence assembly (Day et al., 2005) showed that AR is similar to conventional paper-based methods but allowed less disruption to work and greater mobility. The authors also stated that the users were positive towards the system. In Doil et al. (2003) the improvement of industrial planning process by the use of AR technology was studied. The study showed that the system significantly improved the interaction with the participants and several field studies demonstrated cost reduction and data quality improvements.

The research results (Schenk et. al., 2009) of the static work assistance system based on AR technology proved that the error rate and the time to execute a task is lower than with paper manuals. In the tests the error rate with the static augmented reality system is almost zero and the time to execute a task is 20% faster. Also the change of strain needs to be remarked when using AR equipments. In one research Tümler et al. (2008) discovered that even though the mobile AR system was not optimal the system caused only a biologically agreeable strain for the user.

In Sääski et al. (2008b) a comprehensive assembly study was conducted. There were altogether 59 assembly events; 30 persons did the assembly using paper instructions and 29 used augmented reality instructions. With paper instructions, the assembly task took 36 min. 34 sec. and with augmented-reality based instructions the performance time was 31 min. 45 sec.; approximately five minutes faster. The usage of wrong tools during the assembly took place six times more often with paper instructions. Furthermore, trying to put a part in a wrong place happened twice as much with paper based instructions. This difference as well as all the other differences reported in this paper are statistically significant; statistical results were obtained with the t-test, performed with independent samples and the level of statistical significance was always at least $p < 0.05$.

The results of the studies described show that AR can have a great added value in assembly work as the work can be done faster, with less failures and less effort.

DISCUSSION AND CONCLUSIONS

In this paper we have described the augmented reality technology and usability and human factors studies dealing with this technology. The challenges of implementation are described below.

Display technology

Analysing the different options available in market (Laster Tecnologies, 2009; Trivisio, 2009; Lumus, 2009; Vuzix, 2009; Intersense, 2009; eMagin, 2009; Inition, 2009), and based on main advantages and disadvantages of different AR implementations, it can be concluded that hands-free support devices with minimal obstruction to the work should be considered within the scope of ManuVAR. The challenges with display technology are insufficient ergonomic design of user interfaces, i.e. heavy weight, display size too small causing possible health & safety problems (Wulz and Walch, 2006; University of Southern California, 2007; Wilson, 2004). Further, the brightness and resolution may cause problems when using HMD outdoor (University of Southern California, 2007). Immersive experience should also be taken into account when designing a suitable solution.

Tracking

Different tracking systems are suited for different applications. The main criteria should be the desired accuracy needed and the location where the tracking will be used. In the context of the ManuVAR project, it is obvious that different tracking methods should be used to achieve the best results. Different tracking systems are suited for different applications. The main criteria should be the desired accuracy needed and the location where the tracking will be used. In ManuVAR, there are indoor, controlled applications where precision is important, optical tracking could be used, to deliver better accuracy. The ManuVAR consortium has also large machinery and outdoor activities, in which a more robust and tolerant tracking or hybrid combinations of different technologies could be used.

Hardware

At the moment, difficulties in system integration (bulks, wires, power) and integration in current workplace technology exist (University of Southern California, 2007) (Wilson, 2004). Next consideration is the performance of mobile hardware - calculation time for feature recognition algorithms. This limits the amount of data which can visualised (Geißel et al., 2008). Furthermore, Walch (2006) lists that investment cost for hardware is high. However, limits of performance and price are probably becoming a minor detail in the near future.

General challenges exist for the implementation of AR technologies:

- Lack of social acceptance (University of Southern California, 2007)
- Combination of computer system with flexibility of humans (wearable computer) (Reif, 2008)
- Data preparation through AR specialists because of missing authoring/modelling tools and standardization to create augmented scenarios.

As a conclusion and in the context of the ManuVAR project, AR technology can be considered as a very advanced tool for providing support to workers and improving performance of manual tasks, although some obstacles should be overcome before AR can be implemented in the industry.

ACKNOWLEGDEMENT

The research leading to these results has received funding from the European Community's Seventh Framework Programme FP7/2007-2013 under grant agreement no. 211548 "ManuVAR".

REFERENCES

Azuma, R., Baillot, Y., Behringer, R., Feiner, S., Julier, S., and MacIntyre, B., (2001), "Recent advances in augmented reality*", IEEE Computer Graphics and Applications*, 21 (2001), no. 6, pp. 34—47.

Bernatchez, M. (2007), *Article: Resolution analysis for HMD helmets* http://vresources.org/HMD_rezanalysis.html [Accessed 16 December 2009]

Bimber, O. (2005) *Spatial Augmented Reality: Merging Real and Virtual Worlds.* Natick, MA, USA: A K Peters, Limited, p 74-75.

Day, P.N. Ferguson, G. O'Brian Holt, P. Hogg, S. Gibson, D. (2005). *Wearable augmented virtual reality for enhancing information delivery in high precision defence assembly: an engineering case study.* Virtual Reality. 8: 177–184

Doil, F., Schreiber, W., Alt, T., and Patron, C. (2003), *Augmented reality for manufacturing planning.* Proceedings of the workshop on Virtual environments 2003, pp 71-76, Zürich, Switzerland

eMagin (2009), *Homepage,* http://www.emagin.com/, [Accessed: 12 November 2009]

Eursch, A. (2007) *Increased safety for manual tasks in the field of nuclear science using the technology of augmented reality.* 2007 IEEE Nuclear Science Symposium Conference Record.

Geißel, O., Longhitano, L. , Katzenbach, A. (2008), *Operativer Einsatz von Mixed Reality Technologien im Baubarkeitsprozess der Fahrzeugentwicklung*, Mercedes-Benz Cars, Daimler AG, pp. 9-11, http://www.katzenbach-web.de/datas/ Veroeffentlichungen/2008-5_HNI_Paper.pdf, [Accessed: 15 November 2009]

Hakkarainen, M., Woodward, C., and Billinghurst, M. (2008), *Augmented Assembly using a Mobile Phone*, ISMAR2008, Cambridge 15-18 September 2008 (http://ismar08.org/)

Inition (2009), *Homepage* http://www.inition.co.uk/, [Accessed: 12 November 2009]

Intersense (2009), *Precise Motion Tracking Technology*, Homepage, http://www.intersense.com/, [Accessed: 3 December 2009]

Intuition (2007), *Report on vision based feature tracking* Intuition NoE Augmented Reality Working group 2.2. 1st draft 31.7.2007

Jeon, S., Shim, H, and Kim, G. J., (2006), "Viewpoint Usability for Desktop Augmented Reality." *The International Journal of Virtual Reality, 2006, 5(3):33-39*

Laster Technologies (2009), *Mobile Informative Eyewear – Solutions for Augmented Reality*, Homepage http://www.laster.fr/, [Accessed: 30 November 2009]

Lumus (2009), *Homepage* http://www.lumus-optical.com/, [Accessed: 4 December 2009]

Ong, S. K.; Yuan, M. L. & Nee, A. Y. C. (2007), Augmented reality applications in manufacturing: a survey *International Journal of Production Research*, 46, pp. 2707 — 2742

ManuVAR D1.1 (2009). *Definition of requirements, analysis of standards and tools.* ManuVAR CP-IP-211548. Fraunhofer IFF: Magdeburg.

724

ManuVAR D2 (2010). *Definition of industrial cases.* ManuVAR CP-IP-211548. TNO: Amsterdam

Reif, R. (2008) *Einsatz der Augmented Reality in der Logistik,* Lehrstuhl für Fördertechnik, Materialfluss und Logistik, Technische Universität München, CeMAT Campus Intralogistik http://www.technikwissen.de/libary/common/logistik/Campus/29.05.08-Einsatz-Augmented-Reality.pdf, [Accessed: 15 November 2009]

Schenk, M.; Grubert, J.; Sauer, S.; Berndt, D.; and Mecke, R. (2009), *Augmented Reality basierte Werkerassistenz.* pp. 341-360; ISBN 978-3-940019-80-6; GITO-Verlag 2009

Siltanen, S., Hakkarainen, M., Korkalo, O., Salonen, T., Sääski, J., Woodward C., Kannetis, T., Perakakis, and M., Potamianos, A. (2007), *Multimodal user interface for augmented assembly.* IEEE International Workshop on Robots and Human Interactive Communication

Sääski, J., Salonen, T., Riitahuhta, A.(2008a) *"Augmented Reality in Mobile Machinery Assembly".* Proceedings of TMCE 2008 Symposium, April 21-25, 2008, Izmir, Turkey, edited by I. Horváth and Z. Rusák, pp-257-268.

Sääski, J., Salonen, T., Liinasuo, M., J., Vanhatalo, M., and Riitahuhta, A. (2008b) *"Augmented Reality Efficiency in Manufacturing Industry: A Case Study."* Proceedings NordDesign 2008, Tallinn, Estonia.

Tang, A., Owen, C., Biocca, F., and Mou, W (2003), "Comparative Effectiveness of Augmented Reality in Object Assembly." *Proceedings of the SIGCHI conference on Human factors in computing systems,* pp73-80.

Trivisio (2009), *Homepage,* http://www.trivisio.com/, [Accessed: 10 December 2009]

Tümler, J., Rüdiger, M., Schenk, M., and Huckauf, A. (2008), *Mobile Augmented Reality in Industrial Applications: Approaches for Solution of User-Related Issues.* Symposium on Mixed and Augmented Reality, Proceedings of the 7th IEEE/ACM International Symposium on Mixed and Augmented Reality

University of Southern California (2007), *Augmented Reality- Linking Real and Virtual Worlds* http://graphics.usc.edu/~suyay/class/AR.pdf, [Accessed: 15 November 2009]

Vuzix (2009), *Homepage,* http://www.vuzix.com/home/index.html, [Accessed: 8 December 2009]

Walch, D. (2006), *Einsatz AR/VR zur Schulung,* Presentation, Lehrstuhl für Fördertechnik, Materialfluss und Logistik Technische Universität München, http://www.fml.mw.tu-muenchen.de/PDF/Vortrag_Walch_Einsatz_ARVR_zur_Schulung.pdf [Accessed: 15 November 2009]

Wilson, J.R. (2004), *Virtual reality at work,* literally, Article from 03.Mai 2004, Institute for Occupational Ergonomics, University of Nottingham, http://cordis.europa.eu/ictresults/index.cfm?section=news&tpl=article&ID=64907, [Accessed: 3 November 2009]

Wulz, J., Walch, D. (2006), *Eintauchen in die digitale Welt,* Presentation from 17.11.2006, Lehrstuhl für Fördertechnik, Materialfluss und Logistik, Technische Universität München, http://www.fml.mw.tum.de/PDF/Eintauchen_in_die_digitale_Welt.pdf [Accessed: 4 December 2009]

CHAPTER 78

ManuVAR: A Framework for Improving Manual Work Through Virtual and Augmented Reality

Boris Krassi[1], Mirabelle D'Cruz[2], Peter Vink[3]

[1]VTT Technical Research Centre of Finland
Tekniikankatu 1, FIN-33101 Tampere, Finland

[2]University of Nottingham
Human Factors Research Group, Faculty of Engineering, NG7 2RD, UK

[3]TNO
2132 Hoofddorp, Polarisavenue 151, The Netherlands

ABSTRACT

This paper introduces a European-Union-funded research project ManuVAR that aims to provide a systematic technological and methodological system to support high knowledge high value manual work throughout product lifecycle. The ManuVAR approach is based on a product lifecycle management approach and virtual and augmented reality technology combined with the ergonomics methods. The paper sets the research framework of ManuVAR by identifying the seven most prominent high level problem areas related to the manual work in Europe. Finally, it overviews the first ManuVAR results reported in other publications in this volume.

Keywords: Manual Work, Virtual Reality, Augmented Reality, Product Lifecycle Management, PLM, ManuVAR

INTRODUCTION

BACKGROUND

Manual work (MW) is a central and expensive component of manufacturing, assembly, testing and maintenance services in Europe. There are many industrial sectors that rely on the knowledge and skills of their manual workers, for example, satellite assembly, maintenance of nuclear reactors, operation of complex machinery, design and manufacturing of highly customized products. Despite the pressure of globalization, decline of the working age population in Europe, and labor cost minimization, these industries cannot offshore or automate their MW because it is the core of their operation that has to be made locally and manually with highly experienced and knowledgeable personnel.

According to Eurostat (2008), there are about 19 million people involved in the high knowledge high value MW in Europe mainly as plant and machine assemblers and operators. Considering the entire product lifecycle (LC) from design to recycling, the number of people related to MW is even larger. The benefits from improving MW can be significant, for example, shortening the maintenance by one day saves 300 thousand Euros for a 500MW power plant and 20-30 thousand Euros for a rock crushing machine (ManuVAR research, internal deliverable D1.1).

The niche of the ManuVAR project – Manual Work Support throughout System Lifecycle by Employing Virtual and Augmented Reality – is high value high knowledge MW that can be neither off-shored not automated (ManuVAR, 2010). This kind of MW primarily corresponds to the "knowledge-based behavior", but also partially to the highly complex "skill-based behavior" (Rasmussen & Goodstein, 1985). ManuVAR aims to provide a systematic technical and methodological solution for supporting the MW in the entire LC.

The ManuVAR approach is based on combining several disciplines: product lifecycle management (PLM), Virtual and Augmented Reality (VR/AR) technology, virtual modeling, and ergonomics methods with the special focus on high knowledge high value MW.

PLM is "an integrated information-driven approach comprised of people, processes/practices, and technology to all aspects of a product's life, from its design through manufacture, deployment and maintenance – culminating in the product's removal from service and final disposal" (Grieves, 2006).

MW is present at all stages of the LC and therefore MW is an important constituent of PLM. The human-machine interaction and synergy (Wilson, 2000) is the subject of ergonomics and, in the context of industrial systems, it is included in the production models such as lean and agile manufacturing (Narasimhan, et al., 2006).

However, due to the flexibility of the human and the difficulty of formalizing knowledge and skills in the same way as information, accounting for the human within PLM is very not easy. Besides, PLM systems are technically very complex and they contain a massive amount of various information. Hence, there is a need for an efficient natural interface to the PLM systems, which support bi-directional communication and feedback between all human actors across the LC. VR and AR are known to be the most promising technologies for building such interfaces.

VR is a "paradigm in computer-human interaction, in which three-dimensional computer-generated worlds ... have the effect of containing objects that have their own location in three-dimensional space" with the user perception of the virtual world being "...similar to the real world" (Bryson, 2004). Virtual environments (VEs) are synthetic, computer-generated 'worlds' build with the VR technology. Within a VE participants can not only engage passively with 3-dimensional (3D) objects and avatars via the senses, but where it is also possible to interact with these features. The ability to navigate around and interact intuitively with objects and avatars in the VE by carrying out actions and receiving an immediate, or almost immediate, sense of feedback; and a sense of being present within the VE often referred to as 'Presence' (Wilson & D'Cruz, 2006). AR means the computer-generated enhancement (text, images, video, 3D models) of a view of the real world. These enhancements are location specific and, if visual in nature, can orient themselves to map correctly onto objects or surfaces in the real world.

RELATED WORK

The subject domain covered by ManuVAR has been in the focus of the scientific community for the past 15 years. The Engineering Village scientific search engine contains references to more than 40 thousand publications on VR, AR, VE. The MicroPat database lists about 2000 patents on the VR-related technologies. Computer-aided design (CAD) and product data management (PDM) have been developed since the early computer era, while the digital engineering – digital mock-up, virtual model, PLM – have became especially active in the 1990's. Our state-of-the-art review covered more than 60 EU R&D projects funded under several European Union Framework Programs (EU FP).

Thus, the ongoing EU FP7 projects LeanPPD and FUTURESME aim to develop new production models for the European industries beyond the lean manufacturing model. These projects focus on the business aspects of the innovation and knowledge-based models. The model of "fit manufacturing" (Pham, et al., 2009) combines lean, agile and sustainable paradigms. ManuVAR is more technical in scope and it pays a special attention to the high knowledge high value MW, which is seen as an opportunity for the European industries to build their knowledge and skill-based processes and to sustain the challenges of globalization.

The EU FP6 network of excellence INTUITION (and its ongoing continuation – the EuroVR association) has been elaborating on the application of VR/AR in a variety of industry sectors. The EU FP6 HILAS project concentrated on creating a human factors framework across the system LC. The EU FP6 project CoSpaces focuses on organizational models and technologies supporting collaborative workspaces within distributed virtual manufacturing enterprises. The EU FP6 VIRTHUALIS considers safety enhancements in the LC using VR technology. The

EU FP6 PISA project proposes the combination of human flexibility, intelligence and skills with the advantages of sophisticated technical systems to support the human worker. The EU FP6 CybermanS project develops a set of methods and instruments to support an innovative approach to the organization of flexible production, in agreement with the ergonomics and safety standards. The EU FP6 DiFAC develops a collaborative manufacturing environment to support group work in an immersive and interactive way, for concurrent product design, prototyping and manufacturing, as well as worker training.

The main ManuVAR feature is that it concentrates on various aspects of the high knowledge high value MW in the industrial context and combines all three aspects: PLM, VR/AR and ergonomics. The hypothesis is that solving MW issues is only possible through such a multidisciplinary approach.

MANUVAR OBJECTIVES

The objective of ManuVAR is to develop an innovative technology platform and a methodology framework to support high value manual work throughout the product lifecycle. ManuVAR aims to develop a new production model for the European industries. It will combine the best potential of the lean and agile models with the knowledge and skill management in the product LC. The ultimate goal of ManuVAR is to go "beyond lean manufacturing" in order to

1. Increase productivity and quality and reduce cost of manual work in the whole lifecycle ("lean");
2. Facilitate adaptation to product customization and changes ("agile");
3. Support efficient knowledge and skill management through the lifecycle ("beyond");

ManuVAR pursues a number of results:

- Technological results: the platform of application tools, system architecture, tested technical prototypes;
- Methodological results: the ManuVAR PLM model, application methodology, framework of ergonomics methods to support MW throughout the product LC;
- Business results: ManuVAR industrial cases, demonstrated ManuVAR implementations in the partner companies, a spin-off business plan;
- Policy results: recommendations and evidence for to the policy makers that high knowledge high value MW can be an opportunity of the European industries.

PURPOSE OF THE PAPER

The purpose of this paper is threefold: (1) to introduce the ManuVAR project; (2) to set the research framework by identifying the seven high level problem areas related to the MW in Europe; (3) to overview the research publications in this volume that describe the first project results.

MANUVAR METHOD OF WORK

PROJECT WORKFLOW

To achieve the goals of ManuVAR, the work is structured into twelve work packages (WPs) corresponding to the general system workflow: input/reference → analysis → synthesis → testing → feedback → output. The work plan includes inductive (from requirements to the generalized solution: platform and methodology) and deductive phases (from generalized solution to testing it in the companies) with a number of iterations and feedback links, see Figure 1:

- Inductive phase: requirements analysis (WP1), definition and alignment of industrial cases (WP2), analysis of the requirements from the technological, human factors and PLM perspective (WP3), and general solution synthesis: the ManuVAR platform and methodology (WP4);

- Deductive phase: adaptation of the platform and methodology to the industrial needs (WP5), development of tools, industrial cases and demonstration prototypes (WP6), trials and assessment in the laboratory environment (WP7), training (WP8), and demonstration in the real factory environment (WP9).

The remaining WP10, 11 and 12 are dedicated to business development, dissemination and management and are executed throughout the project.

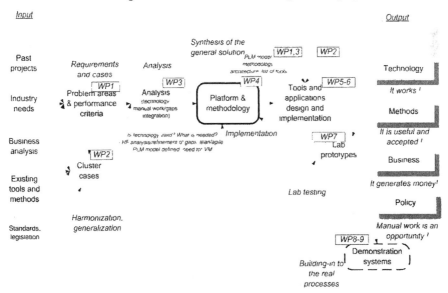

FIGURE 1. ManuVAR workflow (Source: ManuVAR project)

MANUVAR CONSORTIUM AND INDUSTRIAL CLUSTERS

The ManuVAR consortium comprises 18 partners representing industry, research and academia with complementary expertise. The research and academic partners build the scientific and know-how backbone of the project and the universities provide the massive resource essential for the ManuVAR system implementation. In addition to this, the companies provide requirements, domain expertise, assessment of results, innovation and result take up.

The high share of companies in the consortium indicates that industrial cases are the key strength of the ManuVAR project. These cases are developed by clusters, one case each. The clusters are formed around partner companies with the links to respective research institutions. The industrial cases in clusters are designed to fit the partner competence areas within the ManuVAR framework. The clusters provide the requirements and feedback, while the research institutions and universities act as mediators between clusters and the rest of the project and make a common platform and methodology, generic methods as well as tools available. Due to this diversity the clusters and cases cover the most important aspects of MW support throughout all major stages of system LC.

Every company covers a different industry sector so that there are five synergistic application clusters in different industrial areas.

Cluster 1. VR/AR assisted procedure compliance in terrestrial satellite assembly. The industrial partner is Thales Alenia Space Italia (TAS-I) supported by the Association for Advancement of Radical Behavior Analysis (AARBA) in Italy.

Cluster 2. Low cost VR system for improving assembly lines in small and medium enterprises. The industrial partners are Cards PLM solutions and IPA Total Productivity supported by TNO – Netherlands Organization for Applied Scientific Research, all in the Netherlands.

Cluster 3. AR/VR-enhanced remote online maintenance support in the railway sector. The industrial partners are participants are NEM Solutions supported by Fatronik-Tecnalia in Spain.

Cluster 4. Use of VR for improving training on nondestructive techniques. The industrial partners are TECNATOM supported by University of Malaga in Spain.

Cluster 5 VR/AR in the heavy machinery productization and maintenance. The industrial partner is Metso Minerals supported by VTT Technical Research Centre of Finland.

MANUVAR PROBLEM FORMULATION

SEVEN PROBLEM AREAS RELATED TO MANUAL WORK

Several interview sessions were conducted in the five ManuVAR clusters among senior and middle management and workers. As a result, we identified the seven most prominent problem areas faced by European industries in the context of high knowledge high value MW. These problem areas are listed below.

1.   *Problems with communication throughout the lifecycle.* This applies to the situations when MW is involved in two or more stages of the LC, but bi-directional communication between them is not efficient. For example, when the documentation or instructions become out-of-date or out-of-sync due to product changes (forward flow) or when a worker has to provide feedback on the MW operation back to the designers, but there is only a verbal or paper-based system to do it (backward flow). This problem area is also related to the organizational barriers within companies, when companies' information is managed by different information systems creating islands of information and knowledge.

2.   *Poor interfaces.* This means when an actor in the LC has either too little or too much information (e.g. documentation, instructions), this information is not contextualized (e.g. instructions are not focused on the current operation, product, part, situation), or when information is delivered by an improper medium (e.g. paper manuals that are hard to use and that quickly get obsolete).

3.   *Inflexible design process.* This refers to the situations when product design (and, therefore, the associated MW) cannot handle quickly changing requirements and frequent re-designs, in other words, the product design process is not agile. It is especially acute when the physical mockups are not available (expensive, inflexible, out-of-date), while virtual prototyping is not used sufficiently. As a result, there is a lack of knowledge of how MW should be made, e.g. what implications a product change has on the assembly instruction or the maintenance procedures.

4.   *Inefficient knowledge management.* This problem area is related to formalization, capture, processing, archiving, bi-directional transmission and training processes for high knowledge high value MW. By absorbing the facts obtainable from this whole system (product, machinery, work, environment, tools), the human accumulates and brings up tacit knowledge and skills. However, it is extremely difficult to capture tacit knowledge and to formalize it so that it could be manipulated in the same way as information.

5.   *Low productivity.* This is manifested in delays, idle time, non productive work, low-quality and re-work, i.e. falls into the conventional non-lean setting.

6.   *Lack of technology acceptance* regarding AR/AR technology, which is perceived as overly complex, expensive, unreliable, and difficult to use.

7.   *Physical and cognitive stress.* This is related to the traditional subject of physical (health and safety, physical stress) and cognitive (mental stress, motivation, team work environment) ergonomics. While cognitive stress is more relevant to the focus of the project, physical stress (e.g. difficult working postured, hazardous work environment) should also be considered because it distracts the attention of the human from the high knowledge operations.

DISCUSSION AND CONCLUSIONS

The identified seven problem areas capture a large portion of the MW problems as they have been collected from five companies operating in different industry sectors that have in common a unique production and a high proportion of high knowledge high value MW. These problem areas are quite high a level, but nevertheless their company-specific manifestations are visible in each of the five cases. The detailed analysis from the technology, human factors, and product lifecycle management perspectives will be published in (Lawson et al. (eds.), 2010).

These problems set the research framework for the project. On the basis of these problems we will propose the overall ManuVAR system that is valid for a number of companies, which are similar to those in ManuVAR. This system will be implemented with the VR/AR and PLM technologies that will be supported by the ergonomics method.

The specific implementation of ManuVAR in each cluster will cover part of the problem areas and LC stages and will be tuned to fit the technological requirements of each cluster. Without the need to build a large monolithic and potentially useless system, the specific implementations in the clusters collectively will demonstrate the validity of the general ManuVAR solution.

Let us overview the first ManuVAR results that are reported in other publications included in this volume.

(Krassi, et al., 2010) presents the ManuVAR overall solution, methodology, system architecture and drafts the application tools. It also formulates the main principle of the virtual modeling.

(Kiviranta, et al., 2010) reveals the role of the virtual environment and the virtual model for supporting the bi-directional seamless and dynamic communication of multiple heterogeneous actors across the product LC.

(Aromaa, et al., 2010) describes the integration framework of the ergonomic methods and the PLM and it proposes a new grouping to harmonize various ergonomics methods supporting MW in the LC.

(Liston, et al., 2010) outlines the ManuVAR training concept in the context of the quality-improvement tools for manual handling.

(Lawson, et al., 2010) reports an investigation of the MW in the ManuVAR clusters using a combination of human factors approaches, including task analysis, ethnography and behavior analysis.

(Tosolin, et al., 2010) analyzes the joined application of behavior analysis and VR/AR to supporting MW in the Thales Alenia Space Italy cluster.

(Leino, et al., 2010) outlines the plans of the Metso Minerals cluster to exploit PLM and VR/AR for supporting MW at the design and productization LC stages.

(Vink, et al., 2010) sets the user requirements to the mixed reality tool Ergomix for design, evaluation and support of MW.

Finally, two papers describe technical solutions that will be further developed in ManuVAR to support MW: a mixed reality system on mobile devices (Frangakis, et al., 2010) and a virtual interactive training application (Strauchmann, et al., 2010).

ACKNOWLEDGEMENTS

The research leading to these results has received funding from the European Commission's Seventh Framework Programme FP7/2007-2013 under grant agreement 211548 "ManuVAR".

This paper has been written on behalf of the ManuVAR consortium:

- AARBA, IT – Fabio Tosolin, Guido Tosolin, Alessandro Valdina;
- Cards PLM Solutions, NL – Gert Nomden;
- Carr Communications, IE – Eddie Shaw;
- Fatronik-Tecnalia, ES – Jon Agirre Ibarbia, Jon Azpiazo;
- Hermia Business Development, FI – Jarno Kolehmainen, Enna Palander-Palonen;
- Institute of Communication and Computer Systems, GR – Ioannis Karaseitanidis, Konstantinos Loupos, Nikos Frangakis;
- Fraunhofer Institute of Factory Operation and Automation, DE – Daniel Reh, Matthias Strauchmann, Corrina Kunert;
- IPA Total Productivity, NL – Hans Totte;
- Metso Minerals, FI – Juhamatti Heikkilä, Sanna Holma, Petri Rantanen;
- NEM Solutions, ES – Tim Smith;
- Tampere University of Technology – Petteri Multanen;
- Thales Alenia Space Italy, IT – Enrico Gaia, Carlo Vizzi;
- TNO, NL – Gu van Rhijn, Sandra Eikhout, Merle Blok, Tim Bosch;
- Trinity College Dublin – Paul Liston, Maria Chiara Leva, Sam Cromie;
- Tecnatom, ES – Susana Flores, Eva Frutos;
- University of Malaga, ES – Matthieu Poyade, Arcadio Reyes Lecuona;
- University of Nottingham – Richard Eastgate, Anne Floyde, Alyson Langley, Glyn Lawson, Shakil Shaikh, Sarah Sharples;
- VTT, FI – Kaj Helin, Susanna Aromaa, Simo-Pekka Leino, Salla Lind, Magnus Simons, Sauli Kiviranta, Juha Saaski, Juhani Viitaniemi.

REFERENCES

Aromaa, S., Viitaniemi, J., Leva, C., Krassi, B. (2010) "Integration of Human Factors Methods to the ManuVAR Product Lifecycle Management Model." *Applied Human Factors and Ergonomics - AHFE 2010*, USA

Bryson, S. (2004) "Virtual reality" *Computer science handbook.* A.B. Tucker (ed.). Chapman & Hall, CRC, p. 42-1

European Union Labour Force Survey – Annual results (2008). Eurostat (N. Massarelli)

Frangakis, N., Karaseitanidis, I., Blach, R., Gleue, T. (2010) "Supporting workers through the combination of Mixed Reality and mobile devices in wide areas" *Applied Human Factors and Ergonomics - AHFE 2010*, USA

Grieves, M. (2006) *Product Lifecycle Management.* McGraw-Hill

Kiviranta, S., Poyade., M., Helin, K. (2010) "Virtual environments for supporting high knowledge , high value manual work design." *Applied Human Factors and Ergonomics - AHFE 2010*, USA

Krassi, B., Kiviranta, S., Liston, P., Leino, S.-P., Strauchmann, M., Reyes-Lecuona, A., Viitaniemi, J., Sääski, J., Aromaa, S., Helin, K. (2010) "ManuVAR PLM model, methodology, architecture, and tools for manual work support throughout system lifecycle" *Applied Human Factors and Ergonomics - AHFE 2010*, USA

Lawson, G., D'Cruz, M., Krassi, B. (eds.) (2010), "High value high knowledge manual work in the European industries: identification of problem areas and their technical, business and ergonomics analysis." (working title) *VTT working papers* (in publication)

Lawson, G., Tosolin, G., Leva, C., Sharples, S., Langley, A., Shaikh, S. (2010) "Investigating Manual Work Using Human Factors Approaches & Behavioral Analysis" *Applied Human Factors and Ergonomics - AHFE 2010*, USA

Leino, S.-P., Kiviranta, S., Rantanen, P., Heikkilä, J., Martikainen, T., Vehviläinen, M., Mäkiranta, A., Nuutinen, P., Hokkanen, I., Multanen, P. (2010) "Collaboration between Design and Production by exploiting VR/AR and PLM – Case Metso Minerals" *Applied Human Factors and Ergonomics - AHFE 2010*, USA

Liston, P., Cromie, S., Leva, C., Helin, K., D'Cruz, M. (2010) "ManuVAR Training: Supporting the Implementation and Design of VR/AR Quality-improvement Tools for Manual Handling" *Applied Human Factors and Ergonomics - AHFE 2010*, USA

ManuVAR project web site (2010), [www.manuvar.eu] Referred in February 2010

Narasimhan, R., Swink, M., Kim, S.W. (2006) "Disentangling leanness and agility: An empirical investigation." *Journal of Operations Management,* p. 440-457

Strauchmann, M., Haase, T., Viitaniemi, J., Kiviranta, S., Vink, P. (2010) "A virtual interactive training application to support manual work" *Applied Human Factors and Ergonomics - AHFE 2010*, USA

Tosolin, G., Valdina, A., Vizzi, C., Gaia, E. (2010) "From teaching machines to the exploitation of virtual and augmented reality: Behavior Analysis supporting manual workers in aerospace industry within the European project ManuVAR" *Applied Human Factors and Ergonomics - AHFE 2010*, USA

Vink, P., Blok, M., van Rhijn, G., Bosch, T., Totté, H., Loupos, K. (2010) "User demands for new mixed reality tools, first results of the ManuVAR project" *Applied Human Factors and Ergonomics - AHFE 2010*, USA

Wilson, J.R. (2000) "Fundamentals of ergonomics in theory and practice." *Applied Ergonomics,* 31, p. 557-567

Wilson, J.R., D'Cruz, M. (2006) "Virtual and interactive environments for work of the future." *International Journal of Human-Computer Studies*. Vol. 64, Is. 3, p. 158-169

Pham, D.T., Adebayo-Williams, O., Thomas, A.J, Barton, R., Ebrahim, Z., Shamsuddin, T. (2009) "Fit Manufacturing: A Strategy for Achieving Economic Sustainability." *IPROMS 2008 Conference*

Chapter 79

Development of 3D Ergonomics Requirement Models for Cabin Development

Anders Sundin[1], Johan Landén[1], Ann Hedberg[1], Jimmy Holler[2]

[1]Semcon AB
SE-401 17 Gothenburg, Sweden

[2]Atlas Copco Rock Drills AB, SDE
SE-701 91 Örebro, Sweden

ABSTRACT

During product development in general, and in automotive product development in particular, the need for accurate specified requirements within ergonomics is a key factor for user centered design. Traditionally, ergonomic features and requirements have been difficult to incorporate and interpret in the product development phases. This is mainly due to the fact that the human aspects within physical ergonomics as well as issues of human machine interaction (HMI), have contained less hard facts compared to e.g. structural analysis and mechanical design. Digital Human Modeling (DHM) tools have for a long time been used to facilitate this work and consider ergonomics early in the product development process. However, even though this tool is used, the product developers have difficulties finding easy access to ergonomic recommendations and requirements needed for a proper study. Usually facts are spread out in paper folders or digital folders around the company. This paper illustrates how the DHM tool Delmia V5 Human has been used to build

3D ergonomic requirements models for a specific automotive cabin development branch. Based on a foundation of international recommendations and regulations, e.g. EN, ISO, SAE, as well as company-specific demands, CAD models were developed describing the ergonomic requirements as envelopes or planes. The models covered outer visibility, maximum and comfort reach and foot reach. Furthermore, inner visibility was also included covering part of the HMI requirements. Benefits of 3D ergonomic requirement models are discussed in this paper giving illustrations on how these types of models can be used as an effective tool in ergonomics requirements handling as well as for time reduction in product development. In a conclusion it is shortly described how these models have been incorporated in the cabin development company.

Keywords: Ergonomics Requirement, Digital Human Modeling, HMI, Cabin Development

INTRODUCTION

During product development in general, and in automotive product development in particular, the need for accurately specified requirements within ergonomics is a key factor for successful user centered design. Beside the user benefits, a major driver is economical as poor design solutions affect ergonomics and the economical results (e.g. Falck, 2009). Traditionally, ergonomics features and requirements have been difficult to incorporate and interpret into the product development phases, mainly due to the fact that human aspects within physical ergonomics as well as issues of Human Machine Interaction (HMI), have contained less hard facts compared to e.g. structure analysis and mechanical design. Furthermore, Digital Human Modeling (DHM) tools have for many years been used early in the development work to facilitate the consideration and prediction of ergonomics in the product development process as well as in the accompanying workplace and production design phases (e.g. Chaffin, 2001, Sundin, 2006, Lämkull, 2006). However, even though these tools are used and a lot of standards exist, the product developers have difficulty finding easy access to the ergonomics recommendations and requirements needed for proper design and a natural way of working with these requirements.

Atlas Copco, Surface Drilling Equipment, SDE, were in a phase where they wanted to improve ergonomics and user interfaces of the cab interior in their ROC-series, surface drill rigs. This paper describes part of a joint project between the consultancy company Semcon, Group Ergonomics & HMI, and Atlas Copco SDE, where development work was carried out compiling and developing requirements within physical ergonomics and HMI, aimed at this cabin project, but also with future cabin development projects in mind. A physical ergonomics and HMI analysis of an existing ROC-series cab concept was also carried out, but it is not covered in this paper.

Central in the project work described in this paper, was the development of 3D ergonomics requirement models, covering both physical ergonomics and HMI. A 3D ergonomics requirement model is a three-dimensional computer model (CAD model) describing and representing requirements for a specific ergonomic area, such as reach demands to controls in a cockpit.

The basic aim for such model development is to facilitate the daily work within ergonomics in the product development process. Generally it is today often problematic to find and use relevant and correct ergonomics requirements in the daily workplace. Instead of searching for and using paper folders and text documents with ergonomics requirements described often in a difficult way, as is often the case today, the developers should instead get easy access to the requirements via 3D models in the same simple way as any other mechanical CAD part imported into the CAD-system.

METHODOLOGY

The development work, towards a set of requirement described both as 3D models and in text, was made by a group of different competencies. Represented were three ergonomists, two being certified European Ergonomists (Eur.Erg.), one HMI specialist, one mechanical design engineer, and two industrial design engineers. The models covered outer visibility, maximum and comfort reach and foot reach. Also inner visibility was included covering part of the HMI requirements.

The 3D requirement model development was based on standardization documents as far as possible (ISO, EN etc). In total, 14 different standards were used as a base (see references). In those cases where the content of the standards were not completely satisfactory for use in the drilling area, information has been gathered from ergonomics literature, especially Dreyfuss (2002), and a number of analyses carried out in the computer manikin software Delmia V5 Human.

When the standard documents were not used as a base, the requirements were discussed, developed and decided together with employees at Atlas Copco SDE System department.

The requirements were developed in accordance with the anthropometrics defined in ISO 3411 where 5%ile, 50%ile and 95%ile operators (small, medium and large) from a worldwide market is stated.

Models were developed in Catia V5 as Iges-files with the SIP (Seat Index Point) as reference. All reach requirement models have been designed with the reference point SIP as defined in ISO 5353.

Finalized models were incorporated into the company computer systems by Atlas Copco SDE personnel.

RESULTS

For physical ergonomics and HMI respectively, resulting requirements were compiled and/or developed for the following areas to support cabin development: Reach, Inner visibility, Outer visibility, Ingress/egress, Seat position, Positioning and layout, Design of controls, Interaction and Design of Graphical User Interface (GUI). This requirements setting resulted in text documents describing the requirements to be followed by Atlas Copco SDE, one document for physical ergonomics and one for HMI.

For the areas above, 3D ergonomics requirement models were developed and designed in CAD for selected parts of the requirements. The selection of requirements to be modeled was made jointly in the project team. In total, thirteen 3D ergonomics requirement models were developed resulting in one Iges-file per area, see Table 1. Of these, the eight considered most important were chosen to be visualized as *Primary* 3D ergonomics requirement models. Beside these eight models, considered to be the primary requirements to be followed within the company, an additional five extra, less demanding *Complementary* 3D ergonomics requirement models were developed. These complementary 3D models could also be used in the daily work as complementary support. The Complementary models covered 50%ile operators while the Primary models covered 5%ile to 95%ile operators.

Table 1. The developed thirteen 3D requirement models.

Primary 3D Requirement Models	Complementary 3D Requirement Models
Reach comfort	-
Reach inner clearance	-
Reach maximum Atlas Copco SDE	Reach maximum
Reach comfort foot zone	-

Visibility inner comfort maximum 5%ile 95%ile	Visibility inner comfort maximum 50%ile
Visibility outer comfort upwards downwards 5%ile 95%ile	Visibility outer comfort upwards downwards 50%ile
Visibility HMI detection 5%ile to 95%ile	Visibility HMI detection 50%ile
Visibility HMI monitoring 5%ile to 95%ile	Visibility HMI monitoring 50%ile

Below, three examples of the physical ergonomics and HMI 3D requirement models are shown: Reach maximum Atlas Copco SDE (Figure 1), Visibility HMI detection 50%ile (Figure 2) and Reach comfort, used for the positioning of primary functions (Figure 3). The manikins are not included in the model when they are in use in the daily work. These models are to be used by the industrial designers (styling), mechanical designers, etc. When loaded into the CAD environment, the model automatically orients itself according to SIP. While in place the developer can e.g. see if a suggested control or a display position is positioned within reach or good visibility.

Figure 1. Requirement model Reach maximum Atlas Copco SDE, shown together with manikin.

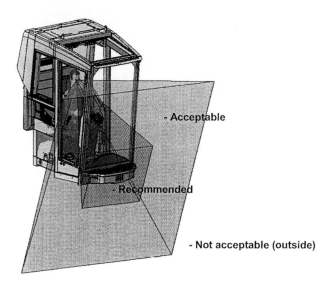

Figure 2. Requirement model Visibility HMI detection 5%ile to 95%ile to be used for detection task, here visualized together with cabin and manikin.

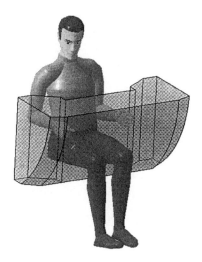

Figure 3. Requirement model Reach comfort, a comfort zone for primary functions here visualized together with manikin.

Beside the text documents and the 3D models describing the requirements, a methodology part was also written, i.e. guidelines, on *how* to use the developed 3D requirement models in the daily work and suggestions on how and when to use the ergonomics and HMI requirement requirements in Atlas Copco development projects. For those ergonomics areas not yet having a 3D requirement model developed, the main text documents with all the requirements described will cover the need to find answers on requirement levels.

CONCLUSIONS

Since this work was carried out at the end of 2008, Atlas Copco SDE has successfully been using the developed 3D ergonomics requirements models in their daily work. Both industrial design engineers and design engineers bring them up into the CAD system to continuously make checks so that the requirements are fulfilled throughout the design changes naturally taking place during the development work. Lately, Atlas Copco SDE has also included the models and procedures into their internal standardization framework. After an evaluation of their use of the thirteen developed models, the aim is to continue with more areas to be modeled.

The benefits of ergonomics 3D requirement models seem to be considerable. Today, almost all product development is made in a 3D-environment. During the last decade also ergonomics has been introduced and is in daily use in this 3D environment, e.g. with computer manikins. However, early requirement and verification work have not been digitalized in the same way. By digitized requirements presented here, the requirement-setting phase in product development can be better supported from the ergonomics and HMI point of view. The most obvious positive factor is the reduction in time required to find, interpret, understand, and finally use ergonomics requirements. Additionally, error detections at an early phase also leads to cost reduction as problematic or faulty design solutions can be found and corrected earlier, clearly brought forward by Falck (2009). However, as in all digital ergonomics work the need of a structural process is vital (e.g. Green, 2000 and Sundin, 2001). In this work, also a company-specific methodology was developed including how Atlas Copco SDE should work with the ergonomics requirements in general and 3D ergonomics requirements in particular in the different product development phases, including e.g. roles and responsibilities.

To summarize, using these types of 3D ergonomics requirements models seems to facilitate how requirements within ergonomics and HMI can be used and followed, also giving ergonomics arguments early in the development process. Stakeholders such as industrial designers, designers and ergonomists also get a common ground where it is easier to communicate demands and requirements, both internally and externally to suppliers etc.

ACKNOWLEDGMENTS

We are grateful to all persons that have been involved in the development work. From Atlas Copco SDE System department; Jimmy Holler, Dennis Hedberg and from Atlas Copco Rocktec Industrial Design; Alex Liebert and Daniel Nadjalin. From Semcon; Ann Hedberg, Niklas Eklund, Johan Landén and Anders Sundin.

REFERENCES

Chaffin, D. (2001). Digital human modeling for vehicle and workplace design. Society of Automotive Engineers, Inc., Warrendale, PA.

Dreyfuss Richard Associates. (2002). The measure of man & woman, Revised edition, Human factors in design.

EN ISO 2867: 2008 Earth moving machinery - Access systems.

Falck, A-C. Örtengren, R. Högberg, D. (2009). The impact of poor assembly ergonomics on product quality: A cost–benefit analysis in car manufacturing. Wiley Periodicals, Inc.

Green, R.F. (2000). A generic process for human modeling analysis. Proc. SAE International. Digital Human Modeling for Design and Engineering 2000, International Conference and Exhibition, June 6-8, Dearborn, Michigan, USA. SAE 2000-01-2167.

ISO 10968: 2004. Earth-moving machinery - Operator's controls.

ISO 15008: 2003. Road vehicles - Ergonomic aspects of transport information and control systems - Specifications and compliance procedures for in-vehicle visual presentation.

ISO 5353: 1995. Earth-moving machinery, and tractors and machinery for agriculture and forestry - Seat index point.

ISO 3411: 2007. Earth-moving machinery - Human physical dimensions of operators and minimum operator space envelope.

ISO 6011: 2003 Earth-moving machinery – Visual display of machine operation.

ISO 6405-1: 2004. Earth-moving machinery – Symbols for operator controls and other displays, Part 1: Common symbols.

ISO 6682: 1986. Earth moving machinery, Zones of comfort and reach for controls.

ISO 9241-10: 1996. Ergonomic requirements for office work with visual display terminals (VDTs), Part 10: Dialogue principles.

ISO 9241-3: 1992. Ergonomic requirement for office work with visual display terminals (VDTs), Part 3 Visual display requirements.

ISO 9241-5: 1998. Ergonomic requirement for office work with visual display terminals (VDTs), Part 5 Workstation and postural layout.

ISO 9355-1 / SS EN 894-1: 1999. Ergonomic requirements for the design of displays and control actuators, Part 1: Human interactions with displays and control actuators.

ISO 9355-2 / SS EN 894-2: 1999. Ergonomic requirements for the design of displays and control actuators, Part 2: Displays.

ISO 9355-3 / SS EN 894-3: 2006. Ergonomic requirements for the design of displays and control actuators, Part 3: Control actuators.

Lämkull, D. (2006). Computer manikins in evaluation of manual assembly tasks. Thesis for the Degree of Licentiate of Technology. Department of Production and Production Development, Production Systems, Chalmers University of Technology, Göteborg, Sweden, ISSN 1652-9243, Report No. 14.

MIL STD 1472F: 1999. Department of defense design criteria, Human engineering SS EN 791: 1995. Drill rigs – Safety.

Sundin, A. (2001). Participatory ergonomics in product development and workplace design. Supported by computerised visualisation and human modeling. Department of Human Factors Engineering, Chalmers University of Technology, Göteborg, Sweden, ISBN 91-7291-046-1.

CONTACT

Ph.D. Anders Sundin, European Ergonomist, Eur.Erg., Technical Area Leader Ergonomics/HMI, Automotive, Semcon, phone +46 733 989635, anders.sundin@semcon.com.

Chapter 80

Usability Evaluation of Target Expansion Schemes in Selecting Targets with a Mouse

Donghun Lee, Sunghyuk Kwon, Min K. Chung

Department of Industrial and Management Engineering,
Pohang University of Science and Technology
San 31, Hyoja-dong, Nam-gu, Pohang, Gyeongbuk, 790-784, South Korea

ABSTRACT

Target expansion, increasing target size according to cursor movement, is a novel scheme to enhance the speed and accuracy of target selection tasks. This study examines the effects of four age groups and eight target expansion methods on target acquisition tasks with multiple targets. Thirty two subjects performed the combinations of four expansion areas and two expansion techniques. Older users took longer to acquire targets than did younger users. The task completion time was shortest when expanding no and one-icon areas with the occlusion technique. The error rate was lowest when expanding group area with the push technique. All age groups were most satisfied with the method of expanding one-icon area with the occlusion technique. We suggest two alternative guidelines in designing target expansion methods for target acquisition tasks using a mouse. In a GUI in which targets expand, it is desirable to expand only the target to which the cursor points. When designing a GUI to simultaneously expand a specific group, it is desirable to push surrounding groups aside. It is expected that usability of related software can be improved when using these target expansion schemes in practical GUI designs.

Keywords: Target Expansion Scheme, Target Acquisition, Graphical User Interface

INTRODUCTION

Graphical user interfaces (GUIs) allow users to interact with personal computers (PCs) by selecting targets such as icons or menus using an input device such as a mouse. Pointing to targets with a mouse is an essential task in GUIs (Blanch et al., 2004). Because GUIs are indispensable in various applications (e.g. MS Office and Adobe Photoshop) and in operating systems such as MS Windows and Mac OS X, usability of GUIs is increasingly important.

Previous studies related to usability of GUIs mostly focused on design factors such as targets' size, color, or shape. Official design guidelines have been also provided. However, usability of GUIs in PC working environments can still be improved. Recent studies on target acquisition tasks have presented methods to ameliorate the limitations of existing GUIs by introducing techniques that simplify the process of putting the cursor on the target.

Several studies have shown that target expansion can improve the usability of GUIs by increasing the size of the target according to the location and movement of a cursor, or the distance between cursor and target (McGuffin & Balakrishnan, 2002, 2005; Zhai et al., 2003). In particular, visual expansion only without enlargement in motor space had a positive effect on performance in target acquisition tasks (Cockburn & Brock, 2006). They also suggested that it may be more useful for practical applications because it allows a wider workspace and offers more information than motor expansion. However, target expansion schemes may spatially disorder nearby targets during cursor movement because they change the layout of toolbars and menus on screen (Blanch et al., 2004). Thus, the effect of fixing or shifting target centers needs to be investigated in detail.

Although there have been some studies on the target expansion methods, several points to be considered remains. First, various interaction designs based on target expansion should be developed, and the comparative evaluation of the designs should be examined. The fish-eye technique can also be considered as one of the target expansion techniques. Second, previous studies have focused on performance measures. Users' subjective assessment as well as performance measures should be considered when evaluating the target expansion methods. Lastly, the usability test should be conducted for the tasks in more realistic two-dimensional (2D) GUIs in which multiple targets occur in groups.

The problem of human-computer interaction may be more critical for older adults than for younger ones because of declines of motor abilities with age. Older people also had difficulty in mouse control tasks because of declines of visual abilities and in cognition (Czaja & Lee, 2002; Fisk et al., 2004). Older users as well as younger users need to be considered simultaneously in designing and evaluating the GUIs operated by a mouse. Thus, this study examines the effects of different age groups and target expansion methods on target acquisition tasks with grouped icons. In particular, this study identifies design alternatives to help improve usability of

GUIs for older adults who have more difficulty performing target acquisition tasks.

METHODS

PARTICIPANTS

A total of 32 volunteers participated in the experiment; they were divided into four age groups: Young (20-44 years old), Middle-aged (45-54 years old), Old (55-64 years old) and Senior (≥65 years old). Each age group consisted of eight individuals. The amount of experience with computers decreased with increasing age (Table 1).

Table 1 Participants' information   mean (SD)

Age group	Age [years]	PC use experience [years]
Young	34.0 (7.4)	10.0 (4.5)
Middle-aged	48.3 (1.9)	9.9 (5.0)
Old	61.3 (3.0)	5.0 (2.6)
Senior	70.1 (3.9)	3.4 (1.9)

APPARATUS

A 15-inch laptop computer and an optical mouse were used in the experiment. An experimental prototype was developed using Microsoft Visual Studio 2003® with icon images used in MS Office 2007 (Figure 1). A total of 114 icons were arranged into six groups at the top of the screen. Each icon was 18 × 18 pixels in size. The distance between the icons was 0 pixels within each group, and the groups were separated by 5 pixels. A black, 20-mm diameter 'Ready' button was displayed at the bottom of the screen. The target that users must select was displayed in red.

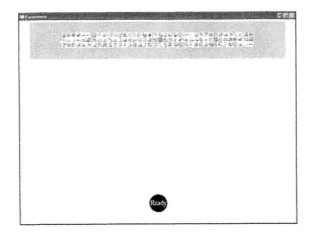

FIGURE 1 Prototype used in the experiment

The scale and timing of the target expansion was based on the method employed in McGuffin & Balakrishnan (2002). The length of the side of each icon or group expanded by a factor of 2 (i.e. a 4 × increase in area). The expansion occurred gradually when the cursor had moved 90% of the distance between the 'Ready' button and the target icon or group. Targets or groups expanded by the cursor's trajectory moving toward a target or group.

EXPERIMENTAL DESIGN

Three independent variables were considered: Age Group (Age), Expansion Area (Area), and Expansion Technique (Technique). Age was divided into four levels: Young, Middle-aged, Old and Senior. Four types of Areas were tested: No, One-icon, Fish-eye and Group. No Area is where the size of target or group never changed. One-icon Area magnifies one target near the cursor by a factor of 4. Fish-eye Area proportionally magnifies both the target near and nearby targets. Group Area magnifies the group nearest cursor. Two Techniques were tested: Occlusion and Push. Occlusion Technique is where the center of the target or group expanded is unchanged, and adjacent targets are partially or totally hidden. Push Technique moves adjacent targets or groups aside, and changes their layout (Figure 2).

	Occlusion	**Push**
No	(a) No Area & Occlusion Technique	(b) No Area & Push Technique

One-icon		
	(c) One-icon Area & Occlusion Technique	(d) One-icon Area & Push Technique
Fish-eye		
	(e) Fish-eye Area & Occlusion Technique	(f) Fish-eye Area & Push Technique
Group		
	(g) Group Area & Occlusion Technique	(h) Group Area & Push Technique

FIGURE 2 Eight experimental conditions

The target expansion methods consist of eight conditions; of these, the combination of No Area and Occlusion Technique (called Normal) is the static condition typically used in applications and web pages. All subjects performed the experiment under all eight conditions.

The dependent variables included two performance measures and a subjective rating. Task completion time (TIME) was the mean interval between clicking the 'Ready' button and clicking the target. Error rate (ERROR) was calculated by the number of clicks used and the number of clicks required for task completion. Preference of Use (PREFERENCE) assessed using a 100-point scale (0: strongly dislike, 100: strongly like) after completion of each experimental condition.

PROCEDURE

Participants first confirmed the location of the target marked in red, clicked the 'Ready' button, and moved the cursor to the target and selected it as accurately and quickly as possible. If the target was selected correctly, another target turned red, and the 'Ready' button reappeared. This process constituted one trial. The participants repeated thirty trials under each experimental condition. Thus, each subject performed a total of 240 trials (4 Areas × 2 Techniques × 30 trials). TIME and ERROR both were logged automatically after each condition was finished. Between the conditions, the participants recorded the score for PREFERENCE.

RESULTS

TASK COMPLETION TIME

Age ($F(3, 28) = 6.35$, $p = .0020$), Area ($F(3, 84) = 32.47$, $p < .0001$) and Technique

(F(1, 28) = 74.10, p < .0001) all significantly affected TIME. TIME was significantly shorter for Young adults than for Old and Senior adults, but the difference among Middle-aged, Old and Senior adults was not significant, nor was difference between Young and Middle-aged adults (Figure 3; left). In addition, TIME was shorter for the No and One-icon Area than for Fish-eye and Group Area, and was longest for Group Area (Figure 3; right). Also, TIME was shorter when using the Occlusion Technique (1.50 s) than when using Push Technique (1.63 s).

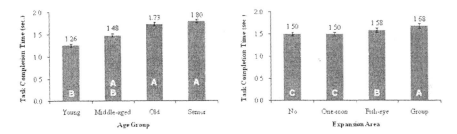

FIGURE 3 TIME for each Age (left) and for each Area (right). Bars show ± 1 standard error of the mean. Means with the same letter are not significantly different

The Area × Technique interaction for TIME was significant (F(3, 84) = 15.12, p < .0001; Figure 4). The Occlusion Technique resulted in shorter time than the Push Technique for No, One-icon and Fish-eye Areas, whereas the difference between the two Techniques for Group Area was not significant.

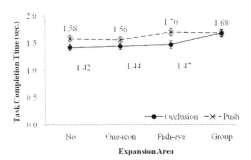

FIGURE 4 Interactions between Area and Technique. Bars show ± 1 standard error of the mean

ERROR RATE

Area (F(3, 84) = 6.20, p = .0007) and Technique (F(1, 28) = 8.70, p = .0064) significantly affected ERROR. ERROR was significantly higher for Fish-eye Area than for No, One-icon and Group Areas, which were not significantly different from each other (Figure 5; left). Using Push Technique (2.60%) resulted in lower error than the Occlusion Technique (3.85%).

The Area × Technique interaction on ERROR was significant (F(2, 56) = 9.83, p = .0009; Figure 5; right). The Push Technique had lower error than the Occlusion Technique when using Fish-eye and Group Areas, whereas difference between Occlusion and Push Techniques was not significant for No and One-icon Areas.

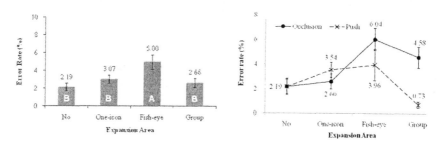

FIGURE 5 ERROR for each Technique (left) and interactions between Area and Technique (right). Bars show ± 1 standard error of the mean. Means with the same letter are not significantly different

PREFERENCE OF USE

Age (F(3, 28) = 2.95, p = .0496) and Area (F(3, 84) = 17.24, p < .0001) significantly affected PREFERENCE. PREFERENCE was higher for Senior adults than for Young adults, and no significant difference was observed among Young, Middle-aged, and Old adults or among Middle-aged, Old, and Senior adults (Figure 6; left). One-icon Area was preferred over other Areas. Among the other Areas, No Area was more preferred than Fish-eye and Group Area, which were not significantly different from each other (Figure 6; right).

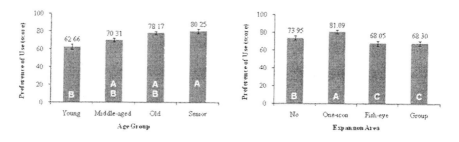

FIGURE 6 PREFERENCE for each Age (left) and for each Area (right). Bars show ± 1 standard error of the mean. Means with the same letter are not significantly different

The Age × Area interaction was significant (F(9, 84) = 2.23, p = .0276; Figure 7). Young adults preferred No and One-icon Areas to Fish-eye and Group Areas, whereas there was no significant difference between No and One-icon Areas, or between Fish-eye and Group Areas. Middle-aged adults preferred One-icon Area to the other Areas, whereas no significant difference was observed among the other Areas. There was no significant difference among Areas for Old and Senior adults.

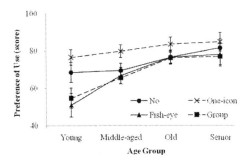

FIGURE 7 Interactions between Age and Area. Bars show ± 1 standard error of the mean

DISCUSSION

Older people had more difficulty selecting targets with a mouse than did younger users. Time to select targets increased with age; in particular, Senior adults took about 1.4 times longer than Young adults. Older adults may reduce movement velocity to compensate for the decline in accuracy that resulted from the decline in their motor abilities. Vercruyssen (1997) indicated a similar cause that older adults' control of mouse movements was inaccurate because of their non-symmetrical velocity trajectories. Fisk et al. (2004) mentioned that increased noise in the movement is a source of the age-related performance decline. Thus, the age-related decrease in motor abilities can cause older people to have difficulty manipulating a mouse, and can result in significant differences in age groups' speeds.

Regardless of Age, when expanding One-icon Area with the Occlusion Technique, the TIME was similar to the Normal condition typically used in various applications and web pages. Whereas the TIMEs of the Young and Middle-age adults were a little longer when using the Method 3 than in the Normal condition, the Method 3 allowed users aged over 55 to select targets more quickly than did the Method 1. However, the difference between the two Methods in speed is negligible. Unlike previous research that showed an improvement of about 6 to 12% in speed with a target expansion technique, we did not find a distinct advantage with the Method 3. A probable cause may be due to the task type tested in a complex, 2D environment with multiple targets. Targets that the subjects should select were evenly in the middle and top rows as well as in the bottom row of the three rows. Thus, the subjects may be distracted more when selecting the targets of the middle or top row rather than in the bottom row.

The layout of icons and groups used in our study may become another problem. Whereas previous studies tested in the 1D environment with a single target (McGuffin & Balakrishnan, 2002; Zhai et al., 2003) or in the 2D environment with sparsely-arranged circular targets from the ISO 9241-9 multi-directional tapping task (Cockburn & Brock, 2006), we performed the experiments in the condition with multiple targets and the groups placed close to each other. Consequently, the subjects may be more interrupted to correctly select a target by adjacent icons.

In contrast, when expanding Group Area with two Techniques (Methods 7 & 8) and expanding Fish-eye Area with the Push Technique (Method 6), the TIME was longer for all Ages than in the Normal condition (an increase of about 20%). The combination of the Fish-eye Area and the Push Technique (Method 6) is almost the same as the scheme used in Mac OS X Dock that may have negative effects on performance in target acquisition tasks (McGuffin & Balakrishnan, 2002, 2005; Zhai et al., 2003; Cockburn & Brock, 2006). A major factor reducing performance in speed may be that the position of the cursor greatly changes the design layout; in particular, expanding Group or Fish-eye Area in the Push Technique (Methods 6 & 8) affected a larger surrounding area than the other techniques. Thus, it is not desirable for designers to employ any GUIs resulting in a large change in the toolbar layout with multiple or grouped targets.

The range of error rates among individuals was larger rather than those between age groups. Accuracy may be a higher priority than speed for older people. We observed that adults aged over 45 often paused before positioning a cursor on a target. Those pauses may be caused by older users' behaviors to overcome the decline in accuracy due to decreased motor abilities with age. Also, the performance may be related to the difficulty of the task. Previous studies showed that older users had more difficulty performing computer mouse tasks than younger ones, particularly with very small (3 × 3 pixels) targets (Walker et al., 1996) or more complex tasks such as double-clicking (Smith et al., 1999). In this study, however, the tasks performed might have been relatively easy because the target size was 18 × 18 pixels, and the tasks included only single-clicking.

When expanding Group Area was combined with the Push Technique (Method 8), the ERROR was lowest regardless of all age groups. Error rate using this combination was about 67% lower than when using the Normal configuration. In particular, the Method 8 improved the accuracy of the adults aged over 45; the Error decreased by 80% for Middle-aged, 50% for Old, and 87.5% for Senior adults, compared with the Method 1. However, expanding Fish-eye Area with the Occlusion Technique (Method 5) increased the error rate of all age groups, especially in adults aged over 55. Thus, designers should be careful when they try to use target expansion techniques in practical interface designs because task accuracy was significantly affected by the Methods used.

Although the combination of expanding One-icon Area and the Occlusion Technique (Method 3) did not give a distinct improvement in speed and accuracy, all age groups gave the most positive assessment of the method. Although younger adults did not obtain any distinct advantage in speed and accuracy, they gave affirmative answers to the expansion method. Experienced users tend to be more interested in the subjective view for the usability of a system than in the performance effect (Nielsen & Levy, 1994). In our study, younger adults, who had more experience with computers and used computers more often, also preferred the combination of expanding One-icon Area with the Occlusion Technique (Method 3). If a target expansion method like the Method 3 is used in a 2D GUI with multiple targets, it will deliver great benefits to younger users as well as to older ones.

Users' subjective assessments may be important factors for designers to consider when designing computer systems and applications. User preference is an important measure in evaluating different UI designs (Nielsen & Levy, 1994). Thus, designers should carefully consider both subjective notions and performance measures when evaluating some GUI designs and developing design guidelines for them.

CONCLUSIONS

A target expansion scheme is a novel technique to improve usability of GUIs on target acquisition tasks using a computer mouse. We examined how four age groups interact with different target expansion methods on target acquisition tasks in 2D usage environments with grouped multiple targets.

Based on these results, we suggest two alternative guidelines in designing target expansion schemes for target acquisition tasks with a computer mouse. First, when designing a GUI to expand a target, it is desirable to expand only the target that the cursor rests on, and leave adjacent targets unchanged. Second, when designing a GUI to simultaneously expand specific targets grouped by function, it is desirable to push surrounding groups aside rather than to occlude them.

The target expansion designs tested may need several refinements to be available for real implementation in various PC environments. If this challenge can be met by appropriate modification, the usability of related software or applications can be improved more than the results of this study, and the benefits may be greater when performing more difficult tasks with smaller targets or longer movement distances than in the tasks of this study.

REFERENCES

Blanch, R., Guiard, Y. and Beaudouin-Lafon, M. (2004) "Semantic Pointing: Improving Target Acquisition with Control-Display Ratio Adaptation", proceedings of the SIGCHI conference on Human factors in computing systems, Vienna, Austria, 519-526.

Cockburn, A. and Brock, P. (2006) "Human On-line Response to Visual and Motor Target Expansion", proceedings of Graphics Interface 2006, Quebec, Canada, 81-87.

Czaja, S.J. and Lee, C.C. (2002) "Designing computer systems for older adults", in: The HCI Handbook, Jacko, J. and Sears, A. (Ed.). pp. 413-427.

Fisk, A.D., Rogers, W.A., Charness, N., Czaja, S.J. and Sharit, J., ed. (2004). Designing for Older Adults: Principles and Creative Human Factors Approaches. CRC Press.

McGuffin, M. and Balakrishnan, R. (2002) "Acquisition of Expanding Targets", proceedings of the SIGCHI conference on Human factors in computing systems, Minneapolis, Minnesota, USA, 57-64.

McGuffin, M. and Balakrishnan, R. (2005). Fitts' Law and Expanding Targets: Experimental Studies and Designs for User Interfaces, ACM Transactions on Computer-Human Interaction Volume 12 No. 4, 388-422.

Nielsen, J. and Levy, J. (1994). Measuring Usability: Preference vs. Performance, Communications of the ACM Volume 37 No. 4, 66-75.

Smith, M.W., Sharit, J. and Czaja, S.J. (1999). Aging, Motor Control, and the Performance of Computer Mouse Tasks, Human Factors Volume 41 No. 3, 389-396.

Vercruyssen, M. (1997) "Movement control and speed of behavior", in: Handbook of Human Factors and the Older Adult, Fisk, A.D. and Rogers, W.A. (Ed.). pp. 55-86.

Walker, N., Millians, J. and Worden, A. (1996) "Mouse accelerations and performance of older computer users", proceedings of the Human Factors and Ergonomics Society 40th Annual Meeting, Philadelphia, Pennsylvania, USA, 151-154.

Zhai, S., Conversy, S., Beaudouin-Lafon, M. and Guiard, Y. (2003) "Human On-line Response to Target Expansion", proceedings of the SIGCHI conference on Human factors in computing systems, Ft. Lauderdale, Florida, USA, 177-184.

Chapter 81

Consumer Perception of Mobile Phone Attributes

Tao Zhang [1], Pei-Luen Patrick Rau [2], Jia Zhou [2]

[1] Department of Electrical Engineering and Computer Science
Vanderbilt University, TN 37211, USA

[2] Department of Industrial Engineering
Tsinghua University, Beijing, 100084, China

ABSTRACT

This paper reported results from a questionnaire survey of consumer perception of product attributes of mobile phones and how they consider the importance of each product attribute. The questionnaire contained 35 items covering brand, physical attributes, functional attributes, and beneficial attributes. All the items were measured with 7-point Likert scale anchored from "strongly disagree" to "strongly agree". The questionnaire was administrated online to a broad sample (N=215) in China. An exploratory factor analysis was conducted to find out the latent constructs of mobile phone product attributes, which also reflect consumers' common idea of attribute category. Eight factors, accounting for 62.88% of the total variation, were identified: common functions, appearance, multimedia functions, connectivity, personal information management functions, body design, brand & country, and product image. These attribute factors can be used as a basis for understanding of product attributes and its influencing factors. Results from the factor analysis also provide information about what consumers are concerned about when they evaluate mobile phones or make purchase decisions.

Keywords: Product Attributes, Perception, Mobile Phones

INTRODUCTION

The perception of various product attributes gives the first impression for consumers, particularly when they have limited time and bandwidth to do a thorough research on a new product. This perception of attributes also has strong relation with consumer judgment of whether the product will satisfy their needs. Studying the structure of product attributions based on consumer perceptions will help practitioners to understand consumers' acceptance of products. This is especially important for mobile phones as more and more technologies and functional designs have been integrated into new mobile phones. Consumer requirements of mobile phone are now not limited to the basic communication functions, but also on functions for other mobile applications, efficiency, ease of use and comfort. Unlike other interactive systems (e.g., desktop computing), consumer criteria of judging benefits expected from mobile phones include factors like function specification, aesthetics, brand image, self-feeling and user experience. These factors altogether add influence on the overall consumer perception. Therefore, the objective of this study was to provide a quantitative and structural assessment of consumer perceptions of product attributes of mobile phones.

LITERATURE REVIEW

PRODUCT ATTRIBUTES

Earlier studies such as Lancaster (1966) held that attributes are the objective physical aspects of a product. He said that people do not acquire products for the sake of the products themselves, but for the utility that is produced by characteristics of the products attributes. However, many of the product characteristics that are important from the point of view of consumers as well as designers are neither physical nor objective. Therefore, according to Grunet (1989), product attribute is "any aspect of the product itself or its use that can be used to compare product alternatives." "Each alternative can (but need not) be characterized by all attributes, that is, using one attribute does not preclude using another" (Nelson, 1970). Attributes may concern concrete product properties, practical consequences the product and its use and possession may cause, or consequences related to consumers' personal values.

Product Attributes are requirement factors placed on a product from a consumer. Every product item is viewed as a combination of correlative product attributes by the consumer. The designed functions of the product, together with the appearance, price, brand, package and after-sale service, are all parts of product attributes. "Attribute is said to be important if a change in the individual's perception of that product attribute leads to a change in the attitude toward the product." (Jaccard, Brinberg, & Ackerman, 1986). Attribute importance is characterized by the salience and determinacy of an attribute. Salience refers to the accessibility of the attribute, and determinacy to the correlation between an attribute and overall preference.

Lefkoff-Hagius and Mason (1990) created a synthesis from a number of earlier studies and divided product attributes into the three categories: characteristics attributes, beneficial attributes and image attributes. Characteristics attributes are related to the physical properties of a product; beneficial attributes refer to benefits or risks that the product may cause; and image attributes are properties of the product that have an ability to define the product owner's relation to other people or self.

INFLUENCE OF PRODUCT ATTRIBUTES

Consumer perception of product attributes can influence the formation and change of attitude, thereby impose its influence on buying behavior. Kotler and Armstrong (2005) said that consumers regard product as the combination of product attributes which are capable to satisfy their interests or requirements. Consumers make evaluation and choice according to certain rules regarding the performance of product attributes, and the importance they put on every attribute. Payne (1976) suggested that consumers attach different importance on product attributes. In certain occasions, only one single product attribute can lead to the buying decision, while in some other occasions there is a complex process when they are making any decision based on product attributes. Bahn et al. (2007) studied how feel of material, elasticity, tactile oneness, shape, and color of the passenger car crash pad influences satisfaction. The results indicated that softness of material was the key affective response factor of satisfaction. Product attributes also influence affect. Seva et al. (2007) found strong relationship between attributes of mobile phone and pre-purchase affect. For example, slimmer phones increase feeling of contentment and encouragement, larger display increase feelings of amazement and encouragement.

Fishbein (1975) proposed the Multi-attribute Attitude Model, in which he stated that the attitude towards certain products will be reflected by the evaluation of important attributes, then attitude can be assessed by measuring the evaluation of these attributes:

$$A_o = \sum_{i=1}^{n} b_i e_i \text{ , where}$$

A_o = Attitude towards the product

b_i = The strength of belief on product attribute i

e_i = The performance rating of product attribute i

n = Number of important product attributes

METHODOLOGY

ITEMS OF PRODUCT ATTRIBUTES GENERATION

Based on examination of prior work on similar products and analysis of the characteristics of mobile phones, an initial pool of items was developed. Items about function specification were based on the official websites of major mobile phone manufacturers. More items were developed by the author to cover all the three classifications of product attributes (characteristics attributes, beneficial attributes and image attributes). All the items were measured with 7-point Likert scale anchored at "strongly disagree" and "strongly agree". Two criteria were used to determine which product attribute information will be presented in the questionnaire: the respondent should understand what the specified attribute is; and the attribute should be relatively important to consumers during their assessment of perceived usability.

Initial items of product attributes were firstly examined by the authors according to the correlation between items, and then critiques of items were sought from a focus group discussion. Three Ph.D. students and two master students majoring in Human-Computer Interaction (HCI) took part in the focus group discussion. They were all familiar with the research topic and questionnaire design. The aim of focus group discussion was to check if there were ambiguities in the wording of items, if there were important missing items regarding mobile phone products, and if there were any inconsistencies in the arrangement of items and overall questionnaire design. Several items were adjusted and wording modifications were made from the critiques. Two master students majoring in Social Science and one master student majoring in Economics were then asked to examine the adjusted questionnaire. The focus of their work was to check if there was overlap between items, if there were suggestive wordings in the item descriptions, and if there were questions that may exceed respondents' knowledge and proficiency. The author took their feedback and incorporated it in the final revision of the questionnaire. There were 35 items in the final version of the questionnaire. These items were categorized into: brand, physical attributes, functional attributes, and beneficial attributes, which were similar to categories in the study by Lefkoff-Hagius and Mason (1990).

DATA COLLECTION

The questionnaire was administrated online to a broad sample in China including undergraduate and graduate students in several universities in Beijing, employees working in telecommunication industry, manufacturing industry, consulting company, and the government. A total of 215 responses were collected. As shown in Table 1, although more than half of the respondents are college students, the sample represents a wide range of mobile phone consumers with different age, education level, and income. The majority of respondents are young people with high level of education and middle level of income in China. The questionnaire began with an introduction of the survey objectives, followed by a survey of demographic information, mobile phone usage and experience. Then the list of 35 items was given. Respondents were asked to choose the extent to which they agree with the descriptive items for the mobile phone product attributes. They were encouraged to put down their email addresses to get a copy of the result analysis.

Table 1 Demographic profile of questionnaire respondents

VARIABLE	CATEGORY	NUMBER OF RESPONSES	PERCENTAGE
Age	<20 years	5	2.3%
	20-25 years	172	80%
	26-30 years	35	16.3%
	>30 years	3	1.4%
Gender	Female	82	38.1%
	Male	133	61.9%
Education Level	High school	3	1.4%
	Associate degree	6	2.8%
	4 years undergraduate	90	41.9%
	Master student	100	46.5%
	Doctorial student	16	7.4%
Occupation	Government official /Company manager	7	3.3%
	Technician/Engineer	42	19.5%
	Business/Service employee	10	4.7%
	Student	136	63.3%
	Others	20	9.4%
Monthly Income	<1000 RMB	140	65.1%
	1000-5000 RMB	52	24.2%
	5000 - 1000 RMB	21	9.8%
	>10000 RMB	2	0.9%

RESULTS

Exploratory factor analysis is a method for reducing the dimensionality of multivariate data and understanding patterns of association among variables, with the underlying common factor model. The model assumes that the observed variance in each measure is attributable to a relatively small number of common factors and a single specific factor. The objective of exploratory factor analysis is to identify the common factors and explain their relationship to the observed data, and the factor solution is derived from the patterns of association in the observations

(Lattin, Carroll, & Green, 2002).

An exploratory factor analysis was conducted in SAS to determine the "latent traits" of the items of product attributes. Principal axis factoring approach and the varimax rotation method was used. After deleting one item causing almost equal loadings on two common factors, the result shows that the first eight common factors account for 62.88% of the total variation in data. Items were then assigned to the common factors, given the criteria that factor loadings of items should be higher than 0.5. The result of exploratory factor analysis is shown in Table 2.

Table 2 Internal consistencies, means, standard deviations and factor loadings of attribute items

FACTORS	ITEMS	MEAN	SD	FACTOR LOADINGS
Common functions (α = 0.89, Variance accounted for = 4.27)	I believe input method is an important feature for me to judge whether the mobile phone is easy to use.	5.87	1.263	0.78
	I think the short message functions (storage and manipulation) are important for me.	5.97	1.199	0.82
	I think the phonebook functions (storage and manipulation) are important to me.	5.95	1.274	0.85
	I think the battery time is important.	6.04	1.087	0.77
Appearance (α = 0.79, Variance accounted for = 3.46)	I will be attracted by the appearance design of certain mobile phone and then want to own it.	5.01	1.448	0.75
	If I like the appearance of certain mobile phone, I will be very interested in it.	4.92	1.450	0.72
	I think the body color is important to me.	5.04	1.395	0.73
	I like the mobile phone appearance with special material.	4.85	1.370	0.65
	I think it is a great pleasure to show the special appearance design of mobile phone to others.	4.61	1.549	0.55
Connectivity (α = 0.70, Variance accounted for = 2.44)	I think the GRRS connection is important.	3.93	1.722	0.55
	I will consider the connectivity (USB/Infrared/Bluetooth/Wifi) when choosing mobile phone.	4.98	1.607	0.75
	I will consider the accessories available when choosing mobile phone.	5.08	1.450	0.57
Body design (Variance accounted	I think one particular body design in more suitable for me.	2.44	1.281	-0.97
	I think the body design should not be	5.56	1.281	0.97

for =2.15)	considered to be important.			
PIM Functions (α = 0.67, Variance accounted for =2.11)	I will ask about the capability to deal with events like temporary note/number taking when buying a mobile phone.	5.07	1.600	0.63
	I think the date planning functions like notepad, to do list, memo are important to me.	5.26	1.436	0.59
	I think the voice recording function is important to me.	4.08	1.488	0.70
Brand & Country (α = 0.69, Variance accounted for =1.92)	I will only use certain brands of mobile phone.	5.13	1.657	0.65
	I will choose mobile phones from certain countries.	3.43	1.625	0.75
	I will exclude mobile phones from certain countries.	4.27	1.939	0.57
Product image (Variance accounted for =1.61)	I believe the mobile phone design partly represents my life style.	4.89	1.457	0.71

CONCLUSIONS

There is no precise definition of product attributes for mobile phones, as consumers generally perceive product attributes in a conceptual way. The notion of product attributes is formed and existing during the perception process. Previous studies on mobile phones mainly considered individual attributes, without a systematic view of how all the attributes influence perception of mobile phones as a whole. This study fills this gap. The exploratory factor analysis was appropriate to find out the latent constructs of mobile phone product attributes, which also reflect consumers' common idea of attribute categories. Eight factors were identified: common functions, appearance, multimedia functions, connectivity, personal information management functions, body design, brand & country, and product image, which accounted for 62.88% of the total variation. Important factors such common functions and appearance are identified. Since people use mobile phones for daily communications, they stress the importance of phonebook and SMS functions. Ease of use of input methods also influences their perception of common functions. The quantitative approach used in the present study can be beneficial to similar studies of consumer digital products. Results of this study can be smoothly implemented into design and marketing practice as a perceptive model. Limited resources can be efficiently allocated to important attributes.

762

REFERENCES

Bahn, S., Lee, C., Lee, J. H., and Yun, M. H. (2007). "A statistical model of relationship between affective responses and product design attributes for capturing user needs." Usability and Internationalization, Pt 2, Proceedings - Global And Local User Interfaces, 4560, 305-313.

Fishbein, M., and Ajzen, I. (1975). Belief, Attitude, Intention, and Behavior: An Introduction to Theory and Research. Reading. MA: Addison-Wesley.

Grunet, K. G. (1989). "Attributes, Attribute Value and Their Characteristics: A Unifying Approach and An Example Involving A Complex Household Investment." Journal of Economic Psychology, 10, 229-251.

Jaccard, J., Brinberg, D., and Ackerman, L. J. (1986). "Assessing Attribute Importance: A Comparison of Six Methods." Journal of Consumer Research, 12(4), 463-468.

Kotler, P., and Armstrong, G. (2005). Principles of Marketing (11 ed.). Prentice Hall.

Lancaster, K. J. (1966). "A New Approach to Consumer Theory." Journal of Political Economy, 74(2), 132-157.

Lefkoff-Hagius, R., and Mason, C. H. (1990). "The Role of Tangible and Intangible Attributes in Similarity and Preference Judgments". In M. E. Goldberg (Ed.), Advances in Consumer Research (Vol. 17, pp. 135-143). Provo, UT: Association for Consumer Research.

Nelson, P. (1970). "Information and Consumer Behavior." Journal of Political Economy, 78(2), 311-329.

Payne, J. W. (1976). "Task Complexity and Contingent Processing in Decision Making: An Information Search and Protocal Analysis." Organizational Behavior and Human Performance, 16, 366-387.

Seva, R. R., Duh, H. B. L., and Helander, M. G. (2007). "The marketing implications of affective product design. " Applied Ergonomics, 38(6), 723-731

Chapter 82

The Human Factor in Lean Software Development

Andrzej Borucki,Leszek Pacholski

Institute of Management Engineering
Poznan University of Technology, Poland

ABSTRACT

The need for quick adjustments of management software to perform current business tasks is a major challenge for IT system designers.The authors have examined 200 projects carried out from 1990 to 2008 and involving the design, deployment and maintenance of business applications. They have implemented the projects in over 30enterprises. Our research focused on defining the extent and form of use of existing IT architecture components with a view to achieving an enterprise's business objectives. Project team leaders prefer the so-called light customization methods:whose design is human-factor-oriented as they allow for software engineer creativity.

Keywords: Project Management,Software Engineering, Requirement Engineering, Lean Software Development.

INTRODUCTION

In an attempt to boost the productivity of project teams, their managers commonly focus on the application of various management support techniques, such as the CASE, which, by definition, are designed to enable project teams to build complex IT systems that are ever more functional, reliable, effective and maintainable and mobile more efficiently. The use of such techniques often leads to the unification of software development procedures which, according to the authors, along with the automation of the generation of source codes based on logic model diagrams for specific IT systems, effectively deprives software design of creativity(Cushman,Rosenberg 1991),(Stallman,Greene 2005).

Having observed many projects, the authors sees the main cause of the problem in the failure to appreciate the intellectual and psychological potential of the human factor.

The experience gathered in the course of many IT projects ranging from the simplest to the very complex, shows that the key adverse factors affecting the productivity of project teams are(Borucki 2008;De Marco2002):

- The stress that accompanies designers in the development and implementation of IT projects. Such stress results mainly from their tremendous liability for any errors in software operation that may be discovered after deployment.
- Poor use of the intellectual capacities of individual project team members.
- Excessive focus on design support tools at the expense of project quality.
- Excessive adherence to a specific method of IT project management.
- The pressures of time and budget restrains on projects .

According to many IT project managers, the treating of people as "spare parts" often jeopardizes project outcome. Managers commonly forget the simple fact that the intellect and personalities of every employee are different and that often neglected in accomplishing project goals by not helping individual staff members to achieve personal satisfaction. In seeking to complete project goals, meet deadlines (especially where clients impose overly short deadlines for the drafting of documentation), managers frequently conclude that good and effective management boils down to persuading personnel to make a greater effort, even at the expense of their leisure time.

A focus on the technical aspects of project management, particularly by applying standard operating procedures, keeps managers from devoting the proper amount of time to think about who best to assign to a given project task. Having observed the performance of various designers I have worked with for over two decades, I conclude that, with only a few exceptions, the majority of working time is devoted to the completion of project tasks rather than finding the best technical solution.

I similar trend can be seen in the selection of IT system design methods. System designers frequently spend time to choose a design method for a specific project. My practice shows that the choice of a design method has little effect on design productivity and often reduces it where the system needs continuous customization to ever new client requirements. The traditional IT project management methodologies(Katonye,Sommervile 1998),(Robertson 1999):

- Turn out to be inadequate to deal with various operating requirements of clients in an environment that has become largely e-business based.
- Restrict the operating freedom on individual designers contributing to less innovative outcomes.
- Extent project lead times.
- Do not lend easily to change making it more of a challenge to develop successive software versions.

ERGONOMIC ASPECTS OF SOFTWARE ENGINEERING

To successfully develop ever more complex software while avoiding the above adverse effects of applying set software design methods, project team managers have taken to using the so called "lean" methodologies .

Such methods tie project success mainly to effectively employing the creativity of all designers and to ensuring creative cooperation with clients through e.g. joint steering committees. As a consequence, the proposed project solutions are easier to accept by the client who feels it has contributed to their development(Laplant 2009).

In all such methods, emphasis is placed on giving considerable leeway to programmers with a focus on accomplishing the top priority goal of satisfying the client by providing it with valuable software

How can this goal be achieved?

By ensuring adherence to certain principles governing the work of project teams and their cooperation with clients.

In my view, the most important principles of cooperation within project teams are:

- Osmotic communication allowing for the transfers of information among members are to be provided with all project information regardless of whether they are directly concerned. Specific information may always come in handy, project team members based on equal access to information. Design team being aware of solutions to problems that do not concern one directly may well help in dealing with other project barriers.

- Being encouraged to freely volunteer one's views on how to approach a project. The comments may concern technical issues or poor project management.with respect to one individual or the entire project team.

- The ability to turn failure to success. The principle will only work if the causes of failure are known (rather than hidden), i.e. if a list is openly compiled of what has been done poorly and well. Since the design process is always repetitive, snags may prompt the development of successive versions that do meet client requirements.
- Awarding courage and assertiveness in designers and promoting software simplicity.

The key principles of cooperating with clients to develop successful software are as follows(Highsmith 2007):

- The priority for all design team members is to satisfy the client by providing it with valuable software that facilitates its work.
- Approaching the client as a partner, not only in business but also in the technical aspects of programming.
- Frequently submitting software fragments to the end user. This allows prompt verifications of models against client requirements, gives the designer team satisfaction with having correctly developed a part of the system while affording the client a chance to check progress.
- Acceptance of changes in client requirements, even if submitted late in the project.

- The client's measure of progress is to see a working piece of software.

The best designer teams are made up of representatives of both the client and the design company. By working hand in hand, client and designer staff find it easier to develop adequate requirements which in turn speeds up system deployment. Client staff end up identifying with the software development project. This also facilitates designer maintenance after implementation

USE OF CUSTOMIZATION TO THE SOFTWARE DEVELOPMENT METHODS

The best software development method is selected by a number of criteria, the most critical ones being project size, project complexity and the degree of involvement of the future user in software design(Borucki 2008;Barenbach 2009).

Project team leaders prefer the so-called light customization methods:
- whose design is human-factor-oriented as they allow for software engineer creativity,
- which are more adaptive then predictive meaning that IT designers go through a *system life-spiral* in the iteration process as they develop successive prototypes and then verify them against updated requirements, as shown in Figure 1,
- which are non-linear meaning that the ultimate structure of a programming project may be defined only partially at project launch and fine-tuned through the feedback loops at successive stages of development.

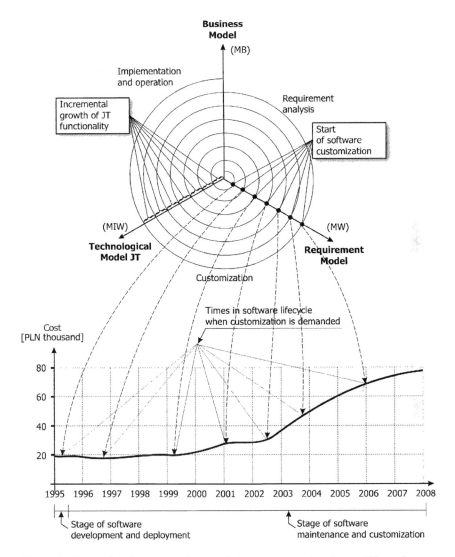

Figure 1. Interaction between software design stages and software life-cycle customization

Our research distinctly demonstrates that the lifecycle of software used by a given client ties closely to designer flexibility and response time in the process of software development. The research findings are partially shown in Figure 1, which presents the software lifecycle at times of software customization. Note how IT infrastructure (hardware and software) gain value from the moment customization projects commence.

In design practice, IT systems are customized by incremental development given in Figure 2.

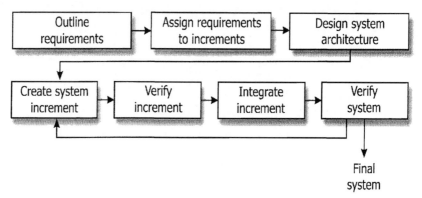

Figure 2. Incremental development.

Incremental development of software is a helpful customization tool as it:
- Helps involve the client in every design stage,
- Allows clients to defer certain software functionality decisions (at least for certain modules) until they gain fuller knowledge and experience from testing the outcome of another development increment.

As indicated by its name, the incremental development method entails design by successive "increments", each of which may be seen as a separate project. In the initial phase of software development, the client shows the designers which functional specifications to incorporate into a given increments. Each increment follows the complete process of cascade development made up of needs definition, design, deployment, testing, integration and tests on the aggregate of all increments. Each increment should deliver an orderly version of the functional requirements previously specified by the future user who should also indicate which functionalities are to be given priority. The critical functional specifications should be delivered in the initial increments. The designer's task here is to work closely with clients helping them to define such increments.

Software customization by incremental development is the most effective approach that guarantees a significantly longer software life cycle. Our studies show that this brand of customization may help extend software life by over a dozen years..customization by incremental development is given in Figure 3.

By the incremental development method, the clients receive, early in the process, the portion of the system they care most about so they can test it to learn about the system, describe their satisfaction with the designer's approach to services and functions and define the next increment, i.e. another list of system functionalities to be delivered in that increment.

When defining the next increment, one should make certain it is not too large, i.e. for the number of functional items not to exceed 500 to 700. Otherwise the future users would find it unmanageable and too much of a challenge to integrate with successive increments.

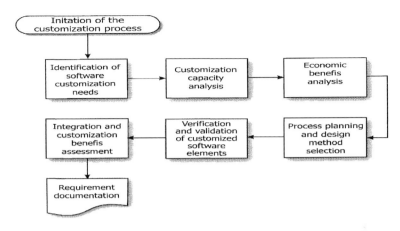

Figure 3. Customization by incremental development.

To ensure consistency and the integration of each increment, it is essential to draw up a preliminary system architecture model made up of the business, requirements and deployment modules. IT architecture model is given in Figure 4. The business model is a central component of any corporate IT architecture. It contains a verbal description of IT structure and the critical economic factors used by all businesses to define their long-term survival strategies(Afuah,Tucci 2003). Software customization is a process in which existing business software is adjusted to meet current company management needs in both the long and the short-term. The underlying purpose is to adjust management support applications to grant managers easier access to software functions that guarantee quick and accurate operational, tactical and strategic decisions to secure a lasting competitive advantage in a given market segment.

The need for quick adjustments of management software to perform current business tasks is a major challenge for IT system designers. The place of customization processes in the software lifecycle is shown in Figure 1.

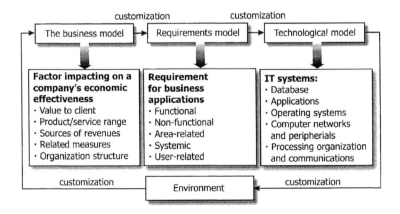

Figure 4. Evolution of the corporate IT architecture model resulting from customization.

The preliminary system model should represent a specific system abstraction level to allow the future user and designer to define individual increments, verify the desired functional specifications and assess the capacity to complete the project within the allocated budget (including spending on the technical infrastructure required for software deployment).

Additionally, software customization by the increment method:

1. allows the clients to benefit from the deployments of successive software increments before the entire system is complete; the effect is that clients gain early experience in software operation and acquire the necessary knowledge (including business knowledge) of the functional requirements needed in subsequent software increments,
2. makes it possible to alleviate the risk of an overall project failure and realistically plan spending.

CONCLUSIONS

In all of the analyzed cases, software customization benefited software users, designers and customizers.

The benefits most valued by clients were:

1. The chance to avoid software replacement which would be highly costly and disturbing to their organizations.
2. A much cheaper way to acquire new software functionalities.
3. Shorter time to deployment, shorter time for staff to learn new software functionalities.
4. Minimal insecurity, reluctance and fear shown by staff in recognizing and learning new software.
5. Shorter time to achieving full functionality in the modernized system.
6. Easier software integration with existing IT infrastructure (customization often helps avoid the replacement of hardware and operating system).
7. Facilitated integration with external IT environments (of clients and suppliers).

The benefits of software customization most treasured by software designer companies are:

1. Significantly longer life of customized software resulting in longer service agreements, new design contracts and more stable income from software functionality development in the long run.
2. Stronger client attachment to the software solidifying a good partner relationship between software user and design team.

REFERENCES

Afuah A.,Tucci Ch.L.(2003),Biznes Internetowy. Strategie i modele.Kraków, Oficyna Ekonomiczna.

Berenbach B., Paulish D., Katzmeir J., Rodorfer A.,(2009)Software and Systems Requirements Engineering : In Practice. New York: MCGraw-Hill Professional. .

Borucki A., (2008)Use of Customization to Enounce the Ergonomic Qualities of Software, Monograph red.L.M.Pacholski, J.M.Marcinkowski, W.M.Horst, Employee Wellness Ergonomics and Occupation Safety, Poznań.

Cushman W.H.,Rosenberg D.J., (1991)Human Factors in Product Design, Elsevier, Amsterdam.

DE Marco.,ListerT.,(2002),Czynnik ludzki,skuteczne przedsięwzięcie i wydajne zespoły.WNT,Warszawa.

Highsmith J., (2007),APM:Agile Project Management, PWN..

Kotonye G., Sommervile J.,(1998) Requirement Engineering:Process and Techinques, John Wiley and Sous.

Laplante P.,(2009),Requirement Engineering for Software and Systems(1st.ed) Redmond, WA:CRC Press..

Robertson S., Robertson J.,(1999)Mastering the Requirements Process, Addison Wesley Longman.

Stallman A., Greene J.,(2005) Applied Software Project Management. Cambridge, MA: O'Reilly Media.

Chapter 83

Ecological Aspects of Macroergonomics

Stanisław Janik, Dominik Grygiel

Institute of Management Engineering
Faculty of Computing Science and Management
Poznan University of Technology
Poznan, Poland

ABSTRACT

The problems of organization and practice of social relationships were analyzed in terms of macroergonomics. Trends were examined in terms of assessing the impact of technological processes and materials on the environment. It was found that the issue of macroergonomics should include environmental aspects in the context of the new achievements of technology and trends in the assessment of technical progress in terms of impact on humans and the environment. The opinions of experts, facilities and scientists from the Committee of Geological Sciences were given. They state that there is no evidence that human activity has any impact on global warming. (www.kngeol.pan.pl). These remarks should be treated in the form of discussion.

The issues were discussed based on examples in the field of mechanical engineering technology.

Keywords: ergonomics, ecology, environmental protection, macroergonomics, the method of calculation, ecological management, social issues, manufacturing, ecological assessment of the manufacturing process, materials, optimization in terms of ecology

INTRODUCTION

Macroergonomics focuses on a various, seemingly not equated with each other, areas of science. The rapid development generates new relationships between different areas of activity.

Analyzing the problems of management and solutions to social problems we find their sources in the activities related to current trends, resulting from the transformation of relations between technology and welfare. Especially evident are social issues on the background of the relations with phenomena in different fields. Welfare being dependent on many factors such as: economic situations, state of mind, security, etc. is in practic function, composed of environmental conditions.

Protection of the environment referred to as the ecology, significantly determines the welfare of society. Taking into account the guidelines and recommendations international institutions, it must be noted that both the ergonomics and the environmental protection are complementary to each other and together with the occupational safety and health, have a huge impact on the health, well-being and welfare of both worker and society. With huge progress (since the 19th century), both in the fields of technology and economy, satisfying meals, there is a problem of choice of not only the materials, but also technology of obtaining and processing the products.

History of technique is the search for materials and technology to fill basic human needs. All at the same time phenomena such as: demography, pollution, etc. are characterized by exponential growth curve and is now approaching asymptotically to a diametrically high results. Inhibition of increase of pollution to our planet has become the main target.

PRESERVATION OF THE ENVIRONMENT AS THE PRIMARY GOAL

Both economic and political strivings of societies (countries) with different aspirations can not guarantee the procedure to preserve the environment. The only possible solution is a joint action under the aegis of the economically dominant societies (countries) in order to protect the environment. All of transnational organizations are involved and the problem is becoming a huge economic imparities. Uneven economic development is the most important obstacle to the preservation of the environment. Military actions in the different places on Earth, blasting (in a decisive majority) in poor and underdeveloped countries, exclude environmentally friendly perceptions of the environment. These countries are unable to make their own efforts to improve the environment. What is more frequent disasters occurring in their areas aggravate unfavorable macroergonomic conditions. At the same time they generally do not employ: monitoring, forecasting, prevention and safety evaluations of the situation in advance. The absence of work, standard of health protection, access to sanitations and instability has a negative impact on social behavior. Pro-ecological policy under guidance of richer countries, going by concern for the environment and not only politico-economic influences, may create a lot of jobs, can lead to awareness through education on environmental protection, reduce the gap separating developing countries from the rest of the world and to stabilize conflict situations. Strict compliance with the principle that it is easier to prevent environmental degradation than to repair damage caused by careless actions should be the primary goal of all the inhabitants of Earth.

INDUSTRIAL AND AGRICULTURAL OPERATION

It is assumed that industry and transport are the main polluters to the environment. Industrial production is characterized by diverse technologies. The use of selected technologies is dependent on modern facilities, the company's financial condition, pressure created by economic partners, demand for products, kinds of raw materials that are used and the perception of the environment. Recent research indicates other important causes of pollution that have so far been ignored. One of them is the production of food for the growing population. All these factors are within the scope of macroergonomics in the broad sense of understanding. In this regard, the choice of used technologies and materials is influenced by a hierarchy of criteria.

The selection criteria are very different and have evolved over the years. Changes and the formation of the new criteria will affect the economic situation, condition, exploring new areas of technology and solutions for the management of industrial and food economy, forcing situation (threat to the environment), groundbreaking inventions, excess or underflow of a particular media, trend and many other factors. For many years, a new leading group is the ecological criteria of environmental protection. It is associated with the growing pollution of the environment and scarcing natural resources. Only plants with a closed production cycle - institutions safe for the environment - can co-exist [Gabryelewicz 2007]. Figure 1. shows a closed production cycle - a cycle that is safe for the environment. [Janik, Grygiel]

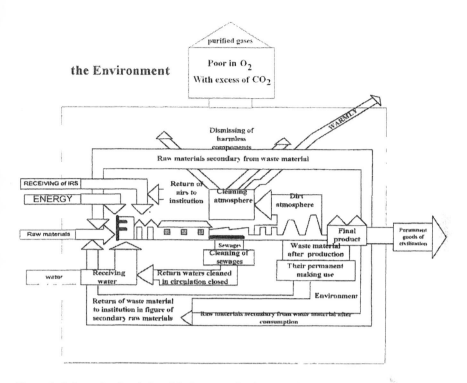

Figure 1. Schematic of an industrial plant completely secured

The concept of environmental selection has to be considered in a comprehensive approach, i.e. beginning on the extraction of raw material, by processing it, production (manufacturing), installation, operating, till destroying the product. The impact of product on the environment is the sum of interactions:

- the process of acquiring raw materials (energy, environmental degradation, gaseous pollutants, waste, sewage, etc.)
- recovery process resulting from environmental changes to the eco-friendly state - what should be the responsibility
- processing raw materials into semi-finished products (energy, gas, waste, sewage
- manufacture of the product (energy, gas, waste, sewage) and cleaning any contamination that arose during the manufacturing process
- use of the product (energy, waste, the media associated with them) - must be conducted in a manner least harmful to the environment
- destruction of used product - the only desirable and correct form is recycling. The main message should be shrinking stock of raw materials and components elimination process (obtaining raw materials) to create a new product which has a direct impact on reducing environmental pollution - rating economic effects higher than recycling is acting to the detriment of humanity. As an example, glass waste and other relatively

low-cost packaging, which transport is economically unprofitable. There is no problem with the alloys, the scrap is relatively expensive and is almost entirely recyclable

Currently, most products can be made from different materials and technologies. The deal to be pursued is storing only bio-degradable waste.

Selective management of municipal waste is a complex process and should seek to improve it. Moreover, in Poland, in some regions there has been tremendous progress in this area.

But there is no excuse for plants irrespective of their profile derogation (even partially) from the conduct of recycling and selective waste management. There is practically no such waste or waste materials which can not be re-used in the technological process [Janik, Grygiel].

Presented here responsibilities show how many jobs (despite the automation of certain processes) can be created with the proper treatment of the problem. It becomes clear that environmental degradation can be reduced by up to a minimum by the use of new technologies and new materials. In general, every thing can be prepared in various ways and new solutions in these areas have emerged. Applied treatment: traditional particle board, contactless (Electro), hybrid, laser, layered hardening and holographic solidification of the liquid.

Environmental contamination by food production (CO_2, methane) can also be reduced by changes in technology, changes in diets and varieties of plants and animals. Absorptivity of CO_2 can be increased by the enforcement of modified plants and trees.

Any activity is related to the use of energy. We are getting it for the whole of non-renewable resources, where stocks are shrinking. This situation, with the ever-growing demand for energy, makes us look for solutions among renewable energy sources. They are wonderful, but their disadvantage is that their power is insufficient to meet growing demand. In addition to energy of water, where tidal power can provide efficiency, wind power stations (windmills), even as farms have no guarantee the wind. Biogas and biofuels need not secure, hence the attempt to secure a variety of ways (worldwide projects). In practice, all energy on our planet is the product of the energy received from the sun. Trend is therefore to use it more. The Figure 2 shows (graphically) the cover of life on Earth by solar energy.

In addition, conventional power stations and internal combustion engines emit both CO_2 and other gaseous pollutants degrading the environment. In this regard, the electricity from conventional power stations can not be regarded as environmentally friendly. In today's technology the only solution are small and safe nuclear power plants. Mankind should not despair because of the depletion of traditional energy. Stocks are sufficient for at least 30 - 50 years. In this era of rapid technological progress during this period people will discover other sources of energy (unknown today or during the preliminary study).

The obligation to comply with the requirements of environmental protection is also implemented in the form of various obligations by all types of organizations (national, global or regional). Used by them system of penalties (if it is not economic policy) is highly indicated. Morally problematic to the planet is a system of resale limits. Earth would be more advantageous for a system of a sustainable progress. Accumulation of environmental impacts (only in such an approach) may be the assessment of both material and technology [Zarzycki, 2007; Janik, 2008].

Both heat and carbon dioxide, regardless of their use and minimization, are spreading into the environment. It is true image and at this stage are carried out only experience of changing circumstances. Industry and running technologies are considered as main source of pollution, this particularly concerns gases called „greenhouse gases". Contrary to the generally prevailing common opinion, the food we eat has a greater impact on the making of so-called „greenhouse gases" than our cars.

This information is not incitement to abandon the food or the achievements of technology, but solutions in the field of ecology. Objective picture may arise when we will consider statistics. Currently prevailing trend is to drop the overall responsibility for global warming on greenhouse gases, especially CO_2. Data on specific materials and technologies, both in technical fields such as economy and food industry(and other areas) could be signpost for example, reducing CO_2 production per year. [Janik, 2005]

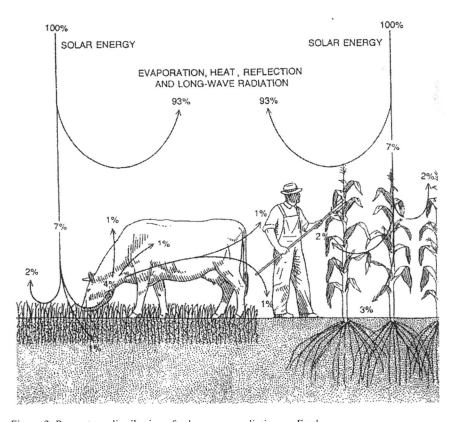

Figure 2. Percentage distribution of solar energy radiating on Earth.

Currently, researches of carbon footprint are carried out (the amount of CO_2 formed, which allowed the emergence of the product) for food products, and more recently for everyday products (shoes, etc.) are carried on. It can be assumed that the carbon footprint of the product (which includes the production of carbon footprint, carbon footprint of raw materials, transport and others) have not deterred customers from products with higher carbon footprint. These studies show that what previously was marginal is becoming an issue important to preserve the environment. People does not take into account the global water pollution, cluttering the nature, the landscape changes caused by bombardments, heat emitted by the conversion of energy (production processes), heat as a result of air pollution is not being reflected from the planet and others. The doctoral thesis "Analysis of manufacturing technology products with the Fe-C alloys in terms of environmental criteria" by Gabrylewicz, I. has brought in technology selection criteria for environmental protection. Environmental hazards were adopted there in the form of:

- energy consumption
- emission of dust into the atmosphere
- waste
- sewage

Table 1 Statement of environmental aspects for press wheel produced with use of various technologies (Gabryelewicz, 2007)

	Machining	Stamping	Molding	Steel sheets
Use of energy [MJ/unit]	339,979	106,774	122,23	241,03
Emission of CO_2 to the atmosphere [kg/unit]	25,441	10,582	6,374	7,3227
Waste [kg/unit]	26,2998	9,6143	7,255	7,0684
Sewage [dm³/unit]	165,542	68,918	51,545	45,055

Not being able to determine the carbon footprint (at the previous time it was not the prime) the most environmentally harmful pollution from manufacturing processes were indicated. Pollutions can be removed using the most severe regimes using innovative solutions. This entails costs. Table 2. shows the conversions of the pollution and energy to bring environment to the state where it was before the impact of pollution. Manufacturing costs were not analyzed here. Environmental costs have been received as components of cost of production. Of course, carbon footprint in terms of energy intensity component is possible to determine. Costs are given in PLN (1 USD = 3 PLN).

The credibility of impact assessment of manufacturing technologies (machine parts based on the differential impact on the environment) is worthy of treatment as a way to improve the environment and is an issue that fall within the scope of macroergonomics.

We believe that the assessment of environmental pollution based on the increase of CO_2, and even the increase of extraction of methane, does not exhaust the subject. Evaluation, which may contribute to the improvement of the environment must take into account all types of pollution persisting for some time and adversely affecting the environment. This assessment must recognize the problem comprehensively. Finding the assessment including a various of sources and physical states of pollutions is temporarily unsolved.

Table 2 Environmental aspects calculated by financial units for various technologies of press wheel production (Gabryelewicz, 2007)

	Machining	Stamping	Molding	Machine cutting
Emission of CO_2 to the atmosphere [kg/unit]	0,025441t × 0,23PLN = 0,00585PLN	0,010582t × 0,23PLN = 0,00243PLN	0,006374t × 0,23PLN = 0,00146PLN	0,0073227t × 0,23PLN = 0,00168PLN
Waste [kg/unit]	0,0262998t × 15,71PLN = 0,413PLN	0,0096143t × 15,71PLN = 0,15PLN	0,007255t × 15,71PLN = 0,114PLN	0,0070684t × 15,71PLN = 0,111PLN
Sewage [dm3/unit]	0,166 m³ × 0,54PLN = 0,0896PLN	0,069m³ × 0,54PLN = 0,0373PLN	0,0515m³ × 0,54PLN = 0,0278PLN	0,045m³ × 0,54PLN = 0,0243PLN
Total:	0,797PLN	0,209PLN	0,782PLN	1,176PLN

ECOLOGICAL FOOTPRINT

It seems that determination and giving to the public information about the carbon footprint of the device is a very important issue. However, in the next step it must be determined by the ecological footprint. For example, it should be the sum of complex trace of independent contaminants - which could not be replaced by a single indicator of pollution. For example:

Eco-footprint = carbon footprint of the product + footprint of water pollution + footprint of pollutions from the residue after the annihilation of the product + footprint of environment pollution during the operation of the product

Eco-footprint must be defined comprehensively as well as carbon footprint. Eco-footprint of product would therefore be "a footprint of damage" to the environment created in the time when the product was manufactured.

Need for such solutions no longer raises doubts. The threat of environment is great and only decisive action can help to improve the situation. We should also find a compromise between the need to narrow specialization in the field and the necessity of interdisciplinary efforts to protect the environment.

It seems highly probable that in the near future products will have the environmental approvals, based on an assessment of environmental impacts throughout the entire lifecycle.

Previous attempt to identify the negative impact of the product in terms of a comprehensive environment are presented in the form of: one parameter, presentation of pollutants from the one physical state, demonstrating one type of pollution, energy intensity in the life cycle of the product, the cost of bringing the environment to the state where it was before the impact of pollution and other indicators.

Marking product with indicator of its negative impact on the environment in terms of complex (from manufacture to disposal) faces significant difficulties. There is no indicator to check the different states of the environmental impact in varying range. Contamination from the same physical state can be reduced with a close approximation to one representative. However, the effects of pollution, physical, chemical and biological, properties of the states of liquid, solid and gas and to the radiation, can not currently be replaced by one indicator.

The trend in recent years towards the prevention of environmental degradation, indicating the main cause (formerly the ozone hole) in the form of CO_2 seems to be the result of political and economic systems.

Calculates, made by experts (including American), many of the world's leading centers of research and expertise, made by scientist from the Committee of Geological Sciences, situated in Polish Academy of Sciences, concludes that "there is no evidence that human activity has an impact on global warming" (www.kngeol.pan.pl).

However, regardless of the main reasons (global warming and the concentration of CO_2 in the atmosphere) we can not underestimate the negative human activities and its impact on the discussed phenomena. We should accept that if the presented expertises were to confirm completeness, it certainly will impact environmental degradation as a result of human influences on global warming at a higher rate than the boundary of error. This impact will grow along with population growth and economic changes (progress in developing countries), and along with progress in technology. In this regard, causing a faster, adverse climate change is not in our interest.

CONCLUSION

Intrigues of politicians, state and business will allow a further fell into investments that will never be accomplished and will not satisfy the hopes placed in them. Promises they proclaim come from the realm of faith, so we can only believe and hope. Unreasonable investment in so-called „renewable energy sources" can bring more damage than good. The balance of CO_2 in the comprehensive approach (from construction to destruction), e.g. of wind turbines is greater than the nuclear power stations. Similarly, necessary for building materials whose manufacturing causes production of CO_2. Projects giant power plants will cause more trouble (whose existence is hidden, and costs for energy, even when the investment subsidy will be so large that only pressure will force customers to use it) than the utility for saving the environment and safeguarding welfare.

These remarks do not negate exploration of energy security, particularly greater use of solar energy, nor any obligation incumbent on every inhabitant of the globe to action to preserve the planet green for generations to come.

REFERENCES

Gabryelewicz, I. (2007) *Analiza technologii wytwarzania wyrobów ze stopów Fe-C w aspekcie kryteriów ekologicznych,* (Promoter: Janik, S.), Zielona Góra

Janik, S. (2008) *Macroergonomic aspects of ecological production* AHFRE International Conference, Las Vegas

Janik, S. and Gabryelewicz, I (2008) *Ocena środowiska procesu wytwórczego elementem układu ergonomicznego.* PTErg, Wrocław, 31 – 36

Janik, S. and Gabryelewicz, I (2008) *Procesy technologiczne – emisja zanieczyszczeń.* 9th International Scientific Conference: NEW WAYS IN MANUFACTURING TECHNOLOGIES, Presov 19-21.06.2008, 397 – 400

Janik, S. and Gabryelewicz, I (2008) *Ergonomiczne aspekty restrukturyzacji,* Opole

Janik, S. and Gabryelewicz, I (2009) *Macroergonomic Assessment of Foundry Production In Reference to Ecological Criteria. Monograph* ed. L.Pacholski Macroergonomics vs Social Ergonomics , Publishing House of Poznan Uniwersity of Technology, Poznań, IBSN 83–7143–841–7

Jasińska, E., .Janik, S. (2009) *The Macroergonomic Leader Surrounding, Monograph* ed. L.Pacholski Macroergonomics vs Social Ergonomics , Publishing House of Poznan University of Technology, Poznań, IBSN 83–7143–841–7

Jasiński, W., Janik S. (2009) *Ergonomic Aspects of Work Conditions in Clothing Manufacturing Monograph* ed. A.Górny, , G. Dahlke - The Formation of Safety in Environment and Space of the Mans Work , Publishing House of Poznan University of Technology, Poznań, IBSN 83–7143–844–8

Jasińska, E., Janik, S. (2009) *Aspekty ergonomiczne okręgu przemysłowego Zastosowania ergonomii,* Wrocław, ISSN 1898–8679

Gabryelewicz, I., Janik, S. (2009*) Problemy środowiskowe w kontekście zdrowia,* Wrocław, ISSN 1898–8679, ISBN 978–83–926630–1

Zarzycki, R., Imbierowicz, M., Stelmachowski, M. (2007) *Wprowadzenie do inżynierii i ochrony środowiska,*Warszawa, ISBN 978–83–204–3142–1

Janik, S., Grygiel, D. – own materials

Chapter 84

Ergonomics' Contributions to a Company's Innovation Strategy

Jan Dul

Rotterdam School of Management, Erasmus University
The Netherlands
E-mail: jdul@rsm.nl; Website: http://www.rsm.nl/jdul

ABSTRACT

Ergonomics can not only contribute to employee health and safety, but also to employee creativity, and therefore to a company's innovation strategy. Employees who work in a creativity stimulating work environment express more novel and useful ideas for the company's products, services, processes, work methods, etc. than employees who work in a less stimulating environment. This paper presents our current research in this new area for ergonomics: (1) The CDQS tool to assess a company's "climate for creativity" and (2) empirical evidence that a climate for creativity leads to more creative behaviour.

Keywords: employee creativity, social-organizational work environment, physical work environment

INTRODUCTION

In traditional productivity driven organizations, employees are placed in work environments with formal structures, time constraints, strict regulations, daily similar tasks, and standardised workplaces. The ergonomics field has contributed to designing such work environments by keeping health and safety risks under control. Although productivity driven work environments may support an organization's cost reduction strategy, in modern economies companies compete primarily on the basis their capability to innovate their products and production processes.

Employees can be a vital resource for an organization's innovation. Employees at any level in the organization can contribute to product and process innovation by producing novel and useful ideas for products, services, processes, systems, work methods, etc. Then, the questions arise which type of work environment is needed to support an organization's innovation strategy though employee creativity, and how the ergonomics field could contribute to such environment.

ASSESSING THE CLIMATE FOR CREATIVITY

Figure 1 presents a conceptual framework that examines the relationship between individual characteristics, organizational and physical work environments and employee creativity for innovation.

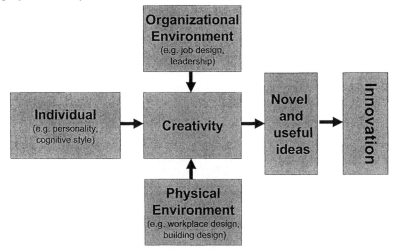

FIGURE 1 The effects of individual factors, organizational environment and physical environment on creativity and innovation (Adapted from Dul and Ceylan, 2006)

Based on this conceptual model we have developed the Creativity Development Quick Scan (CDQS) which is a tool (checklist) for measuring the work environment for creativity at individual, group and organizational level (Dul et al. 2007).

The checklist includes 21 characteristics of the employee's work environment that could foster or hinder creativity. Both social-organizational and physical characteristics are included. These characteristics relate to job design, leadership, interior design, and building design. The checklist has to be filled by the employee; for each characteristic the checklist measures on a 7-point scale the *provided* and *desired* work environment for creativity. Data can be benchmarked against data from more than 1300 employees from 6 countries (Netherlands, Japan, Australia, Brazil, Italy, and Turkey). If employees work in the same organisation the individual data can be aggregated to obtain the work environment for creativity of an organization. Presently data are available from more than 50 companies. The database shows that companies can differ considerably regarding the support they give to the creativity of their employees. CDQS can provide useful information for employees, managers and other stakeholders to start specific improvements.

One main value of the tool is that on the basis of data, discussions can be started with management and employees about the present situation, and possibilities for improvements, similarly to any other participatory ergonomics approach. Many problems and solutions regarding work environments for creativity, identified with this tool are very similar to the problems and solutions for work environments for health and safety. Hence, existing ergonomics knowledge could be readily applied. For other problems the required knowledge may not be readily available (yet) in the current ergonomics domain, although it may be available in other domains (organizational behaviour, business administration, environmental psychology, architecture, etc.). Yet for other problems further studies are needed to develop solutions.

EMPIRICAL RESULTS

A preliminary validation study with 379 knowledge workers from 31 small and medium sized business-to-business service organisations, which were providing financial, advertising, IT, and other services, showed that the work environment is related to employee creativity (Dul et al. 2009).

Figure 2 shows that if more support provided by the work environment (higher "creative climate") is associated with higher levels of creativity (higher level of "creative behavior"). Creative behavior also depends on the employee's desired level of support ("importance"). The highest levels of creativity are realized when the support for creativity that is *provided* by the work environment fits the support that is *desired* by the employee. This perfectly agrees with the ergonomics principle of "fitting the work to the man".

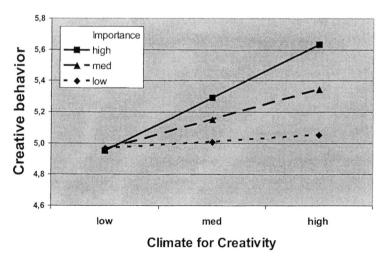

FIGURE 2 The effect of a creative work environment ("climate for creativity") on employee creativity ("creative behavior") (Adapted from Dul et al. 2009)

CONCLUSIONS

The results show that "ergonomics for creativity" is a potential new area of research and practice in ergonomics that can position ergonomics as a field that contributes to a company's innovation strategy, beyond its contributions to health and safety (Ceylan et al. 2008; Dul 2009; Dul and Neumann 2009).

REFERENCES

Ceylan, C., Dul, J. and Aytac, S. (2008), Can the office environment stimulate a manager's creativity? *Human Factors and Ergonomics in Manufacturing* 18 (6), 589-602

Dul, J. (2009), *Business ergonomics beyond health and safety: Work environments for employee productivity, creativity and innovation.* In: P.D. Bust. (ed.). Contemporary Ergonomics 2009, CRC Press, pp 16-23

Dul, J. and Ceylan, C. (2006), *Enhancing organizational creativity from an ergonomics perspective: The Creativity Development model.* Paper presented at the 16th World Congress on Ergonomics (IEA 2006), Maastricht, The Netherlands, July 10–14

Dul , J., Ceylan, C., and Hendriks, H. (2007), *A practical instrument to measure the creativity potential of the work environment.* In: Proceedings of the 10th European conference on Creativity and Innovation, Copenhagen, Denmark.

Dul, J, Ceylan, C. and Jaspers, F. (2009) *Environmental effects on creative behavior and the role of person-environment fit.* Paper presented at the 9th European Academy of Management conference EURAM 2009: Renaissance and Renewal in Management Studies, 10-14 May 2009, Liverpool, UK.

Dul, J. and Neumann, W. P. (2009), "Ergonomics Contributions to Company Strategies", Applied *Ergonomics* 40, 745–752.

Chapter 85

The Social Role of Ergonomics and Material Environment Design

Katarzyna Lis

Department of Labour and Social Policy
Poznań University of Economics
Al. Niepodległości 10, 61-875 Poznań, Poland

ABSTRACT

The aim of this article is to analyze the changes in the nature of work and the importance of ergonomic. With regard to these issues the paper has been divided into four parts. The introduction characterizes demographic changes occurring in society. The second part of the article describes change in the nature of work. In the third section presents the need for the development of material environment design for future generations. The conclusions of the abovementioned subject are in last part of the paper.

Keywords: Labour Market, Ergonomic

INTRODUCTION

Current demographic projections clearly indicate that the global ageing process is well under way and the world population will soon be faced with all its consequences. A recent report by DESA (UN Department of Economic and Social

Affairs) presents an overview of the projected changes in the entire human population over the next 40 years. According to the report, by 2050, the global population will have grown from 6,7 to 9,3 billion. By this time, the over-60s age group will number more than 2 billion and for the first time in history the elderly population will surpass the population of young children. The great majority of elderly people will be living in Asia and Europe (Papiernik, 2007).

In European countries, analyses of the ageing process are considered basic for further functioning of the European Union. According to evaluation studies conducted by the UN and EU (World Population Prospects and Eurostat), by 2050 the population of France will have declined from 60 to 55 million. It may seem only a slight decrease, but it is going to bring about a considerable decline in the number of the country's native citizens. A similar shift is expected in Great Britain, where by 2050 the population will have decreased from 60 to 53 million. Projections for the other European countries are even more disturbing. By 2050, the population of Germany (currently 82 million) will have declined by 21 million. In Italy, the population is expected to have decreased from 57 to 37 million. As for Spain, the projected decline is from 39 to 28 million by 2050 to 12 million by the end of the century. All these projections do not, however, allow for migrations over the next few decades. The projected decrease of the population of Central and Eastern Europe is even more disastrous: Ukraine: 43%; Bulgaria: 34%; Latvia and Lithuania: 25-27%; Russia: 22%; Croatia and Poland: 20%; Hungary: 18% and the Czech Republic: 17%. According to the same sources, by 2050, only Cyprus, Luxembourg, Malta and possibly Sweden will have positive population growth rates (Laqueur, 2008).

The projections show that women will comprise a significant majority of the elderly population because their life expectancy is greater than men's. On the other hand, since women retire earlier (at the age of 55-60) than men (65 for most countries), their share in the labour market will be substantially lower. The decreasing general share of older persons in the labour market within countries with high GDP is also an issue of growing concern. For most well-off regions, the ratio of professionally active men over 60 is 22% and 50% for developing regions. For women, the corresponding ratio is 11% and 19% respectively (Laqueur, 2008).

By 2050, the population of Poland will have declined by 20%. The number of people in their production age is currently growing and by 2050 will almost have doubled that of 2005. The population in their production age had been growing by 2009 and then started to decline. Now it is projected to have decreased by 15% by 2030. By 2050, it is expected to have declined by over one-third of the 2005 rate. The ratio of the entire Polish population in their post-production age is projected to have increased from 15,4% (2005) to nearly 27% by 2030 and over 28% by 2050. The ratio of Poles in their production age has increased from 64% in 2005 to approximately 65,1% in 2010, and is now projected to be decreasing. By 2030, it is expected to have declined to about 58% to a little over 50% by 2050 (ZUS, 2006).

In many countries, demographic changes will require special socioeconomic solutions. One of the most significant changes will involve later retirement, which means that the productive-age group will comprise a large number of elderly persons. It is vital, then, to adjust the working environment to their needs.

STRUCTURAL LABOUR MARKET CHANGES

The labour market is a mechanism within which economic transactions take place: employers offer jobs or work and – on the other hand – employees offer their skills (show their willingness to work) (Mortimer-Szymczak, 1995). Issues concerning the labour market are hence discussed on the assumption that it is internally heterogeneous. In literature on the subject, the labour market is classified by several criteria, most frequently: by market sector, geographic location, by profession, qualifications and type of employees (cf. Table 1).

Table 1: Traditional Classification of the Labour Market (Unolt, 1999),

(Oleksyn, 2001)

Criterion	Types of labour market
Market sector	Labour market: I – Agriculture, Forestry and Fishery
	II – Industry and Construction
	III - Services
Geography	local, regional, domestic, cross-border
Occupation	individual job labour markets
Employee qualifications	Labour markets:
	1) potential employees (trainees, students);
	2) qualified labour force;
	3) unskilled labour force;
Employee types	the labour market of women, ethnical groups, people who often change their workplace, graduates, etc.

Table 1. presents the traditional Fischer-Clark 1930s sectoral division of the labour market. Sector 1 includes Agriculture, Forestry and Fishery; Sector 2 – Industry and Construction; and Sector 3 – Services. It is worth mentioning that with the increase of national income per capita, Sector 2 workers outnumbered those of Sector 1. Growing employment rate in Sector 2 brought about an increase in productivity, but also higher accident occurrence. Currently, Sector 2 is estimated to be employing

20%-30% of the entire working population and employment continues to grow in Sector 3, which is now taking over (Orczyk, 2004).

This traditional division translates into employment statistics, which distinguish three stages of the absorption of the labour force: Stage 1, with the predominant absorptive power of Agriculture (agrarisation), Stage 2 with the predominant absorptive power of Industry and Construction (industrialisation) and stage 3 with the prevailing absorptive power of the Service Sector (servicisation) (Unolt, 1999).

With a view to the globalisation of the labour market, geographic segmentation of the market is currently undergoing radical changes . Every year, 6 million new companies are set up throughout the world, and in consequence 30-40 million new jobs are created. People benefit from the reduction of domestic and cross-border barriers (Oberhänsli and Vera, 2009). However, analyses of globalisation processes also reveal some negative aspects of the situation. The enhanced cross-border economic integration is not equally beneficial to individual countries, industries or even to individual business entities and workers. In particular, developing countries reap smaller benefits than the highly developed ones.

Owing to technological changes, especially concerning IT services, and the emergence of the IT sector (Sector 4), classification by occupational and qualification critera is gaining importance. It is estimated that by 2010, 95% of jobs in the USA will involve gaining, processing and distribution of knowledge and information (Krzysztofek and Szczepański, 2000). Due to the ageing process, the European market is employing workers at more and more advanced ages and the ratio of working women and migrant workers (both legal and unregistered) is growing. Part-time employment and increasing use of new technologies become commonplace.

THE NEED FOR DEVELOPMENT OF ERGONOMIC STANDARDS

The changing labour market and the ageing process ought to keep workplace designers awake at night. Currently, the main trend in workplace design is to reduce factors which may have negative effects on employee health, especially physical health.

Table 2. EU Employees: Exposure to One or More Health Risk Factors in
Individual Sectors (%) (De Norre, 2009)

Sector	Men	Wome
Private households with employed persons	54.47	30.23
Other community, social and personal service activities	39.46	32.62
Health and social work	41.61	44.66
Education	29.74	28.05
Public administration and defence; compulsory social security	37.23	22.68
Real estate, renting and business activities	27.20	22.02
Financial intermediation	12.30	12.91
Transport, storage and communication	55.30	29.22
Hotels and restaurants	39.72	37.46
Wholesale retail trade, repair	44.70	29.02
Construction	66.19	18.00
Electricity, gas and water supply	46.76	19.85
Manufacturing	53.12	38.14
Mining and quarrying	73.77	-
Fishing	69.83	-
Agriculture, hunting and forestry	57.62	48.00

Statistics (cf. Table 2) reveal that sectors with considerable risk factors include
Mining, Fishery and Construction. However, in the changing structure of the labour
market, the number of workers in these sectors will be decreasing. It is crucial, then,
to eliminate risk factors in sectors which are going to take over. With a view to the
extending span of active professional life of elderly persons, it is vital to customize
workplaces to their psycho-physical abilities – a real challenge to workplace
designers.

CONCLUSIONS

The recognition of social needs constitutes one of the fundamental principles of
ergonomic design. The article presents data which illustrate demographic and social
changes, mainly the ageing process and the increasing number of elderly workers
on the labour market. This trend should already be taken into consideration by
workplace designers.

The development of the Service and IT sectors involves extensive application of new technologies, especially IT. In the information society, hardware and software are basic working tools. For this reason, both hardware and software producers, as well as institutions of science and education should focus on hardware and software ergonomy. Ergonomic computer stands and desktops, adjusted to the habits and reasoning of various workers, ought to become commonplace if we want the older population to participate in the labour market.

REFERENCES

Anticipated Receipts and Expenses of the Retirement Fund by 2050. ZUS (Social Insurance Institution), Department of Statistics, Warsaw 2006.

De Norre B. (2009). *Population and social conditions*. Eurostat. Statistic in focus. No. 63.

Górska E. (2002). Projektowanie stanowisk pracy dla osób niepełnosprawnych. Oficyna Wydawnicza Politechniki Warszawskiej. Warszawa.

Grzywiński W., Lis K., Mederski P.S. (2008). *Ergonomiczna analiza oprogramowania harwestera*. Przegląd leśniczy. No. 5. p. 4-5.

Jasiak A. Swereda D. (2005). Ergonomia osób niepełnosprawnych. Wyd. Politechniki Poznańskiej. Poznań.

Krzysztofek K., Szczepański M. S. (2000). *Zrozumieć rozwój. Od społeczeństw tradycyjnych do informacyjnych*. Wydawnictwo Uniwersytetu Śląskiego. Katowice. s. 172-196.

Laqueur W. (2.05.2008). Ostatni spacer po starej Europie. http://www.dziennik.pl/dziennik/europa/article165938/Ostatni_spacer_po_starej_Euro pie.html

Lis K. (2007) *Ergonomia oprogramowania księgowo-finansowego a efektywność.* [w:] Wymiana doświadczeń w dziedzinie ergonomii z państwami ościennymi. - Polska Akademia Nauk (PAN). oddział w Poznaniu - Komisja Ergonomii. p. 151-159.

Mazur T., Nasiński P. (2006). Prognoza wpływów i wydatków funduszu emerytalnego do 2050 roku, Zakład Ubezpieczeń Społecznych, Departament Statystyki, Warszawa.

Mortimer-Szymczak H. (1995). *Rynek pracy i bezrobocie. Acta Universitatis Lodziensis.* Folia Oeconomica No. 135.

Oberhänsli H., Vera O. (01.08.2009) *Globalisation: opportunities and conserns for the people of the developing world.* www.humanglobalisation.org/facts/factsChapters.htm

Oleksyn T. (2001). Praca i płaca. Międzynarodowa Szkoła Menedżerów. Warszawa.

Orczyk J. (2004). *Postęp cywilizacyjny a praca.* w: Przyszłość pracy w XXI wieku. pod red. S. Borkowskiej. IPiSS. Warszawa.

Papiernik J. (19.03.2007) W 2050 r. seniorzy liczniejsi niż dzieci. http://www.egospodarka.pl/20848,W-2050-r-seniorzy-liczniejsi-niz-dzieci,1,39,1.html

Prussak W. (2006). *Ergonomiczne zasady projektowania oprogramowania komputerowego.* [w:] Ergonomia produktu. pod red. J. Jabłońskiego. Wydawnictwo Politechniki Poznańskiej. Poznań.

Unolt J. (1999). *Ekonomiczne problemy rynku pracy.* BPS - Śląsk. Katowice.

Webster F. (1995) *Theories of the information Societ.* Routledge. London/New York. p. 7-27.

Chapter 86

The Leader Impact on Proergonomic Activities

Elżbieta Jasińska, Waldemar Jasiński, Stanisław Janik,

Institute of Management Engineering,
Faculty of Computing Science and Management,
Poznan University of Technology
Poznan, Poland

ABSTRACT

The development of present economical business, social, proergonomic processes caused the situation in which there appeared full-shape clusters. These ones, as it is described in literature on this subject, were classified referring to development or influence on the shape of proergonomic activities in a region. The example is the Lower Silesia Source Cluster in Poland. There is a specific group of firms with the leader in the head together with macroergonomic business surrounding. The need of cooperation and competitiveness (as an attributive necessity of a cluster existence but not a holding in which there is a identifiable dominant firm) of the leader with the direct and indirect surrounding influences the proergonomic relations in the region Lubin-Polkowice in Poland (within the model-structure of the triple helixes: B+R, public sector, business sector). Thanks to the leader's specifics and its multi-track character as a dominant firm in the region (as regards its size, job providing capacity and its macroergonomic surrounding) there are planned and realized characteristic proergonomic activities in the region of Poland described above. These activities become the example to follow for the macrergonomic business surrounding. The ergonomical quality created by the leader and the firms around ought to be the priority in the development of the region (starting from ergonomical education for the needs of the cluster to permanent skill – lifting activities for staff and ergonomical aspects of the first, second and third education).

Keywords: cluster, relation, leader, business surrounding, ergonomics, macroregonomics, ergonomical quality, regional development, proegonomic activities, ergonomical relations among the leader and the surrounding, proergonomic and ergonomic education of staff for the needs of cluster, Polish clusters

INTRODUCTION: THE INTENSITY OF GLOBAL PROCESSES BUT THE SCALE OF PROERGONOMIC IMPACT OF A LEADER IN THE REGION

Quickly progressing process of globalization and intense transformations, on every latitude, caused the course of an intense progress of civilization.. Hence, the progress has developed and, most of all woke in the contemporary man the urge to aspire to obtain comfort in simple everyday life activities, as well as in activities of his professional life. Those phenomena has finally presented the urgent need of initiating actions concerning: first – deep analysis of relations occurring between man (who constitutes an individual unit or a group of people) and the influence in the process of work (taking under consideration means of work), and the environment that constitutes the space of those relations and in which elements mentioned above affect each other and enter in continuous processes [Wykowska M.]. Second – actions that are very much intensified for technological, computer, technical or social processes, in which contemporary man „not only is using benefits and goods of these processes , but also, In result of the tempo of those achievements, is becoming more and more tangled in what he created or he wants to produce in the future [Jasińska 2009]", the more that his technical, natural or social environment has become possibly extremely dominated by him . Therefore the rush of the of civilization development shows clearly that a human being equally fast is losing a hierarchy of importance for her subjectivity in order to get the peak effectiveness of modern technologies, by the cost of the safety of the human health or life, omitting not to say throwing the safety to further plans for better effects of this race. Determined phenomenon is observable in each part of the Word, independently from the micro- or macro-environment, industrial distractor [Figuła 2008] or cluster, region, territories or countries [Jasińska 2009]. The problem becomes the more serious for the leading company in the region that the universalism of the phenomenon is increased by processes of among others [Jasińska 2009; Pietrulewicz 2000]: precipitating changes in every field of social and economical life, including populations living in the surrounding of the leader, who initiates proergonomic actions in the region; appearance of indications of the crisis of the extravocational and professional education inappropriate to needs or for performing future professional roles that are necessary or specific for the development of the dominant and his proergonomic actions for own business environment; creator of techniques don't keep up with the tempo of designing ergonomic workstations equipped and machines and devices built in accordance with ergonomic needs (including goods used in everyday life), which are necessary for optimizing technologies implemented by the leader in the cluster considering proergonomic actions in the region, in which it functions; shaping the population into information society, creating new styles of operating, including constantly changing conditions of work and life of man in the environment consisting the enterprise playing the dominative role, which, through its proergonomic operations, imposes a trend of modern education within the growing structures of the cluster in the region.

TERMINOLOGY, TYPOLOGY AND ATTRIBUTE'S CONDITIONINGS FOR THE LOWER SILESIA RAW MATERIAL CLUSTER IN POLAND

Available world literature is describing enough extensively the comprehending of cluster, with special regard on elaborations of M.E. Porter, A. Marchall, D. Herrtok, Marossini, who are pioneers and "fathers" of the term "cluster" and their concept of development that determines its definition as: "geographical agglomeration of enterprises (mainly small and medium) acting in similar sectors, their suppliers and other organizations (branch societies, centers of knowledge – universities and research centers), who simultaneously compete with each other and cooperate together in the process of production (or services) and in the chain of creation of values, between which they exist (based on trust and cooperation), network connections, for which the participation in clusters can be an important factor affecting their individual competitiveness. (So, economical cluster are open agglomerations, each participant can enter them or go out of them in any moment. Generally they have a local leader, who is accepted by members of the cluster)" [Pacholski 2008; Jasińska 2008]

Accepting terminological decisions of the cluster in the sphere of the rank of the leader and his proergonomic.impact in the region (thanks to which there is a maximum stimulation of the economical growth, increase of local wealth, development concerning modernization of the local economy and its superstructure), it is also important to characterize precisely the typology of clusters that, in the context of the leader's impact of proergonomic actions in the region, constitutes a rich source of knowledge about clusters. It is the same knowledge that indicates the criterion in accordance to [Porter 2001; Jasińska 2008]: localization of the agglomeration of economic units (this is why it is possible to talk about clusters with a local, regional and national scale); age of the cluster (this is why it is possible to talk about embryonic, growing, mature and descending clusters); horizontal connections between sectors (this is why it is possible to talk about narrow or wide clusters); taking under consideration stages of the production chain (this is why it is possible to talk about shallow and deep cluster); competitive position of the cluster sectors (this is why it is possible to talk about cluster being worldwide leaders, national leaders, clusters with an average competitive position and those with weak position); meaning of innovations, technology sectors (this is why it is possible to talk about highly innovative clusters, clusters with low level of innovation and technology).

The same literature determines very strictly attribute conditioning of a cluster, which consist in following list:[Pacholski 2008; Skawińska 2008]

Spatial concentration, which favors easiness of initiating partnership of the leader with the environment and implementing proergonomic actions In reference with the geographical closeness of participants of the cluster. Sector concentration (business environment surrounding the enterprise leading in the region, which can operate in same or similar branches of industry or services), independently from the monolith activity of the dominant from one side, which despite its sector variety can succumb proergonomic influence of the leader of the region from the other side; his results with a maximum economical success of not only low, cheap and specialized factor of production or shortened way from the stock to a final product, but also through simultaneous influence of proergonomic phenomena constituting a space for horizontal and multidimentional structure of the cluster, in which the leader initiates and cares about the proper scale of enumerated phenomena.

Specific specializations and competence (which priority are not only specific competences of the sector, but also which are shaped in the ergonomic education and in Polish system of education or on different stages of education and professional self-education of employees hired by the leader or/and enterprises from its environment.

BUSINESS, ECONOMICAL AND SOCIAL PROCESSES AS DETERMINANTS OF PROERGONOMIC ACTIONS OF THE LEADER OF THE CLUSTER IN THE REGION

The development of economical, business and social processes has contributed from one side to singling out cluster in many regions of the world, from the other side to: [Jasińska 2009]

Change of trends: from individual economical actions to complex operations, such as starting modern technologies, creating innovative network connections, products and services provided in specific circles

Change of bilateral relations, which were preferred until now in the economic reality to new, trilateral, which perfectly fit on the level of a region (triple helix) [Pacholski 2008], especially in the context of proergonomic actions of the cluster's leader.

Changes in itself, which has caused that the enterprise dominating in the region, which was functioning perfectly well, still, it additionally embosoms with companies depending from it. [Jasińska 2008]

So, his progress and those changes initiated the growth and development of the idea of clusters; while their structure has created conditions for perceiving the leader and his business environment as an entity with simultaneous maintenance of the autonomy and freedom of action and the multiplicity of sectors of this environment, which often is very different from the sector of the dominant (which often has monosubjective, leading production in the cluster). The Lower Silesia Raw Material Cluster in Poland is the classic example of such state and it is fulfilling all typological, characterological requirements and attribute's conditioning resulting from the definition of a cluster. It is also determined on a world map of clusters. Within this cluster there exists a specific cluster with a leader of the mining industry and its business environment, which is subjected to proergonomic influences in the Polish region of Lubin-Polkowice. This leader is the KGHM Polska Miedź S.A. The Figure 1 presents the local dominant and his business environment.

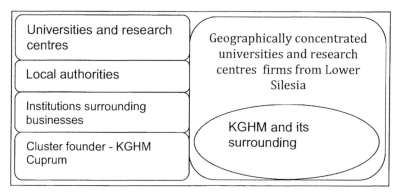

Figure 1. Dominant Holistic Firm: 'Przymus' macroergonomic conditions in the specifications of the Lower Silesia Source Cluster in Lubin – Polkowice region [Jasińska 2009]

Taking under consideration the impact of the leader on proergonomic actions In the region it is possible to initiate deliberations about various ergonomic issues, still one of most important in this subject would be the aspect of actions resulting from the role of employer played by the dominant. Moreover, it should be underlined that the basic feature of such company, as the biggest employer in the mining cluster, is his interdisciplinary character. It results from the fact that, first: as an employer it enters in a wide spectrum of relations with workers and other participants of the work process; second: it seeks optimization in providing comfort of work for own employees, but also for those, who work in its environment and for whom it initiates proergonomic activities in purpose of providing industrial security and simultaneously carrying about obtaining expected economical results of manufacturing processes that take place in the mining cluster. Third: existence of the rigor of guaranteeing means for work [Jasiński 2009] and posts for various and multisectoral business environment of the cluster, in which directly or directly it employs people (taking under consideration proergonomic influences that appear at, e.g. determining high standards and requirements for them, determining "how to do it?" or warnings from threats, as well as model patterns of avoiding them).

Moreover, the same leader, in the context of planned and realized proergonomic actions, must take care about knowing well his employees under direction of employing them in the structure of the cluster (he should initiate correct diagnosis and analysis in order to know better their needs and take under special consideration their needs from the area of ergonomics). Then he should contribute to a rational, functional shaping and accepting specific and standard work posts (including ergonomic standards, also the global ones), on which employees and workers from outside of the company (but operating within the cluster) will be able to work in maximum secure and ergonomic space and with minimum biological input. The table 1 illustrates the widely understood ergonomic quality created by the leader in reference to proergonomic actions initiated in the region [Jasiulewicz 2008]

Table 1 The widely understood ergonomic quality created by the leader in reference to proergonomic actions initiated in the region

Ergonomic quality shaped by influence of the leader with use of proergonomic affects in the region	
Space of the work in structures of the Lower Silesia Raw material Cluster	Proper shaping of the space structure, machines and workstations specific for the cluster
Leader's work environment and his business environment with special consideration of ergonomic needs for his multisectoral surroundings	Indicators formulating the material operating environment of the man in the Lower Silesia Raw Material Cluster
Workstation (maximum spectrum of Workstation In reference to the diversification character of the activity of the leader and other participants of the Lower Silesia Row Material Cluster	Wide spectrum of equipping workstations at considering ergonomic general and peculiar standards
Organization of work in structures of the Lower Silesia Row Material Cluster for the dominant and his environment in the cluster	Planning, documentary service, workshop assistance, distribution also at preserving ergonomic standards intensifying proergonomic actions
Work burden caused by the specificity and the profile of activity of the KGHM as the leader In the mining cluster within Lower Silesia Row Material Cluster (mining, metallurgy)	System man – machine. Burden of hands and legs and the static and dynamic load
Security in copper exploitation and processing (mining, metallurgy) as the leading activity in Lower Silesia Row Material Cluster	Legal regulations, security system including the widely comprehended ergonomics (first, second, third generation), analysis and evaluation of the occupational risk (taking the ergonomics of a workstation into account).

The leader and other units of the Lower Silesia Row Material Cluster unequivocally refer to achievements of many disciplines of science in their proergonomic actions in the region; those reference facilitate them processes of planning, designing (which take place on the position of the dominant's management), creating diversification structure that are necessary and specific for mining and metallurgy, as well as highly standardized workstations. The same workstations, thanks to proergonomic actions, are further subjected to next evolutions, consisting in correcting existing states of reality, which become unwelcome (the aspect of correcting ergonomics) [Wykowska]; as well as upgrading and improving future states (the aspect of conceptual ergonomics) [Wykowska] already expected thanks to new technologies and innovations. It happens so because the dominant has possibilities, thanks to which he gains and implements new, better technological offers in his production branch and better innovations enabling planning, implementation, monitoring, control of effects of initiated operations. [Kowal 2008; Jasińska 2009]

In addition, the essential fact should be taken under account, that for planning, implementing and managing proergonomic programs it is necessary to be considered by the leader a suitably high level of his own awareness about the necessity of ergonomic correcting and designing. Levels of this awareness can shape in two scales: [Stolarek] micro (including microergonomic planning resulting from achievements and states of ergonomic of the first and second generation) and macro (resulting from state and achievements of ergonomics of the third generation). Finally, accepting that the level of awareness of the leader in the region is sufficiently high, he will be able to participate the realization of, among others, planned proergonomic actions, as the biggest and the most important employer in the region, with following actions: [Jasińska 2009]

Study the range of perception of his employees or of all employees from this
Examine anthropometric issues
Analyze and design relatively isolated systems (within the scope of leading employee or the environment) and technical objects
Study cognitive and decisive processes (especially in the scope of ergonomics and proergonomic actions) of his employees and workers from the region.
Examine interactions in the system man -- computer (e.g. ergonomic aspects of a workstation equipped with computers)
Study complex systems (in which the multiobjective subject of the design, like „organization" in Lower Silesia Row Material Cluster will be treated as inseparable component and fragment of the external environment and standard of a bigger space, i.e. mining and metallurgy sector), so that the proergonomic attribute of this examination could treat actions of the leader in the aspect of the "work system" from one side, and in the aspect of "system of own design of management") [Rabenda 2008, Karwowski http..., Jasińska 2009],. Next it can become an integral part of the decisive process of own general board, which is responsible for the addend value of the company, of streamlining in the production, optimal use of intellectual capital employed in structures of the Lower Silesia Row Material Cluster.

Summing up it is important to underline that the leader in the Lower Silesia Row Material Cluster, represented by KGHM Polska Miedź S.A., thanks to his awareness and the fact that he is the biggest employer in the region, "he cares so that man would produce in his company values that will not be lower, even by the cost of loosing higher values; this might reveal itself in mines and mills in Poland and abroad. It is the work of a miner, miller; that requires from the leader a high level of design, organization and management of his workstation and of physical, psychical and emotional security of the worker. Moreover, this work becomes the more valuable that the work environment of a miner is shaped mainly by forces of nature [Jasińska 2009]

THE EXPECTED PATTERN FOR PROERGONOMIC ACTIONS IN THE REGION – FINAL CONCLUSIONS

As it has been already mentioned, thanks to the specificity of the leader and his multidisciplinary character as a dominator in the region; operations are being planned, realized and, most of all determined by him. Those actions, thanks to the rank the leading enterprise has in the cluster, become the example to be followed for other leaders and most of all to other units from the business environment. The "ergonomic quality" enumerated before along presented deliberations, must be created by the leader and the environment in the first place and it should constitute the second most important goal and priority of development of the region. It should develop starting from the level of ergonomic education responding needs of the environment or workstations in the cluster (it hasn't been elaborated issues from the scope of ergonomic education in Poland – it is a wide problem; hence it should constitute an individual subject of analysis and considerations), ending with constant and permanent lifelong learning of human resources in the subject and with further development of fundaments and premises of ergonomics of the first, second and third generation of the example of the Lower Silesia Row Material Cluster in Poland with a peculiar mining circle and its leader – KGHM Polska Miedź S.A.

REFERENCES

Figuła J., (2008), *Dystrakty przemysłowe. Małe i średnie firmy włoskie w dobie globalizacji (Industrial Areas. Small and Medium Italian Firms in the Time of Globalisation)*, Polkowice

Jasińska E., Janik S.,(2009) *The Macroergonomic Leader Surrounding, [in]: Macroergonomics* VS Social Ergonomics. (ed.) L.Pacholski, Monograph, Publishing House of Poznan Univerity of Technology, Poznań, 2009 p.99

Jasińska E., Janik S., (2009), *Aspekty ergonomiczne okręgu przemysłowego, [w]: Wybrane kierunki badań ergonomicznych w 2009 roku*, (ed) J. Charytonowicz,

800

Wrocław, p. 39-40 (*Ergonomical Aspetcs of an Industrial Area. Selected Directions of Ergonomics Research in 2009*).

Jasińska E., Jasiński W., Janik S., (2008) *Otoczenie przedmiotów gospodarczych wokół wiodącego w regionie przedsiębiorstwa*, [in], J. Olszewski, M. Słodowa– Hełpa, (ed.), Koncepcje oraz czynniki rozwoju regionalnego i lokalnego w warunkach funkcjonowania Polski w strukturach zintegrowanej Europy, Poznań 2008, p.43-50; (*The Firms Surrounding the Regional Leader*) (Concepts and Regional Development in Poland's Integrated Structures of United Europe).

Jasińska E., Janik.S, Jasiński M., (2009), *Oddziaływanie lidera na otoczenie*, w materiały konferencyjne, Międzynarodowy Kongres Górnictwa Rud Miedzi, Lubin, Wrzesień (Research Paper for The International Congress of Copper Ore Mining)

Jasińska E.,(2008) *Kompetencje w zakresie bezpieczeństwa i higieny pracy kadry zarządzającej placówką oświatową* [in:] E. Kowal, Zarządzanie warunkami pracy, Zielona Góra 2008, p. 25-28, (The Competences in Work Safety and Hygiene for School Management) (The Management of Work Conditions),

Jasiński W., Janik S.,(2009) *Komputerowe narzędzia do projektowania oświetlenia* [in:] J. Charytonowicz, Wybrane kierunki badań ergonomicznych w 2009r. (The Computer Tools for Light Designing – The Selected Research Directions in 2009).

Jasiulewicz-Kaczmarek M.,Wieczorek R. (2008), *Ocena warunków pracy podczas eksploatacji systemów projakościowych* w [ed.] Edward Kowal, Zarządzanie warunkami pracy, Uniwersytet Zielonogórski, Zielona Góra 2008, p.29-35) (The Assessment of Work Conditions During Quality System Exploitation, Work Condition Management)

Karwowski W., *Zarządzanie systemami pracy*, (Work System Managing) portal.wsiz.rzeszów.pl

Kowal E., (2008), *Zarządzanie warunkami pracy*, Zielona Góra, and E. Kowal, *Ergonomia w zarządzaniu warunkami pracy*, Zielona Góra, (Work Condition Managing, Ergonomics in Work Condition Management)

Pacholski L., (2008) *Kształtowanie sfer wysokiej produktywności w Wielkopolsce*, referat wygłoszony na konferencji Poznań 20.02.2008r., (The Shaping of High Production Spheres, research paper given at the conference in Poznań)

Pietrulewicz B., Pietrulewicz U.,(2000) *Czynniki determinujące przygotowanie ergonomiczne*, [in]: Czynniki determinujące aktywność zawodową pracowników, (ed) B. Pietrulewicz, Zielona Góra, vs.273-274 (The Agents Determining the Preparation of Ergonomics, the Agents Determining Activity of Workers) p.273-274

Porter M., (2001), *Porter o konkurencji*, Warszawa strona 246-249 (*Porter on Competitiveness*)

Skawińska E., Zalewski R. I.,(2009) *Klastry biznesowe w rozwoju konkurencyjności i innowacyjności regionów*, Świat-Europa-Polska, Polskie Wydawnictwo Ekonomiczne, Warszawa; Business Cluster in Competitiveness Development and Region Innovations)

Rabenda A., (2008) *Elementy systemu zarządzania bezpieczeństwem i higieną pracy* [in] E. Kowal, Zarządzanie warunkami pracy, Zielona Góra, p.51-57; (The Elements of Work Safety and Hygiene of Work) (Work Condition Management)

Stolarek M., Czyż M, *Elementy ergonomii, fizjologii i higieny pracy*, (The Elements of Ergonomics, Physiology and Work Hygiene http://www.kultbezp.ciop.pl/.,

Trzcieliński S., (2008) *Przesłanki organizacyjno-strukturalne budowy klastrów w Wielkopolsce*, referat wygłoszony na konferencji Poznań (Organisational and Structural Background in Cluster Building in Wielkopolska Region, research paper given at the conference in Poznań).

Wykowska M., (1994) *Ergonomia- wydanie internetowe*. (Ergonomics – on-line publishing http://www.ergonomia.agh.edu.pl/Skrypt_Ergonomia-M.Wykowska/ergonomia /index.htm

Chapter 87

Macroergonomic Development of Industrial Production Processes

Leszek Pacholski, Bogna Mateja

Institute of Management Engineering
Poznan University of Technology
60-965 Poznań, POLAND

ABSTRACT

The necessity of modification of contemporary macroergonomic approaches results from the process orientation of enterprieses, which at present gains recognition in theory and practice of the organizational design and management (ODAM).

The elaboration presents the course of realization of the development of industrial manufacturing processes, with special consideration of macroergonomic conditionings. At present it includes a chain of operations: starting from making decisions and ending with establishing effects of implementation of designed changes.

There has been also signalized the possibility of practical application of the connection between the aspiration to constant improvement of industrial manufacturing processes and the functioning of systems of security management and industrial security.

Keywords: macroergonomics, process approach

INTRODUCTION

Industrial manufacturing processes are constantly improved; this causes progress and development. There are various premises motivating innovations; they can be traditionally classified as: economical and technical. However, consideration of macroergonomic conditions is characteristic for modern, interdisciplinary approaches. They assume existence of relations between the human factor and material and organizational factors of the work environment, as well as with elements of the technical equipment, with help of which the industrial manufacturing process is realized.

The traditional approach has put impact on technical inventions as factors of realization of the progress and development in production processes. The role of the work organization has been appreciated only in the last century. Nowadays the significance of the human factor for the progress is still considered in a limited scale. While man – the designer and the person realizing and controlling industrial manufacturing processes – is the central point of interest of modern theories describing those processes. Therefore it is inevitable to consider postulates of raising the rank of macroergonomic premises of the development in reference to the theory and practice of functioning of industrial enterprises. The process orientation of enterprises can be a factor contributing to this change; at present it gains wide application in the theory and actions of the organizational design and management.

The process approach forces the realization of organizational innovations out, In reference to the structural and also to the procedural aspect.

ORGANIZATIONAL INNOVATION OF ENTERPRISES

The organizational innovation of enterprises, just as it has been mentioned above; can be perceived in two aspects: structural and procedural. Each aspect has two dimensions: external and internal (see Figure 1).

The external structural organizational innovation assumes the openness of the company to cooperation. So, it includes:

- activity in trade organizations,
- cooperating with institutions from the business environment,
- stimulating the development of cooperation between partners,
- providing companies a complex support of initiating cooperation Since the start of such by maintaining his partnership and ending with examining the efficiency and the range of initiated cooperation,
- maintaining initiated cooperation,
  other actions

Aspect / Dimension	Structural	Procedural
External	External structural organizational innovation ⬇ Openess to cooperation	External procedural organizational innovation ⬇ Innovative forms of organization of logistic processes
Internal	Internal structural organizational innovation ⬇ Systems of organization and distribution of work	Internal procedural organizational innovation ⬇ Qualitative systems of protection of environment and security of work

FIGURE 1. Matrix of organizational innovation of enterprises

The internal structural organizational innovation manifests itself in systems of organization and division of team work, such as:

- organizing group work,
- optimizing powers,
- flattening hierarchical structures,
- systems and techniques of telecommuting,
- electronic model of functioning,
- other actions.

The transition from the industrial era to the era of the processing of information and dynamic technological development has caused nowadays an enormous demand and interest concerning fast access to knowledge and information in response to:

- of the processing of information
- growing requirements of the client,
- shortening cycle of life of the product,
- constant changeability and incertitude of the market requiring actions implementing changes and innovations,
- changes requiring flexibility of actions and fast adaptation to new conditions occurring on the market,
- development of IT technologies,
- globalization of labor markets,
- change of the character of competition on the market of the knowledge-based economy (the value of the company is being created on basis of immaterial assets of the enterprise),
- change of attitude of firms: from concentration on the product to concentration on the client (it requires a systematic canvassing knowledge),

- development of information society and supporting the traditional economy by the knowledge-based economy,
- increasing profits resulting from investments into intellectual assets – In comparison with investments in material assets,
- increasing meaning of the intellectual capital in a company,
- revolutionary development of information techniques, which allow processing and controlling knowledge assets.

From the other hand, procedural innovations concern quality systems, systems of protection of the environment and industrial security (e.g. TQM, Kaizen, ISO, TPM, Concurrent Engineering). They are being applied in following form of activity:

- widening the knowledge about quality systems,
- exchange of experiences concerning implementation and realization of systems of quality,
- concurrent leading tasks,
- improvement of the control,
- other.

However external procedural innovations are innovative forms of organization of logistic processes, such as:

- optimization of supply management,
- synchronization of supplies with current production,
- initiating cooperation on the level of supply chains (suppliers),
- management of logistic localization of the partner / cooperate,
- finding multiple sources of supplies,
- audit of suppliers,
- other actions.

In practice it is possible to present following techniques as examples of such sort of solutions: JUST IN TIME, KANBAN, Supply Chain Management. Multiobjective manufacturing systems are places, in which the development of procedural and structural organizational innovations concerning industrial manufacturing processes occurs. In the same time they constitute the object of interest of macroergonomics (Jasiak, Misztal, 2004). This leads to a finding of authors that there is a necessity of analyzing macroergonomic aspects of transitions of those processes. The macroergonomic, humanocentric and process approach can guarantee a correct direction of effects of management of manufacturing processes. Therefore, the role of macroergonomics can consist in stimulating the progress in many scientific disciplines and engineering operations (including organizational innovations) and, in result, economical development.

DEVELOPMENT OF INDUSTRIAL MANUFACTURING PROCESSES

Since the beginning of the history of all human actions, of individuals, as well as of social groups, success was the power of changes. The pressure of achieving success and, what is inseparably connected with it, more intensive research of methods and ways of acting that would guarantee the defined result has increased under the influence of dynamic development of the modern world. Under the influence of this tendency processes of production are being constantly change along ages. A so-called technical revolution, being an affirmation of the civilization's development has caused that during the life of one generation we deal with a bigger number of important, even revolutionary processes of innovation. They are conditioned by construction, technology, materials, organization, economy or the market.

In way of achieving the effect of development, following actions are initiated in a company oriented on the process (unlike In the case of functional orientation):

- the impact is put on the quality of performer work,
- it is assumed to implement coordination between functions,
- ideas of collective work are being popularized,
- a system approach to the whole system is in effect,
- the orientation on the customer is dominating (McCormack, Johnson, 2001).

Mentioned characteristics are coinciding with aspects quoted in the former chapter concerning manifestations of the organizational innovation.

Innovations of all types include two different situations: designing processes for new manufacturing systems and redesigning, improving and modernizing processes or parts of processes that are already in use in functioning systems. In the first case actions are being directed by standards and theoretical guidelines. In the second case one should rely on observations, results of diagnosis and parameters of current functioning and relations between the human factor, direct environment and external environment and the technological factors of the process. The reconstruction of a network takes place in result of a real evaluation of a macroergonomic situation. This allows obtaining the effect of complex changes and significantly upgrading various aspects of processes being in realization. So, one should avoid ineffective (i.e. expensive, labor-absorbing) interventions in fragments of processes realized in individual systems man – to – machine and regulations of selected parameters, for example concerning the work environment.

IMPLEMENTATIONS OF INNOVATIONS INTO PROCESSES - APPLICATIONS

Decisions concerning implementation of industrial innovations to manufacturing processes are taken under influence of various conditions and they cause different effects.

As it has been assumed in the introduction of present paper, traditional premises of innovation are economical, market and technical motives, while macroergonomic premises have rather a humanistic character, so they are aimed at a good of the man, who has the role of the direct producer, but also of: of employee of support-service processes, recipient or user of products, not to say the inhabitant of the area living in the neighborhood of the enterprise and exposed to its environmental influence.

Therefore it is necessary to take under consideration macroergonomic premises as most important conditions for making decisions. In the same time it should be underlined that the importance of macroergonomic premises is dependent from the stage of actions related with the innovation and from the situation, in which the process is being realized (Mateja, 2009). Presented statement is simultaneously the content of the first principle (see Figure 2) that refers to all stages of described chain of operations. It puts impact on the fact that the consideration of macroergonomic premises in different phases of upgrading the process is proportional to the rank of the problem.

Observations of the rank of innovations realized along last 10 years in 40 selected Polish enterprieses of following sectors of industry: machine, wood processing, construction, food, shoes, electronic and fuel-energy allowed formulating observations that can be summarized in form of a scheme presented in the figure 2.

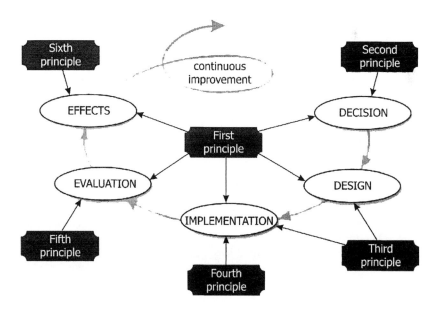

FIGURE 2. Scheme of relations between stages of innovative actions with principles of realizations

Some dissertations concerning relations between stages of innovative actions with rules of realization of the development of industrial manufacturing processes that have been elaborated earlier (Mateja, Pacholski, 2008),(Mateja, 2009), are presented below.

The stage of taking a decision is the most essential and important moment, in which the significance of macroergonomic premises should be especially taken under consideration. It also gives the possibility of first confrontation between the traditional approach and the modern – humanistic one.

The decision-maker most often deals with the tangle of a few kinds of conditioning, affecting one another. The dominating type of causes provokes a need, or even a necessity of taking others under consideration. Technical, economical and market premises appear in various configurations in decisive situations as specific "packages" (Mateja, Pacholski, 2008). Macroergonomic conditions play the leading role while justifying taking the innovation too rarely. Only the tangle of occurrence of traditional and humanistic conditions constitutes a convincing decisive situation.

The second principle refers to this stage (see Figure 2), its content sounds as follows: Management of the development of industrial manufacturing processes requires taking decisions concerning innovations in the process with the appropriate advance (Mateja, 2009). Noticing this principle causes a suitably early initiation of innovative operations, preventing threats for the security, health and ergonomic comfort for workers.

Designing forms, methods and conditions of realization of innovations also reflects the relation of persons taking decisions (managers or directors) to the issue of directives formed by macroergonomics. In this stage takes place the dilemma between financial and realization possibilities and the assumed level of final macroergonomic results of designed innovations.

On his stage it is important to apply the third principle (see Figure 2), which refers to the following phase of actions: Industrial manufacturing processes should be designed and realized in accordance with macroergonomic directives (Mateja, 2009).

Fulfilling assumptions of this principle predicts making the production processes' management aware of the importance of the role of macroergonomic aspects and educating engineers of all specialties with consideration of the issues associated with the human factor included in disciplines, like ergonomics, environmental management, psychology and sociology of work. We understand that "macroergonomic directives" are all ergonomic standards and guidelines, legal regulations related with use of the natural environment and security of work, as well as other directives elaborated by psychology and sociology, organization and management. The macroergonomically correct course of a process in a certain period of time after the initiation is guaranteed thanks to the application of the quoted principle. So, obeying this principle retards the moment of appearance of significant inaccuracies in the course of the process.

Similarly to the former one, the stage of implementation of innovation gives possibilities of verification of the attitude of investors and executors to the issue of macroergonomic conditioning and their realization skills within the range of shaping ergonomic, ecologic and psychosocial conditions. Implementation of innovations related with improvement of industrial manufacturing processes can have a technical, organizational and psychosocial character.

Problems from his stage have initiated formulation of the fourth principle (see Figure 2), which claims that: during the implementation of the innovation to processes being already in realization it is important to use their relation with actions initiated because of other reasons (Mateja, 2009). This principle is justified by tending to avoid stoppages and disturbance in supplies of products and negative financial consequences caused by actions related with implementation of innovations. The coordination of terms of realization of those activities with technologically natural or accordant with marketing motives of breaks of the process also gives a possibility of avoiding multiplying similar operations (realization or result).

The next stage that we can name a control stage or evaluation consists in identifying aspects connected with the human factor in the realization of the functioning process and, in result, to conduction of a monitoring, cyclical and complex diagnosis and evaluation of selected parameters according to unequivocal criteria. The role of this stage in emphasizing the importance of macroergonomic premises is uncontested. It even enables widening the list their list with new premises that have been underestimated until today. Fifth principle (see Figure 2) the premise related to it is saying that: it requires identification and implementation of a monitoring, diagnosis and evaluation – according to explicit criteria of all aspects connected with the human factor (Mateja, 2009). It signalizes the necessity of constant recognizing humanization problems concerning the industrial manufacturing process being subjected to innovations. The identification of cause-and-effect relations for this type of conditionings allows realizing the course of the process without disturbances in the future. It minimizes the probability of occurrence of any unexplained causes of the irregularity in the process.

The last stage of described sequence of actions results directly from the sixth principle (see Figure 2), which is formed as follows: it is necessary to aim to establishing effects, especially humanization ones, of the improvement of industrial manufacturing processes (Mateja, 2009). The evaluation of macroergonomic effects of innovations is difficult to present in measurable and explicitly countable parameters. Macroergonomic premises also have an important role in this stage because they determine directions of criteria enabling a certain type of assessment of effects.

The efficiency of some operations of the enterprise can be evaluated with use of economical, market and financial criteria. The assessment of organizational and humanization innovations is often perceived as subjective or dispersed in time because their consequences have a long term character. It is the long term horizon that presents positive results of humanization effects in big social groups (i.e. improvement of the quality of life, health, measures of satisfaction resulting from work, natural environment). In the scale of medium time horizons one can observe a decrease of measures of numbers of accidents, absence and fluctuation of the labor force, as well as upgrade of the efficiency and quality of work. From the other side, positive results noted in this stage cause stronger motivation of general management of enterprises to initiate innovations conditioned by macroergonomics.

CONCLUSIONS

Both, literature of the problem and experience gathered in practice of numerous enterprises point at the frequent linkage between innovative actions with the implementation of quality management systems, security of work, environmental security or, as it has been happening often in last years, integrated systems. This causes occurrence of a whole range of positive results of the idea of constant improvement applied in those systems as a principle. The so-called Deming circle has a shape of a spiral, which means the necessity of repeating the cycle of actions on a higher level each time. In reference to macroergonomic premises it can mean that the development of industrial manufacturing processes will take place within frames of the mechanism of repeating activities: making decisions, designing, implementing, verifying humanization aspects and evaluating parameters and estimating effects of innovation related with the human factor.

Management of the development of the enterprise is using a range of models (e.g. 7S, McKinsey's, QCDSM Chart, comprehensive models), which allow concentrating actions of the firm in the scope of diagnosing, analyzing and designing its development on determined factors of success (Pascale, Athos, 1986). They are very helpful in the range of revealing or forming initial points of the state of the enterprise and the essential of its development (Wyrwicka, 2003).

Just like models for enterprises presented above, the elaboration of a macroergonomic scheme of development of industrial manufacturing processes can give an analogical practical effect in reference to processes. However systems of industrial security management are some sort of preparation to occurrence of macroergonomic models of management of the development of industrial manufacturing processes.

REFERENCES

Grajewski, P. (2003). *Koncepcja struktury organizacji procesowej.* Towarzystwo Naukowe Organizacji i Kierownictwa. Toruń.

Jasiak, A. (2008) „Determinants of development of contemporery ergonomics", in: *Foundations of Control and Management Sciences,* 11/2008, Publishing House of Poznan University of Technology, Poznan.

Jasiak, A., Misztal, A. (2004). *Makroergonomia i projektowanie makroergonomiczne.* Wydawnictwo Politechniki Poznańskiej. Poznań.

Mateja, B. (2009) „Macroergonomic's principles of advance management of industrial production processes", in: *Macroergonomics vs social ergonomics,* Pacholski Leszek (Ed.). Publishing House of Poznan University of Technology, Poznan.

Mateja, B.E., Pacholski, L.M. (1998) „Implementation current ergonomic renewal undertaking in multiagent manufacturing systems", in: *Global Ergonomics,* Scott, P. et el. (Ed.). pp. 941-944

Mateja, B., Pacholski, L. (2008) "Macroergonomic prerequisites for manufacturing systems modernization", in: *Conference Proceedings of 2nd International Conference on Applied Human Factors and Ergonomics,* (CD ROM), Las Vegas, Nevada.

McCormack, K.P., Johnson W.C. (2001). *Business Process Orientation. Gaining the e-business competitive advantage.* St. Lucie Press, London.

Pacholski, L. (2000) "Macroergonomic paradox of entrepreneurship and economic renewal", in: *Ergonomics for the new millennium, Human Factors and Ergonomics Society,* Santa Monica-San Diego, vol. 2, pp. 185-188

Pacholski, L. (2003) "Macroergonomic circumstances of the manufacturing company development", in: *Ergonomics in the digital age, Ergonomics Society of Korea,* Seoul, vol. 6, pp. 556-559

Pacholski, L., Cempel, W., Pawlewski, P. (2009). *Reengineering. Reformowanie procesów biznesowych i produkcyjnych w przedsiębiorstwie.* Wydawnictwo Politechniki Poznańskiej, Poznań.

Pascale, R.T., Athos, A.G., (1986). *The Art of Japanese Management.* Penguin Books, London.

Wyrwicka, M.K. (2003). *Endogenne przesłanki organizacyjne rozwoju przedsiębiorstwa. Rozprawy,* Wydawnictwo Politechniki Poznańskiej, Poznań.

Chapter 88

The Guidelines for the Sociotechnical Subject: The Practical Results About the Subject

Giles Balbinotti[1], Leila Gontijo[1], André Trautvein[1], Arlete Motter[1], Michelle Robertson[2]

[1]Universidade Federal de Santa Catarina, UFSC
Florianópolis, SC, CEP Brasil

[2]Center for Behavioral Science
Liberty Mutual Research Institute for Safety
71 Frankland Rd.
Hopkinton, MA 01748 USA

ABSTRACT

The present work originated from an establishment need of an administration model that assisted the needs of a relative industry to sociotechnical subjects. If, one side, companies that act at competitive markets, they need administration forms that strengthen them before the competitors, increasing your organization efficiency, on the other hand, the worker desires for better life quality in the work, and aspects of cognitive order, factors which will result motivation and efficiency. That research proposes a model for the unfolding of guidelines in ergonomics, area of the production engineering, considered essential as critic for the efficiency of the organization systems.

Keywords: Improvement of the work conditions; productivity; sociotechnical guidelines; costs and benefits.

INTRODUCTION

Concept of Culture "is a set of beliefs and values shared by all or nearly all members of a group and it is from these beliefs and values that are established behavior and perception of the world of the people" (Coelho and MAGALHAES, 2001, p. 43). Organizational culture is the result of a continuous process of a trial of strength between psychic reality and the reality organizational aspects influenced by unconscious, instinctual and defensive (MENDES, 2004).

The components of the culture, myths, stories, rituals and values, while they may be room to fulfill the wishes and needs of human beings, can be a source of suffering, when the individual identifies himself fully with the organization, and is not to question the conditions under which the work is done. Then, the individual's health can be maintained when there are forces of gratification and frustration of the drive given by the culture (MENDES, 2004).

Usually, the more the organization of work is hard, plus the division of labor is marked, the lower the content of the significant work and less healthy the chances of changing it.A free organization of work becomes an essential part of the psychosomatic balance and satisfaction (DEJOURS, 1992, p. 128). Therefore, the workload increases, so that lower operating alternatives in the changing work situations. However, you can not always say that the relationship between workload and health, expansion of scope of operators coincides with the maintenance of health, as described Dejours (1994, p. 29):

The rearrangement of the organization of work is no longer possible when the relations between workers and the organization of work is blocked, the pain begins, instinctive energy that does not think downloading to their work accumulates in the psychic apparatus, causing a feeling of displeasure and tension. Mergener, and Kehrig Traebert (2008), argue that the musculoskeletal symptoms is a health problem has increased and stood out, especially among workers in jobs with high degree of repetition of movements related to the automation of tasks, which occurs most strongly in computer services.

The RSI is damage from excessive use, imposed on the musculoskeletal system, and lack of time for recovery. They are generally characterized by the occurrence of various symptoms, concomitant or not, of insidious onset, predominantly in the upper limbs, such as pain, numbness, heaviness and fatigue. Cover clinical musculoskeletal system purchased by the employee subject to certain conditions. Constantly, are causes of disability temporary or permanent employment (PICOLOTO; SILVEIRA, 2008; SALIM, 2003).

COMPANY STRATEGIES

In this context of dysfunctional, methods as QC Story and MASP (Method of Analysis and Solutions of Problems) exist to identify and to diagnose problems that affect health, the well-being and people's performance in their work, as well as the existent anomalies that influence the efficiency and the survival of organizations. However, in the phase of resolution of problems we see the technical, human and organizational difficulties appearing, so to seek the necessary transformation actions. The importance of the purpose in the ergonomic intervention is the control of those problems, and so to improve the efficiency of the system. For such intervention to succeed, it is necessary that all involved get together in the solution of the problems, from the leaders to the production auxiliaries, being such involvement an extremely relevant point. Managers, responsible for the administration of production means, starting from the identification of a problem that interferes in the results of his/her department, have the role of promoting, lead and implement actions, using objective and clear methods to guarantee the execution of the established goals, assuring the purposes of the company.

As for the specific objectives, it is intended: - to demonstrate the importance of treating the ergonomic subject in strategic level; - to evaluate the application of the tools of Quality in the improvement of work conditions; - to evidence the practical applicability of the methodology proposed in a cosmetic industry and the analyze of a automotive industry and the obtained results.

The concern with the ergonomic subject and its proposed improvement, contributing directly in the improvement of efficiency, in the reliability and in the quality of organizational operations. (GUÉRIN, 2001). It is important to emphasize that ergonomics seeks to guarantee the health and welfare of the workers and how it can contribute for the reaching of expected financial results to top administration. Workers' health can be affected by several factors, being in relative issues as in work conditions (temperature, noise, illumination), physiologic subjects (posture, effort), or organizational aspects of the work (monotony, fatigue and motivation lack in the execution of tasks) and also in the safety at work (accidents, diseases). Those factors, affect greatly the quality and productivity commitment with such events and as well the financial health of the company, reducing its competitiveness in the market and risking its own survival.

According to Gontijo (2001), "the ergonomic study of the work, seeks the maintenance of physical and mental health of workers, besides better productivity through detailed analyses." Studies from the International Organization of the Work (OIT), show that the bond between safety lack in the work and the fall in competitiveness, places Brazil among the countries with larger problems in that area (Mercantil Gazette, 30 abr. 2003). In such context, ergonomics should be taken seriously, and for that to happen there is a large need to create and promote a methodology of easy implementation in guidelines unfolding so that it might assure the coherence of strategic actions, in the extent of the ergonomic subject, as well as the social and technical problem resolutions, as in production, quality and costs.

It is important to emphasize that the players benefited with the application of that methodology, which brings a new form of organizing the work, will be the employees of the organization, so that their roles and goals inside of the organization become clear and thus benefiting the company, and also possessing a methodology that assures the coherence between the strategic actions and the operational actions developed in the plant grounds, and therefore reaching the goals of the organization. Universities and colleges are also benefited through the explanation of that methodology in organizational administration. It is still a subject of study for the OIT - Work International Organization, "the financial impact of the social practices of companies, including their acting in areas of health and security and their consequent increase!"

METHODOLOGY

The elaboration of this work was accomplished starting from the bibliographical revision of ergonomics concepts and their relationship with work, in the Administration for the Guidelines (*hoshin kanri),* where it is stated that there's a chapter approaching the planning based in the PDCA cycle, the resolution of problems and the transport of the unfolding guidelines. Starting with an abbreviated description of ergonomics, purposes of work, human and technical aspects considered in the administration process, and the motivational and technological aspects that sustain every organizational system. The bibliographical revision is based on the analysis of two models used in the solution of problems, through the administration of guidelines and the ergonomic analysis of the work, being concluded in the composition of the methodology used in this dissertation. The bibliographical research was accomplished with the use of books, articles, master degree dissertations, doctorate theory (developed in CNAM - Paris), laws, norms, documents and organizational procedures.

The first case study was accomplished in a company in the cosmetics industry. This study was based in the application of the proposed model, mentioned above. In agreement with model 1 - administration for the guidelines (FIELDS, 1996, SHIBA, 1997 and AKAO, 1997), where they capitalized the stages of definitions in the annual goal "survival" of the organization, definition of the annual guideline and the process of unfolding of guidelines basically. The model AET - Ergonomic Analysis of the Work (GUÉRIN, 2001 and GONTIJO, 1993), subsidized the composed model, mainly through the consistency of the process of work analysis, in other words, the analysis of the demand, the analysis of the task, analysis of the activity and research action and, finally, the diagnosis of the work activity. The second case was about ergonomics management program in a automotive case.

Besides discussions done with all of the people in the company, interviews were accomplished with the management, employees of several departments, through local observations on the spot , systematic, observations done with audio-visual resources (tape recorder, camcorder and camera), besides brainstorming sessions and questionnaires, seeking to prioritize the raised opinions for the group. Therefore, this research was well worth the analysis of documents, interviews with key people of the researched company, such as, the bottling department responsible, laboratory responsible and industrial manager, as well as free and systematic observations. Eventually, the obtained results are presented with the implementation of the methodology of unfolding guidelines, for the sociotechnical subject. This work was limited to show and analyze the application of unfolding methodology of guidelines, for ergonomic adaptation in a cosmetics company with specific applications in the bottling department.

MODEL COMPOSITION FOR ERGONOMIC ADMINISTRATION

This chapter shows the composition of models (Administration by the Guidelines and Ergonomic Analysis of Work) presented in conceptual base in the previous chapter, for the proposed model, unfolding the guidelines, with focus in this study for the ergonomic subject. The idea here is to capitalize the strong points of those models shown in the diagram below, and to propose a model of easy application, keeping a scientific approach, needed to reach the objectives of the research.

The proposed methodology is structured in twelve main stages, divided in three phases: the first begins with the definition of strategic planning, identification of improvement projects, and a second phase by the administration for the guidelines-GPD,where there is the definition of guidelines and their unfolding, and the last and third phase concludes with daily administration through the analysis and diagnosis of the work, seeking the administration of the ergonomic subject: The first phase of the methodology, the strategic planning - SP, is composed of stage 1, where it is defined the values, the vision and the mission of the organization, and stage 2, that explores the strategic diagnosis through the internal and external sceneries analysis. These aspects were discussed in the previous chapter. The strategic planning serves as a subsidy for the methodology development of the second phase, the administration of guidelines - GPD which is composed by stages 3, 4 and 5. Stage 3 has as objective to define the annual goal of the organization that is the president's goal, which should be vital and inter serviceable, as seen previously. After definition of the annual goal, we establish the annual guidelines - stage 4 – and it is done through the definition of more goals, plus measures to reach that goal (picture 3). The last stage of GPD, stage 5, is the unfolding of the ergonomic guideline, that starting from the establishment of annual guidelines, unfolds it to all hierarchical levels, goals related to the ergonomic subject, and its associated plan of measures. Figure 14 demonstrates this process.

The last phase of the methodology, through their 7 stages, deals with the operational issue, which runs from the execution of defined measures in GPD, until the report (stage 12). Stages 6 and 7 define the form of managing the execution of measures and the verification of goals achievement. Stages 8 and 9, analysis of work and the diagnosis of the working activity are fundamental in the identification and precision of ergonomic problems through ergonomic analysis of the work, and later in the establishment of measures that can eliminate those problems. This approach was made in the ergonomic analysis of the work item. Finally, stages 11 and 12, deal with respectively the incorporation and standardization of the results, and report.

THE RESULTS - CONSIDERATIONS OF THE METHODOLOGY APPLICABILITY

THE CASE OF COSMETIC INDUSTRY

The evaluation done in the application of a methodology which unfolds the guidelines for the ergonomic subject in an industry of cosmetics, indicates that it was applied with success, or in other terms the operation logistic, when compared to the expected results and reached ones in each phase and stage, indicates success in its conception and application. In the associate's survey, we concluded it to be positive, since its beginning, after the explanations and explanations of what would be done, which was the objective, leaving it clear the concern with everyone's welfare inside the company.

That is evidenced starting from the results obtained in each stage of the methodology and that you/they allowed the establishment and implementation of the plan of measures for the process of unfolding of the guidelines, as it proceeds:

Stage 1 - Values were defined, also the vision and the mission of the company through meetings and seminars with higher level administration and employees' involvement as a whole. The company had not yet defined the principles to be used inside the organization, as well as the inexistence of duty, or mission of PICCO, regarding the society and also which way to pursue, in other words the president's strategic vision in which direction to take the organization.

Stage 2 - Strategic diagnosis was defined for top administration, through internal study of sceneries (survey of strong and weak points) and external (opportunities and threats) in which the company is placed. We observed that this stage alerted company administrators of the next steps that the company should look for.

Stage 3 - The establishment of a "survival" annual goal was defined with the president of the company, taking into account the financial data of the previous year. The defined goal is vital for the survival of a company and passes through all sections, from the administrative to the productive. Such goal gives the direction to the organization, and it starts from the goal defined in the organizational guidelines. Starting from the vision in medium and long terms, and taking into account given historical financial data, the annual goal is defined with the president. That goal is financial, exactly to reinforce people's importance in the company, working integrated in the search of a vital result for PICCO.

Stage 4 - THE establishment of the annual guidelines are defined starting from the definition of "survival", goal, where it is complemented with the president's measures, a value (percentile, amount) and periods, resulting in the unfolded and not unfolded guidelines. This stage was important for the definition of great action axes, so to assure the achievement of the company's annual goal.

Stage 5 - The unfolding of the ergonomic guideline, through the definition of goals and measures is done exactly to assure the actions in the ergonomic issue. This point is extremely relevant, and it was done with plenty of energy, mainly for the expectation of looking forward to a better organizational atmosphere and consequently to reach the production and financial results without affecting the moral and people's health.

Stage 6 - The defined actions in stage 5 were executed by me with all the employees' contribution, though some contributing more than others.

Stage 7 - The stage of evaluation of results is important to verify the achievement of expected results. If the result is bad, or the goal was not reached in the defined period, the problem is analyzed; otherwise, if the result is good, it is standardized and employees trained.

Stage 8 - Here we made a work analysis through three analyses: Analysis of the demand, which defined the problem to be analyzed after agreement of several involved parts. The analysis of tasks where we tried to identify what the worker should accomplish, as well as the environmental, technical and organizational conditions for this accomplishment; and the analysis of the activities, where we verified the real work, or as the work is being accomplished. In each stage, besides the brainstorming, we made a research where it was identified the dissatisfaction in relation to the high risks of accidents in the bottling section and throughout the process informing the results to each accomplished stage. Some actions were implemented by the company, in the sense of improving the identified demand initially, as it was in the case of shoes and the jackets. Principles as clarity of objectives at work, methods and used instruments, respect of industrial secrets, collaborators' authorization for observations were in accord with all involved in the process.

Stage 9 - In the diagnosis of the work activity, we analyzed the information obtained in stage 8, correlating the main determinant in the work activity with the demand. Soon afterwards, we created a hierarchy of the main causes in the initial demand, so to be treated.

Stage 10 - Establishment and implementation of measures: Some established measured were implemented after the beginning of our study, seeking the improvement of the organizational atmosphere in the bottling section:

People's larger involvement in the definition of EPI´s: Safety shoes and protection mask case; People's larger awareness of the importance of EPI´s use; Training of all involved in new processes (i.e. package assembly); Use of a white board (visual administration) in the definition of daily tasks; Training practices on hygiene (personal and in relation to the product); Weekly programming (for standard lots).

In our last visits to the company, through conversations with the associates, we observed that the dissatisfaction degree in relation to the risks of accidents decreased.

Stage 11 - In the standardization stage, the objective was to enhance the positive facts that come to assure the effectiveness of future actions. Linked procedures to the production process were elaborated, seeking to assure the coherence in the development of the tasks.

Stage 12 – Establish an information administration through reports (quarterly, half-yearly and annual), so to create a history of actions taken in the company.

It was also concluded that the methodology of unfolding guidelines is in this case, for the ergonomic subject, a tool that assists managers in large companies as well as for small and medium entrepreneurs, evidencing clear ways for that process, seeking coherence to the actions developed in the company, aligning and integrating the actions between departments, involving everybody in the search process towards goal "survival" and to assuring a dynamic and agile process. A direct feature is the external customer's satisfaction, because of clear internal actions and so getting a competitive edge for the organization. Such model allows the company a larger understanding of the internal operation ways, mirrored in the survival of the organization in our highly competitive working atmosphere.

THE CASE OF AUTOMOTIVE INDUSTRY

In case of the Automotive Industry, a Management Program of Ergonomics and Working Conditions, very similar to the example of PICCO, was implemented since

2002. This program is based on the picture 1, where we can see that the main focus is the workplace. The start of any program is the definition of a Common Health, Safety and Working Conditions, that to create a culture in the company. After that we have several steps, such as the PICCO, to manage the working conditions of the industry.

For this case, we don't go into detail of each stage of the program (see Figure 2) because it has already been explained in the case of PICCO, but we are showing the results over the years after the implementation of the Program Management.

The chart below shows the total amount of employees (number of employees) of the company against the amount of employees who passed by the Medical Service (registrations) and we had confirmation that your pain / complaint was related to poor working condition. What we see is that over the years the company had a significant increase in the number of employees, mainly between 2007 and 2008 when they started the second round of production.

On the other hand, a decrease in the number of operators who visited the medical service of the company. In the Figure 3 we can see better what they represent in the Figure 2, we show that the number of employees who passed the medical service (actual amount) and also the same relationship as a percentage, ie, the total crossing the Clinic for the total amount of employees each year

When we move this relationship between the Clinic versus the total number of employees and we see the positive results of implementation of the Program Management, Health and Safety Conditions of work.

Finally, we have a chart (Figure 4) that showed the increased production of the company over the years, it is mean that, we see that the improvement of working conditions can increase performance and capacity of the company.

Figure 1 - Ergonomic Management Program

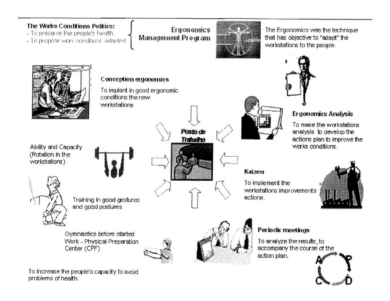

We note that in 2007 we had an increase in production volume of 30% and did not observe this increase in the number of people

Figure 2 – Registrations versus number of employees

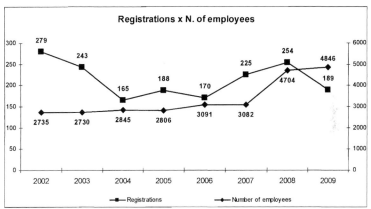

Figure 3 – Registrations versus percentual of employees

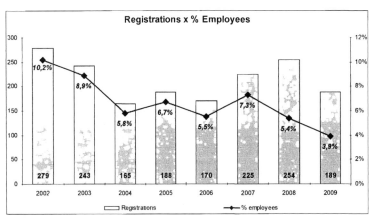

Therefore, we can say that having a management of working conditions is not an expense for business and investment but a medium to long term.

822

Figure 4 – Production of company

	2002	2003	2004	2005	2006	2007	2008	2009
Production	54076	71088	75310	77024	81731	118212	128968	140394

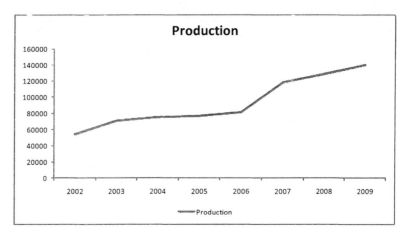

DISCUSSION

This work researched the contribution of a methodology of unfolding guidelines for the ergonomic subject in a production system. Initially a bibliographical revision of the terms related to the research was accomplished, as support to the establishment of that model.

For the verification of the applicability and consistence of the methodology, the same was applied to a system of production of cosmetics. The conclusion is based on the reach of specific objectives proposed in the work that lead to the general objective of this research.

a) To demonstrate the importance of treating the ergonomic subject in strategic level.

We verified that the easiness on the ergonomic subject is as much economical as social problem. An atmosphere with favorable work conditions reduces the working absenteeism increasing the profitability, and so guaranteeing a maintainable economical growth.

After showing the importance of the ergonomic subject in the success of the organization we showed and proved that such subject should be considered by the high administration of PICCO. In the analysis of the case, one of the defined measures to reach the survival goal that is financial was the improvement of the organizational atmosphere.

The viability of the unfolding tasks was emphasized through the involvement of all

PICCO´s associates, the knowledge of the actions to be executed, and the role of each one in the measure plans.

Here we applied the methodology of unfolding the ergonomic guidelines, composed of 12 stages, which adapted (enhanced) some steps of the model presented by Campos and Akao, GPD and the model presented by Guérin, AET. This methodology was composed of the strategic planning, and of the operational use of GPD and the daily administration in the cosmetics field.

b) To evaluate the application of the tools of Quality in the improvement of the work conditions.

The deviations or existent problems in relation to the ideal situation are solvable, otherwise it is not problem. And those solutions will be effective and possible, when a scientific method is used.

The tendency in thinking we know the solution of the problems based only in our experience, in the common sense and in the empiricism, is a not confirmed hypothesis. To solve the problems, it is necessary the application of tools that allow the analysis of processes through facts and data in a systematic and sequential scientific way.

We concluded that the understanding and acceptance of PDCA as base for the planning, as well as the understanding and use of some tools of the quality (graph of Paretto, diagram of Ishikawa). During our study in an Automobile Industry, were very important in the accomplishment of the work and in the identification of the relative problems to health and people's safety. The challenge from now on, is to adopt a culture in the company of resolution of problems based in PDCA, and in the use of those tools, always having the people involved in the process.

c) to evidence the practical applicability of the methodology proposed in a cosmetic industry and the analyse of a automotive industry and the obtained

The application of the methodology was accomplished in an Cosmetic Industry production, a company located in São José of Pinhais, PR / Brazil. The practical applicability of the methodology was evidenced in the results reached in each stage, being compared with the expected results, and in the adoption of this scientific methodology. The positive experiences deal with the definitions of vision and mission, up to the clarity and knowledge of all people involved, and the survival goal of the company. The results were reached, contemplating the improvement of the company's atmosphere. There were results such as improvement of the lay out, change of uncomfortable EPI´s, reduction of physical risks with the installation of attendances in the process of provisioning the bottling machines. The definition and communication of the functional organization chart was important in the improvement of the environment, since it defined the roles and responsibility of each section responsible. A program for managers' development was prepared for the high administration of the company. For old and new associates, the conception of training plans and integration was initiated, looking for the acquisition of competences. And the application of the methodology was accomplished in an **Automobile Industry** production, a company located in São José of Pinhais, PR / Brazil. The practical applicability of the methodology was evidenced in the results reached in each stage, being compared with the expected results, and in the adoption of this scientific methodology. The positive experiences deal with the definitions of vision and mission, up to the clarity and knowledge of all people involved, and the survival goal of the company.

The results were reached, contemplating the improvement of the company's atmosphere. There were results such as improvement of the lay out, change of uncomfortable EPI's, reduction of physical risks with the installation of attendances in the process of provisioning the bottling machines. The definition and communication of the functional organization chart was important in the improvement of the environment, since it defined the roles and responsibility of each section responsible. A program for managers' development was prepared for the high administration of the company. For old and new associates, the conception of training plans and integration was initiated, looking for the acquisition of competences.

CONCLUSION

The administration of the ergonomic guideline is a key to improve the financial health of organizations, and today, the ergonomics is constituted of an administration tool for the organizations. The challenge is to look for synergy among the technical and social systems, assuring an anthrop-metric vision.

As said previously, the relevance of this research is related to people's motivation and to life quality at work that can be understood as welfare related to the individual's work and the extension in his/her work experience being compensatory, satisfactory and free from stress and other negative consequences. The ergonomic subject can be understood as a direct result in the combination of several basic dimensions of the task and other dimensions not directly dependent from the task, capable of producing motivation and satisfaction in different levels, and besides resulting in several types of activities and behavior from individuals' belonging to an organization.

The understanding of the several factors of improvement of work conditions can constitute a solid base to guarantee the survival of organizations so that the work can happen without affecting workers' health, and assure a larger satisfaction in their work and consequently reduction in the absenteeism indexes and turn-over.

Nowadays, a larger attention has been given to QVT – Quality of Life at Work, (ergonomic subject), in the hope of promoting an involvement and motivation of the work atmosphere, providing ways for an increase in productivity.

Having an approach related to ergonomics, high productivity means motivation, dignity and larger personal participation in the project and performance of the work, and that implicates an approach integrated to life quality in the work. The concern with the ergonomic subject affects positively the productivity in an indirect way. The correspondence between productivity and life quality is interwoven and directly proportional, that is, quality of life high, productivity values also high; low life quality will provoke low index of productivity.

The present study presented the results reached with the application of a methodology of unfolding guidelines, which was composed using the model Administration by guidelines - GPD and for the Ergonomic Analysis of the Work - AET, methodological model widely used in Ergonomics.

It is concluded then finally, that the concern with the ergonomic subject is an essential condition for the success of the company that longs for prosperity, and also the administration of strategic actions for the ergonomic subject and its unfolding in all levels of the organization, and that it is possible through the proposed methodology, which assures the coherence in the activities developed by the organization. We owe to the administration, the executions of relative goals for the financial health of the company and also it is workers'.

It is also important the involvement of all in the process of the mission and proposition of improvements of the work conditions definition. It is important for the collaborators' understanding concerning the difficulties of the external atmosphere of the company, which we consider our participation as a researcher, with strong involvement in the work, using our experience and previous knowledge obtained in national and multinational companies.

That methodology is not something that the president of the company can apply and forget, it will take years, from 5 or 10, to accomplish all the fundamental changes that will transform it in a company of national or global importance, in other words, as in every administration model, there is a demand for constant attention and accountability to guarantee the expected performance. For this transformation to happen, it is necessary to raise strategic maturity and thus obtain a competitive advantage.

ACKNOWLEDGEMENTS

For top management of the companies mentioned in this article for believing and investing in a management program of ergonomics as a way to improve the performance and conditions of work and workers' life quality.

REFERENCES

AKAO, Yoji. Desdobramento das diretrizes para o sucesso do TQM. Porto Alegre: Artes Médicas,1997

BALBINOTTI, A Ergonomia como Princípio e Prática nas Empresas. Curitiba: Editora Gênesis, 2003.

CAMPOS, V. Falconi. **Gerenciamento de processos do trabalho do dia-a-dia**. Belo Horizonte: UFMG - Fundação Christiano Ottoni; Rio de Janeiro: Bloch Editores,1994.

CARAYON Pascale. Robertson, M., kleiner, B Hoonakker, P.L.T, 2005. Human Factors in Organizational Design and management – VIII IEA Press, Santa Monica, CA.

DANIELLOU, F. **Le statut de la pratique et des connaissances dans l'intervention ergonomique de conception**. Texte d'habilitation à diriger des recerches. Paris: Université Toulose Le Mirail, 1992.

DEJOURS, C. (1949). **A loucura do trabalho**: estudo de psicopatologia do trabalho. Tradução: Ana Isabel Paraguay e Lúcia Leal Ferreira. 5.ed. ampliada. São Paulo: Cortez, Oboré, 1992. p.42-47.

DEJOURS, C. L'ergonomie en quête de ses principes. Dèbats épistémologiques. Em: F. Daniellou (Org.). **Épistémologie concrète et ergonomie**. Paris: Octares Edition, 1996. p.201-217.

DEJOURS, C. A Carga psíquica do trabalho. In: DEJOURS, C.; ABDOUCHELI, E.; JAYET, C. Coordenação: Maria Irene Stocco Betiol. Tradução de Ideli Domingues.

Psicodinâmica do Trabalho: Contribuições da Escola Dejouriana à análise da relação prazer, sofrimento e trabalho. São Paulo: Atlas, 1994. p. 21- 32.

GONTIJO, Leila. A.; SOUZA, R. J. Macoergonomia e análise do trabalho. In: **II Congresso Latino Americano e VI Seminário Brasileiro de Ergonomia**, Florianópolis, 1993.

GONTIJO, Leila. A.; ULBRICHT, Leandra. Ergonomia aplicada ao estudo dos distúrbios ostemomusculares relacionados ao trabalho dos ordenhadores. **Anais**. ABERGO, 2001. Gramado, RS, 2001.

GUÉRIN, F. et al. **Comprendre le travail pour le transformer** – la pratique de l'ergonomie. São Paulo: Edgar Blücher, 2001.

HENDRICK H. W. **Human factors in ODAM**: the future. Humam factors in organizational design and management. [s.l], 1994. p.6-9.

IIDA, I. **Ergonomia** - projeto e produção. São Paulo: Edgard Blucher, 1990. 465p.

MENDES, A. M. Cultura organizacional e prazer-sofrimento no trabalho: uma abordagem psicodinâmica. In: TAMAYO, A. e colaboradores. Cultura e saúde nas organizações. Porto Alegre: Artmed, 2004, p. 59 – 76.

MERGENER, C.R.; KEHRIG, R.T.; TRAEBERT, J. Sintomatologia Músculo-Esquelética Relacionada ao Trabalho e sua Relação com Qualidade de Vida em Bancários do Meio Oeste Catarinense. **Saúde Soc.** São Paulo, v.17, n.4, p.171-181, 2008.

PICOLOTO, D.; SILVEIRA, E. Prevalência de sintomas osteomusculares e fatores associados em trabalhadores de uma indústria metalúrgica de Canoas - RS. **Ciênc. saúde coletiva**, Rio de Janeiro, v. 13, n. 2, Apr. 2008

SALIM, C. A. Doenças do trabalho: exclusão, segregação e relações de gênero. **São Paulo Perspec.**, São Paulo, v. 17, n. 1, mar. 2003 .

SHIBA Shoji. **TQM**: quatro revoluções na gestão da qualidade. Porto Alegre: Artes Médicas; Bookman, 1997

WISNER, A. Questions épistémológiques en ergonomie et analyse du travail. Em F. Daniellou (Org). **L'ergonomie en quête de ses principes** - dèbats épistemologiques. Toulouse: Octares Éditions, 1996.

CHAPTER 89

Ergonomics in Industrial Design Education Process: A Conceptual and Methodological Experience

Luz Mercedes Sáenz

Universidad Pontificia Bolivariana
Circ.1 # 70-01 Bloque 10 Campus Laureles
Medellín, Colombia

ABSTRACT

This proposal shows the development and implementation of education in Ergonomics and design in the Faculty of Industrial Design in the Universidad Pontificia Bolivariana in Medellín, Colombia through a line of research on the subject. The goal of this line is to create a culture on the subject by introducing courses into the Industrial design undergraduate study program, creating undergraduate research groups, sponsoring research projects as well as applications to everyday university life, as strategies oriented towards articulating both disciplines and strengthening the concepts which lead to a more human design.

Keywords: Ergonomics –Design relationship, education in Ergonomics and Design, Methodology for the design process.

INTRODUCTION

In the faculty of Industrial Design of the Universidad Pontificia Bolivariana (UPB) in Medellin, Colombia, the Ergonomics line of research of the Design Studies

Investigation Group (GED) has made it its goal to establish a relationship between Ergonomics and Design by means of a practice which defines the following:

- Different subjects for the study plan (thematic and conceptual conception).
- A procedural proposal (methodological conception) which allows students of the Faculty of Industrial Design to approach the subject from an anthropocentric, systemic, and interdisciplinary point of view from the very beginning of the design process.
-An alternative to the current education and research in Design (pedagogic conception). This consists of developing activities which encourage the creation and transmission of knowledge such as undergraduate research groups and activities in which ergonomics can be introduced into everyday University life. In this way the participation of students, faculty and other University dependencies help promote a culture on the subject.

This proposal intends to articulate Ergonomics and Design into the design study program and research and application in the development of products. This contributes to the knowledge base and is presented as a methodological alternative to both disciplines whose main goal is to understand the user and his or her specific needs in diverse contexts of use.

ERGONOMICS IN THE EDUCATION PROCESS OF AN INDUSTRIAL DESIGNER. THEMATIC CONCEPTION

Ergonomics in Colombia was introduced into the academic context in the mid 1960's thanks to a proposal put forth by Spanish Engineer Jorge Forcadas. An elective course for students of Industrial Engineering was created in the Universidad Nacional de Colombia in Medellin and it quickly spread to other universities around the country (Estrada, 2005).

Ergonomics has been most frequently included in educational programs related to occupational health and has only recently begun to become a part of Industrial design curricula with enough time intensity so as to permanently introduce the concept into a designer's academic and professional lives(Saenz, 2006).

Today, design faculties around the country which include Ergonomics into their study plans consider not only subjects which teach specific scientific content which supports ergonomics,, such as anthropometrics and biomechanics, but also integrate knowledge from other disciplines in order to turn out an applicable skill.

An example of this are the courses offered by universities such as Pontificia Universidad Javeriana: Ergonomics for Industrial Design, Ergonomics and information, Physical Ergonomics for Design, Ergonomic Analysis, Cognitive Ergonomics, and Transergonomics; the Universidad Industrial de Santander: Physical Ergonomics, Cognitive and environmental Ergonomics, Product Ergonomics, Usability; Universidad Nacional de Colombia in Bogota: Human Factors I, II and III; among others.

In the Faculty of Industrial Design of the Universidad Pontificia Bolivariana (UPB), a more integral education is achieved by introducing the required knowledge in the form of components which are included in a designer's pedagogic and disciplinary development (Faculty of Industrial Design, School of Architecture and Design. 2009). One of these is the Functional – Operational component which represents "the object as useful". From based on this an object's actual "value of use" and its relationship with the user in a given context or activity is defined. From this point of view, function is an object's axis of configuration which determines the technical and operational efficiency, its usefulness, as well as its relationship with the user in both the physical and perceptive. (Saenz, 2005, 3-64).

In this component, in order to approach a product's functionality, subjects such as the physical attributes of form and the material it is made of as well as an object's values of use, understood as the ways in which a function can be performed effectively (Fornari, 1989); subjects which define a product's technical and use functions, according to the proposal of the disciplinary model of the Industrial Design Faculty of UPB (Faculty of Industrial Design of the UPB, 2009), are considered essential. At the same time these subjects must contribute to an effective User – Product relationship (operational), keeping in mind not only the physical requirements and characteristics, but also perceptive and cognitive aspects.

In the functional – operational component, the courses, Functionality of an Object, Dynamic and Static and Technical Functionality consider an object's technical aspects and usefulness. The Ergonomics and Design courses consider the function of the User – Object relationship which allows a detailed analysis of a situation of use from a human perspective (Ergonomics) and the object's perspective (Design). This is done by including physical, cognitive and social characteristics as well as user requirements, technical aspects and shape and material characteristics which may influence a product's usefulness in the configuration process of an industrial object. These courses must also consider the nature of the environment where the object is to be used.

In this way the subjects associated to Ergonomics are a part of the functional – operational component and help not only approach the required content but also to form concepts on the optimization of the User – Product- Context relationship. Several courses are included in the undergraduate study program: Ergonomics and Design 1 (A person as a living being), Ergonomics and Design 2 (A person as the user of an object: Conditions which optimize the User – Product – Context interfaces), elective courses such as Inclusive design, among others.

Also included in the study program are the courses of the Graduation Projects System, supported by the Ergonomics Line of Research, which include the following courses: Investigation and design 1, 2 and 3 and Graduation Project. These courses guide the student along the process of developing the project applying Ergonomics and Design to subjects such as: Recreational objects for visually impaired children, Backpacks and Children, Child Recreation and Sedentary lifestyle, biomedical equipment design, furniture for elderly persons, vascular access system design, and hand rehabilitation objects, military footwear, among others.

In postgraduate education, Ergonomics and Design are a part of the Specialization in Interior Architecture Projects study plan in the content modules about Habitability and Comfort and Interior Workspace Design. They are also included in the Biomedical Engineering Specialization in the Ergonomics and Design module which emphasizes on an interdisciplinary vision.

ERGONOMICS IN THE EDUCATION PROCESS OF AN INDUSTRIAL DESIGNER. METHODOLOGICAL CONCEPTION

The fundamentals of ergonomics were initially conceived as the study of human activity when at work, striving to maintain adequate health conditions during work activities.

From the very beginning this concept has been oriented towards optimizing the Person – Machine – Environment System. These components establish an anthropocentric vision, since its origin and focus are on the person, and a systemic vision which analyses each variable in order to define characteristics and requirements which can be observed and analyzed through methodological proposals which have been developed by this discipline and which also can be used to diagnose and establish applications which maintain health and safety standards in the occupational context.

With time, other areas of knowledge in areas different to health, such as Industrial design, have intervened and through particular experiences inherent to their practice have broadened the view of Ergonomics and thus have extended its reach to other aspects of human life. In this way, when a person uses a product for every day life activities, be it domestic, recreational or other, he or she establishes an analogous relationship to that which is established in the User – Product- Context System in which other disciplines may contribute in order to optimize the relationship.
The methods give Ergonomics a structure which can be used for analysis and

evaluation of specific situations which occur while a person is performing an activity and also a model for scientific practice. It also provides a base from which theories on human performance can be proposed, activities questioned, data collected rigorously for later analysis, and findings and results can be communicated. (Stanton, 2006, 1-3).

Design, on the other hand, originally focused especially on the aesthetic values of a product, in other words appearance and fashion were the motivational elements both for designers and users. The problem of design was reduced to beautifying a product common to industrial society by presenting a style which represented the moment (Campis, 2007, 3-124). In the history of design, it is easier to find testimonies on results rather than processes and/or methods of product configuration(Campis, 2007, appendix 1).

However, design has increasingly shown other methodological alternatives which, by means of procedures, techniques, aids or tools represent the different activities required for an object's configuration process and whose intent is to introduce procedures within a logical framework (Cross, 2002, 3-43).

In order to articulate Ergonomics and design and to assist in the process of an object's configuration which truly considers a users characteristics and requirement, the Ergonomics Research Line GED in UPB has developed a methodological proposal which establishes the relationship between both disciplines. This can be applied as a qualitative analysis tool which is supported by existing quantitative instruments and thus may become a part of the product design process, both in the undergraduate program as well as research and application projects which require a creative process, use and situation analysis and product design proposals.

This procedure also allows, through teaching and research, for the generation of conceptual bases for designers and other professionals, thus aiding in the recognition of all components involved in a specific situation (a person carrying out an activity) and establishes connections necessary for the evaluation, analysis and application of ergonomic conditions (Saenz, 2006).

The methodology includes: basic **Subject units** relative to the User, **the product and the context:** specific and fundamental components of the ergonomic system. These subject units, are individually analyzed and are articulated creating a network which, according to the project's area, establish requirements and/or hierarchies for its development. As seen in figure 1. Subject units from the conceptual proposal presented by the Ergonomics Research Line of the Studies Design Group (GED)from UPB.

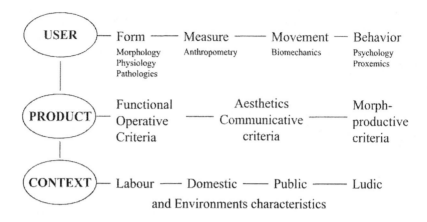

Figure 1. Subject units from the conceptual proposal presented by the
Ergonomics Research Line of the GED from UPB.

It also includes **activities and/or moments** which can be carried out according to
the subject bases which are related: **Identify:** a users characteristics and
requirements. This is a phase in which social and physical/cognitive dimensions are
identified and the product is characterized as well as the context in which it will be
used.

Evaluate: A users activities prior and during the use of the product, critical aspects,
advantages observed, opportunities based on other research related to the subject,
existing legislature. It is a moment which other methodological alternatives offered
by ergonomics can be put to good use, in order to systemically observe a human
being during the activity while using the necessary products. It is a stage for
analysis in which the designer must observe the functional – operational, aesthetic –
communicational and morph – productive (current situation) characteristics and
also it is the moment to observe the aspects of the context.

Integrate: the elements which were identified and evaluated (needs, requirements,
and characteristics) are made tangible by means of the product design proposal. Or,
from an ergonomic perspective, recommendations are suggested in order to adjust
the form of execution (Use/Procedure) and/or the context in which to perform the
activity (environment in which the activity takes place).

Subsequently comes the **Elaborate** stage in which ideas are materialized in the
form of a model or prototype from a design perspective or in the form of
recommendations and/or adjustments from an ergonomic perspective. These must
lead to the development of a plan to **validate** the product with those who will be its
users. It is the moment for applying methodologies and tools available to
Ergonomics and Design and to define if the proposal or the recommendations are
correct.

Next is the construction process (**Production**). During this stage the designer has relinquished responsibilities to the manufacturers which must guarantee all conditions established by the process and keep all quality criteria.

Alter a stage of commercialization, a product is ready to be used or from an Ergonomic perspective, this is the stage in which adjustments and/or recommendations are introduced which guarantee an optimal User – Product – Context relationship, striving to always maximize welfare, health and safety conditions. (Saenz, 2008).

The methodological proposal is used by undergraduate students in order to analyze specific situations in the product design process and in order to analyze determined situations in which the product will be used for the graduation Project. The proposal is also used by researcher of the Ergonomics Line of research for the analysis, evaluation and application of design projects (Saenz 2006) and activities which apply to everyday University life, for example: the Dimension criteria for the new furniture of the UPB school project, anthropometric and dimensional adjustment verification. See figure 2. Activities and moments in the methodological proposal of the Ergonomics line of research in UPB.

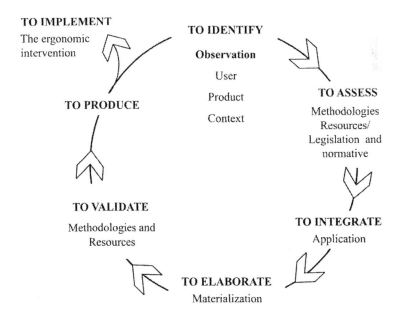

Figure 2. Activities and moments in the methodological proposal of the Ergonomics line of research in UPB.

ERGONOMICS IN THE EDUCATION PROCESS OF AN INDUSTRIAL DESIGNER. PEDAGOGICAL CONCEPTION

The general guidelines of the pedagogic model of the UPB privilege learning, a student's active participation in the construction of knowledge and a teacher's position as mediator; research; and duly supervised experimental and practical independent activities (Universidad Pontificia Bolivariana, 2009).

The faculty of Industrial Design, through the GED's lines of research and, for this specific proposal, through the Ergonomics line of research is developing teaching activities in undergraduate courses, research seedbed, postgraduate courses (teaching), undergraduate research groups, participating in the development of graduation projects (research education) and carrying out applied research and consulting projects which respond to a practical educative model, which among other things, emphasizes on research as an alternative for finding and creating new knowledge.

In this way, students can attain the following during their academic process:

- Apply information on human characteristics which is pertinent: a body's anthropometric and biomechanical data relative to the objects projected and the work environment.
- Define an industrial objects formal attributes seeking to mold it to a user's psycho-physical attributes. Analyze an object's use, verifying how understandable, easy to use and adequate it is to a human being.
- Establish criteria for analysis (based on the functional requirements of an object). Classify the information (from an Ergonomic perspective, seeking to optimize the User – Product –Context relationship).
- Applying methodological proposals to the design process which include ergonomic criteria.
- Recognize the design process as an activity which includes a research component which should, among other things, endow an object with characteristics which respond to a user's capacities and limitations, striving always to improve welfare, health and safety.
- Visualize alternatives and methodologies which can be applied in the design process based on ergonomic criteria.

In this way by using this pedagogic concept during the education of industrial designers, students participate in theoretical-practical activities of an experimental nature and in academic events which improve their skills reaching towards a more human design. See figure 3. Activities for learning, research and practice in Ergonomics and Product Design in UPB.

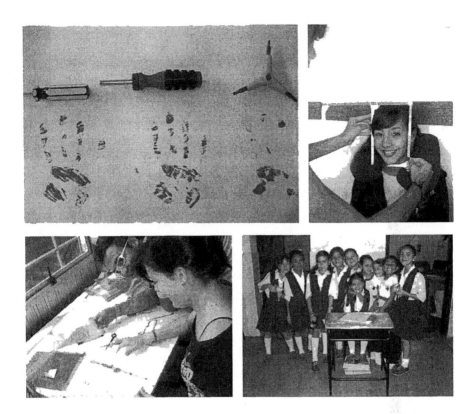

Figure 3. Activities for learning, research and practice in Ergonomics and Product Design in UPB.

CONCLUSIONS

Ergonomics and Design consider subjects which can be articulated and support the creation of knowledge on the configuration process of products focusing on the user.

Ergonomics/Human Factors must be included in the undergraduate study program and are to be used as tools to create concepts (theory), analysis and application (Methodology) in the process of Design, as well as to conceive the product and the relationship of use

Ergonomics and design include contents, tools and procedures which facilitate the paractice of experimentation in the process of education of an Industrial Designer.

The faculty of design of the UPB in Medellín, Colombia integrate Ergonomics and design by implementing courses from the study program which help to get to know and apply the related subjects, strengthen investigative skills, the development of basic research projects (in order to generate knowledge) and applied research (projects carried out with different companies) as well as the participation in University activities which require support from the point of view of Ergonomics and design thus encouraging the creation of a culture on the subject.

REFERENCES

Campis, Isabel (2007). La idea y la materia.Volume 1: El diseño de producto en sus orígenes. Barcelona: Editorial Gustavo Gili.

Cross, Nigel. 2002. Métodos de diseño. Estrategias para el diseño de productos. México, D.F.: Editorial Limusa, S.A. de C.V.

Estrada, Jairo. 2005) "La ergonomía laboral en Colombia, Una aproximación a la identificación de perspectivas" proceedings of Primer Congreso Internacional de Ergonomía, Cali, Colombia.

Facultad de Diseño Industrial, Escuela de Arquitectura y Diseño. 2009. Modelo disciplinar de la facultad de diseño industrial. Universidad Pontificia Bolivariana. Medellín, Colombia.

Fornari, Tulio. 1989. Las funciones de la forma. México: Universidad Autónoma Metropolitana Azcapotzalco.Tilde Editores.

Sáenz, Luz M. 2005. Ergonomía y diseño de productos, criterios de análisis y aplicación. Medellín: Editorial Universidad Pontificia Bolivariana.

Sáenz, Luz M.) 2006. "Creating a culture in ergonomics, an interdisciplinary experience" Proceedings of IEA 2006 Congress. Meeting Diversity in Ergonomics. Elsevier Ltd.

Sáenz, Luz M. 2008. En el proceso de diseño: alternativa metodológica para la concepción de productos. Iconofacto V4, N5:170-182.

Stanton, Neville. et al(2006). Handbook of human factors and ergonomics methods. Taylor & Francis E-Library

Universidad Pontificia Bolivariana. 2009. Modelo pedagógico integrado. Medellín: Editorial Universidad Pontificia Bolivariana.

Chapter 90

Ergonomic Aspect of Urban and Social Dynamics

Agata Bonenberg

Faculty of Architecture, Poznan University of Technology
Nieszawska 13C, 61-021 Poznan, Poland

ABSTRACT

Ergonomics research, investigating the adaptation of living and working spaces to the abilities of human body and mind, develops in the direction set by technology progress and civilisation changes. Increased dynamic economic migration, and work mobility of societies observed in last decades, makes both architecture and design become more flexible and responsive in terms of the new challenges they are supposed to meet. The purpose of this article is to provide an introduction to the research apparatus of ergonomics in such themes as: reconfiguration, adaptation, negotiation, expansion, inversion, and intensification as important elements of human interaction with the environment.

Keywords: flexibility, responsiveness, negotiation space

INTRODUCTION

Ergonomics research explores the adaptation of living and working spaces to the abilities of human body and mind. Its concepts and fields of research develop in the direction set by technology progress and civilisation changes, focusing on the needs and abilities of an individual. This principle has been applied since the late 19th century industrialisation and the time of heavy industry development in the fifties and sixties. In last decades, technological revolution and wide access to information caused substantial social, economic, and cultural transformations, triggering a rapid increase in dynamics of changes in all areas of life. Common use of teleinformation systems – networking of the information flow – becomes a fundamental attribute of social and cultural activities associated with the 'society of knowledge'.

Dynamic external conditions, economic migration, and high work mobility resulted in a fact, that both architecture and everyday use objects get more flexible and responsive to the challenges they are supposed to meet. The author, analysing concepts of flexibility in the development of space, divides them into two groups:
- qualitative changes in the architectural space: reconfiguration, adaptation, and negotiation;
- quantitative changes of urban space: expansion, inversion and intensification.
How much will the changes in the way space is used influence the scope of tasks the ergonomics has to deal with? How much wider will this discipline of knowledge based on the relation between a human being and post-industrial living environment become?
The purpose of this article is to provide an introduction to the research apparatus of ergonomics ideas such as: reconfiguration, adaptation, negotiation, expansion, inversion, and intensification as important elements of human interaction with the environment.

SOCIETY OF KNOWLEDGE

New means of communication and transport, as well as computerisation of many areas of human activity have an impact on contemporary interactions between people. The above mentioned processes are related to the development of the civilisation and influence the shape and spatial qualities of the environment. In *The Rise of the Network Society* (1996) Manuel Castells, Spanish sociologist, created a description of a new area of social interaction which was a reaction to the complexity of these contemporary sociological phenomena. He created a concept of the 'space of flows' describing the flow of information, technology, images, and symbols. According to Castells, the flow not only constitutes a part of social organisation, but also it is an expression of predominating processes in contemporary political, economic, and social lives. The qualities of the flow dictate the way information is received and processed, the way we work and communicate. Its presence influences changes in economic and social conditions which take place in the environment we live in. This constant technological development, changes in attitudes, trends, and need for innovations have a major impact on contemporary culture. The culture is reflected in art and architecture. Objectives of artists and people's expectations have changed: apart from the shapes and images in contemporary forms of artistic expression, aspects related to the dimension of time and ability to interact became important. The same route has been followed by contemporary architecture – new conditions allow it to go far beyond the creating shapes of buildings. Architecture should mainly focus on the analysis of the purpose of the space in terms of ideas such as: social, cultural, economic needs and ergonomics requirements. Flexibility and responsiveness can protect architecture from the devaluation in the condition of constant changes in consumer environment - never before being so important in the space build by people.
This attitude in creative design gives ergonomics an important goal. Ability to adjust space or an object according to the way they are used is far more difficult if

they are multifunctional. The design therefore needs to meet the expectations in various situations, conditions and surroundings - sometimes hard to foresee.

SPATIAL RESPONSIVENESS IN ARCHITECTURE AND STRATEGIES OF ACTIONS TARGETED AT THE IMPROVEMENT OF THE ERGONOMICS OF LIVING SPACE

Manuel Castells foresees in 'The Rise of the Network Society' a gradual increase in the dynamics of changes in social and economical areas. The form of space – contemporary architecture – is adapting to the requirements of these circumstances. The passing decades are first to widely analyse concepts of the architecture performing dynamically on time – related basis. With rapidly changing environment, the potential future ways of using the space need to be taken under consideration already at the design stage. The flexibility of function, negotiability of space and its ability to adapt are subjects of academic research and experimental works. We foresee possible methods of expansion and intensification of existing architectural and urban compositions, and strategies of coping with abandoned architectonical and urban tissues.

Table 1.1. Systematics of dynamics in architectural and urban spaces. acc. to A. Bonenberg

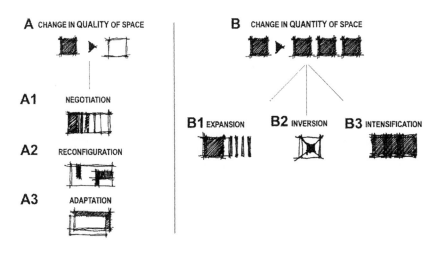

The strategy of actions which are meant to improve the ergonomics of living space of the 'society of knowledge' is related to physical and psychological areas. The physical aspect is associated with working environments typical for the post-industrial era: man – computer arrangement, and performance of interactive systems. The psychological aspect of ergonomics is frequently related to the abilities of memory and perception of computer software interfaces.

Each method of flexible use of space (negotiation, reconfiguration, adaptation, expansion, inversion, intensification) entails the obligation to specify a strategy of actions meant to optimise the ergonomics of space. Every urban, architectural and interior design should be planed and executed in accordance with all aspects of the ergonomics: both psychological and physical characteristics of a man, and technologies present in his environment. In some cases however, certain areas of ergonomics are particularly important. [Table 1.1]

A. CHANGES OF QUALITY IN SPATIAL COMPOSITION

The composition of architectural space may undergo modifications which, although do not necessarily influence the size of occupied area, can result in the optimisation of the way the space is used. (Thackara J., 2006) The changes can progress on different time – related basis, being classified as: long-term (several years), moderately long (few months) or short-term (the change can occur overnight). The time scale therefore constitutes one of the parameters which facilitate the classification of flexible architectural composition. The time scale and the way space is used allow to classify the changes as: negotiation (pt. A1), reconfiguration (pt. A2), and adaptation (pt. A3).

A1. Negotiation

Negotiation spaces are created around static compositional arrangements, designed in a way which facilitates multifunctional usage of the space. In this context, we may talk about the 'negotiation of space' in terms of situation when the useable space does not have one single function, but it may be used for one or more purposes. It is characterised by an ability to change the proportions of territory assigned to the use of a certain function over a short or comparatively short period of time.

Unlike recomposition changes (A2) – it does not generate physical modifications in the architectural form. In practice, the design of negotiation space is related with an arrangement of open plan areas surrounding certain 'immobile' functions, such as staircases, bathrooms, and kitchens.

Classical examples of this solution include: Farnsworth House from 1950 by Mies van der Rohe, Glass House by Philipe Johnson from 1947, Shigeru Ban, Wall-less House, 1998. These historical examples caused popularisation of 'negotiable' spaces in newly designed structures, particularly due to the fact that the concepts of spatial negotiation are in many ways similar to the ideas of minimalism. The acceptance of anthropometrics as a foundation of developed 'immobile' elements of architecture and their adjustments to multifunctional usage is an important element of ergonomics strategy for shaping the negotiation space. Anthropometrical dimensions facilitate design process helping to specify the size of negotiable spaces, appropriate size of the area assigned to the specific elements and guarantees the best internal arrangement. Because anthropometrical dimensions vary within the population, it is difficult to create one universal space - suitable for every user. The project should have very individual character and personalized measures of composition.

Table 1.2. Strategy of actions meant to optimise the ergonomics of designed spaces

	Type of responsiveness of the architectural space	Strategy of actions meant to optimise the ergonomics of designed spaces
A. Qualitative changes of the architectural space	A1. negotiation	The scope of interest: anthropometry in designing immobile elements of the composition
	A2. reconfiguration	The scope of interest: the arrangement of dynamic elements of the interior composition
	A3. adaptation	Adjusting existing space to new functional requirements by anthropometrical analysis and arrangements
B. Quantitative changes in the urban space	B1. expansion	The city ergonomics – determination of limits of expansion by understanding the impact on the environment: microclimate, lighting, noise, vibrations, static perception, dynamic perception
	B2. inversion	Psychological consequences of using an 'abandoned space'
	B3. intensification	Determination of intensification limits by understanding the impact on the environment: microclimate, lighting, noise, vibrations, anthropometry, arrangement

A2. Reconfiguration

Reconfiguration can be understood as modification of the parameters of architectural space in a short time span. Reconfiguration of elements can be observed in the scale of an interior, of a building form and an urban composition. The latter case relates to innovative concepts yet to be realised, whereas kinetic architecture is an example building - scale reconfiguration. Reconfiguration interiors are widely used in open – plan office interiors, and are getting popular in the living spaces. One of first classic examples in residential architecture is a historical villa Schröder by Geritt Rietveld (1924). The composition of the first-floor is based on the staircase, kitchen and bathroom, while other spaces are separated depending on users' needs. This classic structure was followed by contemporary realisations. Internal flexible composition of the space is often seen also from the outside, if mobile facades have been composed according to the rule that the building exterior should reflect its internal order.

The ergonomics strategy in terms of reconfiguration is based on the assessment of dynamic arrangement of composition elements. The localisation and the way these elements are rearranged within the space constitute an elementary design problem connected with reconfiguration space. Traditionally, the assessment of the equipment arrangement in terms of man-working place relationship is based on the criterion of proximity in order to facilitate mobile operation and adequate distance from the equipment to avoid its accidental dislocation. The same elements may be used to evaluate effective reconfiguration of residential interiors. Here, important factors are frequency of use, easy access and intuitive operation of a mobile element.

The classification of ways, in which architectural space change helps to apply solutions for new arrangements suiting ergonomics requirements of the users. There are examples of projects that combine elements of reconfiguration and negotiation. One of such examples is a conceptual design of a temporary residency building in the Isle of Iburgh in Amsterdam, designed by A. Bonenberg. The philosophy of composition was connected with the private – public use of spaces. The work therefore was preceded by preparation of 'privacy' diagram, showing activities which a typical future user is keen to perform accompanied by other people, and these which require maximum privacy. The results of the study show three basic groups of space:

100% flexible – 'public' activities performed in fully negotiable spaces

50% flexible – informal social and family meetings, work from home – recomposition spaces

0% flexible (basically inflexible) – traditionally separated spaces assigned for relaxation and family life

This classification has been reflected in the building architecture, which was based on the interpretation of results of the aforementioned study.

There were three usable floors in the building:

First level – open space (negotiation)

Second level – flexible 50 % (recomposition)

Third level – traditional closed private areas.

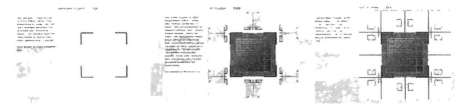

FIGURE 1.1. Residential building, Iburgh in Amsterdam, plans, designed by A. Bonenberg

FIGURE 1.2. Residential building, Iburgh in Amsterdam, designed by A. Bonenberg

A3. Adaptation

Adaptation allows to change the function of a building and adjust it to a new way of use. It is a common procedure which lies within the field of changeable architectural spaces. In terms of time scale adaptation can be classified as a long-term operation. A popular group of realizations in architecture is constituted by lofts. Aside from these, in Poland realizations include difficult adaptations of blocks of flats made of prefabricated concrete sections. Here ergonomics strategies depend on many factors already in situ – often unfavourable spatial conditions. In terms of an existing space, application of new ergonomic solutions can refer mainly to newly designed internal elements.

B. ERGONOMICS AND CHANGES OF THE QUANTITY OF URBAN SPACE

In the build environment occur tendencies of changing the quantity of space used for various functions. They are particularly significant when discussing city – scale transformations. (Bonenberg, W., Baczynski, K., 2006) Building tissue expands, covering larger territory or it can be more intensively developed, remaining in same boundaries. Otherwise – some cities tend to disappear, shrink, 'collapse' in the space. The ergonomics of the city and urban infrastructures has been a subject of many studies on visibility, safety, and stress factors having an impact on an individual in highly developed areas.

B1. Expansion

Expansion is an enlargement (increasing in volume) of building tissue by incorporating new territories. Concepts of linear cities and many vanguard architectural structures had an 'open composition', the expansion of which could theoretically be continued infinitely. The Endless House designed by Frederick Kiesler, an architect and artist who worked on the idea of 'endless space' from 1922 to the 1960s may be an example of such 'continuous' architecture. According to him, an optimal residency model is to create biomorphic, endless space which combines painting, sculpture and architecture as well as the surrounding environment. Linear designs of cities by Alexander Milutin and Arturo Soria y Mata of late 1920s are crucial for the discussion on the expansion of urban system which on the theoretical and compositional level seemed to be unlimited. The analysis of the rules of ergonomics applied to the scale of a city helps to set boundaries of the expansion by identifying environmental possibilities: microclimate changes caused by density of buildings, limited daylight, increased impact of noise and vibrations. The material environment conditions have an impact on the level of residents' satisfaction and their health.

B2. Inversion

Inversion relates to a decreasing need for usable areas. It can be associated with a 'collapse' of a function and changing former arrangement to an opposite one - with opposite characteristics. Shrinking city can be defined from the point of view of demography (people moving out of a city), economy (deindustrialisation, unemployment), space (degradation of residential tissue, abandonment of an area by its users). Search for concepts and ideas on redevelopment of the wasteland gives possibilities of changing the old profile of use. Solutions can be pro-ecological, related to the renaturalisation – returning to the natural state of the environment, possibly close to the initial state before it was altered by a man.
Converting the 'shrinking city' areas to the physical and psychological abilities of residents includes reduction of stress – the feeling of dismissal, abandonment, elapsing time and danger. On the other hand, positive psychological aspects of such spaces are connected with potential for improving the quality of the environment, noise and vibration reduction, etc.

B3. Intensification

Intensification is related to increased usage of space (increasing in volume) within unchanged boundaries. Frequently it is equivalent with a vertical growth – replacing existing structures with higher developments. Introducing high rise buildings in the city structure is one of methods of intensification of the way land is used. Compared to this, theoretical projects created at the prestigious London Architectural Association stand out, with concepts that are not based on the 'drive upwards'. They include, for instance, placing complex urban functions above and around areas of roads crossing the city. This leads to the fusion of earlier separated districts and smooth continuation of the urban organism above the division of roads. Physical factors of the environment which determine appropriate intensification of the development are, similarly to spatial expansion: microclimate, noise, vibrations, and light. With the intensification of the urban tissue the microclimate of the environment significantly changes. Air temperature is raised, the humidity drops down, the air flow changes its direction. All components of the microclimate influence people's dispositions, their physical and mental abilities, work efficiency, and keeping good health. Some of the consequences of spatial expansion and development intensification can be reduced. Suitably prepared design can reduce noise to a large degree, lessen its travelling, it can also partially muffle it.
Intensification of the development influences sociological factors of stress related to the structure of a group and organisational stress, which the situations of highly intensive use of land are accompanied by.

SUMMARY

Ability to employ spatial recombination, adaptation and negotiation is a chance for contemporary architecture. Changing external messages and interior flexibility can facilitate meeting requirements, which the architecture will be facing in the future. At the same time urban tissues demonstrate increasing tendencies of changing the quantity of space used for their functions. The belief that future purposes of buildings cannot be foreseen makes investors, engineers and architects design projects for a fixed period of time – till the moment of technological death of a structure. Ergonomics while setting the strategies of action in terms of living and working space reorganisation can become an instrument facilitating introduction of successive innovative and long-term spatial solutions.

REFERENCES

Bańka A., (1997) *Architektura psychologicznej przestrzeni życia,* Gemini S.C., Poznan
Bonenberg, W., Baczynski, K. (2006), *Urban regeneration,* vol. II., Faculty of Architecture, Poznan University of Technology, Poznan

846

Castells, M., (1996) *The Rise of the Network Society*, Blackwell Publishers Ltd, Oxford, UK
Thackara J., (2006) *In the bubble – designing in a complex world*, MIT Press, Cambridge MA

For Product Safety Concerns and Information please contact our EU
representative GPSR@taylorandfrancis.com Taylor & Francis Verlag GmbH,
Kaufingerstraße 24, 80331 München, Germany

Printed and bound by CPI Group (UK) Ltd, Croydon, CR0 4YY
11/05/2025
01866583-0001